Forward and Inverse Scattering Algorithms based on Contrast Source Integral Equations

Forward and Inverse Scattering Algorithms based on Contrast Source Integral Equations

Peter M. van den Berg
Applied Sciences
Delft University of Technology
Delft, Netherlands

WILEY

This edition first published 2021

Registered Office
John Wiley & Sons, Inc., 111 River Street, Hoboken, NJ 07030, USA

Editorial Office
111 River Street, Hoboken, NJ 07030, USA

For details of our global editorial offices, customer services, and more information about Wiley products visit us at www.wiley.com.

Wiley also publishes its books in a variety of electronic formats and by print-on-demand. Some content that appears in standard print versions of this book may not be available in other formats.

Library of Congress Cataloging-in-Publication Data

Names: van den Berg, Peter. M., author.
Title: Forward and inverse scattering algorithms based on contrast source integral equations / Peter M. van den Berg.
Description: Hoboken, NJ : Wiley, [2020] | Includes bibliographical references and index.
Identifiers: LCCN 2020030605 (print) | LCCN 2020030606 (ebook) | ISBN 9781119741541 (hardback) | ISBN 9781119741572 (adobe pdf) | ISBN 9781119741565 (epub)
Subjects: LCSH: Wave equation. | Scattering (Mathematics) | Integral equations. | Fourier transformations. | Optical data processing–Mathematical models. | Computer-aided design–Mathematical models.
Classification: LCC QC174.26.W28 B464 2020 (print) | LCC QC174.26.W28 (ebook) | DDC 515/.353–dc23
LC record available at https://lccn.loc.gov/2020030605
LC ebook record available at https://lccn.loc.gov/2020030606

Cover Design: Wiley
Cover Image: © tatianazaets/Getty Images

Set in 9.5/12.5pt STIXTwoText by SPi Global, Chennai, India

SKY10024058_011421

To Corrie van Ravesteijn

Contents

Preface *xiii*
Introduction *xv*
About the Companion Website *xxv*

Part I Forward Scattering Problem *1*

1 **Scalar Waves** *3*
1.1 Three-Dimensional Scattering by a Bounded Contrast *5*
1.1.1 Radiation in an Unbounded Homogeneous Embedding *5*
1.1.2 Scattering by a Bounded Contrast *6*
1.1.3 Domain-Integral Equation in the s-Domain *8*
1.1.4 The Born Approximation in the s-Domain *8*
1.1.5 Contrast-Source Integral Equation *9*
1.2 Two-Dimensional and One-Dimensional Scattering *9*
1.3 Numerical Solution of the Integral Equations (1D, 2D, 3D) *11*
1.4 Matlab Input and Output Functions *17*
1.5 Matlab Code for Field Integral Equations (1D, 2D, 3D) *22*
1.6 Matlab Code for Contrast-Source Integral Equation *30*
1.6.1 Performance Analysis *37*
1.6.2 Matlab Built-in Functions for Iterative Solution of the Contrast-Source Integral Equation *41*
1.7 Time-Domain Solution of Contrast-Source Integral Equation *42*
1.A Representation for Homogeneous Green Functions *52*
1.A.1 1D Green Function *52*
1.A.2 2D Green Function *54*
1.A.2.1 Cylindrical Polar Coordinates *55*
1.A.3 3D Green Function *56*
1.A.3.1 Spherical Polar Coordinates *58*
1.B Scattering by a Simple Canonical Configuration *58*
1.B.1 1D Scattering by a Slab *58*
1.B.2 2D Scattering by a Circular Cylinder *60*
1.B.3 3D Scattering by a Sphere *62*
1.C Matlab Codes for Scattering by Canonical Objects *64*
1.C.1 Matlab Code for Slab (1D) *64*
1.C.2 Matlab Code for Circular Cylinder (2D) *65*

1.C.3 Matlab Code for Sphere (3D) *70*
1.C.4 Scattered-Field Computations Canonical Objects *76*

2 Acoustic Waves *79*
2.1 Three-Dimensional Scattering by a Bounded Contrast *82*
2.1.1 Radiation in an Unbounded Homogeneous Embedding *82*
2.1.2 Scattering by a Bounded Contrast *83*
2.1.3 Contrast Source Domain Integral Equation *86*
2.1.4 Numerical Solution and Operators Involved (1D, 2D, 3D) *87*
2.1.4.1 Analytic Differentiation *88*
2.1.4.2 Numerical Differentiation *90*
2.1.4.3 Conjugate Gradient Method *92*
2.1.4.4 Incident Acoustic Wave Field *95*
2.1.4.5 Scattered Acoustic Wave Field *95*
2.1.4.6 Weak Form of the Spatial Derivative of the Green Function *95*
2.2 Matlab Codes Integral Equations: Pressure and Particle Velocity *96*
2.3 Single Integral Equation in Terms of Contrast in Wave Speed and Gradient of Mass Density *112*
2.3.1 Contrast Source Formulation *114*
2.3.2 Conjugate Gradient Iterative Solution and Operators Involved *115*
2.3.2.1 Analytic Differentiation *115*
2.3.2.2 Numerical Differentiation *115*
2.3.2.3 Scattered Acoustic Wave Field *118*
2.4 Matlab Codes Integral Equations: Wave Speed and Gradient of Mass Density *119*
2.4.1 Performance Analysis *130*
2.5 Solution of Integral Equation: Interface Contrast Sources *132*
2.5.1 Contrast-Source Integral Equation *133*
2.5.2 Numerical Solution of Interface Integral Equations (2D) *134*
2.6 Numerical Solution Integral Equations: Volume and Interface Contrast Sources *136*
2.6.1 Discrete Representations in 3D *136*
2.6.2 Discrete Representations in 2D *141*
2.6.3 Discrete Representations in 1D *142*
2.6.4 Conjugate Gradient Iterative Solution and Operators Involved *142*
2.6.4.1 Scattered Acoustic Wave Field *144*
2.7 Matlab Codes Integral Equations: Volume and Interface Contrast Sources *147*
2.7.1 Performance Analysis *158*
2.7.2 Matlab BiCGSTAB Built-in Functions for Iterative Solution of the Contrast Source Integral Equation *160*
2.8 Time-Domain Solution of Contrast Source Integral Equation *163*
2.A Scattering by a Simple Canonical Configuration *170*
2.A.1 1D Scattering by a Slab *170*
2.A.2 2D Scattering by a Circular Cylinder *172*
2.A.2.1 No Contrast in Wave Speed *174*
2.A.3 3D Scattering by a Sphere *177*
2.A.4 Scattered-Field Computations Canonical Objects *179*

3 Electromagnetic Waves *181*
3.1 Three-Dimensional Scattering by a Bounded Contrast *184*
3.1.1 Radiation in an Unbounded Homogeneous Embedding *184*
3.1.2 3D Incident Electromagnetic Field *186*
3.1.3 2D Incident Electromagnetic Field *187*
3.1.4 Scattering by a Bounded Contrast *187*
3.2 Contrast Source (E-field) Integral Equations: Permittivity Contrast Only *190*
3.2.1 2D Contrast Source (E-field) Integral Equations: Permittivity Contrast Only *191*
3.2.2 Conjugate Gradient Iterative Solution and Operators Involved *192*
3.2.2.1 Conjugate Gradient Method *193*
3.2.2.2 Scattered Electromagnetic Wave Field *194*
3.2.3 Matlab Codes E-field Integral Equations: Permittivity Contrast Only *195*
3.2.3.1 Matlab BiCGSTAB Built-in Function *209*
3.3 E-field Equation for Volume and Interface Contrast Sources: Permittivity Contrast Only *211*
3.3.1 Numerical Solution with Volume and Interface Contrast Currents: Permittivity Contrast Only *214*
3.3.1.1 Discrete Representations in 3D *215*
3.3.1.2 Discrete Representations in 2D *219*
3.3.2 Iterative Solution and Operators Involved *219*
3.3.3 Matlab Codes E-field Integral Equations: Volume and Interface Contrast Sources *222*
3.3.4 Performance Analysis *228*
3.4 Contrast Source Integral Equations for Both Permittivity and Permeability Contrast *237*
3.4.1 Numerical Solution and Operators Involved *238*
3.4.1.1 Scattered Electromagnetic Wave Field *240*
3.4.2 Matlab Codes Integral Equations for Both Permittivity and Permeability Contrast: Special Case of Zero Wave-Speed Contrast *242*
3.5 E-field Integral Equation for Zero Wave-Speed Contrast *249*
3.5.1 Numerical Solution for Interface Contrast Source Integral Equation: Zero Wave-Speed Contrast *253*
3.5.1.1 Discrete Representations in 3D *253*
3.5.2 Matlab Codes Integral Equations for Zero Wave-Speed Contrast *256*
3.6 Time-Domain Solution of Contrast Source Integral Equation *268*
3.A Scattering by a Simple Canonical Configuration *279*
3.A.1 2D Scattering by a Circular Cylinder *280*
3.A.1.1 TM Green Function of the Circular Cylinder *280*
3.A.1.2 Electromagnetic Field Strengths *282*
3.A.1.3 Matlab Codes for Circular Cylinder (2D) *284*
3.A.2 3D Scattering by a Sphere *292*
3.A.2.1 TM Green Function of the Sphere *293*
3.A.2.2 Electromagnetic Field Strengths *295*
3.A.2.3 Matlab Codes for Sphere (3D) *298*
3.A.3 Scattered-Field Computations Canonical Objects *306*

Part II Inverse Scattering Problem *307*

4 Scalar Wave Inversion *309*
4.1 Notation *309*
4.2 Synthetic Data *311*
4.3 Nonlinear Inverse Scattering Problem *320*
4.4 Inverse Contrast Source Problem *322*
4.5 Contrast Source Inversion *323*
4.5.1 Discretization of Green's Operators and Norms *324*
4.5.2 Updating the Contrast Sources *325*
4.5.2.1 Gradient Directions *325*
4.5.2.2 Calculation of the Step Length *327*
4.5.3 Updating the Contrast *327*
4.5.4 Initial Estimate *328*
4.5.5 Matlab Codes for the CSI Method *329*
4.6 Multiplicative Regularized Contrast Source Inversion *339*
4.6.1 Regularization Function for the Contrast Update *339*
4.6.2 Updating the Contrast with Multiplicative Regularization *341*
4.6.3 Numerical Implementation of the Regularization *342*
4.6.4 Numerical Solution of Regularization Equation *343*
4.6.5 Matlab Codes for the MRCSI Method *344*
4.7 CSI Method for Reconstruction of a Few Parameters *355*
4.7.1 Gauss–Newton Method for the Contrast Update *358*
4.7.2 Matlab Codes for the Gauss–Newton Type Contrast Updating *359*
4.8 CSI Methods for Phaseless Data *366*
4.8.1 CSI Method for Measured Intensity Data *367*
4.8.2 CSI Method for Measured Amplitude Data *368*
4.9 Gauss–Newton Inversion *369*
4.9.1 Matlab Codes for Gauss–Newton Inversion *372*

5 Acoustic Wave Inversion *377*
5.1 Notation *377*
5.1.1 Compressibility Contrast Only *378*
5.1.2 Mass-density Contrast Only *379*
5.2 Synthetic Data for Zero Compressibility Contrast *379*
5.3 Mass-density Contrast Source Inversion *386*
5.3.1 Updating the Contrast Sources *391*
5.3.1.1 Gradient Directions *391*
5.3.1.2 Calculation of the Step Length *392*
5.3.2 Updating the Contrast *393*
5.3.3 Initial Estimate *393*
5.3.4 Updating the Contrast with Multiplicative TV Regularization *394*
5.3.5 Matlab Codes for the Acoustic MRCSI Method *395*
5.4 Mass-density Interface Model for Zero Wave-Speed Contrast *404*
5.4.1 Synthetic Data for Zero Wave-speed Contrast *409*
5.5 Mass-density Interface Contrast Source Inversion *416*
5.5.1 Updating the Interface-contrast Sources *417*

5.5.1.1 Gradient Directions *417*
5.5.1.2 Calculation of the Step Length *418*
5.5.2 Updating the Interface Contrast *418*
5.5.3 Initial Estimate *419*
5.5.4 Regularization by Resetting Small Interface-contrast Variation to Zero *420*
5.5.5 Matlab Codes for the MICSI Method *420*
5.5.6 Kirchhoff Type of Approximations *430*

6 Electromagnetic Wave Inversion *439*
6.1 Notation *439*
6.1.1 Permittivity Contrast Only *440*
6.2 Synthetic Data for Zero Permeability Contrast *441*
6.3 Data Modeled with Volume and Interface Contrast Sources *453*
6.4 Electromagnetic Contrast Source Inversion *459*
6.4.1 Updating the Contrast Sources *460*
6.4.1.1 Gradient Directions *461*
6.4.1.2 Calculation of the Step Length *462*
6.4.2 Updating the Contrast *462*
6.4.3 Initial Estimate *463*
6.4.4 Updating the Contrast with Multiplicative TV Regularization *464*
6.4.5 Matlab Codes for the MRCSI Method *464*
6.5 Electromagnetic Gauss–Newton Inversion *476*
6.5.1 Matlab Codes for Gauss–Newton Inversion *477*
6.6 Electromagnetic Defects Metrology *486*
6.6.1 Data with Phase Information *490*
6.6.2 Phaseless Data *493*
6.6.3 Focused Data *493*

Matlab Scripts *497*
References *499*
Biography *503*
Index *505*

Preface

As Professor at Delft University of Technology in the period from 1981 to 2014, I learned that university students have great difficulty implementing numerical methods to solve the forward and inverse scattering problem. PhD students need at least one year to understand the theory of wave-field scattering in complex configurations and to cast the problem into an effective numerical formulation. Currently, the scientific community is putting increasing pressure not only to publish the theory and the numerical results but also to store codes and numerical data. In current practice, many scientific works explain the physical problem and the relevant mathematics, and list numerical codes in printed form in an appendix or online on the World Wide Web. I believe that in textbooks, theoretical formulation, numerical discretization, and digital coding should be presented as a coherent package.

Therefore in 2015, I started to work on the book *Forward and Inverse Scattering Algorithms Based on Contrast Source Integral Equations*. The book provides a link between the fundamental scattering theory, as educated at universities, and its discrete counterpart. Next, I convert the discrete formulation into numerical codes and I include them in the book. This is of extreme importance for the development of graduate education, doctoral, and post-doctoral research. The examples are made as simple as possible and serve as canonical problems throughout the book. It can be used either as a starting point or as a benchmark for existing or new methods. I chose Matlab[1] as a software platform because more and more universities use Matlab in teaching. Students are trained to use Matlab as an interactive tool not only to perform the assigned exercises but also to understand the essence of the mathematical problem. It stimulates creative thinking.

The goal I set to myself is that every theoretical problem is formulated in such a way that all physical configurations are discretized in a uniform way. The associated computer code must be executed as quickly as possible with minimal storage requirements, while remaining user-friendly. I consider it the most important requirement. Students should recognize the common thread running through the various Matlab codes. In this book, I organized a Matlab structure so that every code listing does not exceed the end of a page. The description of a Matlab function is paged as close as possible to the theoretical and numerical formulations, enabling easy comparison of the two formulations. Furthermore, the choice for Matlab facilitates a simple conversion to C and C++, also for calling Matlab from C or C++.

The codes for solving the forward scattering problem have been developed exclusively to serve the purpose of the present book. All forward codes are written in such a way they can be used for 1D, 2D, and 3D configurations. To compare the numerical results with analytical results, this book only discusses the canonical examples of a homogeneous scatterer, namely the slab in 1D, the circle

1 Throughout this book Matlab refers to MATLAB® as registered trademark of The MathWorks, Inc.

cylinder in 2D, and the sphere in 3D. For these types of scatterers, the analytical solution for the wave-field presentations in terms of plane waves (1D), radial waves (2D), and spherical waves (3D) are given with the associated codes.

A similar numerical strategy is used for the inverse scattering problem. In the absence of real measurement data, for software development purposes, synthetic data must be used, but with a different calculation model than the discrete numerical model. For that reason, I use the analytical data determined by the canonical objects.

The user of this book can easily change the input configuration to an inhomogeneous scatterer and compare the inversion results with those obtained with a different solution approach, for example a finite-difference technique. I hope that this book will help to reduce the time and energy that students now spend to be introduced in forward and inverse scattering theory and especially in its numerical implementation; often more than half of their research project.

I am very grateful to Dr. Rob Remis who read the entire manuscript for many valuable discussions and suggestions. Thanks also to Dr. Joost van der Neut, who worked with the Matlab codes; his comments resulted in many improvements. A special note of thanks is given to my friend and colleague, professor Jacob Fokkema, who acted as a sounding board and encouraged me to finish the manuscript and publish it as it is. Furthermore, I am indebted to Dr. Aria Abubakar and Dr. Tarek Habashy from Schlumberger-Doll Research. We have been working together for about 20 years and have jointly published around 50 articles on topics in this book; without Aria and Tarek this book would not exist.

May 2020

Peter M. van den Berg
Delft University of Technology

Introduction

The purpose of this book is to describe efficient methods to compute wave fields in an inhomogeneous medium. We consider wave fields of increasing complexity, viz., scalar waves, acoustic waves, and electromagnetic waves. The material media is considered to be isotropic. For scalar waves, the medium is a characterized by a single parameter, the wave speed c. The scalar wave field is represented by the scalar quantity u. For acoustic waves, we deal with two medium parameters: the mass density ρ and compressibility κ. The wave field is represented by the scalar pressure p and the particle-velocity vector v. The acoustic wave speed is given by $c = (\rho\kappa)^{-1/2}$. For electromagnetic waves, the electric permittivity ε and magnetic permeability μ are the two medium parameters, while the electric-field vector E and the magnetic-field vector H represent the wave field. The electromagnetic wave speed is given by $c = (\varepsilon\mu)^{-1/2}$.

The subject area complies with the two-volume book on *Scattering*, edited by Pike and Sabatier [38]. This book yields an overwhelming theoretical overview of *Scattering and Inverse Scattering in Pure and Applied Science*, including some numerical results. Further, a comprehensive overview of the set of partial differential equations that govern the various wave-field descriptions can be found in De Hoop's *Handbook of Radiation Scattering of Waves* [14]. The emphasis is mainly on general principles and theorems that can serve to check numerical results rather than on highly specialized configurations for which more or less complicated analytical answers can be obtained. The system of differential equations customarily serves as the starting point for the computation of the wave field via a numerical discretization procedure applied to the pertaining differential equations, such as finite-difference and finite-element methods. The advantage of these *local methods* is that it leads to the inversion of a linear set of equations where the coefficient matrix is sparse. Presently, the finite-difference technique is very popular, because it admits an easy numerical implementation with the help of staggered rectangular grids in space.

De Hoop [14] replaced the partial differential equations by so-called source-type integral equations. The integrands are the products of known Green functions and unknown contrast sources. For *scalar waves*, the Green function is the field from a point source in an infinite domain with a suitable chosen wave speed c_0. Evidently, the simplest choice is a homogeneous medium with constant wave speed. It is assumed that inside a bounded (scattering) domain $\mathbb{D}_{sct}$, the wave speed c differs from c_0, while outside this domain the wave speed is equal to c_0. Therefore, $\mathbb{D}_{sct}$ represents a contrasting object that disturbs the wave field in an embedding medium with wave speed c_0. For convenience, the total wave field and the incident wave field are introduced. The total wave field is the actual wave field in whole space with wave speed c, and the incident wave field is the wave field in whole space with wave speed c_0. The difference between total and incident wave field is denoted as the scattered wave field. The contrast sources inside the scattering domain $\mathbb{D}_{sct}$ generate the scattered wave field in the whole space. The contrast source is defined as the product of the contrast function and the wave field. In fact, this relation serves as a constitutive relation. The contrast function is given by $1 - (c_0/c)^2$, and its nonzero value defines the scatterer. The contrast source vanishes outside the domain $\mathbb{D}_{sct}$. Note that at this stage of the analysis, the contrast sources are unknown, since the wave field is yet unknown. When we take the observation points inside the scatterer, we arrive at an integral equation for the unknown wave field. Once the wave field is determined, we calculate the contrast-source distribution. Finally, the source-type integral representation provides the wave field in whole space. The major advantage is that the spatial support of the set of integral equations is limited to points where the contrast function does not vanish.

In this book, we consider a slightly different approach. Instead of solving the *field integral equation*, we multiply the field equation by the contrast, and we replace the product of field and contrast by the contrast source, both in the integrand and outside the integrand. We then end up with a *contrast-source integral equation*, where now the contrast source is the unknown function.

In general, the integral equation has to be solved with the aid of numerical methods. However, the integrand of both the field integral equation and the contrast-source integral equation is not defined at points where the Green function is singular. Usually, Galerkin methods are used to overcome this problem and convert the continuous integral equation into a discrete problem. In mathematics, it is referred to as a weak formulation of the continuous operator equation. In principle, it leads to the inversion of a linear set of equations where the coefficient matrix is fully filled. Obviously, this solution method is a *global method*.

This book is intended to provide an easy entry into the numerical methods for an effective solution of the contrast-source integral equations. Similar to the use of rectangular grids in the finite-difference technique, we also employ rectangular grids. On each grid point, our intention is to obtain a global wave function, where its value is replaced by its average over a small symmetrical domain $\mathbb{D}_\delta$ around this grid point. This is achieved automatically when we replace the Green function by its mean (integrated) value over $\mathbb{D}_\delta$. Assuming that the width of each domain is small enough, we replace the value of the embedding medium inside a domain by its value at the grid point. The mean of the Green function, also denoted as the weak form of the Green function, can be calculated analytically. In this book, it is shown that this weak formulation results into second order accurate evaluations.

Since the embedding medium is homogeneous, the Green function only depends on the spatial distance between observation and integration point. The discrete integral operator becomes a discrete convolution and, for known contrast sources, it is calculated efficiently with a fast Fourier technique (FFT). Since the contrast source is unknown, the integral equation is conveniently solved with iterative methods.

A similar analysis applies in principle to *acoustic and electromagnetic waves*, but its becomes more complex due to the vector character of these wave fields. In addition, in the contrast-source integral equations, we have an extra term with spatial derivatives operating on the Green function.

Part I Forward Scattering Problem

In Part I of our book, we discuss the forward scattering problem based on the contrast-source integral equations. In this case, the contrast function is known. For a given incident wave, we show that the contrast sources inside the scatterer are the fundamental unknowns. After solving the integral equations for the contrast sources, the scattered wave field follows directly from its representation. The forward scattering problem is linear and uniquely solvable. In Chapters 1–3, we outline the theoretical formulation and numerical implementation for scalar, acoustic and electromagnetic waves, respectively.

Chapter 1

In this chapter, we discuss **scalar waves**. We start with the wave equation in space and time. We apply a Laplace transformation with respect to time, and then our analysis is performed in the complex s-domain. Causality in time is taken into account by requiring that Re(s) is positive. In Section 1.1, the theory of 3D scattering of waves by a bounded contrast is discussed. First, the radiation from a known source distribution is calculated, resulting into a source-type integral representation. This is the expression for the known incident field. Second, for an unknown contrast-source distribution in a bounded domain, the scattered field is written as a contrast-source integral representation. Third, the contrast-source integral equation is obtained by confining the position of observation to the interior of the scatterer. The unknown contrast-source distribution is the product of the wave-speed contrast and the interior field. In Section 1.2, we present 2D and 1D versions of the contrast-source representations. In Section 1.3, we discuss the numerical solution of the domain integral equations, for 1D, 2D, and 3D configurations. We define the mean of each wave function and we give the expressions of the weak form of the Green function, for all points in 1D, 2D, and 3D, respectively. We show that the discretized integral operators are circular convolutions in space, which can be carried out efficiently using FFTs. We discuss how the Conjugate Gradient (CG) method solves the discrete system of equations in an iterative way. In Section 1.4, we describe the Matlab input and output functions. The computations are carried out for a single frequency f of operation and common time factor $\exp(-i2\pi ft)$. In order to deal with causal signals, we assign a small positive real part of the Laplace parameter, viz. $s = 10^{-16} - i2\pi f$. In Sections 1.5 and 1.6, we present Matlab codes for solving the field integral equation and for solving the contrast-source integral equation, respectively. A performance analysis shows that the contrast-source method outperforms the field method. For the contrast-source method, it is further shown that the use of the BiCGSTAB iterative method reduces the number of iterations by almost a factor of two. In Section 1.7, we describe the contrast-source solution in space time. We solve the contrast-source distributions for a number of discrete frequencies, and we use an inverse Fourier transform to obtain the results in the time domain. Snapshots of the wave propagation in a discrete model of a circular cylinder are shown. In the appendix of Chapter 1, we derive analytical expressions for the 1D, 2D, and 3D Green functions. These expressions play a fundamental role in any analytical wave-field analysis. We present the analytical solution for the wave fields in the canonical configurations, namely the 1D scattering by a slab, the 2D scattering by a circular cylinder, and the 3D scattering by a sphere. The associated Matlab code can also be used to check the convergence of series of radial wave functions (2D) and spherical wave functions (3D).

Chapter 2

This chapter is devoted to **acoustic waves**. We start with the acoustic wave equations in space and time. In the complex s-domain, we obtain a set of four equations for the acoustic pressure and the three components of the particle velocity. In Sections 2.1 and 2.2, the theory of the 3D scattering by bounded acoustic contrasts is discussed and the corresponding Matlab codes are given, with simplifications for the 1D and 2D cases. We define the compressibility and the mass-density contrast. The scalar compressibility-contrast source is defined as the product of compressibility contrast and pressure field. The mass-density contrast source is a vector quantity and is defined as the product of mass-density contrast and particle-velocity field. The source-type integral representations and the contrast-source integral equations are derived. The latter is a set of four contrast-source equations for the four types of contrast sources. When the mass-density contrast vanishes, this system of integral equations reduces to a single integral equation for the compressibility contrast only. This integral equation is identical to the one discussed in Chapter 1. In the full set of integral equations, partial derivatives of the Green functions are present. The consistent theme throughout the discretization method is that with finite spatial discretization, we cannot model the spatial high frequencies. Therefore, special attention is paid to the discretization of spatial derivatives on the Green function. In Sections 2.3 and 2.4, we show that the integral representation for the pressure field can also be written in terms of the wave-speed contrast and the gradient of the mass density. For continuously varying mass density, the corresponding integral equations including some numerical implementations are presented. In Sections 2.5–2.7, we discuss how these integral equations can be adjusted for a discontinuous mass density. The result is that the pressure-field representation is a superposition of a domain integral over the scattering domain, in which the wave-speed contrast sources occur, and a surface integral representation over all interfaces where the mass density jumps. In these interface integrals, interface contrast sources are defined that represent the jumps of mass density. When the point of observation is on an interface, the normal derivative of the Green function is not defined, and we introduce the Cauchy principle value of the relevant integral. To arrive at a system of integral equations, we take the observation point in the interior of the scatterer, both outside all interfaces and on an interface. Using a discretization of a 3D regular grid, we obtain a set of four linear equations. The observations points are located at the centers of the subdomains and at the three interface points that are symmetrically positioned around the center points. In Section 2.8, we describe the contrast-source solution in space time. We solve the contrast-source distributions for a number of discrete frequencies, and we use an inverse Fourier transform to obtain the results in the time domain. Snapshots of the acoustic wave propagation in a discrete model of a circular cylinder are shown. In the appendix of Chapter 2, we present the analytical solution for the acoustic wave fields in the canonical configurations, namely the 1D scattering by a slab, the 2D scattering by a circular cylinder, and the 3D scattering by a sphere. The corresponding Matlab code can also be used to check the convergence of the series of radial wave functions (2D) and spherical wave functions (3D).

Chapter 3

In this chapter, we discuss **electromagnetic waves**. We start with Maxwell's equations in the complex s-domain. In Section 3.1, the theory of the 3D scattering by a bounded electromagnetic contrast is discussed. We define the electric-permittivity contrast and the magnetic-permeability contrast. The 3D electromagnetic field integral equations are presented as a starting point. For both the contrast in permittivity and in permeability, we are dealing with a set of six coupled integral equations for the unknown electric-field vector and unknown magnetic-field vector. In principle, this system of equations allows for a numerical solution for the problem of electromagnetic scattering. In Section 3.2, for zero permeability contrast, we derive the contrast-source type of integral representation for the electric-field vector. The electric-contrast source is a vector quantity and is defined as the product of the permittivity contrast and the electric-field vector. The contrast-source integral equations are derived, which leads to a set of three electric-contrast-source equations for the three unknown components of the electric contrast source. The associated Matlab codes and the simplification for the 2D case are included. In Section 3.3, we then show that it is possible to rewrite the integral representation for the electric wave field as a superposition of an integral over the scattering domain where the permittivity is continuous, and integrals over the interfaces where the permittivity is discontinuous. In the discrete problem, on a rectangular grid, we obtain a set of six integral equations, in which the electric contrast sources occur at the center of each subdomain and the jump of the permittivity takes place on each interface element. In Section 3.4, for both contrast in permittivity and permeability, we discuss the contrast-source integral equations in terms of volume- and interface-contrast sources. In Section 3.5, the special case of vanishing wave-speed contrast is discussed. Then, the electric permittivity and magnetic permeability are inverses of each other. In Section 3.6, we describe the contrast-source solution in space time. We solve the contrast-source distributions for a number of discrete frequencies, and we use an inverse Fourier transform to obtain the results in the time domain. Snapshots of the electromagnetic wave propagation in a discrete model of a circular cylinder are shown. Strong refraction phenomena occur during the propagation along the interface. We also show the wave propagation for vanishing wave-speed contrast. As expected, there is no refraction along the interface, only some interference phenomena occur. In the appendix of Chapter 3, we present the analytical solution for the electromagnetic wave fields in the canonical configurations, namely, the 2D scattering by a circular cylinder and the 3D scattering by a sphere. For a radially oriented electric dipole source in our canonical configuration, we derive the electric Debye potential as spatial derivatives of a scalar potential. We expand it in terms of radial wave functions (2D) and spherical wave functions (3D). The corresponding Matlab code can also be used to check the convergence of the series of radial wave functions (2D) and spherical wave functions (3D).

Part II Inverse Scattering Problem

In Part II, we discuss the inverse scattering problem. It is the problem of determining the material properties of a scattering object based on measured data. In fact these measurements report how this object scatters the incident wave field. The theoretical approach is the inverse of the forward scattering problem that determines how radiation is scattered based on the characteristics of the scatterer. For a suite of incident waves and wave-field measurements, we show that both the contrast function and the contrast sources in the object are the fundamental unknowns. The inverse scattering problem is nonlinear and often not uniquely solvable. In Chapters 4–6, for the single-frequency case, we use synthetic data with respect to a set of incident wave fields. We discuss the 3D theoretical formulation. The description of the numerical discretization and implementation in computer codes are in 2D. Using the 3D forward codes of Part I, the implementation of 3D inversion codes is straightforward; however, it requires much more computer storage and computer time. In this book, we restrict ourselves to the reconstruction of a single contrast function. For scalar waves, it is the wave-speed contrast; for acoustic waves, it is the mass-density contrast; for electromagnetic waves, it is the electrical permittivity.

Chapter 4

In this chapter, we consider **scalar wave inversion**. In Sections 4.1 – 4.4, we discuss the generation of synthetic data and the inverse contrast-source problem. In Section 4.5, the contrast source inversion (CSI) is introduced. The CSI method minimizes a cost functional that consists of two terms. The first term measures the residual error in the data equation defined in the domain $\mathbb{M}$, while the second term measures the residual error in the object equation of domain $\mathbb{D}$. Domain $\mathbb{D}$ includes the unknown scatterer $\mathbb{D}_{sct}$. The first term is linear in the contrast source w, and the second term is nonlinear in the material contrast function w and χ. To circumvent this nonlinearity, the CSI method employs an alternate updating scheme for w and χ. In each iteration, we first update the contrast sources w using a conjugate gradient step for fixed χ and second we update the contrast χ for fixed w. In Section 4.6, we discuss the regularization strategy. Although the inclusion of the object equation in the CSI cost functional can be considered as a physical regularization of the ill-posed data equation, the inversion results may be improved by taking *a priori* information about the contrast profile. The standard method is the addition of a regularization term in the cost functional. In this book, we discuss the multiplicative regularized contrast source inversion (MRCSI) method in which regularization is included as a multiplicative factor in the cost functional. We employ a factor that minimizes the total variation (TV) of the contrast function. This is how we include *a priori* information. In the MRCSI cost functional, the nonlinear minimization of the TV factor is carried out by a conjugate gradient step. To further simplify this procedure, we linearize the related Euler–Lagrange equation, which is solved by a single Jacobi iteration. In Section 4.7, we consider the case that the contrast χ is a function of a few parameters. Then, the regularization is not necessary, and we use the analytical contrast update of the CSI method as starting value for a single iteration of a Gauss–Newton method. The updates for the contrast parameters follow from a linear system of equations for these parameters. In Section 4.8, we show how the CSI method is extended for phaseless data, and we derive the gradient of the data equation. For intensity data, the back-projection operator is based on the known phase information from the previous CSI iteration. The step length of the CG iteration is obtained as the root of a cubic equation. However, for amplitude data, the step length is found as the root of a transcendental equation. In Section 4.9, we again consider the Gauss–Newton method for a contrast that can be described by a few parameters, but we operate in a different way. Let us assume that an approximation of the parameter set is known from the previous iteration. Then, the contrast function χ is approximately known, and we solve the forward problem to arrive at an approximation for the contrast sources. These contrast sources are substituted in the data equation, and we compute its residual. Carrying out a linear Taylor approximation of this residual, we arrive at a system of linear equations for the updates of the contrast parameters.

Chapter 5

In this chapter, we discuss **acoustic wave inversion**. In Sections 5.1 and 5.2, we formulate the contrast-source integral equations for both contrast in compressibility and mass density. To keep the overview of the inverse problem as simple as possible, we assume a vanishing compressibility contrast. For a nonzero mass-density contrast, we discuss the generation of synthetic data and the inverse contrast-source problem. In Section 5.3, the acoustic version of the MRCSI method is presented in detail, including the Matlab codes. In Sections 5.4 and 5.5, we consider the special case that the wave-speed contrast vanishes. From Chapter 2, we have learned that after discretizing the space in rectangular subdomains with constant mass density, we arrive at integral representations in which only integrals over the interfaces between adjacent subdomains occur. In these interface-integral representations, only interface contrasts and interface sources are the actual unknowns. A modified version of the CSI method is proposed, in which the interface contrast and the interface sources are updated in an alternating manner. Without taking any measures, the reconstructed interface contrast shows ringing artifacts. The updating procedure is regularized by setting the small variations to zero. This is consistent with the assumption of piecewise homogeneous wave-speed contrast. This inversion method is denoted as the mass-density interface contrast source inversion (MISCI) method. In order to reduce computation time, the usability of a Kirchhoff type of approximation for a flat piece of interface is discussed.

Chapter 6

In this chapter, we discuss **electromagnetic wave inversion**. In Sections 6.1 – 6.3, we formulate the contrast-source integral equations for both contrast in electric permittivity and magnetic permeability. To keep the overview of the inverse problem as simple as possible, we assume a vanishing permeability contrast. In Section 6.4, the electromagnetic version of the MRCSI method is explained and the Matlab codes are included. In Section 6.5, we discuss the electromagnetic version of the Gauss–Newton method for a permittivity contrast that can be described by a few parameters. As forward solver in this inversion method, we use either (i) the integral equation for volume contrast sources or (ii) the integral equation for volume-contrast sources and interface-contrast sources. We show that method (ii) converges better than method (i), although at the expense of increased computation time. In many technical applications, there is an increasing interest for electromagnetic inspection of small defects in a material structure with a known shape and composition. It is an almost impossible job to subtract any information of the location of the defects in the data domain. Therefore, in Section 6.6, we focus the change of scattered field data due to the presence of the defects, by back-projection of the change of the data from the source/receiver domain into the spatial domain of the scatterer $\mathbb{D}$. As focussing operators, we take the inverses of the scalar Green function in the data operator and the Green function in the incident-field expression. This procedure leads to electric (dipole-type) contrast sources in the domain $\mathbb{D}$. The resolution is enhanced by plotting the divergence of these contrast sources.

Notation

To register the position, we use a Cartesian reference frame with an origin O and three base vectors $\{\boldsymbol{i}_1, \boldsymbol{i}_2, \boldsymbol{i}_3\}$ that are mutual perpendicular oriented and each have a unit length. The property that each base vector specifies geometrically a length and an orientation makes it a vectorial quantity, or a vector; notationally, vectors will be represented by bold-face symbols.

Let $\{x_1, x_2, x_3\}$ denote the three numbers that are needed to specify the spatial position of an observer, then the vectorial position of the observer $\boldsymbol{x}$ is the linear combination $\boldsymbol{x} = x_1\boldsymbol{i}_1 + x_2\boldsymbol{i}_2 + x_3\boldsymbol{i}_3$. The numbers $\{x_1, x_2, x_3\}$ are denoted as the orthogonal Cartesian coordinates of the point of observation. The time coordinate is denoted by t. We employ the International System of Units (Système International d'Unités), abbreviated to SI.

In this book, we employ the summation convention as a shorthand notation to indicate the sum of products of arithmetic arrays. The arrays under consideration have either the same or different dimensions, but the bounds on their subscripts are all the same. The subscripts $\{i, j, k, l\}$ are then to be assigned the values 1, 2, and 3. The convention prescribes that to every repeated subscript in a product of two or more arrays, the values 1, 2, and 3 are assigned successively, while after each assignment, the result is added to the previous one.

In the two-dimensional (2D) space, the summation convention only applies for lowercase subscripts 1 and 2, while in the one-dimensional (1D) space, the summation convention is superfluous.

As regards the differentiation of a vector, two cases have to be distinguished: differentiation with respect to a parameter, e.g. the time coordinate, and differentiation with respect to the spatial (Cartesian) coordinates of the space in which the vector is defined.

Let $\boldsymbol{v}$ be a vector function and assume that $\boldsymbol{v}$ is a differentiable function of the time coordinate t. Let v_k denote the components of $\boldsymbol{v}$, then the derivative $\partial_t \boldsymbol{v}$ of $\boldsymbol{v}$ with respect to t is a vector whose components are given by $\partial_t v_k$.

Let $\boldsymbol{v}$ be a vector function and assume that $\boldsymbol{v}$ is a differentiable function of the spatial (Cartesian) coordinates $\{x_1, x_2, x_3\}$. Let v_l denote the components of $\boldsymbol{v}$, then for each k ($k = 1, 2, 3$), the derivative $\partial_k \boldsymbol{v}$ of $\boldsymbol{v}$ with respect to the spatial coordinate x_k is a vector function. For each k, its components

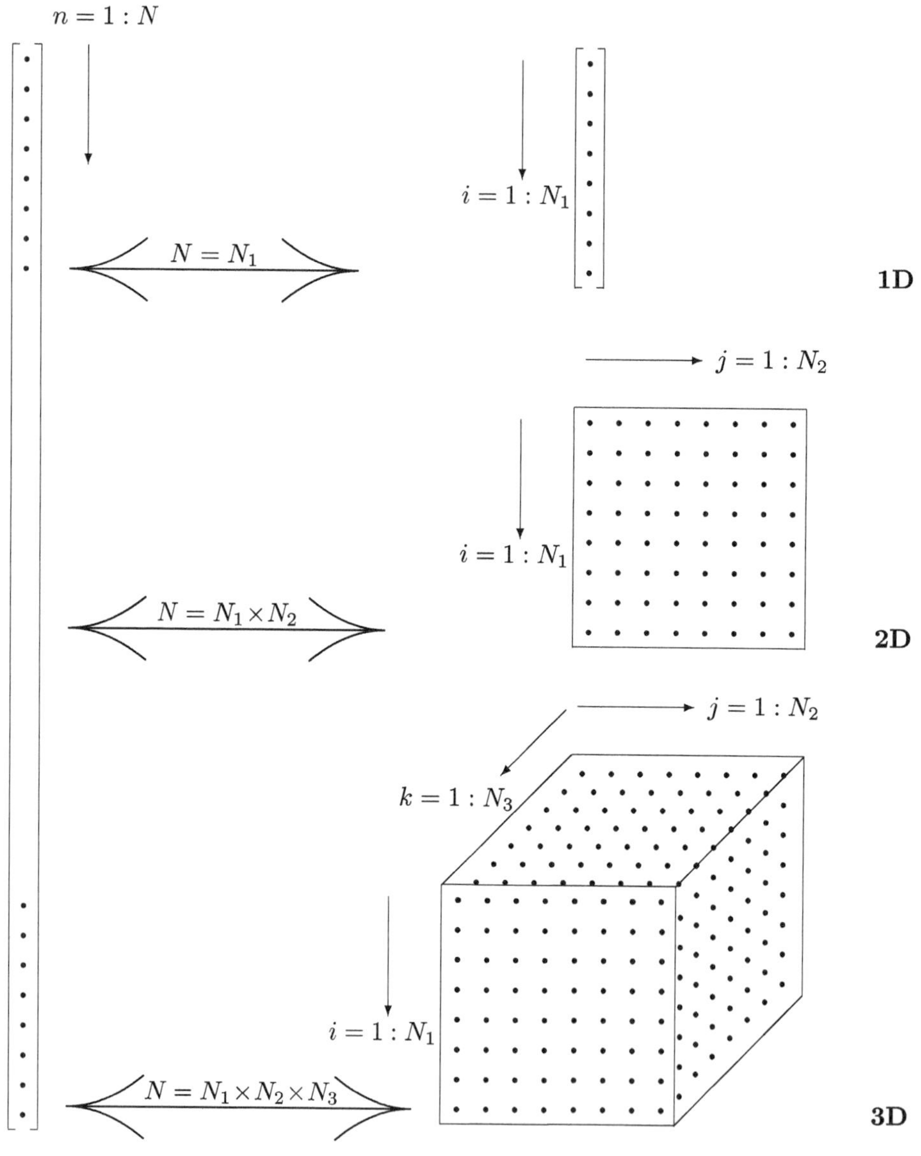

Figure I.1 The multidimensional numbering of spatial position, $x_{i,j,k}$, and its one-dimensional equivalent, x_n, $n = 1, 2, \ldots, N$.

are given by $\partial_k v_l$, where ∂_k denotes the partial derivative with respect to x_k. Derivatives of a higher order are defined in a similar manner.

The gradient of a scalar function ϕ of position $\boldsymbol{x}$ is introduced as grad $\phi = \partial_k \phi$ and it is a vector function of position $\boldsymbol{x}$. The divergence of a vector function $\boldsymbol{v}$ of position $\boldsymbol{x}$ is introduced as div $\boldsymbol{v} = \partial_k v_k = \partial_1 v_1 + \partial_2 v_2 + \partial_3 v_3$ and it is a scalar function of position $\boldsymbol{x}$.

The divgrad operator is denoted as $\partial_k \partial_k$, which stands for $\partial_1^2 + \partial_2^2 + \partial_3^2$. The graddiv operator is denoted as $\partial_j \partial_k$ and it is a tensor function of rank 2.

As far as the numerical formulation is concerned, a scalar function $\phi(\boldsymbol{x}) = \phi(x_1, x_2, x_3)$ is replaced by the array $\phi_{i,j,k} = \phi(x_{1,i}, x_{2,j}, x_{3,k})$, with $i = 1, 2, \ldots, N_1, j = 1, 2, \ldots, N_2$, and $k = 1, 2, \ldots, N_3$.

For a comprehensive formulation of the discretization both in 3D, 2D, and 1D, we introduce the 1D array ϕ_n, for $n = 1, 2, \ldots, N$, where $N = N_1 \times N_2 \times N_3$. This is consistent with the position numbers of the discretization grid generated by the Matlab function `ndgrid` and the lexicographical ordering of Matlab arrays as a single column vector, see Figure I.1. For example, a 3D array A stored in computer memory may be used either as a full 3D array by invoking $A(:,:,:)$ or as a column vector $A(:)$. In mathematics, this procedure is known as vectorization of a matrix A. It is denoted as vec(A), where the column vector is obtained by stacking the columns of the matrix on top of each other.

In effect we employ the notation $\boldsymbol{x}$ for the position vector, while its discrete counterpart is denoted as x_m, independently of the spatial dimension at hand. This notation facilitates a smooth transition from a continuum theory to a numerical notation, and as a consequence, a computer algorithm which follows the theory as well as possible. To denote the difference between two positions $\boldsymbol{x} - \boldsymbol{x}'$, we use the notation $\boldsymbol{x}_m - \boldsymbol{x}'_n$. When no confusion can occur, we take the simple notation $x_m - x'_n$. Note that for repeated subscripts m and n the summation convention does not hold.

About the Companion Website

This book is accompanied by a companion website:

www.wiley.com/go/vandenBerg/ScatteringAlgorithms

The Matlab codes for solving the forward and inverse scattering problems are designed solely to serve the purpose of this book. These codes are stored in a number of folders on this accompanying website. Each folder contains all the Matlab scripts and functions needed to solve a particular forward and/or inverse problem, in 1D, 2D or 3D space.

After downloading a folder, the user can run the codes in their own computing environment and customize them to their own needs.

In the Matlab Scripts chapter at the end of this book, we list the names of the scripts that initiate the calculation of a particular problem and the names of the folders on the Companion website where they are stored.

Part I

Forward Scattering Problem

The forward scattering problem consists of the determination of scattered wave fields based on knowledge of the incident fields and the material contrast χ in the scattering domain $\mathbb{D}_{sct}$. We are interested in the unknown contrast sources inside $\mathbb{D}_{sct}$ that generate the scattered wave field in whole space. This problem is linear, well-posed, and uniquely solvable. We formulate it in terms of contrast-source integral equations. In almost all situations encountered in practice, the integral equations can only be solved approximately with the aid of numerical techniques. How good an approximate solution is can only be quantified after a certain quantitative error has been chosen. For this reason, we recast this problem as an optimization problem of finding contrast sources that minimize the least-square error in the integral equations inside $\mathbb{D}_{sct}$. Once the contrast sources are solved, we can compute the scattered wave fields outside the scattering domain. In this book, we use canonical scattering configurations to validate the resulting scattered wave fields against the analytical results of the canonical configuration at hand. Then, we find ourselves in the ideal situation, that our canonical objects are modeled both analytically and numerically, but in a very different way. Moreover, the configurations are different, viz, an object with smooth boundaries in the analytical model and an object with staircase-shaped boundaries in the numerical model. We use the calculated analytical data for the canonical configurations to serve as synthetic data in the inverse scattering problem. In view of the inverse scattering problem, it is important to understand the essence of the forward scattering model. Hopcraft and Smith [24] made the statement: "*Any inverse theory is only as good as the forward scattering model from which it is derived.*" Data generated by a forward model must have enough sensitivity for small changes in the scattering configuration.

Forward and Inverse Scattering Algorithms based on Contrast Source Integral Equations, First Edition. Peter M. van den Berg.

Companion website: www.wiley.com/go/vandenBerg/ScatteringAlgorithms

1

Scalar Waves

In any problem associated with wave propagation, the wave speed depends on the actual medium parameters. Therefore, the simplest problem is that of a scalar wave that is spread by a change in wave speed. The theory of scalar waves goes back to d'Alembert and Euler, see e.g. [61] with references and discussions of their work in the mid-1700s. Although the wave propagation in inhomogeneous media is more complex, it is common practice to convey the fundamental concepts to the reader in the form of a scalar wave problem. In this case, the scalar amplitude u depends solely on a single medium parameter, the wave speed c.

The quantities, that describe this wave field, depend on position and on time. Their time dependence in the domain, where the source is acting, is impressed by the excitation mechanism of the source. The subsequent dependence on position and time is determined by propagation and scattering laws. The basic scalar wave equation for the wave field $u = u(\boldsymbol{x}, t)$ in a lossless medium is given by

$$\partial_k \partial_k u - c^{-2} \partial_t^2 u = -q, \tag{1.1}$$

where $c = c(\boldsymbol{x})$ is the wave speed and $q = q(\boldsymbol{x}, t)$ is the volume source density.

In those domains where the wave speed changes continuously with the position, the wave field is a continuously differentiable function of position and satisfies the wave equation (1.1). Through an interface, the wave speed may exhibit a jump discontinuity. Let S denote the interface and assume that S has everywhere a unique tangent plane. Let, further, $\boldsymbol{\nu} = \nu_k$ denote the unit vector along the normal to S such that upon traversing S in the direction of ν_k, we pass from the domain $\mathbb{D}_2$ to the domain $\mathbb{D}_1$, $\mathbb{D}_1$ and $\mathbb{D}_2$ being located at either side of S (Figure 1.1). Suppose that the wave speed jumps across S. In the direction parallel to S, the wave field still varies in a continuously differentiable manner, and therefore, the partial derivatives parallel to S give no problem in Eq. (1.1). The partial derivatives perpendicular to S, on the other hand, meet functions that show a jump discontinuity across S; these give rise to interface Dirac distributions (interface impulse functions) located on S. Such distributions, however, would be physically representative of interface sources located on S. In the absence of such sources, the absence of interface impulse functions in the partial derivative perpendicular to S should be enforced. The latter is done by requiring that these normal derivatives only meet functions that are continuous across S. This procedure leads to the continuity condition that u and $\nu_k \partial_k u$ are continuous across S, viz., for $\boldsymbol{x} \in$ S and $h > 0$, we have

$$\lim_{h \to 0} u(\boldsymbol{x} + h\boldsymbol{\nu}, t) = \lim_{h \to 0} u(\boldsymbol{x} - h\boldsymbol{\nu}, t), \tag{1.2}$$

$$\lim_{h \to 0} \nu_k \partial_k u(\boldsymbol{x} + h\boldsymbol{\nu}, t) = \lim_{h \to 0} \nu_k \partial_k u(\boldsymbol{x} - h\boldsymbol{\nu}, t). \tag{1.3}$$

Forward and Inverse Scattering Algorithms based on Contrast Source Integral Equations, First Edition. Peter M. van den Berg.

Companion website: www.wiley.com/go/vandenBerg/ScatteringAlgorithms

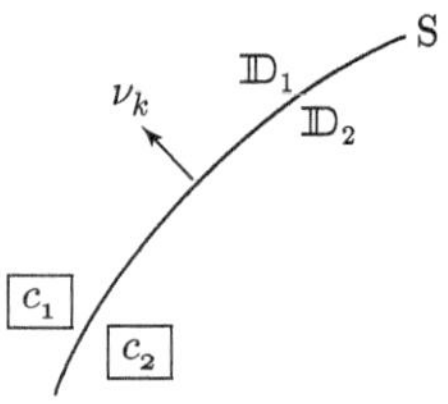

Figure 1.1 Interface between two media with different wave-speed properties.

In a large number of cases, we are interested in the behavior of the wave field in **linear configurations**. Mathematically, one can take advantage of this situation by performing a Laplace transformation with respect to time, over the interval $T = \{t \in \mathbb{R}; t > t_0\}$,

$$\hat{u}(\boldsymbol{x}, s) = \int_{t_0}^{\infty} \exp(-st)u(\boldsymbol{x}, t)\mathrm{d}t, \tag{1.4}$$

and considering the equations governing the wave field in the corresponding Laplace-transform domain or s-domain. In the s-domain relations, the time coordinate has been eliminated, and a wave-field problem in space remains in which the transformation parameter s occurs. Causality of the wave field is taken into account by taking $\mathrm{Re}(s) > 0$, and requires that all causal wave field quantities are analytic functions of s in the right half $0 < \mathrm{Re}(s) < \infty$ of the complex s-plane. For the sake of completeness, we allow a nonvanishing wave field to be present at $t = t_0$, although in the majority of cases we are interested in the causal wave field generated by sources that are switched on at the instant $t = t_0$, in which case the initial values of the wave field are considered zero. Integration by parts, we arrive at

$$\int_{t_0}^{\infty} \exp(-st)\partial_t u(\boldsymbol{x}, t)\mathrm{d}t = -u(\boldsymbol{x}, t_0)\exp(-st_0) + s\hat{u}(\boldsymbol{x}, s), \tag{1.5}$$

and subsequently

$$\int_{t_0}^{\infty} \exp(-st)\partial_t^2 u(\boldsymbol{x}, t)\mathrm{d}t = -[\partial_t u(\boldsymbol{x}, t_0) + su(\boldsymbol{x}, t_0)]\exp(-st_0) + s^2\hat{u}(\boldsymbol{x}, s) \tag{1.6}$$

and

$$\partial_k\partial_k\hat{u} - (s/c)^2\hat{u} = -\hat{q} - c^{-2}[\partial_t u(\boldsymbol{x}, t_0) + su(\boldsymbol{x}, t_0)]\exp(-st_0). \tag{1.7}$$

From Eq. (1.7) it follows that, in the s-domain, one can take into account the influence of a nonvanishing initial wave field by correctly including it in the s-domain volume densities of external volume source rate. In the remainder of our analysis, it will implicitly be understood that nonzero initial acoustic wave field values have been accounted for in this manner. Our wave equation in the s-domain has then the final form

$$\partial_k\partial_k\hat{u} - (s/\hat{c})^2\hat{u} = -\hat{q}. \tag{1.8}$$

We remark that this equation also holds for lossy media, in which the wave speed depends on s as well. We therefore use the notation $\hat{c} = \hat{c}(\boldsymbol{x}, s)$.

The continuity conditions of Eqs. (1.2) and (1.3) in the time-domain leads to continuity conditions in the s-domain:

$$\lim_{h\to 0} \hat{u}(\boldsymbol{x} + h\boldsymbol{\nu}, s) = \lim_{h\to 0} \hat{u}(\boldsymbol{x} - h\boldsymbol{\nu}, s), \tag{1.9}$$

$$\lim_{h\to 0} \nu_k\partial_k\hat{u}(\boldsymbol{x} + h\boldsymbol{\nu}, s) = \lim_{h\to 0} \nu_k\partial_k\hat{u}(\boldsymbol{x} - h\boldsymbol{\nu}, s). \tag{1.10}$$

The behavior of scalar waves of interest is often characterized by the results of a **steady-state analysis**. In such an analysis, the wave quantities are taken to depend sinusoidally on time with a common angular frequency, ω say. Each purely real quantity can then be associated with a complex counterpart and a common time factor $\exp(-i\omega t)$. In doing so, the original quantities in the time domain are found from its complex counterparts as

$$\{u(\boldsymbol{x},t), q(\boldsymbol{x},t)\} = \mathrm{Re}\,[\{\hat{u}(\boldsymbol{x},-i\omega), \hat{q}(\boldsymbol{x},-i\omega)\}\exp(-i\omega t)]. \tag{1.11}$$

Substitution of these complex representations in the time-domain wave equation yields, except for the common time factor $\exp(-i\omega t)$, the Helmholtz equation

$$\partial_k\partial_k\hat{u} + (\omega/\hat{c})^2\hat{u} = -\hat{q}. \tag{1.12}$$

This equation is identical to Eq. (1.8) with $s = -i\omega$. Hence, we interpret the steady-state analysis as a limiting case of the time Laplace-transform analysis in which $s \to -i\omega$ via $\mathrm{Re}(s) > 0$. In this way, we have ensured that causality remains satisfied.

As far as the medium parameter is concerned, we sometimes assume that the wave speed is a real quantity, independent of the frequency ω. We then deal with a lossless medium. If lossy media must be taken into account, the theory and its associated computer codes are straightforward by introducing a complex value $\hat{c}(\boldsymbol{x},s)$ for $s \to -i\omega$ via $\mathrm{Re}(s) > 0$. Only in the calculation of the adjoint of an operator, one should replace the wave speed $\hat{c}$ by its complex conjugate $\overline{\hat{c}}$.

This concludes the discussion of the general framework of the wave equation together with the interface conditions.

1.1 Three-Dimensional Scattering by a Bounded Contrast

We start our analysis with the wave equation in the s-domain. For convenience, we omit the explicit s-dependence of the various field quantities.

1.1.1 Radiation in an Unbounded Homogeneous Embedding

The problem of calculating the wave field generated by a known source distribution $\hat{q} = \hat{q}(\boldsymbol{x})$ in a lossless and homogeneous embedding with real and constant wave speed c_0 is usually denoted as the source problem. The associated wave field is denoted as the incident field $\hat{u}^{inc} = \hat{u}^{inc}(\boldsymbol{x})$. From Eq. (1.8), it follows that this incident field satisfies

$$\partial_k\partial_k\hat{u}^{inc} - \hat{\gamma}_0^2\hat{u}^{inc} = -\hat{q}(\boldsymbol{x}), \tag{1.13}$$

where

$$\hat{\gamma}_0 = s/c_0 \tag{1.14}$$

is the propagation coefficient of waves in a medium with wave speed c_0.

The solution of this equation is obtained by considering the same problem, but now for a point source with unit strength at position $\boldsymbol{x}'$ (Figure 1.2). Then, for each source position $\boldsymbol{x}'$, the corresponding wave field is denoted as $\hat{G} = \hat{G}(\boldsymbol{x} - \boldsymbol{x}')$ and satisfies the equation

$$\partial_k\partial_k\hat{G} - \hat{\gamma}_0^2\hat{G} = -\delta(\boldsymbol{x} - \boldsymbol{x}'), \tag{1.15}$$

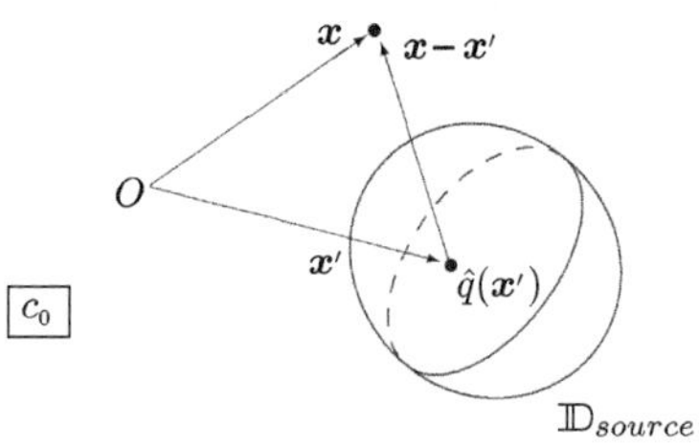

Figure 1.2 Source domain in an unbounded embedding with constant wave speed c_0.

where $\delta(\boldsymbol{x})$ denotes the unit impulse Dirac distribution. Its solution is the well-known 3D Green function, cf. Eq. (1.A.39) of Appendix 1.A.3,

$$\hat{G}(\boldsymbol{x}-\boldsymbol{x}') = \frac{\exp[-\hat{\gamma}_0|\boldsymbol{x}-\boldsymbol{x}'|]}{4\pi|\boldsymbol{x}-\boldsymbol{x}'|}. \tag{1.16}$$

In view of linearity of the problem and the property that

$$\hat{q}(\boldsymbol{x}) = \int_{\boldsymbol{x}' \in \mathbb{D}_{source}} \delta(\boldsymbol{x}-\boldsymbol{x}')\, \hat{q}(\boldsymbol{x}')\, \mathrm{d}V, \tag{1.17}$$

we observe that the solution of Eq. (1.13) is given by

$$\hat{u}^{inc}(\boldsymbol{x}) = \int_{\boldsymbol{x}' \in \mathbb{D}_{source}} G(\boldsymbol{x}-\boldsymbol{x}')\, \hat{q}(\boldsymbol{x}')\mathrm{d}V. \tag{1.18}$$

This equation is the desired source representation in the s-domain.

For simplicity, in our analysis, we assume that the incident field is generated by a point source at position $\boldsymbol{x}^S$, viz. $\hat{q}(\boldsymbol{x}) = \hat{Q}\,\delta(\boldsymbol{x}-\boldsymbol{x}^S)$. Then, the incident wave field at a given observation point $\boldsymbol{x}$ is given by

$$\boxed{\hat{u}^{inc}(\boldsymbol{x}) = \hat{Q}\,\hat{G}(\boldsymbol{x}-\boldsymbol{x}^S) = \hat{Q}\,\frac{\exp[-\hat{\gamma}_0|\boldsymbol{x}-\boldsymbol{x}^S|]}{4\pi|\boldsymbol{x}-\boldsymbol{x}^S|},} \tag{1.19}$$

where $\hat{Q}$ is the Laplace-transformed quantity of the so-called wavelet $Q(t)$.

1.1.2 Scattering by a Bounded Contrast

In this subsection, the scattering of waves by a contrasting domain of bounded extent, present in a homogeneous embedding of infinite extent (Figure 1.3), is investigated in more detail. When we view the scattering object as a volume scatterer, a domain-integral equation formulation is the most versatile technique, while for homogeneous scatterers the boundary-integral equation formulation is most suitable when we treat the scatterer as a surface scatterer [31]. We restrict ourselves to the domain-integral equation formulation of the scattering problem.

Let $\mathbb{D}_{sct}$ be the bounded scattering domain occupied by the scatterer. The domain exterior to $\mathbb{D}_{sct}$ is denoted by $\mathbb{D}'_{sct}$. In $\mathbb{D}'_{sct}$, the wave speed is for simplicity reasons considered to be equal to the real constant c_0. In $\mathbb{D}_{sct}$, we assume the s-dependent wave speed is given by $\hat{c}_{sct} = \hat{c}_{sct}(\boldsymbol{x})$. For real s, this wave speed is real, while for $s = -\mathrm{i}\omega$, it becomes complex, with the real part and the imaginary part satisfying Kramers–Kronig relations, see e.g. section 7.10 of Jackson [25]. When $s = -\mathrm{i}\omega$, the imaginary part of $\hat{c}_{sct}$ should not be negative.

The total wave field in the configuration $\hat{u} = \hat{u}(\boldsymbol{x})$ is decomposed into the incident wave field $\hat{u}^{inc}$ and the scattered wave field $\hat{u}^{sct}$. The incident wave field is the wave field that would be present in

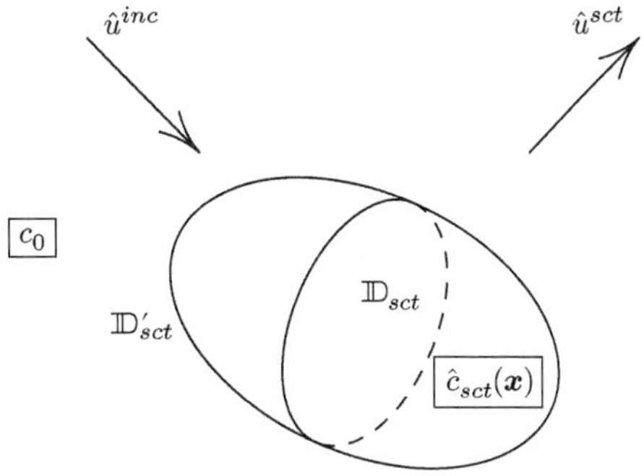

Figure 1.3 The scattering domain.

the entire configuration if the domain $\mathbb{D}_{sct}$ showed no contrast with the embedding. The total wave field is generated by sources that are located outside the scattering domain. Since these sources remain present even if the scattering domain is thought to be absent, they also serve as sources for the incident wave field. The incident wave field quantities can be calculated with the representations obtained in Section 1.1.1.

The scattered wave field is defined as

$$\hat{u}^{sct} = \hat{u} - \hat{u}^{inc}. \tag{1.20}$$

Through a particular reasoning, we now want to express that the scattered wave field depends linearly on a contrast source that represents the presence the scattering object. First, we observe that, since the total wave field is source free in $\mathbb{D}_{sct}$,

$$\partial_k \partial_k \hat{u} - \hat{\gamma}_{sct}^2 \hat{u} = 0, \quad \boldsymbol{x} \in \mathbb{D}_{sct}, \tag{1.21}$$

with

$$\hat{\gamma}_{sct} = s/\hat{c}_{sct}. \tag{1.22}$$

Second, the incident wave field has no sources in $\mathbb{D}_{sct}$, while the wave speeds inside and outside the object have the same value, namely c_0. Hence,

$$\partial_k \partial_k \hat{u}^{inc} - \hat{\gamma}_0^2 \hat{u}^{inc} = 0, \quad \boldsymbol{x} \in \mathbb{D}_{sct}. \tag{1.23}$$

Upon rewriting Eq. (1.21) as

$$\partial_k \partial_k \hat{u} - \hat{\gamma}_0^2 \hat{u} = -(\hat{\gamma}_0^2 - \hat{\gamma}_{sct}^2)\hat{u}, \quad \boldsymbol{x} \in \mathbb{D}_{sct}, \tag{1.24}$$

and subtracting Eq. (1.23) from Eq. (1.24) and using Eq. (1.20), we arrive at

$$\partial_k \partial_k \hat{u}^{sct} - \hat{\gamma}_0^2 \hat{u}^{sct} = -(\hat{\gamma}_0^2 - \hat{\gamma}_{sct}^2)\hat{u}, \quad \boldsymbol{x} \in \mathbb{D}_{sct}. \tag{1.25}$$

Third, we observe that the scattered wave field is source free in the embedding (where the total wave field and the incident wave field have the same sources), and hence

$$\partial_k \partial_k \hat{u}^{sct} - \hat{\gamma}_0^2 \hat{u}^{sct} = 0, \quad \boldsymbol{x} \in \mathbb{D}'_{sct}. \tag{1.26}$$

Equations (1.25) and (1.26) are now combined to

$$\partial_k \partial_k \hat{u}^{sct} - \hat{\gamma}_0^2 \hat{u}^{sct} = -\hat{q}^{sct}, \quad \boldsymbol{x} \in \mathbb{R}^3. \tag{1.27}$$

where the scattering source $\hat{q}^{sct}$ is given by

$$\hat{q}^{sct} = \hat{\gamma}_0^2 \, \hat{\chi}^c \, \hat{u}, \tag{1.28}$$

while the contrast function $\hat{\chi}^c = \hat{\chi}^c(\boldsymbol{x})$ is defined as

$$\boxed{\hat{\chi}^c = 1 - (c_0/\hat{c}_{sct})^2.} \tag{1.29}$$

Note that $\hat{\chi}^c = 0$ outside the scattering domain $\mathbb{D}_{sct}$. If $\hat{q}^{sct}$ were known, Eq. (1.27) would constitute a radiation problem with known sources in an unbounded, homogeneous embedding with wave speed c_0. This problem has been discussed in Section 1.1.1, where we have obtained a solution for $\hat{u}^{inc}$ from Eq. (1.13) as an integral representation in the form of Eq. (1.18). Similarly, we now obtain

$$\boxed{\hat{u}^{sct}(\boldsymbol{x}) = \hat{\gamma}_0^2 \int_{\boldsymbol{x}' \in \mathbb{D}_{sct}} \hat{G}(\boldsymbol{x}-\boldsymbol{x}')\, \hat{\chi}^c(\boldsymbol{x}')\, \hat{u}(\boldsymbol{x}')\, \mathrm{d}V, \quad \boldsymbol{x} \in \mathbb{R}^3.} \tag{1.30}$$

Equation (1.30) is the desired integral representation for the scattered wave field. Once the total wave-field quantity $\hat{u}$ in $\mathbb{D}_{sct}$ is known, this integral representation enables us to calculate the wave field quantities in all space.

So far, our analysis holds for 3D configurations. The changes in 2D and 1D are given in Section 1.2.

1.1.3 Domain-Integral Equation in the *s*-Domain

The domain-integral equation is obtained by confining in Eq. (1.30) the position of observation to the interior of the scatterer, i.e. $\boldsymbol{x} \in \mathbb{D}_{sct}$. Using Eq. (1.20) to express $\hat{u}^{sct}$ in terms of the known values of $\hat{u}^{inc}$ and the as yet unknown values of $\hat{u} \in \mathbb{D}_{sct}$, we arrive at the domain-integral equation,

$$\boxed{\hat{u}^{inc}(\boldsymbol{x}) = \hat{u}(\boldsymbol{x}) - \hat{\gamma}_0^2 \int_{\boldsymbol{x}' \in \mathbb{D}_{sct}} \hat{G}(\boldsymbol{x}-\boldsymbol{x}')\, \hat{\chi}^c(\boldsymbol{x}')\, \hat{u}(\boldsymbol{x}')\, \mathrm{d}V,} \tag{1.31}$$

for $\boldsymbol{x} \in \mathbb{D}_{sct}$. From this Fredholm integral equation of the second kind, see Kantorovich and Krylov [28], the wave field $\hat{u} \in \mathbb{D}_{sct}$ can, in principle, be determined. This integral equation is known as the Lippmann–Schwinger equation [20]. In mathematical physics, this integral equation has been discussed extensively, e.g. [35, p. 1069].

In practice, this integral equation associated with the direct scattering problem must be solved using numerical methods. In special cases, analytical approximations to the values of the wave field in the interior of the scatterer can be made. In particular, the integrals contain a singular point $\boldsymbol{x} = \boldsymbol{x}'$ and around the singular point analytical computations must be performed. The iterative solution of this integral equation will be discussed later. Once the solution of the integral equation has been obtained, the scattered wave field in all space follows by evaluating the right-hand side of Eq. (1.30).

1.1.4 The Born Approximation in the *s*-Domain

For small values of $\int_{\boldsymbol{x}' \in \mathbb{D}_{sct}} |\hat{\gamma}_0^2 \hat{\chi}^c|\, \mathrm{d}V$, the integral equation (1.31) can be solved successively, see e.g. Kress [31, pp. 16 and 17]. The first term of this expansion is

$$\hat{u}^{(0)}(\boldsymbol{x}) \cong \hat{u}^{inc}(\boldsymbol{x}), \quad \boldsymbol{x} \in \mathbb{D}^{sct}. \tag{1.32}$$

This is the so-called Born approximation (see Born and Wolf [8, p. 452]), and it is named after Max Born who presented this approximation in early days (1926) of quantum theory development. Although this approximation is used for optical applications [8], it is actually a low-frequency approximation for scatterers of small size and contrast. The integral representation for observation points outside the scattering domain, Eq. (1.30), is then approximated as

$$\hat{u}^{sct}(\boldsymbol{x}) \cong \hat{\gamma}_0^2 \int_{\boldsymbol{x}' \in \mathbb{D}_{sct}} \hat{G}(\boldsymbol{x}-\boldsymbol{x}')\ \hat{\chi}^c(\boldsymbol{x}')\ \hat{u}^{inc}(\boldsymbol{x}')\ \mathrm{d}V, \quad \boldsymbol{x} \in \mathbb{D}_{sct}. \tag{1.33}$$

An update for the wave field in the scattering domain is successively obtained as,

$$\hat{u}^{(n)}(\boldsymbol{x}) = \hat{u}^{inc}(\boldsymbol{x}) + \hat{\gamma}_0^2 \int_{\boldsymbol{x}' \in \mathbb{D}_{sct}} \hat{G}(\boldsymbol{x}-\boldsymbol{x}')\ \hat{\chi}^c(\boldsymbol{x}')\ \hat{u}^{(n-1)}(\boldsymbol{x}')\ \mathrm{d}V, \tag{1.34}$$

where $n = 1, 2, 3, \ldots$. The solution after n iterations can be written as a summation of repeated operators, the so-called Neumann series, see e.g. Arfken [4, pp. 737–739]. If the Neumann series converges, it converges to the unique solution, see Kress [31]. Because the Neumann method is too restrictive in terms of frequency, wave speed, and size of scattering object, we focus on an efficient numerical solution of the domain integral equation of Eq. (1.31). Before we discuss this, we first show that this integral equation for the field $\hat{u}$ can be converted to an integral equation for the so-called contrast sources.

1.1.5 Contrast-Source Integral Equation

From Eq. (1.30), we observe that the scattered field depends both on the contrast $\hat{\chi}^c$ and the wave field $\hat{u}$. Therefore, in the inverse scattering problem it is advantageous to introduce the contrast source $\hat{w} = \hat{w}(\boldsymbol{x})$ as

$$\boxed{\hat{w} = \hat{\chi}^c \hat{u}.} \tag{1.35}$$

Then, Eq. (1.30) transfers into a contrast-source integral representation

$$\boxed{\hat{u}^{sct}(\boldsymbol{x}) = \hat{\gamma}_0^2 \int_{\boldsymbol{x}' \in \mathbb{D}_{sct}} \hat{G}(\boldsymbol{x}-\boldsymbol{x}')\ \hat{w}(\boldsymbol{x}')\ \mathrm{d}V, \quad \boldsymbol{x} \in \mathbb{R}^3.} \tag{1.36}$$

Multiplying the two sides of Eq. (1.31) by $\hat{\chi}^c$ and using Eq. (1.35), we arrive at the so-called contrast-source integral equation

$$\boxed{\hat{\chi}^c(\boldsymbol{x})\hat{u}^{inc}(\boldsymbol{x}) = \hat{w}(\boldsymbol{x}) - \hat{\chi}^c(\boldsymbol{x})\ \hat{\gamma}_0^2 \int_{\boldsymbol{x}' \in \mathbb{D}_{sct}} \hat{G}(\boldsymbol{x}-\boldsymbol{x}')\ \hat{w}(\boldsymbol{x}')\mathrm{d}V.} \tag{1.37}$$

Once we have solved this integral equation, the solution $\hat{w}(\boldsymbol{x}')$ for $\boldsymbol{x}' \in \mathbb{D}_{sct}$ can be substituted into Eq. (1.36) to arrive at the scattered wave field in whole space, in particular at the receiver points $\boldsymbol{x}^R$. So far, our analysis holds for 3D configurations. The changes in 2D and 1D are summarized in Section 1.2.

1.2 Two-Dimensional and One-Dimensional Scattering

In the **two-dimensional scattering problem**, the scattering object is infinitely long in the x_3-direction, and the wave speed is independent of the x_3-coordinate. Further, in this case, we assume that the sources generate a two-dimensional wave field independent of the x_3-direction. Hence, we only deal with the two-dimensional position vector $\boldsymbol{x}_T = (x_1, x_2)$. The subscript T denotes the transversal component. The spatial derivative ∂_3 of the wave field vanishes. In our 2D analysis, we may replace ∇ by $\nabla_T = \{\partial_1, \partial_2\}$. Consequently, $\nabla_T \cdot \nabla_T = \partial_1\partial_1 + \partial_2\partial_2$.

The 2D Green function $\hat{G}(\boldsymbol{x}_T - \boldsymbol{x}'_T)$ satisfying the wave equation

$$\partial_1\partial_1\hat{G} + \partial_2\partial_2\hat{G} - \hat{\gamma}_0^2\hat{G} = -\delta(\boldsymbol{x}_T - \boldsymbol{x}'_T), \tag{1.38}$$

is obtained as, cf. Eq. (1.A.23),

$$\boxed{\hat{G}(\boldsymbol{x}_T - \boldsymbol{x}'_T) = \frac{1}{2\pi}K_0(\hat{\gamma}_0|\boldsymbol{x}_T - \boldsymbol{x}'_T|),} \tag{1.39}$$

where K_0 is the modified Bessel function of the second kind and zero order.

Note that in the limiting case of either $s \to -\mathrm{i}\omega$, or $s \to \mathrm{i}\omega$ we arrive at

$$\hat{G}(\boldsymbol{x}_T - \boldsymbol{x}'_T) = \begin{cases} \frac{\mathrm{i}}{4}H_0^{(1)}[(\omega/c_0)|\boldsymbol{x}_T - \boldsymbol{x}'_T|], & s \to -\mathrm{i}\omega, \\ -\frac{\mathrm{i}}{4}H_0^{(2)}[(\omega/c_0)|\boldsymbol{x}_T - \boldsymbol{x}'_T|], & s \to \mathrm{i}\omega, \end{cases} \tag{1.40}$$

in which $H_0^{(1)}$ and $H_0^{(2)}$ are the Hankel function of zero order, and of the first kind and the second kind, respectively. To avoid the sometimes confusing use of the time dependency $\exp(\pm\mathrm{i}\omega t)$, we prefer the expression with the modified Bessel function. The choice of the time dependence is determined by choosing either $s \to -\mathrm{i}\omega$ or $s \to \mathrm{i}\omega$.

In 2D, cf. Eq. (1.19), the incident wave field at some observation point $\boldsymbol{x}$, generated by a line source at position $\boldsymbol{x}_T^S$, is given by

$$\boxed{\hat{u}^{inc}(\boldsymbol{x}_T) = \hat{Q}\ G(\boldsymbol{x}_T - \boldsymbol{x}_T^S) = \frac{\hat{Q}}{2\pi}K_0\big(\hat{\gamma}_0|\boldsymbol{x}_T - \boldsymbol{x}_T^S|\big)\,.} \tag{1.41}$$

In 2D, the integral representation for the scattered wave field at a receiver position $\boldsymbol{x}_T^R$ is obtained as, cf. Eq. (1.30),

$$\boxed{\hat{u}^{sct}(\boldsymbol{x}_T^R) = \hat{\gamma}_0^2\int_{\boldsymbol{x}'_T\in\mathbb{D}_{sct}} \hat{G}(\boldsymbol{x}_T^R - \boldsymbol{x}'_T)\ \hat{\chi}^c(\boldsymbol{x}'_T)\ \hat{u}(\boldsymbol{x}'_T)\ \mathrm{d}A} \tag{1.42}$$

and the as yet unknown values of $\hat{u} \in \mathbb{D}_{sct}$ follow from the solution of the 2D domain-integral equation, cf. Eq. (1.31),

$$\boxed{\hat{u}^{inc}(\boldsymbol{x}_T) = \hat{u}(\boldsymbol{x}_T) - \hat{\gamma}_0^2\int_{\boldsymbol{x}'_T\in\mathbb{D}_{sct}} \hat{G}(\boldsymbol{x}_T - \boldsymbol{x}'_T)\ \hat{\chi}^c(\boldsymbol{x}'_T)\ \hat{u}(\boldsymbol{x}'_T)\ \mathrm{d}A,} \tag{1.43}$$

for $\boldsymbol{x}_T \in \mathbb{D}_{sct}$.

In the **one-dimensional scattering problem**, the scattering configuration only changes in the x_1-direction. In addition, we assume that the sources generate a one-dimensional incident wave field, viz. a plane wave that travels in the x_1-direction. Therefore, we only deal with the one-dimensional position x_1. The spatial derivatives ∂_2 and ∂_3 of the wave field vanish. In our 1D analysis, we may replace ∇ by ∂_1. Consequently, $\nabla \cdot \nabla = \partial_1\partial_1$.

The 1D Green function $\hat{G}(x_1 - x'_1)$ satisfying the wave equation

$$\partial_1\partial_1\hat{G} - \hat{\gamma}_0^2\hat{G} = -\delta(x_1 - x'_1), \tag{1.44}$$

is obtained as, cf. Eq. (1.A.9),

$$\boxed{\hat{G}(x_1 - x'_1) = \frac{\exp(-\hat{\gamma}_0|x_1 - x'_1|)}{2\hat{\gamma}_0},} \tag{1.45}$$

In 1D, cf. Eq. (1.19), the incident wave field at some observation point x_1, generated by a plane source at position x_1^S, is given by

$$\boxed{\hat{u}^{inc}(x_1) = \hat{Q}\; G(x_1 - x_1^S) = \hat{Q}\frac{\exp(-\hat{\gamma}_0|x_1 - x_1^S|)}{2\hat{\gamma}_0}.} \tag{1.46}$$

In 1D, the scattered wave field at a receiver position x_1^R is obtained as, cf. Eq. (1.30),

$$\boxed{\hat{u}^{sct}(x_1^R) = \hat{\gamma}_0^2 \int_{x_1' \in \mathbb{D}_{sct}} \hat{G}(x_1^R - x_1')\; \hat{\chi}^c(x_1')\; \hat{u}(x_1')\; \mathrm{d}x_1'} \tag{1.47}$$

and the as yet unknown values of $\hat{u} \in \mathbb{D}_{sct}$ follow from the solution of the 1D domain-integral equation, cf. Eq. (1.31),

$$\boxed{\hat{u}^{inc}(x_1) = \hat{u}(x_1) - \hat{\gamma}_0^2 \int_{x_1' \in \mathbb{D}_{sct}} \hat{G}(x_1 - x_1')\; \hat{\chi}^c(x_1')\; \hat{u}(x_1')\; \mathrm{d}x_1',} \tag{1.48}$$

for $x_1 \in \mathbb{D}_{sct}$.

1.3 Numerical Solution of the Integral Equations (1D, 2D, 3D)

For convenience, we introduce one notation for the integral equation in the 1D case, see Eq. (1.48), the integral equation in the 2D case, see Eq. (1.43), and the integral equation in the 3D case, see Eq. (1.31), viz.,

$$\boxed{\hat{u}^{inc}(x) = \hat{u}(x) - \hat{\gamma}_0^2 \int_{x' \in \mathbb{D}_{sct}} \hat{G}(x - x')\; \hat{\chi}^c(x')\; \hat{u}(x')\; \mathrm{d}x', \quad x \in \mathbb{D}_{sct}.} \tag{1.49}$$

Here, in 1D x stands for x_1, in 2D x stands for $\boldsymbol{x}_T$, and in 3D x stands for $\boldsymbol{x}$, with similar notation for x'. This integral equation is defined for all $x \in \mathbb{D}_{sct}$. In a numerical world, this means that we deal with an infinite number of linear equations with an infinite number of unknown values $\hat{u}(x)$. To limit our number of equations, Galerkin methods are used to convert our continuous operator problem into a discrete problem. In electromagnetics, it is known as the method of moments. Typically, all of these methods use a finite set of basis functions to constrain the solution space. Therefore, in mathematics, it is referred to as a weak formulation of the continuous operator equation.

In the diffraction theory, the concept of the spherical mean of a wave function plays an important role Wilcox [62]. We use this concept for the solution of the integral equation with a singular kernel (Green function). In 3D, we consider a spherical domain with center at x and a radius δ. Then, the spherical mean of the wave function $\hat{u}(x)$ is defined as

$$\langle \hat{u} \rangle(x) = \frac{1}{\Delta V} \int_{|x''|<\delta} \hat{u}(x + x'')\; \mathrm{d}x'', \tag{1.50}$$

where

$$\Delta V = \int_{|x''|<\delta} \mathrm{d}x''. \tag{1.51}$$

In 2D, the spherical mean is the circular mean, while in 1D it becomes the linear mean. In all dimensions, it is an averaging over a domain symmetrically located around the point of observation. In 1D, the averaging is along the Cartesian coordinate. We may apply a Cartesian averaging over a square in 2D and a cube in 3D as well, but then the pertaining integrations cannot be obtained analytically.

We subsequently apply the averaging to both sides of Eq. (1.49). This results into

$$\langle \hat{u}^{inc} \rangle(x) = \langle \hat{u} \rangle(x) - \hat{\gamma}_0^2 \int_{x' \in \mathbb{D}_{sct}} \langle \hat{G} \rangle(x - x')\, \hat{\chi}^c(x')\, \hat{u}(x')\, \mathrm{d}x', \quad x \in \mathbb{D}_{sct}, \tag{1.52}$$

where the weak form of the incident wave field and the weak form of the Green function are defined as

$$\langle \hat{u}^{inc} \rangle(x) = \frac{1}{\Delta V} \int_{|x''|<\delta} \hat{u}^{inc}(x + x'')\, \mathrm{d}x'' \tag{1.53}$$

and

$$\langle \hat{G} \rangle(x) = \frac{1}{\Delta V} \int_{|x''|<\delta} \hat{G}(x + x'')\, \mathrm{d}x'', \tag{1.54}$$

respectively. Since the incident wave field is represented by Eqs. (1.19), (1.41), and (1.46), the weak form of the incident wave field is expressed in terms of the weak form of the Green functions, viz.,

$$\langle \hat{u}^{inc} \rangle(x) = \int_{x' \in \mathbb{D}_{source}} \langle \hat{G} \rangle(x - x')\, \hat{q}(x')\, \mathrm{d}x' . \tag{1.55}$$

The weak form of the Green functions can be computed analytically (for the 3D case we refer to Lecture notes of A.T. de Hoop, see Appendix of [2]). If $|x| \geq \delta$, the results are collected in Table 1.1. If $0 \leq |x| \leq \delta$, the results are given in Table 1.2.

Note that, if $\delta \to 0$, the function $f(\hat{\gamma}_0\delta) = 1$ (see Table 1.1) and for $|x| \neq 0$ the weak Green function $\langle \hat{G} \rangle(x)$ tends to the strong form of the Green function $\hat{G}(x)$.

Until now, the integral equation (1.52) has not been discretized, but the left- and right-hand sides have been filtered in such a way that the spatial variations are reduced. In other words, the amplitudes of the high-frequency spatial spectrum are reduced. This makes sense, because with a finite spatial discretization, we cannot model the spatial high frequencies.

The next part toward our discretization strategy is how to relate the unknown mean $\langle \hat{u} \rangle(x)$ to the wave field $\hat{u}(x')$ in the integral on the right-hand side of the integral equation. At this point, the discretization has to be defined. We have the advantage that, due to the reduction of the spatial high frequencies in the Green function, the minimum mesh size to model the wave field to a certain order of accuracy has increased. For simplicity, we define a regular grid of a number of n equally-sized

Table 1.1 Strong and weak forms of the Green functions, if $|x| \geq \delta$.

	1D	2D	3D
$\hat{G}(x) =$	$\dfrac{1}{2\hat{\gamma}_0} \exp(-\hat{\gamma}_0 \lvert x_1 \rvert)$	$\dfrac{1}{2\pi} K_0(\hat{\gamma}_0 \lvert \boldsymbol{x}_T \rvert)$	$\dfrac{1}{4\pi \lvert \boldsymbol{x} \rvert} \exp(-\hat{\gamma}_0 \lvert \boldsymbol{x} \rvert)$
$f(\hat{\gamma}_0\delta) =$	$\langle \cosh(\hat{\gamma}_0 x_1) \rangle(0)$	$\langle I_0(\hat{\gamma}_0 \lvert \boldsymbol{x}_T \rvert) \rangle(\boldsymbol{0}_T)$	$\left\langle \dfrac{\sinh(\hat{\gamma}_0 \lvert \boldsymbol{x} \rvert)}{\hat{\gamma}_0 \lvert \boldsymbol{x} \rvert} \right\rangle(\boldsymbol{0})$
$=$	$\dfrac{1}{\hat{\gamma}_0\delta} \sinh(\hat{\gamma}_0\delta)$	$\dfrac{2}{\hat{\gamma}_0\delta} I_1(\hat{\gamma}_0\delta)$	$\dfrac{3}{\hat{\gamma}_0^2\delta^2} \left[\cosh(\hat{\gamma}_0\delta) - \dfrac{\sinh(\hat{\gamma}_0\delta)}{\hat{\gamma}_0\delta}\right]$
$\langle \hat{G} \rangle(x) =$	$f(\hat{\gamma}_0\delta)\, \hat{G}(x_1)$	$f(\hat{\gamma}_0\delta)\, \hat{G}(\boldsymbol{x}_T)$	$f(\hat{\gamma}_0\delta)\, \hat{G}(\boldsymbol{x})$
$\langle\langle \hat{G} \rangle\rangle(x) =$	$[f(\hat{\gamma}_0\delta)]^2\, \hat{G}(x_1)$	$[f(\hat{\gamma}_0\delta)]^2\, \hat{G}(\boldsymbol{x}_T)$	$[f(\hat{\gamma}_0\delta)]^2\, \hat{G}(\boldsymbol{x})$

Table 1.2 Weak forms of the Green functions, if $0 \leq |x| \leq \delta$.

	1D	2D	3D
$\hat{G}(x) =$	$\frac{1}{2\hat{\gamma}_0}\exp(-\hat{\gamma}_0\|x_1\|)$	$\frac{1}{2\pi}K_0(\hat{\gamma}_0\|\boldsymbol{x}_T\|)$	$\frac{1}{4\pi\|\boldsymbol{x}\|}\exp(-\hat{\gamma}_0\|\boldsymbol{x}\|)$
$\Delta V =$	$2\,\delta$	$\pi\,\delta^2$	$\frac{4\pi}{3}\delta^3$
$\hat{\gamma}_0^2\Delta V\langle\hat{G}\rangle(x) =$	$1-\exp(-\hat{\gamma}_0\delta)$ $\times\cosh(\hat{\gamma}_0 x_1)$	$1-\hat{\gamma}_0\delta\,K_1(\hat{\gamma}_0\delta)$ $\times I_0(\hat{\gamma}_0\|\boldsymbol{x}_T\|)$	$1-(1+\hat{\gamma}_0\delta)\exp(-\hat{\gamma}_0\delta)$ $\times\frac{1}{\hat{\gamma}_0\|\boldsymbol{x}\|}\sinh(\hat{\gamma}_0\|\boldsymbol{x}\|)$
$\hat{\gamma}_0^2\Delta V\langle\langle\hat{G}\rangle\rangle(0) =$	$1-\exp(-\hat{\gamma}_0\delta)$ $\times f(\hat{\gamma}_0\delta)$	$1-\hat{\gamma}_0\delta\,K_1(\hat{\gamma}_0\delta)$ $\times f(\hat{\gamma}_0\delta)$	$1-(1+\hat{\gamma}_0\delta)\exp(-\hat{\gamma}_0\delta)$ $\times f(\hat{\gamma}_0\delta)$

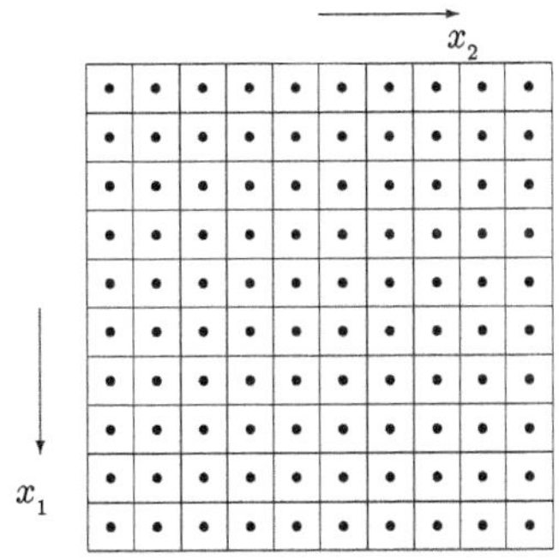

Figure 1.4 Regular grid in 2D with a mesh size of $\Delta x \times \Delta x$. The dots denote the locations of the midpoints on each subdomain $\mathbb{D}_n$.

subdomains, denoted as $\{\mathbb{D}_n,\ n = 1 : N\}$. In 1D, we deal with $N = N_1$ subdomains, denoted as $\{\mathbb{D}_i,\ i = 1 : N_1\}$. Each subdomain has the volume $\Delta V = \Delta x$. In 2D, we deal with $N = N_1 \times N_2$ subdomains, denoted as $\{\mathbb{D}_{i,j},\ i = 1 : N_1,\ j = 1 : N_2\}$. Each subdomain has the volume $\Delta V = (\Delta x)^2$. In 3D, we deal with $N = N_1 \times N_2 \times N_3$ subdomains, denoted as $\{\mathbb{D}_{i,j,k},\ i = 1 : N_1,\ j = 1 : N_2,\ k = 1 : N_3\}$. Each subdomain has the volume $\Delta V = (\Delta x)^3$. We assume that the grid (see Figure 1.4) includes the scattering domain $\mathbb{D}_{sct}$ completely.

On each subdomain $\mathbb{D}_n$, we assume that the contrast is constant and is denoted as $\hat{\chi}^c_n$. If we replace each subdomain by a spherical domain of the same volume, assuming that the wave field $\hat{u}(x)$ is a continuous function of x, this wave field may be replaced by its spherical mean $\langle\hat{u}\rangle(x)$. When we assign x_n as the midpoint of each Cartesian subdomain $\mathbb{D}_n$ and approximate the integral over $\mathbb{D}_n$ as an integral over a spherical domain with a radius δ, so that the volume remains unchanged, the radius δ must be selected as

$$
\begin{aligned}
\delta &= (1/2)\Delta x\,, && \text{in 1D,}\\
\delta &= \pi^{-\frac{1}{2}}\Delta x\,, && \text{in 2D,}\\
\delta &= (4\pi/3)^{-\frac{1}{3}}\Delta x\,, && \text{in 3D.}
\end{aligned}
\tag{1.56}
$$

Then, the integral expression in the integral equation may be approximated as

$$
\begin{aligned}
\int_{x'\in\mathbb{D}_n} \langle\hat{G}\rangle(x-x')\,\hat{\chi}^c(x')\,\hat{u}(x')\,\mathrm{d}x' &:= \hat{\chi}^c_n\,\langle\hat{u}\rangle(x_n)\int_{|x''|<\delta}\langle\hat{G}\rangle(x-x_n+x'')\,\mathrm{d}x''\\
&= \Delta V\,\langle\langle\hat{G}\rangle\rangle(x-x_n)\,\hat{\chi}^c_n\,\langle\hat{u}\rangle(x_n).
\end{aligned}
\tag{1.57}
$$

The integral on the right-hand side is equal to ΔV times the repeated spherical mean of the Green function, which is defined as

$$\langle\langle \hat{G} \rangle\rangle(x) = \frac{1}{\Delta V}\int_{|x''|<\delta} \langle \hat{G}\rangle(x + x'')\,\mathrm{d}x''. \tag{1.58}$$

The resulting expressions for this repeated weak form for various spatial dimensions are given in Tables 1.1 and 1.2. If we now substitute the expression of Eq. (1.57) into the integral equation (1.52) and we require the resulting equation to be satisfied at a discrete spatial set of x_m, identical to the set x_n, we arrive at a system of N linear equations for the N unknowns $\langle \hat{u}\rangle(x_n)$, viz.,

$$\boxed{\langle \hat{u}^{inc}\rangle(x_m) = \langle \hat{u}\rangle(x_m) - \hat{\gamma}_0^2 \Delta V \sum_{n=1}^{N} \langle\langle \hat{G}\rangle\rangle(x_m - x_n)\, \hat{\chi}_n^c\, \langle \hat{u}\rangle(x_n), \quad m = 1 : N.} \tag{1.59}$$

Subsequently, this set of equation is written in operator notation. Without further confusion, we omit the hat-symbols above the s-domain quantities and we write the system of equations as

$$u^{inc} = u - \mathcal{K}\{\chi^c u\}. \tag{1.60}$$

Here, we have

$$\{\chi^c u\}_n = \chi_n^c\, \langle \hat{u}\rangle(x_n), \quad n = 1 : N, \tag{1.61}$$

while the operator $\mathcal{K}$ acting on v is given by

$$[\mathcal{K}v]_m = \sum_{n=1}^{N} \mathcal{K}_{m,n}\, v_n, \quad m = 1 : N, \tag{1.62}$$

where $\mathcal{K}_{m,n}$ is given by

$$\mathcal{K}_{m,n} = \hat{\gamma}_0^2 \Delta V \langle\langle \hat{G}\rangle\rangle(x_m - x_n). \tag{1.63}$$

Let us define the circulant matrix, symbolically written in a 1D world, as

$$\mathcal{M}_{m,n} = |m-n| = \begin{bmatrix} 0 & 1 & \ddots & N-1 & N & N-1 & \ddots & 1 \\ 1 & 0 & 1 & \ddots & N-1 & N & N-1 & \ddots \\ \ddots & 1 & 0 & 1 & \ddots & N-1 & N & N-1 \\ N-1 & \ddots & 1 & 0 & 1 & \ddots & N-1 & N \\ N & N-1 & \ddots & 1 & 0 & 1 & \ddots & N-1 \\ N-1 & N & N-1 & \ddots & 1 & 0 & 1 & \ddots \\ \ddots & N-1 & N & N-1 & \ddots & 1 & 0 & 1 \\ 1 & \ddots & N-1 & N & N & \ddots & 1 & 0 \end{bmatrix}. \tag{1.64}$$

If m and n apply to a certain coordinate direction, $\mathcal{K}_{m,n}$ is a function of $\mathcal{M}_{m,n}$; hence, it is a circulant matrix as well. The operator $\mathcal{K}v$ can be written as the circular convolution

$$\mathcal{K}v = \mathcal{G} \star v, \tag{1.65}$$

where

$$\mathcal{G}_m = \langle\langle \hat{G}\rangle\rangle(x_m) \tag{1.66}$$

is the first column of $\mathcal{K}$ and the vector v of N elements is extended by zero padding to arrive at a vector of $2N$ elements.

The circular convolution of Eq. (1.62) is computed efficiently using the Fast Fourier Transform (FFT). The Matlab functions `fftn` and `ifftn` are the implementations of a n-dimensional forward and inverse discrete Fourier transform. The dimensions of the input function determine the dimension of the space under consideration. The size $N_{1\text{fft}}$ of the FFT grid in the x_1-direction must be at least $2N_1$, and in view of the computation time required, it should be equal to powers of 2. Similar rules apply for the FFT grid in the x_2- and x_3-directions. The computations are written symbolically as

$$C_{1\,:\,2N} = \text{FFT}^{-1}\{\text{FFT}\{\mathcal{G}_{1\,:\,2N}\}\ \text{FFT}\{v_{1\,:\,N}\,;\ \texttt{zeros}_{1\,:\,N}\}\}, \tag{1.67}$$

and we truncate the result to the first N points, resulting into

$$[\mathcal{K}v]_{1\,:\,N} = C_{1\,:\,N}. \tag{1.68}$$

Let us return our attention to the operator equation

$$\boxed{u - \mathcal{K}\{\chi^c u\} = u^{inc},} \tag{1.69}$$

which needs to be solved.

In order to take advantage of fast computation of the operator using FFTs, we opt for an iterative method that minimizes some error criterion. To discuss this aspect, we introduce a discrete norm and inner product:

$$\|u\|^2 = \langle u,\ u\rangle, \quad \langle u,\ v\rangle = \sum_{n=1}^{N} u_n\,\overline{v}_n, \tag{1.70}$$

where the overbar denotes complex conjugate. For further considerations, we introduce the adjoint operator, $\mathcal{K}^\star$ through

$$\langle v,\ \mathcal{K}u\rangle = \langle \mathcal{K}^\star v,\ u\rangle. \tag{1.71}$$

From this it follows from Eq. (1.65) that $K^\star v$ is the circular convolution

$$\mathcal{K}^\star v = \overline{\mathcal{G}} \star v \tag{1.72}$$

Note that in our case the operator $\mathcal{K}$ acts on the product of χ and u as

$$\langle v,\ \mathcal{K}\{\chi^c u\}\rangle = \langle \overline{\chi}^c \overline{\mathcal{K}} v,\ u\rangle. \tag{1.73}$$

From Eqs. (1.71) and (1.73), it follows that

$$\mathcal{K}^\star\{\chi^c v\} = \overline{\chi}^c\ \overline{\mathcal{G}} \star v. \tag{1.74}$$

If u does not solve Eq. (1.69), we define the residual error and a normalized cost functional as

$$\boxed{r = u^{inc} - [u - \mathcal{K}\{\chi^c u\}]} \tag{1.75}$$

and

$$\boxed{F(u) = \frac{\|r\|^2}{\|u^{inc}\|^2},} \tag{1.76}$$

with the property that $F = 1$ for $u = 0$ and $F = 0$ for the exact solution.

In the first instance, an update direction of the wave field in iteration n will be the gradient of the cost functional F with respect to changes in the wave field u, evaluated at the previous step $n-1$, i.e.

$$g^{(n)} = \frac{\partial}{\partial u} F(u)|_{u=u^{(n-1)}}. \tag{1.77}$$

Because the gradient is the update direction with the maximum variant of the cost functional, it is most easily computed by considering the Fréchet derivative [27],

$$\partial_\epsilon F(u^{(n-1)}) = \lim_{\epsilon \to 0} \frac{F(u^{(n-1)} + \epsilon\, v^{(n)}) - F(u^{(n-1)})}{\epsilon} = -2\,\mathrm{Re}\frac{\langle g^{(n)},\, v^{(n)}\rangle}{\|u^{inc}\|^2}, \quad \text{real } \epsilon. \tag{1.78}$$

The Fréchet derivative measures the variation of the functional F as function of the variation of ϵ along a direction $v^{(n)}$ of the function $u^{(n-1)}$. Since the gradient provides the direction in which the variation of a functional is maximal, we may identify $v^{(n)}$ with the gradient of F with respect to the function $u^{(n-1)}$. The gradient is the first member of the resulting inner product with $v^{(n)}$ as second member. In particular, we arrive at

$$g^{(n)} = r^{(n-1)} - \overline{\chi}^c \mathcal{K}^\star r^{(n-1)}. \tag{1.79}$$

Minimization of the cost functional of Eq. (1.76) in each iteration leads to the steepest-descent method. The convergence of this method is substantially improved by enforcing orthogonality of the successive gradients. To solve our linear operator equation, we take the simple Hestenes–Stiefel conjugate gradient directions [22]. Defining

$$A^{(n)} = \|g(n)\|^2, \tag{1.80}$$

these conjugate gradient directions are

$$\begin{aligned} v^{(1)} &= g^{(1)}, \\ v^{(n)} &= g^{(n)} + (A^{(n)}/A^{(n-1)})\, v^{(n-1)}, \quad n = 2, 3, \dots\ . \end{aligned} \tag{1.81}$$

Note that the latter direction follows immediately from a second minimization step [50] in the steepest-descent method. In each iteration, the update for wave field u is obtained as

$$u^{(n)} = u^{(n-1)} + (A^{(n)}/B^{(n)})\, v^{(n)}, \tag{1.82}$$

with

$$B^{(n)} = \left\| v^{(n)} - \mathcal{K}\{\chi^c v^{(n)}\} \right\|^2. \tag{1.83}$$

With this direction, we discussed all the ingredients to employ the conjugate gradient method that minimizes the cost functional F iteratively. It is listed in Table 1.3.

We note that this form of the CG method is convergent for all invertible non-selfadjoint operators. A similar scheme for positive (selfadjoint) operators does not need the computation of the adjoint operator and converges much faster. A convergence criterion is given in terms of the norm of the operator by Kleinman and van den Berg [30]. However, our operator $\mathcal{K}\chi^c \neq \overline{\chi}^c\mathcal{K}^\star$ for complex values of s, which means that the operator is non-selfadjoint. But the major advantage is that the operator $\mathcal{K}$ is circulant in 1D notation. The final method is known as the CGFFT method [51, 63].

To complete this section, we also present the expression for the incident field, defined in Eqs. (1.46), (1.41), and (1.19), viz.,

$$\boxed{\langle \hat{u}^{inc}\rangle(x_n) = \hat{Q}\, f(\hat{\gamma}_0\delta)\langle \hat{G}\rangle(x_n - x^S),} \tag{1.84}$$

where we have used Table 1.1. After the numerical solution of the integral equation, defined in Eqs. (1.47), (1.42), and (1.30), the scattered field is obtained as

$$\boxed{\langle \hat{u}^{sct}\rangle(x^R) = \hat{\gamma}_0^2 \Delta V \sum_{n=1}^{N} \langle \hat{G}\rangle(x^R - x_n)\ \hat{\chi}_n^c\ \langle \hat{u}\rangle(x_n),} \tag{1.85}$$

where we have used Table 1.1.

Table 1.3 The conjugate gradient method for solving the wave field.

Initial estimate		$u^{(0)}$	$=$	0	
		$r^{(0)}$	$=$	$u^{inc} - [u^{(0)} - \mathcal{K}\{\chi^c u^{(0)}\}] = u^{inc}$	
for $n = 1, 2, \cdots$		$g^{(n)}$	$=$	$r^{(n-1)} - \overline{\chi}^c \mathcal{K}^{\star} r^{(n-1)}$	
		$A^{(n)}$	$=$	$\lVert g^{(n)} \rVert^2$	
	if $n = 1$	$v^{(n)}$	$=$	$g^{(n)}$	
	else	$v^{(n)}$	$=$	$g^{(n)} + \dfrac{A^{(n)}}{A^{(n-1)}} v^{(n-1)}$	end
		$B^{(n)}$	$=$	$\lVert v^{(n)} - \mathcal{K}\{\chi^c v^{(n)}\} \rVert^2$	
		$u^{(n)}$	$=$	$u^{(n-1)} + \dfrac{A^{(n)}}{B^{(n)}} v^{(n)}$	
		$r^{(n)}$	$=$	$r^{(n-1)} - \dfrac{A^{(n)}}{B^{(n)}} [v^{(n)} - \mathcal{K}\{\chi^c v^{(n)}\}]$	
	if $\dfrac{\lVert r^{(n)} \rVert}{\lVert u^{inc} \rVert} < \epsilon$ or $n > n_{\max}$ stop				end
end					

1.4 Matlab Input and Output Functions

Before we describe the scattering objects at hand, we assign the specific wave field parameters. We consider values that correspond to the scattering by acoustic wave fields for geophysical applications. We actually take a constant mass density throughout the space. Then, our scalar formulation of this chapter applies. The source wavelet is chosen as $\hat{Q} = 1$. We have selected an example, where the size of the object is of the order of a few wavelengths, and the contrast has moderate values. In this case, both low- and high-frequency approximations, and small contrast approximations (Born approximation) are not useful.

The wave-field parameters are initialized by the `init` function (Matlab Listing I.1). We first define a global parameter `nDIM` that assigns the dimension of space to be used in the full computation sequence. Subsequently, the constant wave speed of the scattering object is $\hat{c}_{sct} = 3000$ m/s, the wave speed c_0 of the embedding is 1500 m/s and the frequency of operation is 50 Hz. In order to deal with causal results, we operate with a small positive real part of s, viz.

$$s = 10^{-16} - \mathrm{i}2\pi f. \tag{1.86}$$

As output, the `init` function displays the operating wavelength. This facilitates the user to check the value $\lambda/\Delta x$. In this example, we take a discretization of 15 samples per wavelength. It is well-known that for the numerical solution of integral equations, we need this discretization to arrive at a mean squared error of about 1%. Each parameter (e.g. `par`) is stored in a structure array `input.par`. This avoids the extensive list of global parameters and this structure array can easily be transferred to a function via de input list. More parameters can be added conveniently. At places where a particular parameter is often used, the long name can be decreased by adding a local redefinition of the following type: `par = input.par`. The `init` function also calls the functions `initSourceReceiver`, `initGrid`, `initFFTGreen`, and `initContrast`.

Matlab Listing I.1 The function `init` to initialize the wave-field parameters.

```
function input = init()
% Time factor = exp(-iwt)
% Spatial units is in m
% Source wavelet  Q = 1

global nDIM;  nDIM = 2;                  % set dimension of space

input.c_0       = 1500;                  % wave speed in embedding
input.c_sct     = 3000;                  % wave speed in scatterer

f               = 50;                    % temporal frequency
wavelength      = input.c_0 / f;         % wavelength
s               = 1e-16 - 1i*2*pi*f;     % LaPlace parameter
input.gamma_0 = s/input.c_0;             % propagation coefficient

disp(['wavelength = ' num2str(wavelength)]);

% add input data to structure array 'input'
  input = initSourceReceiver(input);   % add location of source/receiver

  input = initGrid(input);             % add grid in either 1D, 2D or 3D

  input = initFFTGreen(input);         % compute FFT of Green function

  input = initContrast(input);         % add contrast distribution

  input.Errcri = 1e-3;
```

They initialize the locations of the source and the receivers, the numerical grid in 1D, 2D, and 3D, and the contrast distribution $\hat{\chi}^c$, respectively.

In the function `initSourceReceiver` (Matlab Listing I.2), the vertical source position is $x_1^S = -170$ m, while in 2D the horizontal coordinate is $x_2^S = 0$ m, and in 3D the horizontal coordinates are $x_2^S = x_3^S = 0$ m. In view of the rotational symmetry of both the circular cylinder and the sphere, there is no need to consider other source locations. The receivers are located around the scattering object. In 1D, we take one receiver that measures the reflection from the scattering object and another one that measures the transmission through the scattering object. In 2D, we take a number of NR receivers in a circle around the scattering object. In 3D, we restrict the locations of the receivers to the plane $x_3 = 0$, and hold the same receiver locations like in the 2D case.

The function `initGrid` (Matlab Listing I.3) defines the spatial grids. In the 1D-space, it defines a spatial grid of N_1 subdomains with equal length $\mathrm{d}x = \Delta x$. The coordinates of their midpoints are given by

$$x_1(i) = -(N_1 + 1)\,\mathrm{d}x/2 + i\,\mathrm{d}x, \quad i = 1,\ 2,\ \ldots,\ N_1. \tag{1.87}$$

In the 2D-space, it defines a rectangular grid of N_1 by N_2 square subdomains with equal area of $(\Delta x)^2$. The x_1-coordinates of their midpoints are defined above, while the x_2-coordinates are given by

$$x_2(j) = -(N_2 + 1)\,\mathrm{d}x/2 + j\,\mathrm{d}x, \quad j = 1,\ 2,\ \ldots,\ N_2. \tag{1.88}$$

Matlab Listing I.2 The function `initSourceReceiver` to initialize the source/receivers locations.

```
function input = initSourceReceiver(input)
global nDIM;

if nDIM == 1        %-----------------------------------------------

    input.xS = -170;                              % source position

    input.NR = 1;                                 % receiver positions
    input.xR(1,1) = -150;
    input.xR(1,2) =  150;

elseif nDIM == 2  %-----------------------------------------------

    input.xS(1) = -170;                           % source position
    input.xS(2) =    0;

    input.NR = 180;                               % receiver positions
    input.rcvr_phi(1:input.NR) = (1:input.NR) * 2*pi/input.NR;
    input.xR(1,1:input.NR)     = 150 * cos(input.rcvr_phi);
    input.xR(2,1:input.NR)     = 150 * sin(input.rcvr_phi);

elseif nDIM == 3  %-----------------------------------------------

    input.xS(1) = -170;                           % source position
    input.xS(2) =    0;
    input.xS(3) =    0;

    input.NR = 180;                               % receiver positions
    input.rcvr_phi(1:input.NR) = (1:input.NR) * 2*pi/input.NR;
    input.xR(1,1:input.NR)     = 150 * cos(input.rcvr_phi);
    input.xR(2,1:input.NR)     = 150 * sin(input.rcvr_phi);
    input.xR(3,1:input.NR)     = 0;

end % if
```

In the 3D space, it defines a rectangular grid of N_1 by N_2 by N_3 cubic subdomains with equal volume of $(\Delta x)^3$. The x_1- and x_2-coordinates of their midpoints are defined above, while the x_3-coordinates are given by

$$x_3(k) = -(N_3 + 1)\,\mathrm{d}x/2 + k\,\mathrm{d}x, \quad k = 1,\ 2,\ \dots,\ N_3. \tag{1.89}$$

In order to get the Matlab array numbering $\{i, j, k\}$ consistent with the orientation of the Cartesian coordinates $\{x_1, x_2, x_3\}$, we have enforced the array `input.X`$_1 = X_1(1 : N_1, 1, 1)$ to be a column vector in 1D. As a consequence, in 2D and 3D, we do not use the Matlab function `meshgrid`, but we use Matlab function `ndgrid`. In 2D, the N_2 columns of the array `input.X`$_1$ are copies of the vector x_1 and the N_1 rows of the array `input.X`$_2$ are copies of the vector x_2. In 3D, the N_2 by N_3 columns of the array `input.X`$_1$ are copies of the vector x_1, the N_3 by N_2 rows of the array `input.X`$_2$ are copies of the vector x_2, and the N_1 by N_2 rows of the array `input.X`$_3$ are copies of the vector x_3. In 3D, the equivalence with the Cartesian coordinates and the modified Matlab array numbering is

Matlab Listing I.3 The function `initGrid` to initialize the grid parameters.

```
function input = initGrid(input)
global nDIM;

if nDIM == 1      % Grid in one-dimensional space -----------------------

    input.N1 = 120;                                  % number of samples in x_1
    input.dx = 2;                                    % meshsize dx
    x1 = -(input.N1+1)*input.dx/2 + (1:input.N1)*input.dx;
    input.X1 = x1';
    % Now X1 is a column vector equivalent with x1 axis pointing downwards

elseif nDIM == 2  % Grid in two-dimensional space -----------------------

    input.N1 = 120;                                  % number of samples in x_1
    input.N2 = 100;                                  % number of samples in x_2
    input.dx = 2;                                    % with meshsize dx
    x1 = -(input.N1+1)*input.dx/2 + (1:input.N1)*input.dx;
    x2 = -(input.N2+1)*input.dx/2 + (1:input.N2)*input.dx;
    [input.X1,input.X2] = ndgrid(x1,x2);
    % Now array subscripts are equivalent with Cartesian coordinates
    % x1 axis points downwards and x2 axis is in horizontal direction
    % x1 = X1(:,1) is a column vector in vertical direction
    % x2 = X2(1,:) is a row vector in horizontal direction

elseif nDIM == 3  % Grid in three-dimensional space ---------------------

    input.N1 = 120;                                  % number of samples in x_1
    input.N2 = 100;                                  % number of samples in x_2
    input.N3 = 80;                                   % number of samples in x_3
    input.dx = 2;                                    % with meshsize dx
    x1 = -(input.N1+1)*input.dx/2 + (1:input.N1)*input.dx;
    x2 = -(input.N2+1)*input.dx/2 + (1:input.N2)*input.dx;
    x3 = -(input.N3+1)*input.dx/2 + (1:input.N3)*input.dx;
    [input.X1,input.X2,input.X3] = ndgrid(x1,x2,x3);
    % Now array subscripts are equivalent with Cartesian coordinates
    % x1 axis points downwards and x2 axis is in horizontal direction
    % x1 = X1(:,1,1) is a column vector in vertical direction
    % x2 = X2(1,:,1) is a row vector in first horizontal direction
    % x3 = X3(1,1,:) is a row vector in second horizontal direction

end %if
```

illustrated (Figure 1.5). The coordinate arrays $\texttt{input.X}_1$, $\texttt{input.X}_2$, and $\texttt{input.X}_3$ are typically used to evaluate functions of our spatial variables.

In addition to the advantage that this procedure allows a very compact vectorial description of the various computations on the numerical grid, Matlab vectorization of the functions of $\texttt{input.X}_1$,

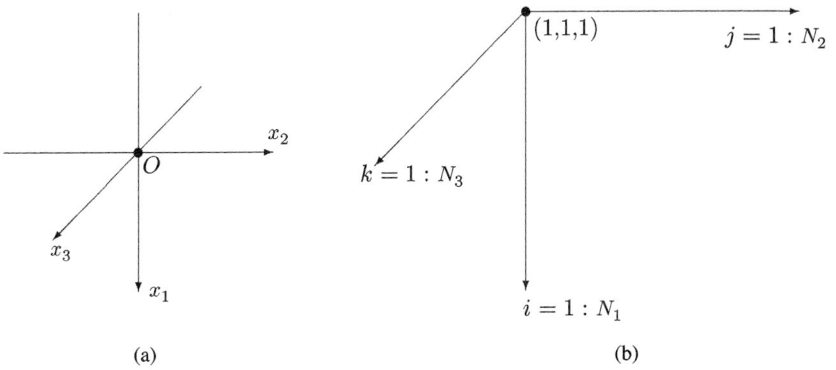

Figure 1.5 The Cartesian coordinates (a) and the modified Matlab array numbering (b).

`input.X`$_2$, and `input.X`$_3$ enables parallel computations of FFTs as well. The latter are widely used for the numerical solution of our integral equations.

The function `initFFTGreen` of Matlab Listing I.4 generates a grid with circulant properties, cf. Eq. (1.64). In 1D, `temp.X`$_{1\text{fft}}$ is the first column of the matrix of Eq. (1.64). In 2D, the $N_{2\text{fft}}$ columns of the array `temp.X`$_{1\text{fft}}$ are copies of the vector x_1, and the $N_{1\text{fft}}$ of rows of `temp.X`$_{2\text{fft}}$ are copies of the vector x_2. In 3D, the $N_{2\text{fft}}$ by $N_{3\text{fft}}$ columns of the array `temp.X`$_{1\text{fft}}$ are copies of the vector x_1, the $N_{3\text{fft}}$ by $N_{2\text{fft}}$ rows of the array `temp.X`$_{2\text{fft}}$ are copies of the vector x_2, and the $N_{1\text{fft}}$ by $N_{2\text{fft}}$ rows of array `temp.X`$_{3\text{fft}}$ are copies of the vector x_3.

For fast computations, the FFT numbers in each direction must have a power of 2. These numbers are a factor of 2 greater than the grid numbers N_1, N_2, and N_3. Although it is not necessary, it is wise to choose the latter numbers also as a factor of 2. At the end of function `initFFTGreen`, there is a call for the Matlab function `Green` to compute the repeated weak form of the Green functions, see Matlab Listing I.5, for either 1D, 2D, or 3D, respectively. The actual expressions were listed in Table 1.2. Note for points, where the Cartesian distance vanishes, the Green function in 2D and 3D becomes singular. Therefore, we replace the zero distance at the ordinal number $(1, 1, 1)$ equal to 1. We get the wrong results for these points, but later we compute the actual repeated Green function at these points correctly. Finally, in function `initFFTGreen`, the output IntG of `Green` is transformed to the discrete Fourier space. The size of each dimension of IntG determines the dimensions of the FFT. The discrete Fourier transformed results are put in the array FFTG, with the same dimension of IntG.

The function `initContrast` (Matlab Listing I.6) initializes the contrast distribution. In our example, it is the discrete distribution of $\hat{\chi}^c(x_1)$ in 1D, $\hat{\chi}^c(x_1, x_2)$ in 2D, and $\hat{\chi}^c(x_1, x_2, x_3)$ in 3D. Inside $\mathbb{D}_{sct}$, the contrast is equal to the constant value given by $1 - (c_0/\hat{c}_{sct})^2$. Note that we employ Matlab matrix inequality, returning a matrix with elements 1 if the test is satisfied and 0 if not.

At last, we initialize the error criterion that should be used as stop criterion in the iterative schemes for solving the integral equations at hand. After all these initializations, before we start with the actual computation of the scattering by our canonical examples, we present a Matlab function `plotWavefield` (Matlib Listing I.8) to plot the total field and the scattered field on the pertaining grid. In order to use the Matlab function `imagesc` to plot two-dimensional distributions consistent with our modified Matlab array numbering, we have to modify this function.

Matlab Listing I.4 The function `initFFTGreen` to compute the discrete Fourier transform of the repeated weak form of the Green function.

```
function input = initFFTGreen(input)
global nDIM;

if nDIM == 1                                   % make one-dimensional FFT grid
    N1fft       = 2^ceil(log2(2*input.N1));
    x1(1:N1fft) = [0 : N1fft/2-1   N1fft/2 : -1 : 1] * input.dx;
    temp.X1fft = x1';  % put it in single column

elseif nDIM == 2                               % make two-dimensional FFT grid
    N1fft       = 2^ceil(log2(2*input.N1));
    N2fft       = 2^ceil(log2(2*input.N2));
    x1(1:N1fft) = [0 : N1fft/2-1   N1fft/2 : -1 : 1] * input.dx;
    x2(1:N2fft) = [0 : N2fft/2-1   N2fft/2 : -1 : 1] * input.dx;
    [temp.X1fft,temp.X2fft] = ndgrid(x1,x2);

elseif nDIM == 3                               % make three-dimensional FFT grid
    N1fft       = 2^ceil(log2(2*input.N1));
    N2fft       = 2^ceil(log2(2*input.N2));
    N3fft       = 2^ceil(log2(2*input.N3));
    x1(1:N1fft) = [0 : N1fft/2-1   N1fft/2 : -1 : 1] * input.dx;
    x2(1:N2fft) = [0 : N2fft/2-1   N2fft/2 : -1 : 1] * input.dx;
    x3(1:N3fft) = [0 : N3fft/2-1   N3fft/2 : -1 : 1] * input.dx;
    [temp.X1fft,temp.X2fft,temp.X3fft] = ndgrid(x1,x2,x3);

end

% compute gam_0^2 * subdomain integrals  of Green function
  [IntG] = Green(temp,input);

% apply n-dimensional Fast Fourier transform
  input.FFTG = fftn(IntG);
```

Therefore, we employ the Matlab function `IMAGESC` of Matlab Listing I.7. For new Matlab users, it is also a quick lesson how to include some essential figure options.

Finally, in this subsection, we also present the Matlab function `displayData.m` of Matlab Listing I.9. For the 1D problem of scattering by a slab, it prints the scattered field amplitudes at receivers above and below the slab. For both the 2D problem of the circular cylinder and the 3D problem of the sphere, it plots the scattered field at the receiver locations. The function `displayData.m` will be used later to plot and compare the results of the integral equation method with the results of the analytic solutions of the slab, the circular cylinder and the sphere (Appendix 1.C).

1.5 Matlab Code for Field Integral Equations (1D, 2D, 3D)

In this section, we present the Matlab codes for the numerical solution of the domain integral equations. In Section 1.3, we have described the CGFFT method to solve the integral equation with a conjugate gradient method, whereby the Green operators are computed by FFTs. We introduced

Matlab Listing I.5 The function `Green` to compute the repeated weak form of the Green function.

```
function [IntG] = Green(temp, input)
global nDIM;
dx = input.dx; gam0 = input.gamma_0;

if nDIM == 1

    X1        = temp.X1fft;
    DIS       = abs(X1);
    G         = 1/(2*gam0) * exp(-gam0.*DIS);
    delta     = 0.5 * dx;
    factor    = sinh(gam0*delta) / (gam0*delta);
    IntG      = (gam0^2 * dx) * factor^2 * G;    % integral includes gam0^2
    IntG(1)   = 1 - exp(-gam0*delta) * factor;

elseif nDIM == 2

    X1        = temp.X1fft;  X2 = temp.X2fft;
    DIS       = sqrt(X1.^2 + X2.^2);
  DIS(1,1)    = 1;                        % avoid Green's singularity for DIS = 0
    G         = 1/(2*pi).* besselk(0,gam0*DIS);
    delta     = (pi)^(-1/2) * dx;              % radius circle with area of dx^2
    factor    = 2 * besseli(1,gam0*delta) / (gam0*delta);
    IntG      = (gam0^2 * dx^2) * factor^2 * G; % integral includes gam0^2
  IntG(1,1)   = 1 - gam0*delta * besselk(1,gam0*delta) * factor;

elseif nDIM == 3

    X1        = temp.X1fft;  X2 = temp.X2fft;  X3 = temp.X3fft;
    DIS       = sqrt(X1.^2 + X2.^2 + X3.^2);
DIS(1,1,1)    = 1;                        % avoid Green's singularity for DIS = 0
    G         = exp(-gam0*DIS) ./ (4*pi*DIS);
    delta     = (4*pi/3)^(-1/3) * dx;         % radius sphere with area of dx^3
    arg       = gam0*delta;
    factor    = 3 * (cosh(arg) - sinh(arg)/arg) / arg^2;
    IntG      = (gam0^2 * dx^3) * factor^2 * G;  % integral includes gam0^2
IntG(1,1,1)= 1 - (1+gam0*delta)*exp(-gam0*delta) * factor;

end
```

a general framework describing the 1D, 2D, and 3D solution with the same operator notation. The specific expression for the Green operators are listed in Tables 1.1 and 1.2. Because Matlab functions allow general n-dimensional arrays as arguments, we take advantage of it, by coding the discrete versions of the 1D, 2D, and 3D integral equations completely parallel. This provides a simple insight into the typical similarities and differences between the different dimensions in space. It is clear that for each dimension, the main Matlab script, the Matlab function with the conjugate gradient scheme, and the Matlab function computing the Green operators using FFTs are completely similar. These functions are determined only by the n-dimensional grid and form the common thread running through the final program code. The actual dimension is determined by the size of the corresponding input argument.

Matlab Listing I.6 The function `initContrast` to initialize the contrast distribution.

```
function [input] = initContrast(input)
global nDIM;

input.a  = 40;   % half width slab / radius circle cylinder / radius sphere

contrast = 1 - input.c_0^2/input.c_sct^2;

if      nDIM == 1;    R = sqrt(input.X1.^2);
elseif nDIM == 2;    R = sqrt(input.X1.^2 + input.X2.^2);
elseif nDIM == 3;    R = sqrt(input.X1.^2 + input.X2.^2 + input.X3.^2);
end % if

input.CHI = contrast .* (R < input.a);
```

Matlab Listing I.7 The function `IMAGESC` to plot 2D functions.

```
function IMAGESC(x1,x2,matrix2D)

% interchange of x1 and x2
  imagesc(x2,x1,matrix2D);

% add some standard options
  xlabel('x_2 \rightarrow');
  ylabel('\leftarrow x_1');
  axis('equal','tight');
  colorbar('hor'); colormap jet;
```

We start our description of the Matlab script and its associated functions with the `Forward-CGFFTu` script listed in Matlab Listing I.10. In this script, we start with the function `init`. It puts in a structured array all the necessary input parameters we need during the whole computation procedure. Subsequently, we carry out the following steps:

(1) We call the last Matlab script `ForwardCanonicalObjects` of Appendix 1.C.4, viz., Matlab Listing I.33, to compute the scattered wave field at a number of receivers, for either the 1D slab, the 2D circle, or the 3D sphere. These data are used to benchmark the scattered field computed from the integral equation method.
(2) We compute the weak form of the incident field either in 1D, 2D, or 3D.
(3) We solve the integral equation for the total wave field with the CGFFT method and plot some results.
(4) We compute the scattered field at the receiver locations. To measure the quality of these data, we benchmark it against the analytical results of step (1).

In step (2), the function `IncWave` of Matlab Listing I.11 computes the weak form of the incident field either in 1D, 2D, or 3D, cf. Eq. (1.84). The particular weak Green functions are listed in Table 1.1. In fact we deal with the weak form of either an incident plane wave, a line source or a point source, respectively.

In step (3), we solve our integral equation. It is really the heart of the program. In the function `ITERCGu` of Matlab Listing I.12, the conjugate gradient scheme is implemented. The code closely

Matlab Listing I.8 The function `plotWavefield` to plot the total and scattered wave fields.

```
function  plotWavefield(u_inc,u,input)
global nDIM;
a = input.a;          u_sct = u - u_inc;

if nDIM == 1        % Plot wave fields in one-dimensional space ------------

   set(figure,'Units','centimeters','Position',[5 5 18 12]);
   x1 = input.X1;  N1 = input.N1;
   x2 = -30 : 30;
   subplot(1,2,1)
      IMAGESC(x1(2:N1-1),x2, abs(u_sct(2:N1-1)) );
      title('\fontsize{13} 1D: abs(u^{sct})');
      hold on;       plot(3*x2,0*x2+a,'--w','LineWidth',2);
                     plot(3*x2,0*x2-a,'--w','LineWidth',2);
  subplot(1,2,2)
      IMAGESC(x1(2:N1-1),x2, abs(u(2:N1-1)) );
      title('\fontsize{13} 1D: abs(u)'); caxis([ 0 max(abs(u_sct))]);
      hold on;       plot(3*x2,0*x2+a,'--w','LineWidth',2);
                     plot(3*x2,0*x2-a,'--w','LineWidth',2);

elseif nDIM == 2  % Plot wave fields in two-dimensional space -------------

    set(figure,'Units','centimeters','Position',[5 5 18 12]);
   x1 = input.X1(:,1);     x2 = input.X2(1,:);
   N1 = input.N1;          N2 = input.N2;
   subplot(1,2,1);
      IMAGESC(x1(2:N1-1),x2(2:N2-1), abs(u_sct(2:N1-1,2:N2-1)));
      title('\fontsize{12} 2D: abs(u^{sct})');
      hold on; phi = 0:.01:2*pi; plot(a*cos(phi),a*sin(phi),'w');
   subplot(1,2,2);
      IMAGESC(x1(2:N1-1),x2(2:N2-1), abs(u(2:N1-1,2:N2-1)));
      title('\fontsize{13} 2D: abs(u)'); caxis([ 0 max(abs(u_sct(:)))]);
      hold on; phi = 0:.01:2*pi; plot(a*cos(phi),a*sin(phi),'w');

elseif nDIM == 3  % Plot wave fields at x3 = 0 or x3 = dx/2 ---------------

   set(figure,'Units','centimeters','Position',[5 5 18 12]);
    N1 = input.N1;          N2 = input.N2;         N3 = input.N3;
    x1 = input.X1(:,1,1);  x2 = input.X2(1,:,1);  N3cross = floor(N3/2+1);
   subplot(1,2,1)
      IMAGESC(x1(2:N1-1),x2(2:N2-1),abs(u_sct(2:N1-1,2:N2-1,N3cross)));
      title('\fontsize{13} 3D:abs(u^{sct})');
      hold on; phi = 0:.01:2*pi; plot(a*cos(phi),a*sin(phi),'w');
   subplot(1,2,2)
      IMAGESC(x1(2:N1-1),x2(2:N2-1),abs(u(2:N1-1,2:N2-1,N3cross)));
      title('\fontsize{13} 3D: abs(u)'); caxis([ 0 max(abs(u_sct(:)))]);
      hold on; phi = 0:.01:2*pi; plot(a*cos(phi),a*sin(phi),'w');

end % if
```

Matlab Listing I.9 The function `displayData` to present the scattered field at the receivers.

```
function displayData(data,input)
global nDIM;

if nDIM == 1    % display scattered data at two receivers
  disp(['scattered wave amplitude above slab = ' num2str(abs(data(1)))]);
  disp(['scattered wave amplitude below slab = ' num2str(abs(data(2)))]);
  if exist(fullfile(cd, 'data1D.mat'), 'file');    load data1D data1D;
     disp(['analytic data above slab = ' num2str(abs(data1D(1)))]);
     disp(['analytic data below slab = ' num2str(abs(data1D(2)))]);
     disp(['error=' num2str(norm(data(:)-data1D(:),1)/norm(data1D(:),1))]);
  end

elseif nDIM == 2 % plot data at a number of receivers -------------------
  if exist(fullfile(cd,'DATA2D.mat'), 'file');    load DATA2D data2D;
     error = num2str(norm(data(:)-data2D(:),1)/norm(data2D(:),1));
     disp(['error=' error]);
  end
  set(figure,'Units','centimeters','Position', [5 5 18 7]);
  angle = input.rcvr_phi * 180 / pi;
  if exist(fullfile(cd,'DATA2D.mat'), 'file')
       plot(angle,abs(data),'--r',angle,abs(data2D),'b')
       legend('Integral-equation method', ...
              'Bessel-function method','Location','NorthEast');
       text(50,0.8*max(abs(data)), ...
            ['Error^{sct} = ' error '  '],'EdgeColor','red','Fontsize',11);
  else plot(angle,abs(data),'b')
       legend('Bessel-function method','Location','NorthEast');
  end
  title('\fontsize{12} scattered wave data in 2D');    axis tight;
  xlabel('observation angle in degrees'); ylabel('abs(data) \rightarrow');

elseif nDIM == 3 % plot data at a number of receivers -------------------
  if exist(fullfile(cd,'DATA3D.mat'), 'file');    load DATA3D data3D;
     error = num2str(norm(data(:)-data3D(:),1)/norm(data3D(:),1));
     disp(['error=' error]);
  end
  set(figure,'Units','centimeters','Position', [5 5 18 7]);
  angle = input.rcvr_phi * 180 / pi;
  if exist(fullfile(cd,'DATA3D.mat'), 'file')
       plot(angle,abs(data),'--r',angle,abs(data3D),'b')
       legend('Integral-equation method', ...
              'Bessel-function method','Location','Best');
       text(50,0.8*max(abs(data)), ...
            ['Error^{sct} = ' error '  '],'EdgeColor','red','Fontsize',11);
  else plot(angle,abs(data),'b')
       legend('Bessel-function method','Location','Best');
  end
  title('\fontsize{12} scattered wave data in 3D');    axis tight;
  xlabel('observation angle in degrees'); ylabel('abs(data) \rightarrow');
end % if
```

Matlab Listing I.10 The `ForwardCGFFTu` script to solve the integral equation.

```
clear all; clc; close all; clear workspace
input = init();

% (1) Compute analytically scattered field data ------------------------
        ForwardCanonicalObjects

% (2) Compute incident field -------------------------------------------
        u_inc = IncWave(input);

% (3) Solve integral equation for wave field with CGFFT method ---------
tic;    u = ITERCGu(u_inc,input);
toc;    plotWavefield(u_inc,u,input);
        w = input.CHI .* u;
        plotContrastSource(w,input);

% (4) Compute scattered field data and plot fields and data ------------
        data = Dop(w,input);
        displayData(data,input);
```

follows the outline in Table 1.3. In this function, we also set the total number of iterations and use the error criterion to stop if the maximum number is reached. Since, the update direction in this conjugate gradient method depends on the exact calculation of the adjoint operator, we include the necessary check of its definition via the inner products of Eq. (1.71). In the development of a numerical code, it is advocated to carry out this check. After validation, one can skip this test. The operator and its adjoint are computed with the help of the functions `Kop` and `AdjKop`, see Matlab Listings I.13 and I.14. The computation of the operator $\mathcal{K}v$ follows Eqs. (1.67) and (1.68). The adjoint operation follows from Eq. (1.74), and the changes in the code for the adjoint are obvious. The Matlab function `plotWavefield` (see Matlab Listing I.8) plots the total field and the scattered field on the pertaining grid. To compare the wave fields for different dimensions, we have collected the plots in Figure 1.6.

After solving for the total wave field, the contrast source is immediately obtained with $w = \chi^c\, u$. Both the contrast and the contrast source distribution are plotted with the help of `PlotContrastSource` (see Matlab Listing I.15). For different dimensions, we have collected the plots in Figure 1.7.

In step (4) of the Matlab script `ForwardCGFFTu`, the scattered field at the selected receiver points are computed. The contrast source distribution $w = \chi^c\, u$ is the input for the function `Dop` of Matlab Listing I.16. The program follows from Eq. (1.85) and Table 1.1, for 1D, 2D, and 3D, respectively.

Finally, these scattered field are compared to the analytical data stored in MAT-files. The function `displayData.m` of Matlab Listing I.9 loads the particular file and for 1D it prints the analytical and numerical data, while for 2D and 3D, it plots the analytical and numerical data. These data

Matlab Listing I.11 The function `IncWave` to compute the incident wave field.

```
function [u_inc] = IncWave(input)
global nDIM;

gam0 = input.gamma_0;
xS   = input.xS;
dx   = input.dx;

if nDIM == 1                     % incident wave on one-dimensional grid

  x1     = input.X1;
  DIS    = abs(x1-xS(1));
  G      = 1/(2*gam0) * exp(-gam0.*DIS);
  delta  = 0.5 * dx;
  factor = sinh(gam0*delta) / (gam0*delta);

elseif nDIM == 2                 % incident wave on two-dimensional grid

  X1     = input.X1; X2 = input.X2;
  DIS    = sqrt( (X1-xS(1)).^2 + (X2-xS(2)).^2 );
  G      = 1/(2*pi).* besselk(0,gam0*DIS);
  delta  = (pi)^(-1/2) * dx;    % radius circle with area of dx^2
  factor = 2 * besseli(1,gam0*delta) / (gam0*delta);

elseif nDIM == 3                 % incident wave on three-dimensional grid

  X1     = input.X1; X2 = input.X2; X3 =input.X3;
  DIS    = sqrt( (X1-xS(1)).^2 + (X2-xS(2)).^2 + (X3-xS(3)).^2 );
  G      = exp(-gam0*DIS) ./ (4*pi*DIS);
  delta  = (4*pi/3)^(-1/3) * dx; % radius sphere with area of dx^3
  arg    = gam0*delta;
  factor = 3 * (cosh(arg) - sinh(arg)/arg) / arg^2;

end

u_inc = factor * G;              % factor for weak form if DIS > delta
```

are presented in Figures 1.8 and 1.9. The visible differences are mainly due to the staircase approximation of the discrete contrast distribution in the integral equation method versus the circular boundary of the circular cylinder and the sphere in the analytical methods.

In these figures, we also display the numerical values of the normalized data error,

$$\text{Error}^{sct} = \frac{\sum_{p=1}^{\text{NR}} |u^{sct}(x_p^R) - u_{exact}^{sct}(x_p^R)|}{\sum_{p=1}^{\text{NR}} |u_{exact}^{sct}(x_p^R)|}. \tag{1.90}$$

For the 3D case, this data error is 30% smaller than the 2D one. This can be explained by observing that the incident field in 3D decays faster than in 2D, so that a smaller part of the stair-cased

Matlab Listing I.12 The function `ITERCGu` to compute the total wave field.

```
function [u] = ITERCGu(u_inc , input)
% CGFFT scheme for wave field integral equation
CHI = input.CHI;   FFTG = input.FFTG;

itmax  = 200;
Errcri = input.Errcri;
it     = 0;                        % initialization of iteration
u      = zeros(size(u_inc));       % first guess for field
r      = u_inc;                    % first residual vector
Norm0  = norm(r(:));               % normalization factor
Error  = 1;                        % error norm initial error

% check adjoint operator [I-conj(CHI)K*] via inner product ----------------
        dummy =  r - conj(CHI).*AdjKop(r,FFTG);
      Result1 = sum(r(:).*conj(dummy(:)));
        dummy =  r - Kop(CHI.*r,FFTG);
      Result2 = sum(dummy(:).*conj(r(:)));
      fprintf('Check adjoint: %e\n',abs(Result1-Result2));
% ---------------------------------------------------------------------------

fprintf('Error =           %g',Error);

while (it <itmax) && (Error > Errcri)
   % determine conjugate gradient direction v
       g  = r - conj(CHI).*AdjKop(r,FFTG);
      AN  = norm(g(:))^2;
     if it == 0
        v = g;
     else
        v = g + (AN/AN_1) * v;
     end
   % determine step length alpha
     Kv    = v - Kop(CHI.*v,FFTG);
     BN    = norm(Kv(:))^2;
     alpha = AN / BN;
   % update field and AN
     u     = u + alpha * v;
     AN_1  = AN;
   % update the residual error r
     r     = r - alpha * Kv;
     Error = sqrt(norm(r(:)) / Norm0);
     fprintf('\b\b\b\b\b\b\b\b%6f',Error);
     it = it+1;
end % while

fprintf('\b\b\b\b\b\b\b\b%6f\n',Error);
disp(['Number of iterations is ' num2str(it)]);
if it == itmax
   disp(['itmax was reached:   Error = ' num2str(Error)]);
end
```

Matlab Listing I.13 The function `Kop` to carry out the $\mathcal{K}$ operator.

```
function [Kv] = Kop(v,FFTG)
global nDIM;
  Cv = zeros(size(FFTG));                              % make fft grid

if nDIM == 1; [N1,~]       = size(v);  Cv(1:N1,1)          = v; end
if nDIM == 2; [N1,N2]      = size(v);  Cv(1:N1,1:N2)       = v; end
if nDIM == 3; [N1,N2,N3]   = size(v);  Cv(1:N1,1:N2,1:N3)  = v; end

Cv = fftn(Cv);    Cv = ifftn(FFTG .* Cv);              % convolution by fft

if nDIM == 1; Kv = Cv(1:N1,1);          end
if nDIM == 2; Kv = Cv(1:N1,1:N2);       end
if nDIM == 3; Kv = Cv(1:N1,1:N2,1:N3);  end
```

Matlab Listing I.14 The function `AdjKop` to carry out the $\mathcal{K}^{\star}$ operator.

```
function [Kf] = AdjKop(f,FFTG)
global nDIM;

Cf = zeros(size(FFTG));                                % make fft grid

if nDIM == 1; [N1,~]       = size(f);  Cf(1:N1,1)          = f; end
if nDIM == 2; [N1,N2]      = size(f);  Cf(1:N1,1:N2)       = f; end
if nDIM == 3; [N1,N2,N3]   = size(f);  Cf(1:N1,1:N2,1:N3)  = f; end

Cf = fftn(Cf);    Cf = ifftn(conj(FFTG) .* Cf);        % convolution by fft

if nDIM == 1; Kf = Cf(1:N1,1);          end
if nDIM == 2; Kf = Cf(1:N1,1:N2);       end
if nDIM == 3; Kf = Cf(1:N1,1:N2,1:N3);  end
```

spherical boundary is “illuminated.” As a consequence, the difference between the amplitudes of the scattered field from such an object and the one from an ideal spherical object is smaller than in the 2D case.

1.6 Matlab Code for Contrast-Source Integral Equation

In our numerical example of this chapter, we have used a computational grid, of which the Cartesian dimensions are much larger than the ones of the scattering object present in the grid. In practice, to minimize the required computer storage and computer time, it is obvious that the grid encloses the scattering object as closely as possible. However, in this book, we want to benchmark the wave fields, not only in the interior of the object and not only in some receiver points at large distance of the object but also in the near region of the scattering object.

An important application of the so-called forward codes of this book is to use them as a model to image the scattering object in terms of position, shape, and contrast. What we designate as a scattering object may consist of a collection of individual scatterers. To make room for the unknown

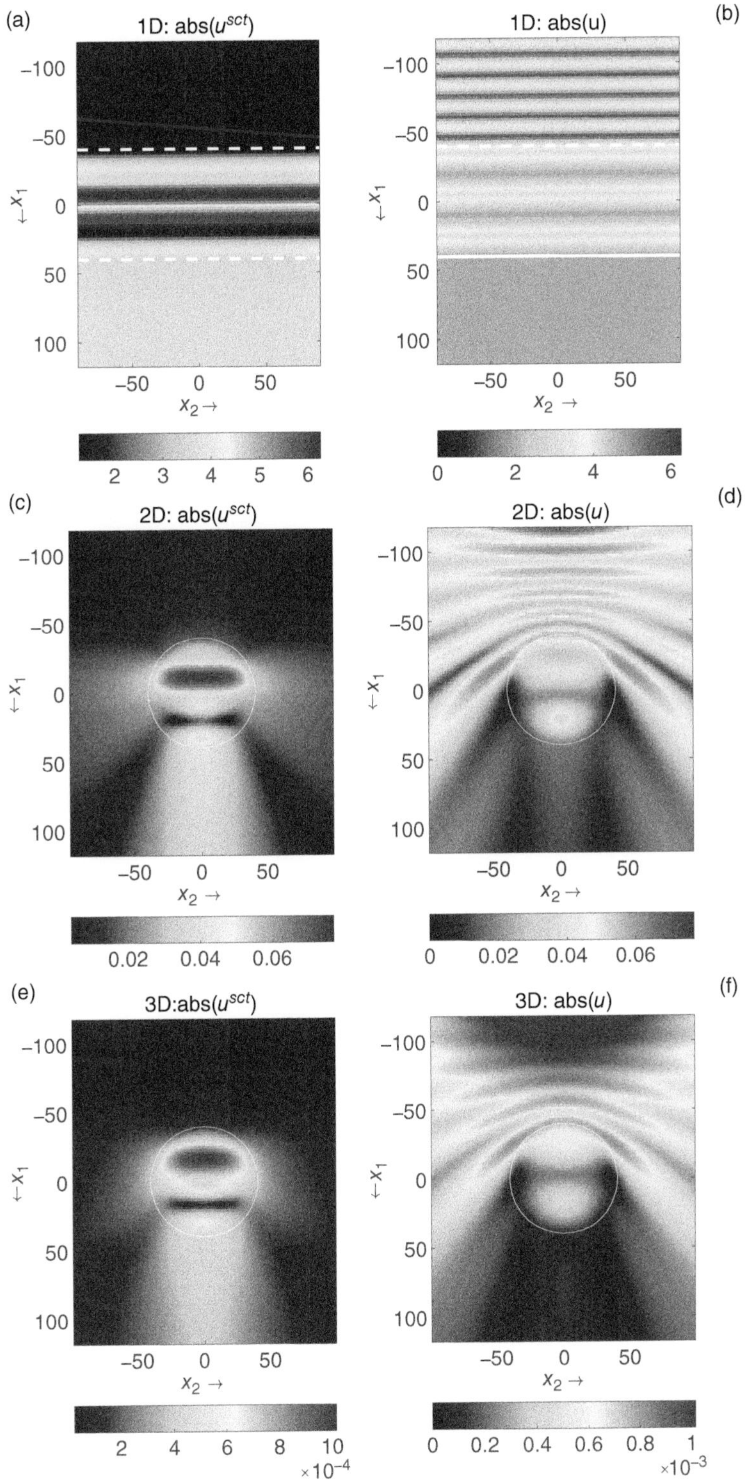

Figure 1.6 Plots of the scattered and total wave field produced by `ForwardCGFFTu`, for 1D (a, b), 2D (c, d), and 3D (e, f), respectively.

Matlab Listing I.15 The function `PlotContrastSource` to plot the contrast distribution and the contrast source distributions.

```
function plotContrastSource(w, input)
global nDIM;
CHI = input.CHI;

if nDIM == 1              % Plot 1D contrast/source distribution ---------------

    x1 = input.X1;
    set(figure,'Units','centimeters','Position',[5 5 18 12]);
    subplot(1,2,1)
       IMAGESC(x1,-30:30, CHI);
       title('\fontsize{13} \chi = 1 - c_0^2 / c_{sct}^2');
    subplot(1,2,2)
       IMAGESC(x1,-30:30, abs(w) )
       title('\fontsize{13} abs(w)');

elseif  nDIM == 2       % Plot 2D contrast/source distribution ---------------

    x1 = input.X1(:,1);    x2 = input.X2(1,:);
    set(figure,'Units','centimeters','Position',[5 5 18 12]);
    subplot(1,2,1)
       IMAGESC(x1,x2, CHI);
       title('\fontsize{13} \chi = 1 - c_0^2 / c_{sct}^2');
       subplot(1,2,2)
       IMAGESC(x1,x2, abs(w) )
       title('\fontsize{13} abs(w)');

elseif nDIM == 3        % Plot 3D contrast/source distribution ---------------
                         % at x3 = 0 or x3 = dx/2

    x1 = input.X1(:,1,1);  x2 = input.X2(1,:,1);
    N3cross = floor(input.N3/2+1);
    set(figure,'Units','centimeters','Position',[5 5 18 12]);
    subplot(1,2,1)
       IMAGESC(x1,x2, CHI(:,:,N3cross));
       title('\fontsize{13} \chi = 1 - c_0^2 / c_{sct}^2');
    subplot(1,2,2)
       IMAGESC(x1,x2, abs(w(:,:,N3cross)));
       title('\fontsize{13} abs(w)');
end % if
```

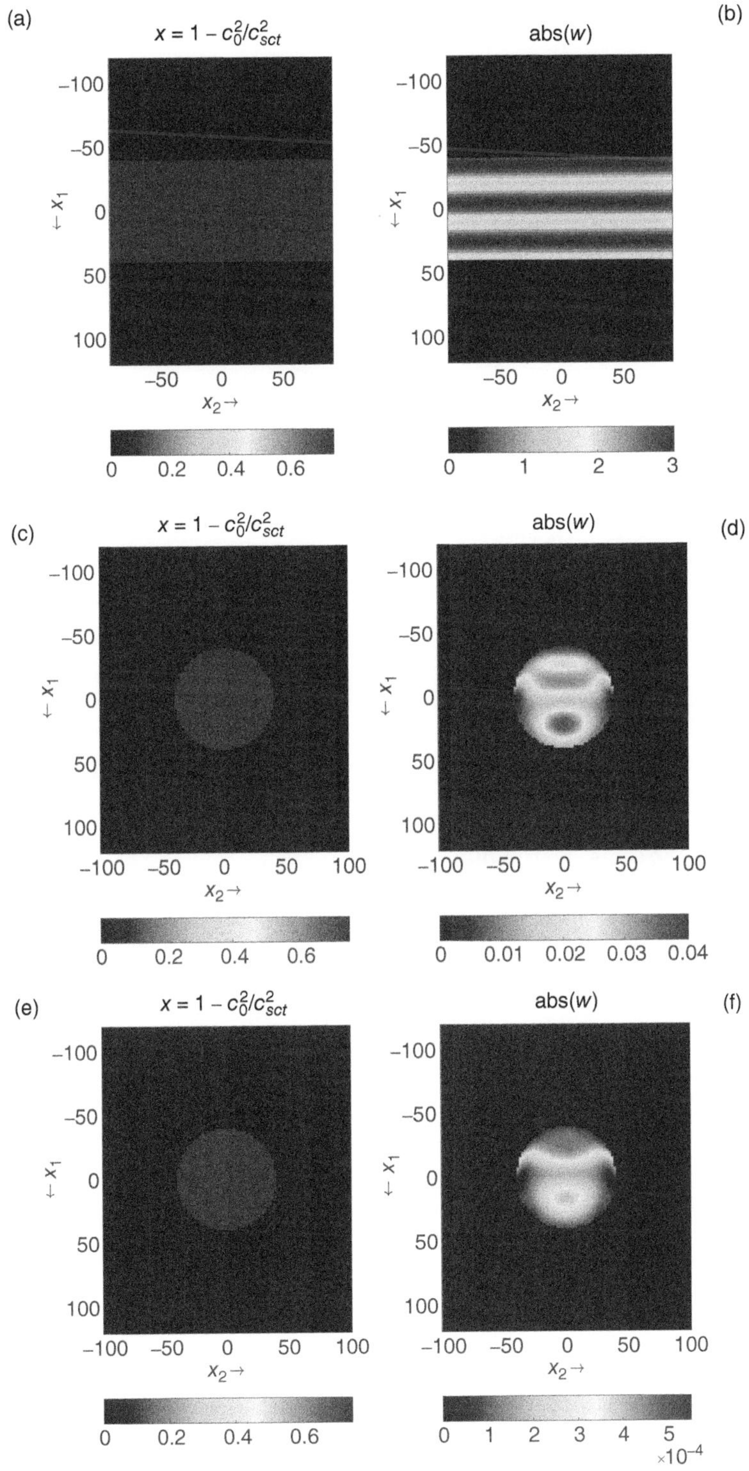

Figure 1.7 Plots of the contrast and the contrast sources produced by `ForwardCGFFTu`, for 1D (a, b), 2D (c, d), and 3D (e, f), respectively.

Matlab Listing I.16 The function Dop to compute the scattered wave field at a number of receivers.

```
function [data] = Dop(w,input)
global nDIM;
gam0 = input.gamma_0;
dx   = input.dx;
xR   = input.xR;

data = zeros(1,input.NR);

if nDIM == 1

  x1       = input.X1;
  delta    = 0.5 * dx;
  factor   = sinh(gam0*delta) / (gam0*delta);
  DIS      = abs(xR(1,1)-x1);                 % receiver for reflected field
  G        = 1/(2*gam0) * exp(-gam0.*DIS);
 data(1,1)= (gam0^2 * dx) * factor * sum(G.*w);
  DIS      = abs(xR(1,2)-x1);                 % receiver for transmitted field
  G        = 1/(2*gam0) * exp(-gam0.*DIS);
 data(1,2)= (gam0^2 * dx) * factor * sum(G.*w);

elseif nDIM == 2

  X1       = input.X1; X2 = input.X2;
  delta    = (pi)^(-1/2) * dx;                % radius circle with area of dx^2
  factor   = 2 * besseli(1,gam0*delta) / (gam0*delta);
  for p = 1 : input.NR
     DIS       = sqrt((xR(1,p)-X1).^2 +(xR(2,p)-X2).^2);
     G         = 1/(2*pi).* besselk(0,gam0*DIS);
   data(1,p) = (gam0(*^*)2 * dx(*^*)2) * factor * sum(G(:).*w(:));
  end % p_loop

elseif nDIM == 3

  X1       = input.X1; X2 = input.X2; X3 = input.X3;
  delta    = (4*pi/3)^(-1/3) * dx;            % radius sphere with area of dx^3
  arg      = gam0*delta;
  factor   = 3 * (cosh(arg) - sinh(arg)/arg) / arg^2;
  for p = 1 : input.NR
     DIS       = sqrt((xR(1,p)-X1).^2 + (xR(2,p)-X2).^2 + (xR(3,p)-X3).^2);
     G         = exp(-gam0*DIS) ./ (4*pi*DIS);
   data(1,p) = (gam0^2 * dx^3) * factor * sum(G(:).*w(:));
  end % p_loop

end % if
```

location of these scatterers, we need a test domain $\mathbb{D}$ that contains the collection of scatterers completely. During this imaging process, which has an iterative character, we cannot apply an iterative resizing of the grid, because of the risk of missing a part of the scattering object, which appears later in the inversion scheme.

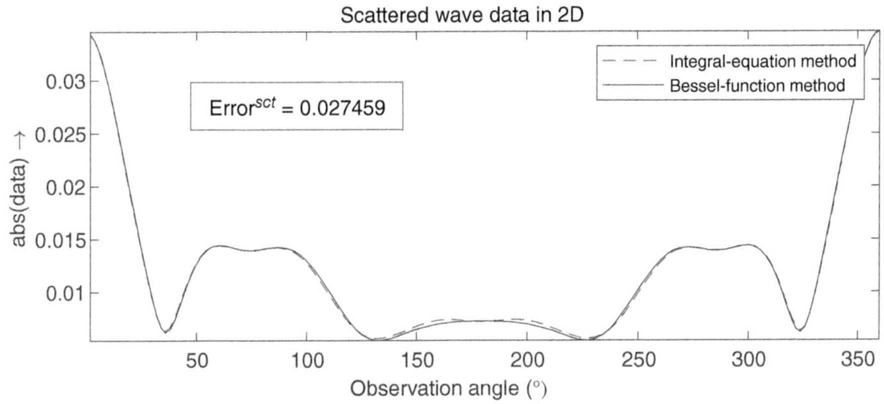

Figure 1.8 Plots of the 2D scattered wave field at the receivers, using either Matlab script `WavefieldSctCircle` (solid line) or Matlab script `ForwardCGFFTu` (dashed line).

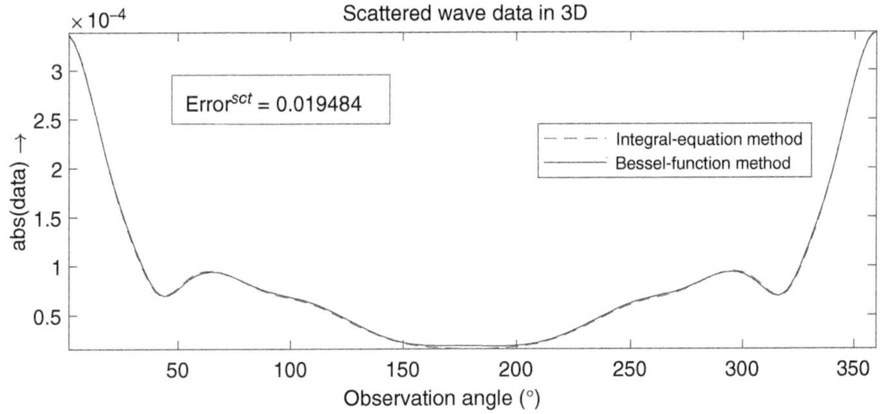

Figure 1.9 Plots of the 3D scattered wave field at the receivers, using either Matlab script `WavefieldSctBessel` (solid line) or Matlab script `ForwardCGFFTu` (dashed line).

However, due to convergence reasons, it is not wise to update some wave-field quantities, where the inverted contrast is negligibly small. Since the wave field u depends in the whole space on each contrast point, the present integral equation solver of `ITERCGu` is difficult to change. Nevertheless, the contrast source w is directly related to the contrast χ^c. Where the contrast vanishes, the contrast source vanishes as well.

We therefore return to our contrast-source integral equation of Eq. (1.37). The operator form of the contrast source integral equation becomes, cf. Eq. (1.69),

$$\boxed{w - \chi^c \mathcal{K} w = \chi^c u^{inc}.} \tag{1.91}$$

Note that we have reached the situation that the contrast-source equation is restricted to a domain where we have nonzero contrast χ^c. Outside this domain, both the contrast χ^c and the contrast source w vanish. On the contrary, in the wave-field integral equation, the wave field u does not vanish at locations where the contrast vanishes. If w does not solve Eq. (1.91), the residual error and the normalized cost functional are now defined as

$$\boxed{r = \chi^c u^{inc} - [w - \chi^c \mathcal{K} w]} \tag{1.92}$$

and

$$\boxed{F(w) = \frac{\|r\|^2}{\|\chi^c u^{inc}\|^2},} \tag{1.93}$$

with the property that $F = 1$ for $w = 0$ and $F = 0$ for the exact solution. The gradient of the cost functional becomes, cf. Eq. (1.79),

$$g^{(n)} = r^{(n-1)} - \mathcal{K}^{\star}\{\chi^c r^{(n-1)}\}. \tag{1.94}$$

With the expressions for the contrast-source operator equation, the residual error and the gradient of the cost functional, the Conjugate Gradient scheme for solving the contrast source w is presented in Table 1.4, cf. Table 1.3 with the CG scheme for the wave field u.

In the modified Matlab script `ForwardCGFFTw` script of Matlab Listing I.17, we have changed step (3), where now the contrast-source integral equation is solved with the CGFFT method. The pertaining Matlab function `ITERCGw` to solve for the contrast source is listed in Matlab Listing I.18.

As discussed, in the conjugate gradient scheme for solving the contrast source, there is no need to update the contrast source at points where the contrast vanishes. In `ITERCGw`, on the line with the three exclamation signs, the gradient $g^{(n)}$ of the cost functional is enforced to zero at those spatial points where the contrast is less than the error criterion imposed. Note that this is allowed if the initial estimate $w^{(0)}$ is zero.

In Figure 1.10, we show the convergence of the three methods:

(1) The CG method `ITERCGu` for the solution of the wave field u;
(2) The CG method `ITERCGw` for the solution of the contrast source w, without windowing the gradient $g^{(0)}$;
(3) The CG method `ITERCGw` for the solution of the contrast source w, with a windowed gradient $g^{(0)}$.

Table 1.4 The Conjugate Gradient FFT method for solving the contrast source.

Initial estimate		$w^{(0)}$	=	0
		$r^{(0)}$	=	$\chi^c u^{inc} - [w^{(0)} - \chi^c \mathcal{K} w^{(0)}] = \chi^c u^{inc}$
for $n = 1, 2, \ldots$		$g^{(n)}$	=	$r^{(n-1)} - \mathcal{K}^{\star}\{\overline{\chi}^c r^{(n-1)}\}$
		$A^{(n)}$	=	$\| g^{(n)} \|^2$
	if $n = 1$	$v^{(n)}$	=	$g^{(n)}$
	else	$v^{(n)}$	=	$g^{(n)} + \frac{A^{(n)}}{A^{(n-1)}} v^{(n-1)}$ end
		$B^{(n)}$	=	$\| v^{(n)} - \chi^c \mathcal{K} v^{(n)} \|^2$
		$w^{(n)}$	=	$w^{(n-1)} + \frac{A^{(n)}}{B^{(n)}} v^{(n)}$
		$r^{(n)}$	=	$r^{(n-1)} - \frac{A^{(n)}}{B^{(n)}} [v^{(n)} - \chi^c \mathcal{K} v^{(n)}]$
	if $\frac{\|r^{(n)}\|}{\|\chi^c u^{inc}\|} < \epsilon$ or $n > n_{\max}$ stop			end
end				

Matlab Listing I.17 The `ForwardCGFFTw` script to solve the contrast source integral equation.

```
clear all; clc; close all; clear workspace
input = init();

% (1) Compute analytically scattered field data ---------------------------
      ForwardCanonicalObjects

% (2) Compute incident field ----------------------------------------------
      u_inc = IncWave(input);

% (3) Solve integral equation for contrast source with CGFFT method -------
      w = ITERCGw(u_inc, input);
      plotContrastSource(w, input);

% (4) Compute synthetic data and plot fields and data --------------------
      data = Dop(w, input);
      displayData(data, input);
```

For these three methods, we plot the normalized mean square error $[F(w^{(n)})]^{\frac{1}{2}}$ and $[F(u^{(n)})]^{\frac{1}{2}}$ as function of the iteration number, n. For a sufficiently small error criterion and for an increasing number of spatial dimensions, `nDim`, the number of iterations in the contrast-source method with windowed gradients does not increase very much, while the number of iterations for the wave-field method increases substantially. It is clear that the simple window function of the gradient in the contrast-source method restricts the number of unknown points and hence reduces the required number of iterations. It outperforms the CG method for the solution of the wave field. On the other hand, we can make some domain restriction for the wave-field equation as well, but for inverse scattering applications, it is not trivial.

1.6.1 Performance Analysis

In this subsection, we report some more details about the computation time and accuracy. In order to investigate the convergence rate of the integral equation results, we consider the results for the scattered data and use Error^{sct} of Eq. (1.90) as our quality criterion. Our aim is to investigate this error as function of the grid size Δx. We begin with $\Delta x = 2$ and reduce it each time with a factor of 2. Specifically, we define a refinement factor $r = 1, 2, \ldots, 6$ and define a new grid size $\Delta x^{(r)} = \Delta x / r$ and new samples numbers $N_1^{(r)} = r\,N_1$, $N_2^{(r)} = r\,N_2$, and $N_3^{(r)} = r\,N_3$.

We have set the error criterion in the Matlab function `ITERCGw` to 10^{-10}. All computations are carried on a Dell computer (64 bit, dual core 3.50 GHz) with installed memory (RAM) of 64 GB. We do this only for the 1D and 2D cases. We have not executed the present convergence test for the 3D case, because it will take too much computation time if we refine the grid discretization.

Starting with the 1D case, we sum in Eq. (1.90) over the two receivers available in the 1D case (NR = 2). We have observed that the error reduces from 2.6% to 0.003%. The 2.6% error is made on a grid discretization of 15 points per wavelength. The 0.03% error is made on a grid with 960 points per wavelength. In Figure 1.11, the results are presented as the dashed line. Note that the horizontal axis represents the refinement factor and the vertical axis represents the function $\frac{1}{2}\log_2(\text{Error}^{sct})$. Then, the 1D error is almost a linear function. This means that the 1D error reduces proportional to $(\Delta x)^2$. This is not surprising, because the local changes made by replacing a wave function by

Matlab Listing I.18 The function `ITERCGw` to compute the contrast-source distribution.

```
function [w] = ITERCGw(u_inc, input)
% CG_FFT scheme for contrast source integral equation
CHI = input.CHI;  FFTG = input.FFTG;

itmax  = 200;
Errcri = input.Errcri;
it     = 0;                                % initialization of iteration
w      = zeros(size(u_inc));               % first guess for contrast source
r      = CHI.*u_inc;                       % first residual vector
Norm0  = norm(r(:));                       % normalization factor
Error  = 1;                                % error norm initial error

% check adjoint operator [I-K*conj(CHI)] via inner product ---------------
          dummy = r - AdjKop(conj(CHI).*r,FFTG);
        Result1 = sum(r(:).*conj(dummy(:)));
          dummy = r - CHI.* Kop(r,FFTG);
        Result2 = sum(dummy(:).*conj(r(:)));
        fprintf('Check adjoint: %e†n',abs(Result1-Result2));
% ------------------------------------------------------------------------

fprintf('Error =          %g',Error);

while (it < itmax) && (Error > Errcri)
   % determine conjugate gradient direction v
        g = r - AdjKop(conj(CHI).*r,FFTG);
        g = (abs(CHI) <= Errcri) .* g; % window for negligible contrast!!!
      AN  = norm(g(:))^2;
      if it == 0
         v = g;
      else
         v = g + (AN/AN_1) * v;
      end
   % determine step length alpha
      Kv    = v - CHI.* Kop(v,FFTG);
      BN    = norm(Kv(:))^2;
      alpha = AN / BN;
   % update contrast source w and AN
      w     = w + alpha * v;
      AN_1  = AN;
   % update the residual error r
      r     = r - alpha * Kv;
      Error =sqrt(norm(r(:)) / Norm0);
      fprintf('†b†b†b†b†b†b†b†b%6f',Error);        it = it+1;
end % while

fprintf('†b†b†b†b†b†b†b†b%6f†n',Error);
disp(['Number of iterations is ' num2str(it)]);
if it == itmax
   disp(['itmax was reached:   err/norm = ' num2str(Error)]);
end
```

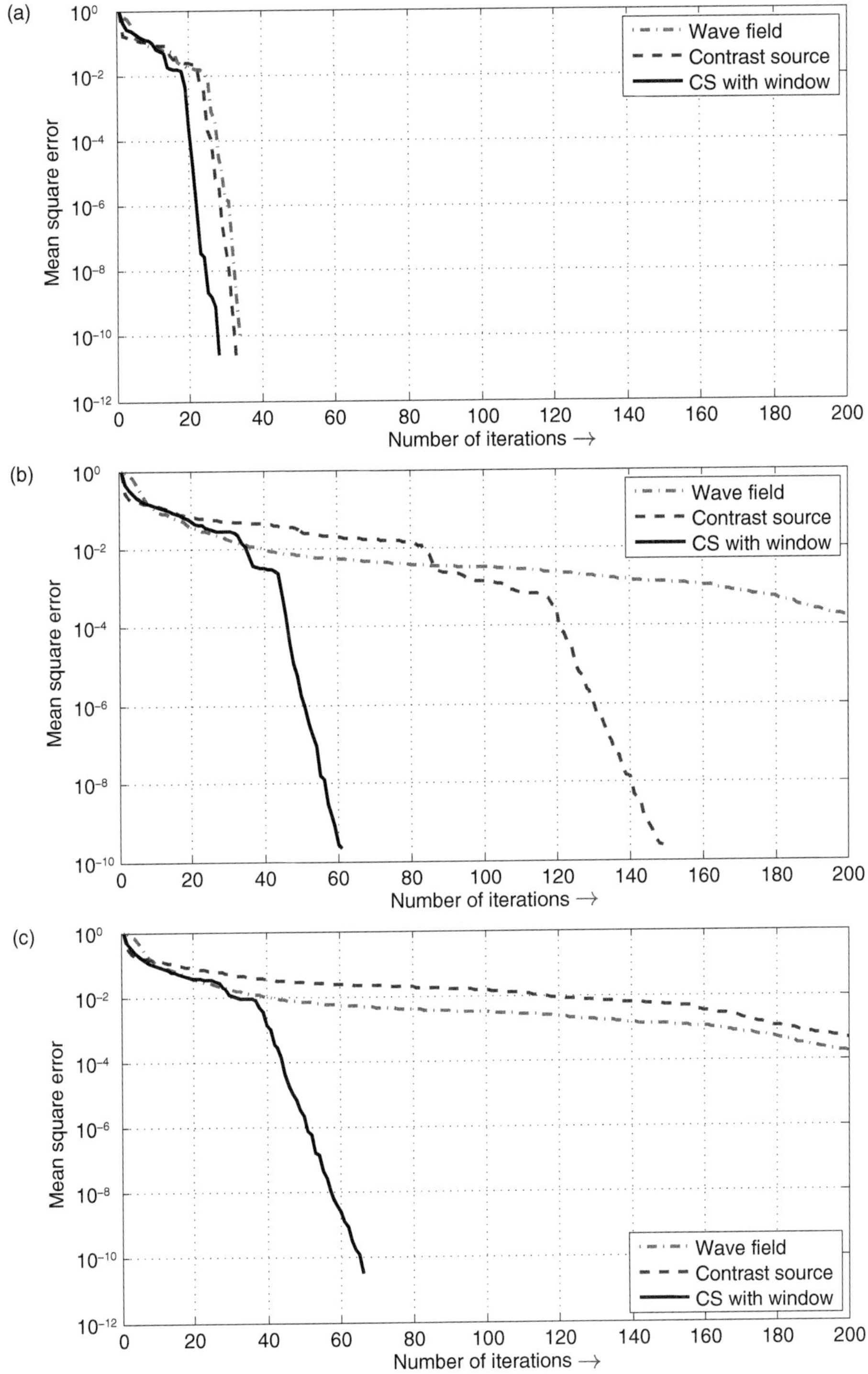

Figure 1.10 The mean square error as function of the number of iterations, for 1D (a), 2D (b) and 3D (c). The dash-dot and dashed lines denote that we have used `ITERCGu` and `ITERCGw`, respectively; the solid lines represent the results using `ITERCGw` with a windowed gradient.

its mean over a domain of length Δx is also of the order of $(\Delta x)^2$. We also note that the number of iterations to reach an error criterion of 10^{-6} in the satisfaction of the integral equation amounts approximately to 22 iterations, when we reduce the grid size from 2 to 0.0625.

Next we consider the 2D case, but now with NR = 180. We have found that the error reduces from 2.7% to 0.01%. In Figure 1.11, the results are displayed as the solid line. Compared with the 1D curve, the 2D curve start with some irregular behavior. This is caused by the staircase to approximate the curved boundary of the circular cylinder, which contributes to some extra errors. Refinement of the grid gives a better approximation of the circular boundary, but a finer staircase approximation is not a systematical refinement of the previous staircase boundary. We will not go into more detail and we will not discuss some ways to correct them because the circular cylinder and the sphere are only considered as a bench mark. On the other hand, one can consider a rectangular cylinder and perform the refining, such that the boundaries always coincide with the grid lines, as in the 1D case. One can do the same error analysis, where the data with respect to the finest grid is considered as the "exact" solution. But anyway in the circular-cylinder case, at the last refinements, the curve behaves more linear, which means that the 2D error also reduces proportional with $(\Delta x)^2$. We further have found that the number of iterations to reach an error criterion of 10^{-6} is almost constant. It changes from 51 to 52 iterations.

For the 3D case, using the computer with 64 GB RAM, we run out of memory when $r = 4$, and therefore, we do not present the 3D convergence results.

For the 2D case, we have registered the computation time needed to solve the integral equation. This is only the time to execute step (3) of the Matlab function ITERCGw. For each refinement, we define this computation time as $t^{(r)}$. In Figure 1.12, we plot the quantity $\log_2(t^{(r)})$ for the refinement $r = 1, 2, \ldots, 7$, respectively. The actual computation time amounts from $t^{(1)} = 0.7$ to $t^{(7)} =$ 1505 seconds. Because in each iteration of the conjugate scheme, the number of operations is of the order $N \log_2 N$ and the number of iterations is almost constant, we consider this latter function as well. For each refinement, we plot the computational complexity function $\log_2(N \log_2 N)$. We notice that the relative computation time per refinement approaches this complexity factor for increasing refinements. The unknown factor is obtained by calibrating it to the computation time at the last refinement. Anyway, Figure 1.12 confirms that the computation cost for the numerical solution of the integral equation is of the order of $N_{it}\ N \log_2 N$, where N_{it} is the number of iterations

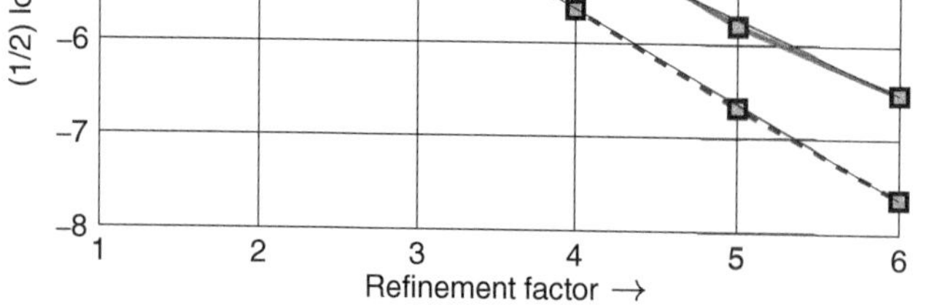

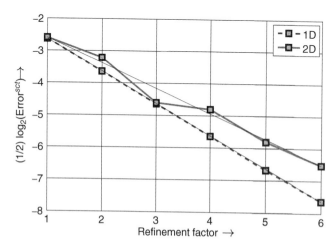

Figure 1.11 Error as function of the refinement factor $r = 1, 2, \ldots, 6$, for the 1D case (dashed line) and the 2D case (solid line). The thin lines are straight lines between the first and the last point of each curve.

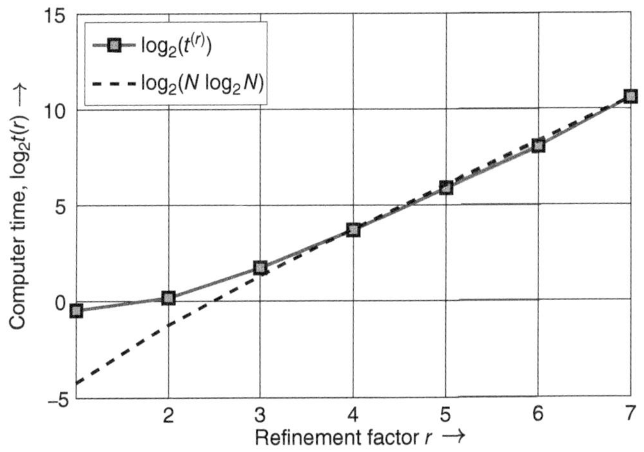

Figure 1.12 Computation time to solve the 2D integral equation as function of the refinement factor r (solid line). For comparison, the computational complexity function is presented as dashed line; $N = N_1 \times N_2$ is the total number of grid cells.

of the Conjugate Gradient scheme. However, it is important to realize that this observation only applies to the numerical solution of the integral equation itself. If you need the scattered field at some receiver points outside the computational domain, the computation of the integral representation cannot take advantage of FFTs. The computational complexity per receiver point is of order N^2. For increasing N, the extra computation time to compute this scattered field can exceed the computation time to solve the integral equation.

1.6.2 Matlab Built-in Functions for Iterative Solution of the Contrast-Source Integral Equation

So far, we have only considered the conjugate gradient method to solve an integral equation with convolutional kernel, with the great advantage that we can employ FFT routines for an efficient computation of the pertaining operator. Because standard routines in Matlab libraries are valid for positive (self-adjoint) operators, symmetrization of the operator through multiplication with its adjoint leads to an error criterion in the adjoint space, which is not always desirable. Therefore, we have given our own Matlab implementation of the CGFFT method, with the adjoint operating on the residual error, where this error is defined in the actual space of the operator itself. However, the multiplication with the adjoint operator increases the norm of the new operator (or in matrix language it squares the condition number). The need for an adjoint operator is circumvented by the full Krylov method [30], by which we mean the general iterative method

$$\begin{aligned} w^{(n)} &= w^{(n-1)} + \alpha^{(n)} v^{(n)}, \\ v^{(n)} &= r^{(n-1)} + \sum_{m=1}^{n-1} \gamma^{(n,m)}\, v^{(m)}, \end{aligned} \tag{1.95}$$

where error minimization of the residual r determines the coefficients $\alpha^{(n)}$ and $\gamma^{(n,m)}$.

In the community for numerical analysis, this error-reducing iterative method is known as GMRES [45]. Although GMRES for our non-self adjoint operators does not require the construction of the adjoint operator, it needs storage of the directions $v^{(m)}$ of all previous iterations. In practice, GMRES is restarted after a certain number of iterations, which number depends on

Matlab Listing I.19 The ForwardBiCGSTABFFTw script to solve the contrast-source integral equation.

```
clear all; clc; close all; clear workspace
input = init();

% (1) Compute analytically scattered field data ----------------
      ForwardCanonicalObjects

% (2) Compute incident field -----------------------------------
      u_inc = IncWave(input);

% (3) Solve integral equation for contrast source with FFT ------
tic;  w = ITERBiCGSTABw(u_inc, input);
toc;  plotContrastSource(w, input);

% (4) Compute synthetic data and plot fields and data ----------
      data = Dop(w, input);
      displayData(data, input);
```

the available computer storage. These restarts significantly reduced the convergence rate. Much attention has been paid to alternative iterative methods for non-self adjoint operators, which has less stringent storage requirements than the full Krylov method of GMRES. In the Matlab library, there are a number of such methods available, e.g. BiCG, BiCGSTAB, and CGS, but these schemes are not always error reducing during the iterations. In view of our goals to focus on the forward and inverse scattering algorithms based on contrast-source integral equations, we do not want to distract readerś attention to these numerical issues that require further numerical research. After many numerical experiments with our contrast-source integral equations, we have found that the BiCGSTAB method [58] shows excellent performance. In Matlab Listing I.19, we show that the Matlab function ITERBiCGSTABw of Matlab Listing I.20 calls Matlab built-in function bicgstab. For the purpose of using FFTs, we do not want to construct a matrix equation $Aw = b$, but we use a function handle that returns the value of $Aw = w - \chi^c \mathcal{K} w$. The function handle is given by the Matlab function Aw. Because Matlab built-in functions operate on vectors, we have to reshape our input matrices to vectors and the output vectors to matrices. This is only necessary for the 2D and 3D cases.

For the 3D case, we have run the Matlab script ForwardCGFFTw and the Matlab script ForwardBiCGSTABFFTw. To reach a criterion of 0.001, function ITERCGw requires 40 iterations with a computation time of 44 seconds, while ITERBiCGSTABw requires 23 iterations with a computation time of 27 seconds. For the two methods, the computation per iteration is roughly the same. The reduction of the number of iterations of almost a factor of 2 is due to the fact that the BiCGSTAB method does not involve products of operator and its adjoint.

1.7 Time-Domain Solution of Contrast-Source Integral Equation

So far, we have considered the solution of the contrast-source equation for a certain value of $s = \epsilon - \mathrm{i}2\pi f \approx -\mathrm{i}2\pi f$, where ϵ is a vanishing positive constant. In our computer codes, we have taken $\epsilon = 10^{-16}$. The time domain solution for the wave function $u = u(\boldsymbol{x}, t)$ is obtained from the frequency

Matlab Listing I.20 The function `ITERBiCGSTABw` to call Matlab's built-in function `bicgstab`.

```
function [w] = ITERBiCGSTABw(u_inc, input)
% BiCGSTAB_FFT scheme for contrast source integral equation Aw = b
itmax  = 1000;
Errcri = input.Errcri;

b = input.CHI(:) .* u_inc(:);                              % known 1D vector
w = bicgstab(@(w) Aw(w,input), b, Errcri, itmax);          % call BiCGSTAB

w = vector2matrix(w,input);                                % output matrix

end %----------------------------------------------------------------------

function y = Aw(w,input)
  w = vector2matrix(w,input);                    % Convert 1D vector to matrix
  y = w - input.CHI .* Kop(w,input.FFTG);
  y = y(:);                                      % Convert matrix to 1D vector
end %----------------------------------------------------------------------

function w = vector2matrix(w,input)
% Modify vector output from 'bicgstab' to matrix for further computations
global nDIM;
  if nDIM == 2
       w = reshape(w,[input.N1,input.N2]);
  elseif nDIM == 3
       w = reshape(w,[input.N1,input.N2,input.N3]);
  end
end
```

domain solutions $\hat{u}(\boldsymbol{x}, s)$ with the help of the Bromwich integral,

$$u(\boldsymbol{x}, t) = \frac{1}{2\pi \mathrm{i}} \int_{\epsilon-\mathrm{i}\infty}^{\epsilon+\mathrm{i}\infty} \exp(st)\; \hat{u}(\boldsymbol{x}, s)\; \mathrm{d}s. \tag{1.96}$$

In the limiting case that $\epsilon \downarrow 0$, we arrive at the inverse Fourier transform, i.e.

$$u(\boldsymbol{x}, t) = \int_{-\infty}^{\infty} \exp(-\mathrm{i}2\pi f t)\; \hat{u}(\boldsymbol{x}, s)\; \mathrm{d}f. \tag{1.97}$$

Discretization of this transform leads us to the inverse discrete Fourier transform,

$$u(\boldsymbol{x}, t_m) = \Delta f \sum_{n=-\frac{1}{2}N+1}^{\frac{1}{2}N} \exp\left(-\mathrm{i}2\pi \frac{m\,n}{N}\right)\; \hat{u}(\boldsymbol{x}, s_n), \tag{1.98}$$

where

$$t_m = m\;\Delta t, \quad f_n = n\;\Delta f, \quad s_n = \epsilon - \mathrm{i}2\pi f_n, \quad \text{and} \quad \Delta t\;\Delta f = \frac{1}{N}, \tag{1.99}$$

with $m = -\frac{1}{2}N + 1 : \frac{1}{2}N$. The forward discrete Fourier transform is given by

$$\hat{u}(\boldsymbol{x}, s_n) = \Delta t \sum_{m=-\frac{1}{2}N+1}^{\frac{1}{2}N} \exp\left(\mathrm{i}2\pi \frac{m\,n}{N}\right) u(\boldsymbol{x}, t_m)\,, \tag{1.100}$$

Matlab Listing I.21 The function `Wavelet` to compute the first derivative of a Gaussian wavelet using Matlab's function `gauswavf`, both in the frequency domain and in the time domain.

```
function [input] = Wavelet(input)

Nfft = 512;            fsamples = Nfft/2;
df = 150 / fsamples;          dt = 1 / (df*Nfft);

% First derivative of Gaussian Wavelet using Matlab function gauswavf
  Nwavel = 21;
  [Waveltime,~] = gauswavf(-5,5,Nwavel,1);

% Transform real function to frequency domain
  Wavelfrq = dt * conj(fft(Waveltime,Nfft));
  Wavelfrq(Nfft/2+1:Nfft) = 0;          % restrict to positive frequencies

% Transform to real function in time domain
  Waveltime = 2*df * real(fft(Wavelfrq,Nfft));

figure;
subplot(2,1,1);
  plot((0:(fsamples-1))*df,abs(Wavelfrq(1:fsamples)),'b','LineWidth',1.5);
  axis('tight');
  xlabel(' frequency  [Hz] \rightarrow ')
  ylabel(' |W(f)|  \rightarrow ');
  legend('frequency amplitude spectrum');
subplot(2,1,2);
  plot((0:fsamples-1)*dt,Waveltime(1:fsamples),'r','LineWidth',1.5);
  axis('tight');
  xlabel(' time  [s] \rightarrow ')
  ylabel(' W(t) \rightarrow ');
  legend('Wavelet in time domain')

input.Nfft      = Nfft;
input.fsamples  = fsamples;
input.Wavelfrq = Wavelfrq(1:fsamples);
input.df        = df;
input.dt        = dt;
```

with $n = -\frac{1}{2}N + 1 : \frac{1}{2}N$. Note that our definitions of the discrete Fourier transforms differ from the standard ones, as far as normalization and numbering are concerned [37]. However, we prefer the situation in which the normalization and numbering of the discrete transform are closely related to the definition and symmetry properties of the continuous form.

In view of causality that $u(\boldsymbol{x}, t_m) = 0$ when $t_m < 0$, we can limit our summation in the forward discrete Fourier transform to values of $m = 0, 1, \ldots, N/2$, viz.,

$$\boxed{\hat{u}(\boldsymbol{x}, s_n) = \Delta t \sum_{m=0}^{\frac{1}{2}N} \exp\left(i2\pi \frac{m\,n}{N}\right) u(\boldsymbol{x}, t_m).} \tag{1.101}$$

Further, $u(\boldsymbol{x}, t_m)$ is a real causal function in the time domain, and the imaginary part of $\hat{u}(\boldsymbol{x}, s_n)$ is an odd function of n. Then, the inverse discrete Fourier transform is obtained as

$$u(\boldsymbol{x}, t_m) = 2\Delta f \ \mathrm{Re}\left[\sum_{n=0}^{\frac{1}{2}N} \exp\left(-\mathrm{i}2\pi\frac{m\,n}{N}\right) \hat{u}(\boldsymbol{x}, s_n)\right]. \tag{1.102}$$

We note that in the last two expressions of our discrete transforms, symmetric domains around $m = 0$ and $n = 0$ must be selected. In applying FFT routines, zero padding for the negative values of n and m is required.

At this point, we also have to select the wavelet $Q(t)$ in Eq. (1.19). The frequency-domain results were calculated for $\hat{Q}(s_n) = 1$. Before computing the time-domain results, we must define the wavelet and multiply the frequency-domain results by $\hat{Q}(s_n)$. We choose the first derivative of the Gaussian as our frequency-domain wavelet. It is computed with the help of the Matlab function `Wavelet` of Matlab Listing I.21. The resulting frequency- and time-domain functions are presented in Figure 1.13.

We change the single-frequency script `ForwardBiCGSTABFFTw` of Matlab Listing I.19 to one for the full frequency spectrum of the wavelet. This script called as `TimedomainBiCGSTABFFTw` is displayed in Matlab Listing I.22. Note that the frequency band ranges from 0 to 120 Hz. To capture the high-frequency range, we reduce our grid size from $\Delta x = 2$ m to $\Delta x = 1$ m. To get a better picture of the wave propagation in the time domain, we take a considerably larger observation domain by choosing $N_1 = N_2 = 600$ in the 2D cross-section. In view of this larger domain and the number of computations for many frequencies, we only consider the 2D situation. For convenience, we have the source position as $(0, -170)$ m; it is located on the left-hand side of the cylinder. Then, the wave propagates from the source in the right-hand direction toward the circular cylinder. We consider

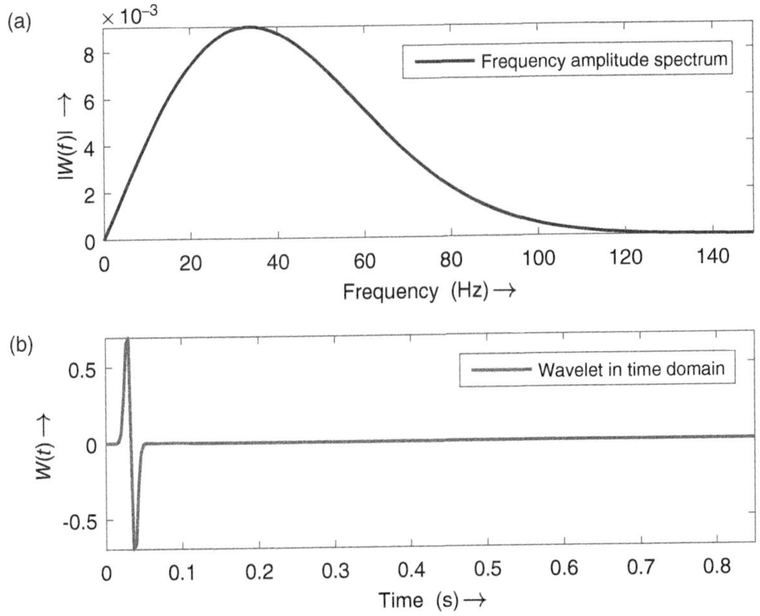

Figure 1.13 The wavelet $\hat{Q}(s)$ in the frequency domain (a) and $Q(t)$ in the time domain (b).

Matlab Listing I.22 The script `TimedomainBiCGSTABFFTw` to compute the time-domain total wave field $u(\mathbf{x}_T, t)$ on a 2D grid.

```
clear all; clc; close all; clear workspace;
input = init();

    input = Wavelet(input);
Wavelfrq = input.Wavelfrq;
Wavelmax  = max(abs(Wavelfrq(:)));

% Redefine frequency-independent parameters
  input.N1 = 600;                          % number of samples in x_1
  input.N2 = 600;                          % number of samples in x_2
  input.dx = 1;                            % grid size
        x1 = -(input.N1+1)*input.dx/2 + (1:input.N1)*input.dx;
        x2 = -(input.N2+1)*input.dx/2 + (1:input.N2)*input.dx;
  [input.X1,input.X2] = ndgrid(x1,x2);
  input.xS     = [0 ,-170];                % source position
  input.c_sct  = 3000;                     % wave speed in scatterer
  input        = initContrast(input);      % contrast distribution

Errcrr_0 = input.Errcri;

ufreq     = zeros(input.N1,input.N2,input.Nfft);
for f = 2 : input.fsamples

%    (0) Make error criterion frequency dependent
        factor = abs(Wavelfrq(f))/Wavelmax;
        Errcri = min([Errcrr_0 / factor, 0.999]);
        input.Errcri = Errcri;

%    (1) Redefine frequency-dependent parameters
        freq = (f-1) * input.df;        disp(['freq sample: ', num2str(f)]);
        s = 1e-16 - 1i*2*pi*freq;          % LaPlace parameter
        input.gamma_0 = s/input.c_0;       % propagation coefficient
        input = initFFTGreen(input);       % compute FFT of Green function

%    (2) Compute incident field ------------------------------------------
        u_inc = IncWave(input);

%    (3) Solve integral equation for contrast source with FFT ------------
        w =ITERBiCGSTABw(u_inc,input);

%    (4) Compute total wave field on grid  and add to freqency components
        u_sct = Kop(w,input.FFTG);
        ufreq(:,:,f) = Wavelfrq(f) .* (u_inc(:,:) + u_sct(:,:));

end; % frequency loop

ut = 2 * input.df * real(fft(ufreq,[],3));                    clear ufreq;
utime(:,:,1:input.fsamples) = ut(:,:,1:input.fsamples);       clear ut;

SnapshotU;                      % Make snapshots for a few time instants
```

Matlab Listing I.23 The function `SnapshotU` to present some snapshots of the time-domain total wave field $u(\mathbf{x}_T, t)$ on a 2D grid.

```
% This script is a continuation of TimedomainBiCGSTABFFTw
clc;

xS(1) = input.xS(1);  xS(2) = input.xS(2);

utimeabs = abs(utime);
set(figure,'Units','centimeters','Position',[0 -1 18 32]);

for n = 1 : 12
  subplot(4,3,n)
  n_t = (20 + n * 7);

% Compute the time domain power in dB
  udB = -20 * log10(utimeabs/max(max(utimeabs(:,:,n_t))));
% Make snapshot
  imagesc(input.X2(1,:),input.X1(:,1),udB(:,:,n_t)); grid on;
  title(['t = ',num2str(n_t * input.dt),' s']);
  hold on;  colormap(hot);  axis('square');  caxis([0, 40]);
  plot(xS(2),xS(1),'ko','MarkerFaceColor',[0 0 0],'MarkerSize',3);
  phi = 0:0.001:2*pi;
  plot(input.a*cos(phi),input.a*sin(phi),'b','LineWidth',1.2);
  pause(0.1)
  hold off;
end
```

two different wave speed values in the scattering object, viz., $\hat{c}_{sct} = 3000$ and 750 m/s, while the wave speed of the embedding, $c_0 = 1500$ m/s, will not change.

To save on computer time, we make the error criterion, to stop the iterative procedure of `ITERBiCGSTABw`, dependent on the frequency amplitude of the wavelet. In step (0) of the frequency loop in `TimedomainBiCGSTABFFTw`, we increase the criterion by a factor inversely proportional to the amplitude of the normalized spectrum of the wavelet. Then, for each frequency, we must call the Matlab functions to recompute (1) the Green function, (2) the incident wave field, and (3) the solution of the integral equation. After solution for the contrast sources, in step (4) the total wave field on the grid is computed. Finally, the total wave fields $\hat{u}(\boldsymbol{x}_T, s_n)$ are transformed to the time domain and only the positive times are retained. To present the wave field at certain time instants t_m, we call the Matlab script `SnapshotU` of Matlab Listing I.23. In fact, we calculate the power in dB of the normalized wave-field values. For each snapshot, the maximum limit of `caxis` is chosen to be 40 dB.

Figure 1.14 displays the snapshots in case the wave speed within the circular cylinder is twice the background wave speed. In the third and fourth picture, it is clear that the wave propagates faster through the object than outside the object. In Figure 1.15, we zoom in on the inner area of the cylinder. For increasing time, we see that the circular cylinder acts as a secondary volume type source. It generates circular types of wave fronts, with maximum amplitudes in the positive x_2-direction and minimum amplitudes in the negative x_2-direction.

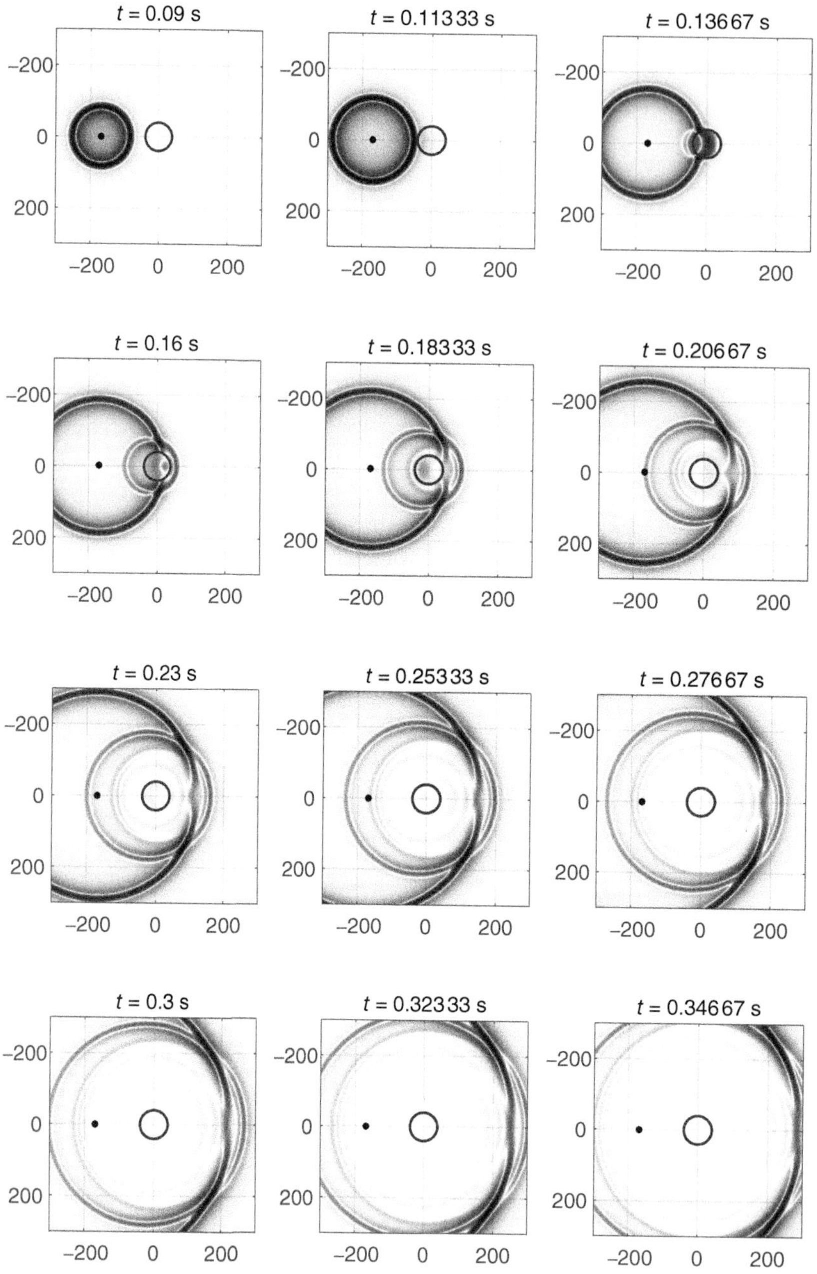

Figure 1.14 Snapshots of the 2D wave propagation in a model of a circular cylinder with $c_0 = 1500$ m and $c_{sct} = 3000$ m/s; the source is located at $x_1^S = 0$ of the vertical axis and $x_2^S = -170$ m of the horizontal axis.

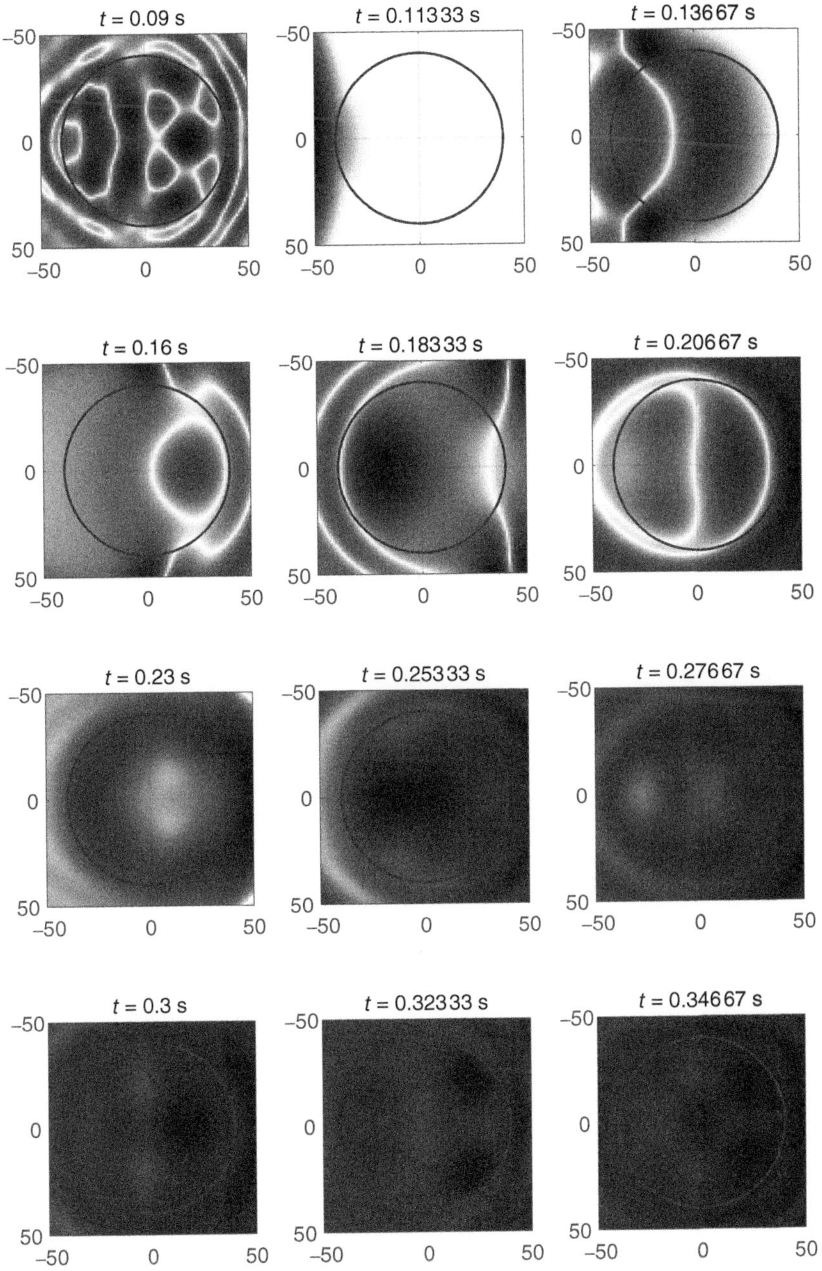

Figure 1.15 Similar to Figure 1.14, but the grid is reduced to 100 m by 100 m. Each image is normalized to the maximum value of the wave-field power (in dB). The wave-field values of the first picture only show the numerical noise.

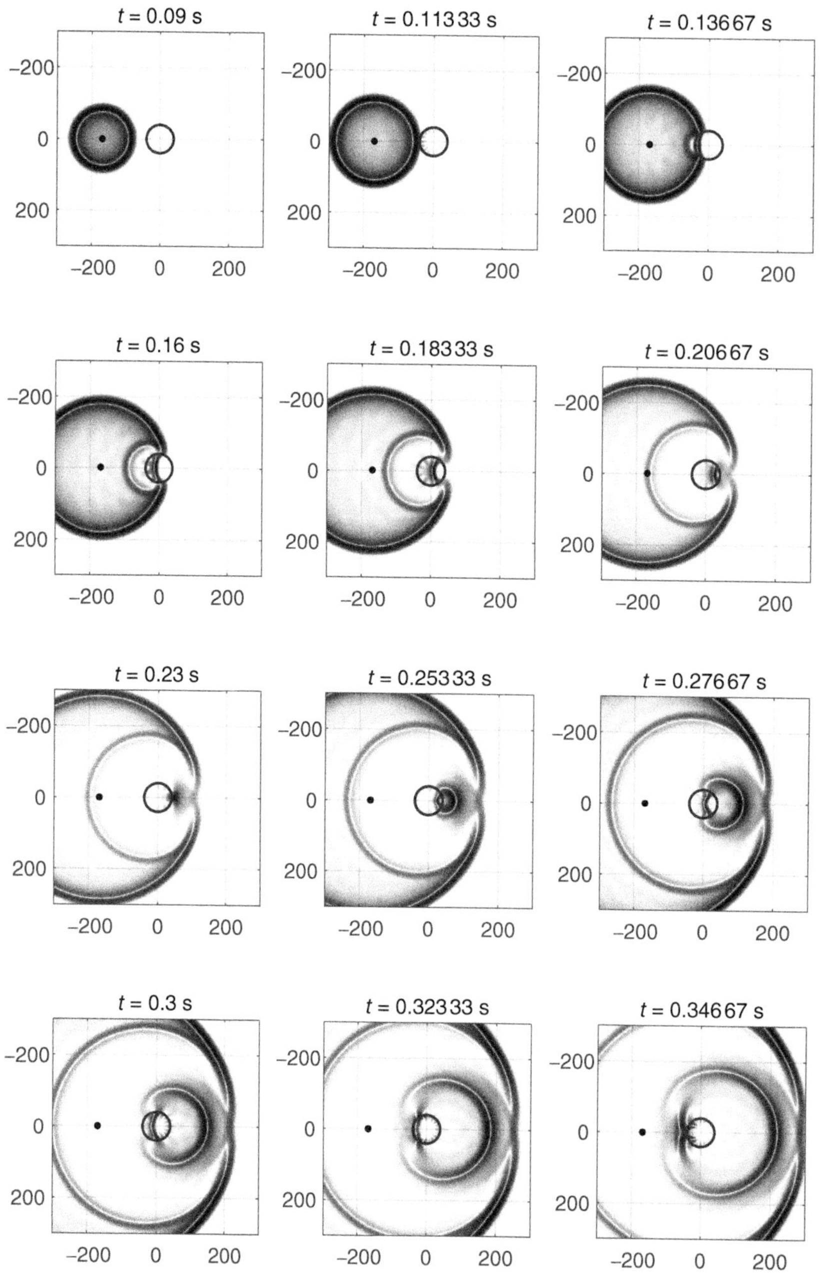

Figure 1.16 Snapshots of the 2D wave propagation in a model of a circular cylinder with $c_0 = 1500$ m and $c_{sct} = 750$ m/s; the source is located at $x_1^S = 0$ of the vertical axis and $x_2^S = -170$ m of the horizontal axis.

Figures 1.16 and 1.17 display the snapshots in case the wave speed within the circular cylinder is half the background wave speed. In the first few pictures of the figure, it is clear that the wave goes slower through the object than outside the object. For larger time instants, we observe that a secondary wave is generated that propagates along both sides of the circular interface of cylinder.

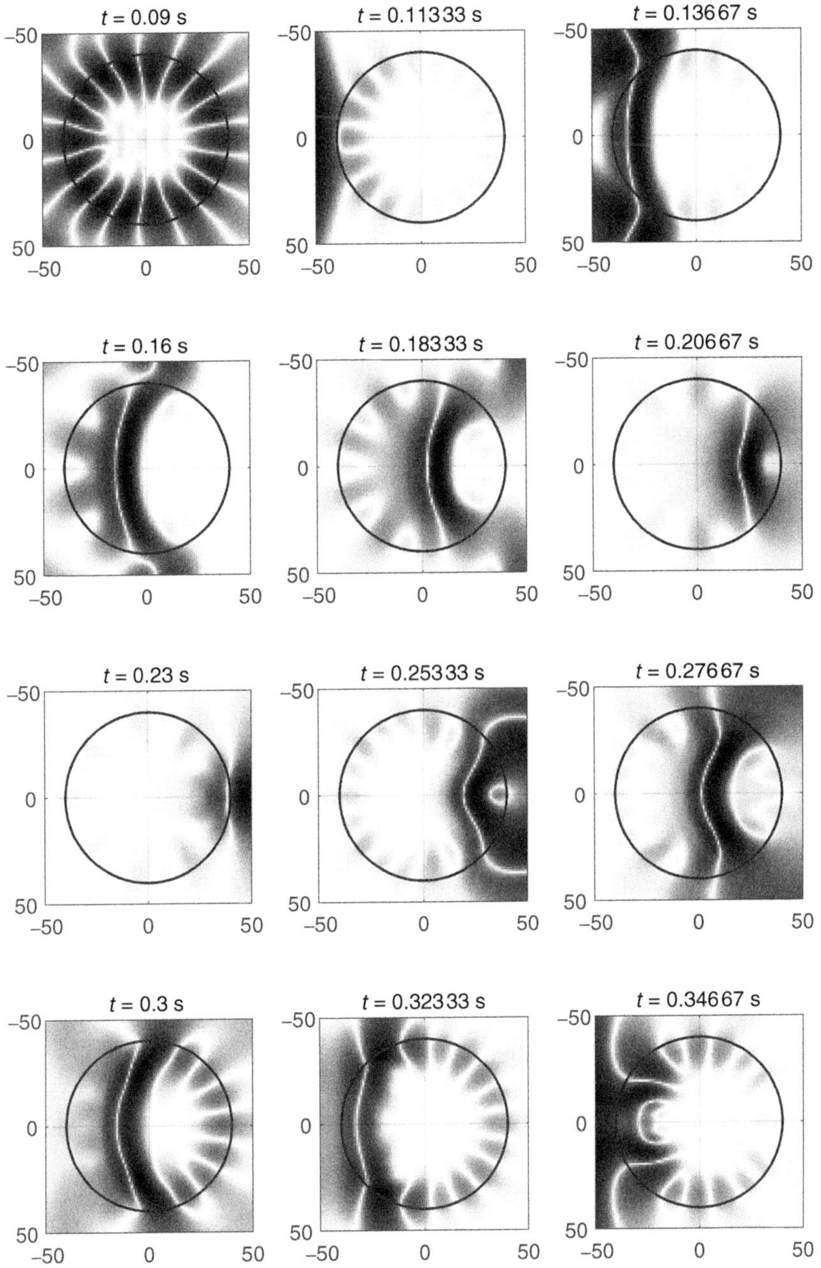

Figure 1.17 Similar to Figure 1.16, but the grid is reduced to 100 m by 100 m. The wave-field values of the first picture only show the numerical noise, although some field structure is present.

When the two wave fronts meet, they form a cardioid wave front together. After further propagation from this point, a new cardioid wave front is formed and is completed at the opposite point of the circular interface. If we investigate Figures 1.16 and 1.17 in more detail, we see that in the last figure there are some strong scattering phenomena inside the circular cylinder. This is definitely the result of a greater difference in wave speed inside and outside the object. From Figures 1.16 and 1.17, it appears that in the case of a smaller wave speed in the object, the circular object acts as a secondary interface type of source.

In the following chapters, we will pay more attention to these interface sources and how to explicitly include them in our theoretical model.

1.A Representation for Homogeneous Green Functions

In this appendix, we present the expressions for the Green function as a solution of the one-, two-, and three-dimensional wave equation. We assume that the wave speed in whole space is determined by the constant value c_0. In the literature, many different derivations have been published to obtain the Green function as the fundamental solution of the wave equation with a Dirac function as source distribution. In this appendix, we will derive the Green functions in such a way that it follows the same line, both in 1D, 2D, and 3D.

In a homogeneous space, we apply a spatial Fourier transformation, so the wave equation becomes algebraic. After solving the corresponding algebraic equation, an inverse spatial Fourier transformation provides the required expressions. In view of our choice that x_1 is the vertical direction, while x_2 and x_3 are the horizontal directions, the x_1-direction is our preferred direction.

Additionally, in each dimension, the Green function $\hat{G}(x|x')$ can be rewritten as a superposition of a product of two types of wave functions. One type of functions satisfy the causality conditions, while the other type does not. For convenience to solve some canonical scattering problems, we shall present them as well.

1.A.1 1D Green Function

Let us start with the one-dimensional wave equation, i.e. Eq. (1.44),

$$\partial_1\partial_1\hat{G} - \hat{\gamma}_0^2\hat{G} = -\delta(x_1 - x_1'), \tag{1.A.1}$$

where $\delta(x_1)$ denotes the unit impulse Dirac distribution. We define the 1D spatial Fourier transform pair as

$$\tilde{u}(k_1) = \int_{x_1 \in \mathbb{R}} \exp(-ik_1x_1)\,\hat{u}(x_1)\,\mathrm{d}x_1, \tag{1.A.2}$$

$$\hat{u}(x_1) = \frac{1}{2\pi}\int_{k_1 \in \mathbb{R}} \exp(ik_1x_1)\,\tilde{u}(k_1)\,\mathrm{d}k_1. \tag{1.A.3}$$

An important feature is that in the spatial domain, the spatial derivative $\partial_1\hat{u}$ is equivalent to an algebraic multiplication $ik_1\tilde{u}$ in the Fourier domain. Further the Fourier transform of $\delta(x_1 - x_1')$ is equal to $\exp(-ik_1x_1')$.

We now apply the 1D Fourier transform to both sides of Eq. (1.A.1) to arrive at

$$-k_1^2\tilde{G} - \hat{\gamma}_0^2\tilde{G} = -\exp(-ik_1x_1'), \tag{1.A.4}$$

with solution

$$\tilde{G} = \frac{\exp(-ik_1x_1')}{k_1^2 + \hat{\gamma}_0^2}. \tag{1.A.5}$$

The inverse Fourier transform yields the 1D Green function,

$$\boxed{\hat{G}(x_1|x_1') = \frac{1}{2\pi}\int_{k_1\in\mathbb{R}} \frac{\exp[ik_1(x_1 - x_1')]}{k_1^2 + \hat{\gamma}_0^2}\,dk_1.} \tag{1.A.6}$$

When $x_1 \geq x_1'$, the integral at the right-hand side of Eq. (1.A.6) is evaluated by continuing the integrand analytically into the complex k_1-plane, supplementing the path of integration by a semicircle situated in the upper half-plane $0 \leq \text{Im}(k_1) < \infty$ and of infinitely large radius, and applying the theorem of residues (see Figure 1.A.1). On account of Jordanś lemma, the contribution from the semicircle at infinity vanishes. Further, the only singularity of the integrand in the upper half of the complex k_1-plane is the simple pole at $k_1 = i\hat{\gamma}_0$, where we have assumed that the real part of $\hat{\gamma}_0 = s/c_0$ has a small positive real part. Taking into account the residue of this pole, we arrive at

$$\hat{G}(x_1|x_1') = \frac{1}{2\hat{\gamma}_0}\exp[-\hat{\gamma}_0(x_1 - x_1')], \quad (x_1 - x_1') \geq 0. \tag{1.A.7}$$

When $x_1 \leq x_1'$, the integral at the right-hand side of Eq. (1.A.6) is evaluated by continuing the integrand analytically into the complex k_1-plane, supplementing the path of integration by a semicircle situated in the lower half-plane $-\infty < \text{Im}(k_1) \leq 0$ and of infinitely large radius, and applying the theorem of residues (see Figure 1.A.1). On account of Jordanś lemma, the contribution from the semicircle at infinity vanishes. Further, the only singularity of the integrand in the lower half of the complex k_1-plane is the simple pole at $k_1 = -i\hat{\gamma}_0$. Taking into account the residue of this pole, we arrive at

$$\hat{G}(x_1|x_1') = \frac{1}{2\hat{\gamma}_0}\exp[\hat{\gamma}_0(x_1 - x_1')], \quad (x_1 - x_1') \leq 0. \tag{1.A.8}$$

Combining the two results, the 1D Green function is obtained as

$$\boxed{\hat{G}(x_1|x_1') = \frac{\exp[-\hat{\gamma}_0|x_1 - x_1'|]}{2\hat{\gamma}_0},} \tag{1.A.9}$$

for all x_1 and x_1' in $\mathbb{R}$. This result is used in Subsection 1.B.1

We remark that this result leads to the **addition theorem** for 1D waves, i.e.

$$\hat{G}(x_1|x_1') = \frac{1}{2\hat{\gamma}_0}\begin{cases} \exp[-\hat{\gamma}_0 x_1]\,\exp(\hat{\gamma}_0 x_1'), & x_1' \leq x_1, \\ \exp[\hat{\gamma}_0 x_1]\,\exp(-\hat{\gamma}_0 x_1'), & x_1' \geq x_1. \end{cases} \tag{1.A.10}$$

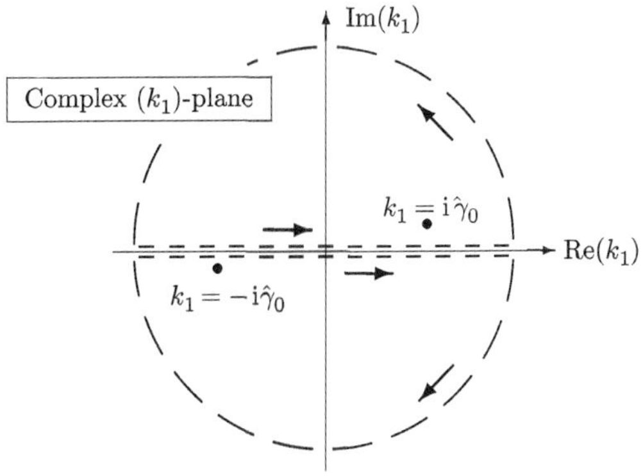

Figure 1.A.1 Contour integration in the complex k_1-plane.

1.A.2 2D Green Function

Let us start with the two-dimensional wave equation, i.e. Eq. (1.38),

$$\partial_1\partial_1\hat{G} + \partial_2\partial_2\hat{G} - \hat{\gamma}_0^2\hat{G} = -\delta(x_1 - x'_1, x_2 - x'_2). \tag{1.A.11}$$

We define the 2D spatial Fourier transform pair as

$$\tilde{u}(k_1,k_2) = \int_{(x,x_2)\in\mathbb{R}^2} \exp(-\mathrm{i}k_1x_1 - \mathrm{i}k_2x_2)\,\hat{u}(x_1,x_2)\,\mathrm{d}x_1\mathrm{d}x_2, \tag{1.A.12}$$

$$\hat{u}(x_1,x_2) = \left(\frac{1}{2\pi}\right)^2 \int_{(k_1,k_2)\in\mathbb{R}^2} \exp(\mathrm{i}k_1x_1 + \mathrm{i}k_1x_1)\,\tilde{u}(k_1,k_2)\,\mathrm{d}k_1\mathrm{d}k_2. \tag{1.A.13}$$

In the spatial domain, the spatial derivatives, $\partial_j\hat{u}$, $j = 1, 2$, are equivalent to algebraic multiplications $\mathrm{i}k_j\tilde{u}$ in the Fourier domain. Further, the Fourier transform of $\delta(x_1 - x'_1, x_2 - x'_1)$ is equal to $\exp(-\mathrm{i}k_1x'_1 - \mathrm{i}k_2x'_2)$.

We now apply the 2D spatial Fourier transform to both sides of Eq. (1.A.11) to arrive at

$$-k_1^2\tilde{G} - k_2^2\tilde{G} - \hat{\gamma}_0^2\tilde{G} = -\exp(-\mathrm{i}k_1x'_1 - \mathrm{i}k_2x'_2), \tag{1.A.14}$$

with solution

$$\tilde{G} = \frac{\exp(-\mathrm{i}k_1x'_1 - \mathrm{i}k_2x'_2)}{k_1^2 + k_2^2 + \hat{\gamma}_0^2}. \tag{1.A.15}$$

The inverse Fourier transform yields

$$\hat{G}(x_1,x_2|x'_1,x'_2) = \frac{1}{4\pi}\int_{(k_1,k_2)\in\mathbb{R}^2} \frac{\exp[-\mathrm{i}k_1(x_1 - x'_1) - \mathrm{i}k_2(x_2 - x'_2)]}{k_1^2 + k_2^2 + \hat{\gamma}_0^2}\,\mathrm{d}k_1\mathrm{d}k_2. \tag{1.A.16}$$

So far, there is no distinction between the two spatial directions. To be consistent with the 1D case, we consider the integration with respect to k_1. Similar as in the 1D case, the integral with respect to k_1 at the right-hand side of Eq. (1.A.16) is evaluated by continuing the integrand analytically into the complex k_1-plane. The only difference is that the poles are now given by $k_1 = \pm\mathrm{i}(\hat{\gamma}_0^2 + k_2^2)^{\frac{1}{2}}$, see Figure 1.A.2. Taking into account the two residues, the 2D Green function is obtained as

$$\boxed{\hat{G}(x_1,x_2|x'_1,x'_2) = \frac{1}{4\pi}\int_{k_2\in\mathbb{R}} \frac{\exp\left[-(\hat{\gamma}_0^2 + k_2^2)^{\frac{1}{2}}|x_1 - x'_1| + \mathrm{i}k_2(x_2 - x'_2)\right]}{(\hat{\gamma}_0^2 + k_2^2)^{\frac{1}{2}}}\,\mathrm{d}k_2,} \tag{1.A.17}$$

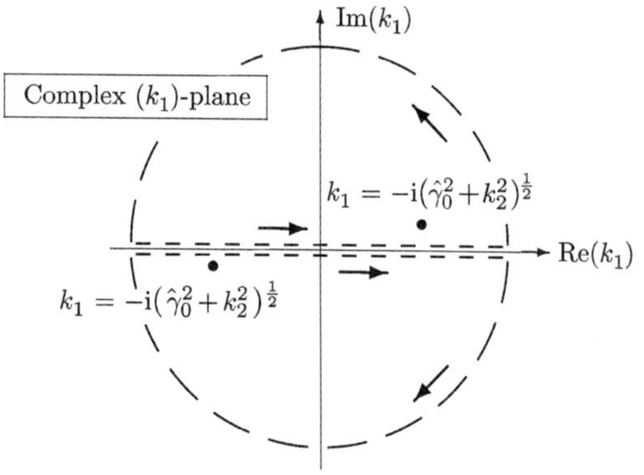

Figure 1.A.2 Contour integration in the complex k_1-plane.

for all (x_1, x_2) and (x'_1, x'_2) in $\mathbb{R}^2$. This plane-wave representation of the 2D Green function is convenient to derive the 2D Green function in a horizontally layered medium parallel to the $x_1 = 0$ plane, where the medium properties are invariant in the x_2-direction.

However, in a homogeneous 2D space, the medium properties are invariant in all directions. Then, the position and the orientation of the Cartesian coordinate system in the spectral domain with coordinates k_1 and k_2 may be chosen freely. This leads to a further simplification of the expression for the 2D Green function. Let $\boldsymbol{x}_T - \boldsymbol{x}'_T = (x_1 - x'_1, x_2 - x'_2)$ be the distance vector. For convenience, we set up a Cartesian system such that k_1 runs parallel to the distance vector and k_2 is perpendicular to the distance vector. Then, the dot product of $\boldsymbol{k}_T$ and $(\boldsymbol{x}_T - \boldsymbol{x}'_T)$, simplifies to

$$k_1(x_1 - x'_1) + k_2(x_2 - x'_2) = \boldsymbol{k}_T \cdot (\boldsymbol{x}_T - \boldsymbol{x}'_T) = \pm k_1 \, |\boldsymbol{x}_T - \boldsymbol{x}'_T|, \tag{1.A.18}$$

where the plus sign holds when the two vectors have the same direction and the minus sign holds when the two vectors have opposite directions. Returning to Eq. (1.A.16), we replace the argument of the exponential function of Eq. (1.A.16) by the right-hand side of Eq. (1.A.18) and carry out the two contour integrations in the complex k_1-plane to arrive at

$$\hat{G}(x_1, x_2|x'_1, x'_2) = \frac{1}{4\pi}\int_{k_2\in\mathbb{R}} \frac{\exp\left[-(\hat{\gamma}_0^2 + k_2^2)^{\frac{1}{2}}\,|\boldsymbol{x}_T - \boldsymbol{x}'_T|\right]}{(\hat{\gamma}_0^2 + k_2^2)^{\frac{1}{2}}}\,\mathrm{d}k_2. \tag{1.A.19}$$

Then, we replace the integration variable k_2 in Eq. (1.A.19) by the variable ψ through

$$k_2 = \mathrm{i}\gamma_0 \cosh(\psi), \tag{1.A.20}$$

we arrive at

$$\hat{G}(x_1, x_2|x'_1, x'_2) = \frac{1}{4\pi}\int_{-\infty}^{\infty} \exp[-\hat{\gamma}_0|\boldsymbol{x}_T - \boldsymbol{x}'_T|\cosh(\psi)]\mathrm{d}\psi. \tag{1.A.21}$$

Introducing the modified Bessel function of the second kind and zero order [1, p. 376],

$$K_0(z) = \int_0^{\infty} \exp[-z \cosh(\psi)]\mathrm{d}\psi, \tag{1.A.22}$$

we obtain the 2D Green function from Eq. (1.A.21) as

$$\boxed{\hat{G}(x_1, x_2|x'_1, x'_2) = \frac{1}{2\pi}K_0(\hat{\gamma}_0|\boldsymbol{x}_T - \boldsymbol{x}'_T|).} \tag{1.A.23}$$

1.A.2.1 Cylindrical Polar Coordinates

For rotationally symmetric configurations, it is convenient to introduce the polar coordinates as

$$x_1 = r\,\cos(\phi), \quad x_2 = r\,\sin(\phi) \quad \text{and} \quad x'_1 = r'\,\cos(\phi'), \quad x_2 = r\,\sin(\phi')\,. \tag{1.A.24}$$

Then, the distance between the position $\boldsymbol{x}$ and $\boldsymbol{x}'$ is written as

$$|\boldsymbol{x}_T - \boldsymbol{x}'_T| = \sqrt{(r^2 + r'^{\,2} - 2\,r\,r'\cos(\phi - \phi')}. \tag{1.A.25}$$

We use the **addition theorem** for the modified Bessel functions, see Watson [60, p. 361],

$$K_0(\hat{\gamma}_0|\boldsymbol{x}_T - \boldsymbol{x}'_T|) = \sum_{m=-\infty}^{\infty}\begin{cases} K_m(\hat{\gamma}_0 r)I_m(\hat{\gamma}_0 r')\exp[im(\phi - \phi')] & r' \leq r, \\ I_m(\hat{\gamma}_0 r)K_m(\hat{\gamma}_0 r')\exp[im(\phi - \phi')] & r' \geq r, \end{cases} \tag{1.A.26}$$

where I_m and K_m are the mth order modified Bessel function of the first kind and second kind, respectively. Then, we observe that the 2D Green function expression of Eq. (1.A.23) can be written as a superposition of the product of causal wave functions $K_m(\hat{\gamma}_0 r)\exp(\mathrm{i}\, m\phi)$ and noncausal wave functions $I_m(\hat{\gamma}_0 r)\exp(\mathrm{i}m\phi)$, viz.,

$$\hat{G}(r,\phi|r',\phi') = \frac{1}{2\pi}\sum_{m=0}^{\infty}\epsilon_m \cos[m(\phi-\phi')]\begin{cases} K_m(\hat{\gamma}_0 r)I_m(\hat{\gamma}_0 r'), & r' \le r,\\ I_m(\hat{\gamma}_0 r)K_m(\hat{\gamma}_0 r'), & r' \ge r, \end{cases} \tag{1.A.27}$$

where we have used that $K_m = K_{-m}$, and $I_m = I_{-m}$, so that the terms of positive and negative m can be combined. Further, $\epsilon_m = 1$ if $m = 0$ and $\epsilon_m = 2$ if $m \neq 0$. This result for the 2D Green function in polar coordinates is used in Subsection 1.B.2.

1.A.3 3D Green Function

Let us start with the three-dimensional wave equation, i.e. Eq. (1.15),

$$\partial_k\partial_k\hat{G} - \hat{\gamma}_0^2\hat{G} = -\delta(\boldsymbol{x}-\boldsymbol{x}'). \tag{1.A.28}$$

We define the 3D spatial Fourier transform pair as

$$\tilde{u}(\boldsymbol{k}) = \int_{\boldsymbol{x}\in\mathbb{R}^3} \exp(-\mathrm{i}k_1x_1 - \mathrm{i}k_2x_2 - \mathrm{i}k_3x_3)\hat{u}(\boldsymbol{x})\,\mathrm{d}x_1\mathrm{d}x_2\mathrm{d}x_3, \tag{1.A.29}$$

$$\hat{u}(\boldsymbol{x}) = \left(\frac{1}{2\pi}\right)^3\int_{\boldsymbol{k}\in\mathbb{R}^3} \exp(\mathrm{i}k_1x_1 + \mathrm{i}k_2x_2 + \mathrm{i}k_3x_3)\,\tilde{u}(\boldsymbol{k})\,\mathrm{d}k_1\mathrm{d}k_2\mathrm{d}k_3. \tag{1.A.30}$$

We use the notation $\boldsymbol{k} = (k_1,\boldsymbol{k}_H) = (k_1,k_2,k_3)$ and $\boldsymbol{x} = (x_1,\boldsymbol{x}_H) = (x_1,x_2,x_3)$, in which $\boldsymbol{k}_H$ and $\boldsymbol{x}_H$ are the horizontal directions $\boldsymbol{k}_H = (k_2,k_3)$ and $\boldsymbol{x}_H = (x_2,x_3)$, respectively. In the spatial domain, the spatial derivatives, $\partial_k\hat{u}$, $k = 1,2,3$, are equivalent to an algebraic multiplication $\mathrm{i}k_k\tilde{u}$ in the Fourier domain. Further, the Fourier transform of $\delta(\boldsymbol{x}-\boldsymbol{x}') = \delta(x_1-x_1',x_2-x_2',x_3-x_3')$ is equal to $\exp(-\mathrm{i}k_1x_1' - \mathrm{i}k_2x_2' - \mathrm{i}k_3x_3')$.

We now apply the 3D Fourier transform to both sides of Eq. (1.A.28) to arrive at

$$-k_1^2\tilde{G} - k_2^2\tilde{G} - k_3^2\tilde{G} - \hat{\gamma}_0^2\tilde{G} = -\exp(-\mathrm{i}k_1x_1' - \mathrm{i}k_2x_2' - \mathrm{i}k_3x_3'), \tag{1.A.31}$$

with solution

$$\tilde{G} = \frac{\exp(-\mathrm{i}k_1x_1' - \mathrm{i}k_2x_2' - \mathrm{i}k_3x_3')}{k_1^2+k_2^2+k_3^2+\hat{\gamma}_0^2}. \tag{1.A.32}$$

The inverse Fourier transform yields

$$\hat{G}(\boldsymbol{x}|\boldsymbol{x}') = \left(\frac{1}{2\pi}\right)^3\int_{\boldsymbol{k}\in\mathbb{R}^3}\frac{\exp[\mathrm{i}k_1(x_1-x_1') + \mathrm{i}\boldsymbol{k}_H\cdot(\boldsymbol{x}_H-\boldsymbol{x}_H')]}{k_1^2+\boldsymbol{k}_H\cdot\boldsymbol{k}_H+\hat{\gamma}_0^2}\,\mathrm{d}k_1\mathrm{d}k_2\mathrm{d}k_3. \tag{1.A.33}$$

So far there is no distinction between the three spatial directions. To be consistent with the 1D case and 2D case, we consider the integration with respect to k_1. Similar as in the 2D case, the integral with respect to k_1 at the right-hand side of Eq. (1.A.33) is evaluated by continuing the integrand analytically into the complex k_1-plane. The only difference is that the poles are now given by $k_1 = \pm\mathrm{i}(\hat{\gamma}_0^2+\boldsymbol{k}_H\cdot\boldsymbol{k}_H)^{\frac{1}{2}}$, see Figure 1.A.3. Taking into account the two residues, the 3D Green function is obtained as

$$\boxed{\hat{G}(\boldsymbol{x}|\boldsymbol{x}') = \frac{1}{8\pi^2}\int_{\boldsymbol{k}_H\in\mathbb{R}^2}\frac{\exp\left[-(\hat{\gamma}_0^2+\boldsymbol{k}_H\cdot\boldsymbol{k}_H)^{\frac{1}{2}}|x_1-x_1'| + \mathrm{i}\boldsymbol{k}_H\cdot(\boldsymbol{x}_H-\boldsymbol{x}_H')\right]}{(\hat{\gamma}_0^2+\boldsymbol{k}_H\cdot\boldsymbol{k}_H)^{\frac{1}{2}}}\mathrm{d}k_2\mathrm{d}k_3,} \tag{1.A.34}$$

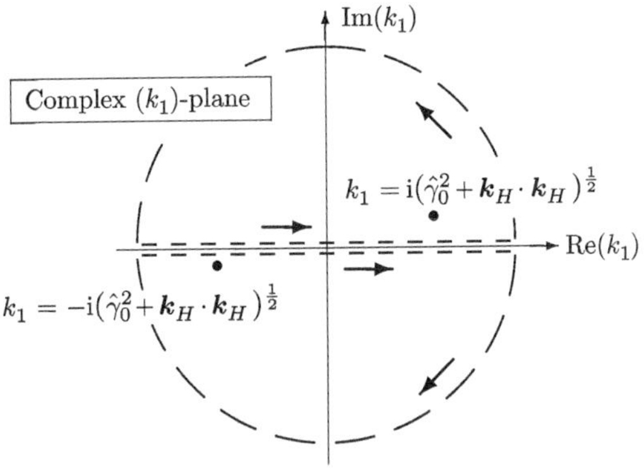

Figure 1.A.3 Contour integration in the complex k_1-plane.

for all $\boldsymbol{x}$ and $\boldsymbol{x}'$ in $\mathbb{R}^3$. This plane-wave representation of the 3D Green function is convenient to derive the 3D Green function in a horizontally layered medium parallel to the $x_1 = 0$ plane. The medium properties are invariant in the x_1- and x_2-directions.

However, in a homogeneous 3D space, the medium properties are invariant in all directions. Then, the position and the orientation of the Cartesian coordinate system in the spectral domain with coordinates k_1, k_2, and k_3 may be chosen freely. This leads to a further simplification of the expression for the 3D Green function. Let $\boldsymbol{x} - \boldsymbol{x}' = (x_1 - x_1', x_2 - x_2', x_3 - x_3')$ be the distance vector. For convenience, we set up a Cartesian system such that k_1 runs parallel to the distance vector and $\boldsymbol{k}_H$ is perpendicular to the distance vector. Then, the dot product of $\boldsymbol{k}$ and $(\boldsymbol{x} - \boldsymbol{x}')$, simplifies to

$$k_1(x_1 - x_1') + k_2(x_2 - x_2') + k_2(x_2 - x_2') = \boldsymbol{k} \cdot (\boldsymbol{x} - \boldsymbol{x}') = \pm k_1 \, |\boldsymbol{x} - \boldsymbol{x}'|, \tag{1.A.35}$$

where the plus sign holds when the two vectors have the same direction and the minus sign holds when the two vectors have opposite directions. Subsequently, we replace the argument of the exponential function of Eq. (1.A.33) by the right-hand side of Eq. (1.A.35) and carry out the two contour integrations in the complex k_3-plane to arrive at

$$\hat{G}(\boldsymbol{x}|\boldsymbol{x}') = \frac{1}{8\pi^2} \int_{\boldsymbol{k}_H \in \mathbb{R}^2} \frac{\exp\left[-(\hat{\gamma}_0^2 + \boldsymbol{k}_H \cdot \boldsymbol{k}_H)^{\frac{1}{2}} |\boldsymbol{x} - \boldsymbol{x}'|\right]}{(\hat{\gamma}_0^2 + \boldsymbol{k}_H \cdot \boldsymbol{k}_H)^{\frac{1}{2}}} \, \mathrm{d}k_2 \mathrm{d}k_3. \tag{1.A.36}$$

Then, we replace the integration variable k_2 and k_3 in Eq. (1.A.17) by the variables κ and ψ through

$$k_2 = \kappa \, \cos(\psi), \quad k_3 = \kappa \, \sin(\psi). \tag{1.A.37}$$

so that

$$\begin{aligned}
\hat{G}(\boldsymbol{x}|\boldsymbol{x}') &= \frac{1}{8\pi^2} \int_0^{2\pi} \mathrm{d}\psi \int_0^{\infty} \frac{\exp\left[-(\hat{\gamma}_0^2 + \kappa^2)^{\frac{1}{2}} |\boldsymbol{x} - \boldsymbol{x}'|\right]}{(\hat{\gamma}_0^2 + \kappa^2)^{\frac{1}{2}}} \, \kappa \, \mathrm{d}\kappa, \\
&= \frac{1}{4\pi} \int_0^{\infty} \frac{\exp\left[-(\hat{\gamma}_0^2 + \kappa^2)^{\frac{1}{2}} |\boldsymbol{x} - \boldsymbol{x}'|\right]}{(\hat{\gamma}_0^2 + \kappa^2)^{\frac{1}{2}}} \, \kappa \, \mathrm{d}\kappa, \\
&= -\frac{1}{4\pi} \left. \frac{\exp\left[-(\hat{\gamma}_0^2 + \kappa^2)^{\frac{1}{2}} |\boldsymbol{x} - \boldsymbol{x}'|\right]}{|\boldsymbol{x} - \boldsymbol{x}'|} \right|_{\kappa=0}^{\kappa=\infty},
\end{aligned} \tag{1.A.38}$$

with final result

$$\boxed{\hat{G}(\boldsymbol{x}|\boldsymbol{x}') = \frac{\exp[-\hat{\gamma}_0 |\boldsymbol{x} - \boldsymbol{x}'|]}{4\pi |\boldsymbol{x} - \boldsymbol{x}'|}.} \tag{1.A.39}$$

1.A.3.1 Spherical Polar Coordinates

For rotationally symmetric configurations, it is convenient to introduce the polar coordinates as

$$\begin{aligned} x_1 &= r\ \sin(\theta)\cos(\phi), & x_2 &= r\ \sin(\theta)\sin(\phi), & x_3 &= r\cos(\theta), \\ x_1' &= r'\ \sin(\theta')\cos(\phi'), & x_2' &= r'\ \sin(\theta')\sin(\phi'), & x_3' &= r'\cos(\theta'). \end{aligned} \tag{1.A.40}$$

Then, the distance between the positions of $\boldsymbol{x}$ and $\boldsymbol{x}'$ can be written as

$$|\boldsymbol{x}-\boldsymbol{x}'| = \sqrt{(r^2 + r'^2 - 2rr'\cos(\boldsymbol{x},\boldsymbol{x}')}, \tag{1.A.41}$$

where $\cos(\boldsymbol{x},\boldsymbol{x}')$ is the cosine of the angle between the vectors $\boldsymbol{x}$ and $\boldsymbol{x}'$, which can be written as

$$\cos(\boldsymbol{x},\boldsymbol{x}') = \cos(\theta)\cos(\theta') + \sin(\theta)\sin(\theta')\cos(\phi-\phi'). \tag{1.A.42}$$

We now use the **addition theorem** for the modified spherical Bessel functions [36, equation 10.60.3],

$$\frac{\exp(-\hat{\gamma}_0|\boldsymbol{x}_T - \boldsymbol{x}_T'|)}{\hat{\gamma}_0|\boldsymbol{x}_T - \boldsymbol{x}_T'|} = \frac{2}{\pi}\sum_{n=0}^{\infty}(2n+1)\begin{cases} k_n(\hat{\gamma}_0 r)\ i_n^{(1)}(\hat{\gamma}_0 r')\ P_n[\cos(\boldsymbol{x},\boldsymbol{x}')], & r' \le r, \\ i_n^{(1)}(\hat{\gamma}_0 r)\ k_n(\hat{\gamma}_0 r')\ P_n[\cos(\boldsymbol{x},\boldsymbol{x}')], & r' \ge r, \end{cases} \tag{1.A.43}$$

where $i_n^{(1)}$ and k_n are the nth order modified spherical Bessel functions of the first kind and second kind, respectively. Further, P_n is the Legendre polynomial of the nth degree. For Legendre polynomials, there exists an addition theorem, which expresses it in terms of products of associated Legendre functions, viz., $P_n^m[\cos(\theta)]\exp(im\phi)$ and $P_n^m[\cos[\theta')]\exp(im\phi')$, cf. Stratton [47, p. 408]. Note that the 3D Green function is a superposition of the product of causal wave functions of the type $k_n(\hat{\gamma}_0 r)P_n^m[\cos(\theta)]\exp(im\phi)$ and noncausal wave functions of the type $i_n^{(1)}(\hat{\gamma}_0 r)P_n^m[\cos(\theta)]\exp(im\phi)$.

Using Eq. (1.A.43), we observe that Eq. (1.A.39) transfers into

$$\hat{G}(\boldsymbol{x}|\boldsymbol{x}') = \frac{\hat{\gamma}_0}{2\pi^2}\sum_{n=0}^{\infty}(2n+1)\begin{cases} k_n(\hat{\gamma}_0 r)\ i_n^{(1)}(\hat{\gamma}_0 r')\ P_n[\cos(\boldsymbol{x},\boldsymbol{x}')], & r' \le r, \\ i_n^{(1)}(\hat{\gamma}_0 r)\ k_n(\hat{\gamma}_0 r')\ P_n[\cos(\boldsymbol{x},\boldsymbol{x}')], & r' \ge r. \end{cases} \tag{1.A.44}$$

This result for the 3D Green function in spherical coordinates is used in Subsection 1.B.3.

1.B Scattering by a Simple Canonical Configuration

In order to have a necessary check on the correctness of the coding of the domain integral equations at hand, we present the analytical expressions for some simple configurations. We begin with the one-dimensional scattering by a homogenous slab. Second, we discuss the two-dimensional scattering by a homogeneous circular cylinder. Third, we consider the three-dimensional scattering by a homogeneous sphere.

1.B.1 1D Scattering by a Slab

In order to generate synthetic data for the one-dimensional configuration, we consider the scattering problem by a homogeneous slab (Figure 1.B.1). The slab has a thickness of $2a$. The medium inside the slab is characterized by the constant wave speed $\hat{c}_{sct}$. The center of the slab is at $x_1 = 0$. The incident wave field is generated by a planar source at x_1^S and is given by

$$\hat{u}^{inc}(x_1|x_1^S) = \frac{\hat{Q}}{2\hat{\gamma}_0}\exp[-\hat{\gamma}_0|x_1 - x_1^S|]. \tag{1.B.1}$$

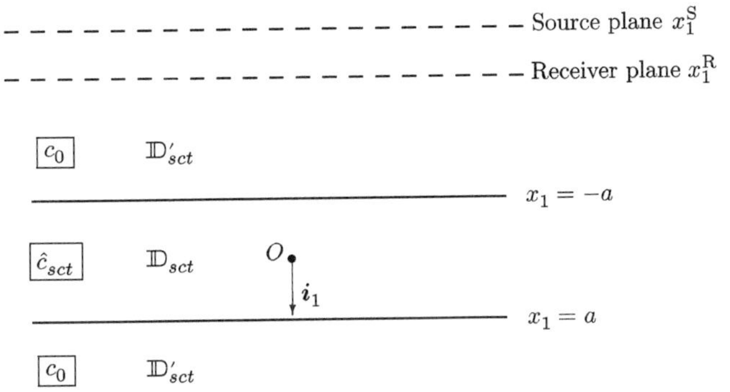

Figure 1.B.1 The 1D configuration with homogeneous slab.

The amplitude of the incident field at $x_1 = -a$ is given by

$$\hat{u}^{inc}(-a|x_1^S) = \frac{\hat{Q}}{2\hat{\gamma}_0} \exp[\hat{\gamma}_0(x_1^S + a)]. \tag{1.B.2}$$

This is a common factor in all wave field quantities.

The different wave constituents are considered to be generated at the interfaces. Taking into account the causality condition for the reflected field, we write the reflected field in the negative x_1-direction as

$$\hat{u}^{rfl}(x_1|x_1^S) = \frac{\hat{Q}}{2\hat{\gamma}_0} \exp[\hat{\gamma}_0(x_1^S + a)]\, \hat{R} \exp[\hat{\gamma}_0(x_1 + a)], \quad x_1 < -a. \tag{1.B.3}$$

The wave field inside the slab consists of two waves propagating in opposite directions, hence, we write this interior wave field as

$$\hat{u}^{int}(x_1|x_1^S) = \frac{\hat{Q}}{2\hat{\gamma}_0} \exp[\hat{\gamma}_0(x_1^S + a\ )]\{\hat{A} \exp[-\hat{\gamma}_{sct}(x_1 + a)] + \hat{B} \exp[\hat{\gamma}_{sct}(x_1 - a)]\}, \tag{1.B.4}$$

when $-a < x_1 < a$. For $x_1 > a$, the field transmitted in the positive x_1-direction is given by

$$\hat{u}^{trm}(x_1|x_1^S) = \frac{\hat{Q}}{2\hat{\gamma}_0} \exp[\hat{\gamma}_0(x_1^S + a)]\, \hat{T} \exp[-\hat{\gamma}_0(x_1 + a)]. \tag{1.B.5}$$

The unknown expansion factors $\hat{R}$, $\hat{A}$, $\hat{B}$, and $\hat{T}$ follow from the interface conditions at $r = -a$ and $r = a$, viz.,

$$\lim_{x_1 \uparrow -a} [\hat{u}^{inc}(\boldsymbol{x}_1 + x_1^S) + \hat{u}^{rfl}(\boldsymbol{x}_1 + x_1^S)] = \lim_{x_1 \downarrow -a} \hat{u}^{int}(\boldsymbol{x}_1 + x_1^S), \tag{1.B.6}$$

$$\lim_{x_1 \uparrow -a} \partial_1 [\hat{u}^{inc}(\boldsymbol{x}_1 + x_1^S) + \hat{u}_k^{rfl}(\boldsymbol{x}_1 + x_1^S)] = \lim_{x_1 \downarrow -a} \partial_1 \hat{u}^{int}(\boldsymbol{x}_1 + x_1^S), \tag{1.B.7}$$

$$\lim_{x_1 \uparrow -a} \hat{u}^{int}(\boldsymbol{x}_1 + x_1^S) = \lim_{x_1 \downarrow -a} \hat{u}^{trm}(\boldsymbol{x}_1 + x_1^S), \tag{1.B.8}$$

$$\lim_{x_1 \uparrow -a} \partial_1 \hat{u}^{int}(\boldsymbol{x}_1 + x_1^S) = \lim_{x_1 \downarrow -a} \partial_1 \hat{u}^{trm}(\boldsymbol{x}_1 + x_1^S), \tag{1.B.9}$$

Substitution of the wave-field expressions into these interface conditions leads to the following set of equations:

$$1 + \hat{R} = \hat{A} + \exp(-2\hat{\gamma}_{sct}a)\ \hat{B}, \tag{1.B.10}$$

$$-\hat{\gamma}_0 + \hat{\gamma}_0 R = -\hat{\gamma}_{sct}\hat{A} + \hat{\gamma}_{sct} \exp(-2\hat{\gamma}_{sct}a)\ \hat{B}, \tag{1.B.11}$$

$$\exp(-2\hat{\gamma}_{sct}a)\,\hat{A}+\hat{B}=\hat{T}\exp(-2\hat{\gamma}_0 a), \tag{1.B.12}$$

$$-\hat{\gamma}_{sct}\exp(-2\hat{\gamma}_{sct}a)\,\hat{A}+\hat{\gamma}_{sct}\hat{B}=-\hat{\gamma}_0\hat{T}\exp(-2\hat{\gamma}_0 a). \tag{1.B.13}$$

Elimination of $\hat{R}$ from the first and the second equation and $\hat{T}$ from the third and the fourth equation leads to two equations for $\hat{A}$ and $\hat{B}$, with solution

$$\hat{A}=\frac{\hat{\tau}}{1-\hat{\rho}^2\exp(-4\hat{\gamma}_{sct}a)},\quad \hat{B}=\frac{-\hat{\tau}\,\hat{\rho}\,\exp(-2\hat{\gamma}_{sct}a)}{1-\hat{\rho}^2\exp(-4\hat{\gamma}_{sct}a)}, \tag{1.B.14}$$

where

$$\hat{\rho}=\frac{\hat{\gamma}_0-\hat{\gamma}_{sct}}{\hat{\gamma}_0+\hat{\gamma}_{sct}},\quad \hat{\tau}=\frac{2\hat{\gamma}_0}{\hat{\gamma}_0+\hat{\gamma}_{sct}}, \tag{1.B.15}$$

are the local reflection and transmission factors. Subsequently, the global reflection and transmission factors are obtained as

$$\hat{R}=\hat{A}+\exp(-2\hat{\gamma}_{sct}a)\,\hat{B}-1, \tag{1.B.16}$$

$$\hat{T}=[\exp(-2(\hat{\gamma}_{sct}a)\,\hat{A}+\hat{B}]\exp(2\hat{\gamma}_0 a). \tag{1.B.17}$$

As far as the scattered field is concerned, we distinguish between a point of observation above the slab and below the slab. When we have a receiver at the point x_1^R above the slab, the scattered wave is denoted as the reflected wave and is given by

$$\hat{u}^{rfl}(x_1^R|x_1^S)=\frac{\hat{Q}}{2\hat{\gamma}_0}\,\hat{R}\,\exp[\hat{\gamma}_0(x_1^R+x_1^S+2a)],\quad x_1^R<-a. \tag{1.B.18}$$

To arrive at the scattered field below the slab, we have to subtract the incident field from the transmitted field. We obtain

$$\hat{u}^{trm}(x_1^R|x_1^S)-\hat{u}^{inc}(x_1^R|x_1^S)=\frac{\hat{Q}}{2\hat{\gamma}_0}\,(\hat{T}-1)\,\exp[-\hat{\gamma}_0(x_1^R-x_1^S)],\quad x_1^R>a. \tag{1.B.19}$$

1.B.2 2D Scattering by a Circular Cylinder

In order to generate synthetic data for the two-dimensional configuration, we consider the scattering problem by a homogeneous, circular cylinder (Figure 1.B.2). We define the spatial position by $\boldsymbol{x}_T=(x_1,x_2)$. The cylinder has a radius a. The medium in the interior of the cylinder is characterized by the constant wave speed $\hat{c}_{sct}$. The center of the cylinder is at $x_1=0,\ x_2=0$. The incident wave field is generated by a monopole line source at $\boldsymbol{x}_T^S=(x_1^S,x_2^S)$ and is given by

$$\hat{u}^{inc}(\boldsymbol{x}_T|\boldsymbol{x}_T^S)=\frac{\hat{Q}}{2\pi}K_0(\hat{\gamma}_0|\boldsymbol{x}_T-\boldsymbol{x}_T^S|). \tag{1.B.20}$$

In order to solve our scattering problem at hand, we introduce polar coordinates adapted to the geometry of the circular cylinder,

$$x_1=r\,\cos(\phi),\quad x_2=r\,\sin(\phi),\quad 0\le\phi<2\pi. \tag{1.B.21}$$

Similarly for the source and receiver coordinates, we introduce

$$x_1^S=r^S\,\cos(\phi^S),\quad x_2^S=r^S\,\sin(\phi^S),\quad 0\le\phi^S<2\pi, \tag{1.B.22}$$

$$x_1^R=r^R\,\cos(\phi^R),\quad x_2^R=r^R\,\sin(\phi^R),\quad 0\le\phi^R<2\pi. \tag{1.B.23}$$

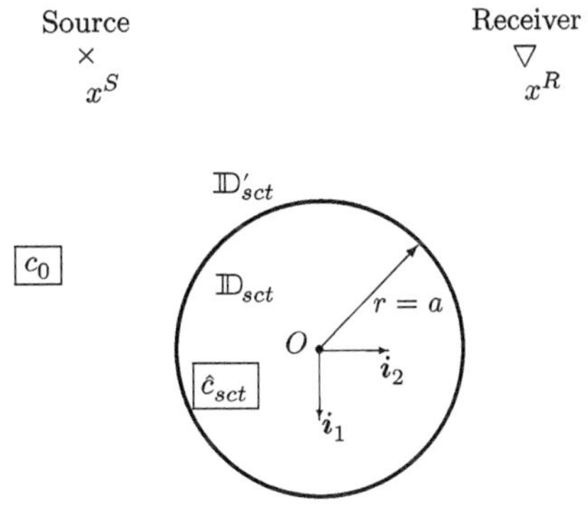

Figure 1.B.2 The 2D configuration with homogeneous circular cylinder.

Then, using the addition theorem for modified Bessel functions, see Watson [60], p. 361], we observe that the incident wave field is represented as an infinite series, in which the spatial dependencies of $\boldsymbol{x}_T$ and $\boldsymbol{x}_T^S$ are degenerated, viz.,

$$\hat{u}^{inc}(\boldsymbol{x}_T|\boldsymbol{x}_T^S) = \frac{\hat{Q}}{2\pi}\sum_{m=0}^{\infty}\epsilon_m \cos[m(\phi-\phi^S)]\begin{cases} I_m(\hat{\gamma}_0 r)\,K_m(\hat{\gamma}_0 r^S), & r \leq r^S, \\ K_m(\hat{\gamma}_0 r)\,I_m(\hat{\gamma}_0 r^S), & r \geq r^S, \end{cases} \tag{1.B.24}$$

where $\epsilon_m = 1$ if $m = 0$ and $\epsilon_m = 2$ if $m \neq 0$. Here, $I_m(\cdot)$ is the modified Bessel function of the first kind and integer order, while $K_m(\cdot)$ is the modified Bessel function of the second kind and integer order. Note that, according to the first expression of Eq. (1.B.24), the incident wave field in $r = 0$ is bounded; according to the second expression of Eq. (1.B.24), the incident wave field satisfies the causality condition when $r \to \infty$. For points on the surface and inside the cylinder with radius a, the first expression of Eq. (1.B.24) applies.

Taking into account the causality condition for the reflected wave field, we write the reflected wave field outside the scattering cylinder as

$$\hat{u}^{rfl}(\boldsymbol{x}_T|\boldsymbol{x}_T^S) = \frac{\hat{Q}}{2\pi}\sum_{m=0}^{\infty}\epsilon_m\hat{A}_m\,K_m(\hat{\gamma}_0 r)\,K_m(\hat{\gamma}_0 r^S)\,\cos[m(\phi-\phi^S)]. \tag{1.B.25}$$

The wave field inside the cylinder has to be bounded at $r = 0$, hence, we write this interior wave field as

$$\hat{u}^{int}(\boldsymbol{x}_T|\boldsymbol{x}_T^S) = \frac{\hat{Q}}{2\pi}\sum_{m=0}^{\infty}\epsilon_m\hat{B}_m\,I_m(\hat{\gamma}_{sct} r)\,K_m(\hat{\gamma}_0 r^S)\,\cos[m(\phi-\phi^S)]. \tag{1.B.26}$$

The unknown expansion factors $\hat{A}_m$ and $\hat{B}_m$ follow from the interface conditions at $r = a$, viz.,

$$\lim_{r\downarrow a}\,[\hat{u}^{inc}(\boldsymbol{x}_T|\boldsymbol{x}_T^S) + \hat{u}^{rfl}(\boldsymbol{x}_T|\boldsymbol{x}_T^S)] = \lim_{r\uparrow a}\,\hat{u}^{int}(\boldsymbol{x}_T|\boldsymbol{x}_T^S), \tag{1.B.27}$$

$$\lim_{r\downarrow a}\,\partial_r[\hat{u}^{inc}(\boldsymbol{x}_T|\boldsymbol{x}_T^S) + \hat{u}^{rfl}(\boldsymbol{x}_T|\boldsymbol{x}_T^S)] = \lim_{r\uparrow a}\,\partial_r\hat{u}^{int}(\boldsymbol{x}_T|\boldsymbol{x}_T^S). \tag{1.B.28}$$

Substitution of the wave-field expressions into these interface conditions leads, for each n to the following set of equations:

$$I_m(\hat{\gamma}_0 a) + \hat{A}_m\,K_m(\hat{\gamma}_0 a) = \hat{B}_m\,I_m(\hat{\gamma}_{sct} a), \tag{1.B.29}$$

$$[\partial_r I_m(\hat{\gamma}_0 r) + \hat{A}_m\partial_r K_m(\hat{\gamma}_0 r)]_{r=a} = \hat{B}_m\,\partial_r I_m(\hat{\gamma}_{sct} r)\Big|_{r=a}. \tag{1.B.30}$$

Solving these equations for the coefficients $\hat{A}_m$ and $\hat{B}_m$ leads to

$$\hat{A}_m = -\frac{\hat{\gamma}_{sct}\partial I_m(\hat{\gamma}_{sct}a)\; I_m(\hat{\gamma}_0 a) - \hat{\gamma}_0 \partial I_m(\hat{\gamma}_0 a)\; I_m(\hat{\gamma}_{sct}a)}{\hat{\gamma}_{sct}\partial I_m(\hat{\gamma}_{sct}a)\; K_m(\hat{\gamma}_0 a) - \hat{\gamma}_0 \partial K_m(\hat{\gamma}_0 a)\; I_m(\hat{\gamma}_{sct}a)} \tag{1.B.31}$$

and

$$\hat{B}_m = (I_m(\hat{\gamma}_0 a) + \hat{A}_m\; K_m(\hat{\gamma}_0 a))/I_m(\hat{\gamma}_{sct}a), \tag{1.B.32}$$

where $\partial I_m(\cdot)$ and $\partial K_m(\cdot)$ denote the derivatives of the functions I_m and K_m with respect to their arguments. They are obtained from the relations

$$\begin{aligned} \partial I_m(z) &= \partial_z I_m(z) = I_{m+1}(z) + \frac{m}{z} I_m(z), \\ \partial K_m(z) &= \partial_z K_m(z) = -K_{m+1}(z) + \frac{m}{z} K_m(z). \end{aligned} \tag{1.B.33}$$

Substituting the reflection factors in the expression for the reflected wave field of Eq. (1.B.25) and taking $\boldsymbol{x}_T = \boldsymbol{x}_T^R$, we obtain

$$\hat{u}^{rfl}(\boldsymbol{x}_T^R|\boldsymbol{x}_T^S) = \frac{\hat{Q}}{2\pi}\sum_{m=0}^{\infty} \epsilon_m \hat{A}_m\; K_m(\hat{\gamma}_0 r^R)\; K_m(\hat{\gamma}_0 r^S)\; \cos[m(\phi^R - \phi^S)]. \tag{1.B.34}$$

With this expression, we are able to compute the synthetic data pertaining to the present example of a domain scatterer. The various expressions to compute the modified Bessel functions can be found in [1, pp. 374–388].

Note that source–receiver reciprocity holds. Interchanging source and receiver does not changes the data.

1.B.3 3D Scattering by a Sphere

In order to generate synthetic data for the three-dimensional scatterer, we consider the scattering problem by a homogeneous sphere (Figure 1.B.3). The sphere has a radius a. The medium in the interior of the sphere is characterized by the constant wave speed $\hat{c}_{sct}$. The center of the sphere is at $\boldsymbol{x} = (0, 0, 0)$. The incident wave field is generated by a monopole source at $\boldsymbol{x}^S = (x_1^S, x_2^S, x_3^S)$ and is given by

$$\hat{u}^{inc}(\boldsymbol{x}|\boldsymbol{x}^S) = \hat{Q}\frac{\exp(-\hat{\gamma}_0|\boldsymbol{x} - \boldsymbol{x}^S|)}{4\pi|\boldsymbol{x} - \boldsymbol{x}^S|}. \tag{1.B.35}$$

In order to solve our scattering problem at hand, we introduce spherical coordinates adapted to the geometry of the sphere,

$$x_1 = r\; \sin(\theta)\; \cos(\phi), \quad x_2 = r\; \sin(\theta)\; \sin(\phi), \quad x_3 = r\; \cos(\theta), \tag{1.B.36}$$

with $0 \le \phi < 2\pi$ and $0 \le \theta < \pi$. Similarly for the source and receiver coordinates, we introduce

$$x_1^S = r^S\; \sin(\theta^S)\; \cos(\phi^S), \quad x_2^S = r^S\; \sin(\theta^S)\; \sin(\phi^S), \quad x_3^S = r^S\; \cos(\theta^S), \tag{1.B.37}$$

$$x_1^R = r^R\; \sin(\theta^R)\; \cos(\phi^R), \quad x_2^R = r^R\; \sin(\theta^R)\; \sin(\phi^R), \quad x_3^R = r^R\; \cos(\theta^R). \tag{1.B.38}$$

Then, using the addition theorem for modified spherical Bessel functions, see [36, equation 10.60.3], we observe that the incident wave field is represented as an infinite series, in which the spatial dependencies of $\boldsymbol{x}$ and $\boldsymbol{x}^S$ are explicitly shown, cf. Eq. (1.A.43),

$$\hat{u}^{inc}(\boldsymbol{x}|\boldsymbol{x}^S) = \frac{\hat{Q}\hat{\gamma}_0}{2\pi^2}\sum_{n=0}^{\infty}(2n+1)\; P_n[\cos(\boldsymbol{x}, \boldsymbol{x}^S)] \begin{cases} i_n^{(1)}(\hat{\gamma}_0 r)\; k_n(\hat{\gamma}_0 r^S), & r \le r^S, \\ k_n(\hat{\gamma}_0 r)\; i_n^{(1)}(\hat{\gamma}_0 r^S), & r \ge r^S. \end{cases} \tag{1.B.39}$$

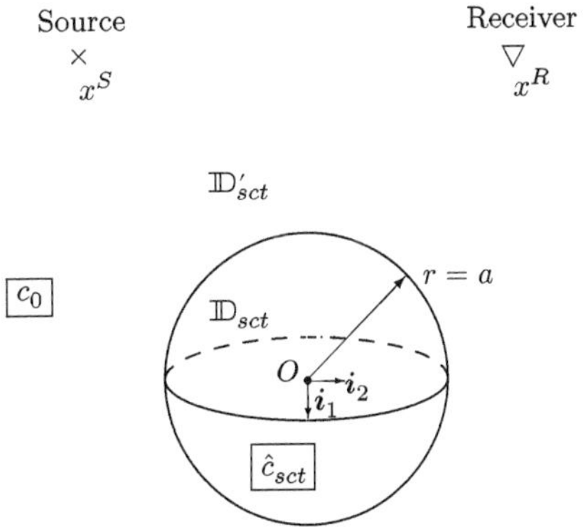

Figure 1.B.3 The 3D configuration with homogeneous sphere.

Here, $i_n^{(1)}(\cdot)$ and $k_n(\cdot)$ are the nth order modified spherical Bessel functions of the first kind and second kind, respectively.

Further, $\cos(\boldsymbol{x}, \boldsymbol{x}^S)$ denotes the cosine of the angle between the vectors $\boldsymbol{x}$ and $\boldsymbol{x}^S$, and $P_n(\cdot)$ is the Legendre polynomial of degree n. It is computed recursively as, see [36, equations (18.9.1) and (18.9.2)],

$$P_n(x) = \frac{2n-1}{n}\, x\, P_{n-1}(x) - \frac{n-1}{n}\, P_{n-2}(x), \tag{1.B.40}$$

where $P_0(x) = 1$ and $P_1(x) = x$.

Continuing with our field analysis, according to the first expression of Eq. (1.B.39), the incident wave field in $r = 0$ is bounded; according to the second expression of Eq. (1.B.39), the incident wave field satisfies the causality condition when $r \to \infty$. For points on the surface of the sphere with radius a, the first expression of Eq. (1.B.39) applies.

Taking into account the causality condition for the reflected wave field, we write the reflected wave field outside the scattering sphere as

$$\hat{u}^{rfl}(\boldsymbol{x}|\boldsymbol{x}^S) = \frac{\hat{Q}\hat{\gamma}_0}{2\pi^2} \sum_{n=0}^{\infty} (2n+1)\, \hat{A}_n\, k_n(\hat{\gamma}_0 r)\, k_n(\hat{\gamma}_0 r^S)\, P_n[\cos(\boldsymbol{x}, \boldsymbol{x}^S)]. \tag{1.B.41}$$

The wave field inside the sphere has to be bounded at $r = 0$, hence, we write this interior wave field as

$$\hat{u}^{int}(\boldsymbol{x}|\boldsymbol{x}^S) = \frac{\hat{Q}\hat{\gamma}_0}{2\pi^2} \sum_{n=0}^{\infty} (2n+1)\, \hat{B}_n\, i_n^{(1)}(\hat{\gamma}_{sct} r)\, k_n(\hat{\gamma}_0 r^S)\, P_n[\cos(\boldsymbol{x}, \boldsymbol{x}^S)]. \tag{1.B.42}$$

The unknown expansion factors $\hat{A}_n$ and $\hat{B}_n$ follow from the interface conditions at $r = a$, where the wave field and its radial derivative are continuous. This leads to

$$\lim_{r \downarrow a} [\hat{u}^{inc}(\boldsymbol{x}|\boldsymbol{x}^S) + \hat{u}^{rfl}(\boldsymbol{x}|\boldsymbol{x}^S)] = \lim_{r \uparrow a} \hat{u}^{int}(\boldsymbol{x}|\boldsymbol{x}^S), \tag{1.B.43}$$

$$\lim_{r \downarrow a} \partial_r[\hat{u}^{inc}(\boldsymbol{x}|\boldsymbol{x}^S) + \hat{u}_k^{rfl}(\boldsymbol{x}|\boldsymbol{x}^S)] = \lim_{r \uparrow a} \partial_r \hat{u}_k^{int}(\boldsymbol{x}|\boldsymbol{x}^S). \tag{1.B.44}$$

Substitution of the wave-field expressions into these interface conditions leads, for each n, to the following set of equations:

$$i_n^{(1)}(\hat{\gamma}_0 a) + \hat{A}_n\, k_n(\hat{\gamma}_0 a) = \hat{B}_n\, i_n^{(1)}(\hat{\gamma}_{sct} a), \tag{1.B.45}$$

$$\left[\partial_r i_n^{(1)}(\hat{\gamma}_0 r) + \hat{A}_n \partial_r k_n(\hat{\gamma}_0 r)\right]_{r=a} = \hat{B}_n\, \partial_r\, i_n^{(1)}(\hat{\gamma}_{sct} r)\Big|_{r=a}. \tag{1.B.46}$$

Solving these equations for the coefficients $\hat{A}_n$ and $\hat{B}_n$ leads to

$$\hat{A}_n = -\frac{\hat{\gamma}_{sct}\ \partial i_n^{(1)}(\hat{\gamma}_{sct}a)\ i_n^{(1)}(\hat{\gamma}_0 a) - \hat{\gamma}_0\ \partial i_n^{(1)}(\hat{\gamma}_0 a)\ i_n^{(1)}(\hat{\gamma}_{sct}a)}{\hat{\gamma}_{sct}\ \partial i_n^{(1)}(\hat{\gamma}_{sct}a)\ k_n(\hat{\gamma}_0 a) - \hat{\gamma}_0\ \partial k_n(\hat{\gamma}_0 a)\ i_n^{(1)}(\hat{\gamma}_{sct}a)} \tag{1.B.47}$$

and

$$\hat{B}_n = \frac{i_n^{(1)}(\hat{\gamma}_0 a) + \hat{A}_n\ k_n(\hat{\gamma}_0 a)}{i_n^{(1)}(\hat{\gamma}_{sct}a)}, \tag{1.B.48}$$

where $\partial i_n^{(1)}(\cdot)$ and $\partial k_n(\cdot)$ denote the derivatives of the functions $i_n^{(1)}(\cdot)$ and $k_n(\cdot)$ with respect to their arguments. The derivatives of the spherical Bessel functions are obtained from the relations

$$\begin{aligned} \partial i_n^{(1)}(z) &= \partial_z i_n^{(1)}(z) = i_{n+1}(z) + \frac{n}{z}\ i_n(z), \\ \partial k_n(z) &= \partial_z k_n(z) = -k_{n+1}(z) + \frac{n}{z}\ k_n(z), \end{aligned} \tag{1.B.49}$$

see [36, equation (10.51.5)].

Substituting the reflection factors $\hat{A}_n$ in the expression for the reflected wave field of Eq. (1.B.41) and taking $\boldsymbol{x} = \boldsymbol{x}^R$, we obtain

$$\hat{u}^{rfl}(\boldsymbol{x}^R|\boldsymbol{x}^S) = \frac{\hat{Q}\hat{\gamma}_0}{2\pi^2}\sum_{n=0}^{\infty}(2n+1)\ \hat{A}_n\ k_n(\hat{\gamma}_0 r^R)\ k_n(\hat{\gamma}_0 r^S)\ P_n[\cos(\boldsymbol{x}^R, \boldsymbol{x}^S)]. \tag{1.B.50}$$

With this expression, we are able to compute the synthetic data pertaining to the present example of a domain scatterer.

As far as the computation of the argument of the Legendre polynomial is concerned, we use

$$\cos(\boldsymbol{x}^R, \boldsymbol{x}^S) = \frac{(r^R)^2 + (r^S)^2 - |\boldsymbol{x}^R - \boldsymbol{x}^S|^2}{2\ r^R\ r^S}. \tag{1.B.51}$$

Note again that Eq. (1.B.50) shows that source–receiver reciprocity holds. Interchanging source and receiver does not change the data.

In numerical computations using Matlab coding, the functions k_n and $i_n^{(1)}$ are not available. We therefore use their expressions in terms of modified Bessel functions of fractional order, viz.,

$$i_n^{(1)}(z) = \sqrt{\frac{1}{2}\pi/z}\ I_{n+\frac{1}{2}}(z) \quad \text{and} \quad k_n(z) = \sqrt{\frac{1}{2}\ \pi/z}\ K_{n+\frac{1}{2}}(z), \tag{1.B.52}$$

see [36, equations (10.47.7)–(10.47.9)].

In Sections 1.C.1–1.C.3 we will present the Matlab scripts to compute these analytic solutions for the scattering of a wave field by the present canonical objects.

1.C Matlab Codes for Scattering by Canonical Objects

1.C.1 Matlab Code for Slab (1D)

For our 1D example, the script `WavefieldTotSlab1D` (nDIM = 1) of Matlab Listing I.24 consists of the following steps:

(1) We compute the incident wave, using Eqs. (1.B.1) and (1.B.2).

Matlab Listing I.24 The `WavefieldTotSlab` script to compute the 1D incident and total wave fields.

```
clear all; clc; close all;
input = init();            X1 = input.X1; xS = input.xS;    a = input.a;
gam0  = input.gamma_0;  gam_sct = input.gamma_0 * input.c_0 / input.c_sct;

% (1) Compute incident wave ------------------------------------------
    u_inc   = 1/(2*gam0) * exp(-gam0*abs(xS-X1));
    u_inc_a = 1/(2*gam0) * exp(gam0*(xS+a));

% (2) Compute coefficients A and B of internal field ------------------
    rho         = (gam0-gam_sct) / (gam0+gam_sct);
    tau         =  2*gam0        / (gam0+gam_sct);
    denominator = 1 - rho * rho * exp(-4*gam_sct*a);
    A           = tau / denominator;
    B           = - rho * A * exp(-2*gam_sct*a);

% (3) Compute reflection and transmision factors ----------------------
    R =  A + exp(-2*gam_sct*a) * B - 1;
    T = (B + exp(-2*gam_sct*a) * A) * exp(2*gam0*a) ;

% (4) Compute total wave field distribution ---------------------------
    u = zeros(input.N1,1);
    for i = 1:input.N1
      if X1(i) < -a;  u(i) = u_inc(i)+R*u_inc_a*exp(gam0*(X1(i)+a)); end
      if abs(X1(i)) < a
         u(i) = u_inc_a * ( A * exp(-gam_sct*(X1(i)+a)) ...
                                  + B * exp(gam_sct*(X1(i)-a)) );
      end
      if  X1(i) > a;  u(i) = T * u_inc_a * exp(-gam0.*(X1(i)+a)); end
    end % i-loop
    plotWavefield(u_inc,u,input)
```

(2) We compute the coefficients A and B of the internal field, using Eqs. (1.B.14) and (1.B.15).
(3) We compute the reflection and transmission factors R and T with the help of Eqs. (1.B.16) and (1.B.17).
(4) We compute the incident and total wavefields. The results are plotted in Figure 1.C.1, using function `plotWavefield` of Matlab Listing I.8.

If we are interested in the scattered field at the two receivers above and below the slab, we use the script `WavefieldSctSlab` of Matlab Listing I.25. For benchmarking needs, the scattered field data are stored in the MAT-file DATA1D.mat.

1.C.2 Matlab Code for Circular Cylinder (2D)

In the analysis of scattering by a circular cylinder, we have introduced polar coordinates and have expanded the wave fields in a series of modified Bessel functions. To determine the number of terms to take into account in this series, we compare the analytic solution of the incident field, Eq. (1.B.20)

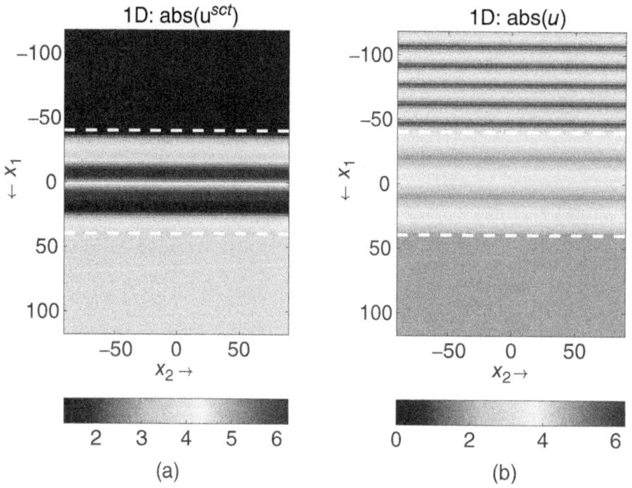

Figure 1.C.1 Plots of the 1D scattered (a) and total (b) wave fields. The dashed lines indicate the upper and lower boundary of the slab.

Matlab Listing I.25 The `WavefieldSctSlab` script to compute the 1D wave fields at the two receivers above and below the slab.

```
clear all; clc; close all;
input = init();          xS = input.xS; a = input.a;
gam0 = input.gamma_0; gam_sct = input.gamma_0 * input.c_0 / input.c_sct;

if exist(fullfile(cd, 'DATA1D.mat'), 'file');    delete DATA1D.mat;   end
% (1) Compute incident wave -----------------------------------------
    u_inc_a = 1/(2*gam0) * exp(gam0*(xS+a));
% (2) Compute coefficents A and B of internal field ------------------
    rho         = (gam0-gam_sct) / (gam0+gam_sct);
    tau         =  2*gam0        / (gam0+gam_sct);
    denominator = 1 - rho * rho * exp(-4*gam_sct*a);
    A           = tau / denominator;
    B           = - rho * A * exp(-2*gam_sct*a);
% (3) Compute reflection and transmision factors ---------------------
    R =  A + exp(-2*gam_sct*a) * B - 1;
    T =  (B + exp(-2*gam_sct*a) * A) * exp(2*gam0*a);
% (4) Compute receiver data for reflection and transmission ----------
         xR = input.xR(1,1);
    data1D(1) = u_inc_a * R * exp(gam0*(xR+a));
         xR = input.xR(1,2);
    data1D(2) = u_inc_a * (T-1) * exp(-gam0*(xR+a));

    displayData(data1D,input);                        save DATA1D data1D;
```

Matlab Listing I.26 The `WavefieldIncCircle` script to compute the 2D incident wave field.

```
clear all; clc; close all;
input = init();  gam0 = input.gamma_0;    a = input.a;

% Transform Cartesian coordinates to polar ccordinates
  xS = input.xS;
  rS = sqrt(xS(1)^2+xS(2)^2);  phiS  = atan2(xS(2),xS(1));
  X1 = input.X1; X2 = input.X2;
  R  = sqrt(X1.^2+X2.^2);  PHI = atan2(X2,X1);

% (1) Compute incident wave in closed form -----------------------------------
      DIS          = sqrt(rS^2+R.^2-2*rS*R.*cos(phiS-PHI));
      u_inc_exact = 1/(2*pi) .* besselk(0,gam0*DIS);

% (2) Compute incident wave as Bessel series with -M:M terms -------------
      M = 50;                                   % increase M for more accuracy
      u_inc = besselk(0,gam0*rS) .* besseli(0,gam0*R);   % zero order term
      for m = 1 : M
        u_inc = u_inc + 2 * besselk(m,gam0*rS) ...
                             .* besseli(m,gam0*R) .* cos(m*(phiS-PHI));
      end % m_loop
      u_inc = 1/(2*pi) * u_inc;

% (3) Determine mean error and plot error in domain -----------------------
      Error = u_inc - u_inc_exact;
      disp(['normalized error = '  ...
                          num2str(norm(Error(:),1)/norm(u_inc_exact(:),1))]);
      set(figure,'Units','centimeters','Position',[5 5 18 10]);
      subplot(1,3,1)
       IMAGESC(X1(:,1),X2(1,:),abs(u_inc_exact))
       title('\fontsize{10} 2D: abs(u^{inc}_{exact})');
       hold on; phi = 0:.01:2*pi; plot(a*cos(phi),a*sin(phi),'w');
      subplot(1,3,2)
       IMAGESC(X1(:,1),X2(1,:),abs(u_inc))
       title('\fontsize{10} 2D: abs(u^{inc})');
       hold on; phi = 0:.01:2*pi; plot(a*cos(phi),a*sin(phi),'w');
     subplot(1,3,3)
      IMAGESC(X1(:,1),X2(1,:),abs(Error))
      title('\fontsize{10} 2D: abs(Error)');
      hold on; phi = 0:.01:2*pi;  plot(a*cos(phi),a*sin(phi),'w');
```

with the results of the series solution, Eq. (1.B.24). The script `WavefieldIncCircle` (nDIM = 2) for this comparison is given in Matlab Listing I.26.

In step (1), we compute the incident field analytically, denoted as $\hat{u}^{inc}_{exact}$, and in step (2), we compute it using the series solution with 50 terms, denoted as $\hat{u}^{inc}$. The results are presented in Figure 1.C.2. From the two plots of the incident fields, it is difficult to estimate the accuracy

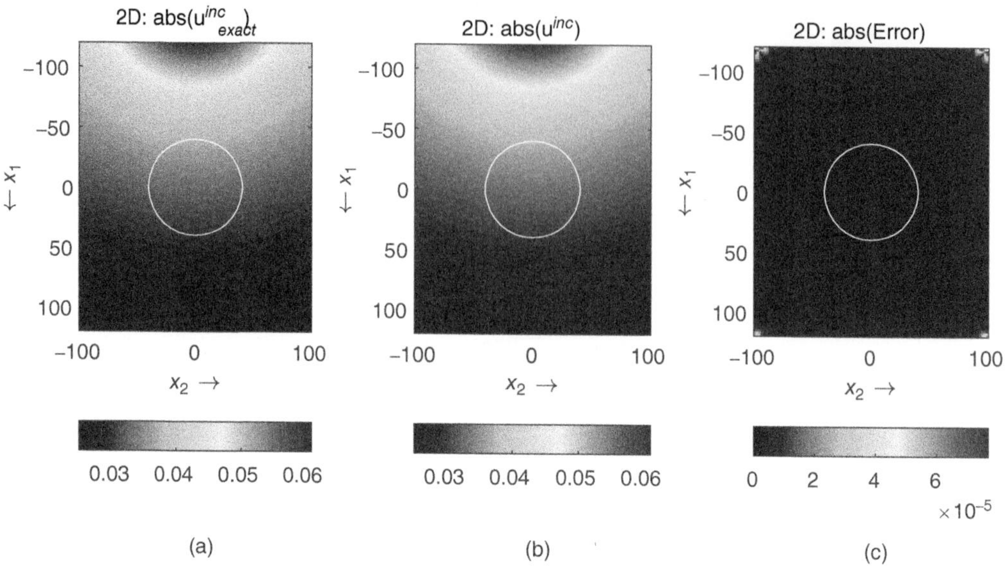

Figure 1.C.2 Comparison of the analytical (a) and the series solution (b) of the incident-wave-field results of `WavefieldIncCircle`, and the resulting error distribution (c).

achieved. Therefore, in step (3) of the Matlab script, we compute the error distribution on the grid, defined as

$$\text{Error}(x_n) = u^{inc}(x_n) - u^{inc}_{exact}(x_n), \tag{1.C.1}$$

and its normalized norm is defined as

$$\text{Mean Error} = \frac{\sum_{n=1}^{N} |\text{Error}(x_n)|}{\sum_{n=1}^{N} |u^{inc}_{exact}(x_n)|}. \tag{1.C.2}$$

The absolute value of the error distribution on the grid is plotted in the right picture of Figure 1.C.2. The error increases with increasing distance from the origin. We observe that the largest error occurs at the corners of the grid. With respect to the largest value of the incident wave field on the grid, the error at the corners are a factor of 10^{-3} smaller. The Mean Error is equal to 8×10^{-6}. For our scattering problem at hand, this is sufficiently small and we may conclude that 50 terms in the series of Bessel functions are sufficient to benchmark the numerical solution of the integral equation.

In the next script `WavefieldTotCircle.m` of Matlab Listing I.27, the total wave field is computed. The following steps are taken:

(1) We compute again the analytic version of the incident wave field.
(2) We compute the coefficients A and B of the reflected wave field and the interior wave field, using Eqs. (1.B.31) and (1.B.32), respectively. Matlab does not have functions for the derivatives of the Bessel functions, but they can directly be computed from the recurrence relation, see last relation of equation (9.6.26) in Abramowitz and Stegun [1].
(3) We compute the scattered wave field and plot the amplitudes of the scattered and total wave fields with the function `plotWavefield` of Matlab Listing 1.8. The results are presented in Figure 1.C.3.

Matlab Listing I.27 The `WavefieldTotCircle` script to compute the 2D total and scattered wave fields.

```
clear all; clc; close all;
input = init();
c_0   = input.c_0;  c_sct = input.c_sct;

gam0    = input.gamma_0;
gam_sct = input.gamma_0 * c_0/c_sct;

% Transform Cartesian coordinates to polar ccordinates
  xS = input.xS; rS = sqrt(xS(1)^2+xS(2)^2);     phiS = atan2(xS(2),xS(1));
  X1 = input.X1; X2 = input.X2; R = sqrt(X1.^2+X2.^2); PHI = atan2(X2,X1);

% (1) Compute incident wave in closed form -----------------------------------
      DIS    = sqrt(rS^2+R.^2-2*rS*R.*cos(phiS-PHI));
      u_inc  = 1/(2*pi) .* besselk(0,gam0*DIS);

% (2) Compute coefficients of series expansion ---------------------------
      arg0 = gam0 * input.a;  args = gam_sct*input.a;
      M = 50;                              % increase M for more accuracy
      A = zeros(1,M+1); B = zeros(1,M+1);
      for m = 0 : M
      Ib0 = besseli(m,arg0);       dIb0 =  besseli(m+1,arg0) + m/arg0 * Ib0;
      Ibs = besseli(m,args);       dIbs =  besseli(m+1,args) + m/args * Ibs;
      Kb0 = besselk(m,arg0);       dKb0 = -besselk(m+1,arg0) + m/arg0 * Kb0;
      A(m+1) = - (gam_sct * dIbs*Ib0 - gam0 * dIb0*Ibs) ...
                 /(gam_sct * dIbs*Kb0 - gam0 * dKb0*Ibs);
      B(m+1) = (Ib0 + A(m+1) * Kb0) / Ibs;
  end

% (3) Compute exterior, interior and total fields --------------------------
      factor = 1/(2*pi) * besselk(0,gam0*rS);
      u_rfl  = A(1) * factor * besselk(0,gam0*R);
      u_int  = B(1) * factor * besseli(0,gam_sct*R);
      for m = 1 : M
        factor = (1/pi) * besselk(m,gam0*rS) .* cos(m*(phiS-PHI));
        u_rfl  = u_rfl + A(m+1) * factor .* besselk(m,gam0*R);
        u_int  = u_int + B(m+1) * factor .* besseli(m,gam_sct*R);
      end
      u_rfl(ceil(input.N1/2),ceil(input.N2/2))=0;% Avoid NaN number at R=0
      u = (u_rfl+u_inc) .* (R >= input.a) +  u_int .* (R < input.a);

      plotWavefield(u_inc,u,input);
```

Comparing Figure 1.C.3 with its 1D version of Figure 1.C.1, we observe that at the vertical line, $x_2 = 0$, the total wave-field behavior is very different. However, at this vertical line, the scattered 2D wave field visibly resembles its 1D counterpart.

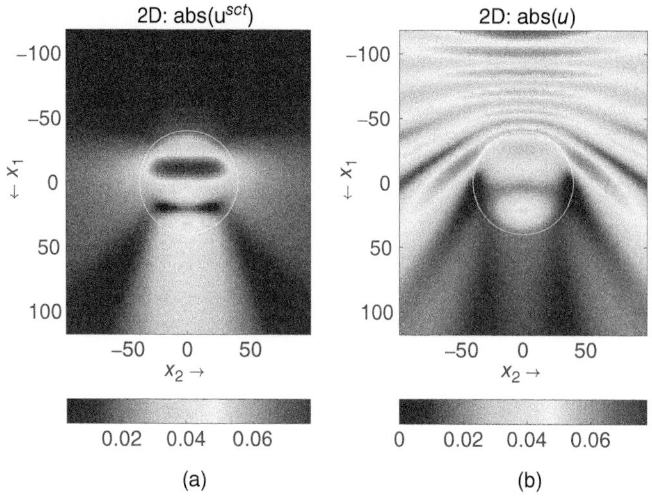

Figure 1.C.3 Plots of the 2D scattered (a) and total (b) wave fields.

From the plots of the scattered 2D wave fields (Figure 1.C.3), we observe that outside the cylinder, the forward scattering (in the direction of propagation of the incident wave) is dominant. It annihilates the incident field in the forward direction.

If one is interested in the scattered field at the receivers outside the cylinder, one can use the script `WavefieldSctCircle` of Matlab Listing I.28. Steps (1) and (2) are the same. Step (3) differs from the previous Matlab script: only a very limited set of observation points outside the grid is selected. In step (3) of the latter script, we use Eq. (1.B.34). The function `displayData` of Matlab Listing I.9 plots the scattered wave-field data at the receiver points around the cylinder, see Figure 1.C.4. In Figure 1.C.4, we observe clearly that the scattered wave is dominant in the forward direction with an angle of 0°.

For benchmarking needs, the scattered field data are stored in the MAT-file DATA2D.mat.

1.C.3 Matlab Code for Sphere (3D)

In the analysis of scattering by a sphere, we have introduced spherical coordinates and have expanded the wave fields in a series of modified spherical Bessel functions and Legendre polynomials.

In the present Matlab versions, there is no function available to compute the Legendre polynomials, except that in the latest releases there is a Matlab function available to compute the associated Legendre functions. We therefore present our own function `Legendre` of Matlab Listing I.29. To determine the number of terms to take into account in the series expansion of the wave fields, we compare the analytic solution of the incident field, Eq. (1.B.35) with the results of the series solution, Eq. (1.B.39).

The script `WavefieldIncSphere` (nDIM = 3) for this comparison is given in Matlab Listing I.30. In step (1) of this script, we compute the incident field analytically, denoted as $\hat{u}^{inc}_{exact}$, and in step (2), we compute it using the series solution with 50 terms, denoted as $\hat{u}^{inc}$. The results are presented in Figure 1.C.5.

Matlab Listing I.28 The `WavefieldSctCircle` script to compute the 2D scattered wave field at the receiver locations around the circular cylinder.

```
gclear all; clc; close all;
input = init();
c_0    = input.c_0;         c_sct   = input.c_sct;
gam0   = input.gamma_0;     gam_sct = input.gamma_0 * c_0/c_sct;

if exist(fullfile(cd, 'DATA2D.mat'), 'file');    delete DATA2D.mat;    end

% (1) Compute coefficients of series expansion -----------------------
     arg0 = gam0 * input.a;   args = gam_sct*input.a;
     M = 100;                                % increase M for more accuracy
     A = zeros(1,M+1);
     for m = 0 : M
       Ib0 = besseli(m,arg0);      dIb0 = besseli(m+1,arg0) + m/arg0 * Ib0;
       Ibs = besseli(m,args);      dIbs = besseli(m+1,args) + m/args * Ibs;
       Kb0 = besselk(m,arg0);      dKb0 =-besselk(m+1,arg0) + m/arg0 * Kb0;
       A(m+1) = - (gam_sct * dIbs*Ib0 - gam0 * dIb0*Ibs) ...
                 /(gam_sct * dIbs*Kb0 - gam0 * dKb0*Ibs);
     end

% (2) Compute reflected field at receivers (data) ----------------------
     xR = input.xR; xS = input.xS;
     rR = sqrt(xR(1,:).^2 + xR(2,:).^2);  phiR = atan2(xR(2,:),xR(1,:));
     rS = sqrt(xS(1)^2+xS(2)^2);            phiS = atan2(xS(2),xS(1));
     data2D = A(1) * besselk(0,gam0*rS).* besselk(0,gam0*rR);
     for m = 1 : M
       factor = 2 * besselk(m,gam0*rS) .* cos(m*(phiS-phiR));
       data2D = data2D + A(m+1) * factor .* besselk(m,gam0*rR);
     end % m_loop
     data2D = 1/(2*pi) * data2D;
     displayData(data2D,input);                        save DATA2D data2D;
```

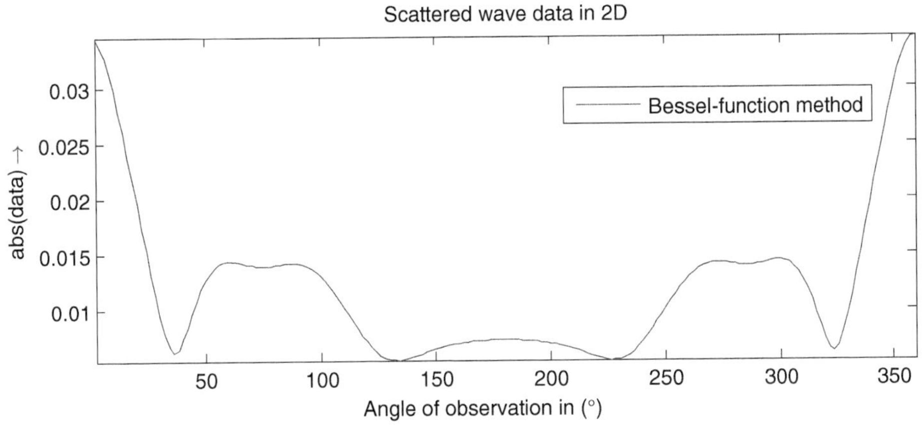

Figure 1.C.4 Plot of the 2D scattered wave field at the receivers around the cylinder.

Matlab Listing I.29 The function `Legendre.m` to compute the Legendre polynomials.

```
function [Pn,Pn_1,Pn_2] = Legendre(n,Z,Pn,Pn_1,Pn_2)
  if n == 0; Pn  = ones(size(Z)); Pn_2 = Pn; end
  if n == 1; Pn  = Z;    end
  if n > 1
     Pn_1 = Pn;
     Pn = (2*n-1)/n * Z .* Pn_1 -(n-1)/n * Pn_2;
     Pn_2 = Pn_1;
  end
```

From the two plots of the incident fields, it is difficult to estimate the accuracy achieved. Therefore, in step (3) of the Matlab script, we compute the error distribution on the grid, defined as

$$\mathrm{Error}(x_n) = u^{inc}_{exact}(\boldsymbol{x}_n) - u^{inc}(\boldsymbol{x}_n) \tag{1.C.3}$$

and its Mean Error, defined as

$$\text{Mean Error} = \frac{\sum_{n=1}^{N} |\mathrm{Error}(x_n)|}{\sum_{n=1}^{N} |u^{inc}(\boldsymbol{x}_n)|}.$$

The absolute value of the error distribution on the grid is plotted in the right picture of Figure 1.C.5. The error increases with increasing distance from the origin. On the grid, the largest error occurs at the corners of the grid. With respect to the largest value of the incident wave field on the grid, the errors at the corners are a factor of 2×10^{-3} smaller. The Mean Error is equal to 2×10^{-4}. For our scattering problem at hand, this is sufficiently small and we may conclude that 50 terms in the series of Bessel functions are sufficient to benchmark the numerical solution of the integral equations.

In the next script `WavefieldTotSphere` of Matlab Listing I.31, the total wave field is computed. The following steps are taken:

(1) We compute again the analytic version of the incident wave field.
(2) We compute the coefficients A and B of the reflected wave field and the interior wave field, using Eqs. (1.B.47) and (1.B.48), respectively. Matlab does not have functions for the derivatives of the Bessel functions, but they can directly be computed from the recurrence relation, see [36, equation (10.51.5)].
(3) We compute scattered wave field and plot the amplitudes of the scattered and total wave fields with the plot function `plotWavefield` of Matlab Listing I.8. The results are presented in Figure 1.C.6.

Comparing Figure 1.C.6 with its 2D version of Figure 1.C.3, we observe that in the plane $x_3 = 0$ the total wave field behavior is very different. From the plots of the scattered wave fields, we observe some similarity, although the amplitudes differ, since the incident wave field in 3D decays faster than in 2D. If we are interested in the scattered field at the receivers outside the sphere, we use the script `WavefieldSctSphere` of Matlab Listing I.32. Steps (1) and (2) are the same. Step (3)

Matlab Listing I.30 The `WavefieldIncSphere` script to compute the 3D incident wave field.

```
clear all; clc; close all;
input = init();
gam0 = input.gamma_0;    a = input.a;

% Transform Cartesian coordinates to spherical coordinates
  xS = input.xS;
  rS = sqrt(xS(1)^2+xS(2)^2+xS(3)^2);
  X1 = input.X1;    X2 = input.X2;    X3 = input.X3;
  R  = sqrt(X1.^2+X2.^2+X3.^2+1e-16);     % add small value to avoid zero R

% (1) Compute incident wave in closed form ----------------------------------
      DIS         = sqrt( (X1-xS(1)).^2 + (X2-xS(2)).^2 + (X3-xS(3)).^2 );
      u_inc_exact = exp(-gam0*DIS)./(4*pi*DIS);

% (2) Compute incident wave as Bessel series with 0:N terms ---------------
      N = 50;                                  % increase N for more accuracy
      COS    = (R.^2 + rS^2 - DIS.^2) ./ (2*rS*R);                % cos(x,xS)
      Pn     = zeros(size(R)); Pn_1 = Pn; Pn_2 = Pn;
      u_inc  = zeros(size(R));
      for n = 0 : N
        [Pn,Pn_1,Pn_2] = Legendre(n,COS,Pn,Pn_1,Pn_2);
        arg = gam0*R; fctr=sqrt(pi/2./arg); Ib0=fctr.* besseli(n+1/2,arg);
        arg = gam0*rS;fctr=sqrt(pi/2 /arg); KbS=fctr * besselk(n+1/2,arg);
        u_inc = u_inc + (2*n+1) * Ib0 .* KbS .* Pn;
      end % n_loop
        u_inc = u_inc * gam0 / (2*pi^2);

% (3) Determine and plot relative error in domain ---------------------------
      Error = u_inc - u_inc_exact;
      disp(['mean_square_error = '   ...
                        num2str(norm(Error(:),1)/norm(u_inc_exact(:),1))]);
    % plot at  cross-section at either x3 = 0 or x3 = dx/2
      N1 = input.N1;  N2 = input.N2;  N3 = input.N3;
      N_cross =floor(N3/2+1);
      set(figure,'Units','centimeters','Position',[5 5 18 10]);
      subplot(1,3,1)
       matrix2D = reshape(abs(u_inc_exact(:,:, N_cross)),N1,N2);
       IMAGESC(X1(:,1,1),X2(1,:,1),matrix2D)
       title('\fontsize{11} 3D: abs(u^{inc}_{exact})');
       hold on; phi = 0:.01:2*pi;   plot(a*cos(phi),a*sin(phi),'w');
      subplot(1,3,2)
       matrix2D = reshape(abs(u_inc(:,:, N_cross)),N1,N2);
       IMAGESC(X1(:,1,1),X2(1,:,1),matrix2D);
       title('\fontsize{11} 3D: abs(u^{inc})');
       hold on; phi = 0:.01:2*pi;   plot(a*cos(phi),a*sin(phi),'w');
      subplot(1,3,3)
       matrix2D = reshape(abs(Error(:,:, N_cross)),N1,N2);
       IMAGESC(X1(:,1,1),X2(1,:,1),matrix2D)
       title('\fontsize{11} 3D: abs({Error})');
       hold on; phi = 0:.01:2*pi;   plot(a*cos(phi),a*sin(phi),'w');
```

Matlab Listing I.31 The `WavefieldTotSphere` script to compute the 3D total and scattered wave fields.

```
clear all; clc; close all;
input = init();
 c_0  = input.c_0;       c_sct   = input.c_sct;
 gam0 = input.gamma_0;  gam_sct = input.gamma_0*c_0/c_sct;

% Transform Cartesian coordinates to spherical coordinates
  xS = input.xS;  rS = sqrt(xS(1)^2+xS(2)^2+xS(3)^2);
  X1 = input.X1;  X2 = input.X2;  X3 = input.X3;
  R  = sqrt(X1.^2+X2.^2+X3.^2+1e-16);    % add small value to avoid zero R

% (1) Compute incident wave in closed form --------------------------------
     DIS   = sqrt( (X1-xS(1)).^2 + (X2-xS(2)).^2 + (X3-xS(3)).^2 );
     u_inc = exp(-gam0*DIS)./(4*pi*DIS);

% (2) Compute coefficients of series expansion ----------------------------
     N = 50;                                 % increase N for more accuracy
     A = zeros(1,N+1);  B = zeros(1,N+1);
     arg0 = gam0 * input.a;  args = gam_sct * input.a;
     for n = 0 : N
      Ib0 = besseli(n+1/2,arg0); dIb0 = besseli(n+3/2,arg0) + n/arg0*Ib0;
      Ibs = besseli(n+1/2,args); dIbs = besseli(n+3/2,args) + n/args*Ibs;
      Kb0 = besselk(n+1/2,arg0); dKb0 =-besselk(n+3/2,arg0) + n/arg0*Kb0;
      A(n+1) = - (gam_sct *dIbs*Ib0 - gam0 *dIb0*Ibs)  ...
                /(gam_sct *dIbs*Kb0 - gam0 *dKb0*Ibs);
  % factor sqrt(pi^2/4/arg0/args) in numerator and denominator is omitted
      B(n+1) = sqrt(args/arg0)* (Ib0 + A(n+1)*Kb0) / Ibs;
     end

% (3) Compute exterior, interior and total fields -------------------------
     DIS = sqrt( (X1-xS(1)).^2 + (X2-xS(2)).^2 + (X3-xS(3)).^2 );
     COS = (R.^2 + rS^2 - DIS.^2) ./ (2*rS*R);            % cos(x,xS)
     Pn = zeros(size(R)); Pn_1 = Pn; Pn_2 = Pn;
     u_rfl = zeros(size(R));
     u_int = zeros(size(R));
     for n = 0 : N
      [Pn,Pn_1,Pn_2] = Legendre(n,COS,Pn,Pn_1,Pn_2);
      arg = gam0*R;    fctr=sqrt(pi/2./arg); Kb0=fctr.*besselk(n+1/2,arg);
      arg = gam0*rS;   fctr=sqrt(pi/2 /arg); KbS=fctr *besselk(n+1/2,arg);
      arg = gam_sct*R; fctr=sqrt(pi/2./arg); Ibs=fctr.*besseli(n+1/2,arg);
      u_rfl = u_rfl + A(n+1)*(2*n+1) * Kb0 .* KbS .* Pn;
      u_int = u_int + B(n+1)*(2*n+1) * Ibs .* KbS .* Pn;
     end
     u_rfl(ceil(input.N1/2),ceil(input.N2/2),ceil(input.N3/2)) = 0;
                                              % Avoid NaN number at R=0
     u_rfl = u_rfl * gam0 / (2*pi^2);
     u_int = u_int * gam0 / (2*pi^2);

     u = (u_rfl+u_inc) .* (R >= input.a) + u_int .* (R < input.a);

     plotWavefield(u_inc,u,input)
```

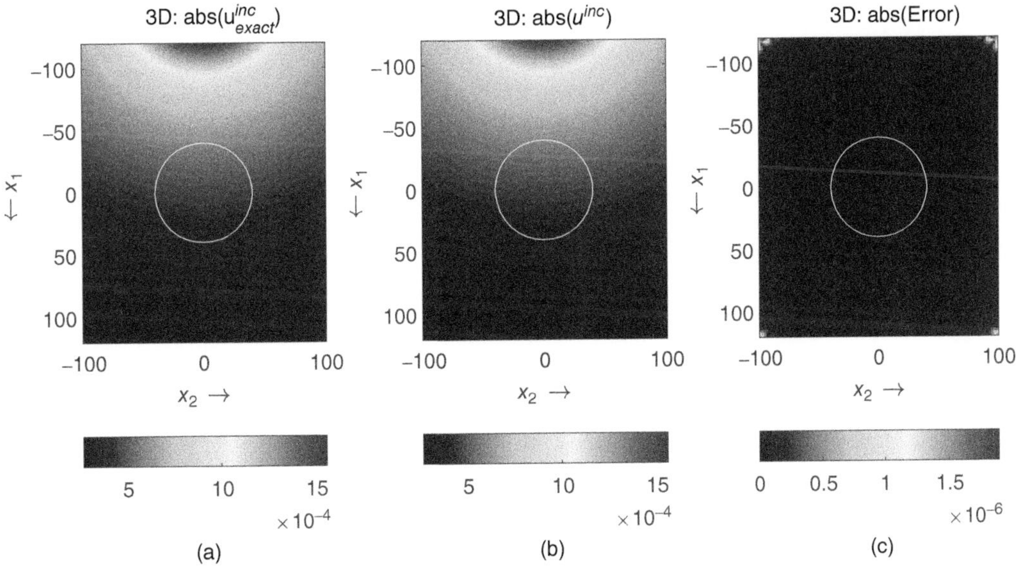

Figure 1.C.5 Comparison of the analytical (a) and the series solution (b) of the incident-wave-field results of `WavefieldIncSphere`, and the resulting error distribution (c).

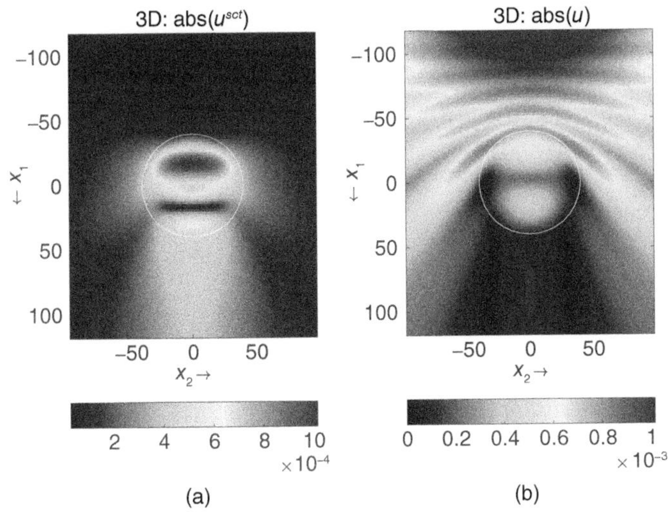

Figure 1.C.6 Plots of the 3D scattered (a) and total (b) wave fields.

differs from the previous Matlab script: a very limited set of receiver points outside the sphere are selected. In step (3) of the latter script, we use Eq. (1.B.50).

The function of Matlab Listing I.9 plots the scattered wave field data at the receiver points around the sphere, see Figure 1.C.7.

In Figure 1.C.7, we observe again that the scattered wave is dominant in the forward direction with an angle of 0°.

For benchmarking needs, the scattered field data are stored in the MAT-file DATA3D.mat.

Matlab Listing I.32 The `WavefieldSctSphere` script to compute the 3D scattered wave field at the receiver locations around the sphere.

```
clear all; clc; close all;
input = init();
c_0  = input.c_0;          c_sct   = input.c_sct;
gam0 = input.gamma_0;      gam_sct = input.gamma_0*c_0/c_sct;

if exist(fullfile(cd, 'DATA3D.mat'), 'file');    delete DATA3D.mat;    end

% (1) Compute coefficients of series expansion -----------------------------
    N = 50;                                     % increase N for more accuracy
    A = zeros(1,N+1);
    arg0 = gam0 * input.a;          args = gam_sct * input.a;
    for n = 0 : N
     Ib0 = besseli(n+1/2,arg0); dIb0 = besseli(n+3/2,arg0) + n/arg0*Ib0;
     Ibs = besseli(n+1/2,args); dIbs = besseli(n+3/2,args) + n/args*Ibs;
     Kb0 = besselk(n+1/2,arg0); dKb0 =-besselk(n+3/2,arg0) + n/arg0*Kb0;
     A(n+1) =   -(gam_sct *dIbs*Ib0 - gam0 *dIb0*Ibs)  ...
               / (gam_sct *dIbs*Kb0 - gam0 *dKb0*Ibs);
  % factor sqrt(pi^2/4/arg0/args) in numerator and denominator is omitted
    end

% (2) Compute reflected field at receivers (data) --------------------------
    xR  = input.xR;  rR = sqrt(xR(1,:).^2 + xR(2,:).^2 + xR(3,:).^2);
    xS  = input.xS; rS = sqrt(xS(1)^2+xS(2)^2+xS(3)^2);
    DIS = sqrt((xR(1,:)-xS(1)).^2+(xR(2,:)-xS(2)).^2+(xR(3,:)-xS(3)).^2);
    COS = (rR.^2 + rS^2 - DIS.^2) ./ (2*rR*rS);             % cos(xR,xS)
    Pn  = zeros(1,input.NR); Pn_1 = Pn; Pn_2 = Pn;
    data3D = zeros(1,input.NR);
    for n = 0 : N
     [Pn,Pn_1,Pn_2] = Legendre(n,COS,Pn,Pn_1,Pn_2);
     arg = gam0*rR; fctr=sqrt(pi/2./arg); Kb0=fctr.* besselk(n+1/2,arg);
     arg = gam0*rS; fctr=sqrt(pi/2 /arg); KbS=fctr * besselk(n+1/2,arg);
     data3D = data3D + A(n+1)*(2*n+1) * Kb0 .* KbS .* Pn;
    end % n_loop
    data3D = data3D * gam0 / (2*pi^2);

    displayData(data3D,input);                          save DATA3D data3D;
```

1.C.4 Scattered-Field Computations Canonical Objects

After completion of the description of the various Matlab scripts to compute the detailed behavior of the scattering by the canonical objects, we finally present a combined script to compute the scattered field at a number of receiver points. This latter script should be used, before we run a particular Matlab script based on a domain integral equation. The resulting field values are used as a benchmark for the wave fields computed with the integral equation methods.

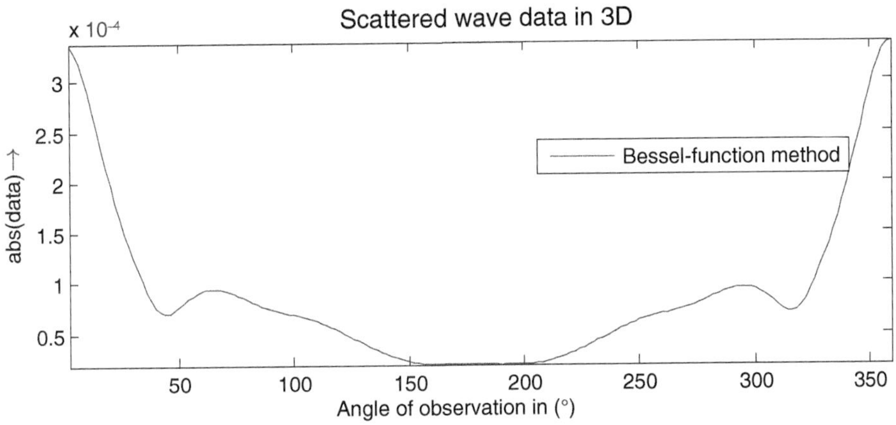

Figure 1.C.7 Plot of the 3D scattered wave field at the receivers around the sphere.

Matlab Listing I.33 The `ForwardCanonicalObjects` script to compute the scattered wave fields for canonical objects.

```
clear all; clc; close all; clear workspace
input = init();
global nDIM;

if nDIM == 1
   % Compute scattered wave field at a receiver above and below the slab
     disp('Running WavefieldSctSlab');      WavefieldSctSlab;

elseif nDIM == 2
   % Compute scattered wave at receivers around the circular cylinder
     disp('Running WavefieldSctCircle');   WavefieldSctCircle;

elseif nDIM == 3
   % Compute scattered wave at a number of receivers around the sphere
     disp('Running WavefieldSctSphere');   WavefieldSctSphere;

end % if
```

2

Acoustic Waves

In this chapter, we now consider the propagation and scattering of acoustic waves in a fluid. In the theory of acoustic waves, the polarization of the wave is taken into account. Instead of a scalar wave field, we now deal with a scalar pressure $p = p(\boldsymbol{x}, t)$ and a particle-velocity vector $v_k = v_k(\boldsymbol{x}, t)$. The theory of acoustic waves goes back to the Theory of Sound by Lord Rayleigh [40]. The basic equations for propagation of acoustic waves in a lossless fluid arise from the equation of motion (Newton's law) and the deformation equation. In the notation of De Hoop [14, Chapter 2] and Fokkema and van den Berg [17], these time domain-equations are

$$\partial_k p + \rho \partial_t v_k = f_k, \tag{2.1}$$

$$\partial_k v_k + \kappa \partial_t p = q, \tag{2.2}$$

where $f_k = f_k(\boldsymbol{x}, t)$ is volume source density of force, $q = q(\boldsymbol{x}, t)$ is the volume source density of injection rate. As far as the material quantities are concerned, $\rho = \rho(x)$ is the volume density of mass and κ is the compressibility. The acoustic wave speed $c = c(\boldsymbol{x})$ is given by $c = 1/\sqrt{\kappa\rho}$. For the special case that the mass density is constant, the particle velocity can be eliminated from these two equations, and we end up with the scalar wave equation of Eq. (1.1), where the wave field u is identified with the pressure p.

In those domains, where mass density and compressibility change continuous with position, the acoustic pressure and the particle velocity are continuously differentiable functions of position and satisfy the differential equations (2.1) and (2.2). In practice, it often occurs that fluids with different material parameters are in contact along the interfaces. Across such an interface, the constitutive parameters ρ and κ show a jump discontinuity. From the equation of motion and the deformation rate equation, it then follows that at least some components of the particle velocity, and possibly the pressure, show a jump discontinuity across the interface. As a result, the pressure and/or the particle velocity are no longer continuously differentiable, and Eqs. (2.1) and (2.2) cease to hold. To interrelate the acoustic wave-field quantities at either side of the interface, a certain set of boundary conditions is needed. To arrive at these boundary conditions, we assume that along the interface the two fluids remain in touch, but do not mix. Then, the component of the particle velocity normal to the interface should be continuous across the interface. Otherwise, the fluid at one side of the interface would either move away from the fluid at the other side of the interface or mix with it. The components of the particle velocity parallel to the interface can be different at each side of the interface because along the interface the two fluids can slide with respect to each other. To arrive at the conditions to be laid upon the acoustic pressure, we proceed as follows:

Let S denote the interface and assume that S has everywhere a unique tangent plane. Let, further, ν_k denote the unit vector along the normal to S such that upon traversing S in the direction of ν_k, we pass from the domain $\mathbb{D}_2$ to the domain $\mathbb{D}_1$, $\mathbb{D}_1$, and $\mathbb{D}_2$ being located at either side of S

Forward and Inverse Scattering Algorithms based on Contrast Source Integral Equations, First Edition. Peter M. van den Berg.

Companion website: www.wiley.com/go/vandenBerg/ScatteringAlgorithms

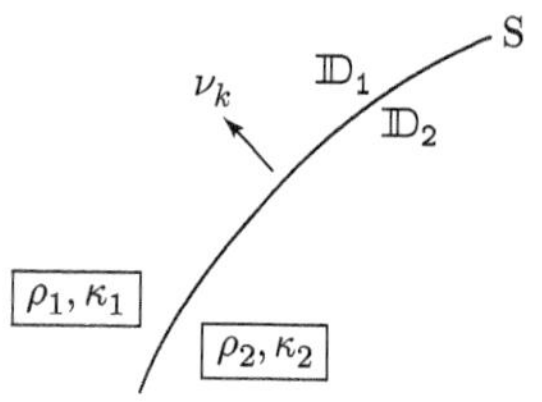

Figure 2.1 Interface between two media with different acoustic properties.

(Figure 2.1). Suppose now that some (or all) acoustic wave quantities jump across S. In the direction parallel to S, all acoustic wave quantities still vary in a continuously differentiable manner, and hence, the partial derivatives parallel to S give no problem in Eqs. (2.1) and (2.2). The partial derivatives perpendicular to S, on the contrary, meet functions that show a jump discontinuity across S; these give rise to interface Dirac distributions (interface impulse functions) located on S. Distributions of this kind would, however, physically be representative of interface sources located on S. In the absence of such sources, the absence of interface impulse functions in the partial derivatives perpendicular to S should be enforced. The latter is done by requiring that these normal derivatives only meet functions that are continuous across S. This procedure leads to the time-domain continuity conditions that p and $\nu_k v_k$ are continuous across S, viz., for $\boldsymbol{x} \in$ S and $h > 0$, we have

$$\lim_{h\to 0} p(\boldsymbol{x} + h\boldsymbol{\nu}, t) = \lim_{h\to 0} p(\boldsymbol{x} - h\boldsymbol{\nu}, t) \tag{2.3}$$

and

$$\lim_{h\to 0} \nu_k v_k(\boldsymbol{x} + h\boldsymbol{\nu}, t) = \lim_{h\to 0} \nu_k v_k(\boldsymbol{x} - h\boldsymbol{\nu}, t). \tag{2.4}$$

In a large number of cases met in practice, we are interested in the behavior of the acoustic wave field in **linear** configurations. Mathematically, one can take advantage of this situation by using a Laplace transformation with respect to time, over the interval $\mathcal{T} = \{t \in \mathbb{R}; t > t_0\}$,

$$\hat{p}(\boldsymbol{x}, s) = \int_{t_0}^{\infty} \exp(-st)p(\boldsymbol{x}, t)\,\mathrm{d}t \tag{2.5}$$

and

$$\hat{v}_k(\boldsymbol{x}, s) = \int_{t_0}^{\infty} \exp(-st)v_k(\boldsymbol{x}, t)\,\mathrm{d}t. \tag{2.6}$$

This results into a set of equations governing the acoustic wave field in the corresponding Laplace-transform domain or s-domain. In the s-domain relations, the time coordinate has been eliminated, and an acoustic wave-field problem in space remains in which the transformation parameter s occurs. Causality of the wave field is taken into account by taking $\mathrm{Re}(s) > 0$, and requiring that all causal wave field quantities are analytic functions of s in the right half $0 < \mathrm{Re}(s) < \infty$ of the complex s-plane.

For linear media, we subject the acoustic wave field equations (2.1) and (2.2) to a Laplace transformation. For completeness, we allow a nonvanishing acoustic wave field to be present at $t = t_0$, although in the majority of cases, we are interested in the causal wave field generated by sources that are switched on at the instant $t = t_0$, in which case the initial values of the acoustic wave field are taken to be zero. Integration by parts, we arrive at

$$\int_{t_0}^{\infty} \exp(-st)\partial_t[v_k(\boldsymbol{x}, t)]\,\mathrm{d}t = -v_k(\boldsymbol{x}, t_0)\,\exp(-st_0) + s\hat{v}_k(\boldsymbol{x}, s) \tag{2.7}$$

and

$$\int_{t_0}^{\infty} \exp(-st)\partial_t[p(\boldsymbol{x}, t)]\,\mathrm{d}t = -p(\boldsymbol{x}, t_0)\exp(-st_0) + s\hat{p}(\boldsymbol{x}, s). \tag{2.8}$$

The acoustic wave equations in the s-domain become

$$\partial_k \hat{p} + s\hat{\rho}\hat{v}_k = \hat{f}_k + \rho(\boldsymbol{x})\, v_k(\boldsymbol{x}, t_0)\, \exp(-st_0), \tag{2.9}$$

$$\partial_k \hat{v}_k + s\hat{\kappa}\hat{p} = \hat{q} + \kappa(\boldsymbol{x})\, p(\boldsymbol{x}, t_0)\, \exp(-st_0). \tag{2.10}$$

We remark that these equations also hold for lossy media in which the mass density and the compressibility depend on s as well. We therefore use the notation $\hat{\rho} = \hat{\rho}(\boldsymbol{x}, s)$ and $\hat{\kappa} = \hat{\kappa}(\boldsymbol{x}, s)$. From Eqs. (2.9) and (2.10) it follows that, in the s-domain, one can take into account the influence of a nonvanishing initial acoustic wave field by properly incorporating it in the s-domain volume densities of external volume force and external volume injection rate. In the remainder of our analysis, it will be tacitly understood that nonzero initial acoustic wave-field values have been accounted for in this manner. Our acoustic wave equations in the s-domain have then the final form

$$\partial_k \hat{p} + s\hat{\rho}\hat{v}_k = \hat{f}_k, \tag{2.11}$$

$$\partial_k \hat{v}_k + s\hat{\kappa}\hat{p} = \hat{q}. \tag{2.12}$$

The continuity conditions of Eqs. (2.3) and (2.4) in the time-domain leads to the continuity conditions in the s-domain:

$$\lim_{h\to 0} \hat{p}(\boldsymbol{x} + h\boldsymbol{\nu}, s) = \lim_{h\to 0} \hat{p}(\boldsymbol{x} - h\boldsymbol{\nu}, s) \tag{2.13}$$

and

$$\lim_{h\to 0} \nu_k \hat{v}_k(\boldsymbol{x} + h\boldsymbol{\nu}, s) = \lim_{h\to 0} \nu_k \hat{v}_k(\boldsymbol{x} - h\boldsymbol{\nu}, s), \tag{2.14}$$

where $\hat{p} = \hat{p}(\boldsymbol{x}, s)$ is the Laplace transform of p and $\hat{v}_k = \hat{v}_k(\boldsymbol{x}, s)$ is the Laplace transform of $\hat{v}_k$.

The behavior of acoustic waves of interest is often characterized by the results of a **steady-state analysis**. In such an analysis, all acoustic wave quantities are taken to depend sinusoidally on time with a common angular frequency, ω say. Each purely real quantity can then be associated with a complex counterpart and a common time factor $\exp(-\mathrm{i}\omega t)$. In doing so, the original quantities in the time domain are found from the complex counterparts as

$$\begin{aligned} &\{p(\boldsymbol{x}, t), v_k(\boldsymbol{x}, t), q(\boldsymbol{x}, t), f_k(\boldsymbol{x}, t)\} \\ &\quad = \mathrm{Re}[\{\hat{p}(\boldsymbol{x}, -\mathrm{i}\omega), \hat{v}_k(\boldsymbol{x}, -\mathrm{i}\omega), \hat{q}(\boldsymbol{x}, -\mathrm{i}\omega), \hat{f}_k(\boldsymbol{x}, -\mathrm{i}\omega)\}\ \exp(-\mathrm{i}\omega t)]\,. \end{aligned} \tag{2.15}$$

Substitution of these complex representations in the acoustic equations in the time domain yields, except for the common time factor $\exp(-\mathrm{i}\omega t)$, a set of equations

$$\partial_k \hat{p} - \mathrm{i}\omega\hat{\rho}\, \hat{v}_k = \hat{f}_k, \tag{2.16}$$

$$\partial_k \hat{v}_k - \mathrm{i}\omega\hat{\kappa}\, \hat{p} = \hat{q}. \tag{2.17}$$

These equations are identical to the one of the Laplace transform domain, viz. Eqs. (2.11) and (2.12) with $s = -\mathrm{i}\omega$. Hence, we interpret the steady-state analysis as a limiting case of the time Laplace transform analysis in which $s \to -\mathrm{i}\omega$ via $\mathrm{Re}(s) > 0$. In this way, we have ensured that causality remains satisfied.

As far as the medium parameters are concerned, sometimes we assume that the mass density and the compressibility are real quantities, independent of the frequency ω. We then deal with a lossless medium. If lossy media must be taken into account, the theory and the pertaining computer codes are straightforward by introducing complex values $\hat{\rho}(\boldsymbol{x}, s)$ and $\hat{\kappa}(\boldsymbol{x}, s)$ for $s \to -\mathrm{i}\omega$ via $\mathrm{Re}(s) > 0$. The wave speed $\hat{c} = 1/\sqrt{\hat{\rho}\hat{\kappa}}$ becomes complex as well. Only in the calculation of the adjoint of an operator, one should replace the the mass density $\hat{\rho}$, the compressibility $\hat{\kappa}$ and the wave speed $\hat{c}$ by their complex conjugates $\overline{\rho}$, $\overline{\kappa}$, and $\overline{c}$.

This concludes the discussion of the general framework of the acoustic wave equations and the acoustic interface conditions.

2.1 Three-Dimensional Scattering by a Bounded Contrast

We start our analysis with the acoustic wave equations in the s-domain. For convenience, we omit the explicit s-dependence of the various field quantities.

2.1.1 Radiation in an Unbounded Homogeneous Embedding

The problem of calculating the acoustic radiation generated by a known volume-source distribution in a known medium is usually denoted as the acoustic source problem. The corresponding acoustic field is denoted as the incident acoustic wave field $\{\hat{p}^{inc}, \hat{v}_k^{inc}\} = \{\hat{p}^{inc}(\boldsymbol{x}), \hat{v}_k^{inc}(\boldsymbol{x})\}$.

We consider a medium of infinite extent with known acoustic properties. In this medium, there is a source that occupies the bounded domain $\mathbb{D}_{source}$ (Figure 2.2). The action of the source is represented by known volume densities of external volume force and/or volume injection $\{\hat{q}, \hat{f}_k\}$.

The s-domain particle velocity and acoustic pressure in a lossless and homogeneous fluid with a real and constant volume density of mass ρ_0 and a real and constant compressibility κ_0 satisfy the s-domain acoustic wave equations

$$\partial_k \hat{p}^{inc} + s\rho_0 \hat{v}_k^{inc} = \hat{f}_k, \tag{2.18}$$

$$\partial_k \hat{v}_k^{inc} + s\kappa_0 \hat{p}^{inc} = Z_0^{-1} \hat{q}'. \tag{2.19}$$

where Z_0 is the acoustic wave impedance of the embedding medium,

$$Z_0 = \rho_0 \, c_0 = (\rho_0/\kappa_0)^{\frac{1}{2}}, \tag{2.20}$$

and $c_0 = (\rho_0 \kappa_0)^{-\frac{1}{2}}$ is the wave speed of the acoustic wave fields in the homogeneous medium at hand. With the definition of the propagation coefficient,

$$\hat{\gamma}_0 = s/c_0, \tag{2.21}$$

we often use the following relations:

$$s\rho_0 = \hat{\gamma}_0 \, Z_0 \quad \text{and} \quad s\kappa_0 = \hat{\gamma}_0 / Z_0. \tag{2.22}$$

The relation between $\hat{q}'$ and $\hat{q}$ is

$$\hat{q}' = Z_0 \, \hat{q}. \tag{2.23}$$

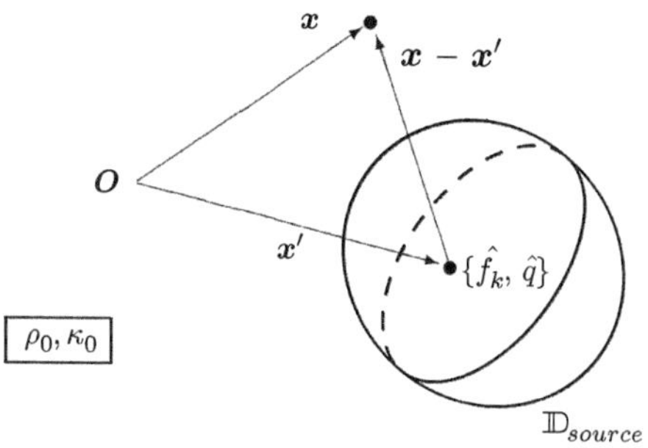

Figure 2.2 Source domain in an unbounded embedding with constant mass density ρ_0 and constant compressibility κ_0.

Note that in numerical calculations, it is preferable to work with field functions of the same unit. In the present case, they are the pressure and the product of acoustic impedance and particle velocity. Therefore, we have included the factor Z_0^{-1} in Eq. (2.19). Then, the acoustic wave equations of (2.18) and (2.19) can be rewritten in terms $\hat{p}^{inc}$ and $Z_0\hat{v}_k^{inc}$ as follows:

$$\partial_k \hat{p}^{inc} + \hat{\gamma}_0\, Z_0 \hat{v}_k^{inc} = \hat{f}_k, \tag{2.24}$$

$$\partial_k Z_0 \hat{v}_k^{inc} + \hat{\gamma}_0\, \hat{p}^{inc} = \hat{q}'. \tag{2.25}$$

We assume that $\hat{f}_k$ and $\hat{q}'$ only differ from zero in some bounded subdomain $\mathbb{D}_{source}$ of $\mathbb{R}^3$ (Figure 2.2). Eliminating $Z_0\hat{v}_k^{inc}$ from these two equations, we obtain the scalar wave equation for the pressure,

$$\partial_k \partial_k \hat{p}^{inc} - \hat{\gamma}_0^2 \hat{p}^{inc} = -\hat{\gamma}_0 \hat{q}' + \partial_k \hat{f}_k. \tag{2.26}$$

The solution for the pressure follows immediately from the scalar radiation problem of the Chapter 1 as, cf. Eqs. (1.13) and (1.18),

$$\hat{p}^{inc}(\boldsymbol{x}) = \int_{\boldsymbol{x}' \in \mathbb{D}_{source}} [\hat{\gamma}_0 \hat{G}(\boldsymbol{x}-\boldsymbol{x}')\, \hat{q}'(\boldsymbol{x}') - \partial_k \hat{G}(\boldsymbol{x}-\boldsymbol{x}')\, \hat{f}_k(\boldsymbol{x}')]\, \mathrm{d}V, \tag{2.27}$$

where we have used the convolutional property of the spatial derivative of the Green function, viz.,

$$\int_{\boldsymbol{x}' \in \mathbb{D}_{source}} \hat{G}(\boldsymbol{x}-\boldsymbol{x}')\, \partial_k' \hat{f}_k(\boldsymbol{x}')\, \mathrm{d}V = \int_{\boldsymbol{x}' \in \mathbb{D}_{source}} \partial_k \hat{G}(\boldsymbol{x}-\boldsymbol{x}')\, \hat{f}_k(\boldsymbol{x}')\, \mathrm{d}V, \tag{2.28}$$

for any function $\hat{f}_k$. Note that we have applied integration by parts and assumed that at boundary of the source domain, the source strength $\hat{f}_k$ vanishes. Further, we use the notation that ∂_k' denotes the spatial derivative with respect to x_k'. The first term in the integrand of Eq. (2.27) represents a monopole contribution with volume density $\hat{q}'$, while the second term represents a dipole contribution with volume density $\hat{f}_k$. Equation (2.27) shows that the acoustic wave field from known sources in a known medium can be calculated in all space, once the wave fields radiated by appropriate point sources have been calculated. The right-hand side of the integral representation expresses the wave field values as a superposition of the wave fields radiated by the elementary volume sources out of which the distributed sources can be envisaged to be composed.

For simplicity, in our analysis, we assume that the incident wave field is generated by a monopole source at $\boldsymbol{x}^S$, viz., $\hat{q}' = Z_0 \hat{Q}\, \delta(\boldsymbol{x}-\boldsymbol{x}^S)$. Then, the incident acoustic pressure at some observation point $\boldsymbol{x}$ is given by

$$\boxed{\hat{p}^{inc}(\boldsymbol{x}|\boldsymbol{x}^S) = \hat{\gamma}_0 Z_0\, \hat{Q}\, \hat{G}(\boldsymbol{x}-\boldsymbol{x}^S).} \tag{2.29}$$

In geophysics, $\hat{\gamma}_0 Z_0 \hat{Q} = s\rho_0 \hat{Q}$ is often denoted as the Laplace transformed quantity of the source signature, the so-called wavelet. The incident particle velocity at some observation point $\boldsymbol{x} \neq \boldsymbol{x}^S$ follows from Eq. (2.24) as

$$\boxed{Z_0 \hat{v}_k^{inc}(\boldsymbol{x}|\boldsymbol{x}^S) = -Z_0\, \hat{Q}\, \partial_k \hat{G}(\boldsymbol{x}-\boldsymbol{x}^S).} \tag{2.30}$$

2.1.2 Scattering by a Bounded Contrast

In this section, the scattering of acoustic waves by a contrasting fluid domain of bounded extent, present in a homogeneous, fluid embedding of infinite extent, is investigated in more detail. The integral-equation formulations of the scattering problem are presented. When we consider the scattering object as a volume scatterer a domain-integral-equation formulation is the most versatile

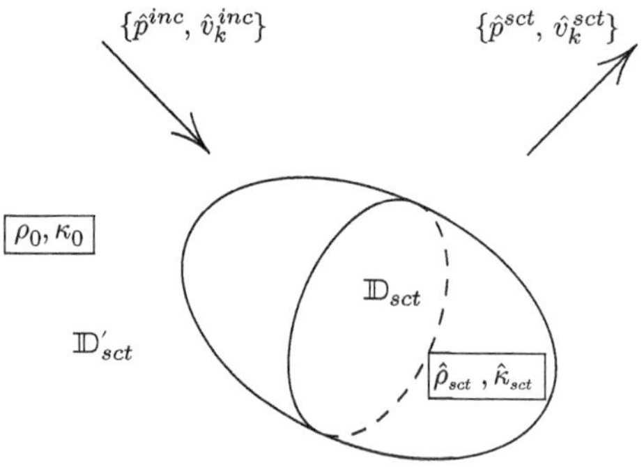

Figure 2.3 The domain of acoustic scattering.

technique, while for homogeneous scatterers the boundary-integral-equation formulation is most convenient when we treat the scatterer as a surface scatterer [12].

Let $\mathbb{D}_{sct}$ be the bounded scattering domain occupied by the fluid scatterer and let $\hat{\rho}_{sct} = \hat{\rho}_{sct}(\boldsymbol{x}, s)$ be its volume density of mass and $\hat{\kappa}_{sct} = \hat{\kappa}_{sct}(\boldsymbol{x}, s)$ its compressibility. The wave speed in $\mathbb{D}_{sct}$ is given by $\hat{c}_{sct} = 1/\sqrt{\hat{\rho}_{sct}\hat{\kappa}_{sct}}$. The domain exterior to $\mathbb{D}_{sct}$ is denoted by $\mathbb{D}'_{sct}$. In $\mathbb{D}'_{sct}$, we assume that in the Embedding, a homogeneous fluid is present with mass density ρ_0 and compressibility κ_0 (Figure 2.3). The total acoustic wave field in the configuration $\{\hat{p}, \hat{v}_k\}$ is decomposed into the incident wave field $\{\hat{p}^{inc}, \hat{v}_k^{inc}\}$ and the scattered wave field $\{\hat{p}^{sct}, \hat{v}_k^{sct}\}$. The incident wave field is the wave field that would be present in the entire configuration if the domain $\mathbb{D}_{sct}$ showed no contrast with the embedding. The total wave field is generated by sources that are located outside the scattering domain. Since these sources remain present even if the scattering domain is thought to be absent, they also serve as sources for the incident wave field. The incident wave field quantities can be calculated with the representations obtained in Section 2.1.1.

The scattered wave field is defined as the difference between the total wave field and the incident wave field. Hence,

$$\{\hat{p}^{sct}, \hat{v}_k^{sct}\} = \{\hat{p} - \hat{p}^{inc}, \hat{v}_k - \hat{v}_k^{inc}\}. \tag{2.31}$$

Through a particular reasoning, we now want to express that the scattered wave field originates from the contrast in acoustic properties that the scattering object shows with respect to its embedding. First, we observe that, since the total wave field is source free in $\mathbb{D}_{sct}$,

$$\partial_k\hat{p} + s\hat{\rho}_{sct}\hat{v}_k = 0, \quad \boldsymbol{x} \in \mathbb{D}_{sct}, \tag{2.32}$$

$$\partial_k\hat{v}_k + s\hat{\kappa}_{sct}\hat{p} = 0, \quad \boldsymbol{x} \in \mathbb{D}_{sct}. \tag{2.33}$$

Second, the incident wave field has no sources in $\mathbb{D}_{sct}$, while inside and outside $\mathbb{D}_{sct}$, the constitutive parameters have the same values. Hence,

$$\partial_k\hat{p}^{inc} + s\rho_0\hat{v}_k^{inc} = 0, \quad \boldsymbol{x} \in \mathbb{D}_{sct}, \tag{2.34}$$

$$\partial_k\hat{v}_k^{inc} + s\kappa_0\hat{p}^{inc} = 0, \quad \boldsymbol{x} \in \mathbb{D}_{sct}. \tag{2.35}$$

Upon rewriting Eqs. (2.32) and (2.33) as

$$\partial_k\hat{p} + s\rho_0\hat{v}_k = s(\rho_0 - \hat{\rho}_{sct})\hat{v}_k, \quad \boldsymbol{x} \in \mathbb{D}_{sct}, \tag{2.36}$$

$$\partial_k\hat{v}_k + s\kappa_0\hat{p} = s(\kappa_0 - \hat{\kappa}_{sct})\hat{p}, \quad \boldsymbol{x} \in \mathbb{D}_{sct}, \tag{2.37}$$

subtracting Eq. (2.34) from Eq. (2.36) and Eq. (2.35) from Eq. (2.37) and using Eq. (2.31), we arrive at

$$\partial_k\hat{p}^{sct} + s\rho_0\hat{v}_k^{sct} = s(\rho_0 - \hat{\rho}_{sct})\hat{v}_k, \quad \boldsymbol{x} \in \mathbb{D}_{sct}, \tag{2.38}$$

$$\partial_k\hat{v}_k^{sct} + s\kappa_0\hat{p}^{sct} = s(\kappa_0 - \hat{\kappa}_{sct})\hat{p}, \quad \boldsymbol{x} \in \mathbb{D}_{sct}. \tag{2.39}$$

Third, we observe that the scattered wave field is source free in the embedding (where the total wave field and the incident wave field have the same sources), and hence,

$$\partial_k \hat{p}^{sct} + s\rho_0 \hat{v}_k^{sct} = 0, \quad \boldsymbol{x} \in \mathbb{D}'_{sct}, \tag{2.40}$$

$$\partial_k \hat{v}_k^{sct} + s\kappa_0 \hat{p}^{sct} = 0, \quad \boldsymbol{x} \in \mathbb{D}'_{sct}. \tag{2.41}$$

Equations (2.38)–(2.41) are now combined to

$$\partial_k \hat{p}^{sct} + s\rho_0 \hat{v}_k^{sct} = \hat{f}_k^{sct}, \quad \boldsymbol{x} \in \mathbb{R}^3, \tag{2.42}$$

$$\partial_k \hat{v}_k^{sct} + s\kappa_0 \hat{p}^{sct} = \hat{q}^{sct}, \quad \boldsymbol{x} \in \mathbb{R}^3, \tag{2.43}$$

where the scattering sources are given by

$$\hat{f}_k^{sct} = s(\rho_0 - \hat{\rho}_{sct})\hat{v}_k \quad \text{and} \quad \hat{q}^{sct} = s(\kappa_0 - \hat{\kappa}_{sct})\hat{p}. \tag{2.44}$$

Similar as for the incident wave-field equations (2.24) and (2.25), we write the scattered field equations as

$$\partial_k \hat{p}^{sct} + \hat{\gamma}_0 \, Z_0 \hat{v}_k^{sct} = \hat{f}_k^{sct}, \tag{2.45}$$

$$\partial_k Z_0 \hat{v}_k^{sct} + \hat{\gamma}_0 \, \hat{p}^{sct} = \hat{q}'^{sct}, \tag{2.46}$$

with $\hat{q}'^{sct} = Z_0 \hat{q}^{sct}$. These scattering sources that generate the scattered field follow from Eqs. (2.22) and (2.44) as

$$\hat{f}_k^{sct} = \hat{\gamma}_0 \left(1 - \frac{\hat{\rho}_{sct}}{\rho_0}\right) Z_0 \hat{v}_k \quad \text{and} \quad \hat{q}'^{sct} = \hat{\gamma}_0 \left(1 - \frac{\hat{\kappa}_{sct}}{\kappa_0}\right) \hat{p}. \tag{2.47}$$

If $\hat{f}_k^{sct}$ and $\hat{q}'^{sct}$ were known, Eqs. (2.45) and (2.46) would constitute an acoustic radiation problem with known sources in an unbounded, homogeneous fluid with the same constitutive parameters as the embedding. This problem has been discussed in Section 2.1.1, cf. Eq. (2.27). Hence, similar as Eq. (2.27), we now arrive at

$$\hat{p}^{sct}(\boldsymbol{x}) = \int_{\boldsymbol{x}' \in \mathbb{D}_{sct}} \left[\hat{\gamma}_0^2 \hat{G}(\boldsymbol{x} - \boldsymbol{x}') \left(1 - \frac{\hat{\kappa}_{sct}}{\kappa_0}\right) \hat{p}(\boldsymbol{x}') - \hat{\gamma}_0 \partial_k \hat{G}(\boldsymbol{x} - \boldsymbol{x}') \left(1 - \frac{\hat{\rho}_{sct}}{\rho_0}\right) Z_0 \hat{v}_k(\boldsymbol{x}') \right] \mathrm{d}V, \tag{2.48}$$

for $\boldsymbol{x} \in \mathbb{D}'_{sct}$. Equation (2.48) is the desired integral representation for the scattered pressure. Once the total wave-field quantities $\hat{p}$ and $Z_0\hat{v}_k$ in $\mathbb{D}_{sct}$ are known, these integral representations enable us to calculate the acoustic wave field quantities in all space. To determine the yet unknown values of $\hat{p}$ and $Z_0\hat{v}_k$, we need a set of integral equations to solve for these acoustic field quantities. By a spatial differentiation of Eq. (2.48) and a substitution of the result into Eq. (2.45), we obtain the integral representation for the scattered particle velocity, $Z_0\hat{v}_k^{sct}$, as

$$Z_0 \hat{v}_j^{sct}(\boldsymbol{x}) = \int_{\boldsymbol{x}' \in \mathbb{D}_{sct}} \left[-\hat{\gamma}_0 \partial_j \hat{G}(\boldsymbol{x} - \boldsymbol{x}') \left(1 - \frac{\hat{\kappa}_{sct}}{\kappa_0}\right) \hat{p}(\boldsymbol{x}') + \partial_j \partial_k \hat{G}(\boldsymbol{x} - \boldsymbol{x}') \left(1 - \frac{\hat{\rho}_{sct}}{\rho_0}\right) Z_0 \hat{v}_k(\boldsymbol{x}') \right] \mathrm{d}V, \tag{2.49}$$

for $\boldsymbol{x} \in \mathbb{D}'_{sct}$. Note that the source term $\hat{f}_k^{sct}$ of Eq. (2.45) vanishes for $\boldsymbol{x} \in \mathbb{D}'_{sct}$, but for points located inside the object, it should be included.

Then, the next step is to let the observation point inside the scattering domain $\mathbb{D}_{sct}$. This procedure yields a set of four integral equations for four unknown components of the wave fields, $\{\hat{p}, Z_0\hat{v}_1, Z_0\hat{v}_2, Z_0\hat{v}_3\}$:

$$\begin{aligned}\hat{p}^{inc}(\boldsymbol{x}) =& \hat{p}(\boldsymbol{x}) - \hat{\gamma}_0^2 \int_{\boldsymbol{x}'\in\mathbb{D}_{sct}} \hat{G}(\boldsymbol{x}-\boldsymbol{x}')\left(1-\frac{\hat{\kappa}_{sct}}{\kappa_0}\right)\hat{p}(\boldsymbol{x}')\,\mathrm{d}V \\ &+ \hat{\gamma}_0\partial_k \int_{\boldsymbol{x}'\in\mathbb{D}_{sct}} \hat{G}(\boldsymbol{x}-\boldsymbol{x}')\left(1-\frac{\hat{\rho}_{sct}}{\rho_0}\right)Z_0\hat{v}_k(\boldsymbol{x}')\,\mathrm{d}V,\end{aligned} \tag{2.50}$$

$$\begin{aligned}Z_0\hat{v}_j^{inc}(\boldsymbol{x}) =& Z_0\hat{v}_j(\boldsymbol{x}) + \hat{\gamma}_0\partial_j \int_{\boldsymbol{x}'\in\mathbb{D}_{sct}} \hat{G}(\boldsymbol{x}-\boldsymbol{x}')\left(1-\frac{\hat{\kappa}_{sct}}{\kappa_0}\right)\hat{p}(\boldsymbol{x}')\,\mathrm{d}V \\ &- \partial_j\partial_k \int_{\boldsymbol{x}'\in\mathbb{D}_{sct}} \hat{G}(\boldsymbol{x}-\boldsymbol{x}')\left(1-\frac{\hat{\rho}_{sct}}{\rho_0}\right)Z_0\hat{v}_k(\boldsymbol{x}')\,\mathrm{d}V - \left(1-\frac{\hat{\rho}_{sct}}{\rho_0}\right)Z_0\hat{v}_j(\boldsymbol{x}),\end{aligned} \tag{2.51}$$

for $\boldsymbol{x} \in \mathbb{D}_{sct}$. Note that the last term of Eq. (2.51) is related to the volume force density of Eqs. (2.45) and (2.47), which is only active inside the scattering domain. In view of the singular behavior of the Green function, we cannot bring the spatial derivatives of the Green function under the integral. If necessary, we can compute the Cauchy principle values of the integrals as a limiting procedure [17, p. 127], in which the observation point approaches the singular point of the Green function.

For completeness, we remark that this set of integral equation has been derived from the acoustic reciprocity theorem by De Hoop [14] and Fokkema and Van den Berg [17]. In principle, one can use this system of equations as the basis to provide a numerical solution of the acoustic scattering problem at hand [59, p. 1539]. However, we have learned from Chapter 1 to opt for a contrast source formulation. This will be discussed in Section 2.1.3.

For contrast in compressibility only, we immediately observe that we end up with a single integral equation for the unknown acoustic pressure only, viz.

$$\hat{p}(\boldsymbol{x})^{inc} = \hat{p}(\boldsymbol{x}) - \int_{\boldsymbol{x}'\in\mathbb{D}_{sct}} \hat{\gamma}_0^2\hat{G}(\boldsymbol{x}-\boldsymbol{x}')\left(1-\frac{c_0^2}{\hat{c}_{sct}^2}\right)\hat{p}(\boldsymbol{x}')\,\mathrm{d}V. \tag{2.52}$$

This integral equation has been discussed in Section 1.1.3.

2.1.3 Contrast Source Domain Integral Equation

To formulate the integral equation in terms of contrast terms, we define the contrast in compressibility,

$$\boxed{\hat{\chi}^{\kappa}(\boldsymbol{x}) = 1 - \frac{\hat{\kappa}_{sct}(\boldsymbol{x})}{\kappa_0},} \tag{2.53}$$

and the contrast in mass density,

$$\boxed{\hat{\chi}^{\rho}(\boldsymbol{x}) = 1 - \frac{\hat{\rho}_{sct}(\boldsymbol{x})}{\rho_0}.} \tag{2.54}$$

Subsequently, the scalar pressure contrast source is defined as

$$\hat{w}^{p}(\boldsymbol{x}) = \hat{\chi}^{\kappa}(\boldsymbol{x})\,\hat{p}(\boldsymbol{x}), \tag{2.55}$$

and the vectorial particle velocity contrast source is defined as

$$\hat{w}_k^{Zv}(\boldsymbol{x}) = \hat{\chi}^{\rho}(\boldsymbol{x})\,Z_0\hat{v}_k(\boldsymbol{x}). \tag{2.56}$$

Introducing these contrast sources in Eqs. (2.50) and (2.51) and multiplying the first equation by $\hat{\chi}^{\kappa}$ and the second equation by $\hat{\chi}^{\rho}$, we arrive at the system of integral equations for pressure and particle-velocity contrast sources

$$\begin{aligned}\hat{\chi}^{\kappa}\hat{p}^{inc}(\boldsymbol{x}) = \hat{w}^{p}(\boldsymbol{x}) - \hat{\chi}^{\kappa}\Bigg[\hat{\gamma}_0^2\int_{\boldsymbol{x}'\in\mathbb{D}_{sct}} \hat{G}(\boldsymbol{x}-\boldsymbol{x}')\hat{w}^{p}(\boldsymbol{x}')\,\mathrm{d}V \\ -\hat{\gamma}_0\partial_k\int_{\boldsymbol{x}'\in\mathbb{D}_{sct}} \hat{G}(\boldsymbol{x}-\boldsymbol{x}')\hat{w}_k^{Zv}(\boldsymbol{x}')\,\mathrm{d}V\Bigg]\end{aligned} \tag{2.57}$$

and

$$\begin{aligned}\hat{\chi}^{\rho}Z_0\hat{v}_j^{inc}(\boldsymbol{x}) = \hat{w}_j^{Zv}(\boldsymbol{x}) - \hat{\chi}^{\rho}\Bigg[-\hat{\gamma}_0\partial_j\int_{\boldsymbol{x}'\in\mathbb{D}_{sct}} \hat{G}(\boldsymbol{x}-\boldsymbol{x}')\hat{w}^{p}(\boldsymbol{x}')\,\mathrm{d}V \\ +\partial_j\partial_k\int_{\boldsymbol{x}'\in\mathbb{D}_{sct}} \hat{G}(\boldsymbol{x}-\boldsymbol{x}')\hat{w}_k^{Zv}(\boldsymbol{x}')\,\mathrm{d}V + \hat{w}_j^{Zv}(\boldsymbol{x})\Bigg].\end{aligned} \tag{2.58}$$

After solving for the contrast sources $\hat{w}^{p}$ and $\hat{w}_k^{Zv}$, the scattered pressure field follows from Eq. (2.48) as

$$\hat{p}^{sct}(\boldsymbol{x}^R) = \int_{\boldsymbol{x}'\in\mathbb{D}_{sct}} [\hat{\gamma}_0^2\hat{G}(\boldsymbol{x}^R-\boldsymbol{x}')\hat{w}^{p}(\boldsymbol{x}') - \hat{\gamma}_0\partial_k^R\hat{G}(\boldsymbol{x}^R-\boldsymbol{x}')\hat{w}_k^{Zv}(\boldsymbol{x}')]\,\mathrm{d}V. \tag{2.59}$$

2.1.4 Numerical Solution and Operators Involved (1D, 2D, 3D)

We now consider the numerical solution of the contrast source-type integral equations, based on the acoustic pressure and the particle-velocity wave fields. In Chapter 1, we have introduced the weak formulation to replace the continuous form of the integral equation by a discrete set of equations. The consistent thread running through the discretization method is that, with a finite spatial discretization, we cannot model the spatial high frequencies. In this section, we continue with the weak formulation of the integral equations by replacing the Green function by its repeated spherical mean. Now there are two procedures: (i) we transfer the spatial differentiations to the Green function and we carry out the differentiation analytically; (ii) we leave the differentiations in front of the integrals and replace the spatial derivatives in Eqs. (2.57) and (2.58) by their central finite difference approximations. To accommodate these central differences, we need extra discretization points outside the discretized scattering domain.

We define a regular grid of a number of N equally sized subdomains, denoted as $\{\mathbb{D}_n,\ n=1:N\}$. In 1D, we are dealing with $N=N_1$ subdomains, denoted as $\{\mathbb{D}_i,\ i=1:N_1\}$. Each subdomain has the volume $\Delta V=\Delta x$. In 2D, we deal with $N=N_1\times N_2$ subdomains, denoted as $\{\mathbb{D}_{i,j},\ i=1:N_1,\ j=1:N_2\}$. Each subdomain has the volume $\Delta V=(\Delta x)^2$. In 3D, we are dealing with $N=N_1\times N_2\times N_3$ subdomains, denoted as $\{\mathbb{D}_{i,j,k},\ i=1:N_1,\ j=1:N_2,\ k=1:N_3\}$. Each subdomain has the volume $\Delta V=(\Delta x)^3$. In addition to assuming that the grid includes the scattering domain $\mathbb{D}_{sct}$, we also have added an extra layer around it to accommodate the central finite-difference approach. The discretization in the (x_1,x_2)-domain is illustrated in Figure 2.4. The dotted domain includes the scattering domain $\mathbb{D}_{sct}$ completely. To this domain, we add extra layer with zero compressibility contrast and zero-mass density contrast. Although the contrast sources vanish at these points, the wave field quantities are nonzero and have to be computed.

On each subdomain $\mathbb{D}_n$, we assume that both the contrast in compressibility and the contrast in mass density are constant and denoted as $\hat{\chi}_n^{\kappa}$ and $\hat{\chi}_n^{\rho}$. Further, in each subdomain, the contrast sources $\hat{w}^{p}$ and $\hat{w}_k^{Zv}$, are replaced by their spherical means $\langle\hat{w}^{p}\rangle$ and $\langle\hat{w}_k^{Zv}\rangle$. The incident wave fields $\hat{p}^{inc}$ and $\hat{v}_j^{inc}$ are replaced by their spherical means $\langle\hat{p}^{inc}\rangle$ and $\langle\hat{v}_j^{inc}\rangle$.

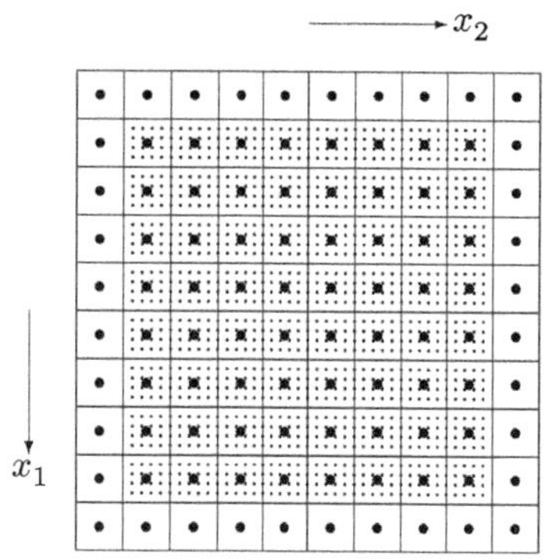

Figure 2.4 Regular grid in the (x_1, x_2)-plane with a mesh size of $\Delta x \times \Delta x$. The center points are denoted as $\boldsymbol{x} = \boldsymbol{x}_n$, $n = 1, 2, \ldots, N$.

2.1.4.1 Analytic Differentiation

We first investigate the method when we interchange integration and differentiation in the integral equations of Eqs. (2.57) and (2.58) and define the associated discrete convolutional operators as

$$\begin{aligned} [\mathcal{K}w_n]_m &= \hat{\gamma}_0^2 \Delta V \sum_{n=1}^{N} \langle\langle \hat{G} \rangle\rangle(\boldsymbol{x}_m - \boldsymbol{x}_n)\ \langle \hat{w}_n \rangle, \\ [\partial_k \mathcal{K} w_n]_m &= \Delta V \sum_{n=1}^{N} \langle\langle \partial_k \hat{G} \rangle\rangle(\boldsymbol{x}_m - \boldsymbol{x}_n)\ \langle \hat{w}_n \rangle, \end{aligned} \tag{2.60}$$

for each contrast-source function $\hat{w}_n$. The convolutional operators on the right-hand sides are computed with the help of function `Kop` of Matlab Listing I.13. Next, we define the adjoint operators

$$\begin{aligned} [\mathcal{K}^\star f_n]_m &= \overline{\hat{\gamma}}_0^2 \Delta V \sum_{n=1}^{N} \langle\langle \overline{\hat{G}} \rangle\rangle(\boldsymbol{x}_m - \boldsymbol{x}_n)\ \langle \hat{f}_n \rangle, \\ [\partial_k \mathcal{K}^\star f_n]_m &= \Delta V \sum_{n=1}^{N} \langle\langle \partial_k \overline{\hat{G}} \rangle\rangle(\boldsymbol{x}_m - \boldsymbol{x}_n)\ \langle \hat{f}_n \rangle, \end{aligned} \tag{2.61}$$

for each function $\hat{f}_n$. The convolutional operators on the right-hand sides are computed with the help of function `AdjKop` of Matlab Listing I.14.

So far, in 3D, we have defined seven convolutions to be carried carry out, viz., four convolutions of Eq. (2.57) and three convolutions of Eq. (2.58). But, in the latter equation, we need also the convolutional operator $\partial_j \partial_k \langle\langle \hat{G} \rangle\rangle$ operating on $\hat{w}_{k,n}^{Zv}$. This means that in 3D, we have to compute an extra set of nine convolutions. This will increase the computation time per iteration by more than a factor of two. In addition special care has to be taken to extract the singularity of the Green function, before we are able to change integration and the differentiations $\partial_j \partial_k$. We therefore leave this grad div operator in front of the integral and choose to replace it by its numerical equivalent:

$$[\partial_j \partial_k \mathcal{K} \hat{w}_{k,n}^{Zv}]_m = \texttt{graddiv}_{j,k} \left[\Delta V \sum_{n=1}^{N} \langle\langle \hat{G} \rangle\rangle(|m-n|\Delta x)\ \langle \hat{w}_{k,n}^{Zv} \rangle \right], \tag{2.62}$$

where we still use the summation convention for the subscript k. As a consequence, we need only three extra convolutions while the increase of computer time by carrying out the discrete graddiv operation is very small. For 1D, 2D, and 3D, it is given by function `graddiv` of Matlab Listing I.34; see formulas 25.3.23 and 25.3.26 of [1, p. 884]. Note that we use Matlab's cell array notation $v\{1\}$, $v\{2\}$ and $v\{3\}$ to denote the vector $\boldsymbol{v} = \{v_1, v_2, v_3\}$. This accommodates an easy transfer of vector functions between various subroutines.

Matlab Listing I.34 The function `graddiv` to compute the graddiv operator.

```
function u = graddiv(v,input)
global nDIM;   dx = input.dx;

if nDIM == 1

N1 = input.N1;    u{1} = zeros(size(v{1}));
% Compute d1d1_v1                    ----------------------------
 u{1}(2:N1-1) = (v{1}(1:N1-2) - 2*v{1}(2:N1-1) + v{1}(3:N1))/dx^2;

elseif nDIM == 2

N1 = input.N1;  N2 = input.N2;  u{1} = zeros(size(v{1}));  u{2} = u{1};
% Compute d1d1_v1, d2d2_v2           ----------------------------
 u{1}(2:N1-1,:) = v{1}(1:N1-2,:) - 2*v{1}(2:N1-1,:) + v{1}(3:N1,:);
 u{2}(:,2:N2-1) = v{2}(:,1:N2-2) - 2*v{2}(:,2:N2-1) + v{2}(:,3:N2);
% Replace the input vector v1 by d1_v and v2 by d2_v2 ----------------
  v{1}(2:N1-1,:) = (v{1}(3:N1,:) - v{1}(1:N1-2,:))/2;   % d1_v1
  v{2}(:,2:N2-1) = (v{2}(:,3:N2) - v{2}(:,1:N2-2))/2;   % d2_v2
% Add d1_v2 = d1d2_v2 to output vector u1 ----------------------------
 u{1}(2:N1-1,:)= u{1}(2:N1-1,:)+(v{2}(3:N1,:)-v{2}(1:N1-2,:))/2;
% Add d2_v1 = d2d1_v1 to output vector u2 ----------------------------
 u{2}(:,2:N2-1)= u{2}(:,2:N2-1)+(v{1}(:,3:N2)-v{1}(:,1:N2-2))/2;

 u{1} = u{1}/dx^2;        u{2} = u{2}/dx^2;             % divide by dx^2

elseif nDIM == 3

N1   = input.N1;              N2   = input.N2;       N3   = input.N3;
u{1} = zeros(size(v{1}));     u{2} = u{1};           u{3} = u{1};
% Compute d1d1_v1, d2d2_v2, d3d3_v3 ----------------------------------
 u{1}(2:N1-1,:,:) = v{1}(1:N1-2,:,:) - 2*v{1}(2:N1-1,:,:) + v{1}(3:N1,:,:);
 u{2}(:,2:N2-1,:) = v{2}(:,1:N2-2,:) - 2*v{2}(:,2:N2-1,:) + v{2}(:,3:N2,:);
 u{3}(:,:,2:N3-1) = v{3}(:,:,1:N3-2) - 2*v{3}(:,:,2:N3-1) + v{3}(:,:,3:N3);
% Replace the input vector v1 by d1_v, v2 by d2_v2 and v3 by d3_v3 -------
  v{1}(2:N1-1,:,:) = (v{1}(3:N1,:,:) - v{1}(1:N1-2,:,:))/2;   % d1_v1
  v{2}(:,2:N2-1,:) = (v{2}(:,3:N2,:) - v{2}(:,1:N2-2,:))/2;   % d2_v2
  v{3}(:,:,2:N3-1) = (v{3}(:,:,3:N3) - v{3}(:,:,1:N3-2))/2;   % d3_v3
% Add d1_v2 = d1d2_v2 and d1_v3 = d1d3_v3  to output vector u1 ----------
 u{1}(2:N1-1,:,:)= u{1}(2:N1-1,:,:)+(v{2}(3:N1,:,:)-v{2}(1:N1-2,:,:))/2 ...
                                   +(v{3}(3:N1,:,:)-v{3}(1:N1-2,:,:))/2;
% Add d2_v1 = d2d1_v1 and dw v3 = d2d3_v3 to output vector u2 -----------
 u{2}(:,2:N2-1,:)= u{2}(:,2:N2-1,:)+(v{1}(:,3:N2,:)-v{1}(:,1:N2-2,:))/2 ...
                                   +(v{3}(:,3:N2,:)-v{3}(:,1:N2-2,:))/2;
% Add d3_v1 = d3d1_v1 and d3_v2 = d3d2_v2 to output vector u3 -----------
 u{3}(:,:,2:N3-1)= u{3}(:,:,2:N3-1)+(v{1}(:,:,3:N3)-v{1}(:,:,1:N3-2))/2 ...
                                   +(v{2}(:,:,3:N3)-v{2}(:,:,1:N3-2))/2;

 u{1} = u{1}/dx^2; u{2} = u{2}/dx^2; u{3} = u{3}/dx^2;
% divide by dx^2
end
```

By now we omit the hat symbols and define the discrete acoustic operator, $\mathcal{K}^p_m$, on the contrast sources as

$$\mathcal{K}^p_m\{w^p_n, w^{Zv}_{k,n}\} = [\mathcal{K}w^p_n]_m - \gamma_0[\partial_k\mathcal{K}w^{Zv}_{k,n}]_m. \tag{2.63}$$

This operator is the discrete version of the expression inside the square brackets on the right-hand side of Eq. (2.57). In fact, it is equivalent to the scattered pressure wave field. The operator $[\mathcal{K}w^p_n]_m$ is obtained from the first convolution of Eq. (2.60) by substituting $w_n := w^p_n$. The convolution $[\partial_k\mathcal{K}w^{Zv}_{k,n}]_m$ is obtained from the second convolution of Eq. (2.60) by substituting $w_n := w^{Zv}_{k,n}$ for $k = 1, 2, 3$, and sum them up for all k. Subsequently, we define the discrete acoustic operator, $\mathcal{K}^{Zv}_{j,m}$, on the contrast sources as follows:

$$\mathcal{K}^{Zv}_{j,m}\{w^p_n, w^{Zv}_{k,n}\} = -\gamma_0[\partial_j\mathcal{K}w^p_n]_m + [\partial_j\partial_k\mathcal{K}w^{Zv}_{k,n}]_m + \langle w^{Zv}_{j,m}\rangle. \tag{2.64}$$

This operator is the discrete version of the expression inside the square brackets on the right-hand side of Eq. (2.58). It is equivalent to the scattered particle-velocity wave field. The operator $[\partial_j\mathcal{K}\hat{w}^p_n]_m$ is obtained from the second convolution of Eq. (2.60) by substituting $w_n := w^p_n$, while operator $[\partial_j\partial_k\mathcal{K}w^{Zv}_{k,n}]_m$ is given by Eq. (2.62). The codes to compute both operators are listed as case `'analytical'` in function `KOPpZv` of Matlab Listing I.37.

For further considerations, we need also the adjoint of the operators, viz.,

$$\begin{aligned}\mathcal{K}^{\star p}_m\{f^p_n, f^{Zv}_{k,n}\} &= [\mathcal{K}^\star f^p_n]_m - \overline{\gamma}_0\,[\partial_k\mathcal{K}^\star f^{Zv}_{k,n}]_m,\\ \mathcal{K}^{\star Zv}_{j,m}\{f^p_n, f^{Zv}_{k,n}\} &= -\overline{\gamma}_0[\partial_j\mathcal{K}^\star f^p_n]_m + [\partial_j\partial_k\mathcal{K}^\star f^{Zv}_{k,n}]_m + \langle f^{Zv}_{j,m}\rangle.\end{aligned} \tag{2.65}$$

These operations are listed as case `'analytical'` in function `AdjKOPZv` of Matlab Listing I.38.

2.1.4.2 Numerical Differentiation

In our previous case `'analytical'`, in 3D, we still need ten convolutions to be carried in each iteration. When the spatial differentiations are carried out numerically, the number of convolutions we need is equal to the four unknown contrast source distributions, viz., w^p_n and $w^{Zv}_{k,n}$ for $k = 1, 2, 3$. The numerical equivalent of Eq. (2.63) becomes

$$\mathcal{K}^p_m\{w^p_n, w^{Zv}_{k,n}\} = [\mathcal{K}w^p_n]_m - \gamma_0^{-1}\mathrm{DIV}_k[\mathcal{K}w^{Zv}_{k,n}]_m, \tag{2.66}$$

where $\mathrm{DIV}_k v_k$ is the finite-difference counterpart of the divergence operator on a vector v_k. Both for 1D, 2D, and 3D, it is given by function `DIV` of Matlab Listing I.35. The convolutions $[\mathcal{K}w^p_n]_m$ and $[\mathcal{K}w^{Zv}_{k,n}]_m$ are obtained from the first convolution of Eq. (2.60), by substituting $w_n := w^p_n$ and $w_n = w^{Zv}_{k,n}$, for $k = 1, 2, 3$.

Subsequently, the numerical equivalent of Eq. (2.64) becomes

$$\mathcal{K}^{Zv}_{j,m}\{w^p_n, w^{Zv}_{k,n}\} = -\gamma_0^{-1}\mathrm{GRAD}_j[\mathcal{K}^p_m\{w^p_n, w^{Zv}_{k,n}\}] + \langle w^{Zv}_{j,m}\rangle, \tag{2.67}$$

where GRAD_j denotes the j-component of the finite difference operator. For 1D, 2D, and 3D, it is given by function `GRAD` of Matlab Listing I.36. The codes to compute both operators are listed as method `numerical` in function `KOPpZv` of Matlab Listing I.37.

For further considerations, we need also the adjoint of the operators. In view of the partial integration rule, we have to take into account that the adjoint of ∂_k changes sign. Then the expressions for the adjoint operations are obtained as

$$\begin{aligned}\mathcal{K}^{\star p}_m\{f^p_n, f^{Zv}_{k,n}\} &= [\mathcal{K}^\star f^p_n]_m + \overline{\gamma}_0^{-1}\mathrm{DIV}_k\,[\mathcal{K}^\star f^{Zv}_{k,n}]_m,\\ \mathcal{K}^{\star Zv}_{j,m}\{f^p_n, f^{Zv}_{k,n}\} &= \overline{\gamma}_0^{-1}\mathrm{GRAD}_j[\mathcal{K}^{\star p}_m\{f^p_n, f^{Zv}_{k,n}\}] + \langle f^{Zv}_{j,m}\rangle.\end{aligned} \tag{2.68}$$

Matlab Listing I.35 The function `DIV` to compute the divergence operator using central finite differences.

```
function u = DIV(v,input)
global nDIM;  dx = input.dx; u = zeros(size(v{1}));

if nDIM == 1;              N1 = input.N1;
  u(2:N1-1) = (v{1}(3:N1) - v{1}(1:N1-2))/(2*dx);                % d1v1

elseif nDIM == 2;          N1 = input.N1;  N2 = input.N2;
  v{1}(2:N1-1,:) =  v{1}(3:N1,:) - v{1}(1:N1-2,:);              % d1v1
  v{2}(:,2:N2-1) =  v{2}(:,3:N2) - v{2}(:,1:N2-2);              % d2v2

  u = (v{1} + v{2}) / (2*dx);

elseif nDIM == 3;          N1 = input.N1;  N2 = input.N2;  N3 = input.N3;
  v{1}(2:N1-1,:,:) = v{1}(3:N1,:,:) - v{1}(1:N1-2,:,:);         % d1v1
  v{2}(:,2:N2-1,:) = v{2}(:,3:N2,:) - v{2}(:,1:N2-2,:);         % d2v2
  v{3}(:,:,2:N3-1) = v{3}(:,:,3:N3) - v{3}(:,:,1:N3-2);         % d3v3

  u = (v{1} + v{2} + v{3}) / (2*dx);
end
```

Matlab Listing I.36 The function `GRAD` to compute the gradient operator $v = \text{grad}\, p$ using central finite differences.

```
function v = GRAD(p,input)
global nDIM;  dx = input.dx;

if nDIM == 1;             N1 = input.N1;
  v{1} = zeros(size(p));
  v{1}(2:N1-1) = (p(3:N1) - p(1:N1-2)) / (2*dx);                 % d1_p

elseif nDIM == 2;         N1 = input.N1;  N2 = input.N2;
  v{1} = zeros(size(p)); v{2} = v{1};
  v{1}(2:N1-1,:) = (p(3:N1,:) - p(1:N1-2,:)) / (2*dx);           % d1_p
  v{2}(:,2:N2-1) = (p(:,3:N2) - p(:,1:N2-2)) / (2*dx);           % d2_p

elseif nDIM == 3;         N1 = input.N1;  N2 = input.N2;  N3 = input.N3;
  v{1} = zeros(size(p)); v{2} = v{1};   v{3} = v{1};
  v{1}(2:N1-1,:,:) = (p(3:N1,:,:) - p(1:N1-2,:,:)) / (2*dx);     % d1_p
  v{2}(:,2:N2-1,:) = (p(:,3:N2,:) - p(:,1:N2-2,:)) / (2*dx);     % d2_p
  v{3}(:,:,2:N3-1) = (p(:,:,3:N3) - p(:,:,1:N3-2)) / (2*dx);     % d3_p
end
```

The convolutions $[\mathcal{K}^{\star} f_n^p]_m$ and $[\mathcal{K}^{\star} f_{k,n}^{Zv}]_m$ are obtained from the first convolution of Eq. (2.61), by substituting $f_n := f_n^p$ and $f_n = w_{k,n}^{Zv}$, for $k = 1, 2, 3$. The codes to compute the operators of Eq. (2.68) are listed as method `numerical` in function `AdjKOPpZv` of Matlab Listing I.38. The adjoint operators $[\mathcal{K}^{\star} f_n^p]_m$ and $[\mathcal{K}^{\star} f_n^{Zv}]_m$ are convolutional operators. They are computed with fast Fourier transforms (FFTs). We use the code of function `AdjKop` of Matlab Listing I.14 (see Chapter 1).

Matlab Listing I.37 The function `KOPpZv` to carry out the $\mathcal{K}^p$ and $\mathcal{K}^{Zv}$ operators.

```
function [Kp,KZv] = KOPpZv(wp,wZv,input)
global nDIM;
gam0=input.gamma_0; FFTG=input.FFTG; FFTdG=input.FFTdG;  KZv=cell(1,nDIM);

switch lower(input.method)
case 'analytical'
   % Acoustic operator using analytical derivatives
   Kp = Kop(wp,FFTG);
   for n = 1:nDIM
       Kp = Kp - gam0 *Kop(wZv{n},FFTdG{n});
       KZv{n} = Kop(wZv{n},FFTG)/gam0^2;
   end
   KZv = graddiv(KZv,input);
   for n = 1:nDIM
       KZv{n} =  - gam0 * Kop(wp,FFTdG{n}) + KZv{n} + wZv{n};
   end

case 'numerical'
   % Acoustic operator using numerical derivatives
   Kp = Kop(wp,FFTG);
   for n=1:nDIM
      KZv{n} = Kop(wZv{n},FFTG);              % use KZv als temporary storage
   end
   Kp = Kp - DIV(KZv,input) / gam0;
   KZv = GRAD(-Kp/gam0,input);
   for n=1:nDIM
       KZv{n} = KZv{n} + wZv{n};
   end

end % switch
```

2.1.4.3 Conjugate Gradient Method

With the defined operators and their pertaining Matlab codes, we observe that, for both case `'analytical'` and case `'numerical'`, the system of integral equations transfers into the following discrete set of equations:

$$
\begin{aligned}
\chi^{\kappa}_m \langle p^{inc}\rangle(x_m) &= w^p_m - \chi^{\kappa}_m \mathcal{K}^p_m\{w^p_n, w^{Zv}_{k,n}\}.\\
\chi^{\rho}_m \langle v^{inc}_j\rangle(x_m) &= w^{Zv}_{j,m} - \chi^{\rho}_m \mathcal{K}^{Zv}_{j,m}\{w^p_n, w^{Zv}_{k,n}\}.
\end{aligned} \tag{2.69}
$$

This system is solved iteratively with the help of a conjugate gradient (CG) method. The conjugate gradient method for solving the contrast sources w^p and w^{Zv}_k is fundamentally the same as the one presented in Table 1.4. We only have to change the scalar w by the vector consisting of $[w^p, w^{Zv}_k]$ and to extend the inner product definition. To discuss it in more detail, we write Eq. (2.69) in the compact form

$$
\boxed{
\begin{aligned}
w^p - \chi^{\kappa}\, \mathcal{K}^p\{w^p, w^{Zv}_k\} &= \chi^{\kappa} p^{inc},\\
w^{Zv}_j - \chi^{\rho}\, \mathcal{K}^{Zv}_j\{w^p, w^{Zv}_k\} &= \chi^{\rho} Z_0 v^{inc}_j.
\end{aligned}} \tag{2.70}
$$

If the contrast sources do not solve Eq. (2.70), the pertaining residual errors are defined as follows:

$$
\boxed{
\begin{aligned}
r^p &= \chi^{\kappa} p^{inc} - [w^p - \chi^{\kappa}\, \mathcal{K}^p\{w^p, w^{Zv}_k\}],\\
r^{Zv}_j &= \chi^{\rho} Z_0 v^{inc}_j - [w^{Zv}_j - \chi^{\rho}\, \mathcal{K}^{Zv}_j\{w^p, w^{Zv}_k\}],
\end{aligned}} \tag{2.71}
$$

Matlab Listing I.38 The function `AdjKOPpZv` to carry out the $\mathcal{K}^{\star p}$ and $\mathcal{K}^{\star Zv}$ operators.

```
function [Kp,KZv] = AdjKOPpZv(fp ,fZv , input)
global nDIM;
gam0=input.gamma_0; FFTG=input.FFTG; FFTdG=input.FFTdG;  KZv=cell(1,nDIM);

switch lower(input.method)
case 'analytical'
   % Acoustic operator using analytical derivatives
   Kp = AdjKop(fp ,FFTG);
   for n = 1:nDIM
       Kp = Kp - conj(gam0) * AdjKop(fZv{n},FFTdG{n});
       KZv{n} = AdjKop(fZv{n},FFTG)/conj(gam0^2);
   end
   KZv = graddiv(KZv, input);
   for n = 1:nDIM
       KZv{n} = - conj(gam0) * AdjKop(fp ,FFTdG{n}) + KZv{n} + fZv{n};
   end

case 'numerical'
   % Acoustic operator using numerical derivatives
   Kp = AdjKop(fp ,FFTG);
   for n=1:nDIM
       KZv{n} = AdjKop(fZv{n},FFTG);          % use KZv as temporary storage
   end
   Kp = Kp + DIV(KZv, input) / conj(gam0);
   KZv = GRAD(Kp/conj(gam0), input);
   for n=1:nDIM
       KZv{n} = KZv{n} + fZv{n};
   end

end % switch
```

and the normalized cost functional as follows:

$$F(w^p, w_j^{Zv}) = \frac{\|r^p\|^2 + \sum_k \|r_k^{Zv}\|^2}{\|\chi^\kappa p^{inc}\|^2 + \sum_k \|\chi^\rho Z_0 v_k^{inc}\|^2}, \tag{2.72}$$

with the property that $F = 1$ for $w^p = w_1^{Zv} = w_2^{Zv} = w_3^{Zv} = 0$ and $F = 0$ for the exact solution. The gradient directions follow from the Fréchet derivative of this cost functional F and are directly related to the adjoint operating on the residual errors in the equations at the previous iteration $n-1$. They are written as follows:

$$\begin{aligned} g^{p(n)} &= r^{p(n-1)} &- \mathcal{K}^{\star p}\{\overline{\chi}^\kappa r^{p(n-1)}, \overline{\chi}^\rho r_k^{Zv(n-1)}\}, \\ g_j^{Zv(n)} &= r_j^{Zv(n-1)} &- \mathcal{K}^{\star Zv}\{\overline{\chi}^\kappa r^{p(n-1)}, \overline{\chi}^\rho r_k^{Zv(n-1)}\}. \end{aligned} \tag{2.73}$$

The Hestenes–Stiefel conjugate gradient directions after the first iteration become

$$\begin{aligned} v^{p(n)} &= g^{p(n)} + (A^{(n)}/A^{(n-1)})\, v^{p(n-1)}, \\ v_j^{Zv(n)} &= g_j^{Zv(n)} + (A^{(n)}/A^{(n-1)})\, v_j^{Zv(n-1)}, \end{aligned} \tag{2.74}$$

where

$$A^{(n)} = \|g^{p(n)}\|^2 + \sum_k \|g_k^{Zv(n)}\|^2. \tag{2.75}$$

In each iteration, the updates for the contrast sources w^p and w_k^{Zv} are obtained as

$$\begin{aligned} w^{p(n)} &= w^{p(n-1)} + (A^{(n)}/B^{(n)})\, v^{p(n)}, \\ w_j^{Zv(n)} &= w_j^{Zv(n-1)} + (A^{(n)}/B^{(n)})\, v_j^{Zv(n)}, \end{aligned} \tag{2.76}$$

with

$$B^{(n)} = \|v^{p(n)} - \chi^\kappa \mathcal{K}^{p(n)}\{v^{p(n)}, v_k^{Zv(n)}\}\|^2 + \sum_j \|v_j^{Zv(n)} - \chi^\rho \mathcal{K}_j^{Zv(n)}\{v^{p(n)}, v_k^{Zv(n)}\}\|^2. \tag{2.77}$$

The complete conjugate gradient scheme, that minimizes the cost functional F iteratively, is presented in Table 2.1. To complete this numerical section, we need the weak forms of the incident and scattered wave fields.

Table 2.1 The conjugate gradient method for solving the contrast sources w^p and w_j^{Zv}

initial estimate		$w^{p(n=0)}$	$=$	0
		$w_j^{Zv(n=0)}$	$=$	0
		$r^{p(n=0)}$	$=$	$\chi^\kappa p^{inc}$
		$r_j^{Zv(n=0)}$	$=$	$\chi^\rho Z_0 v_j^{inc}$
for $n = 1, 2, \ldots$		$g^{p(n)}$	$=$	$r^{p(n-1)} - \mathcal{K}^{\star p}\{\overline{\chi}^\kappa r^{p(n-1)}, \overline{\chi}^\rho r_k^{Zv(n-1)}\}$
		$g_j^{Zv(n)}$	$=$	$r_j^{Zv(n-1)} - \mathcal{K}_j^{\star Zv}\{\overline{\chi}^\kappa r^{p(n-1)}, \overline{\chi}^\rho r_k^{Zv(n-1)}\}$
		$A^{(n)}$	$=$	$\|g^{p(n)}\|^2 + \sum_j \|g_j^{Zv(n)}\|^2$
	if $n = 1$	$v^{p(n)}$	$=$	$g^{p(n)}$
		$v_j^{Zv(n)}$	$=$	$g_j^{Zv(n)}$
	else	$v^{p(n)}$	$=$	$g^{p(n)} + \frac{A^{(n)}}{A^{(n-1)}} v^{p(n-1)}$
		$v_j^{Zv(n)}$	$=$	$g_j^{Zv(n)} + \frac{A^{(n)}}{A^{(n-1)}} v_j^{Zv(n-1)}$ end
		$B^{(n)}$	$=$	$\|v^{p(n)} - \chi^\kappa \mathcal{K}^{p(n)}\{v^{p(n)}, v_k^{Zv(n)}\}\|^2 + \sum_j \|v_j^{Zv(n)} - \chi^\rho \mathcal{K}_j^{Zv(n)}\{v^{p(n)}, v_k^{Zv(n)}\}\|^2$
		$w^{p(n)}$	$=$	$w^{p(n-1)} + \frac{A^{(n)}}{B^{(n)}} v^{p(n)}$
		$w_j^{Zv(n)}$	$=$	$w_j^{Zv(n-1)} + \frac{A^{(n)}}{B^{(n)}} v_j^{Zv(n)}$
		$r^{p(n)}$	$=$	$r^{p(n-1)} - \frac{A^{(n)}}{B^{(n)}}[v^{p(n)} - \chi^\kappa \mathcal{K}^{p(n)}\{v^{p(n)}, v_k^{Zv(n)}\}]$
		$r_j^{Zv(n)}$	$=$	$r_j^{Zv(n-1)} - \frac{A^{(n)}}{B^{(n)}}[v_j^{Zv(n)} - \chi^\rho \mathcal{K}_j^{Zv(n)}\{v^{p(n)}, v_k^{Zv(n)}\}]$
	if $\left[\frac{\|r^{p(n)}\|^2 + \sum_j \|r_j^{Zv(n)}\|^2}{\|\chi^\kappa p^{inc}\|^2 + \sum_j \|\chi^\rho Z_0 v_j^{inc}\|^2}\right]^{\frac{1}{2}} < \epsilon$ or $n > n_{max}$ stop end			
end				

2.1.4.4 Incident Acoustic Wave Field

In 3D, the pressure and the particle velocity of the incident acoustic wave field are given by Eqs. (2.29) and (2.30). Their weak counterparts are expressed in terms of the weak forms $\langle\hat{G}\rangle$ of Table 1.1 as

$$\begin{aligned}\langle\hat{p}^{inc}\rangle(\boldsymbol{x}) &= [\hat{\gamma}_0 Z_0\hat{Q}]\ \langle\hat{G}\rangle(\boldsymbol{x}-\boldsymbol{x}^S),\\ \langle Z_0\hat{v}_j^{inc}\rangle(\boldsymbol{x}) &= -\frac{1}{\hat{\gamma}_0}[\hat{\gamma}_0 Z_0\hat{Q}]\ \partial_j\langle\hat{G}\rangle(\boldsymbol{x}-\boldsymbol{x}^S),\quad j=1,2,3.\end{aligned} \tag{2.78}$$

In 2D, these quantities are

$$\begin{aligned}\langle\hat{p}^{inc}\rangle(\boldsymbol{x}_T) &= [\hat{\gamma}_0 Z_0\hat{Q}]\ \langle\hat{G}\rangle(\boldsymbol{x}_T-\boldsymbol{x}_T^S),\\ \langle Z_0\hat{v}_j^{inc}\rangle(\boldsymbol{x}_T) &= -\frac{1}{\hat{\gamma}_0}[\hat{\gamma}_0 Z_0\hat{Q}]\ \partial_j\langle\hat{G}\rangle(\boldsymbol{x}_T-\boldsymbol{x}_T^S),\quad j=1,2.\end{aligned} \tag{2.79}$$

In 1D, these quantities are

$$\begin{aligned}\langle\hat{p}^{inc}\rangle(x_1) &= [\hat{\gamma}_0 Z_0\hat{Q}]\ \langle\hat{G}\rangle(x_1-x_1^S),\\ \langle Z_0\hat{v}_1^{inc}\rangle(x_1) &= -\frac{1}{\hat{\gamma}_0}[\hat{\gamma}_0 Z_0\hat{Q}]\ \partial_1\langle\hat{G}\rangle(x_1-x_1^S).\end{aligned} \tag{2.80}$$

The spatial derivatives of the weak form of the Green function are listed below.

2.1.4.5 Scattered Acoustic Wave Field

In the 3D case, the scattered pressure is given by Eq. (2.59). Its weak counterpart is expressed in terms of the weak forms $\langle\hat{G}\rangle$ of Table 1.1 as

$$\langle\hat{p}^{sct}\rangle(\boldsymbol{x}^R) = (\Delta x)^3\sum_{m=1}^{N}\left[\hat{\gamma}_0^2\langle\hat{G}\rangle(\boldsymbol{x}^R-\boldsymbol{x}_m)\ \langle\hat{w}_m^p\rangle - \sum_{k=1}^{3}\hat{\gamma}_0\partial_k^R\langle\hat{G}\rangle(\boldsymbol{x}^R-\boldsymbol{x}_m)\ \langle\hat{w}_{k,m}^{Zv}\rangle\right]. \tag{2.81}$$

In the 2D case, this quantity is

$$\langle\hat{p}^{sct}\rangle(\boldsymbol{x}_T^R) = (\Delta x)^2\sum_{m=1}^{N}\left[\hat{\gamma}_0^2\langle\hat{G}\rangle(\boldsymbol{x}_T^R-\boldsymbol{x}_{T,m})\langle\hat{w}_m^p\rangle - \sum_{k=1}^{2}\hat{\gamma}_0\partial_k^R\langle\hat{G}\rangle(\boldsymbol{x}_T^R-\boldsymbol{x}_{T,m})\langle\hat{w}_{k,m}^{Zv}\rangle\right]. \tag{2.82}$$

In the 1D case, this quantity is

$$\langle\hat{p}^{sct}\rangle(x^R) = \Delta x\sum_{m=1}^{N}[\hat{\gamma}_0^2\langle\hat{G}\rangle(x_1^R-x_{1,m})\ \langle\hat{w}_m^p\rangle - \hat{\gamma}_0\partial_1^R\langle\hat{G}\rangle(x_1^R-x_{1,m})\ \langle\hat{w}_{1,m}^{Zv}\rangle]. \tag{2.83}$$

The spatial derivatives of the weak form of the Green function are listed below.

2.1.4.6 Weak Form of the Spatial Derivative of the Green Function

For $|\boldsymbol{x}| > \frac{1}{2}\Delta x$, the spatial derivatives of the weak Green function $\langle\hat{G}\rangle(\boldsymbol{x})$ follow immediately from Table 1.1. In 3D, these derivatives are

$$\partial_k\langle\hat{G}\rangle(\boldsymbol{x}) = -\left(\frac{1}{|\boldsymbol{x}|}+\hat{\gamma}_0\right)\frac{x_k}{|\boldsymbol{x}|}\langle\hat{G}\rangle(\boldsymbol{x}),\quad k=1,2,3. \tag{2.84}$$

In 2D, they are

$$\partial_k\langle\hat{G}\rangle(\boldsymbol{x}_T) = -\hat{\gamma}_0\frac{x_k}{|\boldsymbol{x}_T|}\langle\partial\hat{G}\rangle,\quad k=1,2, \tag{2.85}$$

where $\langle\partial\hat{G}\rangle = f(\hat{\gamma}_0\delta)\ \frac{1}{2\pi}K_1(\hat{\gamma}_0|\boldsymbol{x}_T|)$, while in 1D, we have

$$\partial_1\langle\hat{G}\rangle(x_1) = -\hat{\gamma}_0\frac{x_1}{|x_1|}\langle\hat{G}\rangle(x_1). \tag{2.86}$$

2.2 Matlab Codes Integral Equations: Pressure and Particle Velocity

In this section, we present the Matlab codes for the numerical solution of the coupled set of integral equations for the acoustic pressure and particle velocity inside the scatterer, and we discuss the Matlab codes to compute the scattered acoustic pressure in a number of receiver points outside the scatterer. To initialize the various acoustic wave parameters, we start with the function `initAC` of Matlab Listing I.39. We take the same parameters as used in Chapter 1. We only have to add the actual values of the mass density. The source wavelet is chosen to be $\hat{\gamma}_0 Z_0 \hat{Q} = s\rho_0 \hat{Q} = 1$. The global parameter `nDIM` assigns the dimension of the space at hand. Similar as in Chapter 1, the wave speeds are $c_0 = 1500$ m/s and $\hat{c}_{sct} = 3000$ m/s. The mass densities are chosen to be $\rho_0 = 3000$ kg/m^3 and $\hat{\rho}_{sct} = 1500$ kg/m^3. This means that the wave impedances Z_0 and Z_{sct} are equal to each other and in the 1D acoustic case, there is no reflected wave field. The source/receiver locations and the grid parameters are initialized by calling the Matlab functions `initSourceReceiver` and `initGrid` of Matlab Listings I.2 and I.3. They do not need any change.

The script `initFFTGreen`, however, is replaced by the `initFFTGreenfun` of Matlab Listing I.40. The latter one needs the script `Greenfun` of Matlab Listing I.41. In addition to the weak form of the Green function itself, the weak form of the spatial derivatives of the Green function are computed as well. They follow from Eqs. (2.84)–(2.86), and they are enforced to be zero at zero distance.

Matlab Listing I.39 The function `initAC` to initialize the acoustic wave-field parameters.

```
function input = initAC()
% Time factor = exp(-iwt)
% Spatial units is in m
% Source wavelet  s rho_0 Q = 1

global nDIM;  nDIM = 2;                       % set dimension of space

input.c_0     = 1500;                         % wave speed in embedding
input.c_sct   = 3000;                         % wave speed in scatterer
input.rho_0   = 3000;                         % mass density in embedding
input.rho_sct = 1500;                         % mass density in scatterer

f             = 50;                           % temporal frequency
wavelength    = input.c_0 / f;                % wavelength
s             = 1e-16 - 1i*2*pi*f;            % LaPlace parameter
input.gamma_0 = s/input.c_0;                  % propagation coefficient
disp(['wavelength = ' num2str(wavelength)]);

% add input data to structure array 'input'
  input = initSourceReceiver(input);       % add location of source/receiver

  input = initGrid(input);                 % add grid in either 1D, 2D or 3D

  input = initFFTGreenfun(input);          % compute FFT of Green function

  input = initAcousticContrast(input);     % add contrast distribution

  input.Errcri = 1e-3;
```

Matlab Listing I.40 The function `initFFTGreenfun` computes the discrete Fourier transform of the Green function and its spatial derivatives.

```
function [input] = initFFTGreenfun(input)
global nDIM;

if nDIM == 1                              % make one-dimensional FFT grid
   N1fft        = 2^ceil(log2(2*input.N1));
   x1(1:N1fft) = [0 : N1fft/2-1   N1fft/2 : -1 : 1] * input.dx;
   temp.X1fft = x1';          % put it in single column
   sign_x1 = [0 ones(1,N1fft/2-1)  0  -ones(1,N1fft/2-1)];
   Sign.X1fft = sign_x1'; % put it in single column

elseif nDIM == 2                          % make two-dimensional FFT grid
   N1fft        = 2^ceil(log2(2*input.N1));
   N2fft        = 2^ceil(log2(2*input.N2));
   x1(1:N1fft) = [0 : N1fft/2-1   N1fft/2 : -1 : 1] * input.dx;
   x2(1:N2fft) = [0 : N2fft/2-1   N2fft/2 : -1 : 1] * input.dx;
   [temp.X1fft,temp.X2fft] = ndgrid(x1,x2);
   sign_x1 = [0 ones(1,N1fft/2-1)  0  -ones(1,N1fft/2-1)];
   sign_x2 = [0 ones(1,N2fft/2-1)  0  -ones(1,N2fft/2-1)];
   [Sign.X1fft,Sign.X2fft] = ndgrid(sign_x1,sign_x2);

elseif nDIM == 3                          % make three-dimensional FFT grid
   N1fft        = 2^ceil(log2(2*input.N1));
   N2fft        = 2^ceil(log2(2*input.N2));
   N3fft        = 2^ceil(log2(2*input.N3));
   x1(1:N1fft) = [0 : N1fft/2-1   N1fft/2 : -1 : 1] * input.dx;
   x2(1:N2fft) = [0 : N2fft/2-1   N2fft/2 : -1 : 1] * input.dx;
   x3(1:N3fft) = [0 : N3fft/2-1   N3fft/2 : -1 : 1] * input.dx;
   [temp.X1fft,temp.X2fft,temp.X3fft] = ndgrid(x1,x2,x3);
   sign_x1 = [0 ones(1,N1fft/2-1)  0  -ones(1,N1fft/2-1)];
   sign_x2 = [0 ones(1,N2fft/2-1)  0  -ones(1,N2fft/2-1)];
   sign_x3 = [0 ones(1,N3fft/2-1)  0  -ones(1,N3fft/2-1)];
   [Sign.X1fft,Sign.X2fft,Sign.X3fft] = ndgrid(sign_x1,sign_x2,sign_x3);
end % if

% compute gam_0^2 * subdomain integrals of Green function
% and interface integrals of derivatives of Green function
  [IntG,IntdG] = Greenfun(temp,Sign,input);

% apply n-dimensional Fast Fourier transforms
  input.FFTG = fftn(IntG);
  for n=1:nDIM
    input.FFTdG{n} = fftn(IntdG{n});
  end
```

Matlab Listing I.41 The function `Greenfun` computes the repeated weak form of the Green function and the spatial derivatives of the strong form.

```
function [IntG, IntdG] = Greenfun(temp, Sign, input)
global nDIM;

gam0 = input.gamma_0;
dx   = input.dx;
if nDIM == 1
    delta    = 0.5 * dx;
    factor   = sinh(gam0*delta) / (gam0*delta);
    X1       = temp.X1fft;
    DIS      = abs(X1);
    G        = 1/(2*gam0) * exp(-gam0.*DIS) * factor^2;
    IntG     = (gam0^2 * dx) * G;                 % integral includes gam0^2
  IntG(1)    = 1 - exp(-gam0*delta) * factor;  %---------------------------------
    d_G      = - gam0 * G;
  d1_G       = Sign.X1fft .* d_G;
  IntdG{1}   = dx * d1_G;

elseif nDIM == 2
    delta     = (pi)^(-1/2) * dx;             % radius circle with area of dx^2
    factor    = 2 * besseli(1,gam0*delta) / (gam0*delta);
    X1        = temp.X1fft;  X2 = temp.X2fft;
    DIS       = sqrt(X1.^2 + X2.^2);
  DIS(1,1)    = 1;                          % avoid Green's singularity for DIS = 0
    G         = 1/(2*pi).* besselk(0,gam0*DIS) * factor^2;
    IntG      = (gam0^2 * dx^2) * G;                 % integral includes gam0^2
  IntG(1,1)   = 1 - gam0*delta * besselk(1,gam0*delta) * factor; %----------
    d_G       = - gam0 * (1/(2*pi)) .* besselk(1,gam0*DIS) * factor^2;
    d1_G      = (Sign.X1fft .* X1./DIS) .* d_G;
    d2_G      = (Sign.X2fft .* X2./DIS) .* d_G;
  IntdG{1}    = dx^2 * d1_G;     IntdG{1}(1,1) = 0;
  IntdG{2}    = dx^2 * d2_G;     IntdG{2}(1,1) = 0;

elseif nDIM == 3    % integral over sphere with equivalent volume of dx^3
    delta    = (4*pi/3)^(-1/3) * dx;        % radius sphere with area of dx^3
    arg      = gam0*delta;
    factor   = 3 * (cosh(arg) - sinh(arg)/arg) / arg^2;
     X1      = temp.X1fft;  X2 = temp.X2fft;  X3 = temp.X3fft;
     DIS     = sqrt(X1.^2 + X2.^2 + X3.^2);
DIS(1,1,1)   = 1;                          % avoid Green's singularity for DIS = 0
   G         = exp(-gam0*DIS) ./ (4*pi*DIS) * factor^2;
   IntG      = (gam0^2 * dx^3) * G;                 % integral includes gam0^2
IntG(1,1,1)= 1 - (1+gam0*delta) * exp(-gam0*delta) * factor;
  d_G        = (-1./DIS - gam0) .* G;
  d1_G       = (Sign.X1fft .* X1./DIS) .* d_G;
  d2_G       = (Sign.X2fft .* X2./DIS) .* d_G;
  d3_G       = (Sign.X3fft .* X3./DIS) .* d_G;
   IntdG{1}  = dx^3 * d1_G;     IntdG{1}(1,1,1) = 0;
   IntdG{2}  = dx^3 * d2_G;     IntdG{2}(1,1,1) = 0;
   IntdG{3}  = dx^3 * d3_G;     IntdG{3}(1,1,1) = 0;
end
```

Matlab Listing I.42 The function `initAcousticContrast` to initialize the acoustic parameters.

```
function [input] = initAcousticContrast(input)
global nDIM;

input.a  = 40;  % half width slab / radius circle cylinder / radius sphere

if      nDIM == 1;    R = sqrt(input.X1.^2);
elseif  nDIM == 2;    R = sqrt(input.X1.^2 + input.X2.^2);
elseif  nDIM == 3;    R = sqrt(input.X1.^2 + input.X2.^2 + input.X3.^2);
end % if

shape = (R < input.a);

% (1) Compute compressibbily contrast (kappa = 1/(rho*c^2) ----------------
kappa_0        = 1 / (input.rho_0   * input.c_0^2);
kappa_sct      = 1 / (input.rho_sct * input.c_sct^2);
contrast_kappa = 1 - kappa_sct /kappa_0;
input.CHI_kap  = contrast_kappa .* shape;

% (2) Compute mass density contrast ----------------------------------------
contrast_rho   = 1 - input.rho_sct / input.rho_0;
input.CHI_rho  = contrast_rho .* shape;
```

Matlab Listing I.43 The `ForwardCGFFTwpZv` script to solve the contrast sources $\hat{w}^p$ and $\partial\hat{w}_k^{Zv}$ from the system of integral equations.

```
clear all; clc; close all; clear workspace
input = initAC();

%  (1) Compute analytically scattered field data ---------------------------
       AcForwardCanonicalObjects

%  (2) Compute incident field ----------------------------------------------
       [p_inc,Zv_inc] = IncAcousticWave(input);

%  (3) Solve contrast source integral equations with CGFFT method ----------
       input.method = 'analytical';   % choose 'analytical' or 'numerical'
       disp(['Derivatives are ' lower(input.method)]);
tic;   [w_p,w_Zv] = ITERCGwpZv(p_inc,Zv_inc,input);
toc
    %  Plot contrast sources, compute and plot the pressure wave fields
       plotContrastSourcewp(w_p,input);
       plotContrastSourcewZv(w_Zv,input);
       [Kf,~] = KOPpZv(w_p,w_Zv,input);   p = p_inc + Kf;
       plotPressureWavefield(p_inc,p,input);

% (4)  Compute synthetic data and plot fields and data ----------------------
       data = DOPwpZv(w_p,w_Zv,input);
       displayData(data,input);
```

Matlab Listing I.44 The function `IncAcousticWave` to compute the incident pressure and particle-velocity wave fields.

```
function [p_inc,Zv_inc] = IncAcousticWave(input)
global nDIM;
gam0 = input.gamma_0;
xS   = input.xS;      dx = input.dx;

if nDIM == 1                    % incident wave on one-dimensional grid

% Weak form
   delta      = 0.5 * dx;
   factor     = sinh(gam0*delta) / (gam0*delta);
   x1         = input.X1;
   DIS        = abs(x1-xS(1));
   p_inc      = factor * 1/(2*gam0) * exp(-gam0.*DIS);
   d_p_inc    = - gam0 * p_inc;
   Zv_inc{1}  = - (1/gam0) * sign(x1-xS(1)) .* d_p_inc;

elseif nDIM == 2                % incident wave on two-dimensional grid

% Weak form
   delta      = (pi)^(-1/2) * dx;        % radius circle with area of dx^2
   factor     = 2 * besseli(1,gam0*delta) / (gam0*delta);
   X1         = input.X1;    X2 = input.X2;
   DIS        = sqrt( (X1-xS(1)).^2 + (X2-xS(2)).^2 );
   p_inc      = factor * 1/(2*pi).* besselk(0,gam0*DIS);
   d_p_inc    = - gam0 * factor * 1/(2*pi).* besselk(1,gam0*DIS);
   Zv_inc{1}  = - (1/gam0) * (X1-xS(1))./DIS .* d_p_inc;
   Zv_inc{2}  = - (1/gam0) * (X2-xS(2))./DIS .* d_p_inc;

elseif nDIM == 3                % incident wave on three-dimensional grid

% Weak form
   delta      = (4*pi/3)^(-1/3) * dx; % radius sphere with area of dx^3
   arg        = gam0*delta;
   factor     = 3 * (cosh(arg) - sinh(arg)/arg) / arg^2;
   X1         = input.X1;    X2 = input.X2;     X3 =input.X3;
   DIS        = sqrt( (X1-xS(1)).^2 + (X2-xS(2)).^2 + (X3-xS(3)).^2 );
   p_inc      = factor * exp(-gam0*DIS) ./ (4*pi*DIS);
   d_p_inc    = - (1./DIS + gam0) .* p_inc;
   Zv_inc{1}  = - (1/gam0) * (X1-xS(1))./DIS .* d_p_inc;
   Zv_inc{2}  = - (1/gam0) * (X2-xS(2))./DIS .* d_p_inc;
   Zv_inc{3}  = - (1/gam0) * (X3-xS(3))./DIS .* d_p_inc;

end % if
```

Matlab Listing I.45 The function ITERCGwpZv to compute the contrast sources w^p and w_k^{Zv}.

```
function[w_p,w_Zv] = ITERCGwpZv(p_inc,Zv_inc,input)
global nDIM;    CHI_kap = input.CHI_kap;    CHI_rho = input.CHI_rho;

w_Zv=cell(1,nDIM); g_Zv=cell(1,nDIM); r_Zv=cell(1,nDIM); v_Zv=cell(1,nDIM);

itmax = 1000;   Errcri = input.Errcri;   it = 0;  % initialization iteration
  w_p = zeros(size(p_inc)); r_p = CHI_kap.*p_inc; Norm0 = norm(r_p(:))^2;
for n = 1:nDIM
   w_Zv{n} = zeros(size(Zv_inc{n}));
   r_Zv{n} = CHI_rho.*Zv_inc{n};          Norm0 = Norm0 + norm(r_Zv{n}(:))^2;
end

CheckAdjointpZv(r_p,r_Zv,input); % Check once ----------------------------
Error = 1;    fprintf('Error =          %g',Error);
while (it < itmax) && ( Error > Errcri) && (Norm0 > eps)
% determine conjugate gradient direction v
  dummy=cell(1,nDIM); for n = 1:nDIM; dummy{n}=conj(CHI_rho).*r_Zv{n}; end
  [Kp,KZv] = AdjKOPpZv(conj(CHI_kap).*r_p, dummy,input);
       g_p =  r_p - Kp;                         % and window for small CHI_kap
       g_p = (abs(CHI_kap) >= Errcri).*g_p;       AN = norm(g_p(:))^2;
  for n = 1:nDIM
   g_Zv{n} = r_Zv{n} - KZv{n};                  % and window for small CHI_rho
   g_Zv{n} = (abs(CHI_rho) >= Errcri).*g_Zv{n}; AN = AN+norm(g_Zv{n}(:))^2;
  end
  if it == 0;  v_p=g_p; for n=1:nDIM; v_Zv{n}=g_Zv{n};  end
  else
   v_p=g_p+AN/AN_1*v_p; for n=1:nDIM; v_Zv{n}=g_Zv{n}+AN/AN_1*v_Zv{n}; end
  end
% determine step length alpha
  [Kp,KZv] = KOPpZv(v_p,v_Zv,input);
   Kp = v_p - CHI_kap .* Kp;                    BN = norm(Kp(:))^2  ;
   for n = 1:nDIM
     KZv{n} = v_Zv{n} - CHI_rho .* KZv{n};      BN = BN + norm(KZv{n}(:))^2;
   end
   alpha = AN / BN;
 % update contrast sources and AN
   w_p   = w_p + alpha * v_p;
   for n = 1:nDIM;  w_Zv{n} = w_Zv{n} + alpha * v_Zv{n}; end
   AN_1  = AN;
 % update residual errors
   r_p   = r_p  - alpha * Kp;              Norm = norm(r_p(:))^2;
   for n = 1:nDIM
     r_Zv{n} = r_Zv{n} - alpha * KZv{n};   Norm = Norm + norm(r_Zv{n}(:))^2;
   end
   Error = sqrt(Norm / Norm0); fprintf('\b\b\b\b\b\b\b\b%6f',Error);
   it = it+1;
end % CG iterations
 fprintf('\b\b\b\b\b\b\b\b%6f\n',Error);
 disp(['Number of iterations is ' num2str(it)]);
 if it == itmax; disp(['itmax reached:  err/norm = ' num2str(Error)]); end
```

The acoustic compressibility and mass-density distributions are computed by the Matlab function `initAcousticContrast` of Matlab Listing I.39. This Matlab function is the replacement of the Matlab function `initContrast` of Chapter 1. The changes are obvious.

After the initialization of the acoustic parameters, we start with the main Matlab script `ForwardCGFFTwpZv` of Matlab Listing I.43. Similar to the main Matlab script of scalar wave scattering we have four steps: (1) Computation of scattering by a canonical object; (2) Computation of the incident wave fields; (3) Solution of the set of integral equations; (4) Computation of scattering data, based on the solution of the integral equations.

In step (1) of `ForwardCGFFTwpZv`, script `AcForwardCanonicalObjects` of Matlab Listing I.79 of Appendix 2.A.4 is called. It computes the acoustic pressure at a number of receivers, for either the 1D slab, the 2D circle, or the 3D sphere. These data are used to benchmark the scattered field based on the numerical solution of our present integral equation.

In step (2) of `ForwardCGFFTwpZv`, we compute the weak form of the incident pressure and particle velocity, either in 1D, 2D, or 3D. The function `IncAcousticWave` is presented Matlab Listing I.44.

In step (3) of `ForwardCGFFTwpZv`, we solve the integral equations for the contrast sources w^p and w_k^{Zv} by calling the function `ITERCGwpZv` of Matlab Listing I.45. Observe that we can switch between either an analytical computation of the spatial derivatives in the operators or a numerical one. The code closely follows the CG method outlined in Table 2.1. We start with the case `'analytical'`. To check the adjoint operator, we include the function `CheckAdjointpZv` of

Matlab Listing I.46 The function `CheckAdjointpZv` to check the adjoint operators.

```
function CheckAdjointpZv(r_p,r_Zv,input)
global nDIM;
CHI_kap = input.CHI_kap;    CHI_rho = input.CHI_rho;
  dummy = cell(1,nDIM);

% Adjoint operator [I - K*conj(CHI)] via inner product ----------------
for n = 1:nDIM
    dummy{n} = conj(CHI_rho).*r_Zv{n};
end
[Kp,KZv] = AdjKOPpZv(conj(CHI_kap).*r_p,dummy,input);
Result1  = sum( r_p(:) .* conj(r_p(:)-Kp(:)) );
for n = 1:nDIM
   dummy{n} = r_Zv{n} - KZv{n};
   Result1  = Result1 + sum( r_Zv{n}(:) .* conj(dummy{n}(:)) );
end

% Operator [I - CHI K] via inner product ------------------------------
[Kp,KZv] = KOPpZv(r_p,r_Zv,input);
Result2  = sum( (r_p(:)-CHI_kap(:).*Kp(:)) .* conj(r_p(:)));
for n = 1:nDIM
   dummy{n} = r_Zv{n} - CHI_rho .* KZv{n};
   Result2  = Result2 + sum( dummy{n}(:) .* conj(r_Zv{n}(:)) );
end

% Print the difference ------------------------------------------------
fprintf('Check adjoint: %e\n',abs(Result1-Result2));
```

Matlab Listing I.46. This check is based on the numerical satisfaction of the inner products, using the initial residual errors r^p and r^{Zv}. The difference between both sides of this equation is printed. A value of the order of the computer precision must be reached. Differences of the order 10^{-6} are not sufficient; for these values, the iterative scheme can slow down considerably, because the conjugate-gradient directions will lose accuracy in their orthogonality requirements. The operator function `KOPpZv` and the adjoint function `AdjKOPpZv` are listed in Matlab Listings I.37 and I.38. In these functions, the scalar operators `Kop` and `AdjKop` of Matlab Listings I.13 and I.14 are used. After validation, one can skip this test.

The function `plotContrastSourcewp` of Matlab Listing I.47 plots the scalar contrast source $\hat{w}^p$, while `plotContrastSourcewZv` of Matlab Listing I.48 plots the tangential components

Matlab Listing I.47 The function `plotContrastSourcewp` to plot the contrast distribution χ^κ and the contrast source distribution w^p.

```
function plotContrastSourcewp(w_p,input)
global nDIM;   CHI_kap = input.CHI_kap;

if nDIM == 1               % Plot 1D contrast/source distribution -------------
    x1 = input.X1;
    set(figure,'Units','centimeters','Position',[5 5 18 12]);
    subplot(1,2,1)
       IMAGESC(x1,-30:30, CHI_kap);
       title('\fontsize{13} \chi^{\kappa} = 1 - \kappa_{sct}/ \kappa_{0}');
    subplot(1,2,2)
       IMAGESC(x1,-30:30, abs(w_p));
       title('\fontsize{13} abs(w^{p})');

elseif  nDIM == 2          % Plot 2D contrast/source distribution -------------
    x1 = input.X1(:,1);    x2 = input.X2(1,:);
    set(figure,'Units','centimeters','Position',[5 5 18 12]);
    subplot(1,2,1)
       IMAGESC(x1,x2, CHI_kap);
       title('\fontsize{13} \chi^{\kappa} = 1 - \kappa_{sct}/ \kappa_{0}');
       subplot(1,2,2)
       IMAGESC(x1,x2, abs(w_p));
       title('\fontsize{13} abs(w^p)');

elseif nDIM == 3         % Plot 3D contrast/source distribution --------------
                         % at x3 = 0 or x3 = dx/2
    x1 = input.X1(:,1,1);   x2 = input.X2(1,:,1);
    N3cross = floor(input.N3/2+1);
    set(figure,'Units','centimeters','Position',[5 5 18 12]);
    subplot(1,2,1)
       IMAGESC(x1,x2,CHI_kap(:,:,N3cross));
       title('\fontsize{13} \chi^{\kappa} = 1 - \kappa_{sct}/ \kappa_{0}');
    subplot(1,2,2)
       IMAGESC(x1,x2, abs(w_p(:,:,N3cross)));
       title('\fontsize{13} abs(w^p)');
end % if
```

Matlab Listing I.48 The function `plotContrastSourcewZv` to plot the contrast distribution χ^{ρ} and the contrast-source distribution w_k^{Zv}.

```
function plotContrastSourcewZv(w_Zv,input)
global nDIM;        CHI_rho = input.CHI_rho;

if nDIM == 1              % Plot 1D contrast/source distribution ----------
    x1 = input.X1;
    set(figure,'Units','centimeters','Position',[5 5 18 12]);
    subplot(1,2,1);
        IMAGESC(x1,-30:30,CHI_rho);
        title('\fontsize{13} \chi^{\rho} = 1 - \rho_{sct}/ \rho_{0}');
    subplot(1,2,2);
        IMAGESC(x1,-30:30, abs(w_Zv{1}));
        title('\fontsize{13} abs(w_1^{Zv})');

elseif nDIM == 2          % Plot 2D contrast/source distribution ----------
    x1 = input.X1(:,1);   x2 = input.X2(1,:);
    set(figure,'Units','centimeters','Position',[5 5 18 12]);
    subplot(1,2,1);
        IMAGESC(x1,x2,abs(w_Zv{1}));
        title('\fontsize{13} abs(w_1^{Zv})');
    subplot(1,2,2);
        IMAGESC(x1,x2,abs(w_Zv{2}));
        title('\fontsize{13} abs(w_2^{Zv})');

elseif nDIM == 3          % Plot 3D contrast/source distribution ----------
                          % at x3 = 0 or x3 = dx/2
    x1 = input.X1(:,1,1);  x2 = input.X2(1,:,1);
    N3cross = floor(input.N3/2+1);
    set(figure,'Units','centimeters','Position',[5 5 18 12]);
    subplot(1,2,1);
        IMAGESC(x1,x2,abs(w_Zv{1}(:,:,N3cross)));
        title('\fontsize{13} abs(w_1^{Zv})');
    subplot(1,2,2);
        IMAGESC(x1,x2,abs(w_Zv{2}(:,:,N3cross)));
        title('\fontsize{13} abs(w_2^{Zv})');
end % if
```

of the contrast source w_k^{Zv}. The function `IMAGESC` is a modified version of the standard Matlab function `imagesc`, see Matlab Listing I.7.

For the 1D case, Figure 2.5 shows the compressibility contrast χ^{κ} and the contrast-source amplitude $|w^p|$, while Figure 2.6 shows the mass-density contrast χ^{ρ} and the contrast-source amplitude $|w_1^{Zv}|$. For our specific choice of wave speeds and mass densities, the compressibility contrast is equal to the mass-density contrast, but the two contrast source amplitudes are very different. However, in view of equal wave impedances through the entire 1D space, the correct combination in the expression for the reflected field should result in a vanishing result.

For the 2D case, Figure 2.7 shows the compressibility contrast χ^{κ} and the contrast-source amplitude $|w^p|$, while Figure 2.8 shows the absolute values of the two contrast-source components $|w_1^{Zv}|$ and $|w_2^{Zv}|$. The mass-density contrast is omitted: for our chosen parameters, it is identical to the compressibility contrast.

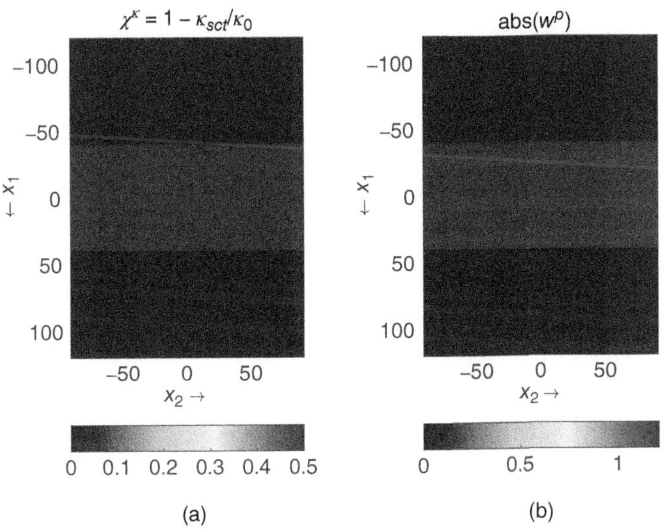

Figure 2.5 Plots of the contrast (a) χ^κ and the contrast-source amplitude (b) $|w^p|$ produced by `ForwardCGFFTwpZv` for the 1D case.

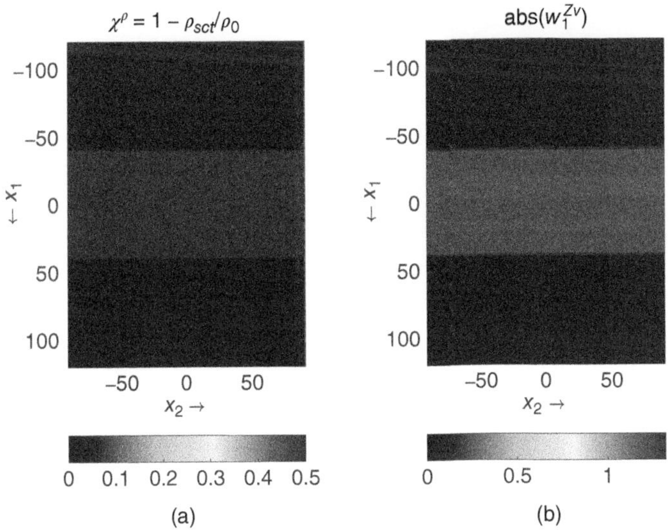

Figure 2.6 Plots of the contrast (a) χ^ρ and contrast-source amplitude (b) $|w_1^{Zv}|$ produced by `ForwardCGFFTwpZv` for the 1D case.

For the 3D case, Figure 2.9 shows the compressibility contrast χ^κ and the contrast-source amplitude $|w^p|$, while Figure 2.10 shows the absolute values of the two tangential components of the contrast sources, $|w_1^{Zv}|$ and $|w_2^{Zv}|$. When, we compare the 2D results and the results in the 2D cross-section of the 3D configuration, we see that there are no arguments to use the 2D contrast sources as an approximation of the 3D ones.

In order to compute the acoustic pressure, in the Matlab script `ForwardCGFFTwpZv` we use again the Matlab function `KOPpZv` to compute the scattered field in our domain of investigation. Subsequently, we compute the total acoustic pressure and call the plot function `plotPressure-Wavefield` of Matlab Listing I.49.

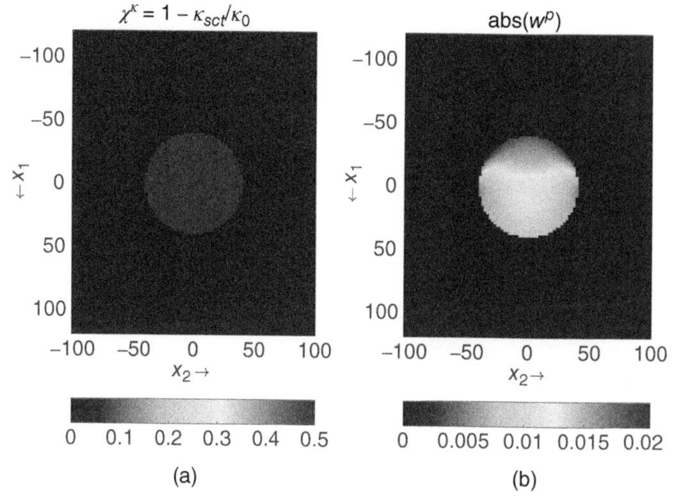

Figure 2.7 Plots of the contrast (a) χ^κ and the contrast-source amplitude (b) $|w^p|$, produced by `ForwardCGFFTwpZv` for the 2D case.

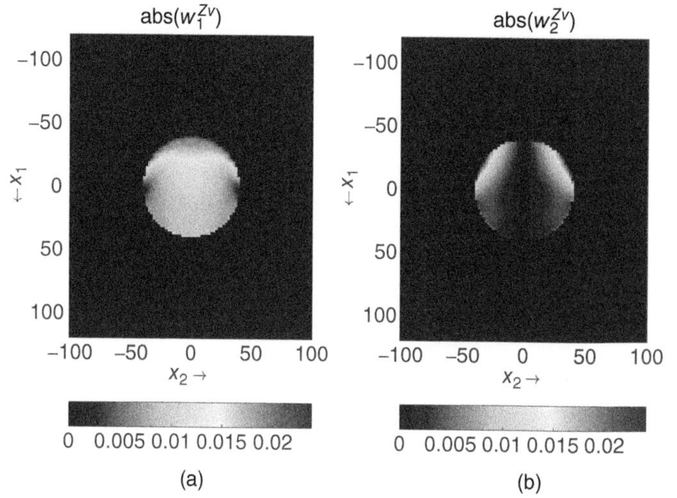

Figure 2.8 Plots of the contrast-source amplitudes (a) $|w_1^{Zv}|$ and (b) $|w_2^{Zv}|$ produced by `ForwardCGFFTwpZv` for the 2D case.

Note that only the field values in the shaded domain of Figure 2.4 are displayed. In Figure 2.11, we show the scattered and total pressure for 1D, 2D, and 3D, respectively. The 1D scattered field plot shows a zero reflected wave above the slab and a transmitted wave below the slab. The 1D total pressure wave field above the slab exhibits an interference pattern of the incident and the reflected wave. Inside this slab, we have a somehow wider interference pattern, due to the change in wave speed. Below the slab, we observe only the propagating transmitted wave. In the pictures of the 2D and 3D cases of Figure 2.11, similar interference effects are still visible. Apart from major differences in the overall amplitude, at first sight the amplitude distributions in 2D and 3D seem to be similar, but further investigation shows considerable local differences.

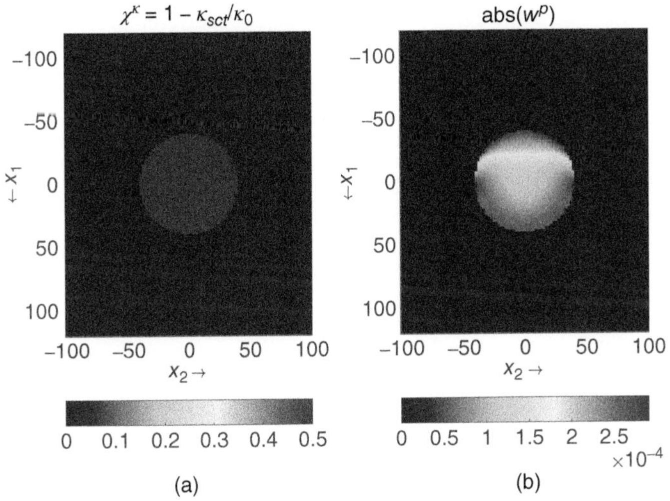

Figure 2.9 Plots of the contrast (a) χ^κ and the contrast source amplitude (b) $|w^p|$ produced by `ForwardCGFFTwpZv` for the 3D case.

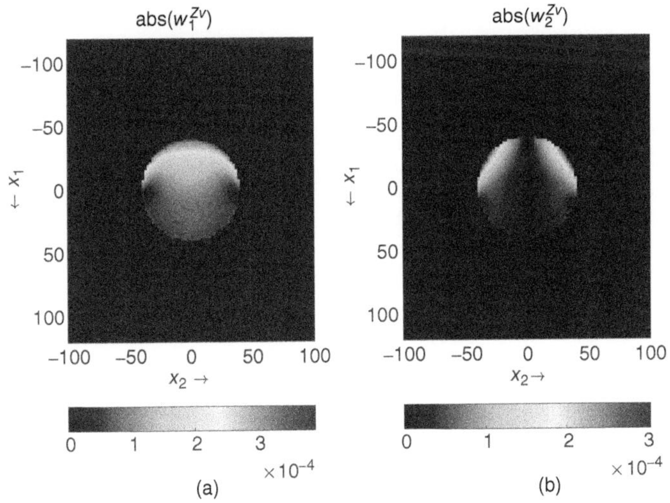

Figure 2.10 Plots of the contrast source amplitudes (a) $|w_1^{Zv}|$ and (b) $|w_2^{Zv}|$ produced by `ForwardCGFFTwpZv` for the 3D case.

In step (4) of the Matlab script `ForwardCGFFTwpZv`, the scattered field at the selected receiver points are computed. The contrast source distributions w^p and w_k^{Zv} are the input for the function `DopwpZv` of Matlab Listing I.50. The program follows from the weak form of the scattered pressure of Eqs. (2.81)–(2.83), the weak form of the spatial derivatives of the weak Green functions of Eqs. (2.84)–(2.86) and the weak Green functions of Table 1.1 for 1D, 2D, and 3D, respectively. These scattered fields are compared to the analytical Bessel function data stored in MAT-files. We use the error norm of Eq. (1.90), with u^{sct} replaced by p^{sct}. The function `displayData.m` of

Matlab Listing I.49 The function `plotPressureWavefield` to plot the scattered pressure and total pressure in the scatterer domain.

```
function  plotPressureWavefield(p_inc,p,input)
global nDIM;
a = input.a;          p_sct = p - p_inc;

if nDIM == 1         % Plot wave fields in one-dimensional space ----------

   set(figure,'Units','centimeters','Position',[5 5 18 12]);
   x1 = input.X1;  N1 = input.N1;
   x2 = -30 : 30;
   subplot(1,2,1)
      IMAGESC(x1(2:N1-1),x2, abs(p_sct(2:N1-1)) );
      title('\fontsize{13} 1D: abs(p^{sct})');
      hold on;      plot(3*x2,0*x2+a,'--w','LineWidth',2);
                    plot(3*x2,0*x2-a,'--w','LineWidth',2);
  subplot(1,2,2)
      IMAGESC(x1(2:N1-1),x2, abs(p(2:N1-1)) );
      title('\fontsize{13} 1D: abs(p)'); caxis([ 0 max(abs(p_sct))]);
      hold on;      plot(3*x2,0*x2+a,'--w','LineWidth',2);
                    plot(3*x2,0*x2-a,'--w','LineWidth',2);

elseif nDIM == 2  % Plot wave fields in two-dimensional space -----------

   set(figure,'Units','centimeters','Position',[5 5 18 12]);
   x1 = input.X1(:,1);    x2 = input.X2(1,:);
   N1 = input.N1;         N2 = input.N2;
   subplot(1,2,1);
      IMAGESC(x1(2:N1-1),x2(2:N2-1), abs(p_sct(2:N1-1,2:N2-1)));
      title('\fontsize{12} 2D: abs(p^{sct})');
      hold on; phi = 0:.01:2*pi; plot(a*cos(phi),a*sin(phi),'w');
   subplot(1,2,2);
      IMAGESC(x1(2:N1-1),x2(2:N2-1), abs(p(2:N1-1,2:N2-1)));
      title('\fontsize{13} 2D: abs(p)'); caxis([ 0 max(abs(p_sct(:)))]);
      hold on; phi = 0:.01:2*pi; plot(a*cos(phi),a*sin(phi),'w');

elseif nDIM == 3  % Plot wave fields at x3 = 0 or x3 = dx/2 --------------

   set(figure,'Units','centimeters','Position',[5 5 18 12]);
    N1 = input.N1;          N2 = input.N2;          N3 = input.N3;
    x1 = input.X1(:,1,1);   x2 = input.X2(1,:,1);   N3cross = floor(N3/2+1);
   subplot(1,2,1)
      IMAGESC(x1(2:N1-1),x2(2:N2-1),abs(p_sct(2:N1-1,2:N2-1,N3cross)));
      title('\fontsize{13} 3D:abs(p^{sct})');
      hold on; phi = 0:.01:2*pi; plot(a*cos(phi),a*sin(phi),'w');
   subplot(1,2,2)
      IMAGESC(x1(2:N1-1),x2(2:N2-1),abs(p(2:N1-1,2:N2-1,N3cross)));
      title('\fontsize{13} 3D: abs(p)'); caxis([ 0 max(abs(p_sct(:)))]);
      hold on; phi = 0:.01:2*pi; plot(a*cos(phi),a*sin(phi),'w');

end % if
```

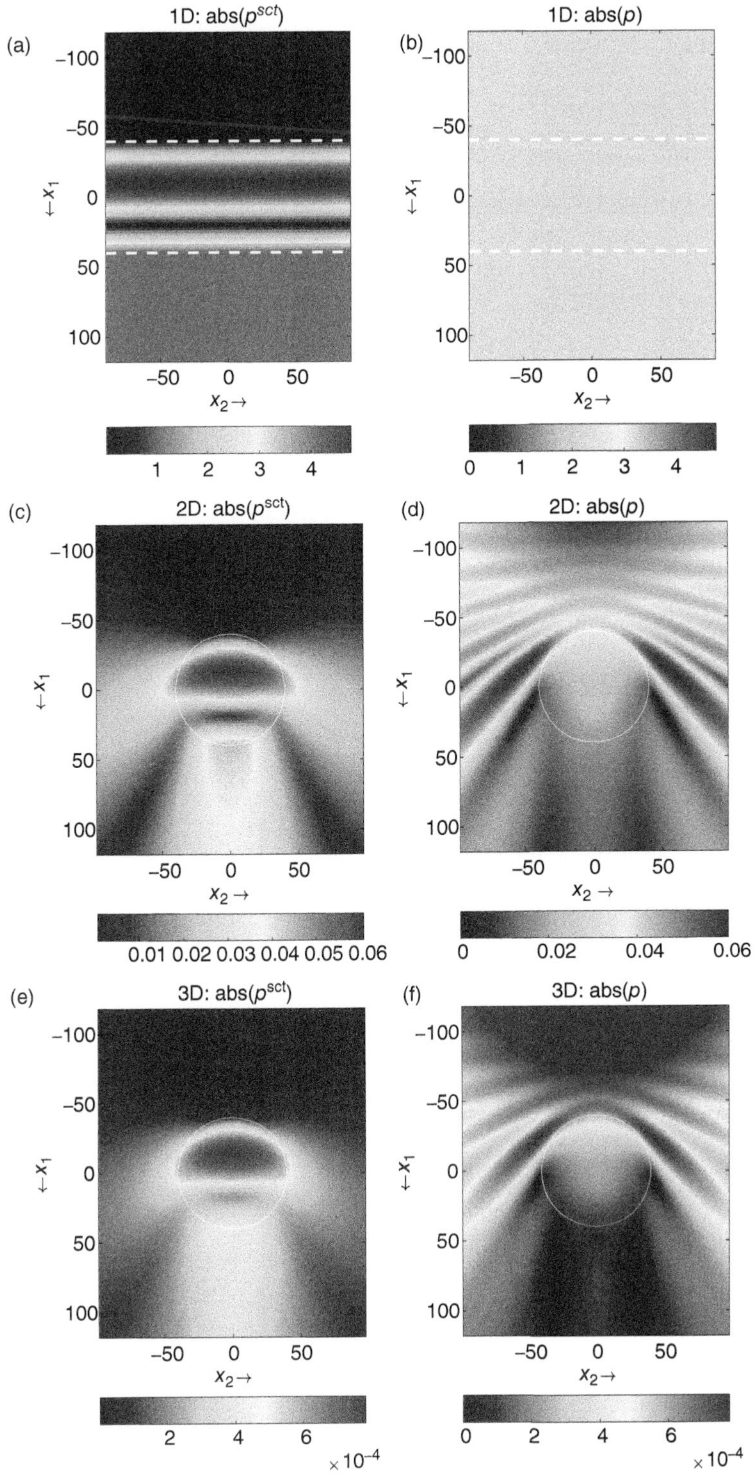

Figure 2.11 Plots of the scattered wave field and the total wave field, produced by `ForwardCGFFTwpZv`, for 1D (a, b), 2D (c, d), and 3D (e, f).

Matlab Listing I.50 The function `DopwpZv` to compute the scattered field at a number of receivers, generated by the volume contrast sources w^p and w_k^{Zv}.

```
function [data] = DOPwpZv(w_p,w_Zv,input)
global nDIM;
gam0 = input.gamma_0;      dx = input.dx;     xR = input.xR;
data = zeros(1,input.NR);

if nDIM == 1
   x1      = input.X1;
   delta   = 0.5 * dx;   factor = sinh(gam0*delta) / (gam0*delta);
   DIS        = abs(x1-xR(1,1));                % receiver for reflected field
   G          = factor /(2*gam0) * exp(-gam0.*DIS);
   d1_G       = - gam0 * sign(xR(1,1)-x1) .* G;
  data(1,1) = (gam0^2 * dx) * sum(G .* w_p) ...
                 - gam0 * dx * sum(d1_G(:) .* w_Zv{1}(:));
   DIS        = abs(x1-xR(1,2));                % receiver for transmitted field
   G          = factor /(2*gam0) * exp(-gam0.*DIS);
   d1_G       = - gam0 * sign(xR(1,2)-x1) .* G;
  data(1,2) = (gam0^2 * dx) * sum(   G(:) .* w_p(:))         ...
               - gam0 * dx   * sum(d1_G(:) .* w_Zv{1}(:));

elseif nDIM == 2
   X1      = input.X1;     X2 = input.X2;
   delta = (pi)^(-1/2)*dx; factor = 2*besseli(1,gam0*delta) / (gam0*delta);
   for p = 1 : input.NR
      DIS        = sqrt((X1-xR(1,p)).^2 +(X2-xR(2,p)).^2);
      G          = factor /(2*pi).* besselk(0,gam0*DIS);
      d_G        = - factor * gam0 * (1/(2*pi)) .* besselk(1,gam0*DIS);
     d1_G        = ((xR(1,p)-X1)./DIS) .* d_G;
     d2_G        = ((xR(2,p)-X2)./DIS) .* d_G;
    data(1,p)    = (gam0^2 * dx^2) * sum(   G(:)  .* w_p(:))       ...
                     - gam0 * dx^2 * sum(d1_G(:)  .* w_Zv{1}(:)) ...
                     - gam0 * dx^2 * sum(d2_G(:)  .* w_Zv{2}(:));
   end % p_loop

elseif nDIM == 3
   X1       = input.X1;    X2 = input.X2;     X3 = input.X3;
   delta    = (4*pi/3)^(-1/3) * dx;    arg       = gam0*delta;
  factor    = 3 * (cosh(arg) - sinh(arg)/arg) / arg^2;
   for p = 1 : input.NR
      DIS        = sqrt((X1-xR(1,p)).^2 + (X2-xR(2,p)).^2 + (X3-xR(3,p)).^2);
      G          = factor * exp(-gam0*DIS) ./ (4*pi*DIS);
      d_G        = (-1./DIS - gam0) .* G;
      d1_G       = ((xR(1,p)-X1)./DIS) .* d_G;
      d2_G       = ((xR(2,p)-X2)./DIS) .* d_G;
      d3_G       = ((xR(3,p)-X3)./DIS) .* d_G;
    data(1,p)    = (gam0^2 * dx^3) * sum(   G(:)  .* w_p(:))       ...
                     - gam0 * dx^3 * sum(d1_G(:)  .* w_Zv{1}(:)) ...
                     - gam0 * dx^3 * sum(d2_G(:)  .* w_Zv{2}(:)) ...
                     - gam0 * dx^3 * sum(d3_G(:)  .* w_Zv{3}(:));
   end % p_loop
end % if
```

Matlab Listing I.9 loads the particular file and for 1D it prints the analytical and numerical data, while for 2D and 3D it plots the analytical data and numerical data. These data are presented in Figures 2.12 and 2.13. The differences are mainly due to the staircase approximation of the discrete contrast distribution of the integral equation method versus the smooth boundary of the circular cylinder and the sphere. Similar to the scattered-field results of Chapter 1, the error in 3D is substantially smaller than in 2D.

At this point, we switch to the case `'numerical'`, where the spatial derivatives in the forward operator and its adjoint are computed numerically. We replace the statement `input.method = 'analytical'` in the script of Matlab Listing I.43 by the statement `input.method = 'numerical'` and run this script again. We do not present the resulting pictures for the contrast sources and wave fields in the interior of the computational domain, because the differences are hardly visible. We only show the results for the scattered field at the receiver points in Figures 2.14 and 2.15. Comparing these results and the ones of Figures 2.12 and 2.13,

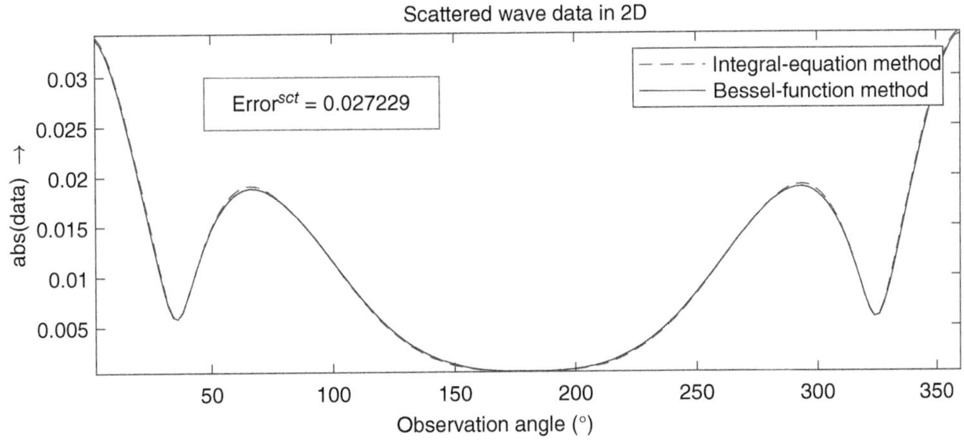

Figure 2.12 Plots of the 2D scattered wave field at the receivers using either Matlab script `WavefieldSctBessel2D` (solid line) or Matlab script `ForwardCGFFTwpZv` (dashed line); case `'analytical'`.

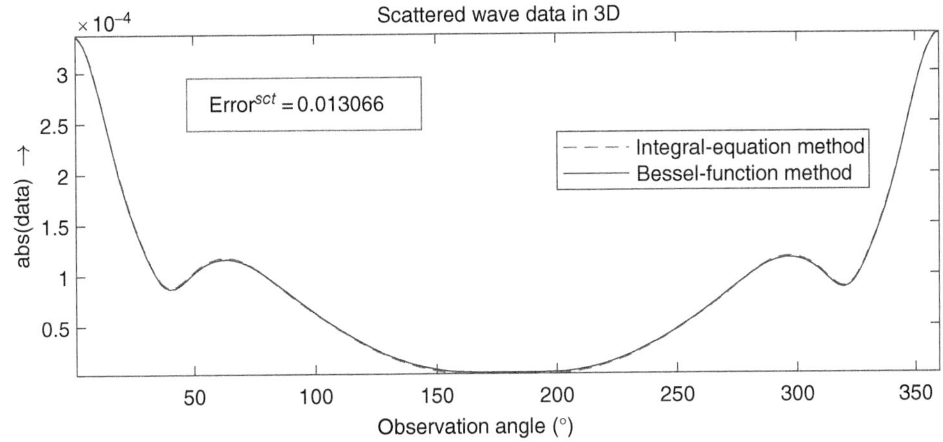

Figure 2.13 Plots of the 3D scattered wave field at the receivers using either Matlab script `WavefieldSctBessel3D` (solid line) or Matlab script `ForwardCGFFTwpZv` (dashed line); case `'analytical'`.

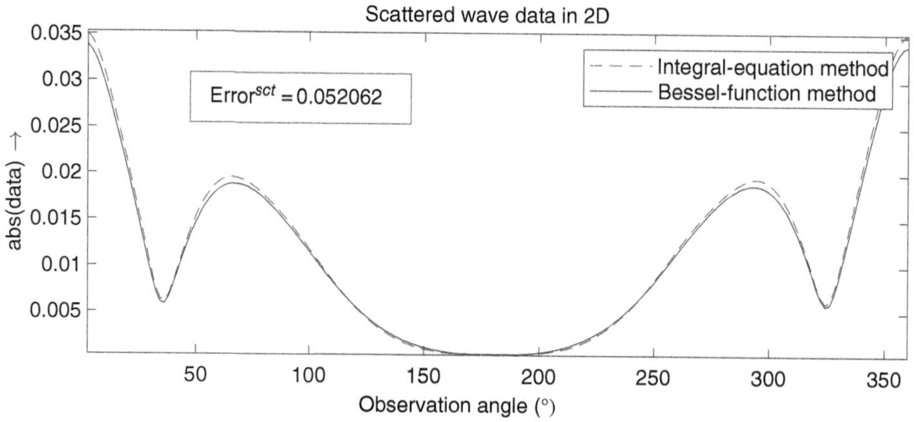

Figure 2.14 Similar as in Figure 2.12, but now for case `'numerical'`.

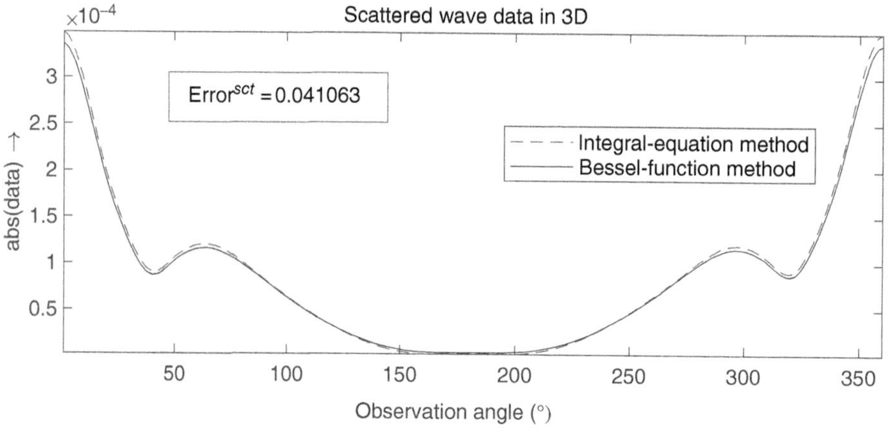

Figure 2.15 Similar as in Figure 2.13, but now for case `'numerical'`.

we immediately see that the errors between the integral equation method and the Bessel function method are now much larger, while the differences in 2D and 3D are now smaller, viz., Error^{sct} is 5.2% in 3D and 4.1% in 2D. In the previous case `'analytical'`, they are 2.7% and 1.3%. Roughly, the numerical differentiation amounts to an extra error of 2–3%. Therefore, the relative difference between the 2D and 3D results in case `'numerical'` is smaller.

2.3 Single Integral Equation in Terms of Contrast in Wave Speed and Gradient of Mass Density

In this section, we show that the coupled system of integral equations for the unknown pressure and unknown particle velocity vector can be transferred to a single equation for the pressure. Let us start again with the integral representation of Eq. (2.48), viz.,

$$\hat{p}^{sct}(\boldsymbol{x}) = \int_{\boldsymbol{x}'\in\mathbb{D}_{sct}} \left[\hat{\gamma}_0^2\hat{G}(\boldsymbol{x}-\boldsymbol{x}')\left(1-\frac{\hat{\kappa}_{sct}}{\kappa_0}\right)\hat{p}(\boldsymbol{x}') - \hat{\gamma}_0\partial_k\hat{G}(\boldsymbol{x}-\boldsymbol{x}')\left(1-\frac{\hat{\rho}_{sct}}{\rho_0}\right)Z_0\hat{v}_k(\boldsymbol{x}')\right]\,\mathrm{d}V, \tag{2.87}$$

for $\boldsymbol{x}' \in \mathbb{D}_{sct}$. Equation (2.87) is the desired integral representation for the scattered wave field. Once the total wave-field quantities $\hat{p}$ and $Z_0\hat{v}_k$ in $\mathbb{D}_{sct}$ are known, this integral representation enables us to calculate the acoustic wave-field quantities in all space. To determine the yet unknown values of $\hat{p}$ and $Z_0\hat{v}_k$, we need a set of integral equations to solve for these acoustic field quantities. In Sections 2.1 and 2.2, we have discussed the integral representation for $Z_0\hat{v}_k^{sct}$ as well, by spatial differentiation of Eq. (2.87). We have shown how direct discretization leads to a set of four integral equations, for the four unknowns $\{\hat{p}, Z_0\hat{v}_1, Z_0\hat{v}_2, Z_0\hat{v}_3\}$. The three extra equations for the particle-velocity vector are substantially more complicated than the first equation for the pressure of Eq. (2.87).

We propose an alternative procedure by substituting the total particle velocity, that follows from Eq. (2.32), into the second term of the integral of Eq. (2.87). The result is

$$\hat{p}^{sct}(\boldsymbol{x}) = \int_{\boldsymbol{x}'\in\mathbb{D}_{sct}} \left[\hat{\gamma}_0^2\hat{G}(\boldsymbol{x}-\boldsymbol{x}')\left(1-\frac{\hat{\kappa}_{sct}}{\kappa_0}\right)\hat{p}(\boldsymbol{x}') - \partial_k\hat{G}(\boldsymbol{x}-\boldsymbol{x}')\left(1-\frac{\rho_0}{\hat{\rho}_{sct}}\right)\partial_k'\hat{p}(\boldsymbol{x}')\right] \,\mathrm{d}V. \tag{2.88}$$

The only unknown function is the acoustic pressure $\hat{p}(\boldsymbol{x})$ and its spatial derivatives $\partial_k\hat{p}(\boldsymbol{x})$ in the domain of the scatterer. Next, using the chain rule, we bring the partial derivative of the pressure to one operating on the mass-density contrast. To achieve this, we rewrite the second term in the integral of Eq. (2.88) as

$$\begin{aligned}
&-\int_{\boldsymbol{x}'\in\mathbb{D}_{sct}} \partial_k\hat{G}(\boldsymbol{x}-\boldsymbol{x}')\left(1-\frac{\rho_0}{\hat{\rho}_{sct}}\right)\partial_k'\hat{p}\,\mathrm{d}V \\
&= -\int_{\boldsymbol{x}'\in\mathbb{D}_{sct}} \partial_k\hat{G}(\boldsymbol{x}-\boldsymbol{x}')\,\partial_k'\left[\left(1-\frac{\rho_0}{\hat{\rho}_{sct}}\right)\hat{p}\right]\mathrm{d}V + \int_{\boldsymbol{x}'\in\mathbb{D}_{sct}} \partial_k\hat{G}(\boldsymbol{x}-\boldsymbol{x}')\left[\partial_k'\left(1-\frac{\rho_0}{\hat{\rho}_{sct}}\right)\right]\hat{p}\,\mathrm{d}V \\
&= -\int_{\boldsymbol{x}'\in\mathbb{D}_{sct}} \partial_k\partial_k\hat{G}(\boldsymbol{x}-\boldsymbol{x}')\left(1-\frac{\rho_0}{\hat{\rho}_{sct}}\right)\hat{p}\,\mathrm{d}V - \int_{\boldsymbol{x}'\in\mathbb{D}_{sct}} \partial_k\hat{G}(\boldsymbol{x}-\boldsymbol{x}')\left[\partial_k'\frac{\rho_0}{\hat{\rho}_{sct}}\right]\hat{p}\,\mathrm{d}V,
\end{aligned} \tag{2.89}$$

where we have used the convolutional property of Eq. (2.28), which states that we have a free choice of placing the differentiation. Note that the first term on the right-hand side of Eq. (2.88) and the first term of the last relation on the right-hand side of Eq. (2.89) can be combined. Next, we use the relation $\partial_k\partial_k\hat{G} = \hat{\gamma}_0^2\,\hat{G} - \delta(\boldsymbol{x}-\boldsymbol{x}')$ and combine the compressibility and mass-density contrast. For observation points $\boldsymbol{x} \neq \boldsymbol{x}'$, the scattered wave-field expression of Eq. (2.88) transfers into

$$\begin{aligned}
\hat{p}^{sct}(\boldsymbol{x}) = &\int_{\boldsymbol{x}'\in\mathbb{D}_{sct}} \hat{\gamma}_0^2\hat{G}(\boldsymbol{x}-\boldsymbol{x}')\,\hat{\chi}^c(\boldsymbol{x}')\,\frac{\rho_0}{\hat{\rho}_{sct}}\hat{p}(\boldsymbol{x}')\,\mathrm{d}V \\
&-\int_{\boldsymbol{x}'\in\mathbb{D}_{sct}} \partial_k\hat{G}(\boldsymbol{x}-\boldsymbol{x}')\left[\partial_k'\frac{\rho_0}{\hat{\rho}_{sct}}\right]\hat{p}(\boldsymbol{x}')\,\mathrm{d}V,
\end{aligned} \tag{2.90}$$

where $\hat{\chi}^c = 1 - c_0^2/\hat{c}_{sct}^2$ is the contrast in wave speed (see Chapter 1). In principle, this is the scattered wave-field expression for observation points in $\mathbb{D}_{sct}'$. However, for integration points $\boldsymbol{x}'$ in the scattered domain $\mathbb{D}_{sct}$, there is always a point that coincides with $\boldsymbol{x}$, so that we have to take into account the delta distribution in the expression of $\partial_k\partial_k\hat{G}$ below Eq. (2.89). Taking into account this extra term $(1-\rho_0/\hat{\rho}_{sct})\hat{p}$ and using $\hat{p} = \hat{p}^{sct} + \hat{p}^{inc}$ we end up with

$$\begin{aligned}
\frac{\rho_0}{\hat{\rho}_{sct}(\boldsymbol{x})}\hat{p}(\boldsymbol{x}) = \hat{p}^{inc}(\boldsymbol{x}) + &\int_{\boldsymbol{x}'\in\mathbb{D}_{sct}} \hat{\gamma}_0^2\hat{G}(\boldsymbol{x}-\boldsymbol{x}')\,\hat{\chi}^c(\boldsymbol{x}')\frac{\rho_0}{\hat{\rho}_{sct}}\hat{p}(\boldsymbol{x}')\,\mathrm{d}V \\
-\partial_k &\int_{\boldsymbol{x}'\in\mathbb{D}_{sct}} \hat{G}(\boldsymbol{x}-\boldsymbol{x}')\left[\partial_k'\frac{\rho_0}{\hat{\rho}_{sct}}\right]\hat{p}(\boldsymbol{x}')\,\mathrm{d}V,
\end{aligned} \tag{2.91}$$

for $\boldsymbol{x} \in \mathbb{D}_{sct}$. Note that the quantity $(\rho_0/\hat{\rho}_{sct})\hat{p}(\boldsymbol{x})$ will jump at discontinuities of $\hat{\rho}_{sct}$. This is consistent with the presence of delta distributions in the second integral of Eq. (2.91).

2.3.1 Contrast Source Formulation

However, with respect to an iterative solution of this integral equation, we have learned from Chapter 1 to opt for a contrast source formulation. We therefore introduce the gradient of the mass density as

$$\hat{\chi}_k^{\partial\rho}(\boldsymbol{x}) = \partial_k \frac{\rho_0}{\rho_{sct}(\boldsymbol{x})} \tag{2.92}$$

and the contrast

$$\hat{\chi}^{c\rho}(\boldsymbol{x}) = \hat{\chi}^{c}(\boldsymbol{x}) \frac{\rho_0(\boldsymbol{x})}{\rho_{sct}}. \tag{2.93}$$

Note that $\hat{\chi}_k^{\partial\rho}$ may be seen as a contrast vector because it vanishes when $\hat{\rho}_{sct} = \rho_0$. The contrast source is defined as follows:

$$\hat{w}^{c\rho}(\boldsymbol{x}) = \hat{\chi}^{c\rho}(\boldsymbol{x})\ \hat{p}(\boldsymbol{x}), \tag{2.94}$$

and the "gradient mass density" source is defined as follows:

$$\hat{w}_k^{\partial\rho}(\boldsymbol{x}) = \hat{\chi}_k^{\partial\rho}(\boldsymbol{x})\ \hat{p}(\boldsymbol{x}). \tag{2.95}$$

We introduce these contrast sources in Eqs. (2.90) and (2.91). Then, we arrive at the integral representation for the scattered field in terms of the unknown contrast sources $\hat{w}^{c\rho}$ and $\hat{w}_k^{\partial\rho}$ as

$$\boxed{\hat{p}^{sct}(\boldsymbol{x}) = \int_{\boldsymbol{x}'\in\mathbb{D}_{sct}} \hat{\gamma}_0^2\ \hat{G}(\boldsymbol{x}-\boldsymbol{x}')\ \hat{w}^{c\rho}(\boldsymbol{x}')\ \mathrm{dV} - \int_{\boldsymbol{x}'\in\mathbb{D}_{sct}} \partial_k \hat{G}(\boldsymbol{x}-\boldsymbol{x}')\ \hat{w}_k^{\partial\rho}(\boldsymbol{x}')\ \mathrm{d}V.} \tag{2.96}$$

After substituting the contrast sources in the integral equation for the pressure, cf. Eq. (2.91), and multiplying both sides of Eq. (2.91) by $\hat{\chi}^c$ and $\frac{\hat{\rho}_{sct}}{\rho_0}\hat{\chi}_j^{\partial\rho}, j = 1, 2, 3$, we obtain a set of integral equations for the unknown contrast sources $\hat{w}^{c\rho}$ and $\hat{w}^{\partial\rho}$ The resulting equations, for $\boldsymbol{x} \in \mathbb{D}_{sct}$, are

$$\boxed{\begin{aligned} \hat{\chi}^c \hat{p}^{inc}(\boldsymbol{x}) &= \hat{w}^{c\rho}(\boldsymbol{x}) - \hat{\chi}^c \left[\hat{\gamma}_0^2 \int_{\boldsymbol{x}'\in\mathbb{D}_{sct}} \hat{G}(\boldsymbol{x}-\boldsymbol{x}')\ \hat{w}^{c\rho}(\boldsymbol{x}')\ \mathrm{dV} \right. \\ &\qquad \left. - \partial_k \int_{\boldsymbol{x}'\in\mathbb{D}_{sct}} \hat{G}(\boldsymbol{x}-\boldsymbol{x}')\ \hat{w}_k^{\partial\rho}(\boldsymbol{x}')\ \mathrm{dV}\right], \\ \hat{\chi}_j^{\rho\partial\rho} \hat{p}^{inc}(\boldsymbol{x}) &= \hat{w}_j^{\partial\rho}(\boldsymbol{x}) - \hat{\chi}_j^{\rho\partial\rho} \left[\hat{\gamma}_0^2 \int_{\boldsymbol{x}'\in\mathbb{D}_{sct}} \hat{G}(\boldsymbol{x}-\boldsymbol{x}')\ \hat{w}^{c\rho}(\boldsymbol{x}')\ \mathrm{dV} \right. \\ &\qquad \left. - \partial_k \int_{\boldsymbol{x}'\in\mathbb{D}_{sct}} \hat{G}(\boldsymbol{x}-\boldsymbol{x}')\ \hat{w}_k^{\partial\rho}(\boldsymbol{x}')\ \mathrm{dV}\right], \end{aligned}} \tag{2.97}$$

where we have introduced the mass density gradient contrast,

$$\boxed{\hat{\chi}_k^{\rho\partial\rho}(\boldsymbol{x}) = \frac{\rho_{sct}}{\rho_0} \partial_k \frac{\rho_0}{\rho_{sct}(\boldsymbol{x})}.} \tag{2.98}$$

It is worth commenting that both in the integral representation of Eq. (2.96) and in the set of integral equations of Eq. (2.97), only the scalar wave-speed contrast $\hat{\chi}^c$ and the vectorial mass-density gradient contrast $\hat{\chi}_k^{\rho\partial\rho}$ occur, while in the equations for the acoustic pressure and the particle velocity the scalar compressibility contrast $\hat{\chi}^{\kappa}$ and scalar mass density $\hat{\chi}^{\rho}$ represent the material parameters of the scatterer at hand. In view of the explicit presence of the gradient of the mass density in the above equations, we expect that the sensitivity of scattered data to variations of the mass density will be greater than operating with the formulation with pressure and particle velocity via the compressibility and mass-density functions.

2.3.2 Conjugate Gradient Iterative Solution and Operators Involved

We now consider the numerical solution of the contrast-source type of integral equation, based on Eq. (2.97). The discretization in the (x_1, x_2)-domain is illustrated in Figure 2.4. The dotted domain includes the scattering domain $\mathbb{D}_{sct}$ completely. To this domain, we add an extra layer with zero compressibility contrast and zero mass-density contrast. Although the contrast sources vanish at these points, the wave-field quantities do not and are computed at these points. On each subdomain $\mathbb{D}_n$, we assume that both the contrast in wave speed and the scaled gradient of the mass density are constant. They are denoted as $\hat{\chi}_n^c$ and $\hat{\chi}_{k,n}^{\rho\partial\rho}$, $k = 1, 2, 3$. Further, in each subdomain, the contrast sources $\hat{w}^{c\rho}$ and $\hat{w}_k^{\partial\rho}$ are replaced by their spherical means $\langle\hat{w}^{c\rho}\rangle$ and $\langle\hat{w}_k^{\partial\rho}\rangle$, respectively. The incident wave field $\hat{p}^{inc}$ is replaced by its spherical mean $\langle\hat{p}^{inc}\rangle$.

Before we describe the various convolutional operators, we note that it does not matter whether we use an analytic differentiation of the Green function inside the operator or a numerical differentiation outside the operator; the total number of convolutions remains the same. We therefore describe the two options.

2.3.2.1 Analytic Differentiation

In this method, we carry out analytical differentiations of the Green function after changing the integration and differentiation in the integral equations. Omitting the hat symbols, the acoustic operator of the pertaining integral equation is written as follows:

$$\mathcal{K}_m^{c\rho}\{w_n^{c\rho}, w_{j,n}^{\partial\rho}\} = [\mathcal{K}w_n^{c\rho}]_m - \gamma_0^{-2}[\partial_k \mathcal{K}w_{k,n}^{\partial\rho}]_m, \tag{2.99}$$

where the convolutional operators $[\mathcal{K}w_n^{c\rho}]_m$ and $[\partial_k \mathcal{K}w_{k,n}^{\partial\rho}]_m$ follow from Eq. (2.60) by substituting $w_n := w_n^{c\rho}$ and $w_n := w_{k,n}^{\partial\rho}$, respectively. The codes to compute both operators are listed as method `analytical` in function `KOPwcdrho` of Matlab Listing I.51. The adjoint of $\mathcal{K}_m^{c\rho}$ is written as

$$\begin{aligned}\mathcal{K}_m^{\star c\rho}\{f_n^{c\rho}, f_{j,n}^{\partial\rho}\} &= \left[\mathcal{K}^\star\left\{f_n^{c\rho} + \sum_k f_{k,n}^{\partial\rho}\right\}\right]_m,\\ \mathcal{K}_{j,m}^{\star\partial\rho}\{f_n^{c\rho}, f_{j,n}^{\partial\rho}\} &= -\left[\partial_j\mathcal{K}^\star\left\{f_n^{c\rho} + \sum_k f_{k,n}^{\partial\rho}\right\}\right]_m,\end{aligned} \tag{2.100}$$

where the convolutional operators $[\mathcal{K}^\star f_n]_m$ and $[\partial_j \mathcal{K}^\star f_n]_m$ follow from Eq. (2.61). These operations are listed as case `'analytical'` in function `AdjKOPwcdrho` of Matlab Listing I.52.

2.3.2.2 Numerical Differentiation

In this method, we leave the differentiation in front of the integral in the integral equations and compute it numerically. Omitting the hat symbols, the acoustic operator of the pertaining integral equation is written as follows:

$$\mathcal{K}_m^{c\rho}\{w_n^{c\rho}, w_{j,n}^{\partial\rho}\} = [\mathcal{K}w_n^{c\rho}]_m - \gamma_0^{-2}\mathrm{DIV}_k[\mathcal{K}w_{k,n}^{\partial\rho}]_m, \tag{2.101}$$

where the convolutional operators $[\mathcal{K}w_n^{c\rho}]_m$ and $[\mathcal{K}w_{k,n}^{\partial\rho}]_m$ follow from the first convolution of Eq. (2.60) by substituting $w_n := w_n^{c\rho}$. The codes to compute both operators are listed as case `'numerical'` in function `KOPwcdrho` of Matlab Listing I.51. The adjoint of $\mathcal{K}_m^{c\rho}$ is written as

$$\begin{aligned}\mathcal{K}_m^{\star c\rho}\{f_n^{c\rho}, f_{j,n}^{\partial\rho}\} &= \left[\mathcal{K}^\star\left\{f_n^{c\rho} + \sum_k f_{k,n}^{\partial\rho}\right\}\right]_m,\\ \mathcal{K}_{j,m}^{\star\partial\rho}\{f_n^{c\rho}, f_{j,n}^{\partial\rho}\} &= \overline{\gamma}_0^{-2}\mathrm{GRAD}_j\left[\mathcal{K}^\star\left\{f_n^{c\rho} + \sum_k f_{k,n}^{\partial\rho}\right\}\right]_m,\end{aligned} \tag{2.102}$$

Matlab Listing I.51 The function `KOPwcdrho` to carry out the $\mathcal{K}$ operator.

```
function [Kwdw] = KOPwcdrho(w,dw,input)
global nDIM;

switch lower(input.method)

case 'analytical'           % Acoustic operator using analytical derivatives
   Kwdw = Kop(w,input.FFTG);
   for n = 1:nDIM
       Kwdw = Kwdw - Kop(dw{n},input.FFTdG{n});
   end

case 'numerical'            % Acoustic operator using numerical derivatives
   gam0  = input.gamma_0;    Kwdw = cell(1,nDIM);
   Kw = Kop(w,input.FFTG);
   for n = 1:nDIM
       Kwdw{n} = Kop(dw{n},input.FFTG);
   end
   Kwdw = Kw  - DIV(Kwdw,input) / gam0^2;

end % switch
```

where we have also used the property that the adjoint of DIV = − GRAD. The convolutional operators $[\mathcal{K}^{\star} f_n]_m$ follow from Eq. (2.61). These operations are listed as case `'numerical'` in function `AdjKOPwcdrho` of Matlab Listing I.52. Surprisingly, this adjoint needs three convolutions, while the analytical version of Eq. (2.100) needs four convolutions.

We further remark that the analytical option with the spatial derivative of the Green function can only be used if the gradient of the mass density is continuous. Using this option for our scattering problem of the canonical problems of homogeneous slab, circular cylinder and sphere, the jump of the mass density at their boundaries may lead to serious convergence problems. The iterative scheme for the numerical option, where all spatial differentiations of the mass density and the Green function are replaced by their numerical counterparts, will converge to the result of a smoothed version of the scattering object at hand.

With the defined operators and their pertaining Matlab codes, we observe that the system of integral equations transfers into the following discrete set of equations:

$$\begin{aligned} \chi_m^c\, \langle p^{inc}\rangle(x_m) &= w_m^{cp} - \chi_m^c\, \mathcal{K}_m^{cp}\{w_n^{cp}, w_{j,n}^{\partial\rho}\}, \\ \chi_{k,m}^{\rho\partial\rho}\langle p^{inc}\rangle(x_m) &= w_{k,m}^{\partial\rho} - \chi_{k,m}^{\rho\partial\rho}\, \mathcal{K}_m^{cp}\{w_n^{cp}, w_{j,n}^{\partial\rho}\}, \end{aligned} \tag{2.103}$$

where the expression of $\mathcal{K}_m^{cp}$ in terms of the unknowns w_m^{cp} and $w_{k,m}^{\partial\rho}$ is given in Eq. (2.99). This system is solved iteratively with the help of a conjugate gradient method. The conjugate gradient method for solving the contrast sources w^{cp} and $w_k^{\partial\rho}$ is presented in Table 2.2. It is very similar to the one of Table 2.1; we only have to change the nomenclature. To discuss it in more detail, we write Eq. (2.103) in the compact form

$$\boxed{\begin{aligned} w^{cp} - \chi^c\, \mathcal{K}^{cp}\{w^{cp}, w_j^{\partial\rho}\} &= \chi^c\, p^{inc}, \\ w_k^{\partial\rho} - \chi_k^{\rho\partial\rho}\, \mathcal{K}^{cp}\{w^{cp}, w_j^{\partial\rho}\} &= \chi_k^{\rho\partial\rho}\, p^{inc}. \end{aligned}} \tag{2.104}$$

Matlab Listing I.52 The function `AdjKOPwcdrho` to carry out the $\mathcal{K}^\star$ operator.

```
function [Kf,dKf] = AdjKOPwcdrho(f,df,input)
global nDIM;          dKf = cell(1,nDIM);

switch lower(input.method)

case 'analytical'      % Acoustic operator using analytical derivatives
    for n = 1:nDIM
        f = f + df{n};
    end
    Kf = AdjKop(f,input.FFTG);

for n = 1:nDIM
  dKf{n} = -AdjKop(f,input.FFTdG{n});
end

case 'numerical'       % Acoustic operator using numerical derivatives
    gam0 = input.gamma_0;
    for n = 1:nDIM
        f = f + df{n};
    end
    Kf = AdjKop(f,input.FFTG);
    dKf = GRAD(Kf/conj(gam0^2),input);

end % switch
```

If the contrast sources do not solve Eq. (2.104), the pertaining residual errors are defined as

$$
\begin{aligned}
r^{c\rho} &= \chi^c\, p^{inc} - [w^{c\rho} - \chi^c\, \mathcal{K}^{c\rho}\{w^{c\rho}, w_j^{\partial\rho}\}], \\
r_j^{\partial\rho} &= \chi_j^{\partial\rho} p^{inc} - [w_j^{\partial\rho} - \chi_j^{\partial\rho}\, \mathcal{K}^{c\rho}\{w^{c\rho}, w_j^{\partial\rho}\}],
\end{aligned}
\tag{2.105}
$$

and the normalized cost functional as

$$
F(w^{c\rho}, w_j^{\partial\rho}) = \frac{\|r^{c\rho}\|^2 + \sum_j \|r_j^{\partial\rho}\|^2}{\|\chi^{c\rho}\, p^{inc}\|^2 + \sum_j \|\chi_j^{\partial\rho} p^{inc}\|^2},
\tag{2.106}
$$

with the property that $F = 1$ for $w^{c\rho} = w_1^{\partial\rho} = w_2^{\partial\rho} = w_3^{\partial\rho} = 0$ and $F = 0$ for the exact solution. The gradient directions follow from the Fréchet derivative of this cost functional F and are directly related to the adjoint operating on the residual errors in the equations at the previous iteration $n - 1$. They are obtained as

$$
\begin{aligned}
g^{c\rho(n)} &= r^{c\rho(n-1)} - \mathcal{K}^{\star c\rho}\{\overline{\chi}^c r^{c\rho(n-1)}, \overline{\chi}_k^{\partial\rho}\, r_k^{\partial\rho(n-1)}\}, \\
g_j^{\partial\rho(n)} &= r_j^{\partial\rho(n-1)} - \mathcal{K}_j^{\star\partial\rho}\{\overline{\chi}^c r^{c\rho(n-1)}, \overline{\chi}_k^{\partial\rho}\, r_k^{\partial\rho(n-1)}\}.
\end{aligned}
\tag{2.107}
$$

The Hestenes–Steifel conjugate gradient directions after the first iteration become

$$
\begin{aligned}
v^{c\rho(n)} &= g^{c\rho(n)} + (A^{(n)}/A^{(n-1)})\, v^{c\rho(n-1)}, \\
v_j^{\partial\rho(n)} &= g_j^{\partial\rho(n)} + (A^{(n)}/A^{(n-1)}) v_j^{\partial\rho(n-1)},
\end{aligned}
\tag{2.108}
$$

where

$$
A^{(n)} = \|g^{c\rho(n)}\|^2 + \sum_j \|g_j^{\partial\rho(n)}\|^2,
\tag{2.109}
$$

while the updates for the contrast sources $w^{c\rho}$ and $w_j^{\partial\rho}$ are obtained as

$$
\begin{aligned}
w^{c\rho(n)} &= w^{c\rho(n-1)} + (A^{(n)}/B^{(n)})v^{c\rho(n)}, \\
w_k^{\partial\rho(n)} &= w_k^{\partial\rho(n-1)} + (A^{(n)}/B^{(n)})v_k^{\partial\rho(n)},
\end{aligned}
\tag{2.110}
$$

with

$$
B^{(n)} = \|v^{c\rho(n)} - \chi^c\,\mathcal{K}^{c\rho(n)}\{v^{c\rho(n)}, v_k^{\partial\rho(n)}\}\|^2 + \sum_j \|v_j^{\partial\rho(n)} - \chi^{\rho\partial\rho}\,\mathcal{K}_j^{\partial\rho(n)}\{v^{c\rho(n)}, v_k^{\partial\rho(n)}\}\|^2.
\tag{2.111}
$$

The complete conjugate gradient scheme, that minimizes the cost functional F iteratively, is presented in Table 2.2. To complete this numerical section, we need the weak forms of the incident and scattered wave fields. The weak form of the incident pressure field is given in Eq. (2.78)

2.3.2.3 Scattered Acoustic Wave Field

In the 3D case, the scattered pressure is given by Eq. (2.96), for $x = x^R \in \mathbb{D}'_{sct}$. Its weak counterpart is expressed in terms of the weak forms $\langle \hat{G} \rangle$ of Table 1.1 as

$$
\langle \hat{p}^{sct} \rangle (x^R) = (\Delta x)^3 \sum_{m=1}^{N} \left[\hat{\gamma}_0^2 \langle \hat{G} \rangle (x^R - x_m)\,\langle \hat{w}_m^{c\rho} \rangle - \sum_{k=1}^{3} \partial_k^R \langle \hat{G} \rangle (x^R - x_m)\,\langle \hat{w}_{k,m}^{\partial\rho} \rangle \right].
\tag{2.112}
$$

Table 2.2 The conjugate gradient method for solving the contrast sources $w^{c\rho}$ and $w_j^{\partial\rho}$

initial estimate	$w^{c\rho(n=0)}$	=	0
	$w_j^{\partial\rho(n=0)}$	=	0
	$r^{c\rho(n=0)}$	=	$\chi^c p^{inc}$
	$r_j^{\partial\rho(n=0)}$	=	$\chi_j^{\rho\partial\rho} p^{inc}$
for $n = 1, 2, \dots$	$g^{c\rho(n)}$	=	$r^{c\rho(n-1)} - \mathcal{K}^{\star c\rho}\{\overline{\chi}^c r^{c\rho(n-1)}, \overline{\chi}^{\rho\partial\rho} r_k^{\partial\rho(n-1)}\}$
	$g_j^{\partial\rho(n)}$	=	$r_j^{\partial\rho(n-1)} - K_j^{\star\partial\rho}\{\overline{\chi}^c r^{c\rho(n-1)}, \overline{\chi}^{\rho\partial\rho} r_k^{\partial\rho(n-1)}\}$
	$A^{(n)}$	=	$\|g^{c\rho(n)}\|^2 + \sum_j \|g_j^{\partial\rho(n)}\|^2$
if $n = 1$	$v^{c\rho(n)}$	=	$g^{c\rho(n)}$
	$v_j^{\partial\rho(n)}$	=	$g_j^{\partial\rho(n)}$
else	$v^{c\rho(n)}$	=	$g^{c\rho(n)} + \dfrac{A^{(n)}}{A^{(n-1)}} v^{c\rho(n-1)}$
	$v_j^{\partial\rho(n)}$	=	$g_j^{\partial\rho(n)} + \dfrac{A^{(n)}}{A^{(n-1)}} v_j^{\partial\rho(n-1)}$ end
	$B^{(n)}$	=	$\|v^{c\rho(n)} - \chi^c \mathcal{K}^{c\rho(n)}\{v^{c\rho(n)}, v_k^{\partial\rho(n)}\}\|^2$
			$+\sum_j \|v_j^{\partial\rho(n)} - \chi^{\rho\partial\rho} \mathcal{K}_j^{\partial\rho(n)}\{v^{c\rho(n)}, v_k^{\partial\rho(n)}\}\|^2$
	$w^{c\rho(n)}$	=	$w^{c\rho(n-1)} + \dfrac{A^{(n)}}{B^{(n)}} v^{c\rho(n)}$
	$w_j^{\partial\rho(n)}$	=	$w_r^{\partial\rho(n-1)} + \dfrac{A^{(n)}}{B^{(n)}} v_r^{\partial\rho(n)}$
	$r^{c\rho(n)}$	=	$r^{c\rho(n-1)} - \dfrac{A^{(n)}}{B^{(n)}}[v^{c\rho(n)} - \chi^c \mathcal{K}^{c\rho(n)}\{v^{c\rho(n)}, v_k^{\partial\rho(n)}\}]$
	$r_j^{\partial\rho(n)}$	=	$r_j^{\partial\rho(n-1)} - \dfrac{A^{(n)}}{B^{(n)}}[v_j^{\partial\rho(n)} - \chi^{\rho\partial\rho} \mathcal{K}_j^{\partial\rho(n)}\{v^{c\rho(n)}, v_k^{\partial\rho(n)}\}]$

$$
\text{if}\ \left[\frac{\|r^{c\rho(n)}\|^2 + \sum_j \|r_j^{\partial\rho(n)}\|^2}{\|\chi^c p^{inc}\|^2 + \sum_j \|\chi^{\rho\partial\rho} p^{inc}\|^2} \right]^{\frac{1}{2}} < \varepsilon\ \text{or}\ n > n_{max}\ \text{stop end}
$$

end

In the 2D case, this quantity is

$$\langle \hat{p}^{sct} \rangle (\boldsymbol{x}_T^R) = (\Delta x)^2 \sum_{m=1}^{N} \left[\hat{\gamma}_0^2 \langle \hat{G} \rangle (\boldsymbol{x}_T^R - \boldsymbol{x}_{T,m}) \, \langle \hat{w}_m^{c\rho} \rangle - \sum_{k=1}^{2} \partial_k^R \langle \hat{G} \rangle (\boldsymbol{x}_T^R - \boldsymbol{x}_{T,m}) \, \langle \hat{w}_{k,m}^{\partial\rho} \rangle \right]. \tag{2.113}$$

In the 1D case, this quantity is

$$\langle \hat{p}^{sct} \rangle (\boldsymbol{x}^R) = \Delta x \sum_{m=1}^{N} [\hat{\gamma}_0^2 \langle \hat{G} \rangle (x_1^R - x_{1,m}) \, \langle \hat{w}_m^{c\rho} \rangle - \partial_1^R \langle \hat{G} \rangle (x_1^R - x_{1,m}) \, \langle \hat{w}_{1,m}^{\partial\rho} \rangle]. \tag{2.114}$$

The weak form of the spatial derivative of the Green function are given in Eqs. (2.84), (2.85), and (2.86), for the 3D case, the 2D case, and the 1D case, respectively.

2.4 Matlab Codes Integral Equations: Wave Speed and Gradient of Mass Density

In this section, we present the Matlab codes for the numerical solution of the coupled set of integral equations for the contrast source related to wave speed contrast and acoustic pressure, and the contrast source related to the gradient of mass-density and the acoustic pressure. After solving these contrast sources, we discuss the Matlab codes to compute the scattered acoustic pressure at a number of receiver points outside the scatterer. To initialize the different acoustic wave parameters, we start with the function `initAC` of Matlab Listing I.39. We do not change the parameters and the Green functions. The source/receiver locations and the grid parameters are initialized by calling the Matlab functions `initSourceReceiver` and `initGrid` of Matlab Listings I.2 and I.3, while the function `initFFTGreenfun` is given in Matlab Listing I.40. The Matlab function `initAcousticContrast` of Matlab Listing I.39 computing the compressibility and the mass density is also not changed. After the initialization of the acoustic parameters, we now start with the main Matlab script `ForwardCGFFTwcdrho` of Matlab Listing I.53. Similar to the main Matlab script of scalar wave scattering, we have four steps: (1) Computation of scattering by a canonical object; (2) Computation of the incident wave fields; (3) Solution of the set of integral equations; (4) Computation of scattering data, based on the solution of the integral equations.

In step (1) of `ForwardCGFFTwcdhro`, script `AcForwardCanonicalObjects` of Matlab Listing I.79 of Appendix 2.A.4 is called. It computes the acoustic pressure at a number of receivers. These data are used to benchmark the scattered field based on the numerical solution of our present integral equation. In the remainder of the program, we need the parameters of the contrast in wave speed χ^c and the mass density gradient contrast $\chi^{\rho\partial\rho}$ of Eq. (2.98). These parameters are initialized in the Matlab function `AcousticModifiedContrast` of Matlab Listing I.54 and added to the structure file `input`. The gradient of the normalized mass density is approximated by a central difference rule, using the function `GRAD` of Matlab Listing I.36. For our present configuration with a discontinuous change of medium parameters at the interface of the scatterer, this approximation is rather crude. This will be discussed later.

In step (2) of `ForwardCGFFTwcdrho`, we compute the weak form of the incident pressure, using the function `IncAcousticWave` presented in Matlab Listing I.44. Note that we do not output the particle velocity, since in the present integral equation formulation we do not need it.

In step (3) of `ForwardCGFFTwpZv`, we solve the integral equations for the contrast sources $w^{c\rho}$ and $w_k^{\partial\rho}$ by calling the function `ITERCGwcdhro` of Matlab Listing I.55. The code closely follows the CG method outlined in Table 2.2. Again, we can switch between either the analytic computation of the spatial derivatives in the operators or the numeric one. We start with the case `analytical`. To check the adjoint operator, we include the function `CheckAdjointwcdrho` of Matlab Listing I.56. This check is based on the numerical satisfaction of the inner product, using the initial

Matlab Listing I.53 The `ForwardCGFFTwcdrho` script to solve the contrast sources $\hat{w}^{c\rho}$ and $\partial\hat{w}_k^{\partial\rho}$ from the system of integral equations.

```
clear all; clc; close all; clear workspace
input = initAC();

%  (1) Compute analytically scattered field data ─────────────────
        AcForwardCanonicalObjects

%        Input modified differentiable contrast functions ────────
        input = AcousticModifiedContrast(input);

%  (2) Compute only the incident pressure field ──────────────────
        [p_inc,~] = IncAcousticWave(input);

%  (3) Solve contrast source integral equations with CGFFT method ─────────
        input.method = 'analytical';  % choose 'analytical' or 'numerical'
        disp(['Derivatives are ' lower(input.method)]);
tic;    [w_crho,w_drho] = ITERCGwcdrho(p_inc,input);
toc;
    %  Plot contrast and contrast sources
        plotContrastSourceWcrho(w_crho,input);
        plotInterfaceSourceWdrho(w_drho,input);

%  (4) Compute synthetic data and plot fields and data ───────────
        data = DOPwcdrho(w_crho,w_drho,input);
        displayData(data,input);
```

Matlab Listing I.54 The function `AcousticModifiedContrast` to initialize the acoustic wave-field parameters.

```
function [input] = AcousticModifiedContrast(input)
global nDIM;

if      nDIM == 1;  R = sqrt(input.X1.^2);
elseif nDIM == 2;  R = sqrt(input.X1.^2 + input.X2.^2);
elseif nDIM == 3;  R = sqrt(input.X1.^2 + input.X2.^2 + input.X3.^2);
end % if

% (1) Compute wave speed contrast CHI^c ─────────────────────────
    ratio_c_sct2 = (1 + (input.c_sct.^2/input.c_0^2 - 1) * (R < input.a));
    input.CHI_c  = 1 - (1./ratio_c_sct2);

% (2) Compute gradient mass density contrast CHI^\rho/drho──────────
    ratio_rho_sct = 1 + (input.rho_sct/input.rho_0-1) * (R < input.a);
    inverse_ratio_rho  = 1./ ratio_rho_sct;
            drho_sct = GRAD(inverse_ratio_rho, input);
    for n =1:nDIM
      input.CHI_drho{n} = ratio_rho_sct.* drho_sct{n};
    end
```

Matlab Listing I.55 The function `ITERCGwcdrho` to compute the contrast sources $w^{c\rho}$ and $w_j^{\partial\rho}$.

```
function [w_c,w_drho]=ITERCGwcdrho(p_inc, input)
global nDIM; CHI_c = input.CHI_c;   CHI_drho = input.CHI_drho;
w_drho = cell(1,nDIM);   Kwdrho = cell(1,nDIM);
r_drho = cell(1,nDIM);   g_drho = cell(1,nDIM);   v_drho = cell(1,nDIM);

itmax = 1000;  Errcri = input.Errcri;  it = 0;  % initialization iteration
w_c = zeros(size(p_inc));  r_c = CHI_c.*p_inc;   Norm0=norm(r_c(:))^2;
for n = 1:nDIM
  w_drho{n} = zeros(size(p_inc));  r_drho{n} = CHI_drho{n}.*p_inc;
      Norm0 = Norm0 + norm(r_drho{n}(:))^2;
end
CheckAdjointwcdrho(r_c,r_drho,input);  % Check once-----------------------
Error = 1;    fprintf('Error =         %g',Error);
while (it < itmax) && ( Error > Errcri) && (Norm0 > eps)
  dummy = cell(1,nDIM);          % determine conjugate gradient direction v
  for n = 1:nDIM; dummy{n} = conj(CHI_drho{n}).*r_drho{n}; end
   [Kc,Kdrho] = AdjKOPwcdrho(conj(CHI_c).*r_c, dummy, input);
          g_c =  r_c - Kc;                    %  and window for small CHI_c
          g_c = (abs(CHI_c) >= Errcri).*g_c;     AN = norm(g_c(:))^2;
  for n = 1:nDIM
   g_drho{n} = r_drho{n} - Kdrho{n};             % and window for small CHI_rho
   g_drho{n} = (abs(CHI_drho{n}) >= Errcri).*g_drho{n};
   AN = AN+norm(g_drho{n}(:))^2;
  end
  if it == 0; v_c=g_c; for n=1:nDIM; v_drho{n}=g_drho{n};  end
  else
  v_c=g_c+AN/AN_1*v_c; for n=1:nDIM; v_drho{n}=g_drho{n}+AN/AN_1*v_drho{n};
  end
  end
% determine step length alpha
  [Kw] = KOPwcdrho(v_c,v_drho,input);
   Kwc = v_c - CHI_c .* Kw;                    BN = norm(Kwc(:))^2;
  for n = 1:nDIM
   Kwdrho{n} = v_drho{n}-CHI_drho{n}.*Kw;  BN = BN + norm(Kwdrho{n}(:))^2;
  end
   alpha = AN / BN;
   w_c   = w_c + alpha * v_c;     % update contrast sources
   for n = 1:nDIM;  w_drho{n} = w_drho{n} + alpha * v_drho{n}; end
   AN_1  = AN;                    % update AN
 % update residual errors
   r_c  = r_c  - alpha * Kwc;               Norm = norm(r_c(:))^2;
  for n = 1:nDIM
   r_drho{n}=r_drho{n}-alpha* Kwdrho{n};  Norm = Norm+norm(r_drho{n}(:))^2;
  end
   Error = sqrt(Norm / Norm0);  fprintf('\b\b\b\b\b\b\b\b%6f',Error);
   it = it+1;
 end % CG iterations
 fprintf('\b\b\b\b\b\b\b\b%6f\n',Error);
 disp(['Number of iterations is ' num2str(it)]);
 if it == itmax; disp(['itmax reached:  err/norm = ' num2str(Error)]); end
```

Matlab Listing I.56 The function `CheckAdjointwcdrho` to check the adjoint operators.

```
function CheckAdjointwcdrho(r_c,r_drho,input)
global nDIM;
CHI_c = input.CHI_c;   CHI_drho = input.CHI_drho;   dummy = cell(1,nDIM);

% Adjoint operator [I - K*conj(CHI)] via inner product ----------------------
for n = 1:nDIM
    dummy{n} = conj(CHI_drho{n}).*r_drho{n};
end
[Kf,dKf] = AdjKOPwcdrho(conj(CHI_c).*r_c,dummy,input);
Result1  = sum( r_c(:) .* conj(r_c(:)-Kf(:)) );
for n = 1:nDIM
   dummy{n} = r_drho{n} - dKf{n};
   Result1  = Result1 + sum( r_drho{n}(:) .* conj(dummy{n}(:)) );
end

% Adjoint operator [I - CHI K] via inner product ----------------------------
[Kwcdrho] = KOPwcdrho(r_c,r_drho,input);
Result2  = sum( (r_c(:)-CHI_c(:).*Kwcdrho(:)) .* conj(r_c(:)));
for n = 1:nDIM
   dummy{n} = r_drho{n} - CHI_drho{n} .* Kwcdrho;
   Result2  = Result2 + sum( dummy{n}(:) .* conj(r_drho{n}(:)) );
end

% Print the differences
fprintf('Check adjoint: %e\n',abs(Result1-Result2));
```

residuals $r^{c\rho}$ and $r_j^{\partial\rho}$. The difference between both sides of this equation is printed. The operator `KOPwcdrho` and the adjoint `AdjKOPwcdrho` are listed in Matlab Listings I.51 and I.52. In these functions, the operators `Kop` and `AdjKop` of Matlab Listings I.13 and I.14 are used. After validation, one can skip this test. The function `plotContrastSourceWcrho` of Matlab Listing I.57 plots the scalar contrast source $w^{c\rho}$, while `plotContrastSourceWdrho` of Matlab Listing I.58 plots the components of the contrast source $w_j^{\partial\rho}$. The function `IMAGESC` is a modified version of the standard Matlab function `imagesc`, see Matlab Listing I.7.

For the 1D case, Figure 2.16 shows the wave-speed contrast and the contrast-source amplitude $|w^{c\rho}|$, while Figure 2.17 shows the mass-density gradient contrast $\chi_1^{\rho\partial\rho}$ and the contrast-source amplitude $|w_1^{\partial\rho}|$. Comparing these figures with Figures 2.5 and 2.6, we see that contrast source $w^{c\rho}$ has a very different structure from that of the source w^{p}. In Figure 2.17, we see that mass-density gradient is zero, except near the interface of the slab. For our chosen mass-density distribution, at the top interface, the gradient exhibits a positive jump and at the bottom interface, it exhibits a negative gradient. It is evident that these jumps lead to contrast sources $|w_1^{\partial\rho}|$ at the interfaces. The consequence is that it also has an effect on the contrast source $|w^{c\rho}|$ at the interfaces.

For the 2D case, Figure 2.18 shows the contrast distributions and the corresponding contrast sources. The pictures on the left-hand side show the wave-speed contrast χ^c and the two components of the mass-density gradient contrast, $\chi_1^{\rho\partial\rho}$, $\chi_2^{\rho\partial\rho}$. At the circular interface, the first component is positive near the top interface and negative near the bottom interface, while the second component is positive near the left interface and negative near the right interface. The consequences for the corresponding contrast-source components are clearly visible in the pictures on the right-hand side.

Matlab Listing I.57 The function `plotContrastSourceWcrho` to plot the contrast distribution χ^c and the contrast source distribution $w_j^{c\rho}$.

```
function plotContrastSourceWcrho(w_crho,input)
global nDIM;    CHI = input.CHI_c;

if nDIM == 1              % Plot 1D contrast/source distribution -----------

   x1 = input.X1;
   set(figure,'Units','centimeters','Position',[5 5 18 12]);
   subplot(1,2,1)
      IMAGESC(x1,-30:30, CHI);
      title('\fontsize{13} \chi^{c}');
   subplot(1,2,2)
      IMAGESC(x1,-30:30, abs(w_crho));
      title('\fontsize{13} abs(w^{c\rho})');

elseif nDIM == 2          % Plot 2D contrast/source distribution -----------

   x1 = input.X1(:,1);    x2 = input.X2(1,:);
   set(figure,'Units','centimeters','Position',[5 5 18 12]);
   subplot(1,2,1)
      IMAGESC(x1,x2, CHI);
      title('\fontsize{13} \chi^{c}');
      subplot(1,2,2)
      IMAGESC(x1,x2, abs(w_crho));
      title('\fontsize{13} abs(w^{c\rho})');

elseif nDIM == 3          % Plot 3D contrast/source distribution -----------
                          % at x3 = 0 or x3 = dx/2

   x1 = input.X1(:,1,1);  x2 = input.X2(1,:,1);
   N3cross = floor(input.N3/2+1);
   set(figure,'Units','centimeters','Position',[5 5 18 12]);
   subplot(1,2,1)
      IMAGESC(x1,x2,CHI(:,:,N3cross));
      title('\fontsize{13} \chi^{c}');
   subplot(1,2,2)
      IMAGESC(x1,x2, abs(w_crho(:,:,N3cross)));
      title('\fontsize{13} abs(w^{c\rho})');

end % if
```

For the 3D case, in the cross-sectional plane $x_3 = 0$, Figure 2.19 shows the contrast distributions and the corresponding contrast sources. At first glance, the results do not seem to differ much from the previous 2D configuration. With a further comparison of Figures 2.18 and 2.19, we see that the amplitude of the 3D contrast sources in the vertical direction decreases faster than the amplitude of the 2D contrast sources. This is understandable because the incident wave field in 3D decays faster than the one in 2D. We have also computed the acoustic pressure field in the scattering domain. The images are hardly different from those of Figure 2.11 and therefore we omit a further presentation.

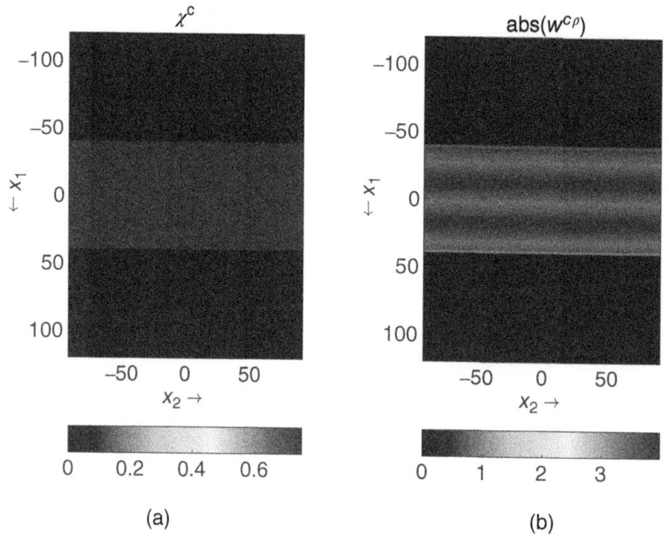

Figure 2.16 Plots of the contrast (a) χ^c and the contrast-source amplitude (b) $|w^{c\rho}|$, produced by `ForwardCGFFTwdcrho` for the 1D case.

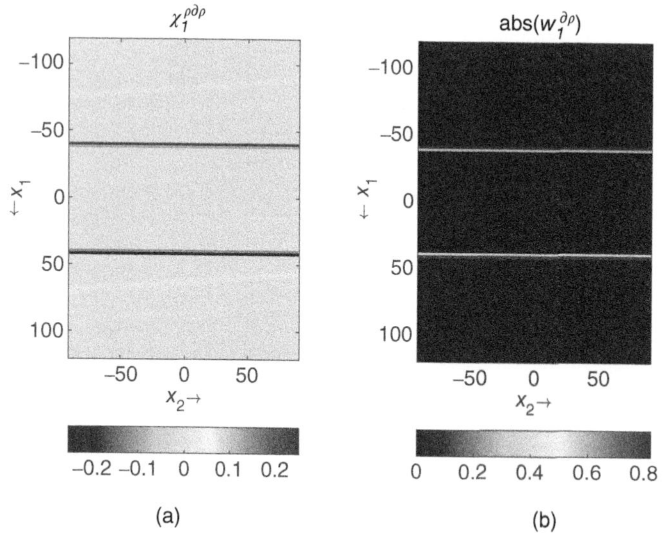

Figure 2.17 Plots of the contrast (a) $\chi_1^{\rho\partial\rho}$ and the contrast-source amplitude (b) $|w_1^{\partial\rho}|$, produced by `ForwardCGFFTwcdrho` for the 1D case.

Matlab Listing I.58 The function `plotInterfaceSourceWdrho` to plot the contrast distribution χ^c and the contrast-source distribution $w_j^{\partial\rho}$.

```
function plotInterfaceSourceWdrho(w_drho, input)
global nDIM;  dx = input.dx;  CHI_drho = input.CHI_drho;

if nDIM == 1              % Plot 1D contrast/source distribution ----------------
   x1 = input.X1;
   set(figure,'Units','centimeters','Position',[5 5 18 12]);
   subplot(1,2,1);
      IMAGESC(x1+dx/2,-30:30,CHI_drho{1});
       title('\fontsize{13} \chi^{\rho\partial\rho}_1');
   subplot(1,2,2);
      IMAGESC(x1+dx/2,-30:30, abs(w_drho{1}));
      title('\fontsize{13} abs(w^{\partial\rho}_1)');

elseif nDIM == 2          % Plot 2D contrast/source distribution ----------------
   x1 = input.X1(:,1);   x2 = input.X2(1,:);
   set(figure,'Units','centimeters','Position',[5 5 18 12]);
   subplot(1,2,1);
      IMAGESC(x1+dx/2,x2,CHI_drho{1});
      title('\fontsize{13} \chi^{\rho\partial\rho}_1');
   subplot(1,2,2);
      IMAGESC(x1+dx/2,x2, abs(w_drho{1}));
      title('\fontsize{13} abs(w^{\partial\rho}_1)');
   set(figure,'Units','centimeters','Position',[5 5 18 12]);
   subplot(1,2,1);
      IMAGESC(x1,x2+dx/2,CHI_drho{2});
      title('\fontsize{13}  \chi^{\rho\partial\rho}_2');
   subplot(1,2,2);
      IMAGESC(x1,x2+dx/2,abs(w_drho{2}));
      title('\fontsize{13} abs(w^{\partial\rho}_2)');

elseif nDIM == 3        % Plot 3D contrast/source distribution ----------------
                          % at x3 = 0 or x3 = dx/2
   x1 = input.X1(:,1,1);  x2 = input.X2(1,:,1);
   N3cross = floor(input.N3/2+1);
   set(figure,'Units','centimeters','Position',[5 5 18 12]);
   subplot(1,2,1);
      IMAGESC(x1+dx/2,x2, CHI_drho{1}(:,:,N3cross));
      title('\fontsize{13} \chi^{\rho\partial\rho}_1');
   subplot(1,2,2);
      IMAGESC(x1+dx/2,x2,abs(w_drho{1}(:,:,N3cross)));
      title('\fontsize{13} abs(w^{\partial\rho}_1)');
   set(figure,'Units','centimeters','Position',[5 5 18 12]);
   subplot(1,2,1);
      IMAGESC(x1,x2+dx/2, CHI_drho{2}(:,:,N3cross));
      title('\fontsize{13}  \chi^{\rho\partial\rho}_2');
   subplot(1,2,2);
      IMAGESC(x1,x2+dx/2,abs(w_drho{2}(:,:,N3cross)));
      title('\fontsize{13} abs(w^{\partial\rho}_2)');

end % if
```

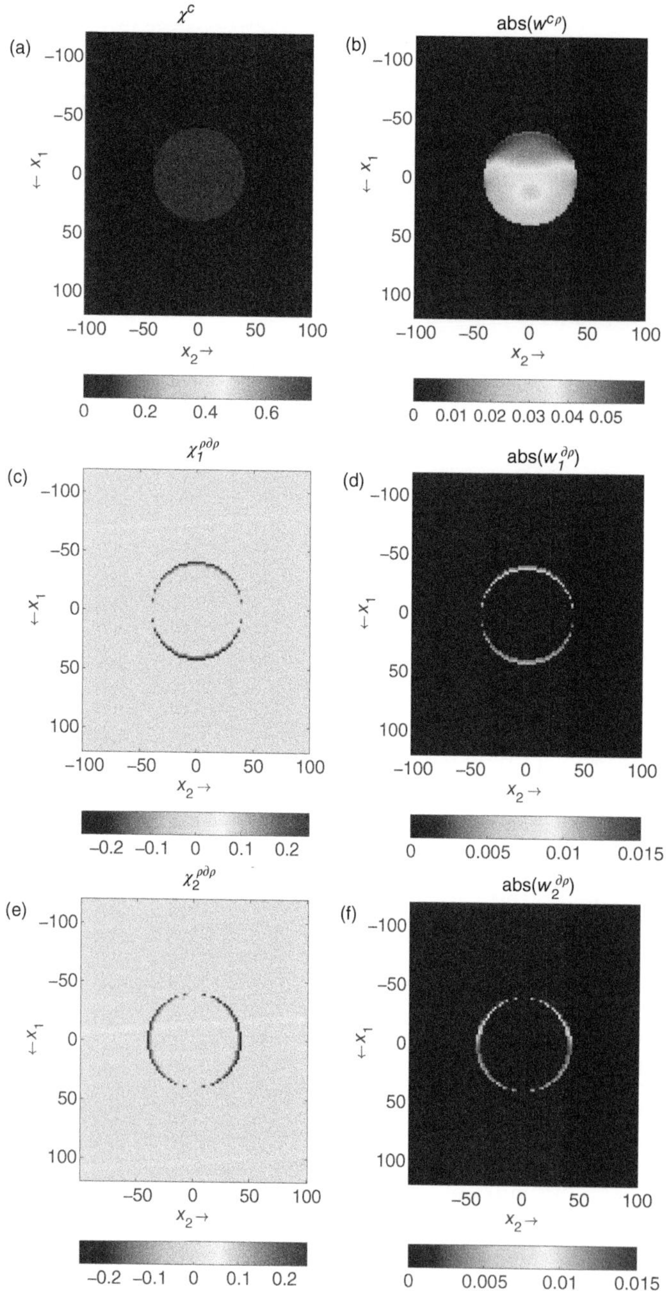

Figure 2.18 Plots of contrast χ^c and contrast source $|w^{c\rho}|$ (a, b), contrast $\chi_1^{\partial\rho_1}$ and contrast source $|w_1^{\partial\rho}|$ (c, d), contrast $\chi_2^{\partial\rho}$ and contrast source $|w_2^{\partial\rho}|$ (e, f), produced by `ForwardCGFFTwdcrho` for the 2D case.

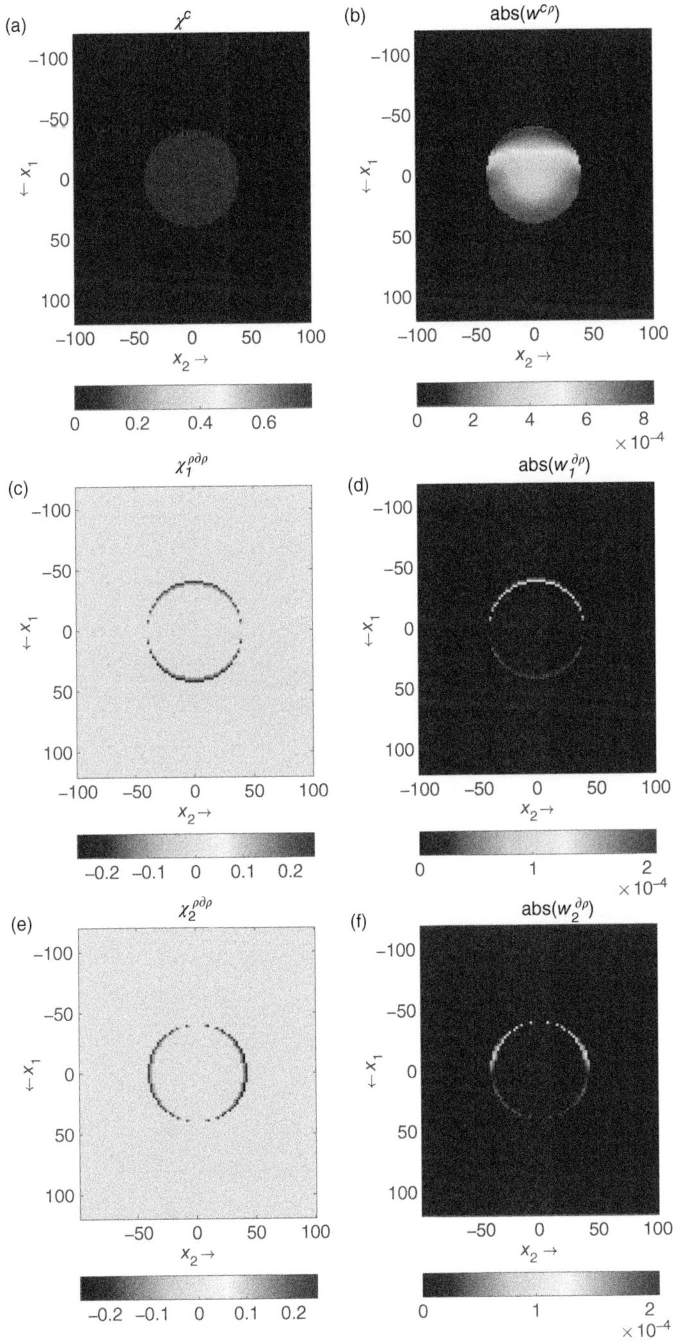

Figure 2.19 Plots of contrast χ^c and contrast source $|w^{c\rho}|$ (a, b), contrast $\chi_1^{\partial\rho_1}$, and contrast source $|w_1^{\partial\rho}|$ (c, d), contrast $\chi_2^{\partial\rho}$ and contrast source $|w_2^{\partial\rho}|$ (e, f), produced by `ForwardCGFFTwdcrho` for the 3D case.

In step (4) of the Matlab script `ForwardCGFFTwcdrho`, the scattered field at the selected receiver points are computed. The contrast-source distributions w^p and $w_j^{\partial\rho}$ are the input for the function `Dopwcdrho` of Matlab Listing I.59. The program follows from the weak form of the scattered pressure of Eqs. (2.112)–(2.114).

The scattered fields are compared to the data obtained from the exact Bessel function series and are stored in MAT-files. The function `displayData.m` of Matlab Listing I.9 loads the specific file and for 1D it prints the analytical data and the data using the integral equation method, while for 2D and 3D it plots these data, see Figures 2.20 and 2.21. Note that a small part of the visible differences between the two curves is caused by the edge effects of the staircase approximation of the boundaries of the circular cylinder and the sphere. A large part of these differences are the result of the finite difference approximation for the gradient of the mass-density contrast. To quantify the

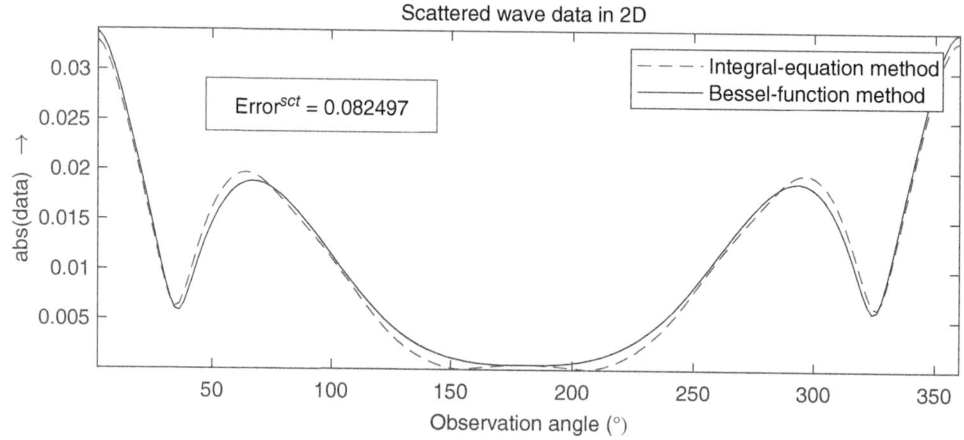

Figure 2.20 Plots of the 2D scattered wave field at the receivers, using either Matlab script `WavefieldSctBessel2D` (solid line) or Matlab script `ForwardCGFFTwcdrho` (dashed line); case `'analytical'`.

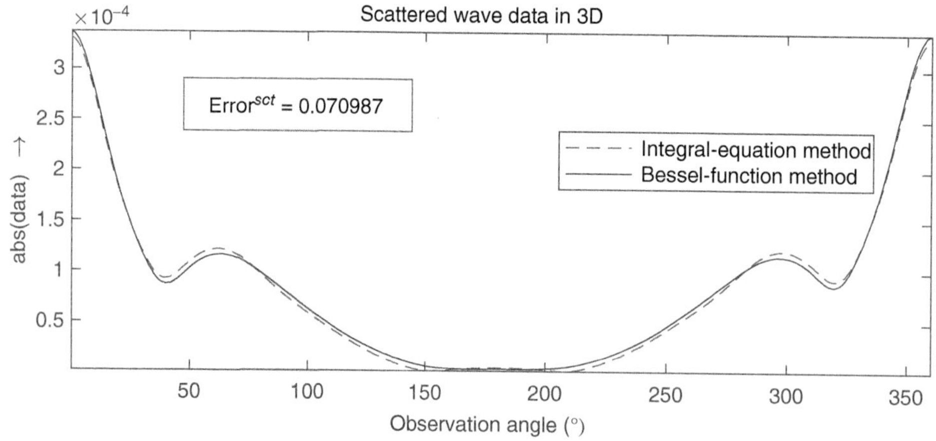

Figure 2.21 Plots of the 3D scattered wave field at the receivers anal, using either Matlab script `WavefieldSctBessel3D` (solid line) or Matlab script `ForwardCGFFTwcdrho` (dashed line); case `'analytical'`.

Matlab Listing I.59 The function `Dopwcdrho` to compute the scattered field at a number of receivers, generated by the volume contrast sources $w^{c\rho}$ and $w_j^{\partial\rho}$.

```
function [data] = DOPwcdrho(w_c,w_drho,input)
global nDIM;
gam0 = input.gamma_0;      dx = input.dx;     xR = input.xR;
data = zeros(1,input.NR);

if nDIM == 1
  x1        = input.X1;
  delta     = 0.5 * dx;   factor = sinh(gam0*delta) / (gam0*delta);
  DIS       = abs(x1-xR(1,1));                  % receiver for reflected field
  G         = 1/(2*gam0) * exp(-gam0.*DIS);
  d1_G      = - gam0 * sign(xR(1,1)-x1) .* G;
 data(1,1) = (gam0^2 * dx) * factor * sum(G .* w_c);
 data(1,1) = data(1,1) - dx * sum(d1_G(:) .* w_drho{1}(:));
  DIS       = abs(x1-xR(1,2));                  % receiver for transmitted field
  G         = 1/(2*gam0) * exp(-gam0.*DIS);
  d1_G      = - gam0 * sign(xR(1,2)-x1) .* G;
 data(1,2) = (gam0^2 * dx) * factor * sum(G(:) .* w_c(:));
 data(1,2) = data(1,2) - dx * factor * sum(d1_G(:) .* w_drho{1}(:));

elseif nDIM == 2
  X1     = input.X1;   X2 = input.X2;
  delta  = (pi)^(-1/2)*dx; factor = 2*besseli(1,gam0*delta) / (gam0*delta);
  for p = 1 : input.NR
     DIS       = sqrt((X1-xR(1,p)).^2 +(X2-xR(2,p)).^2);
     G         = 1/(2*pi).* besselk(0,gam0*DIS);
     d_G       = - gam0 * (1/(2*pi)) .* besselk(1,gam0*DIS);
    data(1,p) = (gam0^2 * dx^2) * factor * sum(G(:) .* w_c(:));
     d1_G      = ((xR(1,p)-X1)./DIS) .* d_G;
    data(1,p) = data(1,p) - dx^2 * sum(d1_G(:) .* w_drho{1}(:));
     d2_G      = ((xR(2,p)-X2)./DIS) .* d_G;
    data(1,p) = data(1,p) - dx^2 * sum(d2_G(:) .* w_drho{2}(:));
  end % p_loop

elseif nDIM == 3
  X1      = input.X1;    X2 = input.X2;    X3 = input.X3;
  delta   = (4*pi/3)^(-1/3) * dx;    arg     = gam0*delta;
 factor   = 3 * (cosh(arg) - sinh(arg)/arg) / arg^2;
  for p = 1 : input.NR
     DIS       = sqrt((X1-xR(1,p)).^2 + (X2-xR(2,p)).^2 + (X3-xR(3,p)).^2);
     G         = exp(-gam0*DIS) ./ (4*pi*DIS);
     d_G       = (-1./DIS - gam0) .* exp(-gam0*DIS) ./ (4*pi*DIS);
    data(1,p) = (gam0^2 * dx^3) * factor * sum(G(:) .* w_c(:));
     d1_G      = ((xR(1,p)-X1)./DIS) .* d_G;
    data(1,p) = data(1,p) - dx^3 * sum(d1_G(:) .* w_drho{1}(:));
     d2_G      = ((xR(2,p)-X2)./DIS) .* d_G;
    data(1,p) = data(1,p) - dx^3 * sum(d2_G(:) .* w_drho{2}(:));
     d3_G      = ((xR(3,p)-X3)./DIS) .* d_G;
    data(1,p) = data(1,p) - dx^3 * sum(d3_G(:) .* w_drho{3}(:));
  end % p_loop
end % if
```

differences, we use the error norm:

$$\text{Error}^{sct} = \frac{\sum_{p=1}^{\text{NR}} |p^{sct}(x_p^R) - p_{exact}^{sct}(x_p^R)|}{\sum_{p=1}^{\text{NR}} |p_{exact}^{sct}(x_p^R)|}. \tag{2.115}$$

The numerical values of this error norm are plotted in the figures as well.

We now switch to case `'numerical'`, where the spatial derivatives in the forward operator and its adjoint are computed numerically, and we show in Figures 2.22 and 2.23 the results for the scattered field at the receiver points. The error estimates are slightly smaller. Therefore, we conclude that the difference between the numerically discretized configuration and that of the circular cylinder (or the sphere) determines largely the error quantities. In Section 2.4.1, we investigate these errors for finer grid sizes.

2.4.1 Performance Analysis

To investigate the convergence rate of the integral equation results, we consider the results for the scattered data and the error norm of Eq. (2.115). Our goal is to investigate this error as function

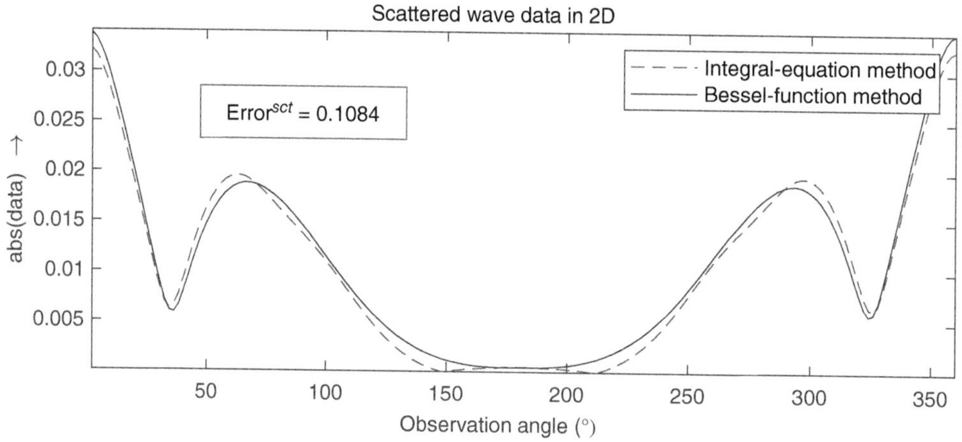

Figure 2.22 Similar as in Figure 2.20, but now for case `'numerical'`.

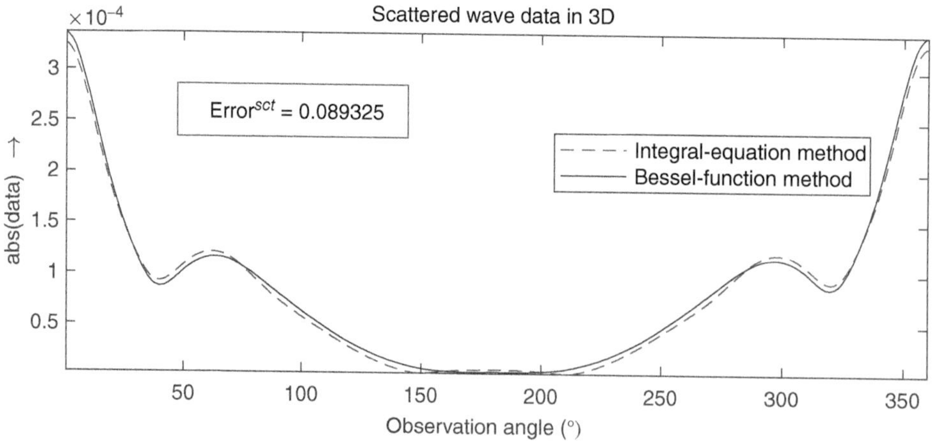

Figure 2.23 Similar as in Figure 2.21, but now for case `'numerical'`.

of the grid size Δx, both for the previous method, the so-called pZv-method and the present method, the so-called $c\partial\rho$-method. We only consider the 2D case. We begin with $\Delta x = 2$ and reduce it each time with a factor of two. Again, we define a refinement factor $r = 1, 2, \ldots, 5$ and define a new grid size $\Delta^{(r)} = \Delta x / r$ and new sample numbers $N_1^{(r)} = r\, N_1$ and $N_2^{(r)} = r\, N_2$. We have set the error criterion in the Matlab functions `ITERCGwpZv` and `ITERCGwcdhro` to 10^{-6}. These computations are performed on a Dell computer (64 bit, dual core 3.50 GHz) with 64 GB of installed memory (RAM).

In Figure 2.24, the results of case `'analytical'` are presented as solid lines, while the results of case `'numerical'` are presented as dashed lines.

For the pZv-method, the error reduces from 2.7% to 0.08% in case `'analytical'`, while the error decreases from 5.2% to 0.18% in case `'numerical'`. For both cases, the number of iterations to reach an error criterion of 10^{-6} is almost constant. It varies between 78 and 80 iterations, when we reduce the grid size from 2 to 0.125 m.

For the $c\partial\rho$-method, the error decreases from 8.2% to 4.33% in case `'analytical'`, while the error decreases from 10.9% to 4.15% in case `'numerical'`.

For these experiments, further grid refining does not provide further error reduction. For case `'analytical'`, the required number of iterations changes from 148 to 82, when we reduces the grid size from 2 to 0.125 m. For case `'numerical'`, this number of iterations changes from 157 to 82. Although the number of unknowns increases with finer grid sizes, the observation tells us that the iterative scheme converges faster, if the jumps in the numerical implementation of the mass-density gradient increases, however the theoretical value of a Dirac distribution can never be obtained and the actual error in the scattered field values will not converge to a result that is small enough. In Sections 2.5–2.7, we discuss a modified version of the present integral equation method, in which at interfaces the jumps in the mass density are taken into account analytically.

Finally, we have registered the computer time needed to solve the integral equation. This is only the time to execute step (3) of the Matlab functions `ITERCGwpZv` and `ITERCGwcdhro`. In Figure 2.25, we have plotted this computer time for a single iteration, when we increase the number of grid cells, $N = N_1 \times N_2$, by refining the grid size. This figure confirms that the computation cost for the numerical solution of the integral equation is roughly of the order of $N \log_2(N)$, where we find that case `'analytical'` of the pZv-method needs the largest computation time. Per iteration, the two cases of the $c\partial\rho$-method take about the same amount of time and, compared to the two cases of the pZv-method, they take the least amount of time.

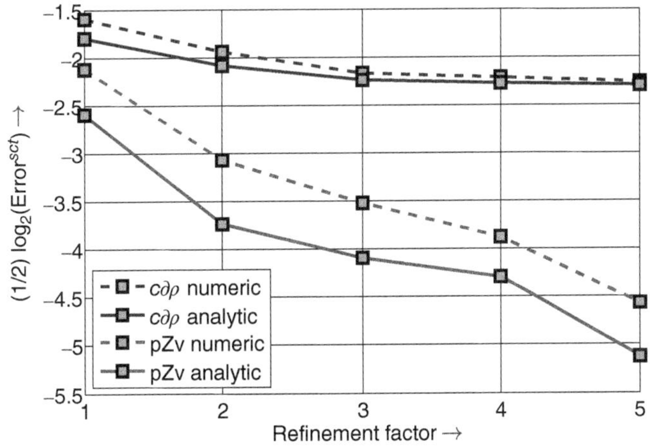

Figure 2.24 Convergence plot of the Error^{sct} as function of the refinement factor r, $n = 1, 2, \ldots, 5$, for the 2D pZv-method and 2D $c\partial\rho$-method.

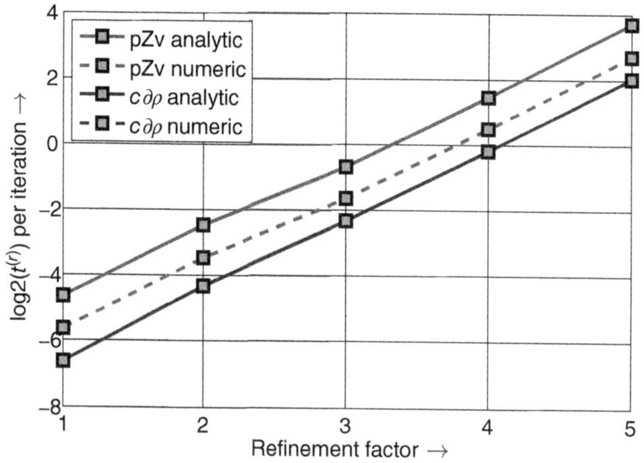

Figure 2.25 Computer time $\log_2(t^{(r)})$ to solve the 2D integral equation as function of the refinement r.

2.5 Solution of Integral Equation: Interface Contrast Sources

In order to have a clear discussion on how to modify the integral equations of Sections 2.3 and 2.4 for discontinuous mass density, we start with the integral equation of Eq. (2.91). For simplicity, we assume that there is **no contrast in wave-speed** ($\hat{\chi}^c = \chi_0$). We then have the scattered field representation

$$\frac{\rho_0}{\hat{\rho}_{sct}(\boldsymbol{x})}\hat{p}(\boldsymbol{x}) = \hat{p}^{inc}(\boldsymbol{x}) - \partial_k \int_{\boldsymbol{x}' \in \mathbb{D}_{sct}} \hat{G}(\boldsymbol{x}-\boldsymbol{x}') \left[\partial'_k \frac{\rho_0}{\hat{\rho}_{sct}}\right] \hat{p}(\boldsymbol{x}')\, \mathrm{d}V. \tag{2.116}$$

Note that the pressure is continuous through medium discontinuities, but the quantity $(\rho_0/\hat{\rho}_{sct})\hat{p}(\boldsymbol{x})$ will jump at discontinuities of $\hat{\rho}_{sct}$. This is consistent with the presence of delta distributions in the integral of Eq. (2.116).

In view of a later consideration regarding numerical discretization, we now assume that the scatterer consists of piecewise homogeneous subdomains. The set of internal interfaces in the domain $\mathbb{D}_{sct}$ are denoted as S_{sct}. It is assumed that the domain $\mathbb{D}_{sct}$ also contains the outer interface. Hence, at the boundary of the domain $\partial\mathbb{D}_{sct}$, there is always a zero contrast. Due to the current assumption of a piecewise homogeneous mass-density distribution, the spatial derivative of the mass density contrast vanishes everywhere, except at the interfaces where the spatial derivative in the normal direction exhibits an interface Dirac distribution. When the observation point is not located at S_{sct}, the integration over the domain $\mathbb{D}_{sct}$ reduces to an integration along all interfaces S_{sct}, viz.,

$$\frac{\rho_0}{\hat{\rho}_{sct}(\boldsymbol{x})}\hat{p}(\boldsymbol{x}) = \hat{p}^{inc}(\boldsymbol{x}) - \partial_k \int_{\boldsymbol{x}' \in \mathrm{S}_{sct}} \hat{G}(\boldsymbol{x}-\boldsymbol{x}') \left[\frac{\rho_0}{\hat{\rho}^+} - \frac{\rho_0}{\hat{\rho}^-}\right] \hat{p}(\boldsymbol{x}')\, \nu_k(\boldsymbol{x}')\, \mathrm{d}A, \tag{2.117}$$

for $\boldsymbol{x} \in \mathbb{D}_{sct}\backslash\mathrm{S}_{sct}$. Further, $\boldsymbol{\nu} = \boldsymbol{\nu}(\boldsymbol{x})$ denotes the normal to an interface through the point $\boldsymbol{x}$, while

$$\hat{\rho}^+ = \lim_{h\to 0}\,[\hat{\rho}_{sct}(\boldsymbol{x}+h\boldsymbol{\nu})] \quad \text{and} \quad \hat{\rho}^- = \lim_{h\to 0}\,[\hat{\rho}_{sct}(\boldsymbol{x}-h\boldsymbol{\nu})]. \tag{2.118}$$

The scattered-field integral representation is written as

$$\frac{\rho_0}{\hat{\rho}_{sct}(\boldsymbol{x})}\hat{p}(\boldsymbol{x}) = \hat{p}^{inc}(\boldsymbol{x}) + \partial_k \int_{\boldsymbol{x}' \in \mathrm{S}_{sct}} \hat{G}(\boldsymbol{x}-\boldsymbol{x}')\, \frac{\hat{\rho}^+ - \hat{\rho}^-}{\hat{\rho}^+\hat{\rho}^-}\rho_0\, \hat{p}(\boldsymbol{x}')\, \nu_k(\boldsymbol{x}')\, \mathrm{d}A, \tag{2.119}$$

for $\boldsymbol{x} \in \mathbb{D}_{sct} \backslash \mathbb{S}_{sct}$. To obtain an integral equation for the unknown $\hat{p}(\boldsymbol{x})$, $\boldsymbol{x} \in \mathbb{S}_{sct}$, we take the point of observation inside the set of homogeneous subdomains of the scatterer, denoted as $\mathbb{D}_{sct} \backslash \mathbb{S}_{sct}$. Subsequently, we let approach the observation point $\boldsymbol{x}$ to a point $\boldsymbol{x}'$ of the set $\mathbb{S}_{sct}$, i.e. $\boldsymbol{x} = \lim_{h \to 0} (\boldsymbol{x}' + h\boldsymbol{\nu})$. In this limiting procedure, the singular point $\boldsymbol{x}$ on the interface is excluded symmetrically, after which the limiting value of the integral has been taken. This leads to

$$\frac{\rho_0}{\hat{\rho}^+}\hat{p}(\boldsymbol{x}) = \hat{p}^{inc}(\boldsymbol{x}) - \frac{1}{2}\frac{\hat{\rho}^+ - \hat{\rho}^-}{\hat{\rho}^-}\frac{\rho_0}{\hat{\rho}^+}\hat{p}(\boldsymbol{x}) + ⨍_{\boldsymbol{x}' \in \mathbb{S}_{sct}} \partial_k \hat{G}(\boldsymbol{x} - \boldsymbol{x}') \, \frac{\hat{\rho}^+ - \hat{\rho}^-}{\hat{\rho}^+ \hat{\rho}^-} \rho_0 \, \hat{p}(\boldsymbol{x}') \, \nu_k(\boldsymbol{x}') \, \mathrm{d}A, \quad (2.120)$$

for $\boldsymbol{x} \in \mathbb{S}_{sct}$. The integral sign $⨍$ denotes the Cauchy principle value of the relevant integral, see [23]. After multiplying both sides of the equation by the factor $2\hat{\rho}^+\hat{\rho}^-/(\hat{\rho}^+ + \hat{\rho}^-)$, we arrive at the interface integral equation,

$$\rho_0 \, \hat{p}(\boldsymbol{x}) = \frac{2\hat{\rho}^+\hat{\rho}^-}{\hat{\rho}^+ + \hat{\rho}^-} \left[\hat{p}^{inc}(\boldsymbol{x}) + ⨍_{\boldsymbol{x}' \in \mathbb{S}_{sct}} \partial_k \hat{G}(\boldsymbol{x} - \boldsymbol{x}') \, \frac{\hat{\rho}^+ - \hat{\rho}^-}{\hat{\rho}^+ \hat{\rho}^-} \rho_0 \, \hat{p}(\boldsymbol{x}') \, \nu_k(\boldsymbol{x}') \, \mathrm{d}A \right] \quad (2.121)$$

for the unknown $\hat{p}(\boldsymbol{x})$, $\boldsymbol{x} \in \mathbb{S}_{sct}$.

2.5.1 Contrast-Source Integral Equation

Similar as in Sections 2.3 and 2.4, we introduce a formulation in terms of contrast sources. However, they are now defined on the interface of the scatterer. Let us denote this contrast source as

$$\partial\hat{w}(\boldsymbol{x}) = \frac{1}{2}\frac{\hat{\rho}^+ - \hat{\rho}^-}{\hat{\rho}^+ \hat{\rho}^-} \, \rho_0 \, \hat{p}(\boldsymbol{x}), \quad \boldsymbol{x} \in \mathbb{S}_{sct}. \quad (2.122)$$

From Eq. (2.119), we obtain the interface contrast-source integral representation

$$\hat{p}^{sct}(\boldsymbol{x}) = \int_{\boldsymbol{x}' \in \mathbb{S}_{sct}} 2 \, \partial_k \hat{G}(\boldsymbol{x} - \boldsymbol{x}') \, \partial\hat{w}(\boldsymbol{x}') \, \nu_k(\boldsymbol{x}') \, \mathrm{d}A, \quad \boldsymbol{x} \in \mathbb{D}'_{sct}, \quad (2.123)$$

where $\hat{\rho}_{sct} = \rho_0$. After multiplying both sides of Eq. (2.121) by the factor $(\hat{\rho}^+ - \hat{\rho}^-)/(2\hat{\rho}^+\hat{\rho}^-)$, we arrive at the interface integral equation,

$$\partial\hat{w}(\boldsymbol{x}) = \frac{\hat{\rho}^+ - \hat{\rho}^-}{\hat{\rho}^+ + \hat{\rho}^-} \left[\hat{p}^{inc}(\boldsymbol{x}) + ⨍_{\boldsymbol{x}' \in \mathbb{S}_{sct}} 2 \, \partial_k \hat{G}(\boldsymbol{x} - \boldsymbol{x}') \, \partial\hat{w}(\boldsymbol{x}') \, \nu_k(\boldsymbol{x}') \, \mathrm{d}A \right], \quad (2.124)$$

for the interface contrast source at the set of interfaces $\mathbb{S}_{sct}$. In particular, this integral equation simplifies to a single integral equation for a homogeneous object with $\hat{\rho}^+ = \rho_0$ and $\hat{\rho}^- = \hat{\rho}_{sct}$.

Let us consider the case that the interfaces are oriented in the Cartesian directions. Then the set of interfaces perpendicular to the x_1-direction is denoted as $\mathbb{S}_1$, the set of interfaces perpendicular to the x_2-direction is denoted as $\mathbb{S}_2$ and the set of interfaces perpendicular to the x_3-direction is denoted as $\mathbb{S}_3$. The set of integral equations may then be written as

$$\boxed{\partial\hat{w}(\boldsymbol{x}) = \frac{\hat{\rho}^+ - \hat{\rho}^-}{\hat{\rho}^+ + \hat{\rho}^-} \left[\hat{p}^{inc}(\boldsymbol{x}) + \sum_{k=1}^{3} ⨍_{\boldsymbol{x}' \in \mathbb{S}_k} 2 \, \partial_k \hat{G}(\boldsymbol{x} - \boldsymbol{x}') \, \partial\hat{w}(\boldsymbol{x}') \, \nu_k(\boldsymbol{x}') \, \mathrm{d}A \right],} \quad (2.125)$$

where $\boldsymbol{\nu}(\boldsymbol{x}) = \boldsymbol{i}_j$ when $\boldsymbol{x} \in \mathbb{S}_j$. After solution for contrast source $\partial\hat{w}$, the scattered wave field at some receiver point $\boldsymbol{x} = \boldsymbol{x}^R$ is obtained as

$$\boxed{\hat{p}^{sct}(\boldsymbol{x}^R) = \sum_{k=1}^{3} \int_{\boldsymbol{x}' \in \mathbb{S}_k} 2 \, \partial_k^R \hat{G}(\boldsymbol{x}^R - \boldsymbol{x}') \, \partial\hat{w}(\boldsymbol{x}') \, \nu_k(\boldsymbol{x}') \, \mathrm{d}A, \quad \boldsymbol{x}^R \in \mathbb{D}'_{sct}.} \quad (2.126)$$

This integral equation is used when we discretize our scattering domain into a set of cubic subdomains with constant material properties.

2.5.2 Numerical Solution of Interface Integral Equations (2D)

So far we have introduced the weak formulation to replace the continuous form of the integral equation by a discrete set of equations. The consistent thread running through the discretization method is that, with a finite spatial discretization, we cannot model the spatial high frequencies. However, for the interface integral equation, both the actual discontinuities of the mass density and the discontinuities due a piecewise constant discretization model must be taken into account in such a way that the spatial high frequencies are also treated consistently. If we can achieve this goal, the solution of the corresponding integral equation may be used as a candidate for a forward solver in an inversion/imaging method to enhance the edge resolution of the image. Then, the current inversion/imaging method based on the wave-speed contrast can be improved by imaging the actual discontinuities in mass density. With this in mind, we choose for a formulation of the interface integral equation.

We demonstrate the solution of the interface integral equation in 2D, which shows all the key features. In 2D, the integral equation (2.125) becomes

$$\partial\hat{w}(\boldsymbol{x}_T) = \frac{\hat{\rho}^+ - \hat{\rho}^-}{\hat{\rho}^+ + \hat{\rho}^-}\left[\hat{p}^{inc}(\boldsymbol{x}_T) + \sum_{k=1}^{2}\int_{\boldsymbol{x}'\in S_k} 2\,\partial_k\hat{G}(\boldsymbol{x}_T - \boldsymbol{x}'_T)\,\partial\hat{w}(\boldsymbol{x}'_T)\,\nu_k(\boldsymbol{x}'_T)\,\mathrm{d}s\right]. \tag{2.127}$$

We have omitted the principle value sign of the integral because in numerical work the spherical mean of the derivative of the Green function vanishes at the singular point $\boldsymbol{x}_T = \boldsymbol{x}'_T$. We start with our regular grid of Figure 2.4. It consists of a number of n equally sized subdomains, denoted as $\{\mathbb{D}_n,\ n = 1 : N\}$. In 2D, we deal with $N = N_1 \times N_2$ subdomains. The mesh size of each square subdomain is equal to $\Delta x \times \Delta x$. In Figure 2.26, we show the grid. The dotted domain includes the scattering domain $\mathbb{D}_{sct}$ completely. To this domain, we add an extra layer with zero mass density contrast. In this domain, the wave-field quantities are computed as well. The midpoints of each subdomain are now defined as

$$\boldsymbol{x}^{(0)}_{T,n} = \{x^{(0)}_{1,n},\ x^{(0)}_{2,n}\}. \tag{2.128}$$

where, for convenience, we have added the superscript $^{(0)}$.

As a next step, we consider the set of points at the horizontal interfaces of the discretized domain (Figure 2.27a)

$$\boldsymbol{x}^{(1)}_{T,n} = \{x^{(0)}_{1,n} + \tfrac{1}{2}\Delta x,\ x^{(0)}_{2,n}\}. \tag{2.129}$$

and a set of points at the vertical interfaces of the discretized domain (Figure 2.27b)

$$\boldsymbol{x}^{(2)}_{T,n} = \{x^{(0)}_{1,n},\ x^{(0)}_{2,n} + \tfrac{1}{2}\Delta x\}. \tag{2.130}$$

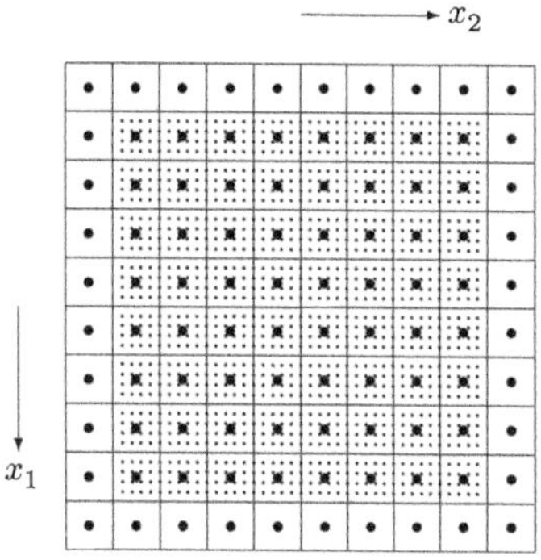

Figure 2.26 2D Regular grid with a mesh size of $\Delta x \times \Delta x$. The center points are denoted as $\boldsymbol{x}_T = \boldsymbol{x}^{(0)}_{T,n}$, $n = 1, 2, \ldots, N$.

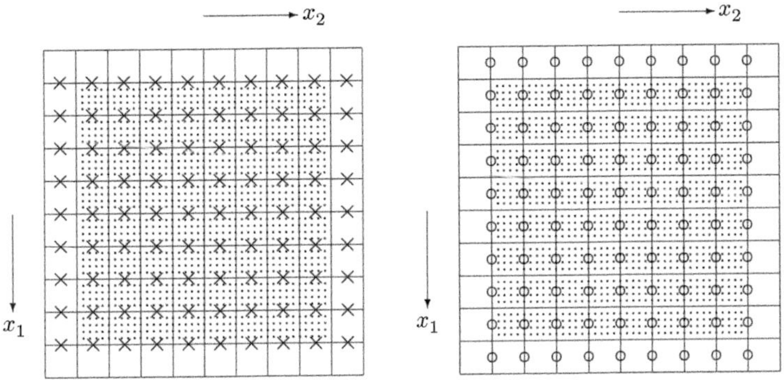

Figure 2.27 Grid shifted $\frac{1}{2}\Delta x$ in positive x_1-direction (a) and grid shifted $\frac{1}{2}\Delta x$ in positive x_2-direction (b). The points indicated by × and o are denoted as $\boldsymbol{x}_n^{(1)} = \boldsymbol{x}_n^{(0)} + \frac{1}{2}\Delta x\, \boldsymbol{i}_1$ and $\boldsymbol{x}_n^{(2)} = \boldsymbol{x}_n^{(0)} + \frac{1}{2}\Delta x\, \boldsymbol{i}_2$, respectively.

The values of the contrast sources at the horizontal interfaces and the vertical interfaces of the discretized domain are denoted as $\partial\hat{w}_n^{(1)} = \partial\hat{w}(\boldsymbol{x}_{T,n}^{(1)})$ and $\partial\hat{w}_n^{(2)} = \partial\hat{w}(\boldsymbol{x}_{T,n}^{(2)})$, respectively. We further assume that these values are the mean values of $\partial\hat{w}$ on each boundary of a subdomain. For points where the argument of the Green function vanishes, we define $\partial_k\hat{G} = 0$, for $\boldsymbol{x}_T - \boldsymbol{x}'_T = 0$ on each boundary of a subsquare. This is consistent with its principal value definition. Subsequently, we approximate each integral over a boundary of a subsquare by its mean value times the integration length Δx. Note that with these mean values of wave functions along an interface, we have left our strong formulation and used some weak form.

At this point, we define our discrete operators

$$
\begin{aligned}
[\mathcal{K}\partial\hat{w}^{(1)}](\boldsymbol{x}_T) &= 2\Delta x \sum_{n=1}^{N} \langle \partial_1 \hat{G}\rangle(\boldsymbol{x}_T - \boldsymbol{x}_{T,n}^{(1)})\ \langle \partial\hat{w}_n^{(1)}\rangle, \\
[\mathcal{K}\partial\hat{w}^{(2)}](\boldsymbol{x}_T) &= 2\Delta x \sum_{n=1}^{N} \langle \partial_2 \hat{G}\rangle(\boldsymbol{x}_T - \boldsymbol{x}_{T,n}^{(2)})\ \langle \partial\hat{w}_n^{(2)}\rangle.
\end{aligned}
\tag{2.131}
$$

These two operators are defined on different grids. But when we take in the first relation of Eq. (2.131), the observation points $\boldsymbol{x}_T$ at a point $\boldsymbol{x}_{T,m}^{(1)}$ of the set $\{\boldsymbol{x}_{T,n}^{(1)},\ n = 1,\ldots,N\}$, the equation becomes a spatial convolution for the ×-points of the shifted grid in Figure 2.27a. On this grid the convolution is computed efficiently using FFTs. This is also valid for the second relation of Eq. (2.131) when we take $\boldsymbol{x}_T$ at a point of the set $\{\boldsymbol{x}_{T,n}^{(2)},\ n = 1,\ldots,N\}$, which are the o-points of the shifted grid in Figure 2.27b. To compute $[\mathcal{K}\partial\hat{w}^{(1)}](\boldsymbol{x}_T)$ for $\boldsymbol{x}_T = \boldsymbol{x}_{T,m}^{(2)}$ and $[\mathcal{K}\partial\hat{w}^{(2)}](\boldsymbol{x}_T)$ for $\boldsymbol{x}_T = \boldsymbol{x}_{T,m}^{(1)}$ we interrelate these two operators to each other by linear interpolation. The result is that the continuous integral equation is replaced by the discrete system of equations:

$$
\begin{aligned}
&\hat{R}_m^{(1)}\hat{p}^{inc}(\boldsymbol{x}_{T,m}^{(1)}) = \partial\hat{w}_m^{(1)} - \hat{R}_m^{(1)}\left[[\mathcal{K}\partial\hat{w}^{(1)}](\boldsymbol{x}_{T,m}^{(1)}) + [\mathcal{K}\partial\hat{w}^{(2)}](\boldsymbol{x}_{T,m}^{(1)})\right] \\
&\hat{R}_m^{(2)}\hat{p}^{inc}(\boldsymbol{x}_{T,m}^{(2)}) = \partial\hat{w}_m^{(2)} - \hat{R}_m^{(2)}\left[[\mathcal{K}\partial\hat{w}^{(1)}](\boldsymbol{x}_{T,m}^{(2)}) + [\mathcal{K}\partial\hat{w}^{(2)}](\boldsymbol{x}_{T,m}^{(2)})\right],
\end{aligned}
\tag{2.132}
$$

where the "reflection coefficients" are given by

$$
\hat{R}_m^{(1)} = \frac{\hat{\rho}(\boldsymbol{x}_{T,m}^{(0)} + \Delta x\, \boldsymbol{i}_1) - \hat{\rho}(\boldsymbol{x}_{T,m}^{(0)})}{\hat{\rho}(\boldsymbol{x}_{T,m}^{(0)} + \Delta x\, \boldsymbol{i}_1) + \hat{\rho}(\boldsymbol{x}_{T,m}^{(0)})}, \quad
\hat{R}_m^{(2)} = \frac{\hat{\rho}(\boldsymbol{x}_{T,m}^{(0)} + \Delta x\, \boldsymbol{i}_2) - \hat{\rho}(\boldsymbol{x}_{T,m}^{(0)})}{\hat{\rho}(\boldsymbol{x}_{T,m}^{(0)} + \Delta x\, \boldsymbol{i}_2) + \hat{\rho}(\boldsymbol{x}_{T,m}^{(0)})},
\tag{2.133}
$$

and where each point $\boldsymbol{x}_{T,m}^{(0)}$ denotes the midpoint of subdomain $\mathbb{D}_m$ (see Figure 2.26). Note that we assume that $\hat{\rho} = \hat{\rho}_{sct}$ for points in the scatterer and $\hat{\rho} = \rho_0$ for points outside the scatterer. The operators $\mathcal{K}$ on the right-hand side of Eq. (2.132) act on $\partial \hat{w}_n^{(1)}$ and $\partial \hat{w}_n^{(2)}$, respectively. Before defining their actual representations, we define

$$
\begin{aligned}
\mathcal{K}_m^{(1,1)} &= [\mathcal{K}\partial\hat{w}^{(1)}](\boldsymbol{x}_{T,m}^{(1)}) = 2\Delta x \sum_{n=1}^{N} \langle\langle\partial_1\hat{G}\rangle\rangle(\boldsymbol{x}_{T,m}^{(0)} - \boldsymbol{x}_{T,n}^{(0)}) \, \langle\partial\hat{w}_n^{(1)}\rangle, \\
\mathcal{K}_m^{(2,2)} &= [\mathcal{K}\partial\hat{w}^{(1)}](\boldsymbol{x}_{T,m}^{(2)}) = 2\Delta x \sum_{n=1}^{N} \langle\langle\partial_2\hat{G}\rangle\rangle(\boldsymbol{x}_{T,m}^{(0)} - \boldsymbol{x}_{T,n}^{(0)}) \, \langle\partial\hat{w}_n^{(2)}\rangle.
\end{aligned}
\tag{2.134}
$$

Note that in the expression of the convolutional operators $\mathcal{K}_m^{(1,1)}$ and $\mathcal{K}_m^{(2,2)}$, we have used the relation $\boldsymbol{x}_m^{(k)} - \boldsymbol{x}_n^{(k)} = \boldsymbol{x}_m^{(0)} - \boldsymbol{x}_n^{(0)}$, for $k = 1,\ 2$. Further, we define

$$
\begin{aligned}
\mathcal{K}_m^{(1,2)} &= [\mathcal{K}\partial\hat{w}^{(2)}](\boldsymbol{x}_{T,m}^{(1)}), \\
\mathcal{K}_m^{(2,1)} &= [\mathcal{K}\partial\hat{w}^{(1)}](\boldsymbol{x}_{T,m}^{(2)}).
\end{aligned}
\tag{2.135}
$$

For computational efficiency, we do not compute $\mathcal{K}_m^{(1,2)}$ and $\mathcal{K}_m^{(2,1)}$ with the help of Eq. (2.131). To compute $\mathcal{K}_m^{(1,2)}$, we perform a simple linear interpolation from the values of $\mathcal{K}_m^{(2,2)}$ at the four adjacent points. Similarly, $\mathcal{K}_m^{(1,2)}$ is obtained from a simple linear interpolation from the values of $\mathcal{K}_m^{(1,1)}$ at the four adjacent points. The results in the extra layer added to the scattering domain are the extrapolated values and are immaterial, since the reflection coefficients vanish in this layer. However, we need this layer to define the correct array numbering of the input and output arrays used in Matlab.

2.6 Numerical Solution Integral Equations: Volume and Interface Contrast Sources

In this section, we present the discrete representations of the integral equations when we deal with contrast in both wave speed and mass density, in a 3D space, 2D space, and 1D space, respectively.

2.6.1 Discrete Representations in 3D

We start with Eq. (2.91),

$$
\begin{aligned}
\frac{\rho_0}{\hat{\rho}_{sct}(\boldsymbol{x})}\hat{p}(\boldsymbol{x}) = \hat{p}^{inc}(\boldsymbol{x}) &+ \int_{\boldsymbol{x}' \in \mathbb{D}_{sct}} \hat{\gamma}_0^2 \hat{G}(\boldsymbol{x} - \boldsymbol{x}') \, \hat{\chi}^c(\boldsymbol{x}') \frac{\rho_0}{\hat{\rho}_{sct}} \hat{p}(\boldsymbol{x}') \, dV \\
&- \partial_k \int_{\boldsymbol{x}' \in \mathbb{D}_{sct}} \hat{G}(\boldsymbol{x} - \boldsymbol{x}') \, [\partial_k' \frac{\rho_0}{\hat{\rho}_{sct}}] \, \hat{p}(\boldsymbol{x}') \, dV,
\end{aligned}
\tag{2.136}
$$

when $\boldsymbol{x} \in \mathbb{D}_{sct} \backslash \sum_k \mathbb{S}_k$.

First, we multiply Eq. (2.136) by the contrast $\hat{\chi}^c$ of Eq. (2.93) and use the volume contrast source $\hat{w}^{c\rho}$ of Eq. (2.94). The result is

$$
\boxed{
\begin{aligned}
\hat{\chi}^c(\boldsymbol{x})\hat{p}^{inc}(\boldsymbol{x}) = \hat{w}^{c\rho}(\boldsymbol{x}) - \hat{\chi}^c(\boldsymbol{x}) &\left[\hat{\gamma}_0^2 \int_{\boldsymbol{x}' \in \mathbb{D}_{sct}} \hat{G}(\boldsymbol{x} - \boldsymbol{x}') \hat{w}^{c\rho}(\boldsymbol{x}') \, dV \right. \\
&\left. + \sum_{k=1}^{3} \int_{\boldsymbol{x}' \in \mathbb{S}_k} 2 \, \partial_k \hat{G}(\boldsymbol{x} - \boldsymbol{x}') \, \partial \hat{w}(\boldsymbol{x}') \, v_k(\boldsymbol{x}') \, dA \right],
\end{aligned}
}
\tag{2.137}
$$

where $\partial \hat{w}$ is defined in Eq. (2.122).

Second, we also require that Eq. (2.136) holds on the interfaces S_k and we carry out the same procedure as in Section 2.5. The result is

$$\frac{\hat{\rho}^+ - \hat{\rho}^-}{\hat{\rho}^+ + \hat{\rho}^-}\hat{p}^{inc}(\boldsymbol{x}) = \partial\hat{w}^{cp}(\boldsymbol{x}) - \frac{\hat{\rho}^+ - \hat{\rho}^-}{\hat{\rho}^+\hat{\rho}^-}\left[\hat{\gamma}_0^2 \int_{\boldsymbol{x}'\in\mathbb{D}_{sct}} \hat{G}(\boldsymbol{x}-\boldsymbol{x}')\hat{w}^{cp}(\boldsymbol{x}')\,\mathrm{d}V + \sum_{k=1}^{3} \fint_{\boldsymbol{x}'\in S_k} 2\,\partial_k\hat{G}(\boldsymbol{x}-\boldsymbol{x}')\,\partial\hat{w}(\boldsymbol{x}')\,\nu_k(\boldsymbol{x}')\,\mathrm{d}A\right], \tag{2.138}$$

where $\boldsymbol{x} \in S_j, j = 1, 2, 3$. Remark that we have now a set of four of integral equations because we require consistency at points of our different spaces, viz., $\mathbb{D}_{sct}\backslash\sum_k S_k$ and $S_j, j = 1, 2, 3$.

In 3D, we define a regular grid of a number of n equally sized subdomains, denoted as $\{\mathbb{D}_n, n = 1 : N\}$. We deal with $N = N_1 \times N_2 \times N_3$ subdomains. The mesh size of each square subdomain is equal to Δx. The volume of a subdomain is $\Delta V = \Delta x \times \Delta x \times \Delta x$ and each side has a surface area of $\Delta A = \Delta x \times \Delta x$. The midpoints $\boldsymbol{x}_n$ of each subdomain are given by

$$\boldsymbol{x}_n^{(0)} = \{x_{1,n}^{(0)}, x_{2,n}^{(0)}, x_{3,n}^{(0)}\}. \tag{2.139}$$

As a next step, we define the shifted grid at the horizontal interfaces of the discretized domain,

$$\boldsymbol{x}_n^{(1)} = \left\{x_{1,n}^{(0)} + \tfrac{1}{2}\Delta x,\ x_{2,n}^{(0)},\ x_{3,n}^{(0)}\right\}, \tag{2.140}$$

and two shifted grids at the vertical interfaces of the discretized domain,

$$\boldsymbol{x}_n^{(2)} = \left\{x_{1,n}^{(0)},\ x_{2,n}^{(0)} + \tfrac{1}{2}\Delta x,\ x_{3,n}^{(0)}\right\}, \tag{2.141}$$

and

$$\boldsymbol{x}_n^{(3)} = \left\{x_{1,n}^{(0)},\ x_{2,n}^{(0)},\ x_{3,n}^{(0)} + \tfrac{1}{2}\Delta x\right\}. \tag{2.142}$$

The values of the volume type contrast sources at the center of a subdomain is denoted as $\hat{w}_n^{(0)} = \hat{w}^{cp}(\boldsymbol{x}_n)$, and it is assumed that these values are the mean values of $\hat{w}$ on the pertaining subdomain. The values of the interface type of contrast sources at the horizontal and vertical interfaces of the discretized domain are denoted as $\partial\hat{w}_n^{(1)} = \partial\hat{w}(\boldsymbol{x}_n^{(1)})$, $\partial\hat{w}_n^{(2)} = \partial\hat{w}(\boldsymbol{x}_n^{(2)})$, and $\partial\hat{w}_n^{(3)} = \partial\hat{w}(\boldsymbol{x}_n^{(3)})$, respectively. We further assume that these values are the mean values of $\partial\hat{w}$ on each boundary of a subdomain. Carrying out the discretization of the volume integral (see Chapter 1) and the interface integrals we obtain the following discrete system of equations. The result is that the continuous integral equation is replaced by the discrete system of equations

$$\begin{aligned}
\hat{\chi}_m^c\,\langle p^{inc}\rangle(\boldsymbol{x}_m^{(0)}) &= \hat{w}_m^{(0)} - \hat{\chi}_m^c\left[[\mathcal{K}\hat{w}^{(0)}](\boldsymbol{x}_m^{(0)}) + \sum_{k=1}^{3}[\mathcal{K}\partial\hat{w}^{(k)}](\boldsymbol{x}_m^{(0)})\right],\\
\hat{R}_m^{(1)}\hat{p}^{inc}(\boldsymbol{x}_m^{(1)}) &= \partial\hat{w}_m^{(1)} - \hat{R}_m^{(1)}\left[[\mathcal{K}\hat{w}^{(0)}](\boldsymbol{x}_m^{(1)}) + \sum_{k=1}^{3}[\mathcal{K}\partial\hat{w}^{(k)}](\boldsymbol{x}_m^{(1)})\right],\\
\hat{R}_m^{(2)}\hat{p}^{inc}(\boldsymbol{x}_m^{(2)}) &= \partial\hat{w}_m^{(2)} - \hat{R}_m^{(2)}\left[[\mathcal{K}\hat{w}^{(0)}](\boldsymbol{x}_m^{(2)}) + \sum_{k=1}^{3}[\mathcal{K}\partial\hat{w}^{(k)}](\boldsymbol{x}_m^{(2)})\right],\\
\hat{R}_m^{(3)}\hat{p}^{inc}(\boldsymbol{x}_m^{(3)}) &= \partial\hat{w}_m^{(3)} - \hat{R}_m^{(3)}\left[[\mathcal{K}\hat{w}^{(0)}](\boldsymbol{x}_m^{(3)}) + \sum_{k=1}^{3}[\mathcal{K}\partial\hat{w}^{(k)}](\boldsymbol{x}_m^{(3)})\right],
\end{aligned} \tag{2.143}$$

where the contrast and the "reflection coefficients" are given by

$$\begin{aligned}
\hat{\chi}^c_m &= \hat{\chi}^c(\boldsymbol{x}^{(0)}_m), & \hat{R}^{(1)}_m &= \frac{\hat{\rho}(\boldsymbol{x}^{(0)}_m + \Delta x\, \boldsymbol{i}_1) - \hat{\rho}(\boldsymbol{x}^{(0)}_m)}{\hat{\rho}(\boldsymbol{x}^{(0)}_m + \Delta x\, \boldsymbol{i}_1) + \hat{\rho}(\boldsymbol{x}^{(0)}_m)}, \\
\hat{R}^{(2)}_m &= \frac{\hat{\rho}(\boldsymbol{x}^{(0)}_m + \Delta x\, \boldsymbol{i}_2) - \hat{\rho}(\boldsymbol{x}^{(0)}_m)}{\hat{\rho}(\boldsymbol{x}^{(0)}_m + \Delta x\, \boldsymbol{i}_2) + \hat{\rho}(\boldsymbol{x}^{(0)}_m)}, & \hat{R}^{(3)}_m &= \frac{\hat{\rho}(\boldsymbol{x}^{(0)}_m + \Delta x\, \boldsymbol{i}_3) - \hat{\rho}(\boldsymbol{x}^{(0)}_m)}{\hat{\rho}(\boldsymbol{x}^{(0)}_m + \Delta x\, \boldsymbol{i}_3) + \hat{\rho}(\boldsymbol{x}^{(0)}_m)},
\end{aligned} \tag{2.144}$$

and where each point $\boldsymbol{x}^{(0)}_m$ denotes the midpoint of the subdomain $\mathbb{D}_m$. Note that we assume that $\hat{\rho} = \hat{\rho}_{sct}$ for points in the scatterer and $\hat{\rho} = \rho_0$ for points outside the scatterer. The operators $\mathcal{K}$ in the matrix on the right-hand side of Eq. (2.143) act on $\partial\hat{w}^{(0)}_n$, $\partial\hat{w}^{(1)}_n$, $\partial\hat{w}^{(2)}_n$, and $\partial\hat{w}^{(3)}_n$, respectively. Before defining their actual representations, we first define the operator expressions

$$\begin{aligned}
[\mathcal{K}\hat{w}^{(0)}](\boldsymbol{x}) &= \hat{\gamma}_0^2 \Delta V \sum_{n=1}^{N} \langle\hat{G}\rangle(\boldsymbol{x} - \boldsymbol{x}^{(0)}_n)\ \langle\hat{w}^{(0)}_n\rangle, \\
[\mathcal{K}\partial\hat{w}^{(1)}](\boldsymbol{x}) &= 2\Delta A \sum_{n=1}^{N} \langle\partial_1\hat{G}\rangle(\boldsymbol{x} - \boldsymbol{x}^{(1)}_n)\ \langle\partial\hat{w}^{(1)}_n\rangle, \\
[\mathcal{K}\partial\hat{w}^{(2)}](\boldsymbol{x}) &= 2\Delta A \sum_{n=1}^{N} \langle\partial_2\hat{G}\rangle(\boldsymbol{x} - \boldsymbol{x}^{(2)}_n)\ \langle\partial\hat{w}^{(2)}_n\rangle, \\
[\mathcal{K}\partial\hat{w}^{(3)}](\boldsymbol{x}) &= 2\Delta A \sum_{n=1}^{N} \langle\partial_3\hat{G}\rangle(\boldsymbol{x} - \boldsymbol{x}^{(3)}_n)\ \langle\partial\hat{w}^{(3)}_n\rangle,
\end{aligned} \tag{2.145}$$

and take their particular values at $\boldsymbol{x}^{(0)}_m$, $\boldsymbol{x}^{(1)}_m$, $\boldsymbol{x}^{(2)}_m$, and $\boldsymbol{x}^{(3)}_m$, respectively. These values are given by

$$\begin{aligned}
\mathcal{K}^{(0,0)}_m = [\mathcal{K}\hat{w}^{(0)}](\boldsymbol{x}^{(0)}_m) &= \hat{\gamma}_0^2 \Delta V \sum_{n=1}^{N} \langle\langle\hat{G}\rangle\rangle(\boldsymbol{x}^{(0)}_m - \boldsymbol{x}^{(0)}_n)\ \langle\hat{w}^{(0)}_n\rangle, \\
\mathcal{K}^{(1,1)}_m = [\mathcal{K}\partial\hat{w}^{(1)}](\boldsymbol{x}^{(1)}_m) &= 2\ \Delta A \sum_{n=1}^{N} \langle\langle\partial_1\hat{G}\rangle\rangle(\boldsymbol{x}^{(0)}_m - \boldsymbol{x}^{(0)}_n)\ \langle\partial\hat{w}^{(1)}_n\rangle, \\
\mathcal{K}^{(2,2)}_m = [\mathcal{K}\partial\hat{w}^{(2)}](\boldsymbol{x}^{(2)}_m) &= 2\ \Delta A \sum_{n=1}^{N} \langle\langle\partial_2\hat{G}\rangle\rangle(\boldsymbol{x}^{(0)}_m - \boldsymbol{x}^{(0)}_n)\ \langle\partial\hat{w}^{(2)}_n\rangle, \\
\mathcal{K}^{(3,3)}_m = [\mathcal{K}\partial\hat{w}^{(3)}](\boldsymbol{x}^{(3)}_m) &= 2\ \Delta A \sum_{n=1}^{N} \langle\langle\partial_3\hat{G}\rangle\rangle(\boldsymbol{x}^{(0)}_m - \boldsymbol{x}^{(0)}_n)\ \langle\partial\hat{w}^{(3)}_n\rangle.
\end{aligned} \tag{2.146}$$

Note that in the expression of the convolutional operators $\mathcal{K}^{(1,1)}_m$, $\mathcal{K}^{(2,2)}_m$, and $\mathcal{K}^{(3,3)}_m$, we have used the relation $\boldsymbol{x}^{(k)}_m - \boldsymbol{x}^{(k)}_n = \boldsymbol{x}^{(0)}_m - \boldsymbol{x}^{(0)}_n$, for $k = 1,\ 2,\ 3$.

After computation of these convolutional operators, the other operators are computed by interpolation and extrapolation of the convolutional results. They are symbolically given by

$$\begin{aligned}
\mathcal{K}^{(1,0)}_m &= \mathcal{K}^{(0,0)}_m(\boldsymbol{x}^{(1)}_m), & \mathcal{K}^{(2,0)}_m &= \mathcal{K}^{(0,0)}_m(\boldsymbol{x}^{(2)}_m), & \mathcal{K}^{(3,0)}_m &= \mathcal{K}^{(0,0)}_m(\boldsymbol{x}^{(3)}_m), \\
\mathcal{K}^{(0,1)}_m &= \mathcal{K}^{(1,1)}_m(\boldsymbol{x}^{(0)}_m), & \mathcal{K}^{(2,1)}_m &= \mathcal{K}^{(1,1)}_m(\boldsymbol{x}^{(2)}_m), & \mathcal{K}^{(3,1)}_m &= \mathcal{K}^{(1,1)}_m(\boldsymbol{x}^{(3)}_m), \\
\mathcal{K}^{(0,2)}_m &= \mathcal{K}^{(2,2)}_m(\boldsymbol{x}^{(0)}_m), & \mathcal{K}^{(1,2)}_m &= \mathcal{K}^{(2,2)}_m(\boldsymbol{x}^{(1)}_m), & \mathcal{K}^{(3,2)}_m &= \mathcal{K}^{(2,2)}_m(\boldsymbol{x}^{(3)}_m), \\
\mathcal{K}^{(0,3)}_m &= \mathcal{K}^{(3,3)}_m(\boldsymbol{x}^{(0)}_m), & \mathcal{K}^{(1,3)}_m &= \mathcal{K}^{(3,3)}_m(\boldsymbol{x}^{(1)}_m), & \mathcal{K}^{(2,3)}_m &= \mathcal{K}^{(3,3)}_m(\boldsymbol{x}^{(2)}_m).
\end{aligned} \tag{2.147}$$

Instead of using the expressions of Eq. (2.145), we compute the numerical values of Eq. (2.147) approximately by interpolation and extrapolation of the results of Eq. (2.146). These interpolation and extra procedures relate the operator values at the different grids to each other.

We carry out the following computing procedure:

(1) Compute $\mathcal{K}^{(0,0)}_m$ using FFTs.
Subsequently, compute $\mathcal{K}^{(1,0)}_m$, $\mathcal{K}^{(2,0)}_m$, and $\mathcal{K}^{(3,0)}_m$ by extrapolating $\mathcal{K}^{(0,0)}_m$ linearly in the x_1-direction, x_2-direction, and x_3-direction, respectively.

Matlab Listing I.60 The function `interpolate` to interpolate the acoustic wave field over four adjacent points.

```
function [v] = interpolate(v,grid_out,grid_in,input)
global nDIM;

if nDIM == 2;             N1 = input.N1;  N2 = input.N2;

  if grid_in     == 1;  v(2:N1-1,:) = (v(1:N1-2,:) + v(2:N1-1,:))/2;
  elseif grid_in == 2;  v(:,2:N2-1) = (v(:,1:N2-2) + v(:,2:N2-1))/2;
  end

  if grid_out     == 1; v(1:N1-1,:) = (v(1:N1-1,:) + v(2:N1,:))/2;
  elseif grid_out == 2; v(:,1:N2-1) = (v(:,1:N2-1) + v(:,2:N2))/2;
  end

elseif nDIM == 3;         N1 = input.N1;  N2 = input.N2;  N3 = input.N3;

  if grid_in     == 1;  v(2:N1-1,:,:) = (v(1:N1-2,:,:) + v(2:N1-1,:,:))/2;
  elseif grid_in == 2;  v(:,2:N2-1,:) = (v(:,1:N2-2,:) + v(:,2:N2-1,:))/2;
  elseif grid_in == 3;  v(:,:,2:N3-1) = (v(:,:,1:N3-2) + v(:,:,2:N3-1))/2;
  end

  if grid_out     == 1; v(1:N1-1,:,:) = (v(1:N1-1,:,:) + v(2:N1,:,:))/2;
  elseif grid_out == 2; v(:,1:N2-1,:) = (v(:,1:N2-1,:) + v(:,2:N2,:))/2;
  elseif grid_out == 3; v(:,:,1:N3-1) = (v(:,:,1:N3-1) + v(:,:,2:N3))/2;
  end
end
```

(2) Compute $\mathcal{K}_m^{(1,1)}$ using FFTs.
Subsequently, compute $\mathcal{K}_m^{(0,1)}$ by extrapolating $\mathcal{K}_m^{(1,1)}$ linearly in the negative x_1-direction, and compute $\mathcal{K}_m^{(2,1)}$ and $\mathcal{K}_m^{(3,1)}$ by interpolating the values of $\mathcal{K}_m^{(1,1)}$ at four adjacent points of $\boldsymbol{x}_m^{(1)}$, in the x_2-direction and x_3-direction, respectively.
(3) Compute $\mathcal{K}_m^{(2,2)}$ using FFTs.
Subsequently, compute $\mathcal{K}_m^{(0,2)}$ by extrapolating $\mathcal{K}_m^{(2,2)}$ linearly in the negative x_2-direction, and compute $\mathcal{K}_m^{(1,2)}$ and $\mathcal{K}_m^{(3,2)}$ from $\mathcal{K}_m^{(2,2)}$ by interpolating the values at four adjacent points in the x_1-direction and x_3-direction, respectively.
(4) Compute $\mathcal{K}_m^{(3,3)}$ using FFTs.
Subsequently, compute $\mathcal{K}_m^{(0,3)}$ by extrapolating $\mathcal{K}_m^{(3,3)}$ linearly in the negative x_3-direction, and compute $\mathcal{K}_m^{(1,3)}$ and $\mathcal{K}_m^{(2,3)}$ from $\mathcal{K}_m^{(3,3)}$ by interpolating the values at four adjacent points in the x_1-direction and x_2-direction, respectively.

The *interpolation* procedure in a certain coordinate system is the simple summation of the pertaining values at four adjacent points (midpoint rule), see Matlab function `interpolate` of Matlab Listing I.60, for example,

$$\begin{aligned}\mathcal{K}_m^{(2,1)}(\boldsymbol{x}_m^{(2)}) = \tfrac{1}{4}\Big[&\mathcal{K}_m^{(1,1)}\left(\boldsymbol{x}_m^{(1)} + \tfrac{1}{2}\Delta x\, \boldsymbol{i}_1 + \tfrac{1}{2}\Delta x\, \boldsymbol{i}_2\right) + \mathcal{K}_m^{(1,1)}\left(\boldsymbol{x}_m^{(1)} + \tfrac{1}{2}\Delta x\, \boldsymbol{i}_1 - \tfrac{1}{2}\Delta x\, \boldsymbol{i}_2\right)\\ &+\mathcal{K}_m^{(1,1)}\left(\boldsymbol{x}_m^{(1)} - \tfrac{1}{2}\Delta x\, \boldsymbol{i}_1 + \tfrac{1}{2}\Delta x\, \boldsymbol{i}_2\right) + \mathcal{K}_m^{(1,1)}\left(\boldsymbol{x}_m^{(1)} - \tfrac{1}{2}\Delta x\, \boldsymbol{i}_1 - \tfrac{1}{2}\Delta x\, \boldsymbol{i}_2\right)\Big] .\end{aligned} \tag{2.148}$$

The *extrapolation* procedure from either the $\boldsymbol{x}^{(0)}$-coordinates to other coordinates or from other coordinates to the $\boldsymbol{x}^{(0)}$-coordinates is based on a linear Taylor expansion in the pertaining direction, in which its spatial derivative is numerically replaced by a two-point central difference, for

Matlab Listing I.61 The function `extrapolate` to extrapolate the acoustic wave field either from or to the $x^{(0)}$ coordinates.

```
function [u] = extrapolate(v,grid_out,grid_in,input)
global nDIM;
u = zeros(size(v));

if nDIM == 1;          N1 = input.N1;
  if grid_out == 1  %-----------------------------forward extrapolation
    u(2:N1-2) = v(2:N1-2) + (v(3:N1-1)-v(1:N1-3))/4;
  elseif grid_in ==1 %----------------------------backward extrapolation
    u(2:N1-2) = v(2:N1-2) - (v(3:N1-1)-v(1:N1-3))/4;
  end

elseif nDIM == 2;      N1 = input.N1;  N2 = input.N2;
  if grid_out == 1  %-----------------------------forward extrapolation
    u(2:N1-2,:) = v(2:N1-2,:) + (v(3:N1-1,:)-v(1:N1-3,:))/4;
  elseif grid_out == 2
    u(:,2:N2-2) = v(:,2:N2-2) + (v(:,3:N2-1)-v(:,1:N2-3))/4;
  end
  if grid_in == 1 %-------------------------------backward extrapolation
    u(2:N1-2,:) = v(2:N1-2,:) - (v(3:N1-1,:)-v(1:N1-3,:))/4;
  elseif grid_in == 2
    u(:,2:N2-2) = v(:,2:N2-2) - (v(:,3:N2-1)-v(:,1:N2-3))/4;
  end

elseif nDIM == 3;   N1 = input.N1;  N2 = input.N2;  N3 = input.N3;
  if grid_out == 1 %------------------------------forward extrapolation
    u(2:N1-1,:,:) = v(2:N1-1,:,:) + (v(3:N1,:,:)-v(1:N1-2,:,:))/4;
  elseif grid_out == 2
    u(:,2:N2-1,:) = v(:,2:N2-1,:) + (v(:,3:N2,:)-v(:,1:N2-2,:))/4;
  elseif grid_out == 3
    u(:,:,2:N3-1) = v(:,:,2:N3-1) + (v(:,:,3:N3)-v(:,:,1:N3-2))/4;
  end
  if grid_in == 1  %------------------------------backward extrapolation
   u(3:N1-2,:,:) = v(3:N1-2,:,:) - (v(4:N1-1,:,:)-v(2:N1-3,:,:))/4;
  elseif grid_in == 2
   u(:,3:N2-2,:) = v(:,3:N2-2,:) - (v(:,4:N2-1,:)-v(:,2:N2-3,:))/4;
  elseif grid_in == 3
   u(:,:,3:N3-2) = v(:,:,3:N3-2) - (v(:,:,4:N3-1)-v(:,:,2:N3-3))/4;
  end
end
```

example

$$\mathcal{K}_m^{(1,0)}\left(\boldsymbol{x}_m^{(0)} + \tfrac{1}{2}\Delta x\, \boldsymbol{i}_1\right) = \mathcal{K}_m^{(0,0)}(\boldsymbol{x}_m^{(0)}) + \tfrac{1}{4}[\mathcal{K}_m^{(0,0)}(\boldsymbol{x}_m^{(0)} + \Delta x\, \boldsymbol{i}_1) - \mathcal{K}_m^{(0,0)}(\boldsymbol{x}_m^{(0)} - \Delta x\, \boldsymbol{i}_1)] \tag{2.149}$$

and

$$\mathcal{K}_m^{(0,1)}\left(\boldsymbol{x}_m^{(1)} + \tfrac{1}{2}\Delta x\, \boldsymbol{i}_1\right) = \mathcal{K}_m^{(1,1)}(\boldsymbol{x}_m^{(1)}) - \tfrac{1}{4}[\mathcal{K}_m^{(1,1)}(\boldsymbol{x}_m^{(1)} + \Delta x\, \boldsymbol{i}_1) - \mathcal{K}_m^{(1,1)}(\boldsymbol{x}_m^{(1)} - \Delta x\, \boldsymbol{i}_1)], \tag{2.150}$$

see Matlab function `extrapolate` of Matlab Listing I.61.

The results in the extra layer added to the scattering domain are extrapolated values and are immaterial because both the wave-speed contrast and the reflection coefficients vanish in this layer. However, we need this layer to define the proper array numbering of the input and output arrays of the Matlab script.

In summary, we note that Eq. (2.143) may symbolically be written as

$$\begin{aligned}
\hat{\chi}^c_m \langle p^{inc}\rangle(\boldsymbol{x}^{(0)}_m) &= \hat{w}^{(0)}_m - \hat{\chi}^c_m[\mathcal{K}^{(0,0)}_m + \mathcal{K}^{(0,1)}_m + \mathcal{K}^{(0,2)}_m + \mathcal{K}^{(0,3)}_m]\,,\\
\hat{R}^{(1)}_m \langle \hat{p}^{inc}(\boldsymbol{x}^{(1)}_m)\rangle &= \partial\hat{w}^{(1)}_m - \hat{R}^{(1)}_m\,[\mathcal{K}^{(1,0)}_m + \mathcal{K}^{(1,1)}_m + \mathcal{K}^{(1,2)}_m + \mathcal{K}^{(1,3)}_m]\,,\\
\hat{R}^{(2)}_m \langle \hat{p}^{inc}(\boldsymbol{x}^{(2)}_m)\rangle &= \partial\hat{w}^{(2)}_m - \hat{R}^{(2)}_m\,[\mathcal{K}^{(2,0)}_m + \mathcal{K}^{(2,1)}_m + \mathcal{K}^{(2,2)}_m + \mathcal{K}^{(2,3)}_m]\,,\\
\hat{R}^{(3)}_m \langle \hat{p}^{inc}(\boldsymbol{x}^{(3)}_m)\rangle &= \partial\hat{w}^{(3)}_m - \hat{R}^{(3)}_m\,[\mathcal{K}^{(3,0)}_m + \mathcal{K}^{(3,1)}_m + \mathcal{K}^{(3,2)}_m + \mathcal{K}^{(3,3)}_m]\,,
\end{aligned} \tag{2.151}$$

where the $\mathcal{K}$-operators act on the volume contrast sources $\hat{w}^{(0)}_m$ and the interface contrast sources $\partial\hat{w}^{(1)}_m$, $\partial\hat{w}^{(2)}_m$, and $\partial\hat{w}^{(3)}_m$. In this symbolic notation, it is clear that the diagonal of this system of linear equations is represented by the operators $\mathcal{K}^{(0,0)}_m$, $\mathcal{K}^{(1,1)}_m$, $\mathcal{K}^{(2,2)}_m$, and $\mathcal{K}^{(3,3)}_m$. The nondiagonal operators represent the interaction of the fields from the four different types of contrast sources.

2.6.2 Discrete Representations in 2D

In 2D, we do not deal with the shifted grid based on the $\boldsymbol{x}^{(3)}_n$ coordinates and similar as in Chapter 1 the incident wave field and the Green functions are different. The 2D volume is denoted as $\Delta V = \Delta x \times \Delta x$, while the 2D interface area is the side length Δx. In summary, we have the discretized equations

$$\begin{aligned}
\hat{\chi}^c_m \langle p^{inc}\rangle(\boldsymbol{x}^{(0)}_{T,m}) &= \hat{w}^{(0)}_m - \hat{\chi}^c_m[\mathcal{K}^{(0,0)}_m + \mathcal{K}^{(0,1)}_m + \mathcal{K}^{(0,2)}_m]\,,\\
\hat{R}^{(1)}_m\, p^{inc}(\boldsymbol{x}^{(1)}_{T,m}) &= \partial\hat{w}^{(1)}_m - \hat{R}^{(1)}_m\,[\mathcal{K}^{(1,0)}_m + \mathcal{K}^{(1,1)}_m + \mathcal{K}^{(1,2)}_m]\,,\\
\hat{R}^{(2)}_m\, p^{inc}(\boldsymbol{x}^{(2)}_{T,m}) &= \partial\hat{w}^{(2)}_m - \hat{R}^{(2)}_m\,[\mathcal{K}^{(2,0)}_m + \mathcal{K}^{(2,1)}_m + \mathcal{K}^{(2,2)}_m]\,,
\end{aligned} \tag{2.152}$$

where

$$\hat{R}^{(1)}_m = \frac{\hat{\rho}(\boldsymbol{x}^{(0)}_{T,m} + \Delta x\,\boldsymbol{i}_1) - \hat{\rho}(\boldsymbol{x}^{(0)}_{T,m})}{\hat{\rho}(\boldsymbol{x}^{(0)}_{T,m} + \Delta x\,\boldsymbol{i}_1) + \hat{\rho}(\boldsymbol{x}^{(0)}_{T,m})}, \quad \hat{R}^{(2)}_m = \frac{\hat{\rho}(\boldsymbol{x}^{(0)}_{T,m} + \Delta x\,\boldsymbol{i}_2) - \hat{\rho}(\boldsymbol{x}^{(0)}_{T,m})}{\hat{\rho}(\boldsymbol{x}^{(0)}_{T,m} + \Delta x\,\boldsymbol{i}_2) + \hat{\rho}(\boldsymbol{x}^{(0)}_{T,m})}. \tag{2.153}$$

The diagonal operators are given by

$$\begin{aligned}
\mathcal{K}^{(0,0)}_m(\boldsymbol{x}^{(0)}_{T,m}) &= \hat{\gamma}_0^2\Delta V\sum_{n=1}^{N}\langle\langle\hat{G}\rangle\rangle(\boldsymbol{x}^{(0)}_{T,m} - \boldsymbol{x}^{(0)}_{T,n})\,\hat{w}^{(0)}_n,\\
\mathcal{K}^{(1,1)}_m(\boldsymbol{x}^{(1)}_{T,m}) &= 2\Delta x\sum_{n=1}^{N}\langle\langle\partial_1\hat{G}\rangle\rangle(\boldsymbol{x}^{(0)}_{T,m} - \boldsymbol{x}^{(0)}_{T,n})\,\partial\hat{w}^{(1)}_n,\\
\mathcal{K}^{(2,2)}_m(\boldsymbol{x}^{(2)}_{T,m}) &= 2\Delta x\sum_{n=1}^{N}\langle\langle\partial_2\hat{G}\rangle\rangle(\boldsymbol{x}^{(0)}_{T,m} - \boldsymbol{x}^{(0)}_{T,n})\,\partial\hat{w}^{(2)}_n.
\end{aligned} \tag{2.154}$$

The off-diagonal operators are computed by interpolation and extrapolation of the resulting values of the diagonal operators as

$$\begin{aligned}
\mathcal{K}^{(1,0)}_m &= \mathcal{K}^{(0,0)}_m(\boldsymbol{x}^{(1)}_{T,m}), & \mathcal{K}^{(2,0)}_m &= \mathcal{K}^{(0,0)}_m(\boldsymbol{x}^{(2)}_{T,m}),\\
\mathcal{K}^{(0,1)}_m &= \mathcal{K}^{(1,1)}_m(\boldsymbol{x}^{(0)}_{T,m}), & \mathcal{K}^{(2,1)}_m &= \mathcal{K}^{(1,1)}_m(\boldsymbol{x}^{(2)}_{T,m}),\\
\mathcal{K}^{(0,2)}_m &= \mathcal{K}^{(2,2)}_m(\boldsymbol{x}^{(0)}_{T,m}), & \mathcal{K}^{(1,2)}_m &= \mathcal{K}^{(2,2)}_m(\boldsymbol{x}^{(1)}_{T,m}).
\end{aligned} \tag{2.155}$$

2.6.3 Discrete Representations in 1D

In 1D, we only deal with a grid based on the midpoint coordinates $x_{1,n}^{(0)}$ and a shifted grid based on the $x_{1,n}^{(1)}$ coordinates. The 1D volume is denoted as ΔV and is equal to Δx. In summary, we have the discretized equations

$$\begin{aligned} \hat{\chi}_m^c \langle p^{inc} \rangle (\boldsymbol{x}_m^{(0)}) &= \hat{w}_m^{(0)} - \hat{\chi}_m^c [\mathcal{K}_m^{(0,0)} + \mathcal{K}_m^{(0,1)}] , \\ \hat{R}_m^{(1)} \, p^{inc}(\boldsymbol{x}_m^{(1)}) &= \partial \hat{w}_m^{(1)} + \hat{R}_m^{(1)} \, [\mathcal{K}_m^{(1,0)} + \mathcal{K}_m^{(1,1)}] . \end{aligned} \tag{2.156}$$

The diagonal operators are given by

$$\begin{aligned} \mathcal{K}_m^{(0,0)}(x_{1,m}^{(0)}) &= \hat{\gamma}_0^2 \Delta x \sum_{n=1}^{N} \langle\langle \hat{G} \rangle\rangle (x_{1,m}^{(0)} - x_{1,n}^{(0)}) \, \hat{w}_n^{(0)} , \\ \mathcal{K}_m^{(1,1)}(x_{1,m}^{(1)}) &= 2 \sum_{n=1}^{N} \langle\langle \partial_1 \hat{G} \rangle\rangle (x_{1,m}^{(0)} - x_{1,n}^{(0)}) \, \partial \hat{w}_n^{(1)} . \end{aligned} \tag{2.157}$$

The off-diagonal operators are computed by interpolation and extrapolation of the resulting values of the diagonal operators as

$$\begin{aligned} \mathcal{K}_m^{(0,1)} &= \mathcal{K}_m^{(1,1)}(\boldsymbol{x}_{T,m}^{(0)}), \\ \mathcal{K}_m^{(1,0)} &= \mathcal{K}_m^{(0,0)}(\boldsymbol{x}_{T,m}^{(1)}). \end{aligned} \tag{2.158}$$

2.6.4 Conjugate Gradient Iterative Solution and Operators Involved

We now consider the iterative solution of the contrast source type of integral equation, based on the volume contrast sources and the interface contrast sources. In Section 2.6, we have defined all discrete operators and their pertaining Matlab codes. We now observe that the system of integral equations listed in Eq. (2.151) is equivalent to the following discrete set of equations:

$$\begin{aligned} \chi_m^c \langle p^{inc} \rangle (x_m^{(0)}) &= w_m^{(0)} - \chi_m^c \mathcal{K}_m^{(0)} \{ \hat{w}_n^{(0)}, \partial \hat{w}_n^{(k)} \}, \\ R_m^{(j)} \langle p^{inc} \rangle (x_m^{(j)}) &= \partial w_m^{(j)} - R_m^{(j)} \mathcal{K}_m^{(j)} \{ \hat{w}_n^{(0)}, \partial \hat{w}_n^{(k)} \}, \end{aligned} \tag{2.159}$$

where

$$\begin{aligned} \mathcal{K}_m^{(0)} \{ \hat{w}_m^{(0)}, \partial \hat{w}_n^{(k)} \} &= \mathcal{K}_m^{(0)} \{ \hat{w}_n^{(0)}, \partial \hat{w}_n^{(1)}, \partial \hat{w}_n^{(2)}, \partial \hat{w}_m^{(3)} \} \\ &= \mathcal{K}_m^{(0,0)} + \mathcal{K}_m^{(0,1)} + \mathcal{K}_m^{(0,2)} + \mathcal{K}_m^{(0,3)}, \\ \mathcal{K}_m^{(j)} \{ w_n^{(0)}, \partial w_n^{(k)} \} &= \mathcal{K}_m^{(j)} \{ \hat{w}_n^{(0)}, \partial \hat{w}_n^{(1)}, \partial \hat{w}_n^{(2)}, \partial \hat{w}_n^{(3)} \} \\ &= \mathcal{K}_m^{(j,0)} + \mathcal{K}_m^{(j,1)} + \mathcal{K}_m^{(j,2)} + \mathcal{K}_m^{(j,3)}, \end{aligned} \tag{2.160}$$

for $j, k = 1, 2, 3$ in 3D, $j, k = 1, 2$ in 2D and $j, k = 1$ in 1D, respectively. The operators $\mathcal{K}_m^{(0)}$ and $\mathcal{K}_m^{(j)}$ act on the unknown volume contrast sources $\hat{w}_n^{(0)}$ and the unknown interface-contrast sources $\partial \hat{w}_n^{(k)}$. The Matlab function `KOPwdw` to compute these operators is given in Matlab Listing I.62.

This system is solved iteratively using the conjugate gradient method. The conjugate gradient method for solving the contrast sources $w^{(0)}$ and $w^{(k)}$ is fundamentally the same as the one shown in Table 2.1. To discuss it in more detail, we write Eqs. (2.159) and (2.160) in the compact form

$$\boxed{\begin{aligned} w^{(0)} - \chi^c \, \mathcal{K}^{(0)} \{ w^{(0)}, \partial w^{(k)} \} &= \chi^c \, p^{inc(0)}, \\ \partial w^{(j)} - R^{(j)} \, \mathcal{K}^{(j)} \{ w^{(0)}, \partial w^{(k)} \} &= R^{(j)} p^{inc(j)}. \end{aligned}} \tag{2.161}$$

Matlab Listing I.62 The function `KOPwdw` to carry out the $\mathcal{K}$ operators.

```
function [Kw,Kdw] = KOPwdw(w,dw,input)
global nDIM;

FFTG = input.FFTG;   FFTdG = cell(1,nDIM);
for n=1: nDIM;   FFTdG{n} = input.FFTdG{n} * 2 / input.dx;   end

if nDIM == 1

    Kdiag = Kop(w,FFTG); % -------------------------------------- % K00
       Kw = Kdiag;  Kdw{1} = extrapolate(Kdiag,1,0,input);        % K10
    Kdiag = Kop(dw{1},FFTdG{1}); % ----------------------------------
                    Kdw{1} = Kdw{1} + Kdiag;                      % K11
       Kw = Kw + extrapolate(Kdiag,0,1,input);                    % K01

elseif nDIM == 2

    Kdiag = Kop(w,FFTG); % -------------------------------------- % K00
       Kw = Kdiag;  Kdw{1} = extrapolate(Kdiag,1,0,input);        % K10
                    Kdw{2} = extrapolate(Kdiag,2,0,input);        % K20
    Kdiag = Kop(dw{1},FFTdG{1}); % ----------------------------------
                    Kdw{1} = Kdw{1} + Kdiag;                      % K11
       Kw = Kw + extrapolate(Kdiag,0,1,input);                    % K01
                    Kdw{2} = Kdw{2} + interpolate(Kdiag,2,1,input); % K21
    Kdiag = Kop(dw{2},FFTdG{2}); % ----------------------------------
                    Kdw{2} = Kdw{2} + Kdiag;                      % K22
       Kw = Kw + extrapolate(Kdiag,0,2,input);                    % K02
                    Kdw{1} = Kdw{1} + interpolate(Kdiag,1,2,input); % K12

elseif nDIM == 3

    Kdiag = Kop(w,FFTG); % ---------------------------------------- K00
       Kw = Kdiag;  Kdw{1} = extrapolate(Kdiag,1,0,input);        % K10
                    Kdw{2} = extrapolate(Kdiag,2,0,input);        % K20
                    Kdw{3} = extrapolate(Kdiag,3,0,input);        % K30
    Kdiag = Kop(dw{1},FFTdG{1}); % ----------------------------------
                    Kdw{1} = Kdw{1} + Kdiag;                      % K11
       Kw = Kw + extrapolate(Kdiag,0,1,input);                    % K01
                    Kdw{2} = Kdw{2} + interpolate(Kdiag,2,1,input); % K21
                    Kdw{3} = Kdw{3} + interpolate(Kdiag,3,1,input); % K31
    Kdiag = Kop(dw{2},FFTdG{2}); % ----------------------------------
                    Kdw{2} = Kdw{2} + Kdiag;                      % K22
       Kw = Kw + extrapolate(Kdiag,0,2,input);                    % K02
                    Kdw{1} = Kdw{1} + interpolate(Kdiag,1,2,input); % K12
                    Kdw{3} = Kdw{3} + interpolate(Kdiag,3,2,input); % K32
    Kdiag = Kop(dw{3},FFTdG{3}); % ----------------------------------
                    Kdw{3} = Kdw{3} + Kdiag;                      % K33
       Kw = Kw + extrapolate(Kdiag,0,3,input);                    % K03
                    Kdw{1} = Kdw{1} + interpolate(Kdiag,1,3,input); % K13
                    Kdw{2} = Kdw{2} + interpolate(Kdiag,2,3,input); % K23

end
```

If the contrast sources do not solve Eq. (2.161), the pertaining residual errors are defined as

$$\boxed{\begin{aligned} r^{(0)} &= \chi^c p^{inc(0)} - [w^{(0)} - \chi^c\ \mathcal{K}^{(0)}\{w^{(0)}, \partial w^{(k)}\}], \\ \partial r^{(j)} &= R^{(j)} p^{inc(j)} - [\partial w^{(j)} - R^{(j)}\ \mathcal{K}^{(j)}\{w^{(0)}, \partial w^{(k)}\}], \end{aligned}} \tag{2.162}$$

and the normalized cost functional

$$\boxed{F(w^{(0)}, \partial w^{(j)}) = \frac{\|r^{(0)}\|^2 + \sum_k \|\partial r^{(k)}\|^2}{\|\chi^c p^{inc(0)}\|^2 + \sum_k \|R^{(k)} p^{inc(k)}\|^2},} \tag{2.163}$$

with the property that $F = 1$ for $w^{(0)} = \partial w^{(1)} = \partial w^{(2)} = \partial w^{(3)} = 0$ and $F = 0$ for the exact solution.

The gradient directions follow from the Fréchet derivative of this cost functional F and are directly related to the adjoint operating on the residual errors in the equations at the previous iteration $n - 1$. They are obtained as

$$\begin{aligned} g^{(0,n)} &= r^{(0,n-1)} &- \mathcal{K}^{\star(0)}\{\overline{\chi}^c r^{(0,n-1)}, \overline{R}^{(k)} \partial r^{(k,n-1)}\}, \\ \partial g^{(j,n)} &= \partial r^{(j,n-1)} &- \mathcal{K}^{\star(j)}\{\overline{\chi}^c r^{(0,n-1)}, \overline{R}^{(k)} \partial r^{(k,n-1)}\}, \end{aligned} \tag{2.164}$$

where the expressions of $\mathcal{K}^{\star(0)}$ and $\mathcal{K}^{\star(j)}$ follow from the inner product definition. Since it is a rather tedious procedure, we present the results directly as Matlab function `AdjKOPwdw` of Matlab Listing I.63. The Hestenes–Stiefel conjugate gradient directions after the first iteration become

$$\begin{aligned} v^{(0,n)} &= g^{(0,n)} + (A^{(n)}/A^{(n-1)})\ v^{(0,n-1)}, \\ \partial v^{(j,n)} &= \partial g^{(j,n)} + (A^{(n)}/A^{(n-1)})\ \partial v^{(j,n-1)}, \end{aligned} \tag{2.165}$$

where

$$A^{(n)} = \|g^{(0,n)}\|^2 + \sum_j \|\partial g^{(j,n)}\|^2. \tag{2.166}$$

In each iteration, the updates for the contrast sources $w^{(0)}$ and $\partial w^{(j)}$ are obtained as

$$\begin{aligned} w^{(0,n)} &= w^{(0,n-1)} + (A^{(n)}/B^{(n)})\ v^{(0,n)}, \\ \partial w^{(j,n)} &= \partial w^{(j,n-1)} + (A_n/B_n)\ \partial v^{(j,n)}, \end{aligned} \tag{2.167}$$

with

$$B^{(n)} = \|v^{(0,n)} - \chi^c \mathcal{K}^{(0,n)}\{v^{(0,n)}, v^{v(k,n)}\}\|^2 + \sum_j \|v^{(j,n)} - R^{(j)} \mathcal{K}^{(j,n)}\{v^{(0,n)}, v^{(k,n)}\}\|^2. \tag{2.168}$$

With these results we have discussed all the ingredients to employ the conjugate gradient method for solving the volume and interface contrast sources $w^{(0)}$ and $\partial w^{(j)}$, see Table 2.3. To complete this numerical section, we need the weak form of the incident pressure wave field at the points $\boldsymbol{x}^{(0)}$ and the shifted points $\boldsymbol{x}^{(j)}$. The weak form of the incident pressure field at some point $\boldsymbol{x}$ is given in Eq. (2.78).

2.6.4.1 Scattered Acoustic Wave Field

After a 1D numerical solution for the volume-contrast source $\hat{w}^0_m$ and the interface-contrast sources $\hat{w}^{(1)}_m$, the numerical value of the scattered wave field at some receiver point $x_1 = x_1^R$ is computed as

$$\langle \hat{p}^{sct} \rangle (x_1^R) = \sum_{m=1}^{N} [\hat{\gamma}_0{}^2 \Delta x \langle \hat{G} \rangle (x_1^R - x_{1,m}^{(0)})\ \hat{w}_m^{(0)} + 2\partial_1^R \hat{G}(x_1^R - x_{1,m}^{(1)})\ \partial \hat{w}_m^{(1)}], \tag{2.169}$$

when $x_1^R \in \mathbb{D}'_{sct}$.

Matlab Listing I.63 The function `AdjKOPwdw` to carry out the $\mathcal{K}^\star$ operator.

```
function [Kf,Kdf] = AdjKOPwdw(f,df,input)
global nDIM;

 FFTG = input.FFTG;  FFTdG = cell(1,nDIM);
 for n=1: nDIM;   FFTdG{n} = input.FFTdG{n} * 2 / input.dx;  end

if nDIM == 1

    arg = f + extrapolate(df{1},0,1,input);        % from x^(1) back to x^(0)
     Kf = AdjKop(arg,FFTG);        % -------------- convolution in x^(0) space

    arg = df{1} + extrapolate(f,1,0,input);        % from x^(0) towards x^(1)
 Kdf{1}= AdjKop(arg,FFTdG{1});  % -------------- convolution in x^(1) space

elseif nDIM == 2

   arg = f    + extrapolate(df{1},0,1,input);      % from x^(1) back to x^(0)
   arg = arg + extrapolate(df{2},0,2,input);       % from x^(2) back to x^(0)
    Kf = AdjKop(arg,FFTG);          % ------------- convolution in x^(0) space

   arg = df{1} + extrapolate(f,1,0,input);        % from x^(0) towards x^(1)
   arg = arg    + interpolate(df{2},1,2,input); % from x^(2) towards x^(1)
 Kdf{1}= AdjKop(arg,FFTdG{1});    % ------------- convolution in x^(1) space

   arg = df{2} + extrapolate(f,2,0,input);        % from x^(0) towards x^(2)
   arg = arg    + interpolate(df{1},2,1,input); % from x^(1) towards x^(2)
 Kdf{2}= AdjKop(arg,FFTdG{2});    % ------------- convolution in x^(2) space

elseif nDIM == 3

   arg = f    + extrapolate(df{1},0,1,input);      % from x^(1) back to x^(0)
   arg = arg + extrapolate(df{2},0,2,input);       % from x^(2) back to x^(0)
   arg = arg + extrapolate(df{3},0,3,input);       % from x^(3) back to x^(0)
    Kf = AdjKop(arg,FFTG);          % ------------- convolution in x^(0) space

   arg = df{1} + extrapolate(f,1,0,input);         % from x^(0) towards x^(1)
   arg = arg    + interpolate(df{2},1,2,input);   % from x^(2) towards x^(1)
   arg = arg    + interpolate(df{3},1,3,input);   % from x^(3) towards x^(1)
 Kdf{1}= AdjKop(arg,FFTdG{1});    % ------------- convolution in x^(1) space

   arg = df{2} + extrapolate(f,2,0,input);         % from x^(0) towards x^(2)
   arg = arg    + interpolate(df{1},2,1,input);   % from x^(1) towards x^(2)
   arg = arg    + interpolate(df{3},2,3,input);   % from x^(3) towards x^(2)
 Kdf{2}= AdjKop(arg,FFTdG{2});    % ------------- convolution in x^(2) space

   arg = df{3} + extrapolate(f,3,0,input);         % from x^(0) towards x^(3)
   arg = arg    + interpolate(df{1},3,1,input);   % from x^(1) towards x^(3)
   arg = arg    + interpolate(df{2},3,2,input);   % from x^(2) towards x^(3)
 Kdf{3}= AdjKop(arg,FFTdG{3});    % ------------- convolution in x^(3) space

end
```

Table 2.3 The CGFFT method for solving the contrast sources $w^{(0)}$ and $\partial w^{(j)}$

initial estimate	$w^{(0,n=0)}$	$=$	0
	$\partial w^{(j,n=0)}$	$=$	0
	$r^{(0,n=0)}$	$=$	$\chi^c p^{inc(0)}$
	$\partial r^{(j,n=0)}$	$=$	$R^{(j)} p^{inc(j)}$
for $n = 1, 2, \ldots$	$g^{(0,n)}$	$=$	$r^{(0,n-1)} - \mathcal{K}^{\star(0)}\{\overline{\chi}^c r^{(0,n-1)}, \overline{R}^{(k)} \partial r^{(k,n-1)}\}$
	$\partial g^{(j,n)}$	$=$	$\partial r^{(j,n-1)} - \mathcal{K}^{\star(j)}\{\overline{\chi}^c r^{(0,n-1)}, \overline{R}^{(k)} \partial r^{(k,n-1)}\}$
	$A^{(n)}$	$=$	$\Vert g^{(0,n)} \Vert^2 + \sum_j \Vert g^{(j,n)} \Vert^2$
if $n = 1$	$v^{(0,n)}$	$=$	$g^{(0,n)}$
	$\partial v^{(j,n)}$	$=$	$\partial g^{(j,n)}$
else	$v^{(0,n)}$	$=$	$g^{(0,n)} + \frac{A^{(n)}}{A^{(n-1)}} v^{(0,n-1)}$
	$\partial v^{(j,n)}$	$=$	$\partial g^{(j,n)} + \frac{A^{(n)}}{A^{(n-1)}} \partial v^{(j,n-1)}$ end
	$B^{(n)}$	$=$	$\Vert v^{(0,n)} - \chi^c \mathcal{K}^{(0)}\{v^{(0,n)}, \partial v^{(k,n)}\} \Vert^2$
			$+\sum_j \Vert \partial v^{(j,n)} - R^{(j)} \mathcal{K}^{(j)}\{v^{(0,n)}, \partial v^{(k,n)}\} \Vert^2$
	$w^{(0,n)}$	$=$	$w^{(0,n-1)} + \frac{A^{(n)}}{B^{(n)}} v^{(0,n)}$
	$\partial w^{(j,n)}$	$=$	$\partial w^{(j,n-1)} + \frac{A^{(n)}}{B^{(n)}} \partial v^{(j,n)}$
	$r^{(0,n)}$	$=$	$r^{(0,n-1)} - \frac{A^{(n)}}{B^{(n)}} [v^{(n)} - \chi^c \mathcal{K}^{(0)}\{v^{(0,n)}, \partial v^{(k,n)}\}]$
	$\partial r^{(j,n)}$	$=$	$\partial r^{(j,n-1)} - \frac{A^{(n)}}{B^{(n)}} [\partial v^{(j,n)} - R^{(j)} \mathcal{K}^{(j)}\{v^{(0,n)}, \partial v^{(k,n)}\}]$
	if $\left[\frac{\Vert r^{(0,n)} \Vert^2 + \sum_k \Vert \partial r^{(k,n)} \Vert^2}{\Vert \chi^c p^{inc} \Vert^2 + \sum_k \Vert R^{(k)} p^{inc(k)} \Vert^2} \right]^{\frac{1}{2}}$		$< \varepsilon$ or $n > n_{\max}$ stop end
end			

After a 2D numerical solution for the volume-contrast source $\partial \hat{w}_m$ and the two interface-contrast sources $\partial \hat{w}_m^{(k)}$, the numerical value of the scattered wave field at some receiver point $\boldsymbol{x}_T = \boldsymbol{x}_T^R$ is computed as

$$\langle \hat{p}^{sct} \rangle(\boldsymbol{x}_T^R) = \sum_{m=1}^{N} \left[\hat{\gamma}_0^{\,2} \Delta V \langle \hat{G} \rangle(\boldsymbol{x}_T^R - \boldsymbol{x}_{T,m}^{(0)})\, \hat{w}_m^{(n)} + 2\Delta x \sum_{k=1}^{2} \partial_k^R \hat{G}(\boldsymbol{x}_T^R - \boldsymbol{x}_{T,m}^{(k)})\, \partial \hat{w}_m^{(k)} \right], \tag{2.170}$$

when $\boldsymbol{x}_T^R \in \mathbb{D}'_{sct}$.

After a 3D numerical solution for the volume-contrast source $\hat{w}$ and the three interface-contrast sources $\partial \hat{w}_m^{(k)}$, the numerical value of the scattered wave field at some receiver point $\boldsymbol{x} = \boldsymbol{x}^R$ is computed as

$$\langle \hat{p}^{sct} \rangle(\boldsymbol{x}^R) = \sum_{m=1}^{N} \left[\hat{\gamma}_0^{\,2} \Delta V \langle \hat{G} \rangle(\boldsymbol{x}^R - \boldsymbol{x}_m^{(0)})\, \hat{w}_m^{(0)} + 2\Delta A \sum_{k=1}^{3} \partial_k^R \hat{G}(\boldsymbol{x}^R - \boldsymbol{x}_m^{(k)})\, \partial \hat{w}_m^{(k)} \right]. \tag{2.171}$$

The weak forms of the spatial derivative of the Green function are given in Eqs. (2.84), (2.85), and (2.86), for the 3D case, the 2D case, and the 1D case, respectively.

2.7 Matlab Codes Integral Equations: Volume and Interface Contrast Sources

The main Matlab script is `ForwardCGFFTwdw` of Matlab Listing I.64, where we start with the Matlab functions `initAC` of Matlab Listing I.39, in which both the source/receiver locations and the grid parameters are initialized by invoking the Matlab functions `initSourceReceiver` and `initGrid` of Matlab Listings I.2 and I.3. Additionally, in `initAC` the Green functions are initialized by `initFFTGreenfun` of Matlab Listing I.40, while the distributions of the acoustic compressibility and the mass density are initialized by `initAcousticContrast` of Matlab Listing I.42. After this initialization, the main Matlab script again consists of four steps.

In step (1) of `ForwardCGFFTwdw`, the script `AcForwardCanonicalObjects` of Matlab Listing I.79 computes the acoustic pressure at a number of receivers, for the three canonical objects. These data serve as a benchmark for the numerical solution of our present integral equation. For the numerical solution of the present integral equations, we need both the wave-speed contrast χ^c at the midpoints of the square subdomains on the unshifted grid, and the reflection coefficients $R^{(k)}$ at the interfaces between adjacent subdomains. Therefore, the main script calls the function `initAcousticContrastIntf` of Matlab Listing I.65. First, the contrast function χ^c on the midpoints of the grid is determined. Second, for a given mass-density distribution, the "reflection coefficients" $R^{(1)}$, $R^{(2)}$, and $R^{(3)}$ on the shifted grids are determined.

In step (2) of `ForwardCGFFTwpZv`, we compute the weak form of the incident pressure, either in 1D, 2D, or 3D. The function `IncPressureWave` of Matlab Listing I.66 is used. The pressure at

Matlab Listing I.64 The `ForwardCGFFTwdw` script to solve the volume contrast sources w and the interface contrast sources ∂w from the system of integral equations.

```
clear all; clc; close all; clear workspace
input = initAC();

% (1) Comput analytically scattered field data
      AcForwardCanonicalObjects;

%     Input interface contrast
      [input] = initAcousticContrastIntf(input);

% (2) Compute incident field at different grids
      [p_inc,Pinc] = IncPressureWave(input);

% (3) Solve contrast source integral equations with CGFFT method ---------
tic;  [w,dw] = ITERCGwdw(p_inc,Pinc,input);
toc
   %  Plot contrast sources, compute and plot the pressure wave fields
      plotContrastSource(w,input);
      plotInterfaceSource(dw,input);

% (4) Compute synthetic data and plot fields and data --------------------
     data = DOPw(w,input);
     data = DOPdw(dw,data,input);
     displayData(data,input);
```

Matlab Listing I.65 The function `initAcousticContrastIntf` to initialize the acoustic wave-field parameters.

```
function [input] = initAcousticContrastIntf(input)
global nDIM;

if      nDIM == 1;  R = sqrt(input.X1.^2);                    [N1,~]  = size(R);
elseif nDIM == 2;  R = sqrt(input.X1.^2 + input.X2.^2); [N1,N2] = size(R);
elseif nDIM == 3;  R = sqrt(input.X1.^2 + input.X2.^2 + input.X3.^2);
                    [N1,N2,N3] = size(R);
end % if

% (1) Compute wave speed contrast mass density contrast ---------------------
           a = input.a;
  c_contrast = 1 - input.c_0^2/input.c_sct^2;
  input.CHI  = c_contrast * (R < a);

% (2) Compute mass density and 'reflection factors' ----------------------
  rho = input.rho_sct * (R < a) + (R >= a) * input.rho_0;

  if nDIM == 1
          Rfl{1} = zeros(N1,1);
  Rfl{1}(1:N1-1) = (rho(2:N1)-rho(1:N1-1))./(rho(2:N1)+rho(1:N1-1));
  elseif nDIM == 2
               Rfl{1} = zeros(N1,N2);
               Rfl{2} = zeros(N1,N2);
      Rfl{1}(1:N1-1,:) = (rho(2:N1,:) - rho(1:N1-1,:)) ...
                         ./(rho(2:N1,:) + rho(1:N1-1,:));
      Rfl{2}(:,1:N2-1) = (rho(:,2:N2) - rho(:,1:N2-1)) ...
                         ./(rho(:,2:N2) + rho(:,1:N2-1));
  elseif nDIM == 3
               Rfl{1} = zeros(N1,N2,N3);
               Rfl{2} = zeros(N1,N2,N3);
               Rfl{3} = zeros(N1,N2,N3);
  Rfl{1}(1:N1-1,:,:) = (rho(2:N1,:,:) - rho(1:N1-1,:,:)) ...
                        ./(rho(2:N1,:,:) + rho(1:N1-1,:,:));
  Rfl{2}(:,1:N2-1,:) = (rho(:,2:N2,:) - rho(:,1:N2-1,:)) ...
                        ./(rho(:,2:N2,:) + rho(:,1:N2-1,:));
  Rfl{3}(:,:,1:N3-1) = (rho(:,:,2:N3) - rho(:,:,1:N3-1)) ...
                        ./(rho(:,:,2:N3) + rho(:,:,1:N3-1));
 end % if

 input.Rfl = Rfl;                    clear input.CHI_rho  input.CHI_kap
```

the midpoints $\boldsymbol{x}^{(0)}$ are denoted as `p_inc` and the pressure at the shifted points $\boldsymbol{x}^{(k)}$ are denoted as `Pinc{k}`. For the latter, we use the Matlab structure array notation.

In step (3) of `ForwardCGFFTwdw`, we solve the integral equations for the contrast sources w and $\partial w^{(j)}$ by calling the function `ITERCGwdw` of Matlab Listing I.67. The code closely follows the CG method outlined in Table 2.3. To check the adjoint operator, we include the function `Check-AdjointwdW` of Matlab Listing I.68. This check is based on the numerical satisfaction of the inner products, using the initial residual errors $r^{(0)}$ and $\partial r^{(j)}$. The difference between both sides

Matlab Listing I.66 The function `IncPressurewave` to compute the incident pressure wave field.

```
function [p_inc,Pinc] = IncPressureWave(input)
global nDIM;
gam0 = input.gamma_0;
xS   = input.xS;     dx = input.dx;

if nDIM == 1                          % incident wave on one-dimensional grid

% Weak form
  delta  = 0.5 * dx;
  factor = sinh(gam0*delta) / (gam0*delta);
  x1     = input.X1;
  DIS    = abs(x1-xS(1));
  p_inc  = factor * 1/(2*gam0) * exp(-gam0.*DIS); % Grid (0) on midpoints
% Strong form
  DIS     = abs(x1+dx/2-xS(1));
  Pinc{1} = 1/(2*gam0) * exp(-gam0.*DIS);       % Grid (1) on shifted points

elseif nDIM == 2                      % incident wave on two-dimensional grid

% Weak form
  delta  = (pi)^(-1/2) * dx;        % radius circle with area of dx^2
  factor = 2 * besseli(1,gam0*delta) / (gam0*delta);
  X1     = input.X1;   X2 = input.X2;
  DIS    = sqrt( (X1-xS(1)).^2 + (X2-xS(2)).^2 );
  p_inc  = factor * 1/(2*pi).* besselk(0,gam0*DIS);% Grid (0) on midpoints
% Strong form
  DIS     = sqrt( (X1+dx/2-xS(1)).^2 + (X2-xS(2)).^2 );
  Pinc{1} = 1/(2*pi).* besselk(0,gam0*DIS);   % Grid (1) on shifted points
  DIS     = sqrt( (X1-xS(1)).^2 + (X2+dx/2-xS(2)).^2 );
  Pinc{2} = 1/(2*pi).* besselk(0,gam0*DIS);   % Grid (2) on shifted points

elseif nDIM == 3                    % incident wave on three-dimensional grid

% Weak form
  delta  = (4*pi/3)^(-1/3) * dx; % radius sphere with area of dx^3
  arg    = gam0*delta;
  factor = 3 * (cosh(arg) - sinh(arg)/arg) / arg^2;
  X1     = input.X1;   X2 = input.X2;    X3 =input.X3;
  DIS    = sqrt( (X1-xS(1)).^2 + (X2-xS(2)).^2 + (X3-xS(3)).^2 );
  p_inc  = factor * exp(-gam0*DIS) ./ (4*pi*DIS); % Grid (0) on midpoints
% Strong form
  DIS     = sqrt( (X1+dx/2-xS(1)).^2 + (X2-xS(2)).^2 + (X3-xS(3)).^2 );
  Pinc{1} = exp(-gam0*DIS) ./ (4*pi*DIS);      % Grid (1) on shifted points
  DIS     = sqrt( (X1-xS(1)).^2 + (X2+dx/2-xS(2)).^2 + (X3-xS(3)).^2 );
  Pinc{2} = exp(-gam0*DIS) ./ (4*pi*DIS);      % Grid (2) on shifted points
  DIS     = sqrt( (X1-xS(1)).^2 + (X2-xS(2)).^2 + (X3+dx/2-xS(3)).^2 );
  Pinc{3} = exp(-gam0*DIS) ./ (4*pi*DIS);      % Grid (3) on shifted points

end % if
```

Matlab Listing I.67 The function `ITERCGwdw` for an iterative solution of the contrast sources w and $\partial w^{(j)}$.

```
function [w,dw] = ITERCGwdw(p_inc, Pinc, input)
global nDIM;      CHI = input.CHI;    Rfl = input.Rfl;
dw = cell(1,nDIM); dr = cell(1,nDIM); dg = cell(1,nDIM); dv = cell(1,nDIM);

itmax = 1000;  Errcri = input.Errcri;  it = 0;  % initialization iteration
    w = zeros(size(p_inc));
    r = CHI.*p_inc;                          Norm0 = norm(r(:))^2;
for n = 1:nDIM
   dw{n} = zeros(size(Pinc{n}));
   dr{n} = Rfl{n}.*Pinc{n};                  Norm0 = Norm0 + norm(dr{n}(:))^2;
end
CheckAdjointwdw(r,dr,input);      % Check once

Error = 1;    fprintf('Error =          %g', Error);
 while (it < itmax) && ( Error > Errcri) && (Norm0 > eps)
 % determine conjugate gradient direction v
   dummy=cell(1,nDIM); for n = 1:nDIM; dummy{n} = conj(Rfl{n}).*dr{n}; end
   [Kf,Kdf] = AdjKOPwdw(conj(CHI).*r, dummy, input);
          g =  r - Kf;      % window for negligible CHI !!
          g = (abs(CHI) >= Errcri) .* g;        AN = norm(g(:))^2;
   for n = 1:nDIM
    dg{n}= dr{n} - Kdf{n}; % window for negligible Rfl !!
    dg{n}= (abs(Rfl{n}) >= Errcri) .* dg{n};    AN = AN + norm(dg{n}(:))^2;
   end
   if it == 0;        v  = g;    else  v   = g    + AN/AN_1 *  v;     end
   for n=1:nDIM
     if it == 0;  dv{n} = dg{n}; else dv{n}= dg{n} + AN/AN_1 * dv{n}; end
   end
% determine step length alpha
  [Kv,Kdv] = KOPwdw(v,dv,input);
   Kv = v - CHI .* Kv;                         BN = norm(Kv(:))^2;
   for n = 1:nDIM
      Kdv{n} = dv{n} - Rfl{n} .* Kdv{n};       BN = BN + norm(Kdv{n}(:))^2;
   end
   alpha = AN / BN;
 % update contrast sources, AN  and update residual errors
   w   = w + alpha *  v;
   r   = r - alpha * Kv;                     Norm = norm(r(:))^2;
   for n = 1:nDIM
      dw{n} = dw{n} + alpha * dv{n};
      dr{n} = dr{n} - alpha * Kdv{n};        Norm = Norm + norm(dr{n}(:))^2;
   end
   AN_1   = AN;
   Error = sqrt(Norm / Norm0); fprintf('\b\b\b\b\b\b\b\b%6f', Error);
   it = it+1;
 end % CG iterations

 fprintf('\b\b\b\b\b\b\b\b%6f\n', Error);
 disp(['Number of iterations is ' num2str(it)]);
 if it == itmax; disp(['itmax reached:  err/norm = ' num2str(Error)]); end;
```

Matlab Listing I.68 The function `CheckAdjointwdw` to check the adjoint operators.

```
function CheckAdjointwdw(r,dr,input)
global nDIM;
  CHI = input.CHI;   Rfl = input.Rfl;
dummy = cell(1,nDIM);

% Adjoint operator [I - K*conj(CHI)] via inner product ---------------------
for n = 1:nDIM
    dummy{n} = conj(Rfl{n}).*dr{n};
end
[Kf,dKf] = AdjKOPwdw(conj(CHI).*r,dummy,input);
Result1  = sum( r(:) .* conj(r(:)-Kf(:)) );
for n = 1:nDIM
   dummy{n} = dr{n} - dKf{n};
   Result1  = Result1 + sum( dr{n}(:) .* conj(dummy{n}(:)) );
end
[Kf,Kdf] = KOPwdw(r,dr,input);

% Operator via [I - CHI K] via inner product ------------------------------
Result2  = sum( (r(:)-CHI(:).*Kf(:)) .* conj(r(:)));
for n = 1:nDIM
   dummy{n} = dr{n} - Rfl{n} .* Kdf{n};
   Result2  = Result2 + sum( dummy{n}(:) .* conj(dr{n}(:)) );
end

% Print the difference ------------------------------------------------------
fprintf('Check adjoint: %e\n',abs(Result1-Result2));
```

of this equation is printed. The operator function `KOPwdw` and the adjoint function `AdjKOP-wdw` are listed in Matlab Listings I.62 and I.63. In these functions, the scalar operators `Kop` and `AdjKop` of Matlab Listings I.13 and I.14 are used. After validation, one can skip this test. The function `plotContrastSourcewp` of Matlab Listing I.47 plots the scalar contrast source $w^{(0)}$, while `plotInterfaceSource` of Matlab Listing I.69 plots the interface contrast sources $\partial w^{(j)}$. The function `IMAGESC` is a modified version of the standard Matlab function `imagesc`, see Matlab Listing I.7.

For the 1D case, Figure 2.28 shows the wave-speed contrast χ^c, the volume contrast source amplitude $|w^{(0)}|$, the "reflection coefficient" $R^{(1)}$, and the interface contrast source amplitude $|w^{(1)}|$. In the bottom-right figure, we observe non zero interface contrast sources at the two interfaces of the slab. The presence of these interface sources is also visible in the volume-contrast source in the top-right figure. Both interface-contrast and volume-contrast sources contribute to the scattered field.

For our choice of equal acoustic compressibility and mass density, the scattered field (reflected field) above the slab vanishes. We did not present figures for the pressure field. But the results are similar to the ones of Figure 2.11.

For the 2D case, Figure 2.29 shows the wave-speed contrast χ^c, the volume-contrast-source amplitude $|w^{(0)}|$, and the absolute values of the two interface-contrast-source components $|\partial w^{(1)}|$ and $|\partial w^{(2)}|$. The mass-density contrast is omitted: for our chosen parameters it is identical to the compressibility contrast. Note that both the reflection coefficients and the interface-contrast

Matlab Listing I.69 The function `plotInterfaceSource` to plot the contrast distribution χ^c and the contrast-source distribution $\partial w^{(j)}$.

```
function plotInterfaceSource(dw, input)
global nDIM;  dx =input.dx;  Rfl = input.Rfl;

if nDIM == 1          % Plot 1D contrast/source distribution ------------
   x1 = input.X1;
   set(figure,'Units','centimeters','Position',[5 5 18 12]);
   subplot(1,2,1);
      IMAGESC(x1+dx/2,-30:30,Rfl{1});
      title('\fontsize{13} R^{(1)}=(\rho^+-\rho^-) / (\rho^++\rho^-)');
   subplot(1,2,2);
      IMAGESC(x1+dx/2,-30:30, abs(dw{1}));
      title('\fontsize{13} abs(dw^{(1)})');

elseif nDIM == 2      % Plot 2D contrast/source distribution ------------
   x1 = input.X1(:,1);   x2 = input.X2(1,:);
   set(figure,'Units','centimeters','Position',[5 5 18 12]);
   subplot(1,2,1);
      IMAGESC(x1+dx/2,x2,Rfl{1});
      title('\fontsize{13} R^{(1)}=(\rho^+-\rho^-) / (\rho^++\rho^-)');
   subplot(1,2,2);
      IMAGESC(x1+dx/2,x2, abs(dw{1}));
      title('\fontsize{13} abs(dw^{(1)})');
   set(figure,'Units','centimeters','Position',[5 5 18 12]);
   subplot(1,2,1);
      IMAGESC(x1,x2+dx/2,Rfl{2});
      title('\fontsize{13} R^{(2)}=(\rho^+-\rho^-) / (\rho^++\rho^-)');
   subplot(1,2,2);
      IMAGESC(x1,x2+dx/2,abs(dw{2}));
      title('\fontsize{13} abs(dw^{(2)})');

elseif nDIM == 3      % Plot 3D contrast/source distribution ------------
                      % at x3 = 0 or x3 = dx/2
   x1 = input.X1(:,1,1);  x2 = input.X2(1,:,1);
   N3cross = floor(input.N3/2+1);
   set(figure,'Units','centimeters','Position',[5 5 18 12]);
   subplot(1,2,1);
      IMAGESC(x1+dx/2,x2, Rfl{1}(:,:,N3cross));
      title('\fontsize{13} R^{(1)}=(\rho^+-\rho^-) / (\rho^++\rho^-)');
   subplot(1,2,2);
      IMAGESC(x1+dx/2,x2,abs(dw{1}(:,:,N3cross)));
      title('\fontsize{13} abs(dw^{(1)})');
   set(figure,'Units','centimeters','Position',[5 5 18 12]);
   subplot(1,2,1);
      IMAGESC(x1,x2+dx/2, Rfl{2}(:,:,N3cross));
      title('\fontsize{13} R^{(2)}=(\rho^+-\rho^-) / (\rho^++\rho^-)');
   subplot(1,2,2);
      IMAGESC(x1,x2+dx/2,abs(dw{2}(:,:,N3cross)));
      title('\fontsize{13} abs(dw^{(2)})');
end % if
```

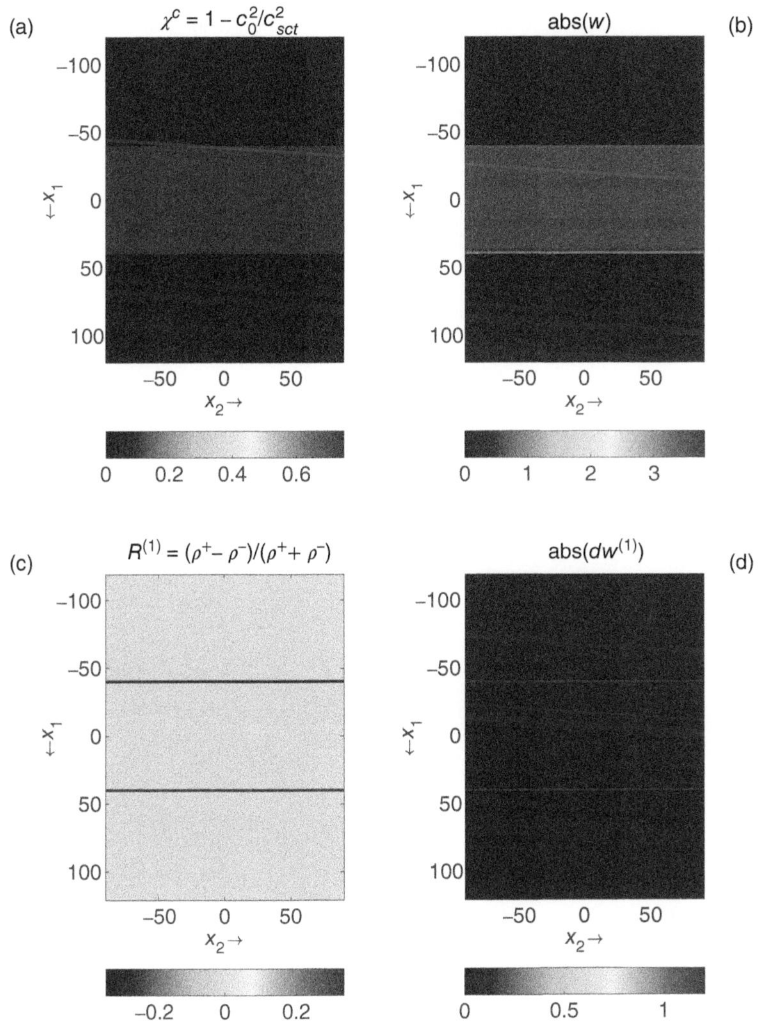

Figure 2.28 Resulting plots of the contrast χ^c and the volume contrast-source amplitude $|w| = |w^{(0)}|$ (a, b), the reflection factor $R^{(1)}$ and the interface contrast-source amplitude $|\partial w^{(1)}|$ (c, d), produced by `ForwardCGFFTwdw` for the 1D case.

sources indicate the staircase approximation of the circular interface. At the circular interface, the reflection coefficients and the interface-contrast sources act in the normal direction and vanish along the tangential direction.

For the 3D case, Figure 2.30 shows, in the vertical cross-section $x_3 = 0$, the wave-speed contrast χ^c, the contrast-source amplitude $|w^{(0)}|$, the absolute values of the two interface-contrast-source components $|\partial w^{(1)}|$ and $|\partial w^{(2)}|$. Comparing the 2D results and the results in this 2D cross-section of the 3D configuration, we observe similar results, but there are no arguments to use the 2D contrast sources as an approximation of the 3D ones.

In step (4) of the Matlab script `ForwardCGFFTwdw`, the scattered field at the selected receiver points are computed. The contrast-source distributions $w^{(0)}$ and $\partial w^{(j)}$ are the input for the function `Dopw` of Matlab Listing I.70. The program follows from the weak form of the scattered pressure

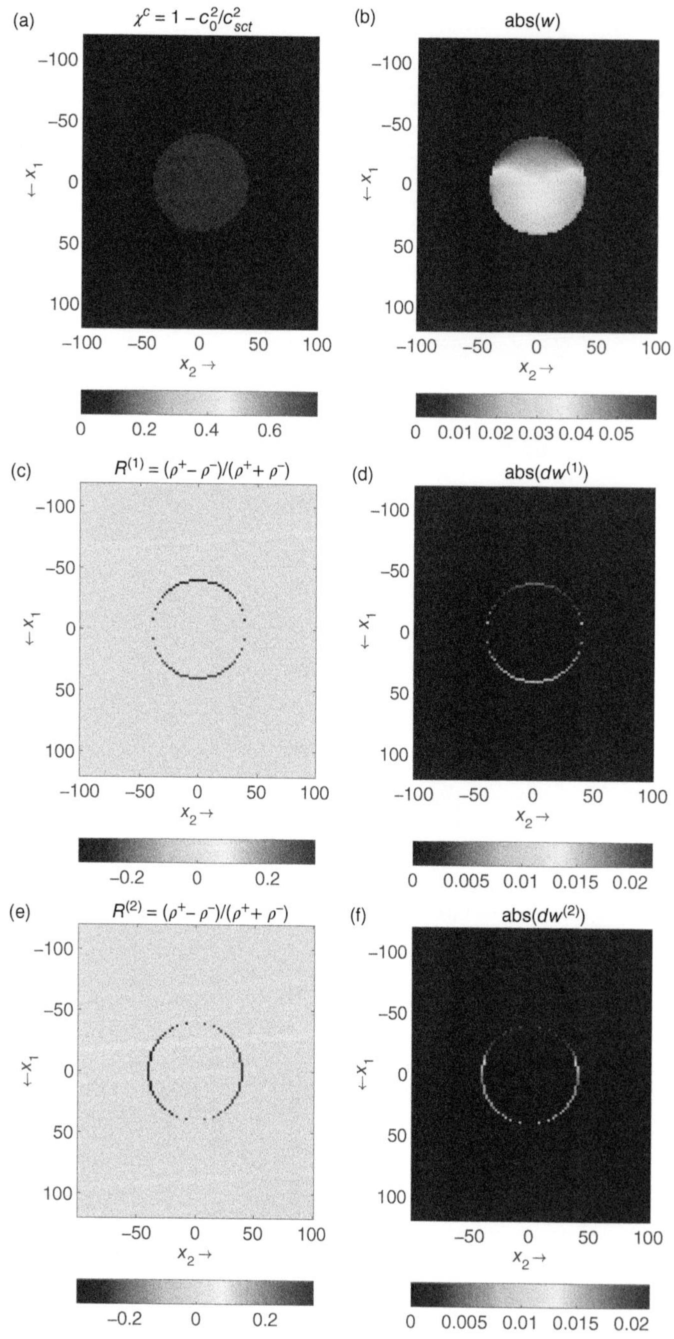

Figure 2.29 Resulting plots of the contrast χ^c and volume-contrast-source amplitude $|w| = |w^{(0)}|$ (a, b), the reflection coefficient $R^{(1)}$ and the interface-contrast-source amplitude $|\partial w^{(1)}|$ (c, d), the reflection coefficient $R^{(2)}$ and the interface-contrast-source amplitude $|\partial w^{(2)}|$ (e, f), produced by `ForwardCGFFTwdw` for the 2D case.

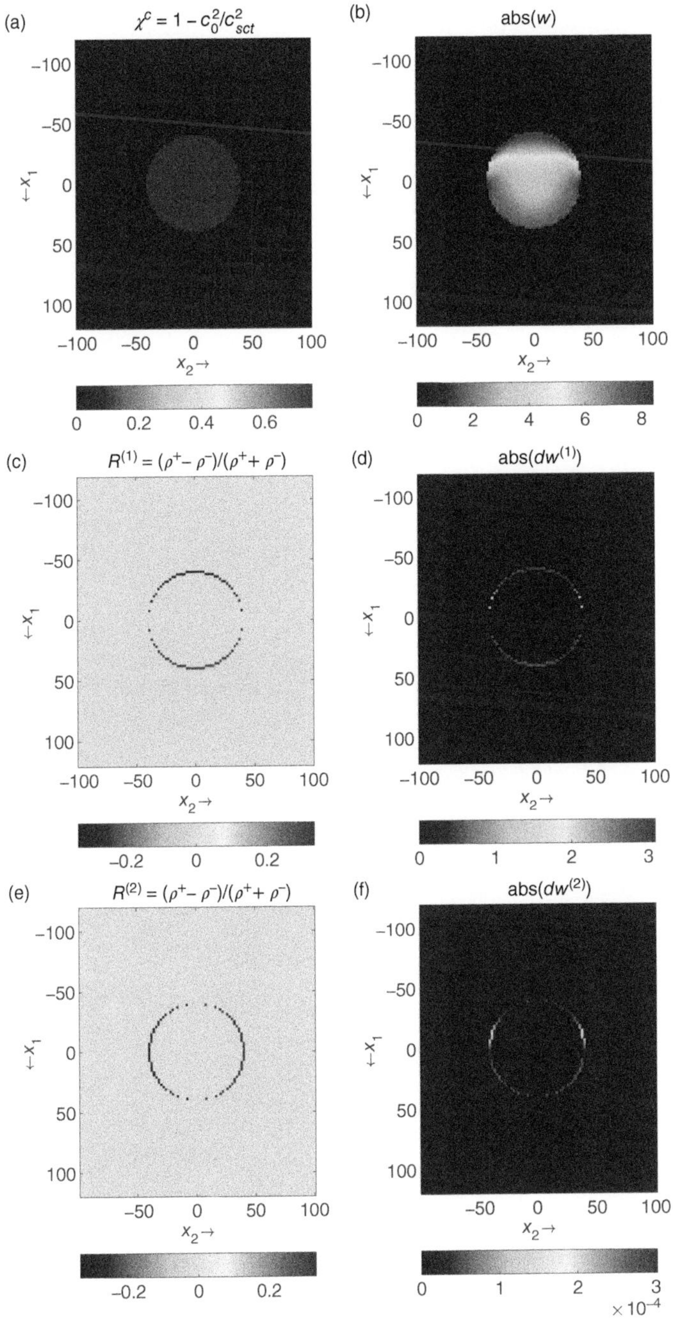

Figure 2.30 Resulting plots of the contrast χ^c and volume-contrast-source amplitude $|w| = |w^{(0)}|$ (a, b), the reflection coefficient $R^{(1)}$ and the interface-contrast-source amplitude $|\partial w^{(1)}|$ (c, d), the reflection coefficient $R^{(2)}$ and the interface-contrast-source amplitude $|\partial w^{(2)}|$ (e, f), produced by `ForwardCGFFTwdw` for the 3D case.

Matlab Listing I.70 The function `Dopw` to compute the scattered field generated by the volume contrast sources at a number of receivers.

```
function [data] = DOPw(w, input)
global nDIM;
gam0 = input.gamma_0;      dx = input.dx;     xR = input.xR;
data = zeros(1,input.NR);

if nDIM == 1
% Weak form
   x1      = input.X1;
   delta   = 0.5 * dx;
   factor  = sinh(gam0*delta) / (gam0*delta);
   DIS     = abs(x1-xR(1,1));                % receiver for reflected field
   G       = 1/(2*gam0) * exp(-gam0.*DIS);
  data(1,1)= (gam0^2 * dx) * factor * sum(G.*w);
   DIS     = abs(x1-xR(1,2));          % receiver for transmitted field
   G       = 1/(2*gam0) * exp(-gam0.*DIS);
  data(1,2)= (gam0^2 * dx) * factor * sum(G(:).*w(:));

elseif nDIM == 2
% Weak form
   X1      = input.X1; X2 = input.X2;
   delta   = (pi)^(-1/2) * dx;                % radius circle with area of dx^2
   factor  = 2 * besseli(1,gam0*delta) / (gam0*delta);
   for p = 1 : input.NR
      DIS     = sqrt((X1-xR(1,p)).^2 +(X2-xR(2,p)).^2);
      G       = 1/(2*pi).* besselk(0,gam0*DIS);
    data(1,p) = (gam0^2 * dx^2) * factor * sum(G(:).*w(:));
   end % p_loop

elseif nDIM == 3
% Weak form
   X1      = input.X1; X2 = input.X2; X3 = input.X3;
   delta   = (4*pi/3)^(-1/3) * dx;            % radius sphere with area of dx^3
   arg     = gam0*delta;
   factor  = 3 * (cosh(arg) - sinh(arg)/arg) / arg^2;
   for p = 1 : input.NR
      DIS     = sqrt((X1-xR(1,p)).^2 + (X2-xR(2,p)).^2 + (X3-xR(3,p)).^2);
      G       = exp(-gam0*DIS) ./ (4*pi*DIS);
    data(1,p) = (gam0^2 * dx^3) * factor * sum(G(:).*w(:));
   end % p_loop

end % if
```

of Eqs. (2.169)–(2.171), the weak form of the spatial derivatives of the weak Green functions of Eqs. (2.84)–(2.86) and the weak Green functions of Table 1.1, for 1D, 2D, and 3D, respectively.

The scattered data at the receivers are computed as the sum of the volume-contrast-source contributions and the interface-contrast-source distributions. They are compared to the analytical Bessel function data stored in MAT-files. We use the error norm of Eq. (2.115). The function `displayData.m` of Matlab Listing I.9 loads the particular file and for 1D it prints them, while for 2D and

Matlab Listing I.71 The function Dopdw to compute the scattered field generated by the interface sources at a number of receivers.

```
function [data] = DOPdw(dw,data,input)
global nDIM;
gam0 = input.gamma_0; xR   = input.xR;   dx = input.dx;

if nDIM == 1

    x1        = input.X1;
    DIS       = abs(xR(1,1)-(x1+dx/2));           % receiver for reflected field
    G         = 1/(2*gam0) * exp(-gam0.*DIS);
    d1_G      = - gam0 * sign(xR(1,1)-(x1+dx/2)) .* G;
  data(1,1) = data(1,1) + 2 * sum(d1_G(:) .* dw{1}(:));
    DIS       = abs(xR(1,2)-(x1+dx/2));           % receiver for transmitted field
    G         = 1/(2*gam0) * exp(-gam0.*DIS);
    d1_G      = - gam0 * sign(xR(1,2)-(x1+dx/2)) .* G;
  data(1,2) = data(1,2) + 2 * sum(d1_G(:) .* dw{1}(:));

elseif nDIM == 2

   X1 = input.X1;    X2 = input.X2;
   for p = 1 : input.NR
     DIS       = sqrt((xR(1,p)-(X1+dx/2)).^2 + (xR(2,p)-X2).^2);
     d_G       = - gam0 * (1/(2*pi)) .* besselk(1,gam0*DIS);
     d1_G      = ((xR(1,p)-(X1+dx/2))./DIS) .* d_G;
    data(1,p) = data(1,p) + 2 * dx * sum(d1_G(:) .* dw{1}(:));
     DIS       = sqrt((xR(1,p)-X1).^2 + (xR(2,p)-(X2+dx/2)).^2);
     d_G       = - gam0 * (1/(2*pi)) .* besselk(1,gam0*DIS);
     d2_G      = ((xR(2,p)-(X2+dx/2))./DIS) .* d_G;
    data(1,p) = data(1,p) + 2 * dx * sum(d2_G(:) .* dw{2}(:));
   end % p_loop

elseif nDIM == 3

   X1 = input.X1;    X2 = input.X2;    X3 = input.X3;
  for p = 1 : input.NR
     DIS      = sqrt((xR(1,p)-(X1+dx/2)).^2+(xR(2,p)-X2).^2+(xR(3,p)-X3).^2);
     d_G      = (-1./DIS-gam0) .* exp(-gam0*DIS) ./ (4*pi*DIS);
     d1_G     = ((xR(1,p)-(X1+dx/2))./DIS) .* d_G;
   data(1,p) = data(1,p) + 2 * dx^2 * sum(d1_G(:) .* dw{1}(:));
     DIS      = sqrt((xR(1,p)-X1).^2+(xR(2,p)-(X2+dx/2)).^2+(xR(3,p)-X3).^2);
     d_G      = (-1./DIS-gam0) .* exp(-gam0*DIS) ./ (4*pi*DIS);
     d2_G     = ((xR(2,p)-(X2+dx/2))./DIS) .* d_G;
   data(1,p) = data(1,p) + 2 * dx^2 * sum(d2_G(:) .* dw{2}(:));
     DIS      = sqrt((xR(1,p)-X1).^2+(xR(2,p)-X2).^2+(xR(3,p)-(X3+dx/2)).^2);
     d_G      = (-1./DIS-gam0) .* exp(-gam0*DIS) ./ (4*pi*DIS);
     d3_G     = ((xR(3,p)-(X3+dx/2))./DIS) .* d_G;
   data(1,p) = data(1,p) + 2 * dx^2 * sum(d3_G(:) .* dw{3}(:));
   end % p_loop

end % if
```

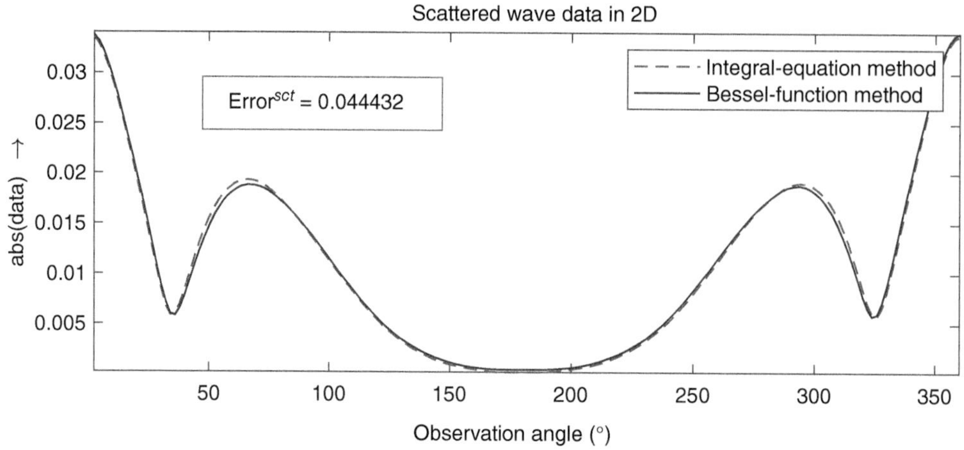

Figure 2.31 Plots of the 2D scattered wave field at the receivers, using either the Matlab script `WavefieldSctBessel2D` (solid line) or the Matlab script `ForwardCGFFTwdw` (dashed line); case `'analytic'`.

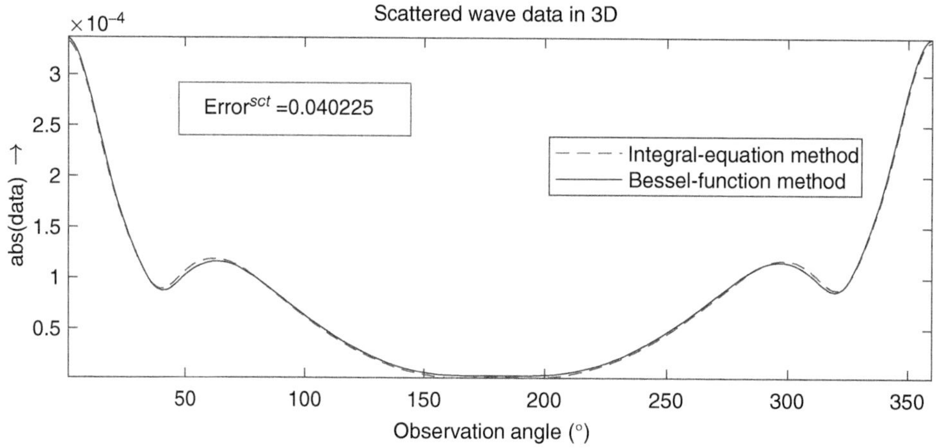

Figure 2.32 Plots of the 3D scattered wave field at the receivers, using either the Matlab script `WavefieldSctBessel3D` (solid line) or the Matlab script `ForwardCGFFTwdw` (dashed line); case `'analytic'`.

3D it plots both the exact data and the numerical data. The scattered field data are presented in Figures 2.31 and 2.32. The differences are mainly due to the staircase approximation of the discrete contrast distribution of the integral equation method versus the smooth boundary of the circular cylinder and the sphere. In contrast to previous observations, we now observe that the error norm Error^{sct} for the 2D and 3D cases are of the same order (see Figures 2.12 and 2.13). In Section 2.7.1, we investigate these errors for finer grid sizes.

2.7.1 Performance Analysis

To investigate the convergence rate of the integral equation results, we consider the results for the scattered data and the error norm of Eq. (2.115). Our goal is to investigate this error as function of

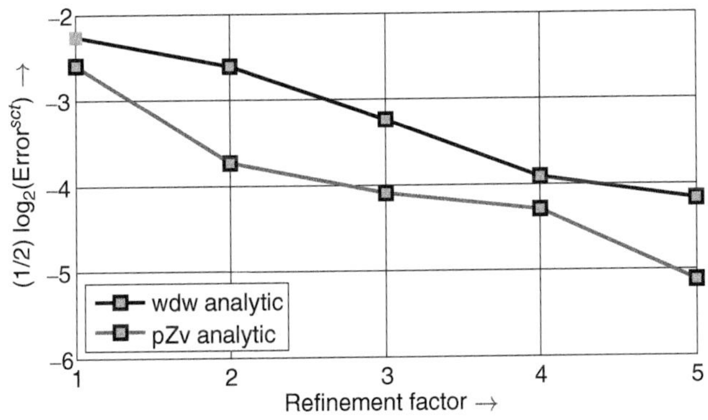

Figure 2.33 Convergence plot of the Error^{sct} as function of the refinement factor r, $n = 1, 2, \ldots, 5$, for the 2D pZv-method and the 2D wdw-method.

the grid size Δx, both for the previous method, the so-called pZv-method and the present method, the so-called wdw-case. We only consider the 2D case. We begin with $\Delta x = 2$ and reduce it each time with a factor of two. Again, we define a refinement factor $r = 1, 2, \ldots, 5$, and define a new grid size $\Delta^{(r)} = \Delta x / r$ and new sample numbers $N_1^{(r)} = r\,N_1$ and $N_2^{(r)} = r\,N_2$. We have set the error criterion in the Matlab functions `ITERCGwpZv` and `ITERCGwdw` to 10^{-6}. These computations are performed on a Dell computer (64 bit, dual core 3.50 GHz, 64 GB RAM).

For the case `'analytical'` of the pZv-method, we have observed that the error reduces from 2.7% to 0.08%. In Figure 2.33, the results are presented as the lower solid line. For the wdw-method, the error reduces from 4.4% to 0.3%. In Figure 2.33, the results are presented as the upper solid line. Contrary to the nonconverging results of the $c\partial\rho$-method (see upper solid line in Figure 2.24), the error now reduces satisfactory. The jumps in the numerical implementation of the mass-density gradient are now treated correctly. The theoretical value of a Dirac distribution is taken into account by replacing it as an interface-contrast-source distribution.

In Figure 2.34, we registered the computation time per iteration. For both the pZv-method and the wdw-method, the computation time as function of the refinement increase in the same way. This figure confirms that the computation costs for the numerical solution of both integral equations are roughly of the order of $N\,\log_2(N)$, where we find that the case `'analytical'` of the pZv-method needs the largest computation time.

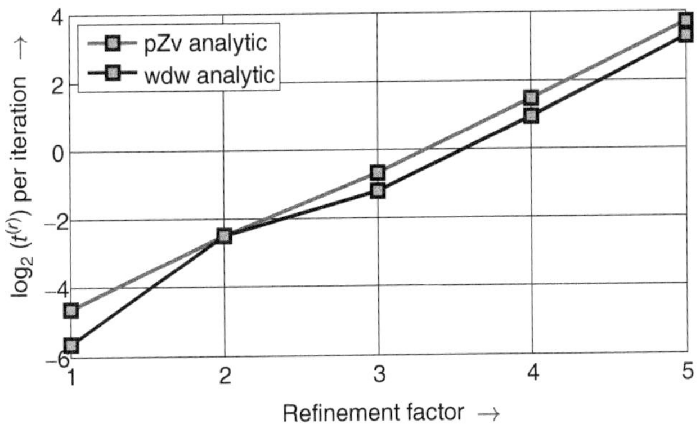

Figure 2.34 Commutation time $\log_2(t^{(r)})$ to solve the 2D integral equation as function of the refinement r, for the 2D pZv-method and 2D wdw-method.

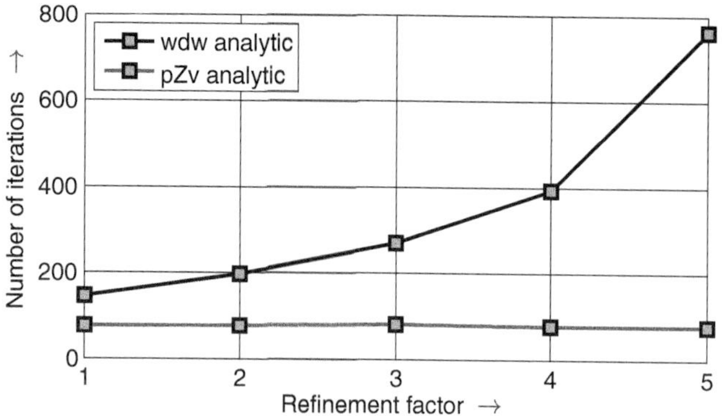

Figure 2.35 Number of iterations to solve the 2D integral equation as function of the refinement r, for the 2D pZv-method and 2D wdw-method.

For, the pZv-method, the number of iterations to reach an error criterion of 10^{-6} is almost constant. It varies between 78 and 80 iterations, when we reduce the grid size from 2 to 0.125 m. However, in the wdw-method, the number of iterations increases from 147 iterations to 762 iterations, see Figure 2.35. This is an indication that the latter method is more sensitive to the jumps in the acoustic mass-density distribution by introducing the interface contrast sources. Reduction of the grid-size leads to a larger collection of individual interface sources. In other words, the number of unknowns increases as a function of the refinement factor, with the consequence that the total number of CG iterations will also increase.

2.7.2 Matlab BiCGSTAB Built-in Functions for Iterative Solution of the Contrast Source Integral Equation

Similar as in Section 1.6.2, we show how the BiCGSTAB built-in Matlab function can be used for the iterative solutions of the acoustic contrast-source integral equations.

In the pZv-method, we have to change function `ITERCGwpZv` in step (3) of the script `ForwardCGFFTwpZv` (see Matlab Listing I.43) by function `ITERBiCGSTABwpZv` of Matlab Listing I.72. In the wdw-method, we have to change function `ITERCGwdw` in step (3) of the script `ForwardCGFFTwdw` (see Matlab Listing I.64) by the function `ITERBiCGSTABwdw` of Matlab Listing I.73. Since the Matlab built-in functions operate on a single column vector, we need to reshape our input matrices to single column vectors and the output column vector of `ITERBiCGSTABw` to matrices. This is only required for the 2D and 3D cases.

We will discuss some results for the 3D case only. These computations are performed on a Dell computer (64 bit, dual core 3.50 GHz, 64 GB RAM).

For the case `'analytical'` of the pZv-method and an error criterion of 0.001, the function `ITERCGwpZv` needs 59 iterations with a computation time of 704 seconds, while function `ITERBiCGSTABpZv` needs 15 iterations with a computation time of 191 seconds. For the case `'numerical'` of the pZv-method, the function `ITERCGwpZv` requires 61 iterations with a computation time of 305 seconds, while function `ITERBiCGSTABwpZv` again requires 15 iterations, but with a computation time of 119 seconds. However, from Figure 2.24, we learn that in the case of `'numerical'`, the error in the scattered data is approximately two to three times as high. Hence, refining the mesh grid with a factor of two, will yield comparable results within the same number of iterations and approximately the same computation time as in the case `'analytical'`.

Matlab Listing I.72 The function `ITERBiCGSTABwpZv` to use Matlab's built-in function BiCGSTAB in `ForwardCGFFTwpZv` of Matlab Listing I.43.

```
function [w_p,w_Zv] = ITERBiCGSTABwpZv(p_inc,Zv_inc,input)
% GMRES_FFT scheme for contrast source integral equation Aw = b
  global nDIM;        [N,~] = size(input.CHI_kap(:));
  itmax  = 1000;   Errcri = input.Errcri;
                                                    % known vector
                  b(1:N,1)        = input.CHI_kap(:)  .* p_inc(:);
                  b(N+1:2*N,1)    = input.CHI_rho(:)  .* Zv_inc{1}(:);
  if nDIM >= 2;   b(2*N+1:3*N,1)  = input.CHI_rho(:)  .* Zv_inc{2}(:); end
  if nDIM == 3;   b(3*N+1:4*N,1)  = input.CHI_rho(:)  .* Zv_inc{3}(:); end

  w = bicgstab(@(w) Aw(w,input), b, Errcri, itmax);   % call BICGSTAB

  [w_p,w_Zv] = vector2matrix(w,input);                % output matrices
end %-------------------------------------------------------------------------


function y = Aw(w,input)
  global nDIM;        [N,~] = size(input.CHI_kap(:));

  [w_p,w_Zv] = vector2matrix(w,input);
  [Kp,KZv]   = KOPpZv(w_p,w_Zv,input);
          Kp = w_p - input.CHI_kap .* Kp;
  for n = 1:nDIM
      KZv{n} = w_Zv{n} - input.CHI_rho .* KZv{n};
  end
                  y(1:N,1)        = Kp(:);
                  y(N+1:2*N,1)    = KZv{1}(:);
  if nDIM >= 2;   y(2*N+1:3*N,1)  = KZv{2}(:);   end
  if nDIM == 3;   y(3*N+1:4*N,1)  = KZv{3}(:);   end
end %-------------------------------------------------------------------------


function [w_p,w_Zv] = vector2matrix(w,input)
% Modify vector output from 'bicgstab' to matrices for further computation
  global nDIM;        [N,~] = size(input.CHI_kap(:));   w_Zv = cell(1,nDIM);

  if nDIM == 2;   DIM = [input.N1,input.N2];              end
  if nDIM == 3;   DIM = [input.N1,input.N2,input.N3];     end

                      w_p = w(1:N,1);
                  w_Zv{1} = w(N+1:2*N,1);
  if nDIM >= 2;   w_Zv{2} = w(2*N+1:3*N,1);   end
  if nDIM == 3;   w_Zv{3} = w(3*N+1:4*N,1);   end

  if nDIM >= 2;       w_p = reshape(w_p,DIM); end
  for n = 1:nDIM
     if nDIM >= 2;   w_Zv{n} = reshape(w_Zv{n},DIM);   end
  end

end
```

Matlab Listing I.73 The function `ITERBiCGSTABwdw` to call Matlab's built-in function BiCGSTAB in `ForwardCGFFTwdw` of Matlab Listing I.64.

```
function [wc,dw] = ITERBiCGSTABwdw(p_inc , Pinc , input)
  % GMRES_FFT scheme for contrast source integral equation Aw = b
  global nDIM;          [N,~] = size(input.CHI(:));
  itmax  = 1000;       Errcri = input.Errcri;
                                                        % known vector
                   b(1:N,1)        = input.CHI(:)      .* p_inc(:);
                   b(N+1:2*N,1)    = input.Rfl{1}(:)   .* Pinc{1}(:);
  if nDIM >= 2;    b(2*N+1:3*N,1)  = input.Rfl{2}(:)   .* Pinc{2}(:);  end
  if nDIM == 3;    b(3*N+1:4*N,1)  = input.Rfl{3}(:)   .* Pinc{3}(:);  end

w = bicgstab(@(w) Aw(w,input), b, Errcri , itmax);     % call BICGSTAB

[wc,dw] = vector2matrix(w,input);                      % output matrices
end %---------------------------------------------------------------------------


function y = Aw(w,input)
  global nDIM;        [N,~] = size(input.CHI(:));

  [wc,dw]   = vector2matrix(w,input);
  [Kw,Kdw]  = KOPwdw(wc,dw,input);
        Kw  = wc - input.CHI .* Kw;
  for n = 1:nDIM
    Kdw{n} = dw{n} - input.Rfl{n} .* Kdw{n};
  end
                   y(1:N,1)        = Kw(:);
                   y(N+1:2*N,1)    = Kdw{1}(:);
  if nDIM >= 2;    y(2*N+1:3*N,1)  = Kdw{2}(:);   end
  if nDIM == 3;    y(3*N+1:4*N,1)  = Kdw{3}(:);   end
end %---------------------------------------------------------------------------


function [wc,dw] = vector2matrix(w,input)
% Modify vector output from 'bicgstab' to matrices for further computation
  global nDIM;        [N,~] = size(input.CHI(:));     dw = cell(1,nDIM);

  if nDIM == 2;      DIM = [input.N1 ,input.N2];               end
  if nDIM == 3;      DIM = [input.N1 ,input.N2 ,input.N3];     end

                       wc = w(1:N,1);
                   dw{1}  = w(N+1:2*N,1);
  if nDIM >= 2;    dw{2}  = w(2*N+1:3*N,1);   end
  if nDIM == 3;    dw{3}  = w(3*N+1:4*N,1);   end

  if nDIM >= 2;        wc = reshape(wc,DIM); end
  for n = 1:nDIM
     if nDIM >= 2; dw{n} = reshape(dw{n},DIM);    end
  end

end
```

For our wdw-method, where we operate with both volume-contrast sources and interface-contrast sources, the function `ITERCGwdw` requires 116 iterations with a computation time of 723 seconds, while function `ITERBiCGSTABwdw` requires only 18 iterations with a computation time of 121 seconds. From these numerical experiments regarding the present type of integral equation, we conclude that it is beneficial to use the BiCGSTAB iterative solver instead of the CG method (including the use of the adjoint of the operators). For the solution of the integral equation for both volume-contrast and interface-contrast sources, it is the preferred method.

2.8 Time-Domain Solution of Contrast Source Integral Equation

So far, we have considered the solution of the contrast source solution for a certain value of $s_n = \varepsilon - i2\pi f_n \approx -i\, 2\pi f_n$, where ε is a vanishing positive constant. In our computer codes, we have taken $\varepsilon = 10^{-16}$. Similar to the scalar case of Chapter 1, the forward discrete Fourier transform for causal pressure wave fields is

$$\hat{p}(\boldsymbol{x}, s_n) = \Delta t \sum_{m=0}^{\frac{1}{2}N} \exp\left(i2\pi \frac{m\,n}{N}\right) p(\boldsymbol{x}, t_m), \tag{2.172}$$

for real function $p(\boldsymbol{x}, t_m)$ in the time domain, The inverse discrete Fourier transform is obtained as

$$p(\boldsymbol{x}, t_m) = \Delta f \sum_{n=0}^{\frac{1}{2}N} \exp\left(-i2\pi \frac{m\,n}{N}\right)\, \hat{p}(\boldsymbol{x}, s_n). \tag{2.173}$$

We note that in the last two expressions of our discrete transforms, symmetric domains around $m = 0$ and $n = 0$ must be selected. In applying FFT routines, zero padding for the negative values of n and m is required.

At this point, we also have to select the wavelet $Q(t)$ in Eq. (2.29). The frequency domain results were calculated for $\hat{Q}(s_n) = 1$. Before computing the time domain results, we must define the wavelet and multiply the frequency domain results by $\hat{Q}(s_n)$. We choose the same wavelet as in Chapter 1, viz., the first derivative of the Gaussian as our frequency domain wavelet. It is computed with the help of the Matlab function `Wavelet` of Matlab Listing I.21. The resulting frequency and time domain functions are presented in Figure (1.13).

We change the single frequency script to one for the full frequency spectrum of the wavelet. This script is called as `TimedomainBiCGSTABFFTwdw`, and it is displayed as Matlab Listing I.74. Note that the frequency band ranges from 0 to 120 Hz. To capture the high-frequency range, we reduce our grid size from $\Delta x = 2$ m to $\Delta x = 1$ m. To get a better picture of the wave propagation in the time domain, we take a considerably larger observation domain by choosing $N_1 = N_2 = 600$ in the 2D cross-section. In view of this larger domain and the number of computations for many frequencies, we only consider the 2D situation. For convenience, we have the source position as $(0, -170)$ m; this is located on the left-hand side of the cylinder. Then, the wave propagates from the source in the right-hand direction toward the circular cylinder. We consider two different set of acoustic parameters, viz., the set $c_0 = 1500$, $\hat{c}_{sct} = 3000$, $\rho_0 = 1500$, $\hat{\rho}_{sct} = 1500$ and the set $c_0 = 1500$, $\hat{c}_{sct} = 750$, $\rho_0 = 750$, $\hat{\rho}_{sct} = 1500$. In both cases, the wave impedance $\hat{Z}_{sct}$ is equal to Z_0. The present script `TimedomainBiCGSTABFFTwdw` is very similar to the script `Timedomain-BiCGSTABFFTw` of Matlab Listing I.22. We omit a further explanation. To present the wave field at certain times instants t_m, we call the Matlab script `SnapshotP` of Matlab Listing I.75. The power

Matlab Listing I.74 The script `TimedomainBiCGSTABFFTwdw` to compute the time-domain total pressure $p(x_T, t)$ on a 2D grid.

```
clear all; clc; close all; clear workspace;
input = initAC();

    input = Wavelet(input);
Wavelfrq = input.Wavelfrq;
Wavelmax  = max(abs(Wavelfrq(:)));

% Redefine frequency-independent parameters
  input.N1 = 600;    N1 = input.N1;              % number of samples in x_1
  input.N2 = 600;    N2 = input.N2;              % number of samples in x_2
  input.dx = 1;                                  % grid size
    x1 = -(input.N1+1)*input.dx/2 + (1:input.N1)*input.dx;
    x2 = -(input.N2+1)*input.dx/2 + (1:input.N2)*input.dx;
  [input.X1,input.X2] = ndgrid(x1,x2);
  input.xS      = [0 ,-170];                     % source position
input.c_0       = 1500;                          % wave speed embedding
input.c_sct     = 3000;                          % wave speed scatterer
input.rho_0     = 3000;                          % mass density embedding
input.rho_sct   = 1500;                          % mass density scatterer
      input     = initAcousticContrast(input);   % contrast distribution
      input     = initAcousticContrastIntf(input); % interface contrast

Errcrr_0 = input.Errcri;
pfreq     = zeros(input.N1,input.N2,input.Nfft);
for f = 2 : input.fsamples
%    (0) Make error criterion frequency dependent
         factor = abs(Wavelfrq(f))/Wavelmax;
         Errcri = min([Errcrr_0 / factor, 0.999]);
         input.Errcri = Errcri;
%    (1) Redefine frequency-dependent parameters
         freq = (f-1) * input.df;       disp(['freq sample: ', num2str(f)]);
         s = 1e-16 - 1i*2*pi*freq;            % LaPlace parameter
         input.gamma_0 = s/input.c_0;         % propagation coefficient
         input = initFFTGreenfun(input);      % compute FFT of Green function
%    (2) Compute incident field at different grids
         [p_inc,Pinc] = IncPressureWave(input);
%    (3) Solve integral equation for contrast source with FFT ---------------
        [wc,dw] = ITERBiCGSTABwdw(p_inc,Pinc,input);
%    (4) Compute total wave field on grid and add to freqency components
         [Kw,Kdw] = KOPwdw(wc,dw,input);
         p_sct = Kw;
         p_sct(2:N1,:)  = p_sct(2:N1,:)+(Kdw{1}(1:N1-1,:)+Kdw{1}(2:N1,:))/2;
         p_sct(:,2:N2)  = p_sct(:,2:N2)+(Kdw{2}(:,1:N2-1)+Kdw{2}(:,2:N2))/2;
         pfreq(:,:,f)   = Wavelfrq(f) .* (p_inc(:,:) + p_sct(:,:));
end  % frequency loop

pt = 2 * input.df * real(fft(pfreq,[],3));                 clear pfreq;
ptime(:,:,1:input.fsamples) = pt(:,:,1:input.fsamples);    clear pt;
save ptime;
SnapshotP;                          % Make snapshots for a few time instants
```

Matlab Listing I.75 The function `SnapshotP` to present some snapshots of the time domain total pressure $p(x_T, t)$ on a 2D grid.

```
% This script is a continuation of TimedomainBiCGSTABFFTwdw
clc;

xS(1) = input.xS(1); xS(2) = input.xS(2);

ptimeabs = abs(ptime);
set(figure,'Units','centimeters','Position',[0 -1 18 32]);
for n = 1 : 12
  subplot(4,3,n)
 n_t = (20 + n * 7);

% Compute the time domain power in dB
  udB = -20 * log10(ptimeabs/max(max(ptimeabs(:,:,n_t))));
% Make snapshot
  imagesc(input.X2(1,:),input.X1(:,1),udB(:,:,n_t)); grid on;
  title(['t = ',num2str(n_t * input.dt),' s']);
  hold on; colormap(hot); axis('square'); caxis([0, 40]);
  plot(xS(2),xS(1),'ko','MarkerFaceColor',[0 0 0],'MarkerSize',3);
  phi = 0:0.001:2*pi;
  plot(input.a*cos(phi),input.a*sin(phi),'b','LineWidth',1.2);
  pause(0.1)
  hold off;

end % if
```

in dB of the normalized acoustic wave fields are presented. For each snapshot, we take the same color distribution, by setting `caxis = ([0,40])`.

Figures 2.36 and 2.37 display the snapshots in case the wave speed within the circular cylinder is twice the background wave speed. In the third and fourth picture, it is clear that the wave propagates faster through the object than outside the object. Comparing these figures with the corresponding figures for scalar waves, see Figures 1.14 and 1.15, where we do not have mass-density contrast, we notice that there is more acoustic forward scattering than scalar forward scattering and the acoustic multiple scattering phenomena are weaker than in the scalar case. This can be explained by our specific choice of mass densities, so that the acoustic wave impedance is constant in space.

Figures 2.38 and 2.39 display the snapshots in case the wave speed within the circular cylinder is half the background wave speed. In the first few pictures of the figure, it is clear that the wave goes slower through the object than outside the object. For larger time instants, we observe that a secondary wave is generated that propagates along both sides of the circular interface of cylinder. When the two wave fronts meet, they form a cardioid wave front together. After further propagation from this point, a new cardioid wave front is formed and is completed at the opposite point of the circular interface. Comparing these figures with the corresponding figures for scalar waves, see Figures 1.16 and 1.17, where we do not have mass-density contrast, we notice that there is more acoustic forward scattering than scalar forward scattering, but the acoustic multiple scattering phenomena do not differ very much. It seems that in the case of a smaller wave speed in the object, the object acts as a secondary interface type source determined by the different wave speeds and not by the wave impedances.

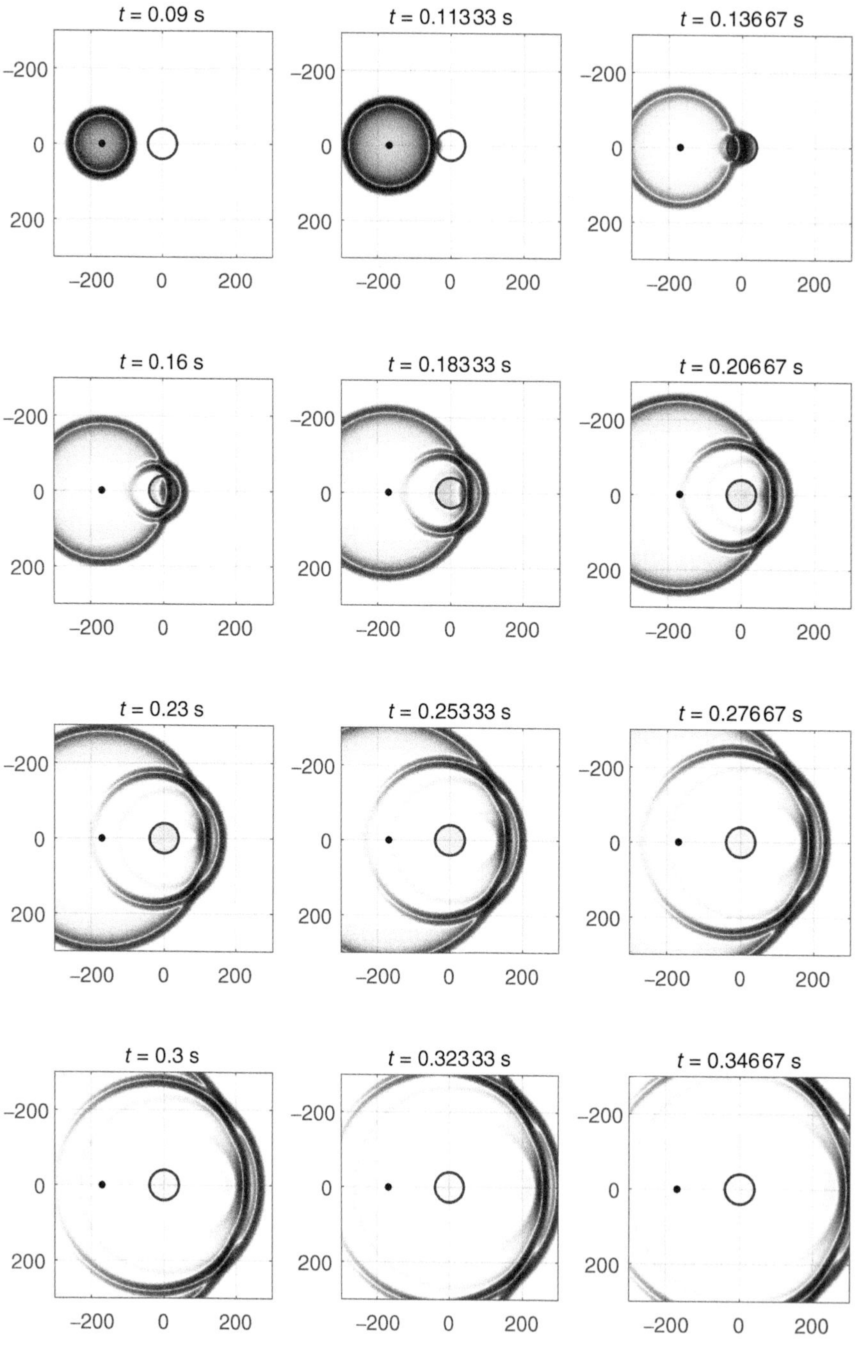

Figure 2.36 Snapshots of the 2D acoustic wave propagation in a model of a circular cylinder with $c_0 = 1500$ m, $\hat{c}_{sct} = 3000$ m/s, $\rho_0 = 3000$ kg/m^3, and $\hat{\rho}_{sct} = 1500$ kg/m^3; the source is located at $x_1^S = 0$ of the vertical axis and $x_2^S = -170$ m of the horizontal axis.

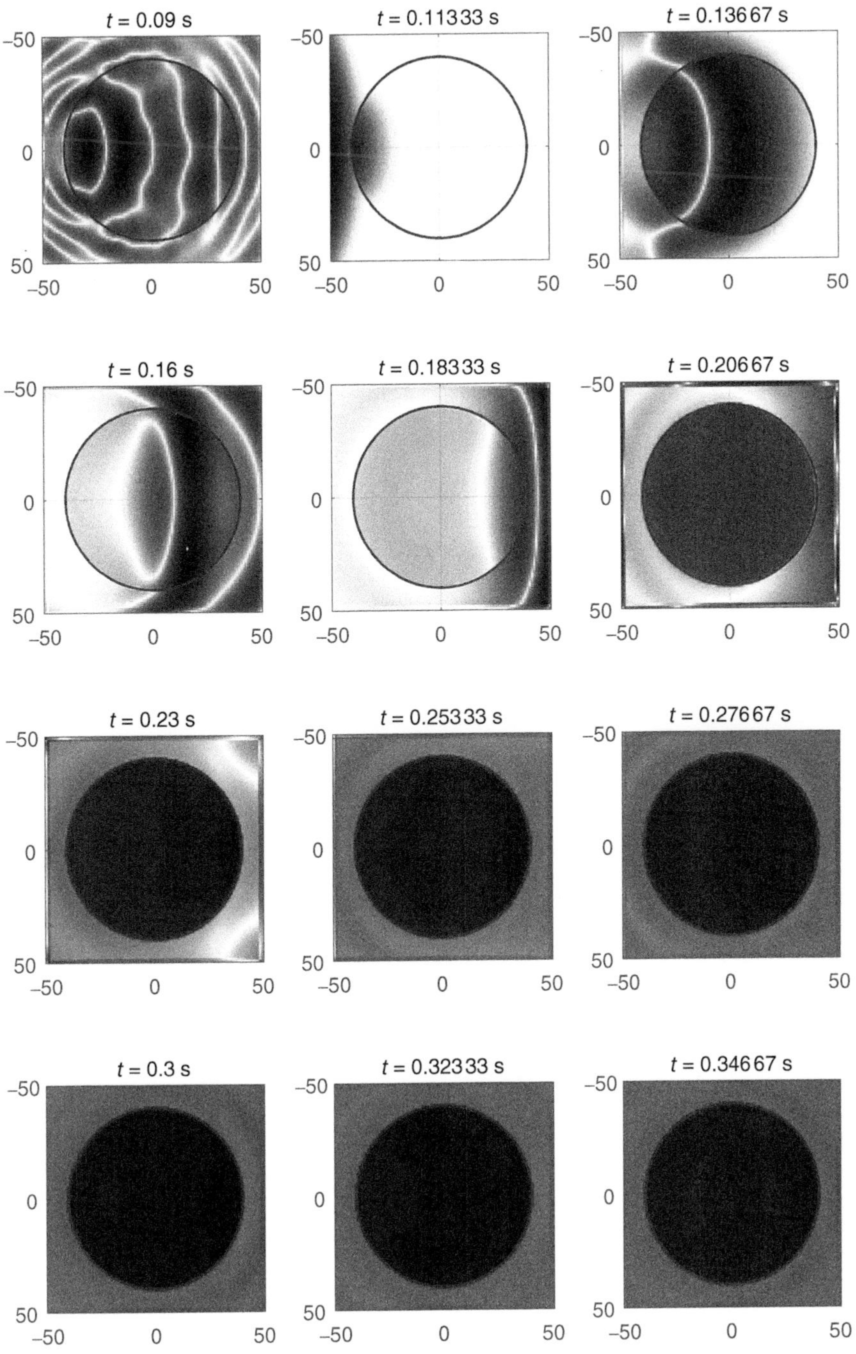

Figure 2.37 Similar to Figure 2.36, but the grid is reduced to 100 m by 100 m. Each image is normalized to the maximum value of the wave-field power (in dB). The wave-field values of the first picture only show the numerical noise.

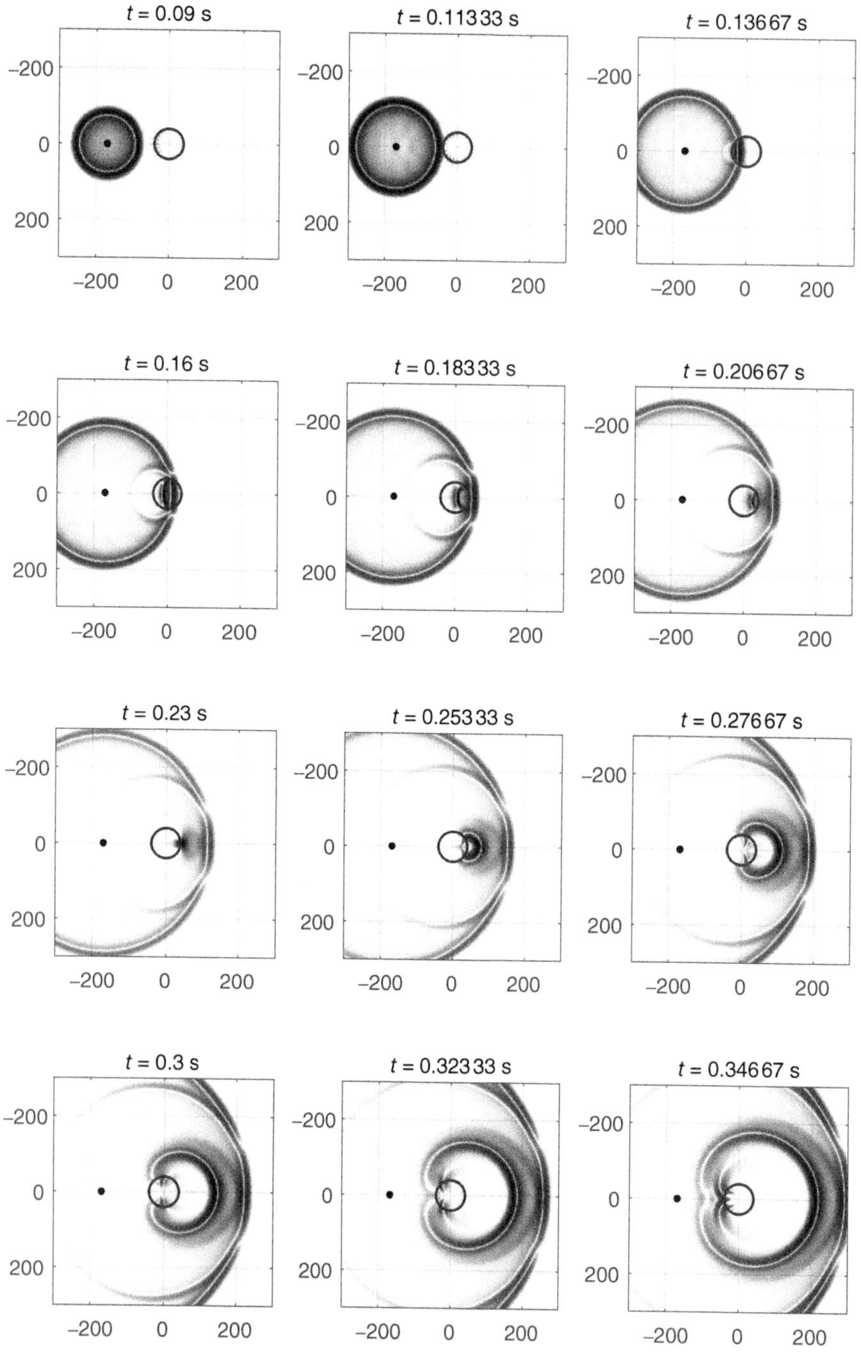

Figure 2.38 Snapshots of the 2D acoustic wave propagation in a model of a circular cylinder with $c_0 = 1500$ m, $\hat{c}_{sct} = 750$ m/s, $\rho_0 = 750$ kg/m^3 and $\hat{\rho}_{sct} = 1500$ kg/m^3; the source is located at $x_1^S = 0$ of the vertical axis and $x_2^S = -170$ m of the horizontal axis.

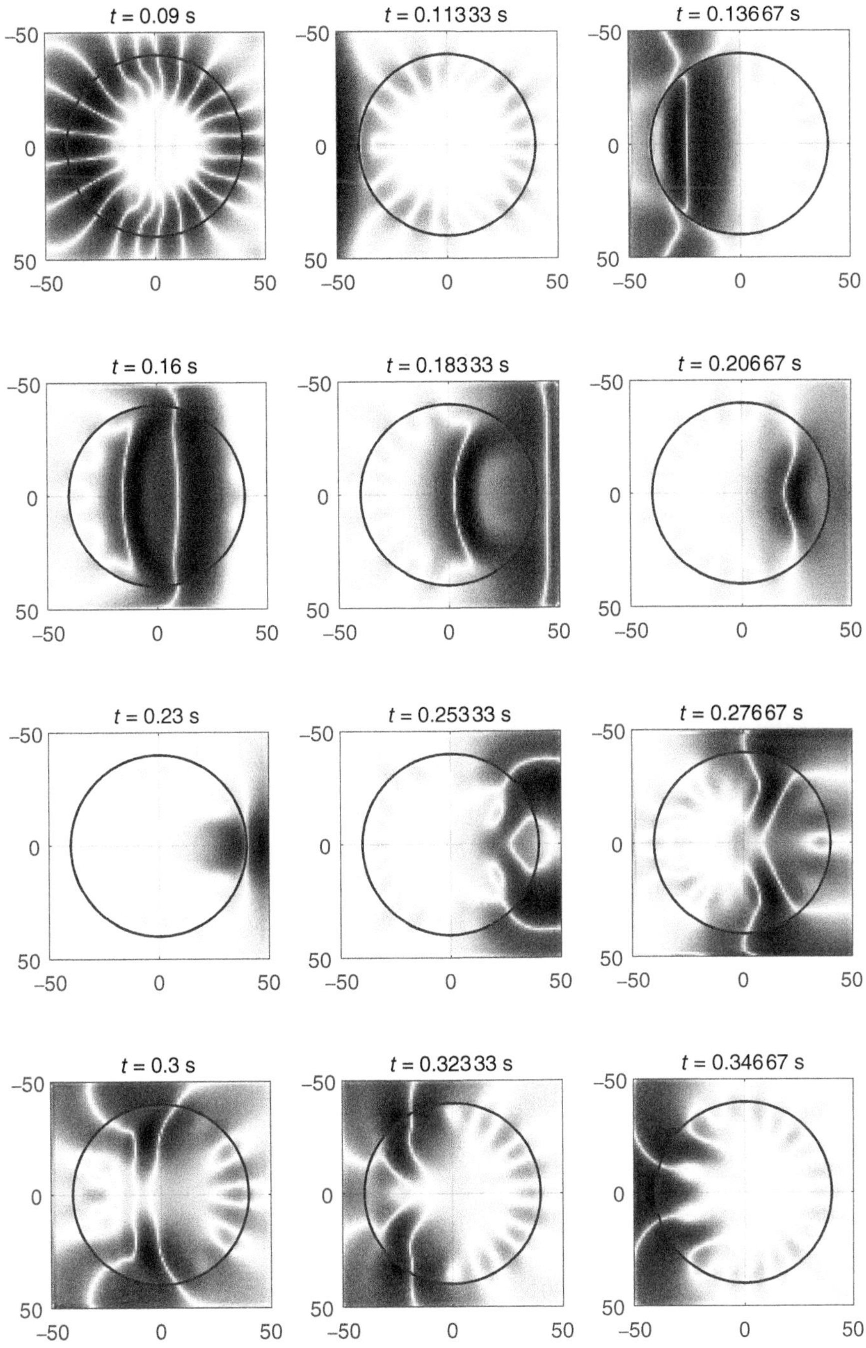

Figure 2.39 Similar to Figure 2.38, but the grid is reduced to 100 m by 100 m. The wave-field values of the first picture only show the numerical noise, although some field structure is present.

2.A Scattering by a Simple Canonical Configuration

In order to have a necessary check on the correctness of the coding of the domain integral equations at hand, we present the analytical expressions for some simple configurations. We first start with the one-dimensional scattering by a homogenous slab. Second, we discuss the two-dimensional scattering by a homogeneous circular cylinder, where we consider the special case of no contrast in wave speed. Third, we consider the 3D scattering by a homogeneous sphere. In addition, we present the Matlab code for computing the scattered field outside the scattering object at hand. Therefore, we need some Matlab scripts to initialize the various parameters. We take the same parameters as used in Chapter 1. We only have to add the actual values of the mass density. The source wavelet is chosen to be $s\hat{Q}\rho_0 = 1$. The pertaining `initAC` function is given in Matlab Listing I.39. The global parameter `nDIM` assigns the dimension of the space at hand. Similar to the wave speeds, we set the mass density $\rho_0 = 1500$ kg/m^3 and the mass density $\hat{\rho}_{sct} = 3000$ kg/m^3. The `initAC` function also calls the functions `initSourceReceiver` and `initGrid`. They need not be changed. They are listed in Chapter 1. The acoustic compressibility and mass density distributions are computed by the Matlab function `initAcousticContrast` as given in Matlab Listing I.42. This Matlab function is the replacement of the Matlab function `initContrast` of Chapter 1.

Subsequently, the script `AcForwardCanonicalObjects` of Matlab Listing I.79 generates the scattered acoustic pressure wave field for scattering in 1D, 2D, and 3D, respectively. The theoretical expressions are derived in Sections 2.A.1–2.A.3.

2.A.1 1D Scattering by a Slab

In order to generate synthetic data for the 1D configuration, we consider the scattering problem by a homogeneous slab (Figure 2.A.1). The slab has a thickness of $2a$. The medium inside the slab is characterized by the constant mass density $\hat{\rho}_{sct}$ and constant compressibility $\hat{\kappa}_{sct}$. Then, the wave speed $\hat{c}_{sct}$ is constant as well. The center of the slab is at $x_1 = 0$. The pressure of the acoustic wave field is generated by a planar source at x_1^S and is given by

$$\hat{p}^{inc}(x_1|x_1^S) = \frac{Z_0\hat{Q}}{2}\exp\left[-\hat{\gamma}_0|x_1 - x_1^S|\right]. \tag{2.A.1}$$

The amplitude of the pressure of the incident field at $x_1 = -a$ is given by

$$\hat{p}^{inc}(-a|x_1^S) = \frac{Z_0\hat{Q}}{2}\exp\left[\hat{\gamma}_0(x_1^S + a)\right]. \tag{2.A.2}$$

This is a common factor in all wave field quantities.

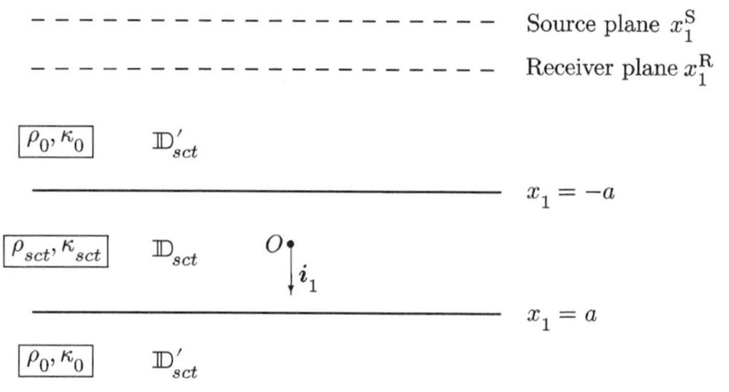

Figure 2.A.1 The 1D configuration with homogeneous slab.

The different wave constituents are considered to be generated at the interfaces. Taking into account the causality condition for the reflected field, we write the reflected field in the negative x_1-direction as

$$\hat{p}^{rfl}(x_1|x_1^S) = \frac{Z_0\hat{Q}}{2}\exp\,[\hat{\gamma}_0(x_1^S + a)]\;\hat{R}\exp[\hat{\gamma}_0(x_1 + a)], \quad x_1 < -a. \tag{2.A.3}$$

The wave field inside the slab consists of two waves propagating in opposite directions, hence, we write the pressure of this interior wave field as

$$\hat{p}^{int}(x_1|x_1^S) = \frac{Z_0\hat{Q}}{2}\exp\,[\hat{\gamma}_0(x_1^S + a\,)]\{\hat{A}\exp[-\hat{\gamma}_{sct}(x_1 + a)] + \hat{B}\exp[\hat{\gamma}_{sct}(x_1 - a)]\}, \tag{2.A.4}$$

when $-a < x_1 < a$. For $x_1 > a$, the field transmitted in the positive x_1-direction is given by

$$\hat{p}^{trm}(x_1|x_1^S) = \frac{Z_0\hat{Q}}{2}\exp\,[\hat{\gamma}_0(x_1^S + a)]\;\hat{T}\exp[-\hat{\gamma}_0(x_1 + a)]. \tag{2.A.5}$$

The unknown factors $\hat{R}$, $\hat{A}$, $\hat{B}$, and $\hat{T}$ follow interface conditions at $r = -a$ and $r = a$, where the pressure is continuous and the inverse of the mass density times the normal derivative of the pressure is continuous. This leads to

$$\lim_{x_1\uparrow -a}\,[\hat{p}^{inc}(x_1|x_1^S) + \hat{p}^{rfl}(x_1|x_1^S)] = \lim_{x_1\downarrow -a}\hat{p}^{int}(x_1|x_1^S), \tag{2.A.6}$$

$$\lim_{x_1\uparrow -a}\frac{1}{s\rho_0}\partial_1[\hat{p}^{inc}(x_1|x_1^S) + \hat{p}_k^{rfl}(x_1|x_1^S)] = \lim_{x_1\downarrow -a}\frac{1}{s\hat{\rho}_{sct}}\partial_1\hat{p}^{int}(x_1|x_1^S), \tag{2.A.7}$$

$$\lim_{x_1\uparrow a}\hat{p}^{int}(x_1|x_1^S) = \lim_{x_1\downarrow a}\hat{p}^{trm}(x_1|x_1^S), \tag{2.A.8}$$

$$\lim_{x_1\uparrow a}\frac{1}{s\hat{\rho}_{sct}}\partial_1\hat{p}^{int}(x_1|x_1^S) = \lim_{x_1\downarrow a}\frac{1}{s\rho_0}\partial_1\hat{p}^{trm}(x_1|x_1^S). \tag{2.A.9}$$

Substitution of the wave-field expressions into these interface conditions leads to the following set of equations:

$$1 + \hat{R} = \hat{A} + \exp(-2\hat{\gamma}_{sct}a)\;\hat{B}, \tag{2.A.10}$$

$$-\frac{1}{Z_0} + \frac{1}{Z_0}R = -\frac{1}{\hat{Z}_{sct}}\hat{A} + \frac{1}{\hat{Z}_{sct}}\exp(-2\hat{\gamma}_{sct}a)\;\hat{B}, \tag{2.A.11}$$

$$\exp(-2\hat{\gamma}_{sct}a)\;\hat{A} + \hat{B} = \hat{T}\exp(-2\hat{\gamma}_0 a), \tag{2.A.12}$$

$$-\frac{1}{\hat{Z}_{sct}}\exp(-2\hat{\gamma}_{sct}a)\;\hat{A} + \frac{1}{\hat{Z}_{sct}}\hat{B} = -\frac{1}{Z_0}\hat{T}\exp(-2\hat{\gamma}_0 a), \tag{2.A.13}$$

where

$$Z_0 = \rho_0\,c_0 \quad \text{and} \quad \hat{Z}_{sct} = \hat{\rho}_{sct}\,\hat{c}_{sct}\,. \tag{2.A.14}$$

Elimination of $\hat{R}$ from the first and the second equation and $\hat{T}$ from the third and the fourth equation leads to two equations for $\hat{A}$ and $\hat{B}$, with solution

$$\hat{A} = \frac{\hat{\tau}}{1 - \hat{\rho}^2\exp(-4\hat{\gamma}_{sct}a)}, \qquad \hat{B} = \frac{-\hat{\tau}\;\hat{\rho}\;\exp(-2\hat{\gamma}_{sct}a)}{1 - \hat{\rho}^2\exp(-4\hat{\gamma}_{sct}a)}, \tag{2.A.15}$$

where

$$\hat{\rho} = \frac{Z_0^{-1} - \hat{Z}_{sct}^{-1}}{Z_0^{-1} + \hat{Z}_{sct}^{-1}}, \qquad \hat{\tau} = \frac{2Z_0^{-1}}{Z_0^{-1} + \hat{Z}_{sct}^{-1}}, \tag{2.A.16}$$

are the local reflection and transmission factors. Subsequently, the global reflection and transmission factors are obtained as

$$\hat{R} = \hat{A} + \exp(-2\hat{\gamma}_{sct}a)\,\hat{B} - 1, \tag{2.A.17}$$

$$\hat{T} = [\exp(-2\hat{\gamma}_{sct}a)\,\hat{A} + \hat{B}]\exp(2\hat{\gamma}_0 a). \tag{2.A.18}$$

Note that, in case the impedance of the medium in the vertical directions does not change, i.e. $\hat{Z}_{sct} = Z_0$, we have $\hat{\rho} = 0$, $\hat{\tau} = 1$, $\hat{A} = 1$, $\hat{B} = 0$, and $\hat{R} = 0$. Then, the slab is reflection free and is invisible for acoustic waves.

As far as the scattered field is concerned, we distinguish between a point of observation above the slab and below the slab. When we have a receiver at the point x_1^R above the slab, the scattered wave is denoted as the reflected wave and is given by

$$\hat{p}^{rfl}(x_1^R|x_1^S) = \frac{Z_0\hat{Q}}{2}\,\hat{R}\,\exp\,[\hat{\gamma}_0(x_1^R + x_1^S + 2a)], \quad x_1^R < -a. \tag{2.A.19}$$

To arrive at the scattered field below the slab, we have to subtract the incident field from the transmitted field. We obtain

$$\hat{p}^{trm}(x_1^R|x_1^S) - \hat{p}^{inc}(x_1^R|x_1^S) = \frac{Z_0\hat{Q}}{2}\,(\hat{T} - 1)\,\exp\,[-\hat{\gamma}_0(x_1^R - x_1^S)], \quad x_1^R > a. \tag{2.A.20}$$

The computation of the pressure wave field is very similar to the scalar wave field. If we require (see Matlab Listing I.39) that the source wavelet $s\rho_0\hat{Q} = 1$, i.e. $Z_0\hat{Q} = 1/\hat{\gamma}_0$, the differences are manifested in the reflection and transmission factors. In particular, the difference is in the acoustic wave impedance that now contains both the wave speed and the mass density. Therefore, we suffice to present the Matlab script `AcousticSctWavefieldSlab`, where the scattered field is computed at a receiver above and below the slab (see Matlab Listing I.76).

2.A.2 2D Scattering by a Circular Cylinder

In order to generate synthetic data for the 2D configuration, we consider the scattering problem by a homogeneous, circular cylinder (Figure 2.A.2). We define the spatial position by $\boldsymbol{x}_T = (x_1, x_2)$. The cylinder has a radius a. The medium in the interior of the cylinder is characterized by the constant wave speed $\hat{c}_{sct}$ and constant mass density $\hat{\rho}_{sct}$. The center of the cylinder is at $x_1 = 0$, $x_2 = 0$. The pressure of the incident wave field is generated by a monopole line source at $\boldsymbol{x}_T^S = (x_1^S, x_2^S)$ and is given by

$$\hat{p}^{inc}(\boldsymbol{x}_T|\boldsymbol{x}_T^S) = \hat{\gamma}_0 Z_0 \frac{\hat{Q}}{2\pi} K_0(\hat{\gamma}_0|\boldsymbol{x}_T - \boldsymbol{x}_T^S|). \tag{2.A.21}$$

In order to solve our scattering problem at hand, we introduce polar coordinates adapted to the geometry of the circular cylinder,

$$x_1 = r\,\cos(\phi), \quad x_2 = r\,\sin(\phi), \quad 0 \le \phi < 2\pi. \tag{2.A.22}$$

Similarly for the source and receiver coordinates, we introduce

$$x_1^S = r^S\,\cos(\phi^S), \quad x_2^S = r^S\,\sin(\phi^S), \quad 0 \le \phi^S < 2\pi, \tag{2.A.23}$$

$$x_1^R = r^R\,\cos(\phi^R), \quad x_2^R = r^R\,\sin(\phi^R), \quad 0 \le \phi^R < 2\pi. \tag{2.A.24}$$

Similar as in Chapter 2, for $r \le r^S$, we write the pressure of the incident wave field as the infinite series

$$\hat{p}^{inc}(\boldsymbol{x}_T|\boldsymbol{x}_T^S) = \hat{\gamma}_0 Z_0 \frac{\hat{Q}}{2\pi} \sum_{m=0}^{\infty} \varepsilon_m\, I_m(\hat{\gamma}_0 r)\, K_m(\hat{\gamma}_0 r^S)\, \cos[m(\phi - \phi^S)], \tag{2.A.25}$$

where $\varepsilon_m = 1$ if $m = 0$ and $\varepsilon_m = 2$ if $m \neq 0$.

Matlab Listing I.76 The `AcousticSctWavefieldSlab` script to compute the 1D acoustic field at the two receivers above and below the slab.

```
clear all; clc; close all; clear workspace
input = initAC();
a      = input.a;                  xS = input.xS;
c_0    = input.c_0;         c_sct   = input.c_sct;
rho_0  = input.rho_0;       rho_sct = input.rho_sct;
gam0   = input.gamma_0;     gam_sct = input.gamma_0 * c_0/c_sct;
Z_0    = rho_0 * c_0;       Z_sct   = rho_sct * c_sct;

if exist(fullfile(cd, 'DATA1D.mat'), 'file');     delete DATA1D.mat;     end
% (1) Compute incident pressure wave field ------------------------------
      p_inc_a = 1/(2*gam0) * exp(gam0*(xS+a));
% (2) Compute coefficents A and B of internal field ----------------------
      Rho = (1/Z_0-1/Z_sct) / (1/Z_0+1/Z_sct);
      Tau =   2/Z_0           / (1/Z_0+1/Z_sct);
      denominator = 1 - Rho * Rho * exp(-4*gam_sct*a);
      A = Tau / denominator;
      B = - Rho * A * exp(-2*gam_sct*a);
% (3) Compute reflection and transmision factors ------------------------
      R = A + exp(-2*gam_sct*a) * B - 1;
      T = (B + exp(-2*gam_sct*a) * A) * exp(2*gam0*a);
% (4) Compute receiver data for reflection and transmission -------------
      xR = input.xR(1,1);
      data1D(1) = p_inc_a * R * exp(gam0*(xR+a));
      xR = input.xR(1,2);
      data1D(2) = p_inc_a * (T-1) * exp(-gam0*(xR+a));
      displayData(data1D,input);                       save DATA1D data1D;
```

Figure 2.A.2 The 2D configuration with homogeneous circular cylinder.

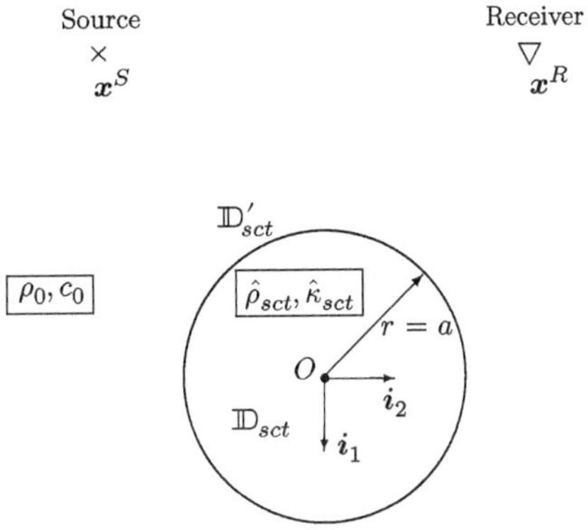

Taking into account the causality condition for the reflected wave field, we write the reflected wave field outside the scattering cylinder as

$$\hat{p}^{rfl}(\boldsymbol{x}_T|\boldsymbol{x}_T^S) = \hat{\gamma}_0 Z_0 \frac{\hat{Q}}{2\pi} \sum_{m=0}^{\infty} \varepsilon_m \hat{A}_m \, K_m(\hat{\gamma}_0 r) \, K_m(\hat{\gamma}_0 r^S) \, \cos\,[m(\phi - \phi^S)]. \tag{2.A.26}$$

The wave field inside the cylinder has to be bounded at $r = 0$, hence, we write this interior wave field as

$$\hat{p}^{int}(\boldsymbol{x}_T|\boldsymbol{x}_T^S) = \hat{\gamma}_0 Z_0 \frac{\hat{Q}}{2\pi} \sum_{m=0}^{\infty} \varepsilon_m \hat{B}_m \, I_m(\hat{\gamma}_{sct} r) \, K_m(\hat{\gamma}_0 r^S) \, \cos\,[m(\phi - \phi^S)]. \tag{2.A.27}$$

The unknown expansion factors $\hat{A}_m$ and $\hat{B}_m$ follow from the interface conditions at $r = a$, where the pressure is continuous and the inverse of the mass density times the radial derivative of the pressure is continuous. This leads to

$$\lim_{r\downarrow a} [\hat{p}^{inc}(\boldsymbol{x}_T|\boldsymbol{x}_T^S) + \hat{p}^{rfl}(\boldsymbol{x}_T|\boldsymbol{x}_T^S)] = \lim_{r\uparrow a} \hat{p}^{int}(\boldsymbol{x}_T|\boldsymbol{x}_T^S), \tag{2.A.28}$$

$$\frac{1}{s\rho_0}\lim_{r\downarrow a} \partial_r[\hat{p}^{inc}(\boldsymbol{x}_T|\boldsymbol{x}_T^S) + \hat{p}^{rfl}(\boldsymbol{x}_T|\boldsymbol{x}_T^S)] = \frac{1}{s\hat{\rho}_{sct}}\lim_{r\uparrow a} \partial_r\hat{p}^{int}(\boldsymbol{x}_T|\boldsymbol{x}_T^S). \tag{2.A.29}$$

Substitution of the wave-field expressions into these interface conditions leads, for each m to the following set of equations:

$$I_m(\hat{\gamma}_0 a) + \hat{A}_m\, K_m(\hat{\gamma}_0 a) = \hat{B}_m\, I_m(\hat{\gamma}_{sct} a), \tag{2.A.30}$$

$$\frac{1}{s\rho_0}[\partial_r I_m(\hat{\gamma}_0 r) + \hat{A}_m \partial_r K_m(\hat{\gamma}_0 r)]_{r=a} = \frac{1}{s\hat{\rho}_{sct}}\hat{B}_m\, \partial_r I_m(\hat{\gamma}_{sct} r)\Big|_{r=a}. \tag{2.A.31}$$

Solving these equations for the coefficients A_m and B_m leads to

$$\hat{A}_m = -\frac{\hat{Z}_{sct}^{-1}\, \partial I_m(\hat{\gamma}_{sct} a)\, I_m(\hat{\gamma}_0 a) - Z_0^{-1}\, \partial I_m(\hat{\gamma}_0 a)\, I_m(\hat{\gamma}_{sct} a)}{\hat{Z}_{sct}^{-1}\, \partial I_m(\hat{\gamma}_{sct} a)\, K_m(\hat{\gamma}_0 a) - Z_0^{-1}\, \partial K_m(\hat{\gamma}_0 a)\, I_m(\hat{\gamma}_{sct} a)} \tag{2.A.32}$$

and

$$\hat{B}_m = (I_m(\hat{\gamma}_0 a) + \hat{A}_m\, K_m(\hat{\gamma}_0 a))/I_m(\hat{\gamma}_{sct} a), \tag{2.A.33}$$

where $\partial I_m(\cdot)$ and $\partial K_m(\cdot)$ denote the derivatives of the functions I_m and K_m with respect to their arguments. Their expressions are given in Eq. (1.B.33). Substituting the factors $\hat{A}_m$ in the expression for the reflected wave field of Eq. (2.A.26) and taking $\boldsymbol{x}_T = \boldsymbol{x}_T^R$, we obtain

$$\hat{p}^{rfl}(\boldsymbol{x}_T^R|\boldsymbol{x}_T^S) = \hat{\gamma}_0 Z_0 \frac{\hat{Q}}{2\pi}\sum_{m=0}^{\infty} \varepsilon_m \hat{A}_m\, K_m(\hat{\gamma}_0 r^R)\, K_m(\hat{\gamma}_0 r^S)\, \cos\,[m(\phi^R - \phi^S)]. \tag{2.A.34}$$

With this expression we are able to compute the synthetic data pertaining to the present example of a domain scatterer.

Similar to the 1D case, the computation of the pressure wave field is very similar to the scalar wave field. The differences are manifested in the reflection and transmission factors, where the acoustic wave impedance, now contains both the wave speed and the mass density. Therefore, we again suffice to present the Matlab script `AcousticSctWavefieldCircle`, where the scattered field is computed at a number of receivers around the circular sphere (see Matlab Listing I.77).

2.A.2.1 No Contrast in Wave Speed

We now consider the special case that the two wave speeds are the same; then,

$$\hat{c}_{sct} = c_0, \quad \hat{\gamma}_{sct} = \hat{\gamma}_0, \quad \text{and} \quad \frac{\hat{Z}_{sct}}{Z_0} = \frac{\hat{\rho}_{sct}}{\rho_0}. \tag{2.A.35}$$

The reflection coefficient of the reflected field simplifies to

$$\hat{A}_m = -\frac{\left(\dfrac{1}{\hat{\rho}_{sct}} - \dfrac{1}{\rho_0}\right) \partial I_m(\hat{\gamma}_0 a)\, I_m(\hat{\gamma}_0 a)}{\dfrac{1}{\hat{\rho}_{sct}}\partial I_m(\hat{\gamma}_0 a)\, K_m(\hat{\gamma}_0 a) - \dfrac{1}{\rho_0}\, \partial K_m(\hat{\gamma}_0 a)\, I_m(\hat{\gamma}_0 a)}. \tag{2.A.36}$$

We now use the Wronskian,

$$-\partial K_m(\hat{\gamma}_0 a) I_m(\hat{\gamma}_0 a) + K_m(\hat{\gamma}_0 a)\partial I_m(\hat{\gamma}_0 a) = \frac{1}{\hat{\gamma}_0 a}, \tag{2.A.37}$$

Matlab Listing I.77 The `AcousticSctWavefieldCircle` script to compute the 1D acoustic scattered field at the receiver locations around the circular cylinder.

```
clear all; clc; close all; clear workspace
input = initAC();
c_0   = input.c_0;         c_sct   = input.c_sct;
rho_0 = input.rho_0;       rho_sct = input.rho_sct;
gam0  = input.gamma_0;     gam_sct = input.gamma_0 * c_0/c_sct;
Z_0   = rho_0 * c_0;       Z_sct   = rho_sct * c_sct;

if exist(fullfile(cd, 'DATA2D.mat'), 'file');     delete DATA2D.mat;    end

% (1) Compute coefficients of series expansion ---------------------------
    arg0 = gam0 * input.a;   args = gam_sct*input.a;
    M = 50;                                     % increase M for more accuracy
    A = zeros(1,M+1); B = zeros(1,M+1);
    for m = 0 : M
      Ib0 = besseli(m,arg0);     dIb0 =  besseli(m+1,arg0) + m/arg0 * Ib0;
      Ibs = besseli(m,args);     dIbs =  besseli(m+1,args) + m/args * Ibs;
      Kb0 = besselk(m,arg0);     dKb0 = -besselk(m+1,arg0) + m/arg0 * Kb0;
      A(m+1) = - ((1/Z_sct) * dIbs*Ib0 - (1/Z_0) * dIb0*Ibs) ...
                /((1/Z_sct) * dIbs*Kb0 - (1/Z_0) * dKb0*Ibs);
      B(m+1) = (Ib0 + A(m+1) * Kb0) / Ibs;
    end

% (2) Compute reflected field at receivers (data) -------------------------
    xR = input.xR; xS = input.xS;
    rR = sqrt(xR(1,:).^2 + xR(2,:).^2);   phiR = atan2(xR(2,:),xR(1,:));
    rS = sqrt(xS(1)^2+xS(2)^2);           phiS = atan2(xS(2),xS(1));
    data2D = A(1) * besselk(0,gam0*rS).* besselk(0,gam0*rR);
    for m = 1 : M
      factor = 2 * besselk(m,gam0*rS) .* cos(m*(phiS-phiR));
      data2D = data2D + A(m+1) * factor .* besselk(m,gam0*rR);
    end % m_loop
    data2D = 1/(2*pi) * data2D;

    displayData(data2D,input);                       save DATA2D data2D;
```

and obtain

$$\hat{A}_m = \frac{\dfrac{\hat{\rho}_{sct}-\rho_0}{\hat{\rho}_{sct}}\,\hat{\gamma}_0 a\;\partial I_m(\hat{\gamma}_0 a)\;I_m(\hat{\gamma}_0 a)}{1-\dfrac{\hat{\rho}_{sct}-\rho_0}{\hat{\rho}_{sct}}\,\hat{\gamma}_0 a\;\partial K_m(\hat{\gamma}_0 a)\;I_m(\hat{\gamma}_0 a)}. \tag{2.A.38}$$

Alternatively, we can start with the 2D form of the integral representation of Eq. (2.119), where we use $\partial_k = -\partial'_k$, $\hat{\rho}^+ = \rho_0$, and $\hat{\rho}^- = \rho_{sct}$. Since the latter two are constant, we obtain the integral representation

$$\hat{p}^{sct}(\boldsymbol{x}_T) = \frac{\hat{\rho}_{sct}-\rho_0}{\hat{\rho}_{sct}} \int_{\boldsymbol{x}'_T\in\mathbb{S}_{sct}} \nu_k(\boldsymbol{x}'_T)\partial'_k\hat{G}(\boldsymbol{x}_T-\boldsymbol{x}'_T)\;\hat{p}(\boldsymbol{x}'_T)\;\mathrm{d}s. \tag{2.A.39}$$

Here, $\boldsymbol{\nu}(\boldsymbol{x})$ denotes the normal to the interface of the scatterer pointing away of this scatterer.

So far the present analysis holds for each homogeneous 2D scatterer. For the circular cylinder with radius a, we introduce polar coordinates, and we then have

$$\hat{p}^{sct}(r,\phi) = \frac{\hat{\rho}_{sct} - \rho_0}{\hat{\rho}_{sct}} \int_0^{2\pi} \partial_r' \hat{G}(r, r', \phi - \phi')|_{r'=a}\ \hat{p}(a, \phi')\ a\ d\phi'. \tag{2.A.40}$$

The derivative of the Green function, for $r \geq a$ is obtained from Eq. (I.A.26) as

$$\partial_r' \hat{G}(r, r', \phi - \phi')|_{r'=a} = \frac{\hat{\gamma}_0}{2\pi} \sum_{m=-\infty}^{\infty} K_m(\hat{\gamma}_0 r)\ \partial I_m(\hat{\gamma}_0 a)\ \exp[im(\phi - \phi')]. \tag{2.A.41}$$

For convenience, we write

$$\begin{aligned} \hat{p}^{inc}(r,\phi) &= \sum_{m=-\infty}^{\infty} \hat{P}_m^{inc}(r) \exp(im\phi), \\ \hat{p}^{sct}(r,\phi) &= \sum_{m=-\infty}^{\infty} \hat{P}_m^{sct}(r) \exp(im\phi), \\ \hat{p}(r,\phi) &= \sum_{m=-\infty}^{\infty} \hat{P}_m(r) \exp(im\phi). \end{aligned} \tag{2.A.42}$$

In view of the convolutional operator with respect to ϕ, Eq. (2.A.40) leads to the angular spectral relation

$$\hat{P}_m^{sct}(r) = \frac{\hat{\rho}_{sct} - \rho_0}{\hat{\rho}_{sct}}\ \hat{\gamma}_0 a\ K_m(\hat{\gamma}_0 r)\ \partial I_m(\hat{\gamma}_0 a) \hat{P}_m(a). \tag{2.A.43}$$

We subsequently require that $\hat{P}_m(a) = \hat{P}_m^{inc}(a) + \hat{P}_m^{sct}(a)$ at the interface $r = a$. This leads to an equation for $\hat{P}_m^{sct}(a)$ only, viz.,

$$\hat{P}_m^{sct}(a) = \frac{\hat{\rho}_{sct} - \rho_0}{\hat{\rho}_{sct}}\ \hat{\gamma}_0 a\ K_m(\hat{\gamma}_0 a)\ \partial I_m(\hat{\gamma}_0 a)\ [\hat{P}_m^{inc}(a) + \hat{P}_m^{sct}(a)], \tag{2.A.44}$$

with solution

$$\hat{P}_m^{sct}(a) = \frac{\dfrac{\hat{\rho}_{sct} - \rho_0}{\hat{\rho}_{sct}}\ \hat{\gamma}_0 a\ K_m(\hat{\gamma}_0 a)\ \partial I_m(\hat{\gamma}_0 a)\ \hat{P}_m^{inc}(a)}{1 - \dfrac{\hat{\rho}_{sct} - \rho_0}{\hat{\rho}_{sct}} \hat{\gamma}_0 a\ K_m(\hat{\gamma}_0 a)\ \partial I_m(\hat{\gamma}_0 a)}. \tag{2.A.45}$$

From the first relation of Eqs. (2.A.25) and (2.A.26), we conclude that

$$\frac{\hat{P}_m^{sct}(a)}{\hat{P}_m^{inc}(a)} = \frac{\hat{A}_m\ K_m(\hat{\gamma}_0 a)}{I_m(\hat{\gamma}_0 a)}, \tag{2.A.46}$$

where we have the reflected field associated with the scattered field. Substituting this result in Eq. (2.A.45) we find that

$$\hat{A}_m = \frac{\dfrac{\hat{\rho}_{sct} - \rho_0}{\hat{\rho}_{sct}}\ \hat{\gamma}_0 a\ \partial I_m(\hat{\gamma}_0 a)\ I_m(\hat{\gamma}_0 a)}{1 - \dfrac{\hat{\rho}_{sct} - \rho_0}{\hat{\rho}_{sct}}\ \hat{\gamma}_0 a\ \partial K_m(\hat{\gamma}_0 a)\ I_m(\hat{\gamma}_0 a)}, \tag{2.A.47}$$

which is identically equal to the expression of Eq. (2.A.38). This proofs that the integral representation at hand yields the same analytic result.

2.A.3 3D Scattering by a Sphere

In order to generate synthetic data for the 3D configuration, we consider the scattering problem by a homogeneous sphere (Figure 2.A.3). The sphere has a radius a. The medium in the interior of the sphere is characterized by the constant wave speed $\hat{c}_{sct}$ and the constant mass density $\hat{\rho}_{sct}$. The center of the sphere is at $\boldsymbol{x} = (0, 0, 0)$. The incident wave field is generated by a monopole source at $\boldsymbol{x}^S = (x_1^S, x_2^S, x_3^S)$ and is given by

$$\hat{p}^{inc}(\boldsymbol{x}|\boldsymbol{x}^S) = \hat{\gamma}_0 Z_0 \hat{Q} \frac{\exp(-\hat{\gamma}_0|\boldsymbol{x} - \boldsymbol{x}^S|)}{4\pi|\boldsymbol{x} - \boldsymbol{x}^S|}. \tag{2.A.48}$$

In order to solve our scattering problem at hand, we introduce spherical coordinates adapted to the geometry of the sphere,

$$x_1 = r\ \sin(\theta)\ \cos(\phi), \quad x_2 = r\ \sin(\theta)\ \sin(\phi), \quad x_3 = r\ \cos(\theta), \tag{2.A.49}$$

with $0 \le \phi < 2\pi$ and $0 \le \theta < \pi$. Similarly, for the source and receiver coordinates, we introduce

$$x_1^S = r^S\ \sin(\theta^S)\ \cos(\phi^S), \quad x_2^S = r^S\ \sin(\theta^S)\ \sin(\phi^S), \quad x_3^S = r^S\ \cos(\theta^S), \tag{2.A.50}$$

$$x_1^R = r^R\ \sin(\theta^R)\ \cos(\phi^R), \quad x_2^R = r^R\ \sin(\theta^R)\ \sin(\phi^R), \quad x_3^R = r^R\ \cos(\theta^R). \tag{2.A.51}$$

Similarly as in Chapter 1, for $r \le r^S$, the incident acoustic pressure is represented as

$$\hat{p}^{inc}(\boldsymbol{x}|\boldsymbol{x}^S) = \frac{\hat{\gamma}_0^2 Z_0 \hat{Q}}{2\pi^2} \sum_{n=0}^{\infty} (2n+1)\ i_n^{(1)}(\hat{\gamma}_0 r)\ k_n(\hat{\gamma}_0 r^S)\ P_n[\cos(\boldsymbol{x}, \boldsymbol{x}^S)]. \tag{2.A.52}$$

Here, $i_n^{(1)}(\cdot)$ and $k_n(\cdot)$ are the nth order modified spherical Bessel functions of the first kind and second kind, respectively.

Taking into account the causality condition for the reflected wave field, we write the reflected wave field outside the scattering sphere as

$$\hat{p}^{rfl}(\boldsymbol{x}|\boldsymbol{x}^S) = \frac{\hat{\gamma}_0^2 Z_0 \hat{Q}}{2\pi^2} \sum_{n=0}^{\infty} (2n+1)\ \hat{A}_n\ k_n(\hat{\gamma}_0 r)\ k_n(\hat{\gamma}_0 r^S)\ P_n[\cos(\boldsymbol{x}, \boldsymbol{x}^S)]. \tag{2.A.53}$$

The wave field inside the sphere has to be bounded at $r = 0$, hence, we write this interior wave field as

$$\hat{p}^{int}(\boldsymbol{x}|\boldsymbol{x}^S) = \frac{\hat{\gamma}_0^2 Z_0 \hat{Q}}{2\pi^2} \sum_{n=0}^{\infty} (2n+1)\ \hat{B}_n\ i_n^{(1)}(\hat{\gamma}_{sct} r)\ k_n(\hat{\gamma}_0 r^S)\ P_n[\cos(\boldsymbol{x}, \boldsymbol{x}^S)]. \tag{2.A.54}$$

Figure 2.A.3 The 3D configuration with homogeneous sphere.

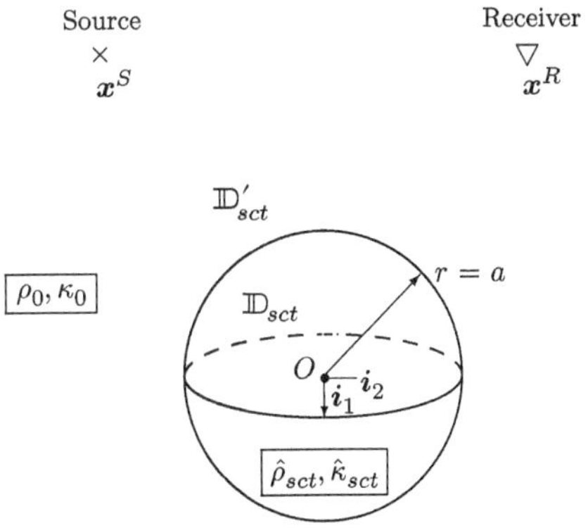

The unknown expansion factors $\hat{A}_n$ and $\hat{B}_n$ follow from the interface conditions at $r = a$, where the wave field and the inverse of the mass density times the radial derivative of the pressure are continuous. This leads to

$$\lim_{r\downarrow a} [\hat{p}^{inc}(\boldsymbol{x}|\boldsymbol{x}^S) + \hat{p}^{rfl}(\boldsymbol{x}|\boldsymbol{x}^S)] = \lim_{r\uparrow a} \hat{p}^{int}(\boldsymbol{x}|\boldsymbol{x}^S), \tag{2.A.55}$$

$$\frac{1}{s\hat{\rho}_0}\lim_{r\downarrow a} \partial_r[\hat{p}^{inc}(\boldsymbol{x}|\boldsymbol{x}^S) + \hat{p}_k^{rfl}(\boldsymbol{x}|\boldsymbol{x}^S)] = \frac{1}{s\hat{\rho}_{sct}}\lim_{r\uparrow a} \partial_r\hat{p}_k^{int}(\boldsymbol{x}|\boldsymbol{x}^S). \tag{2.A.56}$$

Substitution of the wave-field expressions into these interface conditions leads, for each n, to the following set of equations:

$$i_n^{(1)}(\hat{\gamma}_0 a) + \hat{A}_n\, k_n(\hat{\gamma}_0 a) = \hat{B}_n\, i_n^{(1)}(\hat{\gamma}_{sct} a), \tag{2.A.57}$$

$$\frac{1}{s\rho_0}[\partial_r i_n^{(1)}(\hat{\gamma}_0 r) + \hat{A}_n \partial_r k_n(\hat{\gamma}_0 r)]_{r=a} = \hat{B}_n\, \frac{1}{s\hat{\rho}_{sct}}\partial_r\, i_n^{(1)}(\hat{\gamma}_{sct} r)\Big|_{r=a}. \tag{2.A.58}$$

Matlab Listing I.78 The `AcousticSctWavefieldSphere` script to compute the 2D scattered wave field at the receiver locations around the sphere.

```
clear all; clc; close all; clear workspace;    input = initAC();
c_0   = input.c_0;         c_sct   = input.c_sct;
rho_0 = input.rho_0;       rho_sct = input.rho_sct;
gam0  = input.gamma_0;     gam_sct = input.gamma_0 * c_0/c_sct;
Z_0   = rho_0 * c_0;       Z_sct   = rho_sct * c_sct;

if exist(fullfile(cd, 'DATA3D.mat'), 'file');   delete DATA3D.mat;   end
% (1) Compute coefficients of series expansion ---------------------------
    N = 50;                                    % increase N for more accuracy
    A = zeros(N+1); B = zeros(N+1);
    arg0 = gam0    * input.a;
    args = gam_sct * input.a;
  for n = 0 : N
    Ib0 = besseli(n+1/2,arg0);   dIb0 = besseli(n+3/2,arg0) + n/arg0*Ib0;
    Ibs = besseli(n+1/2,args);   dIbs = besseli(n+3/2,args) + n/args*Ibs;
    Kb0 = besselk(n+1/2,arg0);   dKb0 =-besselk(n+3/2,arg0) + n/arg0*Kb0;
    denominator = (1/Z_sct) * dIbs*Kb0 - (1/Z_0) * dKb0*Ibs;
    A(n+1) = -((1/Z_sct) * dIbs*Ib0 - (1/Z_0) * dIb0*Ibs) / denominator;
    B(n+1) = (Ib0 + A(n+1)*Kb0)/Ibs;
  end

% (2) Compute reflected field at receivers (data) --------------------------
    xR  = input.xR;  rR = sqrt(xR(1,:).^2 + xR(2,:).^2 + xR(3,:).^2);
    xS  = input.xS; rS = sqrt(xS(1)^2+xS(2)^2+xS(3)^2);
    DIS = sqrt((xR(1,:)-xS(1)).^2+(xR(2,:)-xS(2)).^2+(xR(3,:)-xS(3)).^2);
    COS = (rR.^2 + rS^2 - DIS.^2) ./ (2*rR*rS);           % cos(xR,xS)
    Pn  = zeros(1,input.NR); Pn_1 = Pn; Pn_2 = Pn;
    data3D = zeros(1,input.NR);
    for n = 0 : N
      [Pn,Pn_1,Pn_2] = Legendre(n,COS,Pn,Pn_1,Pn_2);
       factor        = (2*n+1)*besselk(n+1/2,gam0*rS).*Pn;
       data3D        = data3D + A(n+1).*factor.* besselk(n+1/2,gam0*rR);
    end % n_loop
    data3D = data3D ./ (4*pi*sqrt(rS*rR));

    displayData(data3D,input);                        save DATA3D data3D;
```

Solving these equations for the coefficients A_n and B_n leads to

$$\hat{A}_n = -\frac{\hat{Z}_{sct}^{-1}\ \partial i_n^{(1)}(\hat{\gamma}_{sct}a)\ i_n^{(1)}(\hat{\gamma}_0 a) - Z_0^{-1}\ \partial i_n^{(1)}(\hat{\gamma}_0 a)\ i_n^{(1)}(\hat{\gamma}_{sct}a)}{\hat{Z}_{sct}^{-1}\ \partial i_n^{(1)}(\hat{\gamma}_{sct}a)\ k_n(\hat{\gamma}_0 a) - Z_0^{-1}\ \partial k_n(\hat{\gamma}_0 a)\ i_n^{(1)}(\hat{\gamma}_{sct}a)} \tag{2.A.59}$$

and

$$\hat{B}_n = \frac{i_n^{(1)}(\hat{\gamma}_0 a) + \hat{A}_n\ k_n(\hat{\gamma}_0 a)}{i_n^{(1)}(\hat{\gamma}_{sct}a)}, \tag{2.A.60}$$

where $\partial i_n^{(1)}(\cdot)$ and $\partial k_n(\cdot)$ denote the derivatives of the functions $i_n^{(1)}(\cdot)$ and $k_n(\cdot)$ with respect to their arguments.

Substituting the reflection factors $\hat{A}_n$ in the expression for the reflected wave field of Eq. (2.A.53) and taking $\boldsymbol{x} = \boldsymbol{x}^R$, we obtain

$$\hat{p}^{rfl}(\boldsymbol{x}^R|\boldsymbol{x}^S) = \frac{\hat{\gamma}_0^2 Z_0 \hat{Q}}{2\pi^2}\sum_{n=0}^{\infty}(2n+1)\ \hat{A}_n\ k_n(\hat{\gamma}_0 r^R)\ k_n(\hat{\gamma}_0 r^S)\ P_n[\cos(\boldsymbol{x}^R, \boldsymbol{x}^S)]. \tag{2.A.61}$$

With this expression we are able to compute the synthetic data pertaining to the present example of a domain scatterer. For the computation of the Bessel functions and the Legendre polynomials, we refer to the pertaining section of Chapter 1. We again suffice to present the Matlab script `AcousticSctWavefieldSphere` of Matlab Listing I.78.

2.A.4 Scattered-Field Computations Canonical Objects

After completing the description of the various Matlab scripts to compute the detailed behavior of the scattering by the canonical objects, we finally present a combined script to compute the scattered field at a number of receivers points, see the script `AcForwardCanonicalObjects` of Matlab Listing I.79. This latter script should be used, before we run a particular Matlab script based on a domain integral equation. The resulting field values are used as a measure for the acoustic wave fields computed with the integral equation methods.

Matlab Listing I.79 The `AcForwardCanonicalObjects` script to compute the scattered wave fields for canonical objects.

```
clear all; clc; close all; clear workspace
input = initAC();
global nDIM;

if nDIM == 1
  % Compute scattered acoustic field at receivers above/below the slab
  disp('Running AcousticSctWavefieldSlab');     AcousticSctWavefieldSlab;

elseif nDIM == 2
   % Compute scattered acoustic field at receivers around circular cylinder
   disp('Running AcousticSctWavefieldCircle'); AcousticSctWavefieldCircle;

elseif nDIM == 3
   % Compute scattered acoustic field at receivers around sphere
   disp('Running AcousticSctWavefieldSphere'); AcousticSctWavefieldSphere;

end % if
```

3

Electromagnetic Waves

In this chapter, we consider the propagation and scattering of electromagnetic waves. Instead of an acoustic wave field with a scalar pressure and a particle-velocity vector, we now deal with two vector quantities, viz., the electric-field strength and the magnetic-field strength. The theory of electromagnetic waves goes back to the Treatise on Electricity and Magnetism by Maxwell [33]. The basic equations for propagation of electromagnetic waves through matter are given by the Maxwell equations

$$-\nabla \times \boldsymbol{H} + \partial_t \boldsymbol{D} = -\boldsymbol{J}, \tag{3.1}$$

$$\nabla \times \boldsymbol{E} + \partial_t \boldsymbol{B} = -\boldsymbol{K}, \tag{3.2}$$

where $\boldsymbol{E} = \boldsymbol{E}(\boldsymbol{x}, t)$ is the electric-field strength, $\boldsymbol{H} = \boldsymbol{H}(\boldsymbol{x}, t)$ is the magnetic-field strength, $\boldsymbol{D} = \boldsymbol{D}(\boldsymbol{x}, t)$ is the electric-flux density, and $\boldsymbol{B} = \boldsymbol{B}(\boldsymbol{x}, t)$ is the magnetic-flux density. Further, $\boldsymbol{J} = \boldsymbol{J}(\boldsymbol{x}, t)$ is volume-source density of electric current and $\boldsymbol{K} = \boldsymbol{K}(\boldsymbol{x}, t)$ is volume-source density of magnetic current. The latter quantity has been introduced for symmetry and mathematical convenience reasons. It is noted that $\boldsymbol{K}$ has no physical meaning.

For an isotropic and lossless medium, we have the constitutive relations

$$\boldsymbol{D}(\boldsymbol{x}, t) = \varepsilon \boldsymbol{E}(\boldsymbol{x}, t), \tag{3.3}$$

$$\boldsymbol{B}(\boldsymbol{x}, t) = \mu \boldsymbol{H}(\boldsymbol{x}, t), \tag{3.4}$$

where $\varepsilon = \varepsilon(\boldsymbol{x})$ is the electric permittivity and $\mu = \mu(\boldsymbol{x})$ is the magnetic permeability.

Further, applying the divergence operator to Eqs. (3.1) and (3.2) leads to

$$\nabla \cdot (\partial_t \boldsymbol{D}) = -\nabla \cdot \boldsymbol{J}, \tag{3.5}$$

$$\nabla \cdot (\partial_t \boldsymbol{H}) = -\nabla \cdot \boldsymbol{K}. \tag{3.6}$$

These equations are not independent of Maxwell's equations and are a kind of compatibility relationships.

In those domains in a medium where the permittivity and the permeability change continuously with position, the electromagnetic field vectors are continuously differentiable functions of position and satisfy the differential equations (3.1)–(3.2) and (3.5)–(3.6). Across certain interfaces in a configuration, the electromagnetic field quantities may exhibit a discontinuous behavior. Because on those positions they are no longer continuously differentiable, the electromagnetic field equations cease to hold. To interrelate the electromagnetic wave-field quantities at either side of the interface, a certain set of boundary conditions is needed. When crossing the interface of two adjacent media that differ in their electromagnetic properties, the electric and the magnetic field strengths are, in

Forward and Inverse Scattering Algorithms based on Contrast Source Integral Equations, First Edition. Peter M. van den Berg.

Companion website: www.wiley.com/go/vandenBerg/ScatteringAlgorithms

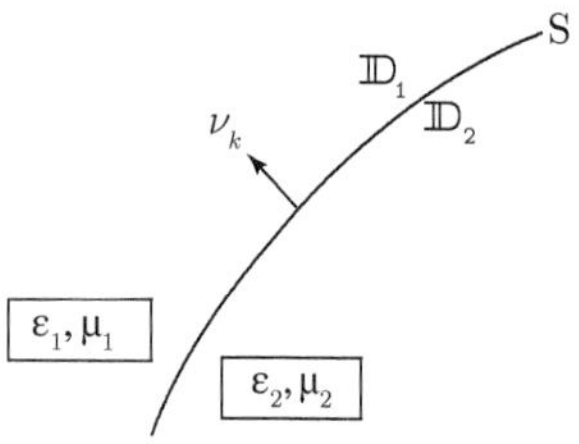

Figure 3.1 Interface between two media with different electromagnetic properties.

general, no longer continuously differentiable in a domain that contains (part of) an interface, and Eqs. (3.1)–(3.6) cease to hold. To solve electromagnetic wave-field problems in domains that contain abrupt boundaries, the electromagnetic field equations must be supplemented by conditions that interrelate the field values at either side of the interface, the so-called boundary conditions.

Let S indicate the interface and assume that S has everywhere a unique tangent plane. Let, further, $\boldsymbol{\nu} = \nu_k$ denote the unit vector along the normal S such that upon traversing S in the direction of ν_k, we pass from the domain $\mathbb{D}_2$ to the domain $\mathbb{D}_1$, $\mathbb{D}_1$ and $\mathbb{D}_2$ being located at either side of S (Figure 3.1). Suppose, now, that some (or all) electromagnetic wave quantities jump across S. In the direction parallel to S, all electromagnetic wave quantities still vary in a continuously differentiable manner, and hence, the partial derivatives parallel to S give no problem in Eqs. (3.1)–(3.2) and (3.5)–(3.6). The partial derivatives perpendicular to S, on the contrary, meet functions that show a jump discontinuity across S; these give rise to interface Dirac distributions (interface impulses) located on S. Distributions of this kind would, however, physically be representative of interface sources located on S. In the absence of such sources, the absence of interface impulses in the partial derivatives perpendicular to S should be enforced. The latter is done by requiring that these normal derivatives only meet functions that are continuous across S. This procedure leads to the time-domain continuity conditions that the tangential components of electromagnetic field strengths $\boldsymbol{E}$ and $\boldsymbol{H}$ and the normal components of the electromagnetic flux densities $\boldsymbol{D}$ and $\boldsymbol{B}$ should be continuous across S, viz., for $\boldsymbol{x} \in$ S and $h > 0$, i.e.

$$\lim_{h\to 0} \boldsymbol{\nu} \times \boldsymbol{E}(\boldsymbol{x} + h\boldsymbol{\nu}, t) = \lim_{h\to 0} \boldsymbol{\nu} \times \boldsymbol{E}(\boldsymbol{x} - h\boldsymbol{\nu}, t), \tag{3.7}$$

$$\lim_{h\to 0} \boldsymbol{\nu} \times \boldsymbol{H}(\boldsymbol{x} + h\boldsymbol{\nu}, t) = \lim_{h\to 0} \boldsymbol{\nu} \times \boldsymbol{H}(\boldsymbol{x} - h\boldsymbol{\nu}, t), \tag{3.8}$$

$$\lim_{h\to 0} \boldsymbol{\nu} \cdot \boldsymbol{D}(\boldsymbol{x} + h\boldsymbol{\nu}, t) = \lim_{h\to 0} \boldsymbol{\nu} \cdot \boldsymbol{D}(\boldsymbol{x} - h\boldsymbol{\nu}, t), \tag{3.9}$$

$$\lim_{h\to 0} \boldsymbol{\nu} \cdot \boldsymbol{B}(\boldsymbol{x} + h\boldsymbol{\nu}, t) = \lim_{h\to 0} \boldsymbol{\nu} \cdot \boldsymbol{B}(\boldsymbol{x} - h\boldsymbol{\nu}, t). \tag{3.10}$$

In a large number of cases met in practice, we are interested in the behavior of electromagnetic fields in **linear configurations**. Mathematically, one can take advantage of this situation by carrying out a Laplace transformation with respect to time,

$$\hat{\boldsymbol{E}}(\boldsymbol{x}, s) = \int_{t_0}^{\infty} \exp(-st)\boldsymbol{E}(\boldsymbol{x}, t)\mathrm{d}t, \tag{3.11}$$

$$\hat{\boldsymbol{H}}(\boldsymbol{x}, s) = \int_{t_0}^{\infty} \exp(-st)\boldsymbol{H}(\boldsymbol{x}, t)\mathrm{d}t, \tag{3.12}$$

and considering the equations governing the wave field in the corresponding Laplace-transform domain or s-domain. In the s-domain relations, the time coordinate has been eliminated, and a wave-field problem in space remains in which the transform parameter s occurs. Causality of the wave field is taken into account by taking $\mathrm{Re}(s) > 0$, and requiring that all causal wave-field quantities are analytic functions of s in the right half $0 < \mathrm{Re}(s) < \infty$ of the complex s-plane.

For linear media, we subject the electromagnetic field equations to a Laplace transformation. For completeness, we allow a nonvanishing electromagnetic field to be present at $t = t_0$, although in the majority of cases we are interested in the causal wave field generated by sources that are switched on at the instant $t = t_0$, in which case the initial values of the electromagnetic field are considered zero. Integrating by parts and using Eqs. (3.3) and (3.4), we arrive at

$$\int_{t=t_0}^{\infty} \exp(-st)\partial_t \boldsymbol{D}(\boldsymbol{x}, t)\mathrm{d}t = -\boldsymbol{D}(\boldsymbol{x}, t_0)\ \exp(-st_0) + s\hat{\varepsilon}(\boldsymbol{x}, s)\hat{\boldsymbol{E}}(\boldsymbol{x}, s), \tag{3.13}$$

$$\int_{t=t_0}^{\infty} \exp(-st)\partial_t \boldsymbol{B}(\boldsymbol{x}, t)\mathrm{d}t = -\boldsymbol{B}(\boldsymbol{x}, t_0) \exp(-st_0) + s\hat{\mu}(\boldsymbol{x}, s)\hat{\boldsymbol{H}}(\boldsymbol{x}, s), \tag{3.14}$$

and subsequently

$$-\boldsymbol{\nabla} \times \hat{\boldsymbol{H}} + s\hat{\varepsilon}\hat{\boldsymbol{E}} = -\hat{\boldsymbol{J}} + \boldsymbol{D}(\boldsymbol{x}, t_0) \exp(-st_0), \tag{3.15}$$

$$\boldsymbol{\nabla} \times \hat{\boldsymbol{E}} + s\hat{\mu}\hat{\boldsymbol{H}} = -\hat{\boldsymbol{K}} + \boldsymbol{B}(\boldsymbol{x}, t_0) \exp(-st_0). \tag{3.16}$$

We remark that these equations also hold for lossy media, in which the permittivity and permeability depend on s as well. We therefore have used the notation $\hat{\varepsilon} = \hat{\varepsilon}(\boldsymbol{x}, s)$ and $\hat{\mu} = \hat{\mu}(\boldsymbol{x}, s)$. From Eqs. (3.15) to (3.16), it follows that in the s-domain, one can take into account the influence of a nonvanishing initial electromagnetic field by properly incorporating it in the s-domain volume densities of external electric current and external magnetic current. In the remainder of our analysis, it will be tacitly understood that nonzero initial electromagnetic field values have been accounted for in this manner. Our electromagnetic wave equations in the s-domain have then the final form

$$-\boldsymbol{\nabla} \times \hat{\boldsymbol{H}} + s\hat{\varepsilon}\hat{\boldsymbol{E}} = -\hat{\boldsymbol{J}}, \tag{3.17}$$

$$\boldsymbol{\nabla} \times \hat{\boldsymbol{E}} + s\hat{\mu}\hat{\boldsymbol{H}} = -\hat{\boldsymbol{K}}. \tag{3.18}$$

The continuity conditions of Eqs. (3.8)–(3.10) in the time-domain leads to continuity conditions in the s-domain:

$$\lim_{h\to 0} \boldsymbol{v} \times \hat{\boldsymbol{E}}(\boldsymbol{x} + h\boldsymbol{v}, s) = \lim_{h\to 0} \boldsymbol{v} \times \hat{\boldsymbol{E}}(\boldsymbol{x} - h\boldsymbol{v}, s), \tag{3.19}$$

$$\lim_{h\to 0} \boldsymbol{v} \times \hat{\boldsymbol{H}}(\boldsymbol{x} + h\boldsymbol{v}, s) = \lim_{h\to 0} \boldsymbol{v} \times \hat{\boldsymbol{H}}(\boldsymbol{x} - h\boldsymbol{v}, s), \tag{3.20}$$

$$\lim_{h\to 0} \boldsymbol{v} \cdot \hat{\boldsymbol{D}}(\boldsymbol{x} + h\boldsymbol{v}, s) = \lim_{h\to 0} \boldsymbol{v} \cdot \hat{\boldsymbol{D}}(\boldsymbol{x} - h\boldsymbol{v}, s), \tag{3.21}$$

$$\lim_{h\to 0} \boldsymbol{v} \cdot \hat{\boldsymbol{B}}(\boldsymbol{x} + h\boldsymbol{v}, s) = \lim_{h\to 0} \boldsymbol{v} \cdot \hat{\boldsymbol{B}}(\boldsymbol{x} - h\boldsymbol{v}, s). \tag{3.22}$$

The behavior of electromagnetic waves of interest is often characterized by the results of a **steady-state analysis**. In such an analysis, all electromagnetic wave quantities are taken to depend sinusoidally on time with a common angular frequency, ω say. Each purely real quantity can then be associated with a complex counterpart and a common time factor $\exp(-\mathrm{i}\omega t)$. In doing so, the original quantities in the time domain are found from the complex counterparts as

$$\begin{aligned}&\{\boldsymbol{E}(\boldsymbol{x}, t), \boldsymbol{H}(\boldsymbol{x}, t), \boldsymbol{J}(\boldsymbol{x}, t), \boldsymbol{K}(\boldsymbol{x}, t)\} \\ &\quad = \mathrm{Re}\,[\{\hat{\boldsymbol{E}}(\boldsymbol{x}, -\mathrm{i}\omega), \hat{\boldsymbol{H}}(\boldsymbol{x}, -\mathrm{i}\omega), \hat{\boldsymbol{J}}(\boldsymbol{x}, -\mathrm{i}\omega), \hat{\boldsymbol{K}}(\boldsymbol{x}, -\mathrm{i}\omega)\}\ \exp(-\mathrm{i}\omega t)].\end{aligned} \tag{3.23}$$

Substitution of these complex representations in the basic equations in the time domain yields, except for the common time factor $\exp(-\mathrm{i}\omega t)$, a set of basic equations

$$-\nabla \times \hat{\boldsymbol{H}} - \mathrm{i}\omega\hat{\varepsilon}\hat{\boldsymbol{E}} = -\hat{\boldsymbol{J}}, \tag{3.24}$$

$$\nabla \times \hat{\boldsymbol{E}} - \mathrm{i}\omega\hat{\mu}\hat{\boldsymbol{H}} = -\hat{\boldsymbol{K}}. \tag{3.25}$$

These equations are identical to the one of the Laplace-transform domain, viz. Eqs. (3.17)–(3.18) with $s = -\mathrm{i}\omega$. Hence, we interpret the steady-state analysis as a limiting case of the time Laplace-transform analysis in which $s \to -\mathrm{i}\omega$ via $\mathrm{Re}(s) > 0$. In this way, we have ensured that the causality remains satisfied.

When we deal with a lossless medium, the electric permittivity and the magnetic permeability are real quantities, independent of the frequency ω. If lossy media must be taken into account, the theory and the pertaining computer codes are straightforward by introducing complex values $\hat{\varepsilon}(\boldsymbol{x}, s)$ and $\hat{\mu}(\boldsymbol{x}, s)$ for $s \to -\mathrm{i}\omega$ via $\mathrm{Re}(s) > 0$. The wave speed $\hat{c} = (\hat{\varepsilon}\hat{\mu})^{-\frac{1}{2}}$ becomes complex as well. Only in the calculation of the adjoint of an operator, one should replace the the electric permittivity $\hat{\varepsilon}$, the magnetic permeability $\hat{\mu}$, and the wave speed $\hat{c}$ by their complex conjugates $\overline{\varepsilon}$, $\overline{\mu}$, and $\overline{c}$.

This concludes the discussion of the general framework of the electromagnetic wave equations and the electromagnetic interface conditions.

3.1 Three-Dimensional Scattering by a Bounded Contrast

We start our analysis with the electromagnetic wave equations in the s-domain. For convenience, we omit the explicit s-dependence of the various field quantities.

3.1.1 Radiation in an Unbounded Homogeneous Embedding

The problem of calculating the electromagnetic radiation generated by a known volume-source distribution in a known medium is often called the electromagnetic source problem. The corresponding electromagnetic wave field is denoted as the incident electromagnetic wave field $\{\hat{\boldsymbol{E}}^{inc}, \hat{\boldsymbol{H}}^{inc}\} = \{\hat{\boldsymbol{E}}^{inc}(\boldsymbol{x}), \hat{\boldsymbol{H}}^{inc}(\boldsymbol{x})\}$.

The configuration consists of a medium of infinite extent with known electromagnetic properties. In this medium, there is a source that occupies the bounded domain $\mathbb{D}_{source}$ (Figure 3.2). The action of the source is represented by known volume densities of external electric and magnetic currents, $\{\hat{\boldsymbol{J}}, \hat{\boldsymbol{K}}\}$.

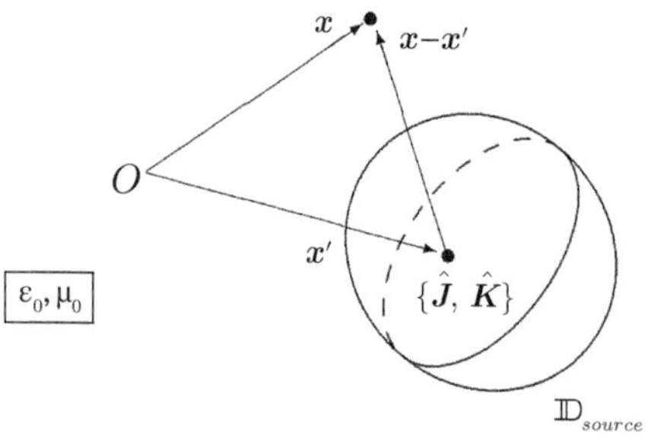

Figure 3.2 Source domain in an unbounded embedding with constant mass density ε_0 and constant compressibility μ_0.

The s-domain electromagnetic field strengths in a lossless and homogeneous medium with real and constant permittivity ε_0 and real and constant permeability μ_0 satisfy the s-domain electromagnetic wave equations

$$-\nabla \times \hat{\boldsymbol{H}}^{inc} + s\varepsilon_0 \hat{\boldsymbol{E}}^{inc} = -Z_0^{-1}\hat{\boldsymbol{J}}', \tag{3.26}$$

$$\nabla \times \hat{\boldsymbol{E}}^{inc} + s\mu_0 \hat{\boldsymbol{H}}^{inc} = -\hat{\boldsymbol{K}}, \tag{3.27}$$

where Z_0 is the electromagnetic impedance of the embedding medium,

$$Z_0 = \mu_0\, c_0 = (\mu_0/\varepsilon_0)^{\frac{1}{2}}, \tag{3.28}$$

and $c_0 = (\varepsilon_0\mu_0)^{-\frac{1}{2}}$ is the wave speed of the electromagnetic wave fields in the homogeneous medium at hand. With the definition of the propagation coefficient,

$$\hat{\gamma}_0 = s/c_0\,, \tag{3.29}$$

we often use the following relations

$$s\mu_0 = \hat{\gamma}_0\, Z_0, \quad \text{and} \quad s\varepsilon_0 = \hat{\gamma}_0/Z_0\,. \tag{3.30}$$

Note that we have included the factor Z_0^{-1} in Eq. (3.26) for computational convenience. The relation between $\hat{\boldsymbol{J}}'$ and $\hat{\boldsymbol{J}}$ is

$$\hat{\boldsymbol{J}}' = Z_0\, \hat{\boldsymbol{J}}\,. \tag{3.31}$$

In numerical practice, it is preferable to work with field functions of the same unit. In the present case, they are the electric field and the product of electromagnetic impedance and magnetic field. The electromagnetic wave equations of (3.26) and (3.27) can be rewritten in terms of the electric-field strength and the product of the electromagnetic impedance and the magnetic-field strength as

$$-\nabla \times Z_0\hat{\boldsymbol{H}}^{inc} + \hat{\gamma}_0\hat{\boldsymbol{E}}^{inc} = -\hat{\boldsymbol{J}}', \tag{3.32}$$

$$\nabla \times \hat{\boldsymbol{E}}^{inc} + \hat{\gamma}_0 Z_0\hat{\boldsymbol{H}}^{inc} = -\hat{\boldsymbol{K}}. \tag{3.33}$$

We assume that $\hat{\boldsymbol{J}}'$ and $\hat{\boldsymbol{K}}$ only differ from zero in some bounded subdomain $\mathbb{D}_{source}$ of $\mathbb{R}^3$ (Figure 3.2).

Eliminating $Z_0\hat{\boldsymbol{H}}^{inc}$ from Eqs. (3.32) and (3.33), we obtain

$$\nabla \times \nabla \times \hat{\boldsymbol{E}}^{inc} + \hat{\gamma}_0^2\hat{\boldsymbol{E}}^{inc} = -\hat{\gamma}_0\hat{\boldsymbol{J}}' - \nabla \times \hat{\boldsymbol{K}}. \tag{3.34}$$

With the identity $\nabla \times \nabla \times \boldsymbol{v} = \nabla\nabla \cdot \boldsymbol{v} - (\nabla \cdot \nabla)\boldsymbol{v}$ and the compatibility relation $\hat{\gamma}_0 \nabla \cdot \boldsymbol{E}^{inc} = -\nabla \boldsymbol{J}'$, we arrive at

$$(\nabla \cdot \nabla)\hat{\boldsymbol{E}}^{inc} - \hat{\gamma}_0^2\hat{\boldsymbol{E}}^{inc} = -\hat{\gamma}_0^{-1}[-\hat{\gamma}_0^2 + \nabla\nabla\cdot]\hat{\boldsymbol{J}}' + \nabla \times \hat{\boldsymbol{K}}. \tag{3.35}$$

Using the Green function formulation, the solution for the electric-field strength becomes

$$\hat{\boldsymbol{E}}^{inc}(\boldsymbol{x}) = \hat{\gamma}_0^{-1}(-\hat{\gamma}_0^2 + \nabla\nabla\cdot)\hat{\boldsymbol{A}} - \nabla \times \hat{\boldsymbol{F}}, \tag{3.36}$$

where the electric-vector potential $\hat{\boldsymbol{A}}$ and the magnetic-vector potential $\hat{\boldsymbol{F}}$ satisfy the equations

$$(\nabla \cdot \nabla)\hat{\boldsymbol{A}} - \hat{\gamma}_0^2\hat{\boldsymbol{A}} = -\hat{\boldsymbol{J}}', \tag{3.37}$$

$$(\nabla \cdot \nabla)\hat{\boldsymbol{F}} - \hat{\gamma}_0^2\hat{\boldsymbol{F}} = -\hat{\boldsymbol{K}}, \tag{3.38}$$

with solutions

$$\hat{\boldsymbol{A}}(\boldsymbol{x}) = \int_{\boldsymbol{x}' \in \mathbb{D}_{source}} \hat{G}(\boldsymbol{x} - \boldsymbol{x}')\, \hat{\boldsymbol{J}}'(\boldsymbol{x}')\mathrm{d}V, \tag{3.39}$$

$$\hat{\boldsymbol{F}}(\boldsymbol{x}) = \int_{\boldsymbol{x}' \in \mathbb{D}_{source}} \hat{G}(\boldsymbol{x} - \boldsymbol{x}')\, \hat{\boldsymbol{K}}(\boldsymbol{x}')\mathrm{d}V. \tag{3.40}$$

The solution for the magnetic-field strength follows from Eq. (3.33) as

$$Z_0 \hat{\boldsymbol{H}}^{inc}(\boldsymbol{x}) = \boldsymbol{\nabla} \times \hat{\boldsymbol{A}} + \hat{\gamma}_0^{-1}(-\hat{\gamma}_0^2 + \boldsymbol{\nabla}\boldsymbol{\nabla}\cdot)\hat{\boldsymbol{F}}. \tag{3.41}$$

These equations show that the electromagnetic wave field from known sources in a known medium can be calculated in all space, once the wave fields radiated by appropriate electric- and magnetic-dipole sources have been calculated. The right-hand sides of the representations for the vector potentials express the wave-field values as a superposition of the wave fields radiated by the elementary volume sources out of which the distributed sources can be envisaged to be composed.

3.1.2 3D Incident Electromagnetic Field

For simplicity, in our 3D analysis we assume that the incident wave field is generated by a vertical electric dipole at $\boldsymbol{x}^S$, viz., $\hat{\boldsymbol{J}} = -\hat{M}\boldsymbol{i}_1\, \delta(\boldsymbol{x} - \boldsymbol{x}^S)$, where $\hat{M}$ denotes the electric-dipole moment, oriented in the negative x_1-direction. Then, the incident electric-field strength at some observation point $\boldsymbol{x}$ is given by

$$\hat{\boldsymbol{E}}^{inc}(\boldsymbol{x}|\boldsymbol{x}^S) = -\frac{Z_0\, \hat{M}}{\hat{\gamma}_0}(-\hat{\gamma}_0^2 + \boldsymbol{\nabla}\partial_1)\hat{G}(\boldsymbol{x} - \boldsymbol{x}^S)\boldsymbol{i}_1, \tag{3.42}$$

with components,

$$\begin{aligned}
\hat{E}_1^{inc}(\boldsymbol{x}|\boldsymbol{x}^S) &= -\frac{Z_0\, \hat{M}}{\hat{\gamma}_0}(-\hat{\gamma}_0^2 + \partial_1\partial_1)\hat{G}(\boldsymbol{x} - \boldsymbol{x}^S),\\
\hat{E}_2^{inc}(\boldsymbol{x}|\boldsymbol{x}^S) &= -\frac{Z_0\, \hat{M}}{\hat{\gamma}_0}\partial_2\partial_1\hat{G}(\boldsymbol{x} - \boldsymbol{x}^S),\\
\hat{E}_3^{inc}(\boldsymbol{x}|\boldsymbol{x}^S) &= -\frac{Z_0\, \hat{M}}{\hat{\gamma}_0}\partial_3\partial_1\hat{G}(\boldsymbol{x} - \boldsymbol{x}^S).
\end{aligned} \tag{3.43}$$

The incident magnetic-field strength at some observation point $\boldsymbol{x}$ is given by

$$Z_0\hat{\boldsymbol{H}}^{inc}(\boldsymbol{x}|\boldsymbol{x}^S) = -\frac{Z_0\, \hat{M}}{\hat{\gamma}_0}\hat{\gamma}_0\ \boldsymbol{\nabla} \times \hat{G}(\boldsymbol{x} - \boldsymbol{x}^S)\boldsymbol{i}_1. \tag{3.44}$$

with components,

$$\begin{aligned}
Z_0\hat{H}_1^{inc}(\boldsymbol{x}|\boldsymbol{x}^S) &= 0,\\
Z_0\hat{H}_2^{inc}(\boldsymbol{x}|\boldsymbol{x}^S) &= -\frac{Z_0\, \hat{M}}{\hat{\gamma}_0}\hat{\gamma}_0\ \partial_3\hat{G}(\boldsymbol{x} - \boldsymbol{x}^S),\\
Z_0\hat{H}_3^{inc}(\boldsymbol{x}|\boldsymbol{x}^S) &= \frac{Z_0\, \hat{M}}{\hat{\gamma}_0}\hat{\gamma}_0\ \partial_2\hat{G}(\boldsymbol{x} - \boldsymbol{x}^S).
\end{aligned} \tag{3.45}$$

The Green function and its various spatial derivatives are computed from

$$\hat{G}(\boldsymbol{x}) = \frac{\exp(-\hat{\gamma}_0|\boldsymbol{x}|)}{4\pi|\boldsymbol{x}|}, \tag{3.46}$$

$$\partial_k\hat{G}(\boldsymbol{x}) = -\left(\hat{\gamma}_0 + \frac{1}{|\boldsymbol{x}|}\right)\frac{x_k}{|\boldsymbol{x}|}\ \hat{G}(\boldsymbol{x}), \tag{3.47}$$

$$\partial_j\partial_k\hat{G}(\boldsymbol{x}) = \left[\left(\frac{3x_jx_k}{|\boldsymbol{x}|^2} - \delta_{j,k}\right)\left(\frac{\hat{\gamma}_0}{|\boldsymbol{x}|} + \frac{1}{|\boldsymbol{x}|^2}\right) + \hat{\gamma}_0^2\frac{x_jx_k}{|\boldsymbol{x}|^2}\right]\ \hat{G}(\boldsymbol{x}), \tag{3.48}$$

where $\delta_{j,k} = 1$, for $j = k$ and $\delta_{j,k} = 0$, for $j \neq k$. Note that $(\partial_k\partial_k - \hat{\gamma}_0^2)\hat{G}(\boldsymbol{x}) = 0$ for $\boldsymbol{x} \neq \boldsymbol{0}$.

3.1.3 2D Incident Electromagnetic Field

In the two-dimensional situation, the source that generate the electromagnetic field is independent of the x_3-coordinate. Consistent to the 3D dipole source, we now assume that we have a vertical electric-dipole line source at $\boldsymbol{x}_T^S$, viz., $\hat{\boldsymbol{J}}_T = -\hat{M}\boldsymbol{i}_1\ \delta(\boldsymbol{x}_T - \boldsymbol{x}_T^S)$, where $\hat{M}$ denotes the electric-dipole moment, oriented in the negative x_1-direction. Then, the Cartesian components of the incident electric-field strength at some observation point $\boldsymbol{x}_T$ are given by

$$\begin{aligned}
\hat{E}_1^{inc}(\boldsymbol{x}_T|\boldsymbol{x}_T^S) &= -\frac{Z_0\ \hat{M}}{\hat{\gamma}_0}(-\hat{\gamma}_0^2 + \partial_1\partial_1)\hat{G}(\boldsymbol{x}_T - \boldsymbol{x}_T^S),\\
\hat{E}_2^{inc}(\boldsymbol{x}_T|\boldsymbol{x}_T^S) &= -\frac{Z_0\ \hat{M}}{\hat{\gamma}_0}\partial_2\partial_1\hat{G}(\boldsymbol{x}_T - \boldsymbol{x}_T^S),\\
\hat{E}_3^{inc}(\boldsymbol{x}_T|\boldsymbol{x}_T^S) &= 0.
\end{aligned} \tag{3.49}$$

The incident magnetic-field strength at some observation point $\boldsymbol{x}_T$ is given by

$$Z_0\hat{\boldsymbol{H}}_T^{inc}(\boldsymbol{x}_T|\boldsymbol{x}_T^S) = -Z_0\ \hat{M}\ \boldsymbol{\nabla}_T \times \hat{G}(\boldsymbol{x}_T - \boldsymbol{x}_T^S)\boldsymbol{i}_1, \tag{3.50}$$

so that the components of the magnetic-field strength become

$$\begin{aligned}
Z_0\hat{H}_1^{inc}(\boldsymbol{x}_T|\boldsymbol{x}_T^S) &= 0,\\
Z_0\hat{H}_2^{inc}(\boldsymbol{x}_T|\boldsymbol{x}_T^S) &= 0,\\
Z_0\hat{H}_3^{inc}(\boldsymbol{x}_T|\boldsymbol{x}_T^S) &= \frac{Z_0\ \hat{M}}{\hat{\gamma}_0}\hat{\gamma}_0\ \partial_2\hat{G}(\boldsymbol{x}_T - \boldsymbol{x}_T^S).
\end{aligned} \tag{3.51}$$

We observe that $\hat{\boldsymbol{H}}_T = \boldsymbol{0}$ and $\hat{H}_3 \neq 0$, hence we deal with a 2D **H-polarized** field in the x_3-direction. Since $\hat{E}_3 = 0$, the electric field is transversal to the x_3-direction. The case of **E-polarization**, where $\hat{\boldsymbol{E}}_T = 0$ and $\hat{E}_3 \neq 0$, only exists in the far-field region of a 3D source, where the electromagnetic field is approximately a plane wave propagating in perpendicular direction to the x_3-axis. In this region, we obtain a scalar scattering problem for $\hat{u} := \hat{E}_3$, and this case has been discussed in Chapter 1.

The 2D Green function is

$$\hat{G}(\boldsymbol{x}_T) = \frac{1}{2\pi}K_0(\hat{\gamma}_0|\boldsymbol{x}_T|), \tag{3.52}$$

where K_0 is the modified Bessel function of the second kind and zero order. The various spatial derivatives are computed as

$$\partial_k\hat{G}(\boldsymbol{x}_T) = -\hat{\gamma}_0\frac{x_k}{|\boldsymbol{x}_T|}\ \frac{1}{2\pi}K_1(\hat{\gamma}_0|\boldsymbol{x}_T|), \tag{3.53}$$

$$\partial_j\partial_k\hat{G}(\boldsymbol{x}_T) = \left(\frac{2x_jx_k}{|\boldsymbol{x}|^2} - \delta_{j,k}\right)\frac{\hat{\gamma}_0}{|\boldsymbol{x}_T|}\ \frac{1}{2\pi}K_1(\hat{\gamma}_0|\boldsymbol{x}_T|) + \hat{\gamma}_0^2\frac{x_jx_k}{|\boldsymbol{x}_T|^2}\hat{G}(\boldsymbol{x}_T), \tag{3.54}$$

where $\delta_{j,k} = 1$, for $j = k$ and $\delta_{j,k} = 0$, for $j \neq k$ and where we have used that the derivative of $K_1(z)$ is obtained from $-\partial K_1(z) = K_0(z) + z^{-1}K_1(z)$. Note that here in 2D, the subscripts j and k are assigned the values 1 and 2, and that $(\partial_k\partial_k - \hat{\gamma}_0^2)\ \hat{G}(\boldsymbol{x}_T) = 0$, for $\boldsymbol{x}_T \neq \boldsymbol{0}$.

3.1.4 Scattering by a Bounded Contrast

In this section, the scattering of electromagnetic waves by a contrasting domain of bounded extent, present in a homogeneous, embedding of infinite extent, is investigated in more detail. The integral equation formulations of the scattering problem are presented. When we consider the scattering object as a volume scatterer, a domain-integral equation formulation is the most versatile

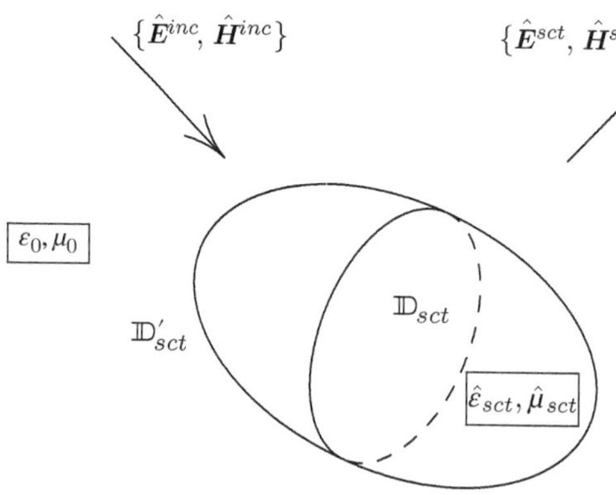

Figure 3.3 The domain of electromagnetic scattering.

technique, while for homogeneous scatterers, the boundary-integral equation formulation is most convenient when we treat the scatterer as a surface scatterer [12].

Let $\mathbb{D}_{sct}$ be the bounded domain occupied by the scatterer and let $\hat{\varepsilon}_{sct} = \hat{\varepsilon}_{sct}(\boldsymbol{x})$ be its permittivity and $\hat{\mu}_{sct} = \hat{\mu}_{sct}(\boldsymbol{x})$ its permeability. The domain exterior to $\mathbb{D}_{sct}$ is denoted by $\mathbb{D}'_{sct}$. In $\mathbb{D}'_{sct}$, we assume that in the embedding, a homogeneous medium is present with permittivity ε_0 and permeability μ_0 (Figure 3.3). The total electromagnetic wave field in the configuration $\{\hat{\boldsymbol{E}}, \hat{\boldsymbol{H}}\}$ is decomposed into the incident wave field $\{\hat{\boldsymbol{E}}^{inc}, \hat{\boldsymbol{H}}^{inc}\}$ and the scattered wave field $\{\hat{\boldsymbol{E}}^{sct}, \hat{\boldsymbol{H}}^{sct}\}$. The incident wave field is the wave field that would be present in the entire configuration if the domain $\mathbb{D}_{sct}$ showed no contrast with the embedding. The total wave field is generated by sources that are located outside the scattering domain. Because these sources remain present, even if the scattering domain is supposed to be absent, they also serve as sources for the incident wave field. The incident wave-field quantities may be calculated with the representations obtained in Section 3.1.1. We start our analysis in the s-domain. The results in the time domain are obtained by applying the standard rules for inversion from the s-domain to the time domain.

The scattered wave field is defined as the difference between the total wave field and the incident wave field. Hence,

$$\{\hat{\boldsymbol{E}}^{sct}, \hat{\boldsymbol{H}}^{sct}\} = \{\hat{\boldsymbol{E}} - \hat{\boldsymbol{E}}^{inc}, \hat{\boldsymbol{H}} - \hat{\boldsymbol{H}}^{inc}\}\,. \tag{3.55}$$

Through a particular reasoning, we now want to express that the scattered wave field originates from the contrast in electromagnetic properties that the scattering object shows with respect to its embedding. First of all, because the total wave field is free of sources in the domain $\mathbb{D}_{sct}$, it satisfies

$$-\boldsymbol{\nabla} \times \hat{\boldsymbol{H}} + s\hat{\varepsilon}_{sct}\hat{\boldsymbol{E}} = \boldsymbol{0}, \quad \boldsymbol{x} \in \mathbb{D}_{sct}\,, \tag{3.56}$$

$$\boldsymbol{\nabla} \times \hat{\boldsymbol{E}} + s\hat{\mu}_{sct}\hat{\boldsymbol{H}} = \boldsymbol{0}, \quad \boldsymbol{x} \in \mathbb{D}_{sct}\,. \tag{3.57}$$

Second, the incident wave field has no sources in $\mathbb{D}_{sct}$, while inside and outside $\mathbb{D}_{sct}$, the constitutive parameters have the same values. Hence,

$$-\boldsymbol{\nabla} \times \hat{\boldsymbol{H}}^{inc} + s\varepsilon_0\hat{\boldsymbol{E}}^{inc} = \boldsymbol{0}, \quad \boldsymbol{x} \in \mathbb{D}_{sct}\,, \tag{3.58}$$

$$\boldsymbol{\nabla} \times \hat{\boldsymbol{E}}^{inc} + s\mu_0\hat{\boldsymbol{H}}^{inc} = \boldsymbol{0}, \quad \boldsymbol{x} \in \mathbb{D}_{sct}\,. \tag{3.59}$$

Upon rewriting Eqs. (3.56) and (3.57) as

$$-\boldsymbol{\nabla} \times \hat{\boldsymbol{H}} + s\varepsilon_0\hat{\boldsymbol{E}} = s(\varepsilon_0 - \hat{\varepsilon}_{sct})\hat{\boldsymbol{E}}, \quad \boldsymbol{x} \in \mathbb{D}_{sct}\,, \tag{3.60}$$

$$\nabla \times \hat{\boldsymbol{E}} + s\mu_0 \hat{\boldsymbol{H}} = s(\mu_0 - \hat{\mu}_{sct})\hat{\boldsymbol{H}}, \quad \boldsymbol{x} \in \mathbb{D}_{sct}\,, \tag{3.61}$$

subtracting Eq. (3.58) from Eq. (3.60) and Eq. (3.59) from Eq. (3.61) and using Eq. (3.55), we arrive at

$$-\nabla \times \hat{\boldsymbol{H}}^{sct} + s\varepsilon_0 \hat{\boldsymbol{E}}^{sct} = s(\varepsilon_0 - \hat{\varepsilon}_{sct})\hat{\boldsymbol{E}}, \quad \boldsymbol{x} \in \mathbb{D}_{sct}\,, \tag{3.62}$$

$$\nabla \times \hat{\boldsymbol{E}}^{sct} + s\mu_0 \hat{\boldsymbol{H}}^{sct} = s(\mu_0 - \hat{\mu}_{sct})\hat{\boldsymbol{H}}, \quad \boldsymbol{x} \in \mathbb{D}_{sct}\,. \tag{3.63}$$

Third, we observe that the scattered wave field is source free in the embedding (where the total wave field and the incident wave field have the same sources), and hence

$$-\nabla \times \hat{\boldsymbol{H}}^{sct} + s\varepsilon_0 \hat{\boldsymbol{E}}^{sct} = \boldsymbol{0}, \quad \boldsymbol{x} \in \mathbb{D}'_{sct}\,, \tag{3.64}$$

$$\nabla \times \hat{\boldsymbol{E}}^{sct} + s\mu_0 \hat{\boldsymbol{H}}^{sct} = \boldsymbol{0}, \quad \boldsymbol{x} \in \mathbb{D}'_{sct}\,. \tag{3.65}$$

Equations (3.62)–(3.65) are now combined to

$$-\nabla \times \hat{\boldsymbol{H}}^{sct} + s\varepsilon_0 \hat{\boldsymbol{E}}^{sct} = -\hat{\boldsymbol{J}}^{sct}, \quad \boldsymbol{x} \in \mathbb{R}^3\,, \tag{3.66}$$

$$\nabla \times \hat{\boldsymbol{E}}^{sct} + s\mu_0 \hat{\boldsymbol{H}}^{sct} = -\hat{\boldsymbol{K}}^{sct}, \quad \boldsymbol{x} \in \mathbb{R}^3\,, \tag{3.67}$$

where

$$\hat{\boldsymbol{J}}^{sct} = s(\hat{\varepsilon}_{sct} - \varepsilon_0)\hat{\boldsymbol{E}}, \quad \text{and} \quad \hat{\boldsymbol{K}}^{sct} = s(\hat{\mu}_{sct} - \mu_0)\hat{\boldsymbol{H}}\,. \tag{3.68}$$

Similar as for the incident wave-field equations (3.32) and (3.33), we write the scattered wave-field equations as

$$-\nabla \times Z_0 \hat{\boldsymbol{H}}^{sct} + \hat{\gamma}_0\, \hat{\boldsymbol{E}}^{sct} = -\hat{\boldsymbol{J}}^{\prime,sct}\,, \tag{3.69}$$

$$\nabla \times \hat{\boldsymbol{E}}^{sct} + \hat{\gamma}_0\, Z_0 \hat{\boldsymbol{H}}^{sct} = -\hat{\boldsymbol{K}}^{sct}\,, \tag{3.70}$$

with $\hat{\boldsymbol{J}}^{\prime sct} = Z_0 \hat{\boldsymbol{J}}^{sct}$. The sources that generate the scattered field follow from Eq. (3.68) as

$$\hat{\boldsymbol{J}}^{\prime sct} = \hat{\gamma}_0 (\frac{\hat{\varepsilon}_{sct}}{\varepsilon_0} - 1)\hat{\boldsymbol{E}}, \quad \text{and} \quad \hat{\boldsymbol{K}}^{sct} = \hat{\gamma}_0 (\frac{\hat{\mu}_{sct}}{\mu_0} - 1) Z_0 \hat{\boldsymbol{H}}\,. \tag{3.71}$$

If $\hat{\boldsymbol{J}}^{\prime sct}$ and $\hat{\boldsymbol{K}}^{sct}$ were known, Eqs. (3.69) and (3.70) would constitute an electromagnetic radiation problem with known sources in an unbounded, homogeneous medium with the same constitutive parameters as the embedding. This problem has been discussed in Section 3.1.1, cf. Eqs. (3.36) and (3.41). Hence, we now arrive at

$$\begin{aligned} \hat{\boldsymbol{E}}^{sct}(\boldsymbol{x}) = \int_{\boldsymbol{x}' \in \mathbb{D}_{sct}} \Big[& (-\hat{\gamma}_0^2 + \nabla\nabla\cdot)\hat{G}(\boldsymbol{x} - \boldsymbol{x}')\, (\frac{\hat{\varepsilon}_{sct}}{\varepsilon_0} - 1)\hat{\boldsymbol{E}}(\boldsymbol{x}') \\ & - \hat{\gamma}_0 \nabla \times \hat{G}(\boldsymbol{x} - \boldsymbol{x}')\, (\frac{\hat{\mu}_{sct}}{\mu_0} - 1) Z_0 \hat{\boldsymbol{H}}(\boldsymbol{x}') \Big]\, \mathrm{d}V, \end{aligned} \tag{3.72}$$

for $\boldsymbol{x} \in \mathbb{D}'_{sct}$. Equation (3.72) is the desired integral representation for the scattered wave field. Once the total wave-field quantities $\hat{\boldsymbol{E}}$ and $Z_0\hat{\boldsymbol{H}}$ in $\mathbb{D}_{sct}$ are known, these integral representations enable us to calculate the electric-field strength in all space. To determine the yet unknown values of $\hat{\boldsymbol{E}}$ and $Z_0\hat{\boldsymbol{H}}$, we need a set of integral equations to solve for these electromagnetic field strengths.

Therefore, we now consider Eq. (3.41). Similarly, we obtain an integral representation for $Z_0\hat{\boldsymbol{H}}^{sct}$ as

$$Z_0\hat{\boldsymbol{H}}^{sct}(\boldsymbol{x}) = \int_{\boldsymbol{x}'\in\mathbb{D}_{sct}} \left[\hat{\gamma}_0\boldsymbol{\nabla}\times\hat{G}(\boldsymbol{x}-\boldsymbol{x}')\left(\frac{\hat{\varepsilon}_{sct}}{\varepsilon_0}-1\right)\hat{\boldsymbol{E}}(\boldsymbol{x}') + (-\hat{\gamma}_0^2+\boldsymbol{\nabla}\boldsymbol{\nabla}\cdot)\hat{G}(\boldsymbol{x}-\boldsymbol{x}')\left(\frac{\hat{\mu}_{sct}}{\mu_0}-1\right)Z_0\hat{\boldsymbol{H}}(\boldsymbol{x}')\right]\,\mathrm{d}V, \tag{3.73}$$

for $\boldsymbol{x}\in\mathbb{D}'_{sct}$.

Then, the next step is to let the observation point inside the scattering domain $\mathbb{D}_{sct}$. This procedure yields a set of six integral equations, for the six unknowns $\{\hat{E}_1,\ \hat{E}_2,\ \hat{E}_3,\ Z_0\hat{H}_1,\ Z_0\hat{H}_2,\ Z_0\hat{H}_3\}$, viz.,

$$\hat{\boldsymbol{E}}^{inc}(\boldsymbol{x}) = \hat{\boldsymbol{E}}(\boldsymbol{x}) - (-\hat{\gamma}_0^2+\boldsymbol{\nabla}\boldsymbol{\nabla}\cdot)\int_{\boldsymbol{x}'\in\mathbb{D}_{sct}}\hat{G}(\boldsymbol{x}-\boldsymbol{x}')\left(\frac{\hat{\varepsilon}_{sct}}{\varepsilon_0}-1\right)\hat{\boldsymbol{E}}(\boldsymbol{x}')\,\mathrm{d}V + \hat{\gamma}_0\boldsymbol{\nabla}\times\int_{\boldsymbol{x}'\in\mathbb{D}_{sct}}\hat{G}(\boldsymbol{x}-\boldsymbol{x}')\left(\frac{\hat{\mu}_{sct}}{\mu_0}-1\right)Z_0\hat{\boldsymbol{H}}(\boldsymbol{x}')\,\mathrm{d}V, \tag{3.74}$$

$$Z_0\hat{\boldsymbol{H}}^{inc}(\boldsymbol{x}) = Z_0\hat{\boldsymbol{H}}(\boldsymbol{x}) - \hat{\gamma}_0\boldsymbol{\nabla}\times\int_{\boldsymbol{x}'\in\mathbb{D}_{sct}}\hat{G}(\boldsymbol{x}-\boldsymbol{x}')\left(\frac{\hat{\varepsilon}_{sct}}{\varepsilon_0}-1\right)\hat{\boldsymbol{E}}(\boldsymbol{x}')\,\mathrm{d}V - (-\hat{\gamma}_0^2+\boldsymbol{\nabla}\boldsymbol{\nabla}\cdot)\int_{\boldsymbol{x}'\in\mathbb{D}_{sct}}\hat{G}(\boldsymbol{x}-\boldsymbol{x}')\left(\frac{\hat{\mu}_{sct}}{\mu_0}-1\right)Z_0\hat{\boldsymbol{H}}(\boldsymbol{x}')\,\mathrm{d}V, \tag{3.75}$$

for $\boldsymbol{x}\in\mathbb{D}_{sct}$. In principle, this system of equations can be used as the basis to provide a numerical solution of the electromagnetic scattering problem at hand. For both contrast in permittivity and in permeability, we are dealing with a set of coupled integral equations for the unknown electric-field vector and unknown magnetic-field vector. For zero contrast in permeability, we have an integral equation for the electric-field vector. For zero contrast in permeability, we have an integral equation for the magnetic-field vector only. Note that the magnetic type of equation is obtained from the electric type of equation, when we replace the permittivity by the permeability and $\hat{E}$ by $Z_0\hat{H}$.

3.2 Contrast Source (E-field) Integral Equations: Permittivity Contrast Only

In case that there is no permeability contrast, $\hat{\mu}_{sct}=\mu_0$, the integral representation of Eq. (3.72) reduces to

$$\hat{\boldsymbol{E}}^{sct}(\boldsymbol{x}) = \int_{\boldsymbol{x}'\in\mathbb{D}_{sct}}(-\hat{\gamma}_0^2+\boldsymbol{\nabla}\boldsymbol{\nabla}\cdot)\hat{G}(\boldsymbol{x}-\boldsymbol{x}')\left(\frac{\hat{\varepsilon}_{sct}}{\varepsilon_0}-1\right)\hat{\boldsymbol{E}}(\boldsymbol{x}')\,\mathrm{d}V, \tag{3.76}$$

or, written out in subscript notation,

$$\hat{E}_j^{sct}(\boldsymbol{x}) = \int_{\boldsymbol{x}'\in\mathbb{D}_{sct}}(-\hat{\gamma}_0^2\delta_{j,k}+\partial_j\partial_k)\hat{G}(\boldsymbol{x}-\boldsymbol{x}')\left(\frac{\hat{\varepsilon}_{sct}}{\varepsilon_0}-1\right)\hat{E}_k(\boldsymbol{x}')\,\mathrm{d}V. \tag{3.77}$$

Let us introduce the permittivity contrast as

$$\hat{\chi}^{\varepsilon}(\boldsymbol{x}) = 1-\frac{\hat{\varepsilon}_{sct}(\boldsymbol{x})}{\varepsilon_0}, \tag{3.78}$$

which is, for $\hat{\mu}_{sct}=\mu_0$, equal to $\chi^c = 1-c_0^2/\hat{c}_{sct}^2(\boldsymbol{x})$. Subsequently, the electric contrast source vector related to the permittivity contrast is introduced as

$$\hat{w}_k^E(\boldsymbol{x}) = \hat{\chi}^{\varepsilon}(\boldsymbol{x})\hat{E}_k(\boldsymbol{x}), \tag{3.79}$$

and we obtain the source type integral representation

$$\hat{E}_j^{sct}(\boldsymbol{x}) = \int_{\boldsymbol{x}'\in\mathbb{D}_{sct}} (\hat{\gamma}_0^2\delta_{j,k} - \partial_j\partial_k)\hat{G}(\boldsymbol{x}-\boldsymbol{x}')\,\hat{w}_k^E(\boldsymbol{x}')\,\mathrm{d}V, \quad \boldsymbol{x}\in\mathbb{D}'_{sct}. \tag{3.80}$$

By letting the point of observation into the scatterer and multiplying the equation through with the permittivity contrast, the contrast-source integral equation is obtained as

$$\hat{\chi}^{\varepsilon}(\boldsymbol{x})\hat{E}_j^{inc}(\boldsymbol{x}) = \hat{w}_j^E(\boldsymbol{x}) - \hat{\chi}^{\varepsilon}(\boldsymbol{x})(\hat{\gamma}_0^2\delta_{j,k} - \partial_j\partial_k)\int_{\boldsymbol{x}'\in\mathbb{D}_{sct}} \hat{G}(\boldsymbol{x}-\boldsymbol{x}')\,\hat{w}_k^E(\boldsymbol{x}')\,\mathrm{d}V, \tag{3.81}$$

when $\boldsymbol{x}\in\mathbb{D}_{sct}$. Once we have solved this integral equation, the solution for $\hat{w}_k^E(\boldsymbol{x}')$ may be substituted into Eq. (3.80). Then, the electric-field strength of the scattered wave field at receiver points $\boldsymbol{x}^R\in\mathbb{D}'_{sct}$ follows from

$$\boxed{\hat{E}_j^{sct}(\boldsymbol{x}^R) = \int_{\boldsymbol{x}'\in\mathbb{D}_{sct}} (\hat{\gamma}_0^2\delta_{j,k} - \partial_j^R\partial_k^R)\hat{G}(\boldsymbol{x}^R-\boldsymbol{x}')\,\hat{w}_k^E(\boldsymbol{x}')\,\mathrm{d}V, \quad \boldsymbol{x}^R\in\mathbb{D}'_{sct}.} \tag{3.82}$$

Similarly, the integral representation for $Z_0\hat{\boldsymbol{H}}^{sct}$ in subscript notation follows from Eq. (3.73) as

$$\boxed{Z_0\hat{H}_j^{sct}(\boldsymbol{x}^R) = \int_{\boldsymbol{x}'\in\mathbb{D}_{sct}} \hat{\gamma}_0\epsilon_{j,k,l}\partial_k^R\hat{G}(\boldsymbol{x}^R-\boldsymbol{x}')\,\hat{w}_l^E(\boldsymbol{x}')\,\mathrm{d}V, \quad \boldsymbol{x}^R\in\mathbb{D}'_{sct}.} \tag{3.83}$$

Here, $\epsilon_{j,k,l}$ is the completely antisymmetrical unit tensor of rank 3 (Levi-Civita tensor) given by

$$\begin{aligned} &\epsilon_{1,2,3} = \epsilon_{2,3,1} = \epsilon_{3,1,2} = +1,\\ &\epsilon_{3,2,1} = \epsilon_{2,1,3} = \epsilon_{1,3,2} = -1,\\ &\epsilon_{j,k,l} = 0, \text{ when not all subscripts are different.}\end{aligned} \tag{3.84}$$

3.2.1 2D Contrast Source (E-field) Integral Equations: Permittivity Contrast Only

When the permittivity contrast is invariant in the x_3-direction,

$$\hat{\chi}^{\varepsilon}(\boldsymbol{x}_T) = 1 - \frac{\hat{\varepsilon}_{sct}(\boldsymbol{x}_T)}{\varepsilon_0}, \tag{3.85}$$

the integral equation of Eq. (3.81) decouples into equations for the transversal contrast sources $\hat{w}_k^E$, for $k = 1, 2$ and an equation for $\hat{w}_3^E$ only. The latter integral equation is the scalar integral equation:

$$\hat{\chi}^{\varepsilon}\hat{E}_3^{inc}(\boldsymbol{x}_T) = \hat{w}_3^E(\boldsymbol{x}^T) - \hat{\chi}^{\varepsilon}\hat{\gamma}_0^2\int_{\boldsymbol{x}'_T\in\mathbb{D}_{sct}} \hat{G}(\boldsymbol{x}_T-\boldsymbol{x}'_T)\,\hat{w}_3^E(\boldsymbol{x}'_T)\,\mathrm{d}A, \quad \boldsymbol{x}_T\in\mathbb{D}_{sct}. \tag{3.86}$$

This equation is equivalent to the contrast-source integral equations of Chapter 1, viz. $\hat{u} := \hat{E}_3$. In addition, the 2D electric-dipole source under consideration has no electric-field component in the $\boldsymbol{i}_3$-direction. Therefore, this case of **E-polarization** is not further discussed.

For **H-polarization**, the set of integral equations for $\hat{w}_1^E$ and $\hat{w}_2^E$ reads

$$\boxed{\hat{\chi}^{\varepsilon}\hat{E}_j^{inc}(\boldsymbol{x}^T) = \hat{w}_j^E(\boldsymbol{x}^T) - \hat{\chi}^{\varepsilon}(\hat{\gamma}_0^2\delta_{j,k} - \partial_j\partial_k)\int_{\boldsymbol{x}'_T\in\mathbb{D}_{sct}} \hat{G}(\boldsymbol{x}_T-\boldsymbol{x}'_T)\,\hat{w}_k^E(\boldsymbol{x}'_T)\,\mathrm{d}A,} \tag{3.87}$$

when $\boldsymbol{x}_T\in\mathbb{D}_{sct}$. After solving this set of integral equations, the electric-contrast sources $\hat{w}_1^E$ and $\hat{w}_2^E$ are substituted in the scattered-field expressions at the receiver points. For $\boldsymbol{x}_T^R\in\mathbb{D}'_{sct}$, the electric-field components are

$$\boxed{\hat{E}_j^{sct}(\boldsymbol{x}_T^R) = \int_{\boldsymbol{x}'_T\in\mathbb{D}_{sct}} (\hat{\gamma}_0^2\delta_{j,k} - \partial_j^R\partial_k^R)\hat{G}(\boldsymbol{x}_T^R-\boldsymbol{x}'_T)\,\hat{w}_k^E(\boldsymbol{x}'_T)\,\mathrm{d}A.} \tag{3.88}$$

For $\hat{w}_3^E = 0$, the magnetic-field components $\hat{H}_1$ and $\hat{H}_2$ are zero, and the integral representation for $Z_0\hat{H}_3^{sct}$ follows from Eq. (3.83) as

$$\boxed{Z_0\hat{H}_3^{sct}(\boldsymbol{x}_T^R) = \int_{\boldsymbol{x}'\in\mathbb{D}_{sct}} \hat{\gamma}_0[\partial_2^R\hat{G}(\boldsymbol{x}_T^R - \boldsymbol{x}_T')\,\hat{w}_1^E(\boldsymbol{x}_T') - \partial_1^R\hat{G}(\boldsymbol{x}_T^R - \boldsymbol{x}_T')\,\hat{w}_2^E(\boldsymbol{x}_T')]\,\mathrm{d}A.} \tag{3.89}$$

In this chapter, we only treat the 2D and 3D cases. In 1D, the grad-div operator $\partial_j\partial_k$ vanishes, and the integral equation becomes scalar. We now consider the numerical solution of the contrast-source integral equations, based on the electric-field strength.

3.2.2 Conjugate Gradient Iterative Solution and Operators Involved

As in Chapter 2, we define a regular grid of a number of n equally sized subdomains, denoted as $\{\mathbb{D}_n,\ n = 1 : N\}$. In 2D, we deal with $N = N_1 \times N_2$ subdomains, referred to as $\{\mathbb{D}_{i,j},\ i = 1 : N_1,\ j = 1 : N_2\}$. Each subdomain has the volume $\Delta V = (\Delta x)^2$. In 3D, we are dealing with $N = N_1 \times N_2 \times N_3$ subdomains, denoted as $\{\mathbb{D}_{i,j,k},\ i = 1 : N_1,\ j = 1 : N_2,\ k = 1 : N_3\}$. Each subdomain has the volume $\Delta V = (\Delta x)^3$. The discretization in the (x_1, x_2)-domain is illustrated in Figure 2.4. The dotted domain includes the scattering domain $\mathbb{D}_{sct}$ completely. We also added an extra layer around it to accommodate the central finite-difference approach. Although the contrast sources vanish at all points inside this extra layer, the wave-field quantities are computed at these points.

Further, in each subdomain $\mathbb{D}_n$, we assume that the permittivity contrast is constant and denote it as $\hat{\chi}_n^\varepsilon$, and consistently the contrast sources $\hat{w}_k^E$ are replaced by their spherical means $\langle\hat{w}_k^E\rangle$. The incident wave fields $\hat{E}_j^{inc}$ are replaced by their spherical means $\langle\hat{E}_j^{inc}\rangle$. For the 3D case, they follow from Section 3.1.2 by replacing the strong form of the Green function by its weak counterpart. In 2D, the expressions of Section 3.1.3 are converted in the same way.

As far as the discretization of the E-field integral equation is concerned, in Chapter 2, from a computational point of view, we have argued that a numerical discretization of the grad-div operator is the most efficient one. Therefore, we follow now the same procedure. At this point, we define our discrete convolutional operators as

$$[\mathcal{K}w_n]_m = \hat{\gamma}_0^2\Delta V\sum_{n=1}^{N}\langle\langle\hat{G}\rangle\rangle(\boldsymbol{x}_m - \boldsymbol{x}_n')\,\langle\hat{w}_n\rangle. \tag{3.90}$$

This convolutional operator is computed with the help of function `Kop` of Matlab Listing I.13 of Chapter 1. Next, we define the adjoint operator

$$[\mathcal{K}^\star f_n]_m = \overline{\hat{\gamma}}_0^2\Delta V\sum_{n=1}^{N}\langle\langle\overline{\hat{G}}\rangle\rangle(\boldsymbol{x}_m - \boldsymbol{x}_n')\,\langle\hat{f}_n\rangle. \tag{3.91}$$

This convolutional operator is computed with the help of function `AdjKop` of Matlab Listing I.14 of Chapter 1.

By now, we omit the hat-symbols and define the discrete electric-field operator

$$[\mathcal{K}_{j,k}^E w_k^E]_m = [\mathcal{K}w_j^E]_m - \gamma_0^{-2}\texttt{graddiv}_{j,k}[\mathcal{K}w_k^E]_m, \tag{3.92}$$

where the finite-difference operator `graddiv` is given by Matlab function I.34 of Chapter 2. The convolution $[\mathcal{K}w_k^E]_m$ is obtained from Eq. (3.90), by substituting $w_n := w_{k,n}^E$, for $k = 1, 2, 3$. The code to compute this operator is given by function `KopE` of Matlab Listing I.80.

Matlab Listing I.80 The function `KopE` to carry out the $\mathcal{K}^E$ operator.

```
function [KwE] = KopE(wE, input)
global nDIM;
KwE = cell(1:nDIM);

for n = 1:nDIM
    KwE{n} = Kop(wE{n}, input.FFTG);
end

dummy = graddiv(KwE, input);             % dummy is temporary storage

for n = 1:nDIM
    KwE{n} = KwE{n} - dummy{n} / input.gamma_0^2;
end
```

Matlab Listing I.81 The function `AdjKopE` to carry out the $\mathcal{K}^{\star E}$ operator.

```
function [Kf] = AdjKopE(f, input)
global nDIM;
Kf = cell(1:nDIM);

for n = 1:nDIM
    Kf{n} = AdjKop(f{n}, input.FFTG);
end

dummy = graddiv(Kf, input);              % dummy is temporary storage

for n = 1:nDIM
    Kf{n} = Kf{n} - dummy{n} / conj(input.gamma_0^2);
end
```

For further considerations, we need also the adjoint of this operator, viz.,

$$[\mathcal{K}^{\star E}_{j,k} f_{k,n}]_m = [\mathcal{K}^{\star} f_{j,n}]_m - \overline{\gamma}_0^{-2} \texttt{graddiv}_{\,j,k}[\mathcal{K}^{\star} f_{k,n}]_m. \tag{3.93}$$

The convolution $[\mathcal{K}^{\star} f_{k,n}]_m$ is obtained from Eq. (3.91), by substituting $f_n := f_{k,n}$, for $k = 1, 2, 3$, respectively. The code to compute this adjoint operator is given by function `AdjKopE` of Matlab Listing I.81. The adjoint operators in Eq. (3.93) are convolutional operators as well. They are computed with FFTs. We use the code `AdjKop` of Matlab Listing I.14.

3.2.2.1 Conjugate Gradient Method

With the defined operators and their pertaining Matlab codes, we observe that the integral equation transfers into the following discrete set of equations:

$$\chi^{\varepsilon}_m \langle E^{inc}_j \rangle (x_m) = w^E_{j,m} - \chi^{\varepsilon}_m [\mathcal{K}^E_{j,k} w^E_k]_m. \tag{3.94}$$

This system is solved iteratively with the help of a Conjugate Gradient method. The method for solving the contrast sources w_k^E is fundamentally the same as the one presented in Table 1.4. We only have to change the scalar w by the vector w_j^E and to extend the inner product definition.

To discuss the iterative procedure in more detail, we write Eq. (3.94) in the compact form

$$\boxed{w_j^E - \chi^\varepsilon [\mathcal{K}_{j,k}^E w_k^E]_m = \chi^\varepsilon \; E_j^{inc}.} \tag{3.95}$$

If the contrast sources do not solve Eq. (3.95), the pertaining residual errors are defined as

$$\boxed{r_j^E = \chi^\varepsilon E_j^{inc} - [w_j^E - \chi^\varepsilon \; \mathcal{K}_{j,k}^E w_k^E],} \tag{3.96}$$

and the normalized cost functional

$$\boxed{F(w^p, w_j^{Zv}) = \frac{\sum_k \| r_k^E \|^2}{\sum_k \| \chi^\varepsilon E_k^{inc} \|^2},} \tag{3.97}$$

with the property that $F = 1$ for $w_1^E = w_2^E = w_3^E = 0$ and $F = 0$ for the exact solution. The gradient directions follow from the Fréchet derivative of this cost functional F and they are directly related to the adjoint operating on the residual errors in the equations at the previous iteration $n - 1$. They are written as

$$g_j^{E(n)} \; = r_j^{E(n-1)} \; - \; \mathcal{K}_{j,k}^{\star E} \; \overline{\chi}^\varepsilon r_k^{E(n-1)}. \tag{3.98}$$

The Hestenes-Stiefel conjugate gradient directions after the first iteration become

$$v_j^{E(n)} \; = g_j^{E(n)} + (A^{(n)}/A^{(n-1)}) \; v_j^{E(n-1)}, \tag{3.99}$$

where

$$A^{(n)} = \sum_k \| \; g_k^{E(n)} \|^2. \tag{3.100}$$

In each iteration, the updates for the contrast sources w_k^E are obtained as

$$w_j^{Zv(n)} \; = w_j^{Zv(n-1)} + (A^{(n)}/B^{(n)}) \; v_j^{Zv(n)}. \tag{3.101}$$

with

$$B^{(n)} = \sum_j \| \; v_j^{E(n)} - \chi^\varepsilon \mathcal{K}_{j,k}^{E(n)} v_k^{E(n)} \|^2. \tag{3.102}$$

The complete Conjugate Gradient scheme, that minimizes the cost functional F iteratively, is presented in Table 3.1. To complete this numerical section, we need the weak forms of scattered wave fields.

3.2.2.2 Scattered Electromagnetic Wave Field

After the numerical solution for the 3D contrast sources $\hat{w}_k^{Zv}(\boldsymbol{x}_m)$, the scattered electric wave field at some receiver point $\boldsymbol{x}^R$ is computed as

$$\langle \hat{E}_j^{sct} \rangle (\boldsymbol{x}^R) = \Delta V \sum_{m=1}^{N} [\hat{\gamma}_0^2 \delta_{j,k} - \partial_j^R \partial_k^R] \langle \hat{G} \rangle (\boldsymbol{x}^R - \boldsymbol{x}_m) \langle \hat{w}_k^E(\boldsymbol{x}_m) \rangle, \tag{3.103}$$

and the scattered magnetic wave field as

$$\langle Z_0 \hat{H}_j^{sct} \rangle (\boldsymbol{x}^R) = \Delta V \sum_{m=1}^{N} \hat{\gamma}_0 \epsilon_{j,k,l} \partial_k^R \langle \hat{G} \rangle (\boldsymbol{x}^R - \boldsymbol{x}_m) \langle \hat{w}_l^E(\boldsymbol{x}_m) \rangle, \tag{3.104}$$

Table 3.1 The CGFFT method for solving the contrast sources w_j^E.

Initial estimate		$w_k^{E(n=0)}$	$=$	0
		$r_j^{E(n=0)}$	$=$	$\chi^{\varepsilon} E_j^{inc}$
for $n = 1, 2, \cdots$				
		$g_j^{E(n)}$	$=$	$r_j^{E(n-1)} - \mathcal{K}^{\star E}_{j,k} \overline{\chi}^{\varepsilon} r_k^{E(n-1)}$
		$A^{(n)}$	$=$	$\sum_j \parallel g_j^{E(n)} \parallel^2$
	if $n = 1$	$v_j^{E(n)}$	$=$	$g_j^{E(n)}$
	else	$v_j^{E(n)}$	$=$	$g_j^{E(n)} + \dfrac{A^{(n)}}{A^{(n-1)}} v_j^{E(n-1)}$ end
		$B^{(n)}$	$=$	$\sum_j \parallel v_j^{E(n)} - \chi^{\varepsilon} \mathcal{K}^E_{j,k} v_k^{E(n)} \parallel^2$
		$w_k^{E(n)}$	$=$	$w_k^{E(n-1)} + \dfrac{A^{(n)}}{B^{(n)}} v_k^{E(n)}$
		$r_j^{E(n)}$	$=$	$r_j^{E(n-1)} - \dfrac{A^{(n)}}{B^{(n)}} \mathcal{K}^E_{j,k} v_k^{E(n)}$
	if $\left[\dfrac{\sum_k \lVert r_k^{E(n)} \rVert^2}{\sum_k \lVert \chi^{\varepsilon} E_k^{inc} \rVert^2}\right]^{\frac{1}{2}}$		$< \epsilon$ or $n > n_{\max}$ stop end	
end				

for $j = 1, 2, 3$, and $k = 1, 2, 3$. The spatial derivatives on the weak Green functions follow from the strong form of the Green function of Section 3.1.2 by replacing the strong form of the Green function by its weak counterpart.

In 2D, the scattered electric wave field becomes

$$\langle \hat{E}_j^{sct} \rangle(\boldsymbol{x}_T^R) = \Delta V \sum_{m=1}^{N} [\hat{\gamma}_0^2 \delta_{j,k} - \partial_j^R \partial_k^R] \langle \hat{G} \rangle(\boldsymbol{x}_T^R - \boldsymbol{x}_{T,m}) \langle \hat{w}_k^E(\boldsymbol{x}_m) \rangle, \tag{3.105}$$

for $j = 1, 2$, and $k = 1, 2$. In 2D, the electric wave field has no component in the x_3-direction and the magnetic wave field has only one component, viz.

$$\langle Z_0 \hat{H}_3^{sct} \rangle(\boldsymbol{x}_T^R) = \Delta V \sum_{m=1}^{N} \hat{\gamma}_0 \left[\partial_2^R \langle \hat{G} \rangle(\boldsymbol{x}_T^R - \boldsymbol{x}_{T,m}) \langle \hat{w}_1^E(\boldsymbol{x}_{T,m}) \rangle - \partial_1^R \langle \hat{G} \rangle(\boldsymbol{x}_T^R - \boldsymbol{x}_{T,m}) \langle \hat{w}_2^E(\boldsymbol{x}_{T,m}) \rangle \right]. \tag{3.106}$$

The spatial derivatives on the weak Green functions follow from the strong form of the Green function of Section 3.1.3 by replacing the strong form of the Green function by its weak counterpart.

3.2.3 Matlab Codes E-field Integral Equations: Permittivity Contrast Only

As in Chapters 1 and 2, we will present the Matlab code for computing the scattered field outside the scattering object at hand. Therefore, we need the Matlab scripts again to initialize the various parameters. The source wavelet is chosen to be $\hat{M}Z_0/\hat{\gamma}_0 = 1$. The corresponding `initEM` function is given in Matlab Listing I.82. The global parameter `nDIM` indicates the dimension of the space at hand. In our case either two or three. The electromagnetic wave speed in the embedding is set to $c_0 = 3 \times 10^8$ m/s, while in the scattering domain, we have set the relative permittivity to

Matlab Listing I.82 The function `initEM` to initialize the electromagnetic wave-field parameters.

```
function input = initEM()
% Time factor = exp(-iwt)
% Spatial units is in m
% Source wavelet M Z_0 / gamma_0  = 1    (Z_0 M = gamma_0)

global nDIM;  nDIM = 2;                 % set dimension of space
if nDIM ==1;  disp('nDIM should be either 2 or 3');  return; end
input.c_0       = 3e8;                  % wave speed in embedding
input.eps_sct = 1.75;                   % relative permittivity of scatterer
input.mu_sct  = 1;                      % relative permeability of scatterer

f               = 10e6;                 % temporal frequency
wavelength      = input.c_0 / f;        % wavelength
s               = 1e-16 - 1i*2*pi*f;    % LaPlace parameter
input.gamma_0 = s/input.c_0;            % propagation coefficient
disp(['wavelength = ' num2str(wavelength)]);

% add input data to structure array 'input'
  input = initSourceReceiver(input);  % add location of source/receiver

  input = initEMgrid(input);          % add grid in either 2D or 3D

  input = initFFTGreen(input);        % compute FFT of Green function

  input = initEMcontrast(input);      % add contrast distribution

  input.Errcri = 1e-3;
```

$\varepsilon_{sct}/\varepsilon_0 = 1.75$ and the relative permeability to $\mu_{sct}/\mu_0 = 1$. Apart from a sign difference, the permittivity contrast is equal to the wave speed contrast of Chapters 1 and 2. The frequency of operation is $f = 10$ MHz. The wave speed and the frequency are chosen such that the wavelength has the same value as that chosen in Chapters 1 and 2. The `initEM` function calls the function `initSourceReceiver`, which does not need to be changed. It initializes the locations of the source and the receivers and it is listed in Chapter 1. The `initEM` function further calls the function `initEMgrid` of Matlab Listing I.83, which defines the numerical grid in 2D and 3D. To show that, in 3D, the component of the contrast-source vector in the x_3-direction vanishes in the vertical plane $x_3 = 0$, we prefer an odd number of discretization points along the x_3-axis. The `initEM` function invokes the function `initFFTGreen` of Chapter 1, which does not have to be changed. The permittivity and permeability contrast distributions are computed by the Matlab function `initEMcontrast` as given in Matlab Listing I.84. This Matlab function is the replacement of the Matlab function `initContrast` of Chapter 1. The changes are obvious.

After the initialization of the electromagnetic parameters, we start the main Matlab script `ForwardCGFFTwE` of Matlab Listing I.85. Similar to the main Matlab script of scalar wave scattering, we have four steps: (1) Computation of scattering by a canonical object; (2) Computation of the incident wave fields; (3) Solution of the set of integral equations; and (4) Computation of scattering data, based on the solution of the integral equations.

Matlab Listing I.83 The function `initEMgrid` to initialize the grid parameters.

```
function input = initEMgrid(input)
global nDIM;

if nDIM == 2        % Grid in two dimensional space ----------------------

    input.N1 = 120;                                 % number of samples in x_1
    input.N2 = 100;                                 % number of samples in x_2
    input.dx = 2;                                   % with meshsize dx
    x1 = -(input.N1+1)*input.dx/2 + (1:input.N1)*input.dx;
    x2 = -(input.N2+1)*input.dx/2 + (1:input.N2)*input.dx;
    [input.X1 ,input.X2] = ndgrid(x1,x2);

elseif nDIM == 3  % Grid in three-dimensional space (odd input.N3) ------

    input.N1 = 120;                                 % number of samples in x_1
    input.N2 = 100;                                 % number of samples in x_2
    input.N3 = 81;                                  % number of samples in x_3
    input.dx = 2;                                   % with meshsize dx
    x1 = -(input.N1+1)*input.dx/2 + (1:input.N1)*input.dx;
    x2 = -(input.N2+1)*input.dx/2 + (1:input.N2)*input.dx;
    x3 = -(input.N3+1)*input.dx/2 + (1:input.N3)*input.dx;
    [input.X1 ,input.X2 ,input.X3] = ndgrid(x1,x2,x3);

end %if
```

Matlab Listing I.84 The function `initEMcontrast` to initialize the electromagnetic wave-field parameters.

```
function [input] = initEMcontrast(input)
global nDIM;
input.a  = 40;        % radius circle cylinder / radius sphere

if      nDIM == 2;    R = sqrt(input.X1.^2 + input.X2.^2);
elseif nDIM == 3;    R = sqrt(input.X1.^2 + input.X2.^2 + input.X3.^2);
end % if

% (1) Compute permittivity contrast ----------------------------------
input.CHI_eps = (1-input.eps_sct) * (R < input.a);

% (2) Compute permeability contrast ----------------------------------
input.CHI_mu  = (1-input.mu_sct)  * (R < input.a);
```

In step (1) of `ForwardCGFFTwE`, script `EMForwardCanonicalObjects` of Matlab Listing I.139 of Appendix 3.A.2.3 is called. It computes the scattered electromagnetic field at a number of receivers, for either the 2D circular cylinder or the 3D sphere. These data are used to benchmark the scattered field based on the numerical solution of our present integral equation. Next, the Matlab function `plotEMcontrast` of Matlab Listing I.86 displays the electric permittivity and magnetic permeability distributions, which are the electromagnetic input parameters. Both for the

Matlab Listing I.85 The `ForwardCGFFTwE` script to solve contrast sources w^E.

```
clear all; clc; close all; clear workspace;
input = initEM();

% (1) Compute analytically scattered field data ------------------------
        EMForwardCanonicalObjects

        plotEMcontrast(input); % plot permittivity / permeability contrast

% (2) Compute incident field -------------------------------------------
        [E_inc,~] = IncEMwave(input);

% (3) Solve integral equation for contrast source with CGFFT method ------
tic;    [w_E] = ITERCGwE(E_inc,input);
        plotContrastSourcewE(w_E,input);
toc
% (4) Compute synthetic data and plot fields and data ------------------
        [Edata,Hdata] = DOPwE(w_E,input);
        displayEdata(Edata,input);
        displayHdata(Hdata,input);
```

Matlab Listing I.86 The function `plotEMcontrast` to plot the electromagnetic permittivity and permeability contrast.

```
function [input] = plotEMcontrast(input)
global nDIM;
set(figure,'Units','centimeters','Position',[5 5 18 12]);

if  nDIM == 2                                   % Plot 2D cross section
    x1 = input.X1(:,1);   x2 = input.X2(1,:);
    subplot(1,2,1)
       IMAGESC(x1,x2, input.CHI_eps);
       title('\fontsize{13}\chi^\epsilon = 1 - \epsilon_{sct}/\epsilon_0');
    subplot(1,2,2)
       IMAGESC(x1,x2, input.CHI_mu);
       title('\fontsize{13} \chi^\mu = 1 - \mu_{sct}/\mu_0');

elseif nDIM == 3                                % Plot 3D cross section
   x1 = input.X1(:,1,1);  x2 =
                  input.X2(1,:,1);
    N3cross = floor(input.N3/2+1);
    subplot(1,2,1)
       IMAGESC(x1,x2, input.CHI_eps(:,:,N3cross));
       title('\fontsize{13} \chi^\epsilon = 1 - \epsilon_{sct}');
    subplot(1,2,2)
       IMAGESC(x1,x2, input.CHI_mu(:,:,N3cross));
       title('\fontsize{13} \chi^\mu = 1 - \mu_{sct}');

end % if
```

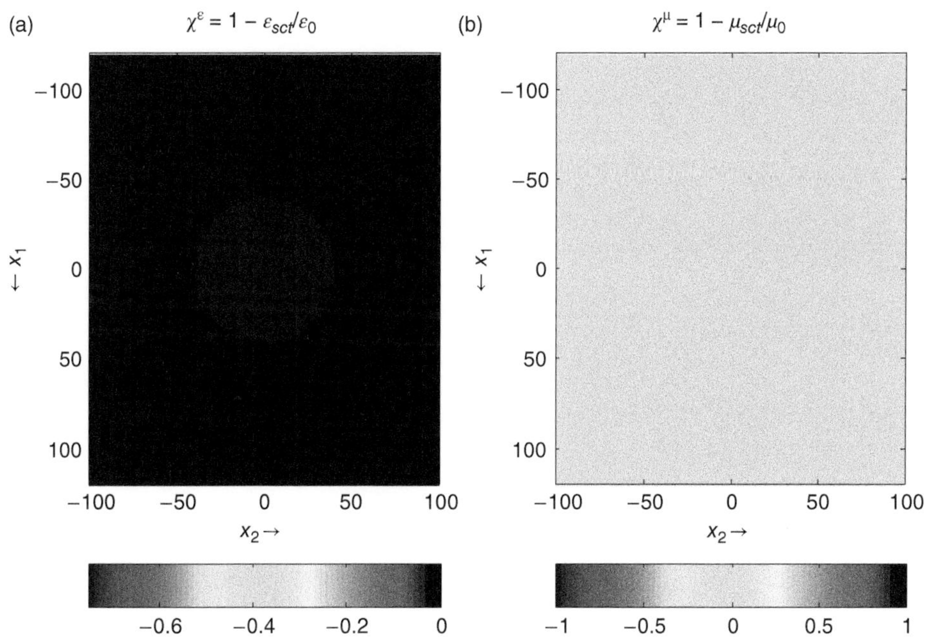

Figure 3.4 Plots of the contrast distributions (a) χ^ε and (b) χ^μ in the 2D cross-section.

2D case and the 3D case, the results in the cross-sectional plane $x_3 = 0$ are shown in Figure 3.4. The function `IMAGESC` is a modified version of the standard Matlab function `imagesc`, see Matlab Listing I.7.

In step (2) of `ForwardCGFFTwE`, we compute the weak form of the electric and the magnetic incident wave fields, either in 2D or 3D. The function `IncEMwave` is presented Matlab Listing I.87.

In step (3) of `ForwardCGFFTwE`, we solve the integral equations for the contrast sources w_j^E by calling the function `ITERCGwE` of Matlab Listing I.88. The code closely follows the CG method outlined in Table 3.1. To check the adjoint operator, we take the function `CheckAdjointE` of Matlab Listing I.89. This check is based on the numerical satisfaction of the inner products, using the initial value of r_j^E. The difference between both sides of this equation is printed. A value of the order of the computer precision should be reached. After validation, one can skip this test.

The operator function `KopE` and the adjoint function `AdjKopE` are listed in Matlab Listings I.80 and I.81. In these functions, the scalar operators `Kop` and `AdjKop` of Matlab Listings I.13 and I.14 are used. The function `plotContrastSourcewE` of Matlab Listing I.90 plots the absolute values of the contrast sources. For the 2D case, Figure 3.5 shows the contrast-source amplitudes $|w_j^E|$, for $j = 1, 2$. For the 3D case, Figure 3.6 shows the contrast-source amplitudes $|w_j^E|$, for $j = 1, 2, 3$. Note that w_3^E is an odd function with respect to the plane $x_3 = 0$ and should vanish in this plane. From the figure it follows that in this plane, w_3^E is five orders smaller than the other components. Furthermore, we see that w_2^E vanishes at the plane $x_2 = 0$.

In step (4) of the Matlab script `ForwardCGFFTwE`, the scattered field at the selected receiver points are computed. The contrast-source distribution w_j^E is the input for function `DopwE` of Matlab Listing I.91. This Matlab function computes the 2D scattered-field strengths and it calls the

Matlab Listing I.87 The function `IncEMwave` to compute the incident electromagnetic field components from a vertically oriented electric dipole.

```
function [E_inc,ZH_inc] = IncEMwave(input)
global nDIM;
gam0 = input.gamma_0;
xS   = input.xS;      dx = input.dx;

% incident wave from electric dipole in negative x_1
if nDIM == 2

  delta  = (pi)^(-1/2) * dx;          % radius circle with area of dx^2
  factor = 2 * besseli(1,gam0*delta) / (gam0*delta);

  X1    = input.X1-xS(1);        X2 = input.X2-xS(2);
  DIS   = sqrt(X1.^2 + X2.^2);
  X1    = X1./DIS;               X2 = X2./DIS;
   G    =   factor *             1/(2*pi).* besselk(0,gam0*DIS);
  dG    = - factor * gam0 .* 1/(2*pi).* besselk(1,gam0*DIS);
  dG11  = (2*X1.*X1 - 1) .* (-dG./DIS) + gam0^2 * X1.*X1 .* G;
  dG21  =  2*X2.*X1        .* (-dG./DIS) + gam0^2 * X2.*X1 .* G;

  E_inc{1} = - (-gam0^2 * G + dG11);
  E_inc{2} = - dG21;
  E_inc{3} = 0;

 ZH_inc{1} = 0;
 ZH_inc{2} = 0;
 ZH_inc{3} = gam0 * X2 .* dG;

elseif nDIM == 3

  delta  = (4*pi/3)^(-1/3) * dx;   % radius sphere with area of dx^3
  arg    = gam0*delta;
  factor = 3 * (cosh(arg) - sinh(arg)/arg) / arg^2;

  X1    = input.X1-xS(1);    X2 = input.X2-xS(2);        X3 = input.X3-xS(3);
  DIS   = sqrt(X1.^2 + X2.^2 + X3.^2);
  X1    = X1./DIS;           X2 = X2./DIS;               X3 = X3./DIS;
   G    = factor * exp(-gam0*DIS) ./ (4*pi*DIS);
  dG    = -(gam0 + 1./DIS) .* G;
  dG11  = ((3*X1.*X1 -1).* (gam0./DIS + 1./DIS.^2) + gam0^2*X1.*X1 ) .* G;
  dG21  = ( 3*X2.*X1     .* (gam0./DIS + 1./DIS.^2) + gam0^2*X2.*X1 ) .* G;
  dG31  = ( 3*X3.*X1     .* (gam0./DIS + 1./DIS.^2) + gam0^2*X3.*X1 ) .* G;

  E_inc{1} = - (-gam0^2 * G + dG11);
  E_inc{2} = - dG21;
  E_inc{3} = - dG31;

 ZH_inc{1} =  zeros(size(DIS));
 ZH_inc{2} = -gam0 * X3 .* dG;
 ZH_inc{3} =  gam0 * X2 .* dG;
end % if
```

Matlab Listing I.88 The function `ITERCGwE` for an iterative solution of w_j^E.

```
function [w_E] = ITERCGwE(E_inc,input)
global nDIM;      CHI_eps = input.CHI_eps;
w_E=cell(1,nDIM);  r_E=cell(1,nDIM);   g_E=cell(1,nDIM);
v_E=cell(1,nDIM);

itmax = 1000; Errcri = input.Errcri;  Norm0 = 0;  it = 0; % initialization
 for n = 1:nDIM
   w_E{n} = zeros(size(E_inc{n}));
   r_E{n} = CHI_eps .* E_inc{n};        Norm0 = Norm0 + norm(r_E{n}(:))^2;
 end
CheckAdjointE(r_E,input);              % Check once

Error  = 1;      fprintf('Error =          %g',Error);
while (it < itmax) && ( Error > Errcri) && (Norm0 > eps)
% determine conjugate gradient direction v
  dummy=cell(1,nDIM);  for n=1:nDIM;  dummy{n}=conj(CHI_eps).*r_E{n}; end
  KE  = AdjKopE(dummy,input);
  AN  = 0;
  for n = 1:nDIM
     g_E{n} =  r_E{n} - KE{n};                % and window for small CHI_eps
     g_E{n} = (abs(CHI_eps) >= Errcri).*g_E{n};    AN=AN+norm(g_E{n}(:))^2;
  end
  if it == 0
     for n = 1:nDIM;  v_E{n} = g_E{n}; end
  else
     for n = 1:nDIM;  v_E{n} = g_E{n} + AN/AN_1 * v_E{n};  end
  end
% determine step length alpha
  KE =  KopE(v_E,input);
  BN = 0;
  for n = 1:nDIM
      KE{n}= v_E{n} - CHI_eps .* KE{n};        BN = BN + norm(KE{n}(:))^2;
  end
  alpha = AN / BN;
% update contrast sources and AN
  for n = 1:nDIM
      w_E{n} = w_E{n} + alpha * v_E{n};
  end
  AN_1  = AN;
% update residual errors
   Norm = 0;
   for n = 1:nDIM
     r_E{n} = r_E{n} - alpha * KE{n};      Norm = Norm + norm(r_E{n}(:))^2;
   end
   Error = sqrt(Norm / Norm0);         fprintf('\b\b\b\b\b\b\b\b%6f',Error);
   it = it+1;
 end % CG iterations

 fprintf('\b\b\b\b\b\b\b\b%6f\n',Error);
 disp(['Number of iterations is ' num2str(it)]);
 if it == itmax; disp(['itmax reached: err/norm = ' num2str(Error)]); end;
```

Matlab Listing I.89 The function `CheckAdjointE` for the iterative CG scheme.

```
function CheckAdjointE(r_E,input)
global nDIM;           CHI_eps = input.CHI_eps;           dummy = cell(1,nDIM);

% Adjoint operator [I-CHI*K] via inner product ----------------------------
for n = 1:nDIM
    dummy{n} = conj(CHI_eps).*r_E{n};
end
[KE] = AdjKopE(dummy,input);
Result1  = 0;
for n = 1:nDIM
   dummy{n} = r_E{n} - KE{n};
   Result1  = Result1 + sum( r_E{n}(:) .* conj(dummy{n}(:)) );
end
% Operator [I-K*CHI] via inner product ------------------------------------
[KE] = KopE(r_E,input);
Result2  = 0;
for n = 1:nDIM
   dummy{n} = r_E{n} - CHI_eps .* KE{n};
   Result2  = Result2 + sum( dummy{n}(:) .* conj(r_E{n}(:)) );
end
fprintf('Check adjoint: %e\n',abs(Result1-Result2));
```

Matlab Listing I.90 The function `plotContrastSourcewE` to plot the contrast-source distributions.

```
function plotContrastSourcewE(w_E,input)
global nDIM;       set(figure,'Units','centimeters','Position',[5 5 18 12]);

if  nDIM == 2        % Plot 2D contrast/source distribution ---------------
    x1 = input.X1(:,1);     x2 = input.X2(1,:);
    subplot(1,2,1);
        IMAGESC(x1,x2,abs(w_E{1}));
        title('\fontsize{13} abs(w_1^E)');
    subplot(1,2,2);
        IMAGESC(x1,x2,abs(w_E{2}));
        title('\fontsize{13} abs(w_2^E)');
elseif nDIM == 3     % Plot 3D contrast/source distribution ---------------
    x1 = input.X1(:,1,1);    x2 = input.X2(1,:,1);
    N3cross = floor(input.N3/2+1);
    subplot(1,3,1);
        IMAGESC(x1,x2,abs(w_E{1}(:,:,N3cross)));
        title('\fontsize{13} abs(w_1^E)');
    subplot(1,3,2);
        IMAGESC(x1,x2,abs(w_E{2}(:,:,N3cross)));
        title('\fontsize{13} abs(w_2^E)');
    subplot(1,3,3);
       IMAGESC(x1,x2,abs(w_E{3}(:,:,N3cross)));
        title('\fontsize{13} abs(w_3^E)');
end % if
```

Matlab Listing I.91 The function `DopwE` to compute the 2D scattered electromagnetic field strengths at a number of receivers.

```
function [Edata,Hdata] = DOPwE(w_E,input)
global nDIM;
gam0 = input.gamma_0;       dx = input.dx;     xR = input.xR;

Edata = zeros(1,input.NR);
Hdata = zeros(1,input.NR);

if nDIM == 2
 % Weak form
 delta  = (pi)^(-1/2)*dx;                  % radius circle with area of dx^2
 factor = 2*besseli(1,gam0*delta) / (gam0*delta);

 for p = 1 : input.NR
     X1  = xR(1,p)-input.X1;       X2 = xR(2,p)-input.X2;
     DIS = sqrt(X1.^2 + X2.^2);  X1 = X1./DIS; X2 = X2./DIS;
     G   =     factor          * 1/(2*pi).* besselk(0,gam0*DIS);

    dG   =  - factor * gam0 * 1/(2*pi).* besselk(1,gam0*DIS);
   d1_G  =   X1 .* dG;       d2_G  =  X2 .* dG;

    dG11 = (2*X1.*X1 - 1) .* (-dG./DIS) + gam0^2*X1.*X1 .* G;
    dG22 = (2*X2.*X2 - 1) .* (-dG./DIS) + gam0^2*X2.*X2 .* G;
    dG21 =   2*X2.*X1       .* (-dG./DIS) + gam0^2*X2.*X1 .* G;

    E1rfl = dx^2 * sum( (gam0^2*G(:) - dG11(:)) .* w_E{1}(:)   ...
                                    - dG21(:)   .* w_E{2}(:) );
    E2rfl = dx^2 * sum(               - dG21(:)   .* w_E{1}(:)   ...
                       +(gam0^2*G(:) - dG22(:)) .* w_E{2}(:) );

   ZH3rfl = gam0 * dx^2 * sum( d2_G(:).*w_E{1}(:) - d1_G(:).*w_E{2}(:) );

   Edata(1,p) = sqrt(abs(E1rfl)^2 + abs(E2rfl)^2);
   Hdata(1,p) = abs(ZH3rfl);

 end; % p_loop


elseif nDIM == 3

 [Edata,Hdata] = DOP3DwE(w_E,input);

end % if
```

Matlab Listing I.92 The function `Dop3DwE` to compute the 3D scattered electromagnetic field strengths at a number of receivers.

```
function [Edata,Hdata] = DOP3DwE(w_E,input)
gam0 = input.gamma_0;      dx = input.dx;    xR = input.xR;

Edata = zeros(1,input.NR);
Hdata = zeros(1,input.NR);

% Weak form
delta  = (4*pi/3)^(-1/3) * dx;          % radius sphere with area of dx^3
   arg = gam0*delta;
factor = 3 * (cosh(arg) - sinh(arg)/arg) / arg^2;

for p = 1 : input.NR
    X1  = xR(1,p)-input.X1;  X2 = xR(2,p)-input.X2; X3 = xR(3,p)-input.X3;
    DIS = sqrt(X1.^2+X2.^2+X3.^2); X1=X1./DIS;  X2=X2./DIS;  X3=X3./DIS;
    G   = factor * exp(-gam0*DIS) ./ (4*pi*DIS);

    dG   =  - (gam0 + 1./DIS) .* G;
   d1_G = X1 .* dG;          d2_G = X2 .* dG;          d3_G = X3 .* dG;

   dG11 = ( (3*X1.*X1 - 1).*(gam0./DIS+1./DIS.^2) + gam0^2*X1.*X1 ) .* G;
   dG22 = ( (3*X2.*X2 - 1).*(gam0./DIS+1./DIS.^2) + gam0^2*X2.*X2 ) .* G;
   dG33 = ( (3*X3.*X3 - 1).*(gam0./DIS+1./DIS.^2) + gam0^2*X3.*X3 ) .* G;
   dG21 = (  3*X2.*X1      .*(gam0./DIS+1./DIS.^2) + gam0^2*X2.*X1 ) .* G;
   dG31 = (  3*X3.*X1      .*(gam0./DIS+1./DIS.^2) + gam0^2*X3.*X1 ) .* G;
   dG32 = (  3*X3.*X2      .*(gam0./DIS+1./DIS.^2) + gam0^2*X3.*X2 ) .* G;

   E1rfl = dx^3 * sum( (gam0^2*G(:) - dG11(:)) .* w_E{1}(:)  ...
                                    - dG21(:)  .* w_E{2}(:)  ...
                                    - dG31(:)  .* w_E{3}(:) );
   E2rfl = dx^3 * sum(              - dG21(:)  .* w_E{1}(:)  ...
                       +(gam0^2*G(:) - dG22(:)) .* w_E{2}(:)  ...
                                    - dG32(:)  .* w_E{3}(:) );
   E3rfl = dx^3 * sum(              - dG31(:)  .* w_E{1}(:)  ...
                                    - dG32(:)  .* w_E{2}(:)  ...
                       +(gam0^2*G(:) - dG33(:)) .* w_E{3}(:) );

  ZH1rfl  = -gam0 * dx^3 * sum( d2_G(:).*w_E{3}(:) - d3_G(:).*w_E{2}(:) );
  ZH2rfl  = -gam0 * dx^3 * sum( d3_G(:).*w_E{1}(:) - d1_G(:).*w_E{3}(:) );
  ZH3rfl  = -gam0 * dx^3 * sum( d1_G(:).*w_E{2}(:) - d2_G(:).*w_E{1}(:) );

  Edata(1,p) = sqrt(abs(E1rfl)^2  + abs(E2rfl)^2  + abs(E3rfl^2));
  Hdata(1,p) = sqrt(abs(ZH1rfl)^2 + abs(ZH2rfl)^2 + abs(ZH3rfl^2));

end % p_loop
```

Matlab Listing I.93 The function `displayEdata` to present the scattered electric-field strengths at a number of receivers.

```
function  displayEdata(Edata, input)
global nDIM;

if nDIM == 2  % plot data at a number of receivers --------------------

    if exist(fullfile(cd,'EDATA2D.mat'), 'file')
      load EDATA2D Edata2D;
      E_error = num2str(norm(Edata(:)-Edata2D(:),1)/norm(Edata2D(:),1));
      disp(['E-error=' E_error]);
   end
   set(figure,'Units','centimeters','Position', [5 5 18 7]);
   angle = input.rcvr_phi * 180 / pi;
   if exist(fullfile(cd,'EDATA2D.mat'), 'file')
      plot(angle,abs(Edata),'--r',angle,abs(Edata2D),'b')
      legend('Integral-equation method', ...
             'Bessel-function method','Location','Best');
      text(50,0.8*max(abs(Edata)), ...
      ['Error(E^{sct}) = ' E_error '  '],'EdgeColor','red','Fontsize',11);
   else
      plot(angle,abs(Edata),'b')
      legend('Bessel-function method','Location','Best');
   end
   title('\fontsize{12} scattered E data in 2D');   axis tight;
   xlabel('observation angle in degrees');
   ylabel('abs(data) \rightarrow');

elseif nDIM == 3  % plot data at a number of receivers ----------------

   if exist(fullfile(cd,'EDATA3D.mat'), 'file')
      load EDATA3D Edata3D ;
      E_error = num2str(norm(Edata(:)-Edata3D(:),1)/norm(Edata3D(:),1));
      disp(['E-error=' E_error]);
   end
   set(figure,'Units','centimeters','Position', [5 5 18 7]);
   angle = input.rcvr_phi * 180 / pi;
   if exist(fullfile(cd,'EDATA3D.mat'), 'file')
      plot(angle,abs(Edata),'--r',angle,abs(Edata3D),'b')
      legend('Integral-equation method', ...
             'Bessel-function method','Location','Best');
      text(50,0.8*max(abs(Edata)), ...
      ['Error(E^{sct}) = ' E_error '  '],'EdgeColor','red','Fontsize',11);
   else
      plot(angle,abs(Edata),'b')
      legend('Bessel-function method','Location','Best');
   end
   title('\fontsize{12} scattered E data in 3D');  axis tight;
   xlabel('observation angle in degrees');
   ylabel('abs(data) \rightarrow');
end % if
```

Matlab Listing I.94 The function `displayHdata` to present the scattered magnetic-field strengths at a number of receivers.

```
function displayHdata(Hdata,input)
global nDIM;

if nDIM == 2 % plot data at a number of receivers -----------------------

   if exist(fullfile(cd,'HDATA2D.mat'), 'file')
     load HDATA2D Hdata2D;
      H_error = num2str(norm(Hdata(:)-Hdata2D(:),1)/norm(Hdata2D(:),1));
      disp(['H-error=' H_error]);
   end
   set(figure,'Units','centimeters','Position', [5 5 18 7]);
   angle = input.rcvr_phi * 180 / pi;
   if exist(fullfile(cd,'HDATA2D.mat'), 'file')
      plot(angle,abs(Hdata),'--r',angle,abs(Hdata2D),'b')
      legend('Integral-equation method', ...
              'Bessel-function method','Location','Best');
      text(50,0.8*max(abs(Hdata)), ...
      ['Error(Z_0H^{sct})=' H_error '  '],'EdgeColor','red','Fontsize',11);
   else
      plot(angle,abs(Hdata),'b')
      legend('Bessel-function method','Location','Best');
   end
   title('\fontsize{12} scattered Z_0H data in 2D');   axis tight;
   xlabel('observation angle in degrees');
   ylabel('abs(data) \rightarrow');

elseif nDIM == 3 % plot data at a number of receivers --------------------

   if exist(fullfile(cd,'HDATA3D.mat'), 'file')
     load HDATA3D Hdata3D;
     H_error = num2str(norm(Hdata(:)-Hdata3D(:),1)/norm(Hdata3D(:),1));
     disp(['H-error=' H_error]);
   end
   set(figure,'Units','centimeters','Position', [5 5 18 7]);
   angle = input.rcvr_phi * 180 / pi;
   if exist(fullfile(cd,'HDATA3D.mat'), 'file')
      plot(angle,abs(Hdata),'--r',angle,abs(Hdata3D),'b')
      legend('Integral-equation method', ...
              'Bessel-function method','Location','Best');
      text(50,0.8*max(abs(Hdata)), ...
      ['Error(Z_0H^{sct})=' H_error '  '],'EdgeColor','red','Fontsize',11);
   else
      plot(angle,abs(Hdata),'b')
      legend('Bessel-function method','Location','Best');
   end
   title('\fontsize{12} scattered Z_0H data in 3D');   axis tight;
   xlabel('observation angle in degrees');
   ylabel('abs(data) \rightarrow');
end % if
```

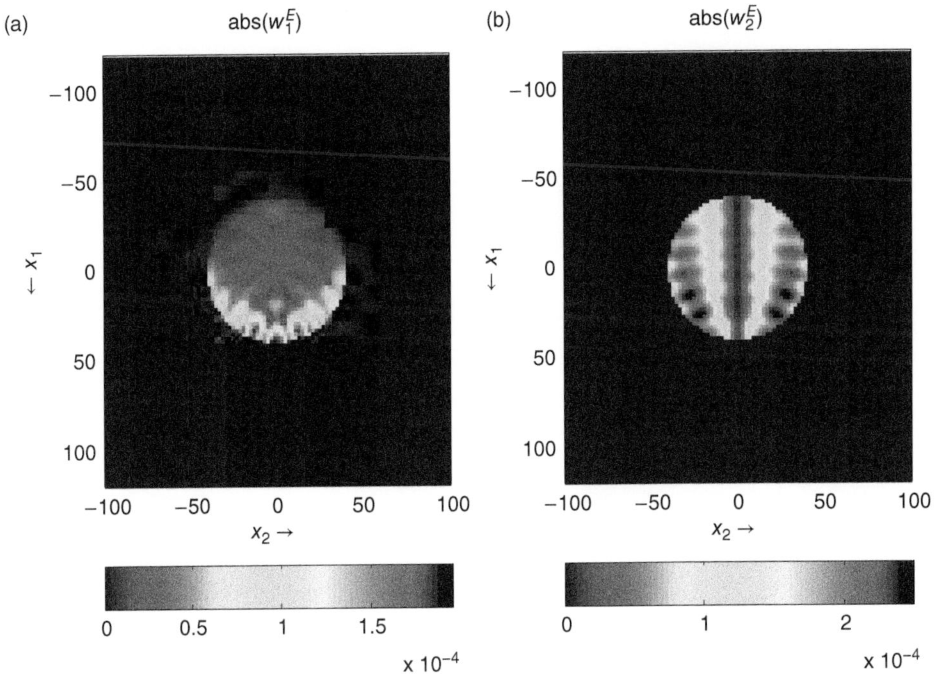

Figure 3.5 Plots of the 2D contrast-source amplitudes (a) $|w_1^E|$ and (b) $|w_2^E|$, produced by `ForwardCGFFTwE`.

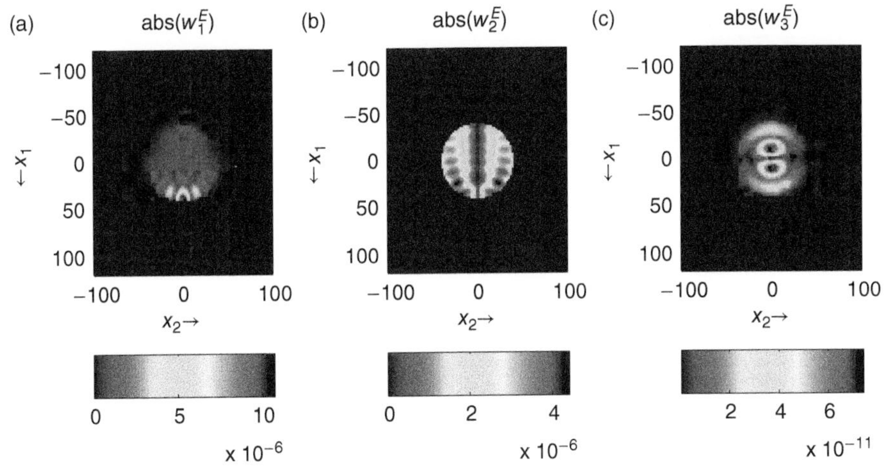

Figure 3.6 Cross-sectional plots of the 3D contrast-source amplitudes (a) $|w_1^E|$, (b) $|w_2^E|$, and (c) $|w_3^E|$, produced by `ForwardCGFFTwE`.

function `Dop3DwE` of Matlab Listing I.92 to compute the 3D scattered-field strengths. The program follows from the weak form of the scattered electromagnetic wave fields of Eqs. (3.103)–(3.106). These scattered fields are compared to the analytical Bessel-function data stored in MAT-files. The functions `displayEdata.m` and `displayHData.m` of Matlab Listings I.93 and I.94 load the

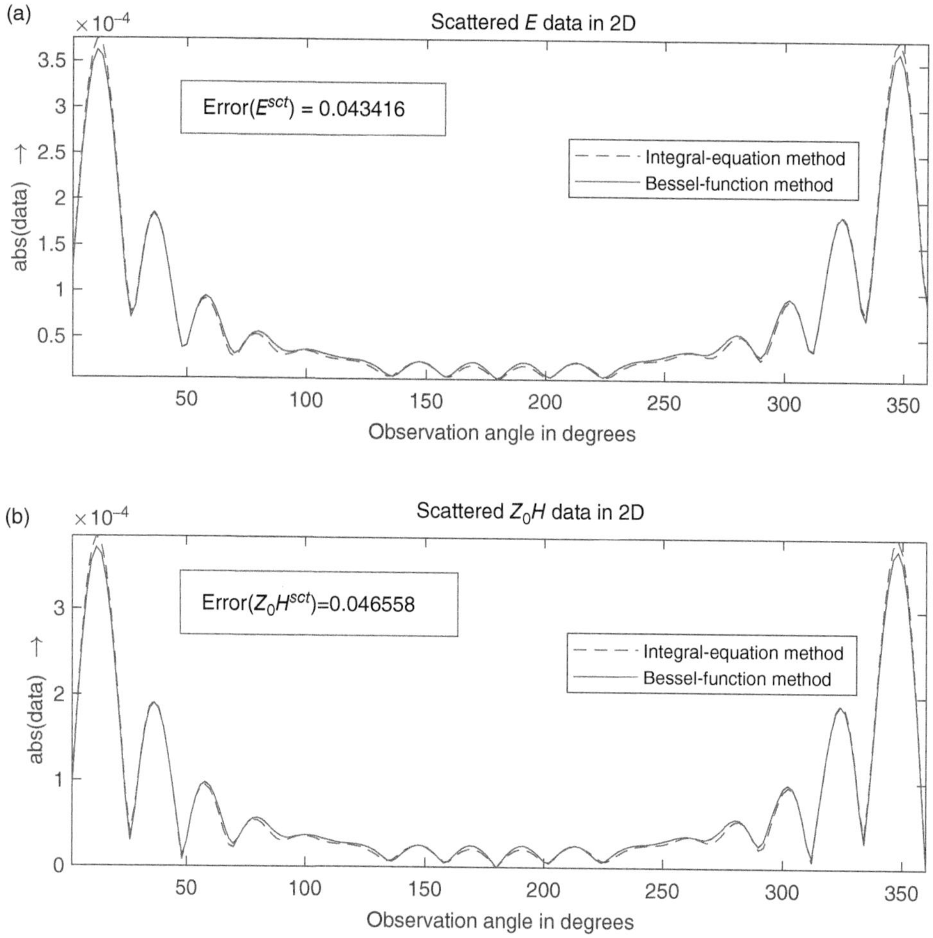

Figure 3.7 The 2D scattered electromagnetic field strengths at the receivers.

particular files and display both the exact data and the numerical data. These data are presented in Figures 3.7 and 3.8. We use the error norms

$$\text{Error}(E^{sct}) = \frac{\sum_{p=1}^{\text{NR}} ||\boldsymbol{E}^{sct}(x_p^R)| - |\boldsymbol{E}_{exact}^{sct}(x_p^R)||}{\sum_{p=1}^{\text{NR}} |\boldsymbol{E}_{exact}^{sct}(x_p^R)|}, \tag{3.107}$$

and

$$\text{Error}(Z_0 H^{sct}) = \frac{\sum_{p=1}^{\text{NR}} ||Z_0\boldsymbol{H}^{sct}(x_p^R)| - |Z_0\boldsymbol{H}_{exact}^{sct}(x_p^R)||}{\sum_{p=1}^{\text{NR}} |Z_0\boldsymbol{H}_{exact}^{sct}(x_p^R)|}. \tag{3.108}$$

Similar to the scattered-field results of Chapter 1, the error in 3D is significantly smaller than in 2D. We also note that in our Chapters 1 and 2, the Error^{sct} is determined by the complex difference

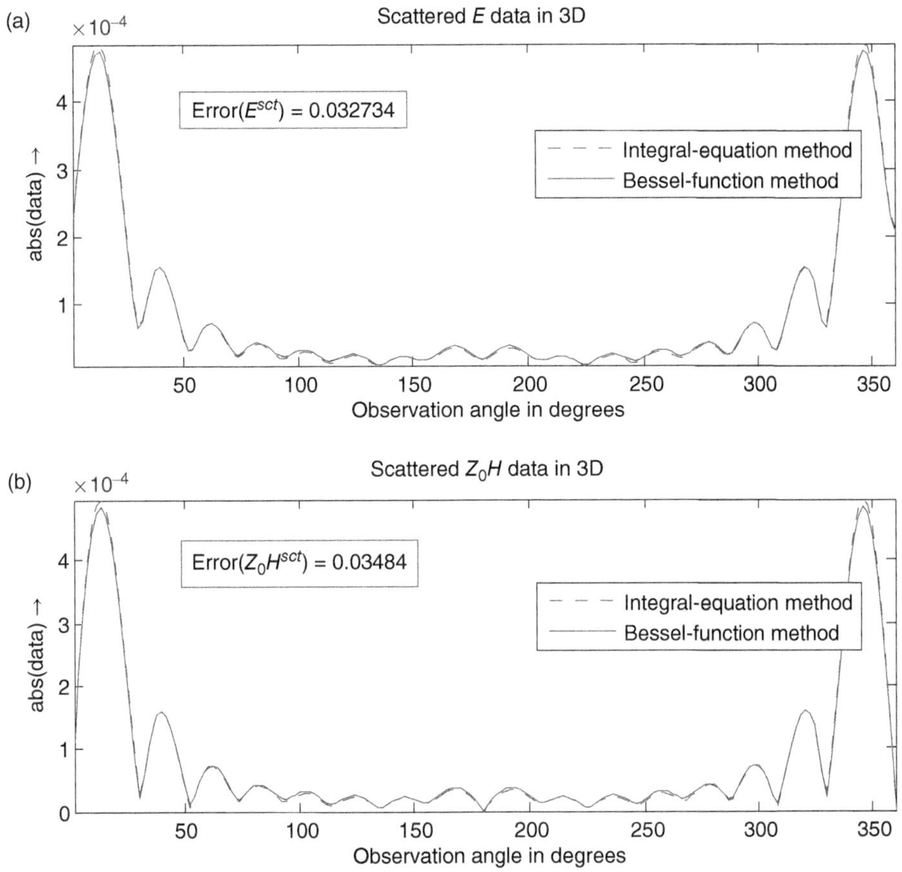

Figure 3.8 The 3D scattered electromagnetic field strengths at the receivers.

between the two scalar results. In the present electromagnetic case, the error is determined by the difference between the amplitudes of the two vectorial results. The error estimate is then a bit too optimistic.

3.2.3.1 Matlab BiCGSTAB Built-in Function

Finally, we replace `ITERCGwE` in step (3) of the script `ForwardCGFFTwE` (see Matlab Listing I.85) by the function `ITERBiCGSTABwE` of Matlab Listing I.95. The modified script is called `Forward-BiCGSTABwE`. The modified script is called `ForwardBiCGSTABwE`. For the 3D case and error criterion of 0.001, the function `ITERCGwE` requires 38 iterations with a computation time of 162 seconds, while function `ITERBiCGSTABwE` requires 14 iterations with a computation time of 66 seconds. The time per iteration is approximately the same: 4.3 and 4.7 seconds. From these numerical experiments regarding the current type of integral equation, we conclude that it is preferable to use the BiCGSTAB iterative solver, instead of the CG method (including the use of the adjoint of the operators).

Matlab Listing I.95 The function `ITERBiCGSTABwE` to use Matlab's built-in function BiCGSTAB in `ForwardCGFFTwE` of Matlab Listing I.85.

```
function [w_E] = ITERBiCGSTABwE(E_inc, input)
% BiCGSTAB scheme for contrast source integral equation Aw = b
  global nDIM;
  itmax   = 1000;   Errcri = input.Errcri;   [N,~] = size(input.CHI_eps(:));

          b(     1:  N,1) = input.CHI_eps(:) .* E_inc{1}(:);
          b(  N+1:2*N,1) = input.CHI_eps(:) .* E_inc{2}(:);

  if nDIM == 3
          b(2*N+1:3*N,1) = input.CHI_eps(:) .* E_inc{3}(:);
  end % if

   w = bicgstab(@(w) Aw(w,input), b, Errcri, itmax);      % call BICGSTAB

  [w_E] = vector2matrix(w,input);                          % output matrices

end %--------------------------------------------------------------------

function y = Aw(w,input)
  global nDIM;        [N,~] = size(input.CHI_eps(:));

  [w_E]  = vector2matrix(w,input);
  [Kw_E] = KopE(w_E,input);

          y(     1:  N,1) = w_E{1}(:) - input.CHI_eps(:) .* Kw_E{1}(:);
          y(  N+1:2*N,1) = w_E{2}(:) - input.CHI_eps(:) .* Kw_E{2}(:);

  if nDIM == 3
          y(2*N+1:3*N,1) = w_E{3}(:) - input.CHI_eps(:) .* Kw_E{3}(:);
  end % if

end %--------------------------------------------------------------------

function [w_E] = vector2matrix(w,input)
% Modify vector output from 'bicgstab' to matrices for further computation
  global nDIM;    [N,~] = size(input.CHI_eps(:));        w_E = cell(1,nDIM);

  if nDIM == 2;    DIM = [input.N1,input.N2];              end % if
  if nDIM == 3;    DIM = [input.N1,input.N2,input.N3];   end % if

              w_E{1} = reshape(w(     1:  N,1),DIM);
              w_E{2} = reshape(w(  N+1:2*N,1),DIM);

  if nDIM == 3
              w_E{3} = reshape(w(2*N+1:3*N,1),DIM);
  end %if

end
```

3.3 E-field Equation for Volume and Interface Contrast Sources: Permittivity Contrast Only

In Chapter 2, we have derived a set of integral equations, where not only the material parameters in contrast sources occur but also their spatial gradients appear as contrast sources. That is why we are investigating the integral representation of Eq. (3.77) in more detail and rewrite it as

$$\hat{E}_j^{sct}(\boldsymbol{x}) = -\hat{\gamma}_0^2 \int_{\boldsymbol{x}' \in \mathbb{D}_{sct}} \hat{G}(\boldsymbol{x}-\boldsymbol{x}') \left(\frac{\hat{\varepsilon}_{sct}}{\varepsilon_0} - 1\right) \hat{E}_j(\boldsymbol{x}')\, \mathrm{d}V + \int_{\boldsymbol{x}' \in \mathbb{D}_{sct}} \partial_j \partial_k \hat{G}(\boldsymbol{x}-\boldsymbol{x}') \left(\frac{\hat{\varepsilon}_{sct}}{\varepsilon_0} - 1\right) \hat{E}_k(\boldsymbol{x}')\, \mathrm{d}V. \tag{3.109}$$

We note that at an interface between different media the normal component of the electric-flux density $\hat{D}_k = \hat{\varepsilon}_{sct}\hat{E}_k$ is continuous. Therefore, we rewrite the second integral in terms of $\hat{D}_k$. In view of the compatibility relation, $\partial_k \hat{D}_k = 0$, that follows from Eq. (3.56) by taking its divergence, the second integral is rewritten as

$$\begin{aligned}\int_{\boldsymbol{x}' \in \mathbb{D}_{sct}} \partial_j \partial_k \hat{G}(\boldsymbol{x}-\boldsymbol{x}') \left(\frac{\hat{\varepsilon}_{sct}}{\varepsilon_0} - 1\right) \hat{E}_k(\boldsymbol{x}')\, \mathrm{d}V &= \int_{\boldsymbol{x}' \in \mathbb{D}_{sct}} \partial_j \hat{G}(\boldsymbol{x}-\boldsymbol{x}')\, \partial_k' \left[\left(\frac{1}{\varepsilon_0} - \frac{1}{\hat{\varepsilon}_{sct}}\right) \hat{D}_k(\boldsymbol{x}')\right] \mathrm{d}V \\ &= \int_{\boldsymbol{x}' \in \mathbb{D}_{sct}} \partial_j \hat{G}(\boldsymbol{x}-\boldsymbol{x}')\, \partial_k' \left[\frac{1}{\varepsilon_0} - \frac{1}{\hat{\varepsilon}_{sct}}\right] \hat{D}_k(\boldsymbol{x}')\, \mathrm{d}V \\ &= -\int_{\boldsymbol{x}' \in \mathbb{D}_{sct}} \partial_j \hat{G}(\boldsymbol{x}-\boldsymbol{x}')\, \partial_k' \left[\frac{1}{\hat{\varepsilon}_{sct}}\right] \hat{D}_k(\boldsymbol{x}')\, \mathrm{d}V.\end{aligned} \tag{3.110}$$

Using this result, the integral representation of Eq. (3.109) becomes

$$\hat{E}_j^{sct}(\boldsymbol{x}) = \hat{\gamma}_0^2 \int_{\boldsymbol{x}' \in \mathbb{D}_{sct}} \hat{G}(\boldsymbol{x}-\boldsymbol{x}')\, \hat{\chi}^{\varepsilon} \hat{E}_j(\boldsymbol{x}')\, \mathrm{d}V - \int_{\boldsymbol{x}' \in \mathbb{D}_{sct}} \partial_j \hat{G}(\boldsymbol{x}-\boldsymbol{x}')\, \partial_k' \left[\frac{1}{\hat{\varepsilon}_{sct}}\right] \hat{D}_k(\boldsymbol{x}')\, \mathrm{d}V, \tag{3.111}$$

where the permittivity contrast, $\hat{\chi}^{\varepsilon} = 1 - \hat{\varepsilon}_{sct}(\boldsymbol{x})/\varepsilon_0$, is introduced in Eq. (3.78). At this point, we use the electric contrast source, $\hat{w}_k^E(\boldsymbol{x}) = \hat{\chi}^{\varepsilon}(\boldsymbol{x})\hat{E}_k(\boldsymbol{x})$, defined by Eq. (3.79). Similarly, we introduce the "gradient permittivity" contrast vector as

$$\hat{\chi}_k^{\partial\varepsilon}(\boldsymbol{x}) = \partial_k \left[\frac{1}{\hat{\varepsilon}_{sct}(\boldsymbol{x})}\right], \tag{3.112}$$

and its electric-flux contrast source as $\hat{w}_j^E$ and $\hat{w}^{\partial D}$ as

$$\hat{w}^{\partial D}(\boldsymbol{x}) = \hat{\chi}_k^{\partial\varepsilon}(\boldsymbol{x})\hat{D}_k(\boldsymbol{x}). \tag{3.113}$$

Then, the scattered-field integral representation of Eq. (3.98) is written in terms of the two types of contrast sources $\hat{w}_j^E$ and $\hat{w}^{\partial D}$ as

$$\boxed{\hat{E}_j^{sct}(\boldsymbol{x}) = \hat{\gamma}_0^2 \int_{\boldsymbol{x}' \in \mathbb{D}_{sct}} \hat{G}(\boldsymbol{x}-\boldsymbol{x}')\, \hat{w}_j^E(\boldsymbol{x}')\, \mathrm{d}V - \int_{\boldsymbol{x}' \in \mathbb{D}_{sct}} \partial_j \hat{G}(\boldsymbol{x}-\boldsymbol{x}')\, \hat{w}^{\partial D}(\boldsymbol{x}')\, \mathrm{d}V,} \tag{3.114}$$

for $\boldsymbol{x} \in \mathbb{D}_{sct}'$.

The contrast-source integral equations are obtained as by multiplying both sides of Eq. (3.114), for $\boldsymbol{x} \in \mathbb{D}_{sct}$, by $\hat{\chi}^{\varepsilon}$ and $\hat{\chi}_j^{\partial\varepsilon}$, respectively. The resulting equations are

$$\begin{aligned}
\hat{\chi}^{\varepsilon}\hat{E}_j^{inc}(\boldsymbol{x}) &= \hat{w}_j^{E}(\boldsymbol{x}) - \hat{\chi}^{\varepsilon}\left[\hat{\gamma}_0^2\int_{\boldsymbol{x}'\in\mathbb{D}_{sct}} \hat{G}(\boldsymbol{x}-\boldsymbol{x}')\,\hat{w}_j^{E}(\boldsymbol{x}')\,\mathrm{d}V \right.\\
&\quad \left. -\partial_j\int_{\boldsymbol{x}'\in\mathbb{D}_{sct}} \hat{G}(\boldsymbol{x}-\boldsymbol{x}')\,\hat{w}^{\partial D}(\boldsymbol{x}')\,\mathrm{d}V\right],\\
\hat{\chi}_j^{\partial\varepsilon}\hat{E}_j^{inc}(\boldsymbol{x}) &= \hat{w}^{\partial D}(\boldsymbol{x}) - \hat{\chi}_j^{\partial\varepsilon}\left[\hat{\gamma}_0^2\int_{\boldsymbol{x}'\in\mathbb{D}_{sct}} \hat{G}(\boldsymbol{x}-\boldsymbol{x}')\,\hat{w}_j^{E}(\boldsymbol{x}')\,\mathrm{d}V \right.\\
&\quad \left. -\partial_j\int_{\boldsymbol{x}'\in\mathbb{D}_{sct}} \hat{G}(\boldsymbol{x}-\boldsymbol{x}')\,\hat{w}^{\partial D}(\boldsymbol{x}')\,\mathrm{d}V\right].
\end{aligned} \tag{3.115}$$

The most interesting of this set of equations is the search for two different types of contrast sources, one source of which contains exclusive information about the gradient of the permittivity distribution. For inverse scattering problems, this is an important feature, but the three gradient components are hidden in the scalar contrast source $\hat{w}^{\partial D}$. Another problem is that these contrast sources behave as Dirac distributions at discontinuities of the permittivity. To test this method with the solution of the scattering problem of our canonical objects cannot be done with sufficient precision, because refinement of the grid does not provide a converged solution. In order to meet this additional difficulty, we follow the procedure of Chapter 2.

We assume that the scatterer consists of piecewise homogeneous subdomains. The set of internal interfaces in the domain $\mathbb{D}_{sct}$ are denoted as S_{sct}. It is assumed that the domain $\mathbb{D}_{sct}$ also contains the outer interface. Hence, at the boundary of the domain $\partial\mathbb{D}_{sct}$, there is always zero contrast. Due to the current assumption of a piecewise homogeneous permittivity distribution, the spatial derivative of the inverse of the permittivity vanishes everywhere, except at the interfaces where the spatial derivative exhibits an interface Dirac distribution in the normal direction. When the observation point is not located at S_{sct}, the integration over the domain $\mathbb{D}_{sct}$ reduces to an integration along all interfaces S_{sct}. Defining

$$\hat{\varepsilon}^{+} = \lim_{h\to 0}\,[\hat{\varepsilon}_{sct}(\boldsymbol{x}+h\boldsymbol{\nu})] \quad \text{and} \quad \hat{\varepsilon}^{-} = \lim_{h\to 0}\,[\hat{\varepsilon}_{sct}(\boldsymbol{x}-h\boldsymbol{\nu})], \tag{3.116}$$

the integral of Eq. (3.110) becomes

$$\begin{aligned}
&-\int_{\boldsymbol{x}'\in\mathbb{D}_{sct}} \partial_j\hat{G}(\boldsymbol{x}-\boldsymbol{x}')\,\partial_k'\left[\frac{1}{\hat{\varepsilon}_{sct}}\right]\hat{D}_k(\boldsymbol{x}')\,\mathrm{d}V\\
&\quad = \int_{\boldsymbol{x}'\in\mathrm{S}_{sct}} \partial_j\hat{G}(\boldsymbol{x}-\boldsymbol{x}')\,\frac{\hat{\varepsilon}^{+}-\hat{\varepsilon}^{-}}{\hat{\varepsilon}^{+}\hat{\varepsilon}^{-}}\,\nu_k(\boldsymbol{x}')\hat{D}_k(\boldsymbol{x}')\,\mathrm{d}A.
\end{aligned} \tag{3.117}$$

Similar as in Chapter 2, we recognize that result of the integral on the right-hand side jumps when the observation point $\boldsymbol{x}$ passes an interface of the scatterer domain. To investigate this, we let the observation point $\boldsymbol{x}$ approach to a point $\boldsymbol{x}'$ of the set S_{sct}, i.e. $\boldsymbol{x} = \lim_{h\to 0}(\boldsymbol{x}' + h\boldsymbol{\nu})$ and consider only the normal derivative of ∂_j, i.e. $\nu_j\partial_j = \nu_j(\boldsymbol{x})\partial_j$. In this limiting procedure, the singular point $\boldsymbol{x}$ on the interface is excluded symmetrically, after which the limiting value of the integral has been taken. The result is

$$\begin{aligned}
&\nu_j(\boldsymbol{x})\int_{\boldsymbol{x}'\in\mathrm{S}_{sct}} \partial_j\hat{G}(\boldsymbol{x}-\boldsymbol{x}')\,\frac{\hat{\varepsilon}^{+}-\hat{\varepsilon}^{-}}{\hat{\varepsilon}^{+}\hat{\varepsilon}^{-}}\nu_k(\boldsymbol{x}')\hat{D}_k(\boldsymbol{x}')\,\mathrm{d}V\\
&\quad = -\frac{1}{2}\frac{\hat{\varepsilon}^{+}-\hat{\varepsilon}^{-}}{\hat{\varepsilon}^{+}\hat{\varepsilon}^{-}}\nu_j(\boldsymbol{x})\hat{D}_j(\boldsymbol{x}) + \fint_{\boldsymbol{x}'\in\mathrm{S}_{sct}} \nu_j\partial_j\hat{G}(\boldsymbol{x}-\boldsymbol{x}')\,\frac{\hat{\varepsilon}^{+}-\hat{\varepsilon}^{-}}{\hat{\varepsilon}^{+}\hat{\varepsilon}^{-}}\nu_k(\boldsymbol{x}')\hat{D}_k(\boldsymbol{x}')\,\mathrm{d}A,
\end{aligned} \tag{3.118}$$

for $\boldsymbol{x} \in \mathbb{S}_{sct}$. The integral sign $⨍$ denotes the Cauchy principle value of the pertaining integral, see [23].

Combining all the results and using that $\hat{E}_j^{sct} = \hat{E}_j - \hat{E}_j^{inc}$, we have

$$\begin{aligned}\hat{E}_j(\boldsymbol{x}) &= \hat{E}_j^{inc}(\boldsymbol{x}) - \hat{\gamma}_0^2 \int_{\boldsymbol{x}' \in \mathbb{D}_{sct}} \hat{G}(\boldsymbol{x}-\boldsymbol{x}') \, (\frac{\hat{\varepsilon}_{sct}}{\varepsilon_0} - 1)\hat{E}_j(\boldsymbol{x}') \, dV \\ &+ \int_{\boldsymbol{x}' \in \mathbb{S}_{sct}} \partial_j \hat{G}(\boldsymbol{x}-\boldsymbol{x}') \, \frac{\hat{\varepsilon}^+ - \hat{\varepsilon}^-}{\hat{\varepsilon}^+ \hat{\varepsilon}^-} \, \nu_k(\boldsymbol{x}')\hat{D}_k(\boldsymbol{x}') \, dA,\end{aligned} \tag{3.119}$$

for $\boldsymbol{x} \in \mathbb{D}_{sct} \setminus \mathbb{S}_{sct}$, while for $\boldsymbol{x} \in \mathbb{S}_{sct}$, we have

$$\begin{aligned}\frac{1}{2} \frac{\hat{\varepsilon}^+ + \hat{\varepsilon}^-}{\hat{\varepsilon}^+ \hat{\varepsilon}^-} \nu_j \hat{D}_j(\boldsymbol{x}) = \nu_j \hat{E}_j^{inc}(\boldsymbol{x}) - \hat{\gamma}_0^2 \nu_j(\boldsymbol{x}) \int_{\boldsymbol{x}' \in \mathbb{D}_{sct}} \hat{G}(\boldsymbol{x}-\boldsymbol{x}') \, (\frac{\hat{\varepsilon}_{sct}}{\varepsilon_0} - 1)\hat{E}_j(\boldsymbol{x}') \, dV \\ + ⨍_{\boldsymbol{x}' \in \mathbb{S}_{sct}} \nu_j \partial_j \hat{G}(\boldsymbol{x}-\boldsymbol{x}') \, \frac{\hat{\varepsilon}^+ - \hat{\varepsilon}^-}{\hat{\varepsilon}^+ \hat{\varepsilon}^-} \nu_k(\boldsymbol{x}')\hat{D}_k(\boldsymbol{x}') \, dA.\end{aligned} \tag{3.120}$$

Approaching the interface from the side with permittivity $\hat{\varepsilon}^+$, we used that $\hat{E}_k = \hat{D}_j/\hat{\varepsilon}^+$ and we combined this term with the first term on the right-hand side of Eq. (3.118). Multiplication with the factor $2\,\hat{\varepsilon}^+\hat{\varepsilon}^-/(\hat{\varepsilon}^+ + \hat{\varepsilon}^-)$, Eq. (3.120) becomes

$$\begin{aligned}\nu_j \hat{D}_j(\boldsymbol{x}) &= \frac{2\,\hat{\varepsilon}^+\hat{\varepsilon}^-}{\hat{\varepsilon}^+ + \hat{\varepsilon}^-} \left[\nu_j \hat{E}_j^{inc}(\boldsymbol{x}) - \hat{\gamma}_0^2 \, \nu_j(\boldsymbol{x}) \int_{\boldsymbol{x}' \in \mathbb{D}_{sct}} \hat{G}(\boldsymbol{x}-\boldsymbol{x}') \left(\frac{\hat{\varepsilon}_{sct}}{\varepsilon_0} - 1 \right) \hat{E}_j(\boldsymbol{x}') \, dV \right. \\ &\left. + \, \nu_j \partial_j ⨍_{\boldsymbol{x}' \in \mathbb{S}_{sct}} \hat{G}(\boldsymbol{x}-\boldsymbol{x}') \, \frac{\hat{\varepsilon}^+ - \hat{\varepsilon}^-}{\hat{\varepsilon}^+ \hat{\varepsilon}^-} \nu_k \hat{D}_k(\boldsymbol{x}') \, dA \right],\end{aligned} \tag{3.121}$$

for $\boldsymbol{x} \in \mathbb{S}_{sct}$. Equations (3.119) and (3.121) represent the system of integral equations for the unknown electric-field strength $\hat{E}$ in $\mathbb{D}_{sct} \setminus \mathbb{S}_{sct}$, and for the normal component of the flux density $\nu_k \hat{D}_k$ in $\mathbb{S}_{sct}$. It is remarked that we have now two sets of integral equations, because we required consistency at points of $\mathbb{D}_{sct} \setminus \mathbb{S}_{sct}$ and at points of $\mathbb{S}_{sct}$.

The next step is to introduce the volume-contrast sources and the interface-contrast sources. The permittivity contrast $\hat{\chi}^\varepsilon = 1 - \hat{\varepsilon}_{sct}/\varepsilon_0$ and the volume-contrast-source vector $\hat{w}_k^E = \hat{\chi}^\varepsilon \hat{E}_k$ are introduced in Eqs. (3.78) and (3.79). The interface-contrast source is defined as

$$\partial \hat{w}^E(\boldsymbol{x}') = \frac{1}{2} \frac{\hat{\varepsilon}^+ - \hat{\varepsilon}^-}{\hat{\varepsilon}^+ \hat{\varepsilon}^-} \nu_k \hat{D}_k(\boldsymbol{x}'). \tag{3.122}$$

With the definitions of the contrast sources, Eqs. (3.119) and (3.121) become

$$\begin{aligned}\hat{E}_j(\boldsymbol{x}) = \hat{E}_j^{inc}(\boldsymbol{x}) + \hat{\gamma}_0^2 \int_{\boldsymbol{x}' \in \mathbb{D}_{sct}} \hat{G}(\boldsymbol{x}-\boldsymbol{x}') \, \hat{w}_j^E(\boldsymbol{x}') \, dV \\ + \int_{\boldsymbol{x}' \in \mathbb{S}_{sct}} 2\, \partial_j \hat{G}(\boldsymbol{x}-\boldsymbol{x}') \, \partial \hat{w}^E(\boldsymbol{x}') \, dA,\end{aligned} \tag{3.123}$$

and

$$\begin{aligned}\partial \hat{w}^E(\boldsymbol{x}) = \frac{2\,\hat{\varepsilon}^+\hat{\varepsilon}^-}{\hat{\varepsilon}^+ + \hat{\varepsilon}^-} \left[\nu_j \hat{E}_j^{inc}(\boldsymbol{x}) + \hat{\gamma}_0^2 \, \nu_j(\boldsymbol{x}) \int_{\boldsymbol{x}' \in \mathbb{D}_{sct}} \hat{G}(\boldsymbol{x}-\boldsymbol{x}') \, \hat{w}_j^E(\boldsymbol{x}') \, dV \right. \\ \left. + \, \nu_j(\boldsymbol{x}) ⨍_{\boldsymbol{x}' \in \mathbb{S}_{sct}} 2\, \partial_j \hat{G}(\boldsymbol{x}-\boldsymbol{x}') \, \partial \hat{w}^E(\boldsymbol{x}') \, dA \right].\end{aligned} \tag{3.124}$$

After multiplying both sides of Eq. (3.123) by the factor $\hat{\chi}^{\varepsilon}$ and multiplying both sides of Eq. (3.124) by the factor $(\hat{\varepsilon}^+ - \hat{\varepsilon}^-)/(2\,\hat{\varepsilon}^+\hat{\varepsilon}^-)$, we obtain the contrast-source integral equations as

$$\boxed{\begin{aligned}\hat{\chi}^{\varepsilon}(\boldsymbol{x})\hat{E}_j^{inc}(\boldsymbol{x}) = \hat{w}_j^E(\boldsymbol{x}) - \hat{\chi}^{\varepsilon}(\boldsymbol{x})\Bigg[\hat{\gamma}_0^2\int_{\boldsymbol{x}'\in\mathbb{D}_{sct}} \hat{G}(\boldsymbol{x}-\boldsymbol{x}')\,\hat{w}_j^E(\boldsymbol{x}')\,\mathrm{d}V \\ + \int_{\boldsymbol{x}'\in\mathbb{S}_{sct}} 2\,\partial_j\hat{G}(\boldsymbol{x}-\boldsymbol{x}')\,\partial\hat{w}^E(\boldsymbol{x}')\,\mathrm{d}A\Bigg]\,,\end{aligned}} \tag{3.125}$$

for $\boldsymbol{x} \in \mathbb{D}_{sct} \setminus \mathbb{S}_{sct}$, and

$$\boxed{\begin{aligned}\frac{\hat{\varepsilon}^+ - \hat{\varepsilon}^-}{\hat{\varepsilon}^+ + \hat{\varepsilon}^-}\nu_j\hat{E}_j^{inc}(\boldsymbol{x}) = \partial\hat{w}^E(\boldsymbol{x}) - \frac{\hat{\varepsilon}^+ - \hat{\varepsilon}^-}{\hat{\varepsilon}^+ + \hat{\varepsilon}^-}\nu_j(\boldsymbol{x})\Bigg[\hat{\gamma}_0^2\int_{\boldsymbol{x}'\in\mathbb{D}_{sct}} \hat{G}(\boldsymbol{x}-\boldsymbol{x}')\,\hat{w}_j^E(\boldsymbol{x}')\,\mathrm{d}V \\ + \fint_{\boldsymbol{x}'\in\mathbb{S}_{sct}} 2\,\partial_j\hat{G}(\boldsymbol{x}-\boldsymbol{x}')\,\partial\hat{w}^E(\boldsymbol{x}')\,\mathrm{d}A\Bigg]\,,\end{aligned}} \tag{3.126}$$

for $\boldsymbol{x} \in \mathbb{S}_{sct}$. Equations (3.125) and (3.126) represent the system of integral equations for the volume-contrast sources $\hat{w}_j^E$ and interface-contrast sources $\partial\hat{w}^E$.

In view of our remarks made below Eq. (3.115), we now observe that the contrast source $\partial\hat{w}^E$ (see Eq. (3.122) contains the difference between the permittivities at both sides of an interface. At each interface, we have the desired information of local components of the permittivity gradient. Further, the issue of the Dirac distributions at the interface is solved by interpreting them as interface sources, which generates a constituent of the scattered wave field.

Once these volume- and interface-contrast sources are solved, the scattered field at the receiver points $\boldsymbol{x}^R$ in $\mathbb{D}_{sct}$ is obtained from the integral representation

$$\boxed{\hat{E}_j^{sct}(\boldsymbol{x}^R) = \hat{\gamma}_0^2\int_{\boldsymbol{x}'\in\mathbb{D}_{sct}} \hat{G}(\boldsymbol{x}^R-\boldsymbol{x}')\,\hat{w}_j^E(\boldsymbol{x}')\,\mathrm{d}V + \int_{\boldsymbol{x}'\in\mathbb{S}_{sct}} 2\,\partial_j^R\hat{G}(\boldsymbol{x}^R-\boldsymbol{x}')\,\partial\hat{w}^E(\boldsymbol{x}')\,\mathrm{d}A.} \tag{3.127}$$

The magnetic field vector follows from $Z_0\hat{\boldsymbol{H}}^{sct}(\boldsymbol{x}^R) = -\hat{\gamma}_0^{-1}\boldsymbol{\nabla}^R \times \hat{\boldsymbol{E}}^{sct}(\boldsymbol{x}^R)$. In vector notation, we get

$$\boxed{Z_0\hat{\boldsymbol{H}}^{sct}(\boldsymbol{x}^R) = -\hat{\gamma}_0\int_{\boldsymbol{x}'\in\mathbb{D}_{sct}} \boldsymbol{\nabla}^R \times \hat{G}(\boldsymbol{x}^R-\boldsymbol{x}')\,\hat{\boldsymbol{w}}^E(\boldsymbol{x}')\,\mathrm{d}V.} \tag{3.128}$$

It is important to see that the electric fields at the receivers are directly related to the interface sources in the scattering domain, but not the magnetic fields. However, the solution for the contrast sources in the scattering domain is also influenced by the interface sources, but it is only a second order effect on the magnetic field. Therefore, in inverse problems to reconstruct the unknown permittivity and its gradient, we advocate to employ electric-field measurements. Magnetic-field measurements add extra information regarding the permittivity, but not to the jumps of the permittivity at the interfaces.

3.3.1 Numerical Solution with Volume and Interface Contrast Currents: Permittivity Contrast Only

We consider the case that the interfaces are oriented in the Cartesian directions. Then, the set of interfaces perpendicular to the x_1-direction is denoted as $\mathbb{S}_1$, the set of interfaces perpendicular to the x_2-direction is denoted as S_2, and the set of interfaces perpendicular to the x_3-direction is denoted as S_3. The set of integral equations (3.125)–(3.126) simplifies. The first equation becomes

$$\hat{\chi}^{\varepsilon}(\boldsymbol{x})\hat{E}_j^{inc}(\boldsymbol{x}) = \hat{w}_j^E(\boldsymbol{x}) \\ -\hat{\chi}^{\varepsilon}(\boldsymbol{x})\left[\hat{\gamma}_0^2\int_{\boldsymbol{x}'\in\mathbb{D}_{sct}}\hat{G}(\boldsymbol{x}-\boldsymbol{x}')\,\hat{w}_j^E(\boldsymbol{x}')\,\mathrm{d}V + \sum_{k=1}^{3}\int_{\boldsymbol{x}'\in\mathrm{S}_k} 2\,\partial_j\hat{G}(\boldsymbol{x}-\boldsymbol{x}')\,\partial\hat{w}^E(\boldsymbol{x}')\,\mathrm{d}A\right], \quad (3.129)$$

for $\boldsymbol{x}\in\mathbb{D}_{sct}\setminus\sum_{k=1}^{3}\mathrm{S}_k$. Using the relation $\boldsymbol{\nu}(\boldsymbol{x}) = \boldsymbol{i}_j$ when $\boldsymbol{x}\in\mathrm{S}_j$, the second equation becomes

$$\frac{\hat{\varepsilon}^+-\hat{\varepsilon}^-}{\hat{\varepsilon}^++\hat{\varepsilon}^-}\hat{E}_j^{inc}(\boldsymbol{x}) = \partial\hat{w}^E(\boldsymbol{x}) - \frac{\hat{\varepsilon}^+-\hat{\varepsilon}^-}{\hat{\varepsilon}^++\hat{\varepsilon}^-} \\ \times\left[\hat{\gamma}_0^2\int_{\boldsymbol{x}'\in\mathbb{D}_{sct}}\hat{G}(\boldsymbol{x}-\boldsymbol{x}')\,\hat{w}_j^E(\boldsymbol{x}')\,\mathrm{d}V + \partial_j\sum_{k=1}^{3}⨍_{\boldsymbol{x}'\in\mathrm{S}_k} 2\,\hat{G}(\boldsymbol{x}-\boldsymbol{x}')\,\partial\hat{w}^E(\boldsymbol{x}')\,\mathrm{d}A\right], \quad (3.130)$$

for $\boldsymbol{x}\in\mathrm{S}_j$, $j = 1, 2, 3$. Remark that we have now two sets of three integral equations because we require consistency at points of different spaces, viz., $\boldsymbol{x}\in\mathbb{D}_{sct}\setminus\sum_{k=1}^{3}\mathrm{S}_k$ and $\boldsymbol{x}\in\mathrm{S}_j, j = 1, 2, 3$.

3.3.1.1 Discrete Representations in 3D

In 3D, we define a regular grid of a number of n equally sized subdomains, denoted as $\{\mathbb{D}_n,\ n = 1 : N\}$. In 3D, we deal with $N = N_1\times N_2\times N_3$ subdomains. The mesh size of each square subdomain is equal to Δx. The volume of a subdomain is $\Delta V = \Delta x\times\Delta x\times\Delta x$, and each side has a surface area of $\Delta A = \Delta x\times\Delta x$. The midpoints $\boldsymbol{x}_n$ of each subdomain are given by

$$\boldsymbol{x}_n^{(0)} = \{x_{1,n}^{(0)},\ x_{2,n}^{(0)},\ x_{3,n}^{(0)}\}. \quad (3.131)$$

As next step, we define a shifted grid at the horizontal interfaces of the discretized domain,

$$\boldsymbol{x}_n^{(1)} = \{x_{1,n}^{(0)} + \frac{1}{2}\Delta x,\ x_{2,n}^{(0)},\ x_{3,n}^{(0)}\}. \quad (3.132)$$

and two shifted grids at the vertical interfaces of the discretized domain,

$$\boldsymbol{x}_n^{(2)} = \{x_{1,n}^{(0)},\ x_{2,n}^{(0)} + \frac{1}{2}\Delta x,\ x_{3,n}^{(0)}\}. \quad (3.133)$$

and

$$\boldsymbol{x}_n^{(3)} = \{x_{1,n}^{(0)},\ x_{2,n}^{(0)}, x_{3,n}^{(0)} + \frac{1}{2}\Delta x\}. \quad (3.134)$$

For convenience, we use the short-hand notation of the mean values of the volume-contrast sources as

$$\langle\hat{w}_{j,n}^{(0)}\rangle = \hat{w}_j^E(\boldsymbol{x}_n^{(0)}),\quad j = 1, 2, 3, \quad (3.135)$$

and the mean values of the interface-contrast sources as

$$\langle\partial\hat{w}_n^{(j)}\rangle = \partial\hat{w}^E(\boldsymbol{x}_n^{(j)}),\quad j = 1, 2, 3. \quad (3.136)$$

By implementing the discretization of the volume integral (see Chapter 2) and the interface integrals, we obtain a discrete system of equations. As a result, the continuous set of integral equations is replaced by the discrete system of 6 m equations for 6 m unknowns. For each m, the equations are written as

$$
\begin{aligned}
\hat{\chi}^{\varepsilon}_m \langle \hat{E}^{inc}_1 \rangle(\boldsymbol{x}^{(0)}_m) &= \hat{w}^{(0)}_{1,m} - \hat{\chi}^{\varepsilon}_m \left[[\mathcal{K} w^{(0)}_1](\boldsymbol{x}^{(0)}_m) + \partial_1 \sum_{k=1}^{3} [\mathcal{K} \partial \hat{w}^{(k)}](\boldsymbol{x}^{(0)}_m) \right], \\
\hat{\chi}^{\varepsilon}_m \langle \hat{E}^{inc}_2 \rangle(\boldsymbol{x}^{(0)}_m) &= \hat{w}^{(0)}_{2,m} - \hat{\chi}^{\varepsilon}_m \left[[\mathcal{K} w^{(0)}_2](\boldsymbol{x}^{(0)}_m) + \partial_2 \sum_{k=1}^{3} [\mathcal{K} \partial \hat{w}^{(k)}](\boldsymbol{x}^{(0)}_m) \right], \\
\hat{\chi}^{\varepsilon}_m \langle \hat{E}^{inc}_3 \rangle(\boldsymbol{x}^{(0)}_m) &= \hat{w}^{(0)}_{3,m} - \hat{\chi}^{\varepsilon}_m \left[[\mathcal{K} w^{(0)}_3](\boldsymbol{x}^{(0)}_m) + \partial_3 \sum_{k=1}^{3} [\mathcal{K} \partial \hat{w}^{(k)}](\boldsymbol{x}^{(0)}_m) \right], \\
\hat{R}^{(1)}_m \hat{E}^{inc}_1 (\boldsymbol{x}^{(1)}_m) &= \partial \hat{w}^{(1)}_m - \hat{R}^{(1)}_m \left[[\mathcal{K} w^{(0)}_1](\boldsymbol{x}^{(1)}_m) + \partial_1 \sum_{k=1}^{3} [\mathcal{K} \partial \hat{w}^{(k)}](\boldsymbol{x}^{(1)}_m) \right], \\
\hat{R}^{(2)}_m \hat{E}^{inc}_2 (\boldsymbol{x}^{(2)}_m) &= \partial \hat{w}^{(2)}_m - \hat{R}^{(2)}_m \left[[\mathcal{K} w^{(0)}_2](\boldsymbol{x}^{(2)}_m) + \partial_2 \sum_{k=1}^{3} [\mathcal{K} \partial \hat{w}^{(k)}](\boldsymbol{x}^{(2)}_m) \right], \\
\hat{R}^{(3)}_m \hat{E}^{inc}_3 (\boldsymbol{x}^{(3)}_m) &= \partial \hat{w}^{(3)}_m - \hat{R}^{(3)}_m \left[[\mathcal{K} w^{(0)}_3](\boldsymbol{x}^{(3)}_m) + \partial_3 \sum_{k=1}^{3} [\mathcal{K} \partial \hat{w}^{(k)}](\boldsymbol{x}^{(3)}_m) \right],
\end{aligned}
\tag{3.137}
$$

where the contrast $\hat{\chi}^c_m$ and the "reflection coefficients' $\hat{R}^{(j)}_m$ are given by

$$
\begin{aligned}
\hat{\chi}^c_m &= \hat{\chi}^c(\boldsymbol{x}^{(0)}_m), & \hat{R}^{(1)}_m &= \frac{\hat{\varepsilon}(\boldsymbol{x}^{(0)}_m + \Delta x\, \boldsymbol{i}_1) - \hat{\varepsilon}(\boldsymbol{x}^{(0)}_m)}{\hat{\varepsilon}(\boldsymbol{x}^{(0)}_m + \Delta x\, \boldsymbol{i}_1) + \hat{\varepsilon}(\boldsymbol{x}^{(0)}_m)}, \\
\hat{R}^{(2)}_m &= \frac{\hat{\varepsilon}(\boldsymbol{x}^{(0)}_m + \Delta x\, \boldsymbol{i}_2) - \hat{\varepsilon}(\boldsymbol{x}^{(0)}_m)}{\hat{\varepsilon}(\boldsymbol{x}^{(0)}_m + \Delta x\, \boldsymbol{i}_2) + \hat{\varepsilon}(\boldsymbol{x}^{(0)}_m)}, & \hat{R}^{(3)}_m &= \frac{\hat{\varepsilon}(\boldsymbol{x}^{(0)}_m + \Delta x\, \boldsymbol{i}_3) - \hat{\varepsilon}(\boldsymbol{x}^{(0)}_m)}{\hat{\varepsilon}(\boldsymbol{x}^{(0)}_m + \Delta x\, \boldsymbol{i}_3) + \hat{\varepsilon}(\boldsymbol{x}^{(0)}_m)},
\end{aligned}
\tag{3.138}
$$

and where each point $\boldsymbol{x}^{(0)}_m$ denotes the midpoint of the subdomain $\mathbb{D}_m$. Note that $\hat{\varepsilon} = \hat{\varepsilon}_{sct}$ for points inside the scatterer and $\hat{\varepsilon} = \varepsilon_0$ for points outside the scatterer. The operators $\mathcal{K}$ in the matrix on the right-hand side of Eq. (3.137) act on $\hat{w}^{(0)}_{1,n}$, $\hat{w}^{(0)}_{2,n}$, $\hat{w}^{(0)}_{3,n}$, $\partial \hat{w}^{(1)}_n$, $\partial \hat{w}^{(2)}_n$, and $\partial \hat{w}^{(3)}_n$, respectively. Before defining their actual representations, we first define the operator expressions

$$
\begin{aligned}
[\mathcal{K} \hat{w}^{(0)}_1](\boldsymbol{x}) &= \hat{\gamma}^2_0 \Delta V \sum_{n=1}^{N} \langle \hat{G} \rangle (\boldsymbol{x} - \boldsymbol{x}^{(0)}_n)\, \langle \hat{w}^{(0)}_{1,n} \rangle, \\
[\mathcal{K} \hat{w}^{(0)}_2](\boldsymbol{x}) &= \hat{\gamma}^2_0 \Delta V \sum_{n=1}^{N} \langle \hat{G} \rangle (\boldsymbol{x} - \boldsymbol{x}^{(0)}_n)\, \langle \hat{w}^{(0)}_{2,n} \rangle, \\
[\mathcal{K} \hat{w}^{(0)}_3](\boldsymbol{x}) &= \hat{\gamma}^2_0 \Delta V \sum_{n=1}^{N} \langle \hat{G} \rangle (\boldsymbol{x} - \boldsymbol{x}^{(0)}_n)\, \langle \hat{w}^{(0)}_{3,n} \rangle, \\
[\mathcal{K} \partial \hat{w}^{(1)}](\boldsymbol{x}) &= 2\, \Delta A \sum_{n=1}^{N} \langle \hat{G} \rangle (\boldsymbol{x} - \boldsymbol{x}^{(1)}_n)\, \langle \partial \hat{w}^{(1)}_n \rangle, \\
[\mathcal{K} \partial \hat{w}^{(2)}](\boldsymbol{x}) &= 2\, \Delta A \sum_{n=1}^{N} \langle \hat{G} \rangle (\boldsymbol{x} - \boldsymbol{x}^{(2)}_n)\, \langle \partial \hat{w}^{(2)}_n \rangle, \\
[\mathcal{K} \partial \hat{w}^{(3)}](\boldsymbol{x}) &= 2\, \Delta A \sum_{n=1}^{N} \langle \hat{G} \rangle (\boldsymbol{x} - \boldsymbol{x}^{(3)}_n)\, \langle \partial \hat{w}^{(3)}_n \rangle.
\end{aligned}
\tag{3.139}
$$

Subsequently, we take their particular values at $\boldsymbol{x}_m^{(0)}$, $\boldsymbol{x}_m^{(1)}$, $\boldsymbol{x}_m^{(2)}$, and $\boldsymbol{x}_m^{(3)}$, respectively. These values are given by

$$\begin{aligned}
\mathcal{K}_{1,m}^{(0,0)} &= [\mathcal{K}\hat{w}_1^{(0)}](\boldsymbol{x}_m^{(0)}) = \hat{\gamma}_0^2 \Delta V \sum_{n=1}^{N} \langle\langle \hat{G} \rangle\rangle(\boldsymbol{x}_m^{(0)} - \boldsymbol{x}_n^{(0)})\, \langle \hat{w}_{1,n}^{(0)} \rangle, \\
\mathcal{K}_{2,m}^{(0,0)} &= [\mathcal{K}\hat{w}_2^{(0)}](\boldsymbol{x}_m^{(0)}) = \hat{\gamma}_0^2 \Delta V \sum_{n=1}^{N} \langle\langle \hat{G} \rangle\rangle(\boldsymbol{x}_m^{(0)} - \boldsymbol{x}_n^{(0)})\, \langle \hat{w}_{2,n}^{(0)} \rangle, \\
\mathcal{K}_{3,m}^{(0,0)} &= [\mathcal{K}\hat{w}_3^{(0)}](\boldsymbol{x}_m^{(0)}) = \hat{\gamma}_0^2 \Delta V \sum_{n=1}^{N} \langle\langle \hat{G} \rangle\rangle(\boldsymbol{x}_m^{(0)} - \boldsymbol{x}_n^{(0)})\, \langle \hat{w}_{3,n}^{(0)} \rangle, \\
\mathcal{K}_m^{(1,1)} &= [\mathcal{K}\partial\hat{w}^{(1)}](\boldsymbol{x}_m^{(1)}) = 2\,\Delta A \sum_{n=1}^{N} \langle\langle \hat{G} \rangle\rangle(\boldsymbol{x}_m^{(0)} - \boldsymbol{x}_n^{(0)})\, \langle \partial\hat{w}_n^{(1)} \rangle, \\
\mathcal{K}_m^{(2,2)} &= [\mathcal{K}\partial\hat{w}^{(2)}](\boldsymbol{x}_m^{(2)}) = 2\,\Delta A \sum_{n=1}^{N} \langle\langle \hat{G} \rangle\rangle(\boldsymbol{x}_m^{(0)} - \boldsymbol{x}_n^{(0)})\, \langle \partial\hat{w}_n^{(2)} \rangle, \\
\mathcal{K}_m^{(3,3)} &= [\mathcal{K}\partial\hat{w}^{(3)}](\boldsymbol{x}_m^{(3)}) = 2\,\Delta A \sum_{n=1}^{N} \langle\langle \hat{G} \rangle\rangle(\boldsymbol{x}_m^{(0)} - \boldsymbol{x}_n^{(0)})\, \langle \partial\hat{w}_n^{(3)} \rangle,
\end{aligned} \tag{3.140}$$

where $\langle\langle \hat{G} \rangle\rangle$ is the repeated spherical mean of the 3D Green function, see Chapter 1. Note that in the expression of the convolutional operators $\mathcal{K}_m^{(1,1)}$, $\mathcal{K}_m^{(2,2)}$, and $\mathcal{K}_m^{(3,3)}$, we have used the relation $\boldsymbol{x}_m^{(k)} - \boldsymbol{x}_n^{(k)} = \boldsymbol{x}_m^{(0)} - \boldsymbol{x}_n^{(0)}$, for $k = 1, 2, 3$.

After computation of these convolutional operators, the other operators are computed by interpolation and extrapolation of the convolutional results. They are symbolically given by

$$\begin{aligned}
\mathcal{K}_m^{(1,0)} &= \mathcal{K}_{1,m}^{(0,0)}(\boldsymbol{x}_m^{(1)}), & \mathcal{K}_m^{(2,0)} &= \mathcal{K}_{2,m}^{(0,0)}(\boldsymbol{x}_m^{(2)}), & \mathcal{K}_m^{(3,0)} &= \mathcal{K}_{3,m}^{(0,0)}(\boldsymbol{x}_m^{(3)}), \\
\mathcal{K}_m^{(0,1)} &= \mathcal{K}_m^{(1,1)}(\boldsymbol{x}_m^{(0)}), & \mathcal{K}_m^{(2,1)} &= \mathcal{K}_m^{(1,1)}(\boldsymbol{x}_m^{(2)}), & \mathcal{K}_m^{(3,1)} &= \mathcal{K}_m^{(1,1)}(\boldsymbol{x}_m^{(3)}), \\
\mathcal{K}_m^{(0,2)} &= \mathcal{K}_m^{(2,2)}(\boldsymbol{x}_m^{(0)}), & \mathcal{K}_m^{(1,2)} &= \mathcal{K}_m^{(2,2)}(\boldsymbol{x}_m^{(1)}), & \mathcal{K}_m^{(3,2)} &= \mathcal{K}_m^{(2,2)}(\boldsymbol{x}_m^{(3)}), \\
\mathcal{K}_m^{(0,3)} &= \mathcal{K}_m^{(3,3)}(\boldsymbol{x}_m^{(0)}), & \mathcal{K}_m^{(1,3)} &= \mathcal{K}_m^{(3,3)}(\boldsymbol{x}_m^{(1)}), & \mathcal{K}_m^{(2,3)} &= \mathcal{K}_m^{(3,3)}(\boldsymbol{x}_m^{(2)}).
\end{aligned} \tag{3.141}$$

Instead of using the expressions of Eq. (3.139), we compute the numerical values of Eq. (3.141) approximately by interpolating and extrapolating the results of Eq. (3.140). These interpolation and extrapolation procedures relate the respective values at the different grids to each other. We perform the following computing procedure:

(1) Compute $\mathcal{K}_{1,m}^{(0,0)}$ using FFTs.
Subsequently, compute $\mathcal{K}_m^{(1,0)}$ by extrapolating $\mathcal{K}_{1,m}^{(0,0)}$ linearly in the x_1-direction.

(2) Compute $\mathcal{K}_{2,m}^{(0,0)}$ using FFTs.
Subsequently, compute $\mathcal{K}_m^{(2,0)}$ by extrapolating $\mathcal{K}_{2,m}^{(0,0)}$ linearly in the x_2-direction.

(3) Compute $\mathcal{K}_{3,m}^{(0,0)}$ using FFTs.
Subsequently, compute $\mathcal{K}_m^{(3,0)}$ by extrapolating $\mathcal{K}_{3,m}^{(0,0)}$ linearly in the x_3-direction.

(4) Compute $\mathcal{K}_m^{(1,1)}$ using FFTs.
Subsequently, compute $\mathcal{K}_m^{(0,1)}$ by extrapolating $\mathcal{K}_m^{(1,1)}$ linearly in the negative x_1-direction, and compute $\mathcal{K}_m^{(2,1)}$ and $\mathcal{K}_m^{(3,1)}$ by interpolating the values of $\mathcal{K}_m^{(1,1)}$ at four adjacent points of $\boldsymbol{x}_m^{(1)}$, in the x_2-direction and x_3-directions, respectively.

(5) Compute $\mathcal{K}_m^{(2,2)}$ using FFTs.
Subsequently, compute $\mathcal{K}_m^{(0,2)}$ by extrapolating $\mathcal{K}_m^{(2,2)}$ linearly in the negative x_2-direction, and compute $\mathcal{K}_m^{(1,2)}$ and $\mathcal{K}_m^{(3,2)}$ from $\mathcal{K}_m^{(2,2)}$ by interpolating the values at four adjacent points in the x_1-direction and x_3-directions, respectively.

(6) Compute $\mathcal{K}_m^{(3,3)}$ using FFTs.
Subsequently, compute $\mathcal{K}_m^{(0,3)}$ by extrapolating $\mathcal{K}_m^{(3,3)}$ linearly in the negative x_3-direction, and compute $\mathcal{K}_m^{(1,3)}$ and $\mathcal{K}_m^{(2,3)}$ from $\mathcal{K}_m^{(3,3)}$ by interpolating the values at four adjacent points in the x_1-direction and x_2-directions, respectively.

The interpolation procedure in a certain coordinate system is the simple summation of the pertaining values at four adjacent points (midpoint rule), see Matlab function `interpolate` of Matlab Listing I.60, for example,

$$\mathcal{K}_m^{(2,1)}(\boldsymbol{x}_m^{(2)}) = \frac{1}{4}\Big[\mathcal{K}_m^{(1,1)}(\boldsymbol{x}_m^{(1)} + \tfrac{1}{2}\Delta x\,\boldsymbol{i}_1 + \tfrac{1}{2}\Delta x\,\boldsymbol{i}_2) + \mathcal{K}_m^{(1,1)}(\boldsymbol{x}_m^{(1)} + \tfrac{1}{2}\Delta x\,\boldsymbol{i}_1 - \tfrac{1}{2}\Delta x\,\boldsymbol{i}_2)$$
$$+ \mathcal{K}_m^{(1,1)}(\boldsymbol{x}_m^{(1)} - \tfrac{1}{2}\Delta x\,\boldsymbol{i}_1 + \tfrac{1}{2}\Delta x\,\boldsymbol{i}_2) + \mathcal{K}_m^{(1,1)}(\boldsymbol{x}_m^{(1)} - \tfrac{1}{2}\Delta x\,\boldsymbol{i}_1 - \tfrac{1}{2}\Delta x\,\boldsymbol{i}_2)\Big] . \qquad (3.142)$$

The extrapolation procedure from either the $\boldsymbol{x}^{(0)}$-coordinates to the other coordinates or from the other coordinates to the $\boldsymbol{x}^{(0)}$-coordinates is based on a linear Taylor expansion in the pertaining direction, in which its spatial derivative is numerically replaced by a two-point central difference, for example

$$\mathcal{K}_m^{(1,0)}(\boldsymbol{x}_m^{(0)} + \tfrac{1}{2}\Delta x\,\boldsymbol{i}_1) = \mathcal{K}_{1,m}^{(0,0)}(\boldsymbol{x}_m^{(0)}) + \tfrac{1}{4}[\mathcal{K}_{1,m}^{(0,0)}(\boldsymbol{x}_m^{(0)} + \Delta x\,\boldsymbol{i}_1) - \mathcal{K}_{1,m}^{(0,0)}(\boldsymbol{x}_m^{(0)} - \Delta x\,\boldsymbol{i}_1)] \qquad (3.143)$$

and

$$\mathcal{K}_m^{(0,1)}(\boldsymbol{x}_m^{(1)} + \tfrac{1}{2}\Delta x\,\boldsymbol{i}_1) = \mathcal{K}_m^{(1,1)}(\boldsymbol{x}_m^{(1)}) - \tfrac{1}{4}[\mathcal{K}_m^{(1,1)}(\boldsymbol{x}_m^{(1)} + \Delta x\,\boldsymbol{i}_1) - \mathcal{K}_m^{(1,1)}(\boldsymbol{x}_m^{(1)} - \Delta x\,\boldsymbol{i}_1)] , \qquad (3.144)$$

see Matlab function `extrapolate` of Matlab Listing I.61.

The results in the extra layer added to the scattering domain are extrapolated values and are immaterial because both the wave-speed contrast and the reflection coefficients in this layer vanish. However, we need this layer to define the correct array numbering of the input and output arrays of the Matlab script.

In summary, we note that Eq. (3.137) may be symbolically written as

$$\begin{aligned}
\hat{\chi}_m^{\varepsilon}\,\langle\hat{E}_1^{inc}\rangle(\boldsymbol{x}_m^{(0)}) &= \hat{w}_{1,m}^{(0)} - \hat{\chi}_m^{\varepsilon}\,[\mathcal{K}_{1,m}^{(0,0)} + \partial_1(\mathcal{K}_m^{(0,1)} + \mathcal{K}_m^{(0,2)} + \mathcal{K}_m^{(0,3)})] ,\\
\hat{\chi}_m^{\varepsilon}\,\langle\hat{E}_2^{inc}\rangle(\boldsymbol{x}_m^{(0)}) &= \hat{w}_{2,m}^{(0)} - \hat{\chi}_m^{\varepsilon}\,[\mathcal{K}_{2,m}^{(0,0)} + \partial_2(\mathcal{K}_m^{(0,1)} + \mathcal{K}_m^{(0,2)} + \mathcal{K}_m^{(0,3)})] ,\\
\hat{\chi}_m^{\varepsilon}\,\langle\hat{E}_3^{inc}\rangle(\boldsymbol{x}_m^{(0)}) &= \hat{w}_{3,m}^{(0)} - \hat{\chi}_m^{\varepsilon}\,[\mathcal{K}_{3,m}^{(0,0)} + \partial_3(\mathcal{K}_m^{(0,1)} + \mathcal{K}_m^{(0,2)} + \mathcal{K}_m^{(0,3)})] ,\\
\hat{R}_m^{(1)}\,\langle\hat{E}_1^{inc}\rangle(\boldsymbol{x}_m^{(1)}) &= \partial\hat{w}_m^{(1)} - \hat{R}_m^{(1)}\,[\mathcal{K}_m^{(1,0)} + \partial_1(\mathcal{K}_m^{(1,1)} + \mathcal{K}_m^{(1,2)} + \mathcal{K}_m^{(1,3)})] ,\\
\hat{R}_m^{(2)}\,\langle\hat{E}_2^{inc}\rangle(\boldsymbol{x}_m^{(2)}) &= \partial\hat{w}_m^{(2)} - \hat{R}_m^{(2)}\,[\mathcal{K}_m^{(2,0)} + \partial_2(\mathcal{K}_m^{(2,1)} + \mathcal{K}_m^{(2,2)} + \mathcal{K}_m^{(2,3)})] ,\\
\hat{R}_m^{(3)}\,\langle\hat{E}_3^{inc}\rangle(\boldsymbol{x}_m^{(3)}) &= \partial\hat{w}_m^{(3)} - \hat{R}_m^{(3)}\,[\mathcal{K}_m^{(3,0)} + \partial_3(\mathcal{K}_m^{(3,1)} + \mathcal{K}_m^{(3,2)} + \mathcal{K}_m^{(3,3)})] ,
\end{aligned} \qquad (3.145)$$

where the $\mathcal{K}$-operators act on the volume-contrast sources $\hat{w}_{1,m}^{(0)}$, $\hat{w}_{2,m}^{(0)}$, and $\hat{w}_{3,m}^{(0)}$, and the interface-contrast sources $\partial\hat{w}^{(1)}$, $\partial\hat{w}^{(2)}$, and $\partial\hat{w}^{(3)}$. We remark that this system of equations is not written in its standard matrix form. In a standard matrix form, the operators $\mathcal{K}_{1,m}^{(0,0)}$, $\mathcal{K}_{2,m}^{(0,0)}$, $\mathcal{K}_{3,m}^{(0,0)}$, $\mathcal{K}_m^{(1,1)}$, $\mathcal{K}_m^{(2,2)}$, and $\mathcal{K}_m^{(3,3)}$ are present in the diagonal of the system of equations. The non-diagonal operators represent the interaction of the fields from the four different types of contrast sources. For convenience of the reader, we rank the various operators in the square brackets on the right-hand side of Eq. (3.145) as the matrix

$$A = \begin{pmatrix}
\mathcal{K}_{1,m}^{(0,0)} & 0 & 0 & \partial_1\mathcal{K}_m^{(0,1)} & \partial_1\mathcal{K}_m^{(0,2)} & \partial_1\mathcal{K}_m^{(0,3)}\\
0 & \mathcal{K}_{2,m}^{(0,0)} & 0 & \partial_2\mathcal{K}_m^{(0,1)} & \partial_2\mathcal{K}_m^{(0,2)} & \partial_2\mathcal{K}_m^{(0,3)}\\
0 & 0 & \mathcal{K}_{3,m}^{(0,0)} & \partial_3\mathcal{K}_m^{(0,1)} & \partial_3\mathcal{K}_m^{(0,2)} & \partial_3\mathcal{K}_m^{(0,3)}\\
\mathcal{K}_m^{(1,0)} & 0 & 0 & \partial_1\mathcal{K}_m^{(1,1)} & \partial_1\mathcal{K}_m^{(1,2)} & \partial_1\mathcal{K}_m^{(1,3)}\\
0 & \mathcal{K}_m^{(2,0)} & 0 & \partial_2\mathcal{K}_m^{(2,1)} & \partial_2\mathcal{K}_m^{(2,2)} & \partial_2\mathcal{K}_m^{(2,3)}\\
0 & 0 & \mathcal{K}_m^{(3,0)} & \partial_3\mathcal{K}_m^{(3,1)} & \partial_3\mathcal{K}_m^{(3,2)} & \partial_3\mathcal{K}_m^{(3,3)}
\end{pmatrix} . \qquad (3.146)$$

3.3.1.2 Discrete Representations in 2D

In 2D, we do not deal with the shifted grid based on the $\boldsymbol{x}_n^{(3)}$ coordinates and similar as in Chapter 2 the incident wave field and the Green functions are different. The 2D volume is denoted as $\Delta V = \Delta x \times \Delta x$, while the 2D interface area is the side length Δx. In summary, we have the discretized equations

$$\begin{aligned}
\hat{\chi}_m^{\varepsilon}\,\langle E_1^{inc}\rangle(\boldsymbol{x}_{T,m}^{(0)}) &= \hat{w}_{1,m}^{(0)} - \hat{\chi}_m^{\varepsilon}[\mathcal{K}_{1,m}^{(0,0)} + \partial_1\,(\mathcal{K}_m^{(0,1)} + \mathcal{K}_m^{(0,2)})]\,,\\
\hat{\chi}_m^{\varepsilon}\,\langle E_2^{inc}\rangle(\boldsymbol{x}_{T,m}^{(0)}) &= \hat{w}_{2,m}^{(0)} - \hat{\chi}_m^{\varepsilon}[\mathcal{K}_{2,m}^{(0,0)} + \partial_2\,(\mathcal{K}_m^{(0,1)} + \mathcal{K}_m^{(0,2)})]\,,\\
\hat{R}_m^{(1)}\,\langle E_1^{inc}\rangle(\boldsymbol{x}_{T,m}^{(1)}) &= \partial\hat{w}_m^{(1)} - \hat{R}_m^{(1)}\,[\mathcal{K}_m^{(1,0)} + \partial_1\,(\mathcal{K}_m^{(1,1)} + \mathcal{K}_m^{(1,2)})]\,,\\
\hat{R}_m^{(2)}\,\langle E_2^{inc}\rangle(\boldsymbol{x}_{T,m}^{(2)}) &= \partial\hat{w}_m^{(2)} - \hat{R}_m^{(2)}\,[\mathcal{K}_m^{(2,0)} + \partial_2\,(\mathcal{K}_m^{(2,1)} + \mathcal{K}_m^{(2,2)})]\,,
\end{aligned} \tag{3.147}$$

where $\hat{\chi}_m^{\varepsilon} = \hat{\chi}^{\varepsilon}(\boldsymbol{x}_{T,m}^{(0)})$ and

$$\hat{R}_m^{(1)} = \frac{\hat{\varepsilon}(\boldsymbol{x}_{T,m}^{(0)} + \Delta x\,\boldsymbol{i}_1) - \hat{\varepsilon}(\boldsymbol{x}_{T,m}^{(0)})}{\hat{\varepsilon}(\boldsymbol{x}_{T,m}^{(0)} + \Delta x\,\boldsymbol{i}_1) + \hat{\varepsilon}(\boldsymbol{x}_{T,m}^{(0)})},\quad \hat{R}_m^{(2)} = \frac{\hat{\varepsilon}(\boldsymbol{x}_{T,m}^{(0)} + \Delta x\,\boldsymbol{i}_2) - \hat{\varepsilon}(\boldsymbol{x}_{T,m}^{(0)})}{\hat{\varepsilon}(\boldsymbol{x}_{T,m}^{(0)} + \Delta x\,\boldsymbol{i}_2) + \hat{\varepsilon}(\boldsymbol{x}_{T,m}^{(0)})}. \tag{3.148}$$

The diagonal operators are given by

$$\begin{aligned}
\mathcal{K}_{1,m}^{(0,0)}(\boldsymbol{x}_{T,m}^{(0)}) &= \hat{\gamma}_0^2\Delta V\sum_{n=1}^{N}\,\langle\langle\hat{G}\rangle\rangle(\boldsymbol{x}_{T,m}^{(0)} - \boldsymbol{x}_{T,n}^{(0)})\,\langle\hat{w}_{1,n}^{(0)}\rangle,\\
\mathcal{K}_{2,m}^{(0,0)}(\boldsymbol{x}_{T,m}^{(0)}) &= \hat{\gamma}_0^2\Delta V\sum_{n=1}^{N}\,\langle\langle\hat{G}\rangle\rangle(\boldsymbol{x}_{T,m}^{(0)} - \boldsymbol{x}_{T,n}^{(0)})\,\langle\hat{w}_{2,n}^{(0)}\rangle,\\
\mathcal{K}_m^{(1,1)}(\boldsymbol{x}_{T,m}^{(1)}) &= 2\Delta x\sum_{n=1}^{N}\langle\langle\partial_1\hat{G}\rangle\rangle(\boldsymbol{x}_{T,m}^{(0)} - \boldsymbol{x}_{T,n}^{(0)})\,\langle\partial\hat{w}_n^{(1)}\rangle,\\
\mathcal{K}_m^{(2,2)}(\boldsymbol{x}_{T,m}^{(2)}) &= 2\Delta x\sum_{n=1}^{N}\langle\langle\partial_2\hat{G}\rangle\rangle(\boldsymbol{x}_{T,m}^{(0)} - \boldsymbol{x}_{T,n}^{(0)})\,\langle\partial\hat{w}_n^{(2)}\rangle.
\end{aligned} \tag{3.149}$$

The off-diagonal operators are computed by interpolation and extrapolation of the resulting values of the diagonal operators as

$$\begin{aligned}
\mathcal{K}_m^{(1,0)} &= \mathcal{K}_m^{(0,0)}(\boldsymbol{x}_{T,m}^{(1)}),\ \mathcal{K}_m^{(2,0)} = \mathcal{K}_m^{(0,0)}(\boldsymbol{x}_{T,m}^{(2)}),\\
\mathcal{K}_m^{(0,1)} &= \mathcal{K}_m^{(1,1)}(\boldsymbol{x}_{T,m}^{(0)}),\ \mathcal{K}_m^{(2,1)} = \mathcal{K}_m^{(1,1)}(\boldsymbol{x}_{T,m}^{(2)}),\\
\mathcal{K}_m^{(0,2)} &= \mathcal{K}_m^{(2,2)}(\boldsymbol{x}_{T,m}^{(0)}),\ \mathcal{K}_m^{(1,2)} = \mathcal{K}_m^{(2,2)}(\boldsymbol{x}_{T,m}^{(1)}).
\end{aligned} \tag{3.150}$$

3.3.2 Iterative Solution and Operators Involved

We now consider the iterative solution of the contrast-source type integral equations, based on the volume-contrast sources and the interface-contrast sources. In Section 3.3.1, we have defined all discrete operators and their pertaining Matlab codes. We notice that the system of integral equations listed in Eq. (3.137) is equivalent to the following discrete set of equations:

$$\begin{aligned}
\chi_m^{\varepsilon}\langle E_j^{inc}\rangle(x_m^{(0)}) &= w_{j,m}^{(0)} - \chi_m^{\varepsilon}\,\mathcal{K}_{j,m}^{(0)}, \quad j = 1,2,3,\\
R_m^{(j)}\langle E_j^{inc}\rangle(x_m^{(k)}) &= \partial w_m^{(j)} - R_m^{(j)}\,\mathcal{K}_m^{(j)}, \quad j = 1,2,3.
\end{aligned} \tag{3.151}$$

The operator $\mathcal{K}_{j,m}^{(0)}$ maps functions defined on the four different grids to functions on the $\boldsymbol{x}^{(0)}$-grid and $\mathcal{K}_m^{(j)}$ maps the functions defined on the four different grids to functions on the grid $\boldsymbol{x}_m^{(j)}, j = 1,2,3$.

Matlab Listing I.96 The function `gradj` to carry out the partial differentiation in the j-direction.

```
function v = gradj(p,j,input)
global nDIM;  dx = input.dx;

if nDIM == 2;
  N1 = input.N1;  N2 = input.N2;

  v = zeros(size(p));
  if j==1;  v(2:N1-1,:) = (p(3:N1,:) - p(1:N1-2,:)) / (2*dx); end;  % d1_p
  if j==2;  v(:,2:N2-1) = (p(:,3:N2) - p(:,1:N2-2)) / (2*dx); end;  % d2_p

elseif nDIM == 3;
  N1 = input.N1;  N2 = input.N2;  N3 = input.N3;

  v = zeros(size(p));
  if j==1; v(2:N1-1,:,:)= (p(3:N1,:,:)-p(1:N1-2,:,:))/(2*dx); end;  % d1_p
  if j==2; v(:,2:N2-1,:)= (p(:,3:N2,:)-p(:,1:N2-2,:))/(2*dx); end;  % d2_p
  if j==3; v(:,:,2:N3-1)= (p(:,:,3:N3)-p(:,:,1:N3-2))/(2*dx); end;  % d3_p
end;
```

They are found as

$$\begin{aligned} \mathcal{K}_{j,m}^{(0)} &= \mathcal{K}_{j,m}^{(0,0)} + \mathtt{grad}_j\mathcal{K}_{j,m}^{(0,1)} + \mathtt{grad}_j\mathcal{K}_{j,m}^{(0,2)} + \mathtt{grad}_j\mathcal{K}_{j,m}^{(0,3)}, \\ \mathcal{K}_{m}^{(j)} &= \mathcal{K}_{m}^{(j,0)} + \mathtt{grad}_j\mathcal{K}_{m}^{(j,1)} + \mathtt{grad}_j\mathcal{K}_{m}^{(j,2)} + \mathtt{grad}_j\mathcal{K}_{m}^{(j,3)}, \end{aligned} \tag{3.152}$$

where $j = 1, 2, 3$. Further, $\mathtt{grad}_j$ is the finite-difference counterpart of the partial derivative with respect to x_j, see Matlab function `gradj` as given by the Matlab Listing I.96.

The operators $\mathcal{K}_{j,m}^{(0)}$ and $\mathcal{K}_{m}^{(j)}$ act on the unknown volume-contrast sources $w_{k,n}^{(0)}$ and the unknown interface-contrast sources $\partial w_n^{(k)}$. The Matlab function `KopEwdw` to compute these operators is given in Matlab Listing I.97.

The expressions of the adjoint follow from the inner-product definition. Because it is a rather tedious procedure and, in addition, a very slow convergence of the pertaining CG scheme was observed, we refrain to present these codes and use the BiCGSTAB method as a more efficient alternative. With this we discussed all the ingredients to solve the integral equation for volume-contrast sources $w_j^{(0)}$ and interface-contrast sources $\partial w^{(j)}$.

In 3D, after a numerical solution for these volume- and interface-contrast sources, the numerical value of the scattered electric-field strength at some receiver point, $\boldsymbol{x}^R \in \mathbb{D}'_{sct}$, is computed as

$$\langle \hat{E}_j^{sct}\rangle(\boldsymbol{x}^R) = \sum_{m=1}^{N} \left[\hat{\gamma}_0{}^2 \Delta V \langle \hat{G}\rangle(\boldsymbol{x}^R - \boldsymbol{x}_m^{(0)})\, \hat{w}_{j,m}^{(0)} + 2\Delta A \sum_{k=1}^{3} \partial_j^R \langle \hat{G}\rangle(\boldsymbol{x}^R - \boldsymbol{x}_m^{(k)})\, \partial\hat{w}_m^{(k)} \right], \tag{3.153}$$

for $j = 1, 2, 3$, while the scattered magnetic field is computed as

$$\begin{aligned} \langle \hat{H}_1^{sct}\rangle(\boldsymbol{x}^R) &= -\hat{\gamma}_0 \Delta V \sum_{m=1}^{N} [\partial_2^R \langle \hat{G}\rangle(\boldsymbol{x}^R - \boldsymbol{x}_m^{(0)})\, \hat{w}_{3,m}^{(0)} - \partial_3^R \langle \hat{G}\rangle(\boldsymbol{x}^R - \boldsymbol{x}_m^{(0)})\, \hat{w}_{1,m}^{(0)}], \\ \langle \hat{H}_2^{sct}\rangle(\boldsymbol{x}^R) &= -\hat{\gamma}_0 \Delta V \sum_{m=1}^{N} [\partial_3^R \langle \hat{G}\rangle(\boldsymbol{x}^R - \boldsymbol{x}_m^{(0)})\, \hat{w}_{1,m}^{(0)} - \partial_1^R \langle \hat{G}\rangle(\boldsymbol{x}^R - \boldsymbol{x}_m^{(0)})\, \hat{w}_{3,m}^{(0)}], \\ \langle \hat{H}_3^{sct}\rangle(\boldsymbol{x}^R) &= -\hat{\gamma}_0 \Delta V \sum_{m=1}^{N} [\partial_1^R \langle \hat{G}\rangle(\boldsymbol{x}^R - \boldsymbol{x}_m^{(0)})\, \hat{w}_{2,m}^{(0)} - \partial_2^R \langle \hat{G}\rangle(\boldsymbol{x}^R - \boldsymbol{x}_m^{(0)})\, \hat{w}_{1,m}^{(0)}]. \end{aligned} \tag{3.154}$$

Matlab Listing I.97 The function KopEwdw to carry out the $\mathcal{K}$ operator.

```
function [Kw,Kdw] = KopEwdw(w,dw,input)
global nDIM;

% Correction factor for interface integrals
factor = 2 / (input.gamma_0^2 * input.dx);

if nDIM == 2;
   Kw{1} = Kop(w{1},input.FFTG);                    %---------------%  K00_1
  Kdw{1} = extrapolate(Kw{1},1,0,input);                            %  K10
   Kw{2} = Kop(w{2},input.FFTG);                    %---------------%  K00_2
  Kdw{2} = extrapolate(Kw{2},2,0,input);                            %  K20
           KOPdw = factor * Kop(dw{1},input.FFTG); %-----------------
   Kw{1} = Kw{1}  + gradj(extrapolate(KOPdw,0,1,input),1,input); % + d1K01
  Kdw{1} = Kdw{1} + gradj(            KOPdw,           1,input); % + d1K11
   Kw{2} = Kw{2}  + gradj(extrapolate(KOPdw,0,1,input),2,input); % + d2K01
  Kdw{2} = Kdw{2} + gradj(interpolate(KOPdw,2,1,input),2,input); % + d2K21
           KOPdw = factor * Kop(dw{2},input.FFTG); %-----------------
   Kw{1} = Kw{1}  + gradj(extrapolate(KOPdw,0,2,input),1,input); % + d1K02
  Kdw{1} = Kdw{1} + gradj(interpolate(KOPdw,1,2,input),1,input); % + d1K12
   Kw{2} = Kw{2}  + gradj(extrapolate(KOPdw,0,2,input),2,input); % + d2K02
  Kdw{2} = Kdw{2} + gradj(            KOPdw,           2,input); % + d2K22

elseif nDIM == 3;
   Kw{1} = Kop(w{1},input.FFTG);                    %---------------%  K00_1
  Kdw{1} = extrapolate(Kw{1},1,0,input);                            %  K10
   Kw{2} = Kop(w{2},input.FFTG);                    %---------------%  K00_2
  Kdw{2} = extrapolate(Kw{2},2,0,input);                            %  K20
   Kw{3} = Kop(w{3},input.FFTG);                    %---------------%  K00_3
  Kdw{3} = extrapolate(Kw{3},3,0,input);                            %  K30
           KOPdw = factor * Kop(dw{1},input.FFTG); %-----------------
   Kw{1} = Kw{1}  + gradj(extrapolate(KOPdw,0,1,input),1,input); % + d1K01
  Kdw{1} = Kdw{1} + gradj(            KOPdw,           1,input); % + d1K11
   Kw{2} = Kw{2}  + gradj(extrapolate(KOPdw,0,1,input),2,input); % + d2K01
  Kdw{2} = Kdw{2} + gradj(interpolate(KOPdw,2,1,input),2,input); % + d2K21
   Kw{3} = Kw{3}  + gradj(extrapolate(KOPdw,0,1,input),3,input); % + d3K01
  Kdw{3} = Kdw{3} + gradj(interpolate(KOPdw,3,1,input),3,input); % + d3K31
           KOPdw = factor * Kop(dw{2},input.FFTG); %-----------------
   Kw{1} = Kw{1}  + gradj(extrapolate(KOPdw,0,2,input),1,input); % + d1K02
  Kdw{1} = Kdw{1} + gradj(interpolate(KOPdw,1,2,input),1,input); % + d1K12
   Kw{2} = Kw{2}  + gradj(extrapolate(KOPdw,0,2,input),2,input); % + d2K02
  Kdw{2} = Kdw{2} + gradj(            KOPdw,           2,input); % + d2K22
   Kw{3} = Kw{3}  + gradj(extrapolate(KOPdw,0,2,input),3,input); % + d3K02
  Kdw{3} = Kdw{3} + gradj(extrapolate(KOPdw,3,2,input),3,input); % + d3K32
           KOPdw = factor * Kop(dw{3},input.FFTG); %-----------------
   Kw{1} = Kw{1}  + gradj(extrapolate(KOPdw,0,3,input),1,input); % + d1K03
  Kdw{1} = Kdw{1} + gradj(interpolate(KOPdw,1,3,input),1,input); % + d1K13
   Kw{2} = Kw{2}  + gradj(extrapolate(KOPdw,0,3,input),2,input); % + d2K03
  Kdw{2} = Kdw{2} + gradj(extrapolate(KOPdw,2,3,input),2,input); % + d2K23
   Kw{3} = Kw{3}  + gradj(extrapolate(KOPdw,0,3,input),3,input); % + d3K03
  Kdw{3} = Kdw{3} + gradj(            KOPdw,           3,input); % + d3K33
end;
```

The spatial derivatives of the weak Green functions follow from the strong form of the Green function of Section 3.1.2 by replacing the strong form of the Green function by its weak counterpart.

In 2D, the nonzero scattered electric-field components are computed as

$$\langle \hat{E}_j^{sct} \rangle(\boldsymbol{x}_T^R) = \sum_{m=1}^{N} \left[\hat{\gamma}_0^{\,2} \Delta V \langle \hat{G} \rangle(\boldsymbol{x}_T^R - \boldsymbol{x}_{T,m}^{(0)})\, \hat{w}_{j,m}^{(0)} + 2\Delta A \sum_{k=1}^{2} \partial_j^R \langle \hat{G} \rangle(\boldsymbol{x}_T^R - \boldsymbol{x}_{T,m}^{(k)})\, \partial \hat{w}_m^{(k)} \right], \tag{3.155}$$

for $j = 1, 2$, while the nonzero scattered magnetic-field components are computed as

$$\langle \hat{H}_3^{sct} \rangle(\boldsymbol{x}_T^R) = -\hat{\gamma}_0 \Delta V \sum_{m=1}^{N} \left[\partial_1^R \langle \hat{G} \rangle(\boldsymbol{x}_T^R - \boldsymbol{x}_{T,m}^{(0)})\, \hat{w}_{2,m}^{(0)} - \partial_2^R \langle \hat{G} \rangle(\boldsymbol{x}_T^R - \boldsymbol{x}_{T,m}^{(0)})\, \hat{w}_{1,m}^{(0)} \right]. \tag{3.156}$$

It is interesting to compare the expressions of Eqs. (3.153)–(3.156) with the equivalent ones of Eqs. (3.103)–(3.106) and observe again that the scattered magnetic-field components are insensitive for the interface sources.

3.3.3 Matlab Codes E-field Integral Equations: Volume and Interface Contrast Sources

In this section, we present the Matlab codes for the numerical solution of the set of coupled equations for the volume- and interface-contrast sources. After solution for the contrast sources, we discuss the Matlab codes to compute the electromagnetic field strengths in a number of receiver points outside the scatterer. To initialize the various electromagnetic parameters, we start with the function `initEM` of Matlab Listing I.82. We do not change these parameters.

After the initialization of the electromagnetic parameters, we start the main Matlab script `ForwardBiCGSTABFFTEwdw` of Matlab Listing I.98. Again we have four steps: (1) Computation of scattering by a canonical object; (2) Computation of the incident wave fields; (3) Solution of the set of integral equations; and (4) Computation of scattering data, based on the solution of the integral equations.

In step (1) of `ForwardBiCGSTABFFTEwdw`, script `EMForwardCanonicalObjects` of Matlab Listing I.139 of Appendix 3.A.2.3 is called. It computes the scattered electromagnetic field at a number of receivers, for either the 2D circular cylinder or the 3D sphere. These data are used to benchmark the scattered field based on the numerical solution of the present set of integral equations. Next, the Matlab function `initEMcontrastIntf` of Matlab Listing I.99 determines the permittivity contrast χ^{ε} at the midpoints of the square subdomains and the reflection coefficients $R^{(j)}$ at the interfaces between adjacent subdomains.

In step (2) of `ForwardBiCGSTABFFTEwdw`, we compute the weak form of the incident electric field, either in 2D or 3D. It computes the incident electric field (E_inc) on the non-shifted grid (0). On the shifted grids (1), (2), and (3), only the normal components of the incident electric field have to be computed. We denote them as dEinc. For the 2D case, the function `IncEfield` is presented in Matlab Listing I.100, while for the 3D case, the function `IncEfield` calls `incEfield3D`. It is presented in Matlab Listing I.101. The notation of Matlab's cell array for the components of E_inc and dEinc is very useful for a compact form of input and output parameters in a function list.

In step (3) of `ForwardBiCGSTABFFTEwdw`, we solve the integral equations for the volume-contrast sources w_E $:= w_j^{(0)}$ and interface-contrast sources dwE $:= \partial w^{(j)}$ by calling `ITERBiCGSTABEwdw` of Matlab Listing I.102. In each iteration of the last method, the two operators $\mathcal{K}_{j,m}^{(0)}$ and $\mathcal{K}_m^{(j)}$ are computed by the function handle `Aw`. In view of the editorial restriction that a Matlab Listing should fit within a single page, the function `vector2matrix`, that handles the reshaping of vectors to matrices, has been listed separately in Matlab Listing I.103. The operator

Matlab Listing I.98 The `ForwardBiCGSTABFFTEwdw` script to solve the volume-contrast sources w_E $:= w_j^{(0)}$ and interface-contrast sources dwE $:= \partial w^{(j)}$.

```
clear all; clc; close all; clear workspace;
input = initEM();

% (1) Compute analytically scattered field data ------------------------
      EMForwardCanonicalObjects

%     Input interface contrast
      [input] = initEMcontrastIntf(input);

% (2) Compute incident field -------------------------------------------
      [E_inc,dEinc] = IncEfield(input);

% (3) Solve integral equation for contrast source with FFT -------------
tic;  [w_E,dwE] = ITERBiCGSTABEwdw(E_inc,dEinc,input);
toc
%     Plot contrast and contrast sources
      plotContrastSourcewE(w_E,input);
      plotEMInterfaceSourceE(dwE,input);

% (4) Compute synthetic data and plot fields and data ------------------
      [Edata,Hdata] = DOPEwdw(w_E,dwE,input);
      displayEdata(Edata,input);
      displayHdata(Hdata,input);
```

function `KopEwdw` is presented in Matlab Listing I.97. In this function, the scalar operator `Kop` of Matlab Listings I.13 is used. The function `plotContrastSourcewE` of Matlab Listing I.90 plots the absolute values of the volume-contrast sources, while the function `plotEMInterface-SourceE` of Matlab Listing I.104 plots the absolute values of the interface-contrast sources. For the 2D case and $j = 1$ and 2, Figure 3.9 shows the volume-contrast-source amplitudes $|w_j^E|$, while for the 3D case and $j = 1, 2$, and 3, these amplitudes are shown in Figure 3.10.

Note that, in 3D, w_3^E is an odd function with respect to the plane $x_3 = 0$ and should vanish in this plane. From Figure 3.10, it follows that in this plane, the nonzero amplitude is only one order smaller than the amplitude of w_1^E. The reason that the discretion error is now visible, while in Figure 3.6 it was not visible, can be explained by the fact that the operators of `ForwardCGFFTwE` follow the symmetry of the operators, whereas in the present case, the extrapolation of the fields interferes with the symmetry. In Figures 3.11 and 3.12, the resulting reflection coefficients and interface-contrast sources are displayed, for the 2D and 3D case, respectively. Note that both the reflection coefficients and the interface-contrast sources indicate the staircase approximation of the circular and spherical boundaries. At the circular and spherical interfaces, the reflections coefficients and the interface-contrast sources act in the normal direction and vanish along the tangential directions.

In step (4) of the Matlab script `ForwardCGFFTwE`, the scattered field at the selected receiver points are computed. The contrast-source distributions are the input for function `DopwE` of Matlab Listing I.91. This Matlab function computes the 2D scattered-field strengths and calls the function `Dop3DwE` of Matlab Listing I.92 to compute the 3D scattered-field strengths. The program follows from the weak form of the scattered electromagnetic wave fields, see Eqs. (3.153)–(3.156).

Matlab Listing I.99 The function `initEMcontrastIntf` to initialize the electromagnetic parameters.

```
function [input] = initEMcontrastIntf(input)
global nDIM;

if      nDIM == 2;
        R = sqrt(input.X1.^2 + input.X2.^2);
        [N1,N2] = size(R);

elseif nDIM == 3;
        R = sqrt(input.X1.^2 + input.X2.^2 + input.X3.^2);
        [N1,N2,N3] = size(R);

end; % if

% Compute permittivity type of 'reflection factors' ----------------------
    a = input.a;
  eps = input.eps_sct * (R < a) + (R >= a) * 1;

if      nDIM == 2;

               Rfl{1} = zeros(N1,N2);
               Rfl{2} = zeros(N1,N2);
     Rfl{1}(1:N1-1,:) = (eps(2:N1,:) - eps(1:N1-1,:)) ...
                                    ./(eps(2:N1,:) + eps(1:N1-1,:));
     Rfl{2}(:,1:N2-1) = (eps(:,2:N2) - eps(:,1:N2-1)) ...
                                    ./(eps(:,2:N2) + eps(:,1:N2-1));

elseif nDIM == 3;

               Rfl{1} = zeros(N1,N2,N3);
               Rfl{2} = zeros(N1,N2,N3);
               Rfl{3} = zeros(N1,N2,N3);
   Rfl{1}(1:N1-1,:,:) = (eps(2:N1,:,:) - eps(1:N1-1,:,:)) ...
                                  ./(eps(2:N1,:,:) + eps(1:N1-1,:,:));
   Rfl{2}(:,1:N2-1,:) = (eps(:,2:N2,:) - eps(:,1:N2-1,:)) ...
                                  ./(eps(:,2:N2,:) + eps(:,1:N2-1,:));
   Rfl{3}(:,:,1:N3-1) = (eps(:,:,2:N3) - eps(:,:,1:N3-1)) ...
                                  ./(eps(:,:,2:N3) + eps(:,:,1:N3-1));

end; % if

  input.Rfl = Rfl;
```

These scattered fields are compared to the analytical Bessel-function data stored in MAT-files. The functions `displayEdata.m` and `displayHData.m` of Matlab Listings I.93 and I.94 load the particular files and plot both the exact data and the numerical data. The absolute values of the 2D scattered electric and magnetic field is presented in Figure 3.13, while the 3D results are presented in Figure 3.14.

Matlab Listing I.100 The `IncEfield` script to compute the 2D incident electric field and to call the function `IncEfield3D` for computation of the 3D incident electric field.

```
function [E_inc,dEinc] = IncEfield(input)
global nDIM;

gam0 = input.gamma_0;  xS = input.xS;    dx = input.dx;

if nDIM == 2;
  delta  = (pi)^(-1/2) * dx;          % radius circle with area of dx^2
  factor = 2 * besseli(1,gam0*delta) / (gam0*delta);

  % Non shifted grid (0) ---------------------------------------------------
    X1     = input.X1 - xS(1);
    X2     = input.X2 - xS(2);
    DIS    = sqrt(X1.^2 + X2.^2);
    X1     = X1./DIS;
    X2     = X2./DIS;
      G    =    factor *           1/(2*pi).* besselk(0,gam0*DIS);
     dG    = - factor * gam0 .* 1/(2*pi).* besselk(1,gam0*DIS);
     dG11 = (2*X1.*X1 - 1) .* (-dG./DIS) + gam0^2 * X1.*X1 .* G;
     dG21 =  2*X2.*X1       .* (-dG./DIS) + gam0^2 * X2.*X1 .* G;
  E_inc{1} = - (-gam0^2 * G + dG11);
  E_inc{2} = - dG21;

  % Shifted grid (1) -------------------------------------------------------
    X1     = input.X1 + dx/2 - xS(1);
    X2     = input.X2 - xS(2);
    DIS    = sqrt(X1.^2 + X2.^2);
    X1     = X1./DIS;
      G    =    factor *           1/(2*pi).* besselk(0,gam0*DIS);
     dG    = - factor * gam0 .* 1/(2*pi).* besselk(1,gam0*DIS);
     dG11 = (2*X1.*X1 - 1)  .* (-dG./DIS) + gam0^2 * X1.*X1 .* G;
  dEinc{1} = - (-gam0^2 * G + dG11);

  % Shifted grid (2) -------------------------------------------------------
    X1    = input.X1 - xS(1);
    X2    = input.X2 + dx/2 - xS(2);
    DIS   = sqrt(X1.^2 + X2.^2);
    X1    = X1./DIS;
    X2    = X2./DIS;
     G    =    factor *           1/(2*pi).* besselk(0,gam0*DIS);
    dG    = - factor * gam0 .* 1/(2*pi).* besselk(1,gam0*DIS);
    dG21 =  2*X2.*X1        .* (-dG./DIS) + gam0^2 * X2.*X1 .* G;
  dEinc{2} = - dG21;

elseif nDIM == 3;

      [E_inc,dEinc] = IncEfield3D(input);

end; % if
```

Matlab Listing I.101 The function `IncEfield3D` to compute the 3D incident electric field.

```
function [E_inc,dEinc] = IncEfield3D(input)
gam0   = input.gamma_0;  xS = input.xS;     dx = input.dx;
delta  = (4*pi/3)^(-1/3) * dx;      % radius sphere with area of dx^3
arg    = gam0*delta; factor = 3 * (cosh(arg) - sinh(arg)/arg) / arg^2;

% Non shifted grid (0) -------------------------------------------------------
   X1    = input.X1-xS(1);
   X2    = input.X2-xS(2);
   X3    = input.X3-xS(3);
   DIS   = sqrt(X1.^2 + X2.^2 + X3.^2);
   X1    = X1./DIS;
   X2    = X2./DIS;
   X3    = X3./DIS;
    G    = factor * exp(-gam0*DIS) ./ (4*pi*DIS);
   dG11 = ((3*X1.*X1 -1).* (gam0./DIS + 1./DIS.^2) + gam0^2*X1.*X1 ) .* G;
   dG21 = ( 3*X2.*X1     .* (gam0./DIS + 1./DIS.^2) + gam0^2*X2.*X1 ) .* G;
   dG31 = ( 3*X3.*X1     .* (gam0./DIS + 1./DIS.^2) + gam0^2*X3.*X1 ) .* G;
E_inc{1} = - (-gam0^2 * G + dG11);
E_inc{2} = - dG21;
E_inc{3} = - dG31;

% Shifted grid (1) -----------------------------------------------------------
   X1    = input.X1+dx/2-xS(1);
   X2    = input.X2-xS(2);
   X3    = input.X3-xS(3);
   DIS   = sqrt(X1.^2 + X2.^2 + X3.^2);
   X1    = X1./DIS;
    G    = factor * exp(-gam0*DIS) ./ (4*pi*DIS);
   dG11 = ((3*X1.*X1 -1).* (gam0./DIS + 1./DIS.^2) + gam0^2*X1.*X1 ) .* G;
 dEinc{1} = - (-gam0^2 * G + dG11);
 % Shifted grid(2) -----------------------------------------------------------
   X1    = input.X1-xS(1);
   X2    = input.X2+dx/2-xS(2);
   X3    = input.X3-xS(3);
   DIS   = sqrt(X1.^2 + X2.^2 + X3.^2);
   X1    = X1./DIS;
   X2    = X2./DIS;
     G   = factor * exp(-gam0*DIS) ./ (4*pi*DIS);
   dG21 = ( 3*X2.*X1     .* (gam0./DIS + 1./DIS.^2) + gam0^2*X2.*X1 ) .* G;
 dEinc{2} = - dG21;
 % Shifted grid(3) -----------------------------------------------------------
  X1    = input.X1-xS(1);
  X2    = input.X2-xS(2);
  X3    = input.X3+dx/2-xS(3);
  DIS   = sqrt(X1.^2 + X2.^2 + X3.^2);
  X1    = X1./DIS;
  X3    = X3./DIS;
    G   = factor * exp(-gam0*DIS) ./ (4*pi*DIS);
  dG31 = ( 3*X3.*X1     .* (gam0./DIS + 1./DIS.^2) + gam0^2*X3.*X1 ) .* G;
  dEinc{3} = - dG31;
```

Matlab Listing I.102 The function ITERBiCGSTABEwdw to compute the volume-contrast sources w_E := $w_j^{(0)}$ and interface-contrast sources dwE := $\partial w^{(j)}$.

```
function [w_E,dwE] = ITERBiCGSTABEwdw(E_inc,dEinc,input)
% BiCGSTAB scheme for contrast source integral equation Aw = b
  global nDIM;

  itmax  = 1000;   Errcri = input.Errcri;    [N,~] = size(input.CHI_eps(:));

  if nDIM == 2;
        b(      1:  N,1) = input.CHI_eps(:) .* E_inc{1}(:);
        b(  N+1:2*N,1) = input.CHI_eps(:) .* E_inc{2}(:);
        b(2*N+1:3*N,1) = input.Rfl{1}(:)   .* dEinc{1}(:);
        b(3*N+1:4*N,1) = input.Rfl{2}(:)   .* dEinc{2}(:);

   elseif nDIM == 3;
        b(      1:  N,1) = input.CHI_eps(:) .* E_inc{1}(:);
        b(  N+1:2*N,1) = input.CHI_eps(:) .* E_inc{2}(:);
        b(2*N+1:3*N,1) = input.CHI_eps(:) .* E_inc{3}(:);
        b(3*N+1:4*N,1) = input.Rfl{1}(:)   .* dEinc{1}(:);
        b(4*N+1:5*N,1) = input.Rfl{2}(:)   .* dEinc{2}(:);
        b(5*N+1:6*N,1) = input.Rfl{3}(:)   .* dEinc{3}(:);
  end; % if

   w = bicgstab(@(w) Aw(w,input), b, Errcri, itmax);          % call BICGSTAB

  [w_E,dwE] = vector2matrix(w,input);

end %----------------------------------------------------------------------


function y = Aw(w,input)
  global nDIM;         [N,~] = size(input.CHI_eps(:));

   [w_E,dwE]   = vector2matrix(w,input);
  [Kw_E,KdwE] = KopEwdw(w_E,dwE,input);

  if nDIM == 2;
        y(      1:  N,1) = w_E{1}(:) - input.CHI_eps(:) .* Kw_E{1}(:);
        y(  N+1:2*N,1) = w_E{2}(:) - input.CHI_eps(:) .* Kw_E{2}(:);
        y(2*N+1:3*N,1) = dwE{1}(:) - input.Rfl{1}(:)   .* KdwE{1}(:);
        y(3*N+1:4*N,1) = dwE{2}(:) - input.Rfl{2}(:)   .* KdwE{2}(:);

  elseif nDIM == 3;
        y(      1:  N,1) = w_E{1}(:) - input.CHI_eps(:) .* Kw_E{1}(:);
        y(  N+1:2*N,1) = w_E{2}(:) - input.CHI_eps(:) .* Kw_E{2}(:);
        y(2*N+1:3*N,1) = w_E{3}(:) - input.CHI_eps(:) .* Kw_E{3}(:);
        y(3*N+1:4*N,1) = dwE{1}(:) - input.Rfl{1}(:)   .* KdwE{1}(:);
        y(4*N+1:5*N,1) = dwE{2}(:) - input.Rfl{2}(:)   .* KdwE{2}(:);
        y(5*N+1:6*N,1) = dwE{3}(:) - input.Rfl{3}(:)   .* KdwE{3}(:);
  end; % if

end
```

Matlab Listing I.103 The function `vector2matrix` required by the function `ITERBiCGSTABEwdw`.

```
function [w_E,dwE] = vector2matrix(w,input)
% Modify vector output from 'bicgstab' to matrices for further computation
  global nDIM;

  [N,~] = size(input.CHI_eps(:));
   w_E = cell(1,nDIM);
   dwE = cell(1,nDIM);

  if nDIM == 2;                          DIM = [input.N1,input.N2];

        w_E{1} = reshape(w(      1:  N,1),DIM);
        w_E{2} = reshape(w(  N+1:2*N,1),DIM);
        dwE{1} = reshape(w(2*N+1:3*N,1),DIM);
        dwE{2} = reshape(w(3*N+1:4*N,1),DIM);

  end; %if

  if nDIM == 3;                         DIM = [input.N1,input.N2,input.N3];

      w_E{1} = reshape(w(      1:  N,1),DIM);
      w_E{2} = reshape(w(  N+1:2*N,1),DIM);
      w_E{3} = reshape(w(2*N+1:3*N,1),DIM);
      dwE{1} = reshape(w(3*N+1:4*N,1),DIM);
      dwE{2} = reshape(w(4*N+1:5*N,1),DIM);
      dwE{3} = reshape(w(5*N+1:6*N,1),DIM);

  end; %if

end
```

We use the error norm for the electric field, defined in Eq. (3.107), and that for the magnetic field, defined in Eq. (3.108). Similar to the scattered field results of Chapter 1, the error in 3D is smaller than in 2D. The errors in the electric fields are smaller than those in the magnetic fields. We also note that in Chapters 1 and 2, the $\text{Error}^{\text{sct}}$ is determined by the complex difference between two scalar results. In the present electromagnetic case, both the errors $\text{Error}(E^{sct})$ and $\text{Error}(Z_0\text{H}^{sct})$ are determined by the difference between the amplitudes of two scattered field vectors. The error estimates are then a bit too optimistic. The error between the differences in phases are not taken into account.

3.3.4 Performance Analysis

In this section, we investigate the convergence rate of the integral equations with volume-contrast sources w_j^E, solved by using function `ITERBiCGSTABwE` and denoted as the w-method. We compare this convergence rate with the one of the so-called wdw-method, in which we solve the integral equations with volume-contrast sources w_j^E and interface-contrast sources $\partial w^{E(j)}$ by using function `ITERBiCGSTABEwdw`. As before, we investigate the data errors $\text{Error}(E^{sct})$ and $\text{Error}(Z_0\text{H}^{sct})$ as function of the grid size. We only consider the 2D case. We start with $\Delta x = 2$ and reduce it each

Matlab Listing I.104 The function `plotEMInterfaceSourceE` to plot the reflection coefficients $R^{(1)}$ and $R^{(2)}$ and the interface-contrast sources $\partial w^{E(1)}$ and $\partial w^{E(2)}$.

```
function plotEMInterfaceSourceE(dwE,input)
global nDIM;
dx = input.dx;          Rfl = input.Rfl;

if  nDIM == 2
    x1 = input.X1(:,1);     x2 = input.X2(1,:);

    set(figure,'Units','centimeters','Position',[5 5 18 12]);
    subplot(1,2,1);
       IMAGESC(x1+dx/2,x2,Rfl{1});
       title('\fontsize{13} R^{(1)}');
    subplot(1,2,2);
       IMAGESC(x1+dx/2,x2, abs(dwE{1}));
       title('\fontsize{13} abs(dwE^{(1)})');

    set(figure,'Units','centimeters','Position',[5 5 18 12]);
    subplot(1,2,1);
       IMAGESC(x1,x2+dx/2,Rfl{2});
       title('\fontsize{13} R^{(2)}');
    subplot(1,2,2);
       IMAGESC(x1,x2+dx/2,abs(dwE{2}));
       title('\fontsize{13} abs(dwE^{(2)})');

elseif nDIM == 3
    N3cross = floor(input.N3/2+1);              % plot at x3 = 0 or x3 = dx/2
    x1 = input.X1(:,1,1);   x2 = input.X2(1,:,1);

    set(figure,'Units','centimeters','Position',[5 5 18 12]);
    subplot(1,2,1);
       IMAGESC(x1+dx/2,x2, Rfl{1}(:,:,N3cross));
       title('\fontsize{13} R^{(1)}');
    subplot(1,2,2);
       IMAGESC(x1+dx/2,x2,abs(dwE{1}(:,:,N3cross)));
       title('\fontsize{13} abs(dwE^{(1)})');

    set(figure,'Units','centimeters','Position',[5 5 18 12]);
    subplot(1,2,1);
       IMAGESC(x1,x2+dx/2, Rfl{2}(:,:,N3cross));
       title('\fontsize{13} R^{(2)}');
    subplot(1,2,2);
       IMAGESC(x1,x2+dx/2,abs(dwE{2}(:,:,N3cross)));
       title('\fontsize{13} abs(dwE^{(2)})');
end; % if
```

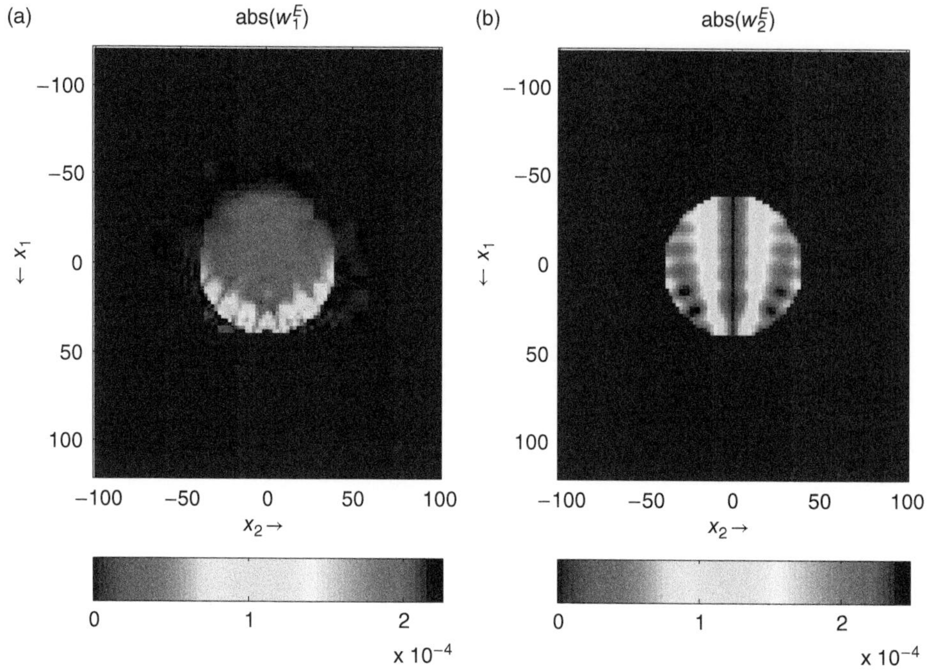

Figure 3.9 Plots of the 2D volume-contrast-source amplitudes (a) $|w_1^E|$ and (b) $|w_2^E|$, produced by `ITERBiCGSTABFFTEwdw`.

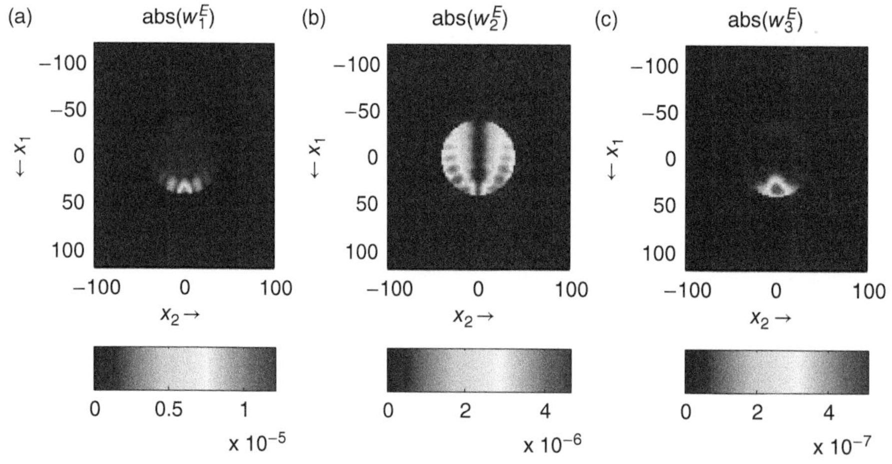

Figure 3.10 Cross-sectional plots of the 3D volume-contrast-source amplitudes (a) $|w_1^E|$, (b) $|w_2^E|$, and (c) $|w_3^E|$, produced by `ITERBiCGSTABFFTEwdw`.

time with a factor of two. We define a refinement factor $r = 1, 2, \ldots, 5$, and a grid size $\Delta^{(r)} = \Delta x/r$ with sample numbers $N_1^{(r)} = r\,N_1$ and $N_2^{(r)} = r\,N_2$. We have set the error criterion to 10^{-6}. These computations are carried on a Dell computer (64 bit, dual core 3.50 GHz, 64 GB RAM). In the wdw-method,

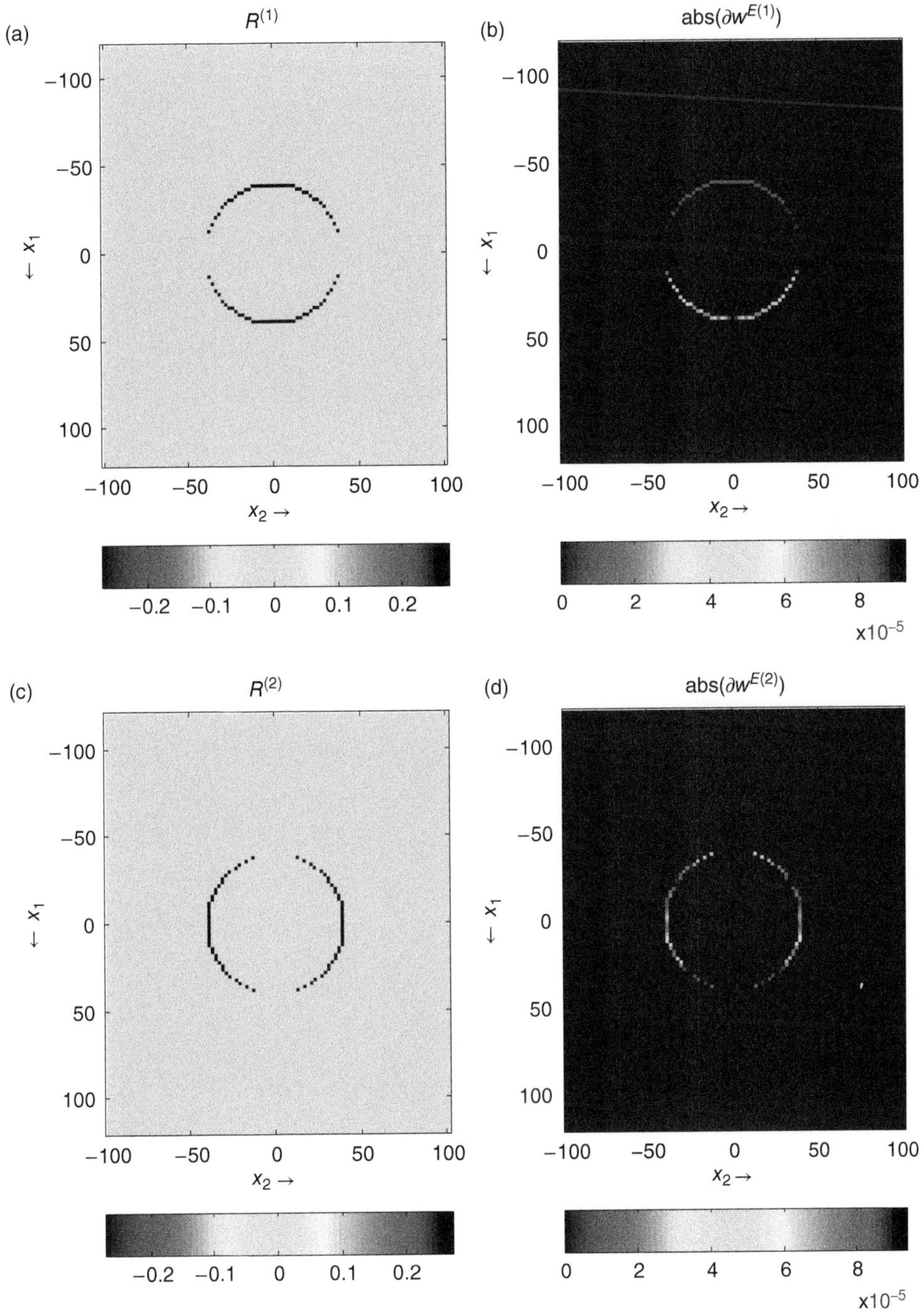

Figure 3.11 Plots of the reflection coefficients (a,c) $R^{(1)}$ and $R^{(2)}$ and the interface-contrast-source amplitudes (b,d) $|\partial w^{E(1)}|$ and $|\partial w^{E(2)}|$, produced by `ITERBiCGSTABFFTEwdw` (2D case).

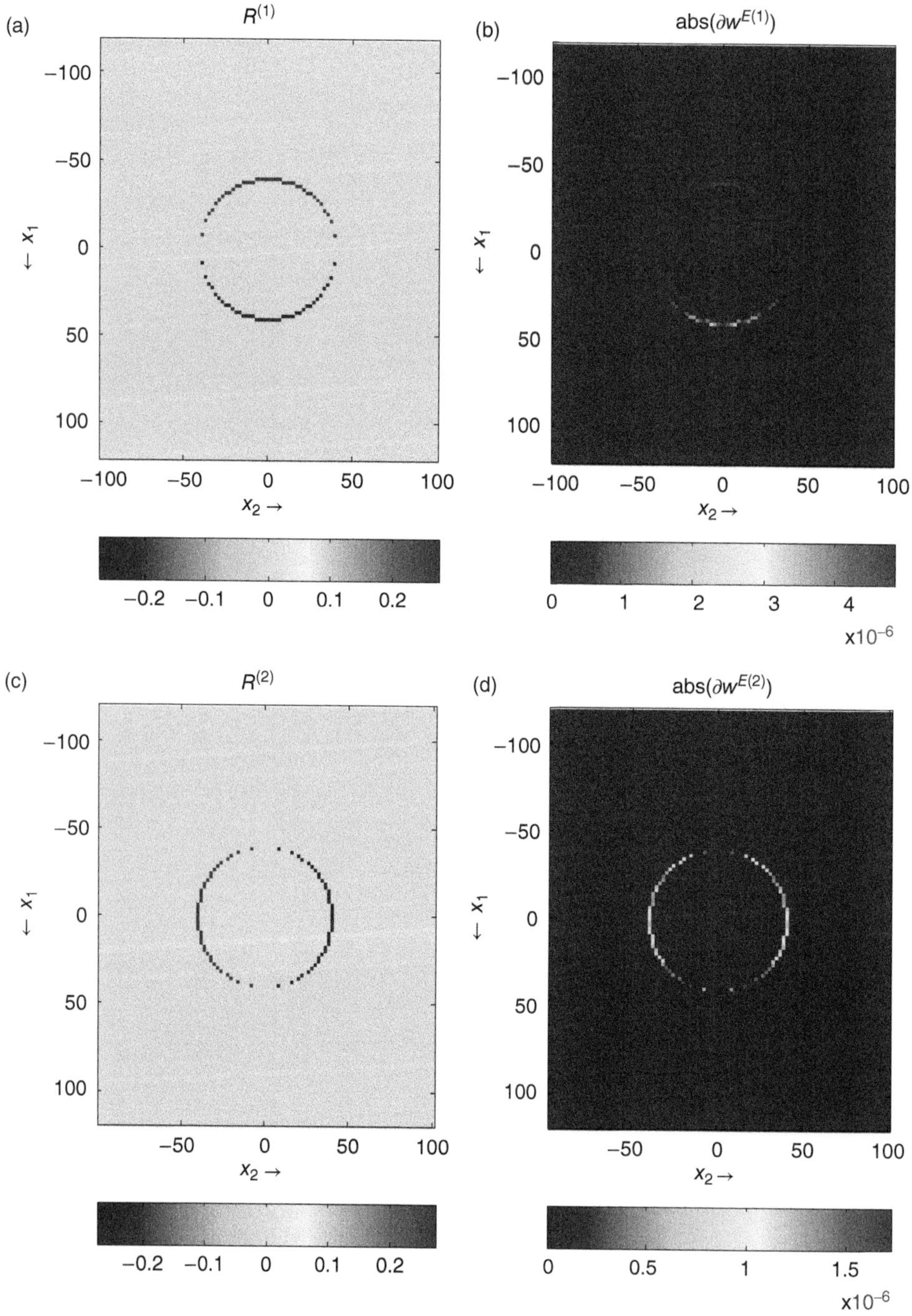

Figure 3.12 Plots of the reflection coefficients (a,c) $R^{(1)}$ and $R^{(2)}$ and the interface-contrast-source amplitudes (b,d) $|\partial w^{E(1)}|$ and $|\partial w^{E(2)}|$, produced by `ITERBiCGSTABFFTEwdw` (3D case).

Matlab Listing I.105 The function DopEwdw to compute the 2D scattered field strengths at a number of receivers.

```
function [Edata,Hdata] = DOPEwdw(w_E,dwE,input)
global nDIM;
gam0 = input.gamma_0;        dx = input.dx;    xR = input.xR;
Edata = zeros(1,input.NR); Hdata = zeros(1,input.NR);

if nDIM == 2;

 delta  = (pi)^(-1/2)*dx;               % radius circle with area dx^2
 factor = 2 * besseli(1,gam0*delta) / (gam0*delta);

 for p = 1 : input.NR
  % Non shifted grid(0)
      X1  = xR(1,p)-input.X1;      X2 = xR(2,p)-input.X2;
     DIS  = sqrt(X1.^2 + X2.^2);   X1 = X1./DIS;  X2 = X2./DIS;
      G   =   factor * 1/(2*pi).* besselk(0,gam0*DIS);
     dG   = - factor * gam0 * 1/(2*pi).* besselk(1,gam0*DIS);
    d1_G  =  X1 .* dG;
    d2_G  =  X2 .* dG;
    E1rfl = gam0^2 * dx^2 * sum( G(:) .* w_E{1}(:) );
    E2rfl = gam0^2 * dx^2 * sum( G(:) .* w_E{2}(:) );
   ZH3rfl = gam0 * dx^2 * sum( d2_G(:).*w_E{1}(:) - d1_G(:).*w_E{2}(:) );

  % Shifted grid(1)
     X1   = xR(1,p) - input.X1 - dx/2;              X2 = xR(2,p)-input.X2;
     DIS  = sqrt(X1.^2 + X2.^2);   X1 = X1./DIS;  X2 = X2./DIS;
     dG   = - factor * gam0 * 1/(2*pi).* besselk(1,gam0*DIS);
    d1_G  =  X1 .* dG;
    d2_G  =  X2 .* dG;
    E1rfl = E1rfl + 2 * dx * sum(d1_G(:) .* dwE{1}(:));
    E2rfl = E2rfl + 2 * dx * sum(d2_G(:) .* dwE{1}(:));

  % Shifted grid(2)
     X1   = xR(1,p) - input.X1;    X2 = xR(2,p) - input.X2 - dx/2;
     DIS  = sqrt(X1.^2 + X2.^2);   X1 = X1./DIS;  X2 = X2./DIS;
     dG   = - factor * gam0 * (1/(2*pi)) .* besselk(1,gam0*DIS);
    d1_G  =  X1 .* dG;
    d2_G  =  X2 .* dG;
   E1rfl  = E1rfl + 2 * dx * sum(d1_G(:) .* dwE{2}(:));
   E2rfl  = E2rfl + 2 * dx * sum(d2_G(:) .* dwE{2}(:));

   Edata(1,p) = sqrt(abs(E1rfl)^2 + abs(E2rfl)^2);
   Hdata(1,p) = abs(ZH3rfl);
 end % p_loop

elseif nDIM == 3;
   delta = (4*pi/3)^(-1/3) * dx;  % radius circle with area dx^2
     arg = gam0*delta;   factor = 3 * (cosh(arg)- sinh(arg)/arg) / arg^2;
 [Edata,Hdata] = DOPE3Dwdw(w_E,dwE,gam0,dx,xR,factor,input);
end % if
```

Matlab Listing I.106 The function `DopE3Dwdw` to compute the 3D scattered electromagnetic field strengths at a number of receivers.

```
function [Edata,Hdata] = DOPE3Dwdw(w_E,dwE,gam0,dx,xR,factor,input)
 Edata = zeros(1,input.NR); Hdata = zeros(1,input.NR);

 for p = 1 : input.NR
  % Non shifted grid (0) -------------------------------------------------------
     X1  = xR(1,p)-input.X1;  X2 = xR(2,p)-input.X2; X3 = xR(3,p)-input.X3;
     DIS = sqrt(X1.^2+X2.^2+X3.^2);
      X1 = X1 ./ DIS;              X2 = X2 ./ DIS;            X3 = X3 ./ DIS;
       G = factor * exp(-gam0*DIS)./(4*pi*DIS);
      dG = -(gam0 + 1./DIS) .* G;
    d1_G = X1 .* dG;           d2_G = X2 .* dG;          d3_G = X3 .* dG;
   E1rfl = gam0^2 * dx^3 * sum(G(:) .* w_E{1}(:));
   E2rfl = gam0^2 * dx^3 * sum(G(:) .* w_E{2}(:));
   E3rfl = gam0^2 * dx^3 * sum(G(:) .* w_E{3}(:));

   ZH1rfl= -gam0 * dx^3 * sum( d2_G(:).*w_E{3}(:) - d3_G(:).*w_E{2}(:) );
   ZH2rfl= -gam0 * dx^3 * sum( d3_G(:).*w_E{1}(:) - d1_G(:).*w_E{3}(:) );
   ZH3rfl= -gam0 * dx^3 * sum( d1_G(:).*w_E{2}(:) - d2_G(:).*w_E{1}(:) );

  % Shifted grid (1) -----------------------------------------------------------
     X1  = xR(1,p)-input.X1-dx/2; X2=xR(2,p)-input.X2; X3=xR(3,p)-input.X3;
     DIS = sqrt(X1.^2 + X2.^2 + X3.^2);
      X1 = X1 ./ DIS;                X2 = X2 ./ DIS;         X3 = X3 ./ DIS;
      dG = - factor * (gam0 + 1./DIS) .* exp(-gam0*DIS)./(4*pi*DIS);
    d1_G = X1 .* dG;    d2_G = X2 .* dG;       d3_G = X3 .* dG;
   E1rfl = E1rfl + 2 * dx^2 * sum(d1_G(:) .* dwE{1}(:));
   E2rfl = E2rfl + 2 * dx^2 * sum(d2_G(:) .* dwE{1}(:));
   E3rfl = E3rfl + 2 * dx^2 * sum(d3_G(:) .* dwE{1}(:));
  % Shifted grid (2) -----------------------------------------------------------
     X1  = xR(1,p)-input.X1; X2=xR(2,p)-input.X2-dx/2; X3=xR(3,p)-input.X3;
     DIS = sqrt(X1.^2 + X2.^2 + X3.^2);
      X1 = X1 ./ DIS;            X2 = X2 ./ DIS;              X3 = X3 ./ DIS;
      dG = - factor * (gam0 + 1./DIS) .* exp(-gam0*DIS)./(4*pi*DIS);
    d1_G = X1 .* dG;    d2_G = X2 .* dG;       d3_G = X3 .* dG;
   E1rfl = E1rfl + 2 * dx^2 * sum(d1_G(:) .* dwE{2}(:));
   E2rfl = E2rfl + 2 * dx^2 * sum(d2_G(:) .* dwE{2}(:));
   E3rfl = E3rfl + 2 * dx^2 * sum(d3_G(:) .* dwE{2}(:));
  % Shifted grid (3) -----------------------------------------------------------
      X1 = xR(1,p)-input.X1; X2=xR(2,p)-input.X2-dx/2; X3=xR(3,p)-input.X3;
     DIS = sqrt(X1.^2 + X2.^2 + X3.^2);
      X1 = X1 ./ DIS;            X2 = X2 ./ DIS;               X3 = X3 ./ DIS;
      dG = - factor * (gam0 + 1./DIS) .* exp(-gam0*DIS)./(4*pi*DIS);
    d1_G = X1 .* dG;         d2_G = X2 .* dG;           d3_G = X3 .* dG;
   E1rfl = E1rfl + 2 * dx^2 * sum(d1_G(:) .* dwE{3}(:));
   E2rfl = E2rfl + 2 * dx^2 * sum(d2_G(:) .* dwE{3}(:));
   E3rfl = E3rfl + 2 * dx^2 * sum(d3_G(:) .* dwE{3}(:));

   Edata(1,p) = sqrt(abs(E1rfl)^2  + abs(E2rfl)^2  + abs(E3rfl^2));
   Hdata(1,p) = sqrt(abs(ZH1rfl)^2 + abs(ZH2rfl)^2 + abs(ZH3rfl^2));
 end % p_loop
```

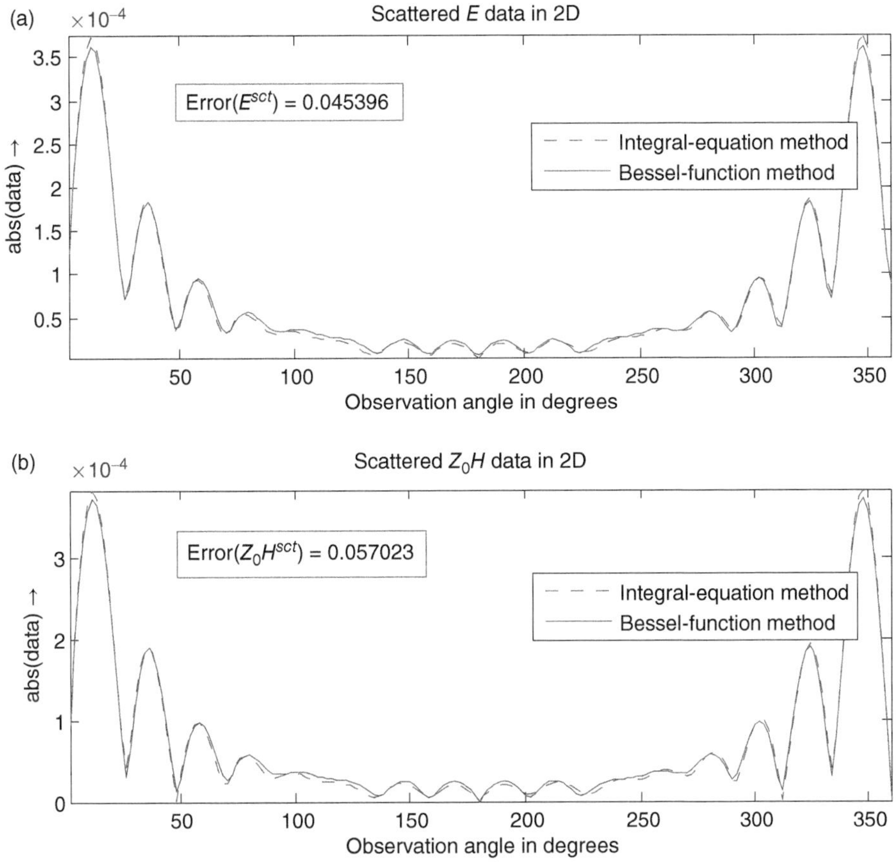

Figure 3.13 The 2D scattered electromagnetic field strengths at the receivers.

the electric-field error decreases from 4.5% to 0.25%, while the magnetic-field error decreases from 5.7% to 0.3%. In the w-method, the electric-field error decreases from 4.3% to 0.13%, while the magnetic-field error decreases from 4.6% to 0.15%.

In Figure 3.15, the E-field errors in the wdw-method and the w-method are displayed. In the first two grid refinements from $r = 1$ to $r = 3$, we see that error curve of the w-method decays almost linearly, indicating that the error is proportional to $(\Delta x)^2$. Assuming that for further refinement, this convergence rate is maintained, we would get the error values indicated by the thin lines. As seen in the scalar case of Chapter 1, where the integral operators are exclusively described by scalar convolutions, this convergence rate is retained for all grid refinements. In the present electromagnetic case, the errors made in the finite differentiations of the operator are visible. Furthermore, these errors are enhanced by the approximation of a smooth interface by a staircase one. This explains that for further refinements $r = 3$ to $r = 5$, the error curve of the w-method begins to differ from the straight line, which indicates that the influence of the interfaces becomes visible. This effect is even more enhanced in the wdw-method. It is surmised that the latter method shows more sensitivity to the location of the interfaces. In principle, this sensitivity issue needs to be further investigated.

In Figure 3.16, the H-field errors are given. Even for grid refinements, from $r = 1$ to $r = 3$, the convergence rate is not a straight line. In principle, the scattered field at the receivers is obtained

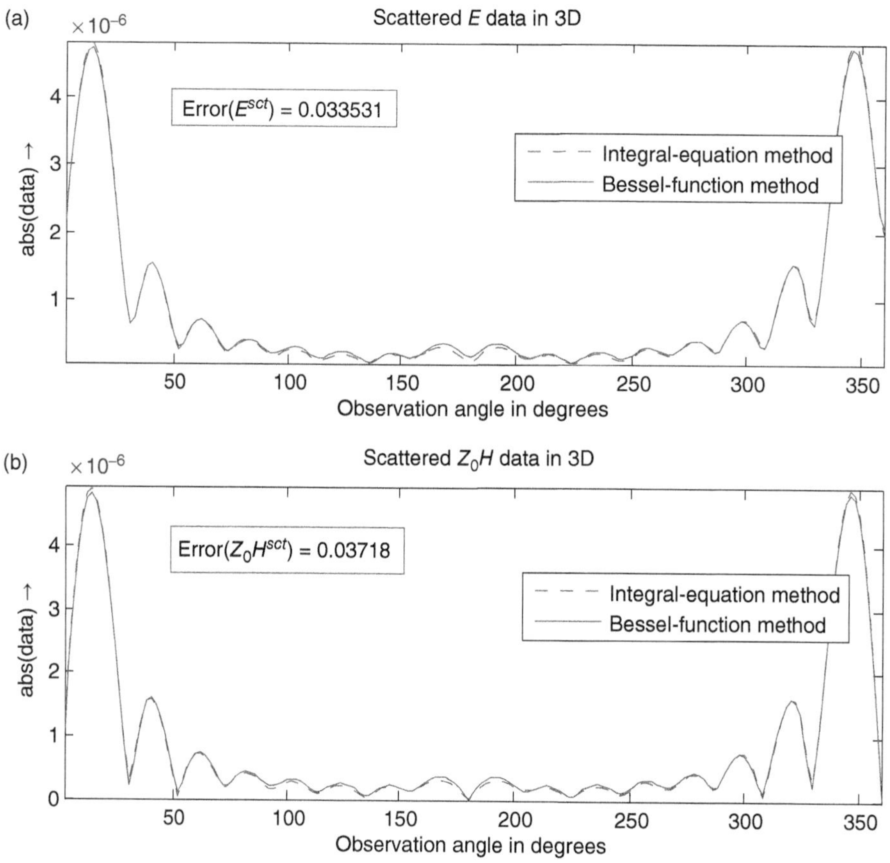

Figure 3.14 The 3D scattered electromagnetic field strengths at the receivers.

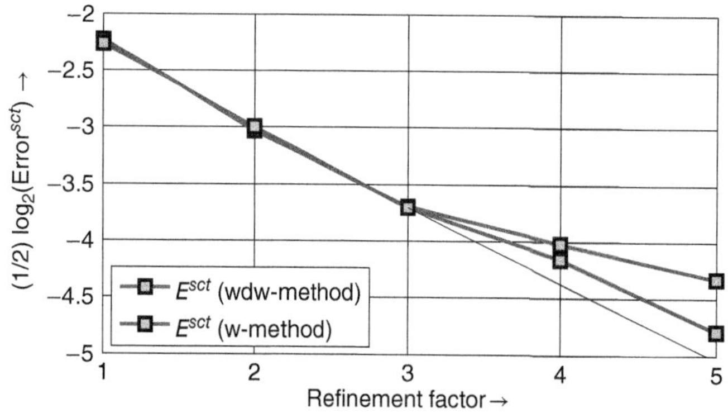

Figure 3.15 Convergence plot of the Error(E^{sct}) as function of the refinement factor r, $n = 1, 2, \ldots, 5$, for the 2D wdw-method and 2D w-method.

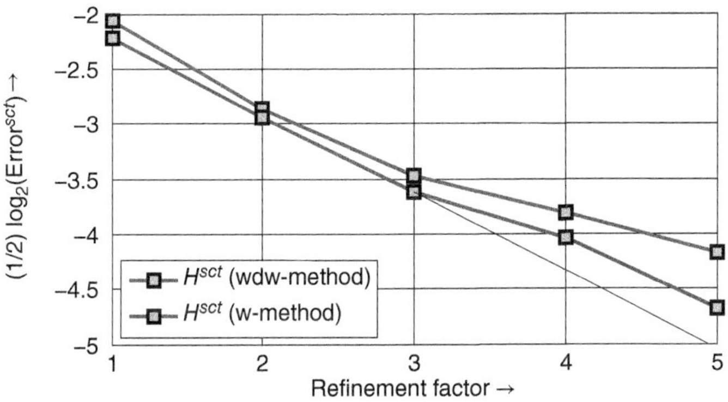

Figure 3.16 Convergence plot of the Error($Z_0\mathrm{H}^{sct}$) as function of the refinement factor r, $n = 1, 2, \ldots, 5$, for the 2D wdw-method and 2D w-method.

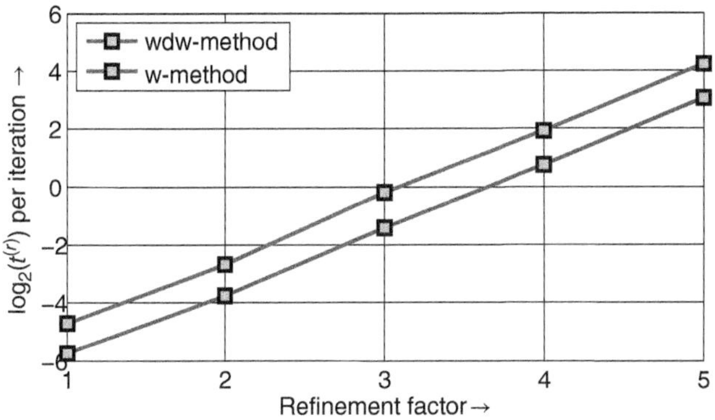

Figure 3.17 Computer time $\log_2(t^{(r)})$ to solve the 2D integral equation as function of the refinement r, for the 2D pZv case and 2D wdw case.

from a spatial differentiation of the scattered electric field, which explains that the H-field errors are always larger than the E-field errors.

In Figure 3.17, we plot the computation time per iteration as function of the grid refinement. This figure confirms that the computation costs for the numerical solution is roughly of the order of $N \log_2(N)$, where $N = N_1 \times N_2$. The wdw-method requires 29 iterations compared to 21 iterations for the w-method, while the computer time per iteration is a factor of three greater for the w-method. For both methods, the number of iterations is invariant for grid refinement.

3.4 Contrast Source Integral Equations for Both Permittivity and Permeability Contrast

In this section, we discuss the electromagnetic scattering problem for scattering objects with both permittivity and permeability contrast. In addition to the introduction of the permittivity contrast $\hat{\chi}^{\varepsilon}$ in Eq. (3.78) and the electric contrast-source vector $\hat{w}_k^E$ in Eq. (3.79), we introduce the permeability contrast as

$$\hat{\chi}^{\mu}(\boldsymbol{x}) = 1 - \frac{\hat{\mu}_{sct}(\boldsymbol{x})}{\mu_0}, \tag{3.157}$$

and the magnetic-contrast-source vector related to the permeability contrast as

$$\hat{\boldsymbol{w}}^H(\boldsymbol{x}) = \hat{\chi}^{\mu}(\boldsymbol{x}) Z_0 \hat{\boldsymbol{H}}(\boldsymbol{x}). \tag{3.158}$$

For $\boldsymbol{x} \in \mathbb{D}'_{sct}$, the source type of integral representations for the electric and magnetic fields $\boldsymbol{E}^{sct} = \boldsymbol{E}^{sct}(\boldsymbol{x})$ and $\boldsymbol{H}^{sct} = \boldsymbol{H}^{sct}(\boldsymbol{x})$ are, cf. Eqs. (3.74) and (3.75),

$$\boxed{\hat{\boldsymbol{E}}^{sct} = \int_{\boldsymbol{x}' \in \mathbb{D}_{sct}} [(\hat{\gamma}_0^2 - \boldsymbol{\nabla}\boldsymbol{\nabla}\cdot)\hat{G}(\boldsymbol{x}-\boldsymbol{x}')\,\hat{\boldsymbol{w}}^E(\boldsymbol{x}') + \hat{\gamma}_0 \boldsymbol{\nabla} \times \hat{G}(\boldsymbol{x}-\boldsymbol{x}')\hat{\boldsymbol{w}}^H(\boldsymbol{x}')]\,\mathrm{d}V} \tag{3.159}$$

and

$$\boxed{Z_0\hat{\boldsymbol{H}}^{sct} = \int_{\boldsymbol{x}' \in \mathbb{D}_{sct}} [-\hat{\gamma}_0 \boldsymbol{\nabla} \times \hat{G}(\boldsymbol{x}-\boldsymbol{x}')\hat{\boldsymbol{w}}^E(\boldsymbol{x}') + (\hat{\gamma}_0^2 - \boldsymbol{\nabla}\boldsymbol{\nabla}\cdot)\hat{G}(\boldsymbol{x}-\boldsymbol{x}')\hat{\boldsymbol{w}}^H(\boldsymbol{x}')]\,\mathrm{d}V.} \tag{3.160}$$

For $\boldsymbol{x} \in \mathbb{D}'_{sct}$, the pertaining integral equations are, cf. Eqs. (3.74) and (3.75),

$$\boxed{\begin{aligned}\hat{\boldsymbol{w}}^E(\boldsymbol{x}) = \chi^{\varepsilon}(\boldsymbol{x}) \Bigg[\hat{\boldsymbol{E}}^{inc}(\boldsymbol{x}) + (\hat{\gamma}_0^2 - \boldsymbol{\nabla}\boldsymbol{\nabla}\cdot)\int_{\boldsymbol{x}' \in \mathbb{D}_{sct}} \hat{G}(\boldsymbol{x}-\boldsymbol{x}')\,\hat{\boldsymbol{w}}^E(\boldsymbol{x}')\,\mathrm{d}V \\ + \hat{\gamma}_0 \boldsymbol{\nabla} \times \int_{\boldsymbol{x}' \in \mathbb{D}_{sct}} \hat{G}(\boldsymbol{x}-\boldsymbol{x}')\,\hat{\boldsymbol{w}}^H(\boldsymbol{x}')\,\mathrm{d}V\Bigg],\end{aligned}} \tag{3.161}$$

$$\boxed{\begin{aligned}\hat{\boldsymbol{w}}^H(\boldsymbol{x}) = \hat{\chi}^{\mu}(\boldsymbol{x}) \Bigg[Z_0\hat{\boldsymbol{H}}^{inc}(\boldsymbol{x}) - \hat{\gamma}_0 \boldsymbol{\nabla} \times \int_{\boldsymbol{x}' \in \mathbb{D}_{sct}} \hat{G}(\boldsymbol{x}-\boldsymbol{x}')\,\hat{\boldsymbol{w}}^E(\boldsymbol{x}')\,\mathrm{d}V \\ + (\hat{\gamma}_0^2 - \boldsymbol{\nabla}\boldsymbol{\nabla}\cdot)\int_{\boldsymbol{x}' \in \mathbb{D}_{sct}} \hat{G}(\boldsymbol{x}-\boldsymbol{x}')\,\hat{\boldsymbol{w}}^H(\boldsymbol{x}')\,\mathrm{d}V\Bigg].\end{aligned}} \tag{3.162}$$

We use this system of equations as the basis to provide a numerical solution of the electromagnetic scattering problem at hand. For both contrast in permittivity and in permeability, we deal with a coupled set of integral equations for the unknown electric contrast-source vector $\hat{\boldsymbol{w}}^E$ and unknown magnetic contrast-source vector $\hat{\boldsymbol{w}}^H$. After a solution of this system of equations, we substitute these contrast sources into Eqs. (3.159) and (3.160) to arrive the scattered field outside the scattering object.

3.4.1 Numerical Solution and Operators Involved

The numerical method for the solution of the set of integral equations follows closely the method of Section 3.2.2, discussing the electric-field integral equation when there is only contrast in permittivity. On each subdomain $\mathbb{D}_n$, we now assume that both the permittivity contrast and the permeability contrast are constant and denote them as $\hat{\chi}_n^{\varepsilon}$ and $\hat{\chi}_n^{\mu}$. Further, in each subdomain, the contrast sources $\hat{w}_k^E$ and $\hat{w}_k^H$ are replaced by their spherical means $\langle \hat{w}_k^E \rangle$ and $\langle \hat{w}_k^E \rangle$.

The incident wave fields $\hat{E}_j^{inc}$ are replaced by their spherical means $\langle \hat{E}_j^{inc} \rangle$. For the 3D case, they follow from Section 3.1.2 by replacing the strong form of the Green function by its weak counterpart. In 2D, the expressions of Section 3.1.3 are converted in the same way.

Similarly, we define our discrete convolutional operators as

$$[\mathcal{K}w_n]_m = \hat{\gamma}_0^2 \Delta V \sum_{n=1}^{N} \langle\langle \hat{G} \rangle\rangle(\boldsymbol{x}_m - \boldsymbol{x}'_n)\,\langle \hat{w}_n \rangle. \tag{3.163}$$

Matlab Listing I.107 The function CURL to carry out the curl operator using finite differences.

```
function w = CURL(v,input)
global nDIM;                                dx = input.dx;

if nDIM == 2;             N1 = input.N1;    N2 = input.N2;
  w{1} = zeros(size(v{1}));  w{2} = w{1};            w{3} = w{1};
  w{1}(:,2:N2-1)        =  (v{3}(:,3:N2) - v{3}(:,1:N2-2));
  w{2}(2:N1-1,:)        = -(v{3}(3:N1,:) - v{3}(1:N1-2,:));
  w{3}(2:N1-1,2:N2-1) =  (v{2}(3:N1,2:N2-1) - v{2}(1:N1-2,2:N2-1)) ...
                        -(v{1}(2:N1-1,3:N2) - v{1}(2:N1-1,1:N2-2));
  w{1} = w{1}/(2*dx);    w{2} = w{2}/(2*dx);    w{3} = w{3}/(2*dx);

elseif nDIM == 3;      N1 = input.N1;      N2 = input.N2;      N3 = input.N3;
  w{1} = zeros(size(v{1}));  w{2} = w{1};            w{3} = w{1};
  w{1}(:,2:N2-1,2:N3-1) = (v{3}(:,3:N2,2:N3-1) - v{3}(:,1:N2-2,2:N3-1)) ...
                         - (v{2}(:,2:N2-1,3:N3) - v{2}(:,2:N2-1,1:N3-2));
  w{2}(2:N1-1,:,2:N3-1) = (v{1}(2:N1-1,:,3:N3) - v{1}(2:N1-1,:,1:N3-2)) ...
                         -(v{3}(3:N1,:,2:N3-1) - v{3}(1:N1-2,:,2:N3-1));
  w{3}(2:N1-1,2:N2-1,:) = (v{2}(3:N1,2:N2-1,:) - v{2}(1:N1-2,2:N2-1,:)) ...
                         -(v{1}(2:N1-1,3:N2,:) - v{1}(2:N1-1,1:N2-2,:));
  w{1} = w{1}/(2*dx);     w{2} = w{2}/(2*dx);     w{3} = w{3}/(2*dx);
end % if
```

This convolutional operator is computed with the help of function Kop of Matlab Listing I.13 of Chapter 1.

By now we omit the hat-symbols and define the discrete electric-field operator

$$[\mathcal{K}_j^E]_m = [\mathcal{K}w_j^E]_m - \gamma_0^{-2}\texttt{graddiv}_{j,k}[\mathcal{K}w_k^E]_m\} + \gamma_0^{-1}\texttt{CURL}_{j,k}[\mathcal{K}w_k^H]_m\}, \tag{3.164}$$

and the discrete magnetic-field operator as

$$[\mathcal{K}_j^H]_m = [\mathcal{K}w_j^H]_m - \gamma_0^{-2}\texttt{graddiv}_{j,k}[\mathcal{K}w_k^H]_m\} - \gamma_0^{-1}\texttt{CURL}_{j,k}[\mathcal{K}w_k^E]_m\}, \tag{3.165}$$

where the finite-difference equivalent graddiv of the graddiv operator is given by the Matlab function I.34 of Chapter 2 and the finite-difference equivalent CURL of the curl operator is given by the Matlab function I.107. The convolutions $[\mathcal{K}w_k^E]_m$ and $[\mathcal{K}w_k^H]_m$ are obtained from Eq. (3.163), by substituting $w_n := w_{k,n}^E$ and $w_n := w_{k,n}^H$ for $k = 1, 2, 3$, respectively. The code to compute $[\mathcal{K}_j^E]_m$ and $[\mathcal{K}_j^E]_m$ is given by function KopEH of Matlab Listing I.108. For the 2D case, we have defined the zero components of KE{3}, KH{1}, and KH{2}. This enables the use of the Matlab function CURL, both for the electric and magnetic wave fields. In 2D, the extra memory needed is not an issue. It also shows, where one has to modify the function for other types of electromagnetic wave fields, e.g. the 2D H-polarized case dealing with the wave field from a vertical magnetic dipole. The electric and magnetic operators of Eqs. (3.164) and (3.165) are used in an iterative solver.

We will use BiCGSTAB method as the iterative solution method, see the function ITERBiCGSTABwEH of Matlab Listing I.109 and the function EHvector2matrix of Matlab Listing I.110.

Matlab Listing I.108 The function KopEH to carry out the $\mathcal{K}^E$ and $\mathcal{K}^H$ operators.

```
function [KE,KH] = KopEH(wE,wH, input)
global nDIM;
gam0  = input.gamma_0;
FFTG  = input.FFTG;

if nDIM == 2
     KE{1} = Kop(wE{1},FFTG);                          KH{1} = zeros(size(wH{3}));
     KE{2} = Kop(wE{2},FFTG);                          KH{2} = zeros(size(wH{3}));
     KE{3} = zeros(size(wH{3}));                       KH{3} = Kop(wH{3},FFTG);
 graddiv_E = graddiv(KE, input);                      curl_E = CURL(KE, input);
                                                      curl_H = CURL(KH, input);
     KE{1} = KE{1} - graddiv_E{1}/gam0^2 + curl_H{1}/gam0;
     KE{2} = KE{2} - graddiv_E{2}/gam0^2 + curl_H{2}/gam0;
     KH{3} = KH{3} - curl_E{3}/gam0;

elseif nDIM == 3
     KE{1} = Kop(wE{1},FFTG);                          KH{1} = Kop(wH{1},FFTG);
     KE{2} = Kop(wE{2},FFTG);                          KH{2} = Kop(wH{2},FFTG);
     KE{3} = Kop(wE{3},FFTG);                          KH{3} = Kop(wH{3},FFTG);
 graddiv_E = graddiv(KE, input);                      curl_E = CURL(KE, input);
 graddiv_H = graddiv(KH, input);                      curl_H = CURL(KH, input);
     KE{1} = KE{1} - graddiv_E{1}/gam0^2 + curl_H{1}/gam0;
     KE{2} = KE{2} - graddiv_E{2}/gam0^2 + curl_H{2}/gam0;
     KE{3} = KE{3} - graddiv_E{3}/gam0^2 + curl_H{3}/gam0;
     KH{1} = KH{1} - graddiv_H{1}/gam0^2 - curl_E{1}/gam0;
     KH{2} = KH{2} - graddiv_H{2}/gam0^2 - curl_E{2}/gam0;
     KH{3} = KH{3} - graddiv_H{3}/gam0^2 - curl_E{3}/gam0;
end
```

3.4.1.1 Scattered Electromagnetic Wave Field

After the numerical solution for the 3D contrast sources $\hat{w}_k^E(\boldsymbol{x}_m)$ and $\hat{w}_k^H(\boldsymbol{x}_m)$, the scattered electric wave field at some receiver point $\boldsymbol{x}^R$ is computed as

$$\langle \hat{E}_j^{sct}\rangle(\boldsymbol{x}^R) = \Delta V \sum_{m=1}^{N} \left[(\hat{\gamma}_0^2\delta_{j,k} - \partial_j^R\partial_k^R)\langle\hat{G}\rangle(\boldsymbol{x}^R - \boldsymbol{x}_m)\langle\hat{w}_k^E(\boldsymbol{x}_m)\rangle \right.$$
$$\left. + \hat{\gamma}_0\epsilon_{j,k,l}\partial_k^R\langle\hat{G}\rangle(\boldsymbol{x}^R - \boldsymbol{x}_m)\langle\hat{w}_l^H(\boldsymbol{x}_m)\rangle\right], \tag{3.166}$$

and the scattered magnetic wave field as

$$\langle Z_0\hat{H}_j^{sct}\rangle(\boldsymbol{x}^R) = \Delta V \sum_{m=1}^{N} \left[- \hat{\gamma}_0\epsilon_{j,k,l}\partial_k^R\langle\hat{G}\rangle(\boldsymbol{x}^R - \boldsymbol{x}_m)\langle\hat{w}_l^E(\boldsymbol{x}_m)\rangle \right.$$
$$\left. + (\hat{\gamma}_0^2\delta_{j,k} - \partial_j^R\partial_k^R)\langle\hat{G}\rangle(\boldsymbol{x}^R - \boldsymbol{x}_m)\langle\hat{w}_k^H(\boldsymbol{x}_m)\rangle\right], \tag{3.167}$$

for $\{j, k, l\} = 1, 2, 3$. The spatial derivatives of the weak Green functions are obtained by replacing the strong form of the Green function (see Section 3.1.2) by its weak counterpart.

Matlab Listing I.109 The function ITERBiCGSTABwEH to use Matlab's built-in function BiCGSTAB.

```
function [w_E,w_H] = ITERBiCGSTABwEH(E_inc, H_inc, input)
% BiCGSTAB scheme for contrast source integral equation Aw = b
 global nDIM;

 itmax  = 1000;   Errcri = input.Errcri;   [N,~] = size(input.CHI_eps(:));

 if nDIM == 2;       b(      1:  N,1) = input.CHI_eps(:) .* E_inc{1}(:);
                     b(  N+1:2*N,1) = input.CHI_eps(:) .* E_inc{2}(:);
                     b(2*N+1:3*N,1) = input.CHI_mu(:)  .* H_inc{3}(:);

 elseif nDIM == 3;   b(      1:  N,1) = input.CHI_eps(:) .* E_inc{1}(:);
                     b(  N+1:2*N,1) = input.CHI_eps(:) .* E_inc{2}(:);
                     b(2*N+1:3*N,1) = input.CHI_eps(:) .* E_inc{3}(:);
                     b(3*N+1:4*N,1) = input.CHI_mu(:)  .* H_inc{1}(:);
                     b(4*N+1:5*N,1) = input.CHI_mu(:)  .* H_inc{2}(:);
                     b(5*N+1:6*N,1) = input.CHI_mu(:)  .* H_inc{3}(:);
 end;

 w = bicgstab(@(w) Aw(w,input), b, Errcri, itmax);        % call BICGSTAB

 [w_E,w_H] = EHvector2matrix(w,input);

end %------------------------------------------------------------------------

function y = Aw(w,input)
 global nDIM;        [N,~] = size(input.CHI_eps(:));

 [w_E,w_H] = EHvector2matrix(w,input);

 [Kw_E,Kw_H] = KopEH(w_E,w_H,input);

 if nDIM == 2;
     y(      1:  N,1) = w_E{1}(:) - input.CHI_eps(:) .* Kw_E{1}(:);
     y(  N+1:2*N,1) = w_E{2}(:) - input.CHI_eps(:) .* Kw_E{2}(:);
     y(2*N+1:3*N,1) = w_H{3}(:) - input.CHI_mu(:)  .* Kw_H{3}(:);

 elseif nDIM ==3;
     y(      1:  N,1) = w_E{1}(:) - input.CHI_eps(:) .* Kw_E{1}(:);
     y(  N+1:2*N,1) = w_E{2}(:) - input.CHI_eps(:) .* Kw_E{2}(:);
     y(2*N+1:3*N,1) = w_E{3}(:) - input.CHI_eps(:) .* Kw_E{3}(:);
     y(3*N+1:4*N,1) = w_H{1}(:) - input.CHI_mu(:)  .* Kw_H{1}(:);
     y(4*N+1:5*N,1) = w_H{2}(:) - input.CHI_mu(:)  .* Kw_H{2}(:);
     y(5*N+1:6*N,1) = w_H{3}(:) - input.CHI_mu(:)  .* Kw_H{3}(:);
 end;
end
```

Matlab Listing I.110 The function `EHvector2matrix` required by the function `ITERBiCGSTABwEH`.

```
function [w_E,w_H] = EHvector2matrix(w, input)
% Modify vector output from 'bicgstab' to matrices for further computation
global nDIM;
[N,~] = size(input.CHI_eps(:));        w_E = cell(1,nDIM); w_H = cell(1,nDIM);

if nDIM == 2;                          DIM = [input.N1,input.N2];

    w_E{1} = reshape(w(     1:  N,1),DIM);
    w_E{2} = reshape(w(  N+1:2*N,1),DIM);
    w_H{3} = reshape(w(2*N+1:3*N,1),DIM);

elseif nDIM == 3;                      DIM = [input.N1,input.N2,input.N3];

    w_E{1} = reshape(w(     1:  N,1),DIM);
    w_E{2} = reshape(w(  N+1:2*N,1),DIM);
    w_E{3} = reshape(w(2*N+1:3*N,1),DIM);
    w_H{1} = reshape(w(3*N+1:4*N,1),DIM);
    w_H{2} = reshape(w(4*N+1:5*N,1),DIM);
    w_H{3} = reshape(w(5*N+1:6*N,1),DIM);
end
```

In 2D, the electric wave field has no component in the x_3-direction and the magnetic wave field has only one component. The scattered electric wave field is computed as

$$\langle \hat{E}_j^{sct} \rangle(\boldsymbol{x}_T^R) = \Delta V \sum_{m=1}^{N} \big[(\hat{\gamma}_0^2 \delta_{j,k} - \partial_j^R \partial_k^R) \langle \hat{G} \rangle(\boldsymbol{x}_T^R - \boldsymbol{x}_{T,m}) \langle \hat{w}_k^E(\boldsymbol{x}_m) \rangle + \hat{\gamma}_0 \epsilon_{j,k,3} \partial_k^R \langle \hat{G} \rangle(\boldsymbol{x}_T^R - \boldsymbol{x}_m) \langle \hat{w}_3^H(\boldsymbol{x}_m) \rangle \big], \tag{3.168}$$

whereas the scattered magnetic wave field becomes

$$\langle Z_0 \hat{H}_3^{sct} \rangle(\boldsymbol{x}_T^R) = \Delta V \sum_{m=1}^{N} \big[- \hat{\gamma}_0 \epsilon_{3,k,l} \partial_k^R \langle \hat{G} \rangle(\boldsymbol{x}_T^R - \boldsymbol{x}_m) \langle \hat{w}_l^E(\boldsymbol{x}_m) \rangle + (\hat{\gamma}_0^2 - \partial_3^R \partial_3^R) \langle \hat{G} \rangle(\boldsymbol{x}_T^R - \boldsymbol{x}_m) \langle \hat{w}_3^H(\boldsymbol{x}_m) \rangle \big], \tag{3.169}$$

for $\{j, k, l\} = 1, 2$. The spatial derivatives of the weak Green functions are obtained by replacing the strong form of the Green function (see Section 3.1.3) by its weak counterpart.

3.4.2 Matlab Codes Integral Equations for Both Permittivity and Permeability Contrast: Special Case of Zero Wave-Speed Contrast

To initialize the electromagnetic wave-field parameters, we start again with the function `initEM` of Matlab Listing I.82. We keep the relative electric permittivity of the scatterer unchanged, viz., $\varepsilon_{sct} = 1.75$ and take the magnetic permeability of the scatterer as $\mu_{sct} = 1/\varepsilon_{sct} = 1/1.75$. For this particular case, we observe that $c_{sct} = c_0$, which means that there is no contrast in wave speed. Later in this chapter, we will show that this special choice leads to a set of integral equations for the unknown interface-contrast sources. There are no contributions of volume-contrast sources. Because we want to compare the numerical results of these two different types of integral equations,

Matlab Listing I.111 The `ForwardBiCGSTABFFTwEH` script to solve contrast sources w^E and w^H.

```
clear all; clc; close all; clear workspace;
input = initEM();

% (1) Compute analytically scattered field data ----------------------
      EMForwardCanonicalObjects

      plotEMcontrast(input); % plot permittivity / permeability contrast

% (2) Compute incident field ----------------------------------------
      [E_inc,H_inc] = IncEMwave(input);

% (3) Solve integral equation for contrast sources with FFT
tic;  [w_E,w_H] = ITERBiCGSTABwEH(E_inc,H_inc,input);
toc
% Plot electric and magnetic contrast sources
      plotContrastSourcewEH(w_E,w_H,input);

% (4) Compute synthetic data and plot fields and data ---------------
      [Edata,Hdata] = DOPwEH(w_E,w_H,input);
      displayEdata(Edata,input);
      displayHdata(Hdata,input);
```

we have made this specific choice for the magnetic permeability. Further, after some numerical experiments, we have seen that the numerical errors in this special case are substantially larger than that in the case for $\mu = 1$. Therefore, we refine our grid and in the function `initEMgrid` of Matlab Listing I.83, we take $N1 = 240$, $N2 = 200$ and $N3 = 161$, with a mesh width $dx = \Delta x = 1$.

After the initialization of the electromagnetic parameters, we start the main Matlab script `ForwardBiCGSTABFFTwEH` of Matlab Listing I.111 with the four steps: (1) Computation of scattering by a canonical object; (2) Computation of the incident wave fields; (3) Solution of the set of integral equations; and (4) Computation of scattering data, based on the solution of the integral equations.

In step (1) of `ForwardBiCGSTABFFTwEH`, script `EMForwardCanonicalObjects` of Matlab Listing I.139 of Appendix 3.A.2 is called. It computes both the electric and magnetic wave fields at a number of receivers, for either the 2D circular cylinder or the 3D sphere. These data are used to benchmark the scattered field based on the numerical solution of our present integral equation. Next, the Matlab function `plotEMcontrast` of Matlab Listing I.86 plots the electric permittivity and magnetic permeability distributions, which are the electromagnetic input parameters. Both for the 2D case and the 3D case, the results in the cross-sectional plane $x_3 = 0$ are shown in Figure 3.18.

In step (2) of `ForwardBiCGSTABFFTwEH`, we compute the weak form of the electric and the magnetic incident wave fields, either in 2D or 3D. The function `IncEMwave` is listed as Matlab Listing I.87 of Section 3.2.3.

In step (3) of `ForwardBiCGSTABFFTwEH`, we solve the integral equations for the contrast sources w_k^E and w_k^H by calling `ITERBiCGSTABwEH` of Matlab Listing I.109. After solution of the integral equations, the function `plotContrastSourcewEH` of Matlab Listing I.112 plots the absolute values of the contrast sources. For the 2D case, Figure 3.19 shows the nonzero amplitudes $|w_1^E|$, $|w_2^E|$, and $|w_3^H|$. For the 3D case, in the cross-section $x_3 = 0$, it shows the contrast-source amplitudes $|w_k^E|$ and $|w_k^H|$ for $k = 1, 2, 3$. Note that in view of the 3D symmetry, the electric field is

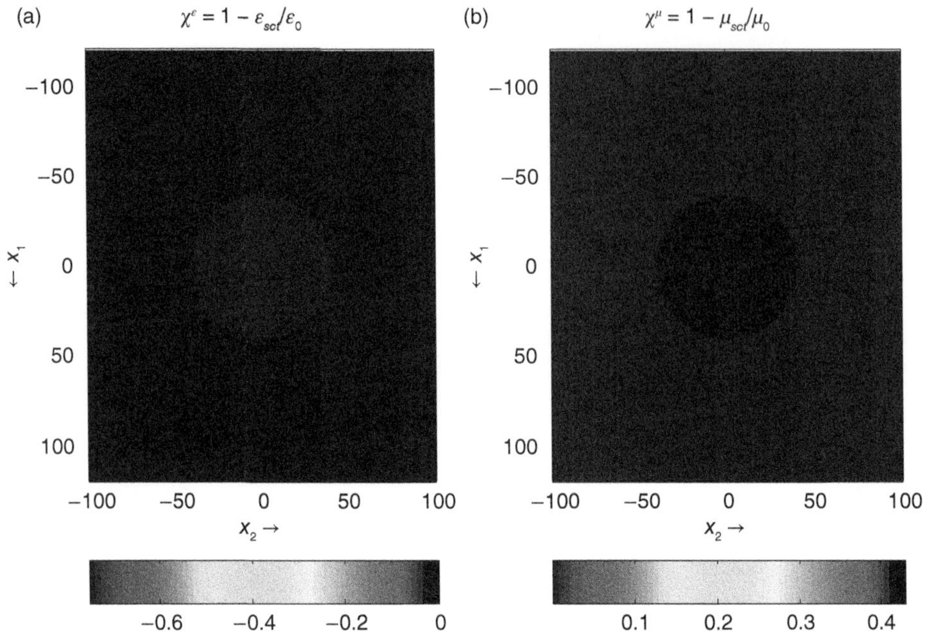

Figure 3.18 Plots of the contrast distributions (a) χ^ε (b) and χ^μ in the 2D cross-section.

even and the magnetic field is odd with respect to the plane $x_3 = 0$, so that at this plane w_3^E, w_1^H, and w_2^H vanish.

In step (4) of the Matlab script `ForwardBiCGSTABFFTFFTwEH`, the scattered electromagnetic field at the selected receiver points are computed. The contrast-source distributions w_k^E and w_k^H are the input for function `DopwEH` of Matlab Listing I.113. This Matlab function computes the 2D electromagnetic field strengths and calls the function `Dop3DwEH` of Matlab Listing I.114 to compute the 3D field components. The programs follow from the weak form of the scattered electromagnetic wave fields of Eqs. (3.166)–(3.169). These scattered field are compared to the analytical Bessel-function data stored in MAT-files. The functions `displayEdata.m` and `displayHData.m` of Matlab Listings I.93 and I.94 load the particular files and plot both the analytical data and the numerical data. These data are presented in Figures 3.20 and 3.21. We use the error norms of Eqs. (3.107) and (3.108).

Although we have refined our mesh size to increase the numerical accuracy, we notice that the error norms are still close to 10%. In both 2D and 3D results, we see a significant error in the backscattering at an angle between −30 and 30°. Hence, the scattering from the staircase interface, which is an approximation of the circular and spherical interfaces now plays a significant role. If we change the magnetic permeability so that the wave-speed contrast does not vanish anymore, the error norms decrease. Further refinement of the mesh size to $\Delta x = 0.5$ reduces the errors norms by almost a factor of four. Remark that the accuracy of the computed electric and magnetic scattered-field amplitudes are roughly the same. Before we draw further conclusions, we first investigate the numerical performance, when we solve a modified set of integral equations based on interface contrast sources.

Matlab Listing I.112 The function `plotContrastSourcewEH` to plot amplitudes of the electromagnetic contrast sources.

```
function plotContrastSourcewEH(w_E,w_H,input)
global nDIM;      set(figure,'Units','centimeters','Position',[5 5 18 12]);

if  nDIM == 2;       % Plot 2D contrast/source distribution ---------------

    x1 = input.X1(:,1);    x2 = input.X2(1,:);
    subplot(1,3,1);
        IMAGESC(x1,x2,abs(w_E{1}));
        title('\fontsize{13} abs(w_1^E)');
    subplot(1,3,2);
        IMAGESC(x1,x2,abs(w_E{2}));
        title('\fontsize{13} abs(w_2^E)');

    subplot(1,3,3);
        IMAGESC(x1,x2,abs(w_H{3}));
        title('\fontsize{13} abs(w_3^H)');


elseif nDIM == 3;    % Plot 3D contrast/source distribution ---------------
    x1 = input.X1(:,1,1);   x2 = input.X2(1,:,1);
    N3cross = floor(input.N3/2+1);
    subplot(1,3,1);
        IMAGESC(x1,x2,abs(w_E{1}(:,:,N3cross)));
        title('\fontsize{13} abs(w_1^E)');
    subplot(1,3,2);
        IMAGESC(x1,x2,abs(w_E{2}(:,:,N3cross)));
        title('\fontsize{13} abs(w_2^E)');
    subplot(1,3,3);
       IMAGESC(x1,x2,abs(w_E{3}(:,:,N3cross)));
        title('\fontsize{13} abs(w_3^E)');

    set(figure,'Units','centimeters','Position',[5 5 18 12]);
    subplot(1,3,1);
        IMAGESC(x1,x2,abs(w_H{1}(:,:,N3cross)));
        title('\fontsize{13} abs(w_1^H)');
    subplot(1,3,2);
        IMAGESC(x1,x2,abs(w_H{2}(:,:,N3cross)));
        title('\fontsize{13} abs(w_2^H)');
    subplot(1,3,3);
       IMAGESC(x1,x2,abs(w_H{3}(:,:,N3cross)));
        title('\fontsize{13} abs(w_3^H)');

end % if
```

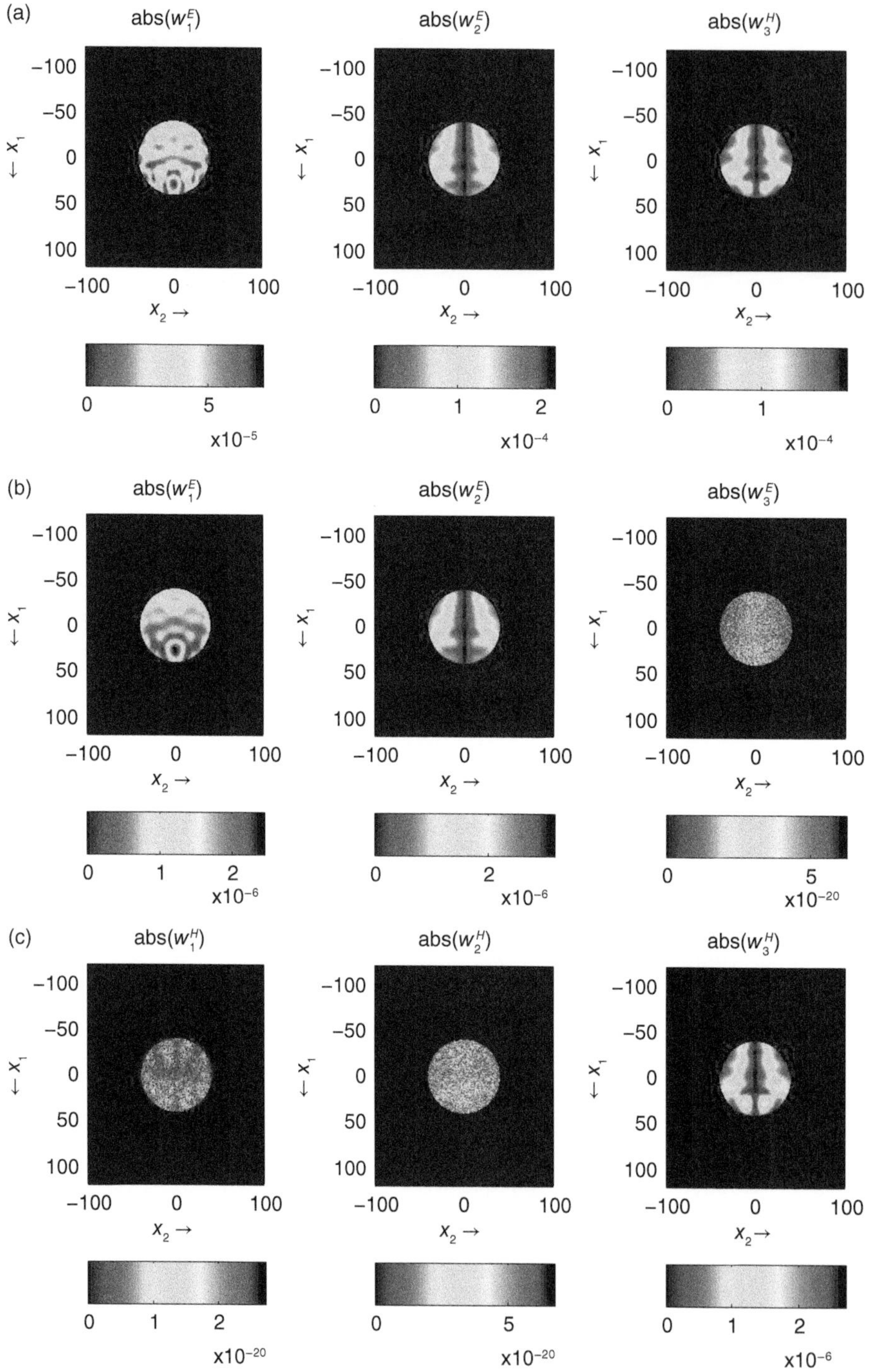

Figure 3.19 Plots of 2D contrast-source amplitudes $|w_1^E|$, $|w_2^E|$ and $|w_3^H|$ (a), and cross-sectional plots of the 3D contrast-source amplitudes $|w_k^E|$ and $|w_k^H|$, $k = 1, 2, 3$, (b,c).

Matlab Listing I.113 The function DopwEH to compute the 2D scattered electromagnetic field strengths at a number of receivers.

```
function [Edata,Hdata] = DOPwEH(w_E,w_H,input)
global nDIM;
gam0 = input.gamma_0;          dx = input.dx;     xR = input.xR;

Edata = zeros(1,input.NR); Hdata = zeros(1,input.NR);

if nDIM == 2;                           % 2D case of H-polarization only

 delta  = (pi)^(-1/2)*dx;
 factor = 2*besseli(1,gam0*delta) / (gam0*delta);
 for p = 1 : input.NR
      X1   = xR(1,p)-input.X1;
      X2   = xR(2,p)-input.X2;
      DIS  = sqrt(X1.^2 + X2.^2);
      X1   = X1./DIS;
      X2   = X2./DIS;
      G    =    factor          * 1/(2*pi).* besselk(0,gam0*DIS);
     dG    =  - factor * gam0 * 1/(2*pi).* besselk(1,gam0*DIS);
    d1_G   =  X1 .* dG;
    d2_G   =  X2 .* dG;
     dG11 = (2*X1.*X1 - 1) .* (-dG./DIS) + gam0^2*X1.*X1 .* G;
     dG22 = (2*X2.*X2 - 1) .* (-dG./DIS) + gam0^2*X2.*X2 .* G;
     dG21 =  2*X2.*X1       .* (-dG./DIS) + gam0^2*X2.*X1 .* G;

 % permittivity contrast distribution ---------------------------------

    E1rfl = dx^2 * sum( (gam0^2*G(:) - dG11(:)) .* w_E{1}(:)   ...
                                      - dG21(:)  .* w_E{2}(:) );
    E2rfl = dx^2 * sum(               - dG21(:)  .* w_E{1}(:)   ...
                      +(gam0^2*G(:) - dG22(:)) .* w_E{2}(:) );

   ZH3rfl = - gam0 * dx^2 * sum(d1_G(:).*w_E{2}(:) - d2_G(:).*w_E{1}(:));

 % permeability contrast distribution ---------------------------------

   ZH3rfl = ZH3rfl +         dx^2 * sum( gam0^2*G(:) .* w_H{3}(:) );
    E1rfl = E1rfl  + gam0 * dx^2 * sum( d2_G(:)  .* w_H{3}(:) );
    E2rfl = E2rfl  - gam0 * dx^2 * sum( d1_G(:)  .* w_H{3}(:) );

   Edata(1,p) = sqrt(abs(E1rfl)^2 + abs(E2rfl)^2);
   Hdata(1,p) = sqrt(abs(ZH3rfl)^2);

 end % p_loop


elseif nDIM == 3;

 [Edata,Hdata] = DOP3DwEH(w_E,w_H,input);

end % if
```

Matlab Listing I.114 The function Dop3DwEH to compute the 3D scattered electromagnetic field strengths at a number of receivers.

```
function [Edata,Hdata] = DOP3DwEH(w_E,w_H,input)
gam0 = input.gamma_0;      dx = input.dx;    xR = input.xR;

Edata = zeros(1,input.NR); Hdata = zeros(1,input.NR);

delta  = (4*pi/3)^(-1/3) * dx;    arg = gam0*delta;
factor = 3 * (cosh(arg) - sinh(arg)/arg) / arg^2;
for p = 1 : input.NR
    X1  = xR(1,p)-input.X1;  X2 = xR(2,p)-input.X2; X3 = xR(3,p)-input.X3;
    DIS = sqrt(X1.^2+X2.^2+X3.^2);  X1=X1./DIS;  X2=X2./DIS;  X3=X3./DIS;
    G   = factor * exp(-gam0*DIS) ./ (4*pi*DIS);
    dG  = -(gam0+1./DIS).*G;  d1_G = X1.*dG;   d2_G=X2.*dG;   d3_G=X3.*dG;
   dG11 = ( (3*X1.*X1 - 1).*(gam0./DIS+1./DIS.^2) + gam0^2*X1.*X1 ) .* G;
   dG22 = ( (3*X2.*X2 - 1).*(gam0./DIS+1./DIS.^2) + gam0^2*X2.*X2 ) .* G;
   dG33 = ( (3*X3.*X3 - 1).*(gam0./DIS+1./DIS.^2) + gam0^2*X3.*X3 ) .* G;
   dG21 = (  3*X2.*X1      .*(gam0./DIS+1./DIS.^2) + gam0^2*X2.*X1 ) .* G;
   dG31 = (  3*X3.*X1      .*(gam0./DIS+1./DIS.^2) + gam0^2*X3.*X1 ) .* G;
   dG32 = (  3*X3.*X2      .*(gam0./DIS+1./DIS.^2) + gam0^2*X3.*X2 ) .* G;

% permittivity contrast contribution -------------------------------------
  E1rfl = dx^3 * sum( (gam0^2*G(:) - dG11(:)) .* w_E{1}(:)  ...
                                   - dG21(:)  .* w_E{2}(:)  ...
                                   - dG31(:)  .* w_E{3}(:) );
  E2rfl = dx^3 * sum(              - dG21(:)  .* w_E{1}(:)  ...
                     +(gam0^2*G(:) - dG22(:)) .* w_E{2}(:)  ...
                                   - dG32(:)  .* w_E{3}(:) );
  E3rfl = dx^3 * sum(              - dG31(:)  .* w_E{1}(:)  ...
                                   - dG32(:)  .* w_E{2}(:)  ...
                     +(gam0^2*G(:) - dG33(:)) .* w_E{3}(:) );
 ZH1rfl = -gam0 * dx^3 * sum( d2_G(:).*w_E{3}(:) - d3_G(:).*w_E{2}(:) );
 ZH2rfl = -gam0 * dx^3 * sum( d3_G(:).*w_E{1}(:) - d1_G(:).*w_E{3}(:) );
 ZH3rfl = -gam0 * dx^3 * sum( d1_G(:).*w_E{2}(:) - d2_G(:).*w_E{1}(:) );

% permeability contrast distribution -------------------------------------
 ZH1rfl = ZH1rfl + dx^3 * sum( (gam0^2*G(:) - dG11(:)) .* w_H{1}(:)   ...
                                            - dG21(:)  .* w_H{2}(:)   ...
                                            - dG31(:)  .* w_H{3}(:) );
 ZH2rfl = ZH2rfl + dx^3 * sum(              - dG21(:)  .* w_H{1}(:)   ...
                              +(gam0^2*G(:) - dG22(:)) .* w_H{2}(:)   ...
                                            - dG32(:)  .* w_H{3}(:) );
 ZH3rfl = ZH3rfl + dx^3 * sum(              - dG31(:)  .* w_H{1}(:)   ...
                                            - dG32(:)  .* w_H{2}(:)   ...
                              +(gam0^2*G(:) - dG33(:)) .* w_H{3}(:) );
 E1rfl  = E1rfl + gam0*dx^3*sum( d2_G(:).*w_H{3}(:) - d3_G(:).*w_H{2}(:) );
 E2rfl  = E2rfl + gam0*dx^3*sum( d3_G(:).*w_H{1}(:) - d1_G(:).*w_H{3}(:) );
 E3rfl  = E3rfl + gam0*dx^3*sum( d1_G(:).*w_H{2}(:) - d2_G(:).*w_H{1}(:) );

 Edata(1,p) = sqrt(abs(E1rfl)^2 +abs(E2rfl)^2 +abs(E3rfl^2));
 Hdata(1,p) = sqrt(abs(ZH1rfl)^2+abs(ZH2rfl)^2+abs(ZH3rfl^2));
end % p_loop
```

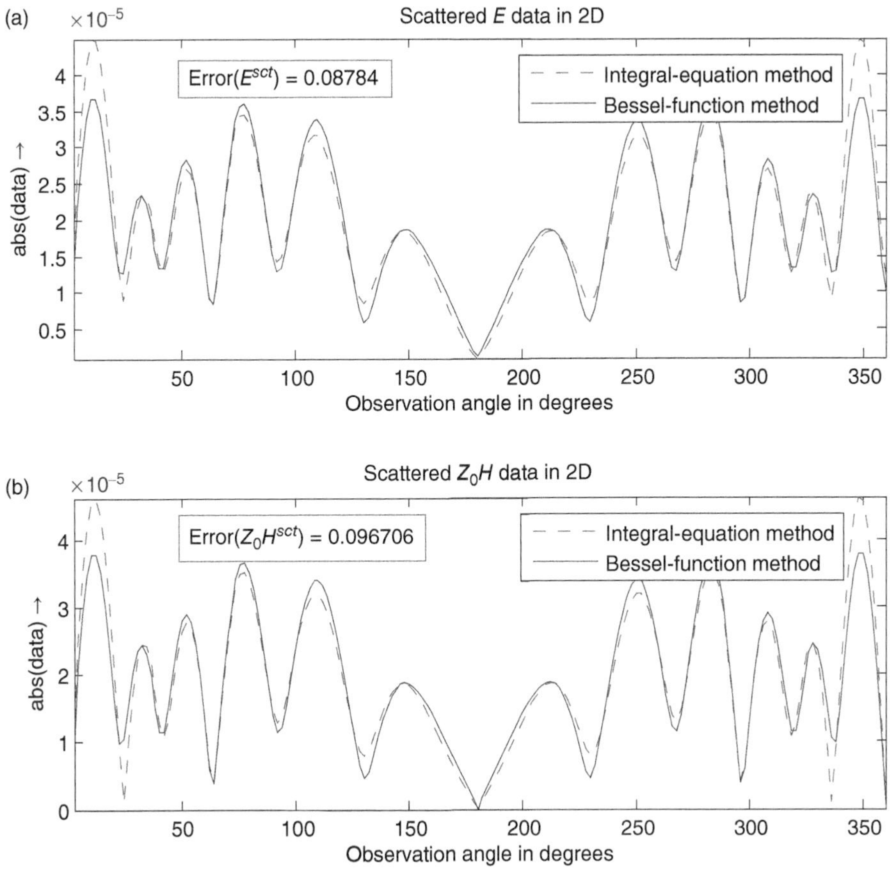

Figure 3.20 The 2D scattered electromagnetic field amplitudes at the receivers.

3.5 E-field Integral Equation for Zero Wave-Speed Contrast

In case that the wave-speed contrast vanishes, the relative permeability is equal to the inverse of the relative permittivity, and the electric-field integral representation of Eq. (3.72) becomes

$$\hat{\boldsymbol{E}}^{sct}(\boldsymbol{x}) = \int_{\boldsymbol{x}'\in\mathbb{D}_{sct}} \left[(-\hat{\gamma}_0^2 + \boldsymbol{\nabla}\boldsymbol{\nabla}\cdot)\hat{G}(\boldsymbol{x}-\boldsymbol{x}') \left(\frac{\mu_0}{\hat{\mu}_{sct}} - 1\right) \hat{\boldsymbol{E}}(\boldsymbol{x}') \right.$$
$$\left. -\hat{\gamma}_0 \boldsymbol{\nabla} \times \hat{G}(\boldsymbol{x}-\boldsymbol{x}') \left(\frac{\hat{\mu}_{sct}}{\mu_0} - 1\right) Z_0 \hat{\boldsymbol{H}}(\boldsymbol{x}') \right] \mathrm{d}V, \tag{3.170}$$

for $\boldsymbol{x} \in \mathbb{D}'_{sct}$. In order to convert this representation into an integral representation for the electric field only, we use Eq. (3.57) to express the magnetic-field vector into the electric-field vector as

$$Z_0\hat{\boldsymbol{H}} = -\frac{Z_0}{s\hat{\mu}_{sct}} \boldsymbol{\nabla} \times \hat{\boldsymbol{E}} = -\hat{\gamma}_0^{-1} \frac{\mu_0}{\hat{\mu}_{sct}} \boldsymbol{\nabla} \times \hat{\boldsymbol{E}}. \tag{3.171}$$

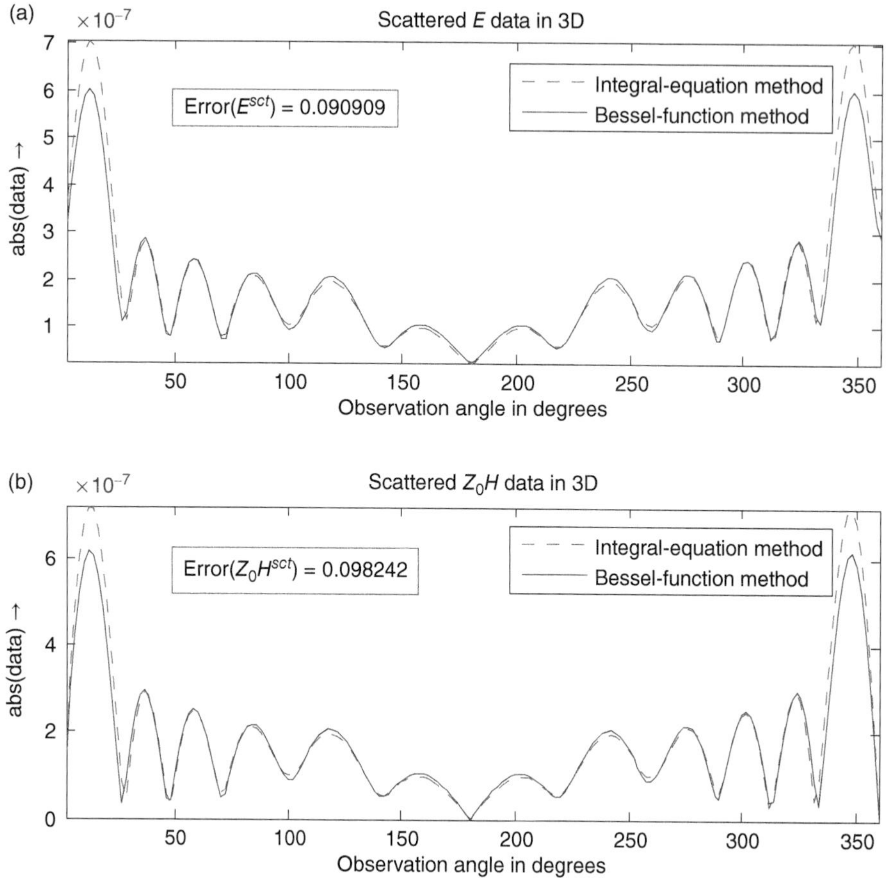

Figure 3.21 The 3D scattered electromagnetic field amplitudes at the receivers.

Then, the integral representation becomes

$$\hat{\boldsymbol{E}}^{sct}(\boldsymbol{x}) = \int_{\boldsymbol{x}' \in \mathbb{D}_{sct}} \left[(-\hat{\gamma}_0^2 + \boldsymbol{\nabla}\boldsymbol{\nabla}\cdot)\hat{G}(\boldsymbol{x}-\boldsymbol{x}') \left(\frac{\mu_0}{\hat{\mu}_{sct}} - 1\right) \hat{\boldsymbol{E}}(\boldsymbol{x}') \right.$$
$$\left. -\boldsymbol{\nabla} \times \hat{G}(\boldsymbol{x}-\boldsymbol{x}') \left(\frac{\mu_0}{\hat{\mu}_{sct}} - 1\right) \boldsymbol{\nabla}' \times \hat{\boldsymbol{E}}(\boldsymbol{x}') \right] \mathrm{d}V. \tag{3.172}$$

Subsequently, we use the relation

$$\left(\frac{\mu_0}{\hat{\mu}_{sct}} - 1\right) \boldsymbol{\nabla}' \times \hat{\boldsymbol{E}}(\boldsymbol{x}') = \boldsymbol{\nabla}' \times \left[\left(\frac{\mu_0}{\hat{\mu}_{sct}} - 1\right) \hat{\boldsymbol{E}}(\boldsymbol{x}')\right] - \left[\boldsymbol{\nabla}' \frac{\mu_0}{\hat{\mu}_{sct}}\right] \times \hat{\boldsymbol{E}}(\boldsymbol{x}'), \tag{3.173}$$

and we obtain

$$\hat{\boldsymbol{E}}^{sct}(\boldsymbol{x}) = \int_{\boldsymbol{x}' \in \mathbb{D}_{sct}} (-\hat{\gamma}_0^2 + \boldsymbol{\nabla}\boldsymbol{\nabla}\cdot)\hat{G}(\boldsymbol{x}-\boldsymbol{x}') \left(\frac{\mu_0}{\hat{\mu}_{sct}} - 1\right) \hat{\boldsymbol{E}}(\boldsymbol{x}')\, \mathrm{d}V$$
$$- \int_{\boldsymbol{x}' \in \mathbb{D}_{sct}} \boldsymbol{\nabla} \times \boldsymbol{\nabla} \times \hat{G}(\boldsymbol{x}-\boldsymbol{x}') \left(\frac{\mu_0}{\hat{\mu}_{sct}} - 1\right) \hat{\boldsymbol{E}}(\boldsymbol{x}')\, \mathrm{d}V$$
$$+ \int_{\boldsymbol{x}' \in \mathbb{D}_{sct}} \boldsymbol{\nabla} \times \hat{G}(\boldsymbol{x}-\boldsymbol{x}') \left[\boldsymbol{\nabla}' \frac{\mu_0}{\hat{\mu}_{sct}}\right] \times \hat{\boldsymbol{E}}(\boldsymbol{x}')\, \mathrm{d}V, \tag{3.174}$$

where we have used the convolutional property of Eq. (2.28), which states that we have a free choice of placing the differentiations. We now use the vector identity $\nabla \times \nabla \times \hat{G}\boldsymbol{v} = (-\nabla \cdot \nabla + \nabla\nabla\cdot)\hat{G}\boldsymbol{v}$, for arbitrary vector function $\boldsymbol{v}$, and $\nabla \cdot \nabla\hat{G} = \hat{\gamma}_0^2\hat{G} - \delta(\boldsymbol{x}-\boldsymbol{x}')$. For observation points $\boldsymbol{x} \in \mathbb{D}'_{sct}$, we observe that the delta function is not active and that the first and second integral of Eq. (3.174) cancel each other. Hence, the E-field integral representation transfers into

$$\boxed{\hat{\boldsymbol{E}}^{sct}(\boldsymbol{x}) = \int_{\boldsymbol{x}'\in\mathbb{D}_{sct}} \nabla \times \hat{G}(\boldsymbol{x}-\boldsymbol{x}') \left[\nabla' \frac{\mu_0}{\hat{\mu}_{sct}}\right] \times \hat{\boldsymbol{E}}(\boldsymbol{x}')\, \mathrm{d}V.} \tag{3.175}$$

This representation holds only for points outside the scattering domain ($\boldsymbol{x} \in \mathbb{D}'_{sct}$). However, for integration points $\boldsymbol{x}'$ in the scattered domain $\mathbb{D}_{sct}$, there is always a point that coincides with $\boldsymbol{x}$, so that we have to take into account the Dirac distribution. Taking into account the extra term $(1 - \mu_0/\hat{\mu}_{sct})\hat{\boldsymbol{E}}$ and using $\hat{\boldsymbol{E}} = \hat{\boldsymbol{E}}^{inc} + \hat{\boldsymbol{E}}^{sct}$, we end up with the integral equation

$$\boxed{\frac{\mu_0}{\hat{\mu}_{sct}}\hat{\boldsymbol{E}}(\boldsymbol{x}) = \hat{\boldsymbol{E}}^{inc}(\boldsymbol{x}) + \nabla \times \int_{\boldsymbol{x}'\in\mathbb{D}_{sct}} \hat{G}(\boldsymbol{x}-\boldsymbol{x}') \left[\nabla' \frac{\mu_0}{\hat{\mu}_{sct}}\right] \times \hat{\boldsymbol{E}}(\boldsymbol{x}')\, \mathrm{d}V.} \tag{3.176}$$

This expression is an exact integral equation for $\boldsymbol{x} \in \mathbb{D}_{sct}$. For a given continuously differentiable permeability $\hat{\mu}_{sct}$, it is an integral equation for the unknown electric field $\hat{E}$. At interfaces, the permeability is not differentiable and we have to take the limiting values of the integral equation at these interfaces. At this point, we assume that the scatterer consists of piecewise homogeneous subdomains. The set of internal interfaces in the domain $\mathbb{D}_{sct}$ are denoted as S_{sct}. It is assumed that the domain $\mathbb{D}_{sct}$ includes the outer interface as well. Hence, at the boundary of the domain $\partial\mathbb{D}_{sct}$, there is always zero contrast. As a consequence of the present assumption of a piecewise homogeneous permeability distribution, the spatial derivative of the inverse permeability vanishes everywhere, except at the interfaces where the spatial derivative in the normal direction shows a interface Dirac distribution. When the observation point is not located at S_{sct}, the integration over the domain $\mathbb{D}_{sct}$ reduces to an integration along all interfaces S_{sct}. Defining

$$\hat{\mu}^+ = \lim_{h\to 0}\,[\hat{\mu}_{sct}(\boldsymbol{x} + h\boldsymbol{\nu})] \quad \text{and} \quad \hat{\mu}^- = \lim_{h\to 0}\,[\hat{\mu}_{sct}(\boldsymbol{x} - h\boldsymbol{\nu})], \quad \boldsymbol{x} \in \mathrm{S}_{sct}, \tag{3.177}$$

we obtain

$$\begin{aligned} \nabla \times \int_{\boldsymbol{x}'\in\mathbb{D}_{sct}} & \hat{G}(\boldsymbol{x}-\boldsymbol{x}') \left[\nabla' \frac{\mu_0}{\hat{\mu}_{sct}}\right] \times \hat{\boldsymbol{E}}(\boldsymbol{x}')\, \mathrm{d}V \\ &= -\nabla \times \int_{\boldsymbol{x}'\in\mathrm{S}_{sct}} \hat{G}(\boldsymbol{x}-\boldsymbol{x}')\, \frac{\hat{\mu}^+ - \hat{\mu}^-}{\hat{\mu}^+\hat{\mu}^-}\mu_0\, \boldsymbol{\nu}(\boldsymbol{x}') \times \hat{\boldsymbol{E}}(\boldsymbol{x}')\, \mathrm{d}A. \end{aligned} \tag{3.178}$$

It is interesting to observe that the interface sources are determined by the interface factor $\eta = \mu_0(\hat{\mu}^+ - \hat{\mu}^-)/(\hat{\mu}^+\hat{\mu}^-)$ and the tangential components of the electric-field vector, $\boldsymbol{\nu}(\boldsymbol{x}') \times \hat{\boldsymbol{E}}(\boldsymbol{x}')$. Therefore, we only have to consider the tangential parts of Eq. (3.176). Further, we observe that the tangential parts of the right-hand side of Eq. (3.178) jumps when the observation point $\boldsymbol{x}$ passes an interface of the scatterer domain. To investigate this, we let the observation $\boldsymbol{x}$ approach to a point $\boldsymbol{x}'$ of the set S_{sct}, i.e. $x_k = \lim_{h\to 0}(x'_k + h\nu_k)$. In this limiting procedure, the singular point $\boldsymbol{x}$ on the interface is excluded symmetrically, after which the limiting value of the integral has been taken. The tangential part of the right-hand side of Eq. (3.178) becomes

$$-\nu \times \nabla \times \int_{x' \in S_{sct}} \hat{G}(x - x')\, \eta(x')\, \nu(x') \times \hat{E}(x')\, dA$$
$$= -\frac{1}{2}\eta(x)\, \nu(x) \times \hat{E}(x) - \nu \times \nabla \times ⨍_{x' \in S_{sct}} \hat{G}(x - x')\, \eta(x')\, \nu(x') \times \hat{E}(x')\, dA, \tag{3.179}$$

for $x \in S_{sct}$. The integral sign $⨍$ denotes Cauchy principle value of the pertaining integral (see [23]). In Eq. (3.179), we have used the expression

$$-\nu(x) \times (\nabla \hat{G}(x - x') \times [\eta(x')\nu(x') \times \hat{E}(x')])$$
$$= (\nu(x) \cdot \nabla \hat{G}(x - x'))\, \eta(x')\nu(x') \times \hat{E}(x')$$
$$-(\nu(x) \cdot [\eta(x')\nu(x') \times \hat{E}(x')])\nabla \hat{G}(x - x'), \tag{3.180}$$

to conclude that the second term on the right-hand side vanishes when x tends to x', because $\nu(x)$ tends to $\nu(x')$ when x tends to x'.

As result, the integral representation of Eq. (3.175) becomes

$$\hat{E}^{sct}(x) = -\nabla \times \int_{x' \in S_{sct}} \hat{G}(x - x')\, \frac{\hat{\mu}^+ - \hat{\mu}^-}{\hat{\mu}^+ \hat{\mu}^-} \mu_0\, \nu(x') \times \hat{E}(x')\, dA, \tag{3.181}$$

and the integral equation (3.176) for the unknown tangential components, $\nu \times \hat{E}$, becomes

$$\frac{\mu_0}{\hat{\mu}^+} \nu(x) \times \hat{E}(x) = \nu(x) \times \hat{E}^{inc}(x) - \frac{1}{2} \frac{\hat{\mu}^+ - \hat{\mu}^-}{\hat{\mu}^+ \hat{\mu}^-} \mu_0\, \nu(x) \times E(x)$$
$$-\nu(x) \times \nabla \times ⨍_{x' \in S_{sct}} \hat{G}(x - x')\, \frac{\hat{\mu}^+ - \hat{\mu}^-}{\hat{\mu}^+ \hat{\mu}^-} \mu_0\, \nu(x') \times \hat{E}(x')\, dA, \tag{3.182}$$

and, after some rearrangements,

$$\mu_0\, \nu(x) \times \hat{E}(x) = \frac{2\hat{\mu}^+ \hat{\mu}^-}{\hat{\mu}^+ + \hat{\mu}^-} \Big[\nu(x) \times \hat{E}^{inc}(x)$$
$$-\nu(x) \times \nabla \times ⨍_{x' \in S_{sct}} \hat{G}(x - x')\, \frac{\hat{\mu}^+ - \hat{\mu}^-}{\hat{\mu}^+ \hat{\mu}^-} \mu_0\, \nu(x') \times \hat{E}(x')\, dA \Big]. \tag{3.183}$$

Similar as before, we introduce a formulation in terms of interface-contrast sources

$$\partial \hat{w} = \frac{1}{2} \frac{\hat{\mu}^+ - \hat{\mu}^-}{\hat{\mu}^+ \hat{\mu}^-} \mu_0\, \hat{E}(x'), \quad x \in S_{sct}. \tag{3.184}$$

Then, we obtain

$$\boxed{\hat{E}^{sct}(x) = -\int_{x' \in S_{sct}} 2\, \nabla \hat{G}(x - x') \times [\nu(x') \times \partial \hat{w}(x')]\, dA.} \tag{3.185}$$

After multiplying the tangential components of both sides of Eq. (3.183) by the factor $(\hat{\mu}^+ - \hat{\mu}^-)/(2\hat{\mu}^+ \hat{\mu}^-)$, we arrive at the interface integral equations,

$$\boxed{\begin{aligned} \frac{\hat{\mu}^+ - \hat{\mu}^-}{\hat{\mu}^+ + \hat{\mu}^-} \nu(x) \times \hat{E}^{inc}(x) &= \nu(x) \times \partial \hat{w}(x) \\ &+ \frac{\hat{\mu}^+ - \hat{\mu}^-}{\hat{\mu}^+ + \hat{\mu}^-} \nu(x) \times \nabla ⨍_{x' \in S_{sct}} 2\hat{G}(x - x') \times [\nu(x') \times \partial \hat{w}(x')]\, dA, \end{aligned}} \tag{3.186}$$

for the unknown interface-contrast sources at the set of interfaces S^{sct}.

3.5.1 Numerical Solution for Interface Contrast Source Integral Equation: Zero Wave-Speed Contrast

We consider the case that the interfaces are oriented in the Cartesian directions. Then, the set of interfaces perpendicular to the x_1-direction is denoted as S_1, the set of interfaces perpendicular to the x_2-direction is denoted as S_2 and the set of interfaces perpendicular to the x_3-direction is denoted as S_3. To express the scattered-field representations and the integral equations explicitly in terms of Cartesian vectors, we use the vector relation

$$\nabla\hat{G}\times[\boldsymbol{\nu}(\boldsymbol{x}')\times\partial\hat{\boldsymbol{w}}(\boldsymbol{x}')]=[\nabla\hat{G}\cdot\partial\hat{\boldsymbol{w}}(\boldsymbol{x}')]\,\boldsymbol{\nu}(\boldsymbol{x}')-[\boldsymbol{\nu}(\boldsymbol{x}')\cdot\nabla\hat{G}]\,\partial\hat{\boldsymbol{w}}(\boldsymbol{x}'),\tag{3.187}$$

where $\hat{G}=\hat{G}(\boldsymbol{x}-\boldsymbol{x}')$. Then, Eq. (3.185) becomes

$$\hat{\boldsymbol{E}}^{sct}(\boldsymbol{x})=2\sum_{k=1}^{3}\int_{\boldsymbol{x}'\in\mathrm{S}_k}\big[[\boldsymbol{\nu}(\boldsymbol{x}')\cdot\nabla\hat{G}]\,\partial\hat{\boldsymbol{w}}(\boldsymbol{x}')-[\nabla\hat{G}\cdot\partial\hat{\boldsymbol{w}}(\boldsymbol{x}')]\,\boldsymbol{\nu}(\boldsymbol{x}')\big]\mathrm{d}A,\tag{3.188}$$

which is equivalent to

$$\begin{aligned}
\hat{E}_1^{sct}(\boldsymbol{x}) &= 2\int_{\boldsymbol{x}'\in\mathrm{S}_2}\partial_2\hat{G}(\boldsymbol{x}-\boldsymbol{x}')\partial\hat{w}_1(\boldsymbol{x}')\,\mathrm{d}A+2\int_{\boldsymbol{x}'\in\mathrm{S}_3}\partial_3\hat{G}(\boldsymbol{x}-\boldsymbol{x}')\partial\hat{w}_1(\boldsymbol{x}')\,\mathrm{d}A\\
&\quad-2\int_{\boldsymbol{x}'\in\mathrm{S}_1}\partial_2\hat{G}(\boldsymbol{x}-\boldsymbol{x}')\partial\hat{w}_2(\boldsymbol{x}')\,\mathrm{d}A-2\int_{\boldsymbol{x}'\in\mathrm{S}_1}\partial_3\hat{G}(\boldsymbol{x}-\boldsymbol{x}')\partial\hat{w}_3(\boldsymbol{x}')\,\mathrm{d}A,\\
\hat{E}_2^{sct}(\boldsymbol{x}) &= 2\int_{\boldsymbol{x}'\in\mathrm{S}_1}\partial_1 G(\boldsymbol{x}-\boldsymbol{x}')\partial\hat{w}_2(\boldsymbol{x}')\,\mathrm{d}A+2\int_{\boldsymbol{x}'\in\mathrm{S}_3}\partial_3\hat{G}(\boldsymbol{x}-\boldsymbol{x}')\partial\hat{w}_2(\boldsymbol{x}')\,\mathrm{d}A\\
&\quad-2\int_{\boldsymbol{x}'\in\mathrm{S}_2}\partial_1\hat{G}(\boldsymbol{x}-\boldsymbol{x}')\partial\hat{w}_1(\boldsymbol{x}')\,\mathrm{d}A-2\int_{\boldsymbol{x}'\in\mathrm{S}_2}\partial_3\hat{G}(\boldsymbol{x}-\boldsymbol{x}')\partial\hat{w}_3(\boldsymbol{x}')\,\mathrm{d}A,\\
\hat{E}_3^{sct}(\boldsymbol{x}) &= 2\int_{\boldsymbol{x}'\in\mathrm{S}_1}\partial_1\hat{G}(\boldsymbol{x}-\boldsymbol{x}')\partial\hat{w}_3(\boldsymbol{x}')\,\mathrm{d}A-2\int_{\boldsymbol{x}'\in\mathrm{S}_2}\partial_2\hat{G}(\boldsymbol{x}-\boldsymbol{x}')\partial\hat{w}_3(\boldsymbol{x}')\,\mathrm{d}A\\
&\quad-2\int_{\boldsymbol{x}'\in\mathrm{S}_3}\partial_1\hat{G}(\boldsymbol{x}-\boldsymbol{x}')\partial\hat{w}_1(\boldsymbol{x}')\,\mathrm{d}A-2\int_{\boldsymbol{x}'\in\mathrm{S}_3}\partial_2\hat{G}(\boldsymbol{x}-\boldsymbol{x}')\partial\hat{w}_2(\boldsymbol{x}')\,\mathrm{d}A.
\end{aligned}\tag{3.189}$$

Similar results are obtained for the integral representation on the right-hand side of Eq. (3.186), where the partial differentiations are located in front of the Cauchy principal value integrals.

3.5.1.1 Discrete Representations in 3D

In 3D, we define a regular grid of a number of n equally sized subdomains, denoted as $\{\mathbb{D}_n,\ n=1:N\}$. In 3D, we deal with $N=N_1\times N_2\times N_3$ subdomains. The mesh size of each square subdomain is equal to Δx. The volume of a subdomain is $\Delta V=\Delta x\times\Delta x\times\Delta x$, and each side has a surface area of $\Delta A=\Delta x\times\Delta x$. The midpoints $\boldsymbol{x}_n$ of each subdomain are given by

$$\boldsymbol{x}_n^{(0)}=\{x_{1,n}^{(0)},\ x_{2,n}^{(0)},\ x_{3,n}^{(0)}\},\tag{3.190}$$

As next step, we employ a shifted grid at the horizontal interfaces of the discretized domain,

$$\boldsymbol{x}_n^{(1)}=\{x_{1,n}^{(0)}+\frac{1}{2}\Delta x,\ x_{2,n}^{(0)},\ x_{3,n}^{(0)}\},\tag{3.191}$$

and two shifted grids at the vertical interfaces of the discretized domain,

$$\boldsymbol{x}_n^{(2)}=\{x_{1,n}^{(0)},\ x_{2,n}^{(0)}+\frac{1}{2}\Delta x,\ x_{3,n}^{(0)}\},\tag{3.192}$$

and

$$\boldsymbol{x}_n^{(3)}=\{x_{1,n}^{(0)},\ x_{2,n}^{(0)},\ x_{3,n}^{(0)}\}.+\frac{1}{2}\Delta x\tag{3.193}$$

The mean values of the contrast sources at the horizontal interfaces and the vertical interfaces of the discretized domain are denoted as $\partial\hat{\boldsymbol{w}}_n^{(1)} = \partial\hat{\boldsymbol{w}}(\boldsymbol{x}_n^{(1)})$, $\partial\hat{\boldsymbol{w}}_n^{(2)} = \partial\hat{\boldsymbol{w}}(\boldsymbol{x}_n^{(2)})$, and $\partial\hat{\boldsymbol{w}}_n^{(3)} = \partial\hat{\boldsymbol{w}}(\boldsymbol{x}_n^{(3)})$, respectively. We further assume that these values are the mean values of $\partial\hat{\boldsymbol{w}}$ on each boundary of a subdomain. Carrying out the discretization of the interface integrals, the weak form of the scattered electric field at the receiver points $\boldsymbol{x}^R$ are obtained as

$$
\begin{aligned}
\langle\hat{E}_1^{sct}\rangle(\boldsymbol{x}^R) &= 2\sum_{m=1}^{N}\Big[\partial_2\langle\hat{G}\rangle(\boldsymbol{x}^R-\boldsymbol{x}_m^{(2)})\,\partial\hat{w}_{1,m}^{(2)}+\partial_3\langle\hat{G}\rangle(\boldsymbol{x}^R-\boldsymbol{x}_m^{(3)})\,\partial\hat{w}_{1,m}^{(3)}\\
&\qquad -\partial_2\langle\hat{G}\rangle(\boldsymbol{x}^R-\boldsymbol{x}_m^{(1)})\,\partial\hat{w}_{2,m}^{(1)}-\partial_3\langle\hat{G}\rangle(\boldsymbol{x}^R-\boldsymbol{x}_m^{(1)})\,\partial\hat{w}_{3,m}^{(1)}\Big],\\
\langle\hat{E}_2^{sct}\rangle(\boldsymbol{x}^R) &= 2\sum_{m=1}^{N}\Big[\partial_1\langle\hat{G}\rangle(\boldsymbol{x}^R-\boldsymbol{x}_m^{(1)})\,\partial\hat{w}_{2,m}^{(1)}+\partial_3\langle\hat{G}\rangle(\boldsymbol{x}^R-\boldsymbol{x}_m^{(3)})\,\partial\hat{w}_{2,m}^{(3)}\\
&\qquad -\partial_1\langle\hat{G}\rangle(\boldsymbol{x}^R-\boldsymbol{x}_m^{(2)})\,\partial\hat{w}_{1,m}^{(2)}-\partial_3\langle\hat{G}\rangle(\boldsymbol{x}^R-\boldsymbol{x}_m^{(2)})\,\partial\hat{w}_{3,m}^{(2)}\Big],\\
\langle\hat{E}_3^{sct}\rangle(\boldsymbol{x}^R) &= 2\sum_{m=1}^{N}\Big[\partial_1\langle\hat{G}\rangle(\boldsymbol{x}^R-\boldsymbol{x}_m^{(1)})\,\partial\hat{w}_{3,m}^{(1)}+\partial_2\langle\hat{G}\rangle(\boldsymbol{x}^R-\boldsymbol{x}_m^{(2)})\,\partial\hat{w}_{3,m}^{(2)}\\
&\qquad -\partial_1\langle\hat{G}\rangle(\boldsymbol{x}^R-\boldsymbol{x}_m^{(1)})\,\partial\hat{w}_{1,m}^{(3)}-\partial_2\langle\hat{G}\rangle(\boldsymbol{x}^R-\boldsymbol{x}_m^{(3)})\,\partial\hat{w}_{2,m}^{(3)}\Big].
\end{aligned}
\tag{3.194}
$$

Carrying out the discretization of Eq. (3.186), the result is that the continuous set of integral equations is replaced by a system of 6 m equations for 6 m unknowns. For each m, the equations are written as

$$
\begin{aligned}
&\hat{R}_m^{(2)}\hat{E}_1^{inc}(\boldsymbol{x}_m^{(2)}) = \partial\hat{w}_{1,m}^{(2)} - \hat{R}_m^{(2)}\,\mathcal{K}_m^{E(1,2)},\\
&\hat{R}_m^{(3)}\hat{E}_1^{inc}(\boldsymbol{x}_m^{(3)}) = \partial\hat{w}_{1,m}^{(3)} - \hat{R}_m^{(3)}\,\mathcal{K}_m^{E(1,3)},\\
&\hat{R}_m^{(3)}\hat{E}_2^{inc}(\boldsymbol{x}_m^{(3)}) = \partial\hat{w}_{2,m}^{(3)} - \hat{R}_m^{(3)}\,\mathcal{K}_m^{E(2,3)},\\
&\hat{R}_m^{(1)}\hat{E}_2^{inc}(\boldsymbol{x}_m^{(1)}) = \partial\hat{w}_{2,m}^{(1)} - \hat{R}_m^{(1)}\,\mathcal{K}_m^{E(2,1)},\\
&\hat{R}_m^{(1)}\hat{E}_3^{inc}(\boldsymbol{x}_m^{(1)}) = \partial\hat{w}_{3,m}^{(1)} - \hat{R}_m^{(1)}\,\mathcal{K}_m^{E(3,1)},\\
&\hat{R}_m^{(2)}\hat{E}_3^{inc}(\boldsymbol{x}_m^{(2)}) = \partial\hat{w}_{3,m}^{(2)} - \hat{R}_m^{(2)}\,\mathcal{K}_m^{E(3,2)},
\end{aligned}
\tag{3.195}
$$

where the "reflection coefficients" are given by

$$
\begin{aligned}
\hat{R}_m^{(1)} &= \frac{\hat{\mu}(\boldsymbol{x}_m^{(0)}+\Delta x\,\boldsymbol{i}_1)-\hat{\mu}(\boldsymbol{x}_m^{(0)})}{\hat{\mu}(\boldsymbol{x}_m^{(0)}+\Delta x\,\boldsymbol{i}_1)+\hat{\mu}(\boldsymbol{x}_m^{(0)})},\\
\hat{R}_m^{(2)} &= \frac{\hat{\mu}(\boldsymbol{x}_m^{(0)}+\Delta x\,\boldsymbol{i}_2)-\hat{\mu}(\boldsymbol{x}_m^{(0)})}{\hat{\mu}(\boldsymbol{x}_m^{(0)}+\Delta x\,\boldsymbol{i}_2)+\hat{\mu}(\boldsymbol{x}_m^{(0)})},\\
\hat{R}_m^{(3)} &= \frac{\hat{\mu}(\boldsymbol{x}_m^{(0)}+\Delta x\,\boldsymbol{i}_3)-\hat{\mu}(\boldsymbol{x}_m^{(0)})}{\hat{\mu}(\boldsymbol{x}_m^{(0)}+\Delta x\,\boldsymbol{i}_3)+\hat{\mu}(\boldsymbol{x}_m^{(0)})}.
\end{aligned}
\tag{3.196}
$$

Before defining the actual representations of $\mathcal{K}_m^{E(j,k)}$, for $j \neq k$, $\{j,k\} = 1, 2, 3$, we first define the operator expressions

$$
\begin{aligned}
[\mathcal{K}\partial\hat{w}_2^{(1)}](\boldsymbol{x}) &= 2\,\Delta A \sum_{n=1}^{N} \langle\hat{G}\rangle(\boldsymbol{x}-\boldsymbol{x}_n^{(1)})\;\partial\hat{w}_{2,n}^{(1)},\\
[\mathcal{K}\partial\hat{w}_3^{(1)}](\boldsymbol{x}) &= 2\,\Delta A \sum_{n=1}^{N} \langle\hat{G}\rangle(\boldsymbol{x}-\boldsymbol{x}_n^{(1)})\;\partial\hat{w}_{3,n}^{(1)},\\
[\mathcal{K}\partial\hat{w}_3^{(2)}](\boldsymbol{x}) &= 2\,\Delta A \sum_{n=1}^{N} \langle\hat{G}\rangle(\boldsymbol{x}-\boldsymbol{x}_n^{(2)})\;\partial\hat{w}_{3,n}^{(2)},\\
[\mathcal{K}\partial\hat{w}_1^{(2)}](\boldsymbol{x}) &= 2\,\Delta A \sum_{n=1}^{N} \langle\hat{G}\rangle(\boldsymbol{x}-\boldsymbol{x}_n^{(2)})\;\partial\hat{w}_{1,n}^{(2)},\\
[\mathcal{K}\partial\hat{w}_1^{(3)}](\boldsymbol{x}) &= 2\,\Delta A \sum_{n=1}^{N} \langle\hat{G}\rangle(\boldsymbol{x}-\boldsymbol{x}_n^{(3)})\;\partial\hat{w}_{1,n}^{(3)},\\
[\mathcal{K}\partial\hat{w}_2^{(3)}](\boldsymbol{x}) &= 2\,\Delta A \sum_{n=1}^{N} \langle\hat{G}\rangle(\boldsymbol{x}-\boldsymbol{x}_n^{(3)})\;\partial\hat{w}_{2,n}^{(3)}.
\end{aligned}
\tag{3.197}
$$

Subsequently, we take their particular values $\boldsymbol{x}$ in such a way that it coincide with the grid points $\boldsymbol{x}_m^{(1)}, \boldsymbol{x}_m^{(2)}$, and $\boldsymbol{x}_m^{(3)}$, respectively. Then, in the expression of these convolutional operators, we can use the relations $\boldsymbol{x}_m^{(k)} - \boldsymbol{x}_n^{(k)} = \boldsymbol{x}_m^{(0)} - \boldsymbol{x}_n^{(0)}$, for $k = 1,\ 2,\ 3$. The results are denoted as

$$
\begin{aligned}
\mathcal{K}_m^{(2,1)} &= [\mathcal{K}\partial\hat{w}_2^{(1)}](\boldsymbol{x}_m^{(1)}) = 2\,\Delta A \sum_{n=1}^{N} \langle\langle\hat{G}\rangle\rangle(\boldsymbol{x}_m^{(0)}-\boldsymbol{x}_n^{(0)})\;\partial\hat{w}_{2,n}^{(1)},\\
\mathcal{K}_m^{(3,1)} &= [\mathcal{K}\partial\hat{w}_3^{(1)}](\boldsymbol{x}_m^{(1)}) = 2\,\Delta A \sum_{n=1}^{N} \langle\langle\hat{G}\rangle\rangle(\boldsymbol{x}_m^{(0)}-\boldsymbol{x}_n^{(0)})\;\partial\hat{w}_{3,n}^{(1)},\\
\mathcal{K}_m^{(3,2)} &= [\mathcal{K}\partial\hat{w}_3^{(2)}](\boldsymbol{x}_m^{(2)}) = 2\,\Delta A \sum_{n=1}^{N} \langle\langle\hat{G}\rangle\rangle(\boldsymbol{x}_m^{(0)}-\boldsymbol{x}_n^{(0)})\;\partial\hat{w}_{n}^{(3)},\\
\mathcal{K}_m^{(1,2)} &= [\mathcal{K}\partial\hat{w}_1^{(2)}](\boldsymbol{x}_m^{(2)}) = 2\,\Delta A \sum_{n=1}^{N} \langle\langle\hat{G}\rangle\rangle(\boldsymbol{x}_m^{(0)}-\boldsymbol{x}_n^{(0)})\;\partial\hat{w}_{1,n}^{(2)},\\
\mathcal{K}_m^{(1,3)} &= [\mathcal{K}\partial\hat{w}_1^{(3)}](\boldsymbol{x}_m^{(3)}) = 2\,\Delta A \sum_{n=1}^{N} \langle\langle\hat{G}\rangle\rangle(\boldsymbol{x}_m^{(0)}-\boldsymbol{x}_n^{(0)})\;\partial\hat{w}_{1,n}^{(3)},\\
\mathcal{K}_m^{(2,3)} &= [\mathcal{K}\partial\hat{w}_2^{(3)}](\boldsymbol{x}_m^{(3)}) = 2\,\Delta A \sum_{n=1}^{N} \langle\langle\hat{G}\rangle\rangle(\boldsymbol{x}_m^{(0)}-\boldsymbol{x}_n^{(0)})\;\partial\hat{w}_{2,n}^{(3)},
\end{aligned}
\tag{3.198}
$$

where $\langle\langle\hat{G}\rangle\rangle$ is the repeated spherical mean of the 3D Green function, defined in Chapter 1. For values of $\boldsymbol{x}_m^{(j)} - \boldsymbol{x}_m^{(k)}$, $j \neq k$, in the arguments of the Green functions, we interpolate the resulting expressions for $j = k$ on the grid (k) with coordinates $\boldsymbol{x}_m^k$ to the grid (j) with coordinates $(\boldsymbol{x}_m^j)$. For example, $\mathcal{K}_m^{(2,1)}(\boldsymbol{x}_m^{(1)\to(2)})$ results into the interpolated values of $\mathcal{K}_m^{(2,1)}$ from grid (1) to grid (2). The interpolation procedure in a certain coordinate system is the simple summation of the pertaining values at four adjacent points (midpoint rule), see Matlab function `interpolate` of Matlab Listing I.60.

With these definitions, we obtain

$$\begin{aligned}
\mathcal{K}_m^{E(1,2)} &= \partial_2[\mathcal{K}_m^{(1,2)} - \mathcal{K}_m^{(2,1)}(\boldsymbol{x}_m^{(1)\to(2)})] + \partial_3[\mathcal{K}_m^{(1,3)}(\boldsymbol{x}_m^{(3)\to(2)}) - \mathcal{K}_m^{(3,1)}(\boldsymbol{x}_m^{(1)\to(2)})],\\
\mathcal{K}_m^{E(1,3)} &= \partial_3[\mathcal{K}_m^{(1,3)} - \mathcal{K}_m^{(3,1)}(\boldsymbol{x}_m^{(1)\to(3)})] + \partial_2[\mathcal{K}_m^{(1,2)}(\boldsymbol{x}_m^{(2)\to(3)}) - \mathcal{K}_m^{(2,1)}(\boldsymbol{x}_m^{(1)\to(3)})],\\
\mathcal{K}_m^{E(2,3)} &= \partial_3[\mathcal{K}_m^{(2,3)} - \mathcal{K}_m^{(3,2)}(\boldsymbol{x}_m^{(2)\to(3)})] + \partial_1[\mathcal{K}_m^{(2,1)}(\boldsymbol{x}_m^{(1)\to(3)}) - \mathcal{K}_m^{(1,2)}(\boldsymbol{x}_m^{(2)\to(3)})],\\
\mathcal{K}_m^{E(2,1)} &= \partial_1[\mathcal{K}_m^{(2,1)} - \mathcal{K}_m^{(1,2)}(\boldsymbol{x}_m^{(2)\to(1)})] + \partial_3[\mathcal{K}_m^{(2,3)}(\boldsymbol{x}_m^{(3)\to(1)}) - \mathcal{K}_m^{(3,2)}(\boldsymbol{x}_m^{(2)\to(1)})],\\
\mathcal{K}_m^{E(3,1)} &= \partial_1[\mathcal{K}_m^{(3,1)} - \mathcal{K}_m^{(1,3)}(\boldsymbol{x}_m^{(3)\to(1)})] + \partial_2[\mathcal{K}_m^{(3,2)}(\boldsymbol{x}_m^{(2)\to(1)}) - \mathcal{K}_m^{(2,3)}(\boldsymbol{x}_m^{(3)\to(1)})],\\
\mathcal{K}_m^{E(3,2)} &= \partial_2[\mathcal{K}_m^{(3,2)} - \mathcal{K}_m^{(2,3)}(\boldsymbol{x}_m^{(3)\to(2)})] + \partial_1[\mathcal{K}_m^{(3,1)}(\boldsymbol{x}_m^{(1)\to(2)}) - \mathcal{K}_m^{(1,3)}(\boldsymbol{x}_m^{(3)\to(2)})].
\end{aligned} \tag{3.199}$$

The Matlab function `KopEdw` to compute these operators is given in Matlab Listing I.116. It computes either the 2D case or the 3D case.

In the 2D case, we deal with a system of equations of 2 m for 2 m unknowns, viz.

$$\begin{aligned}
\hat{R}_m^{(2)}\hat{E}_1^{inc}(\boldsymbol{x}_{T,m}^{(2)}) &= \partial\hat{w}_{1,m}^{(2)} - \hat{R}_m^{(2)}\,\mathcal{K}_m^{E(1,2)},\\
\hat{R}_m^{(1)}\hat{E}_2^{inc}(\boldsymbol{x}_{T,m}^{(1)}) &= \partial\hat{w}_{2,m}^{(1)} - \hat{R}_m^{(1)}\,\mathcal{K}_m^{E(2,1)},
\end{aligned} \tag{3.200}$$

where

$$\begin{aligned}
\mathcal{K}_m^{E(1,2)} &= \partial_2[\mathcal{K}_m^{(1,2)} - \mathcal{K}_m^{(2,1)}(\boldsymbol{x}_m^{(1)\to(2)})],\\
\mathcal{K}_m^{E(2,1)} &= \partial_1[\mathcal{K}_m^{(2,1)} - \mathcal{K}_m^{(1,2)}(\boldsymbol{x}_m^{(2)\to(1)})].
\end{aligned} \tag{3.201}$$

Note that we have used Matlab's notation for a cell array met two indices. For example dw{2, 1} stands for $\partial\hat{w}_2^{(1)}$ and denotes the component of the interface-contrast source $\hat{\boldsymbol{w}}$ in the $\boldsymbol{i}_2$ direction of the grid (1).

To compute the 3D operators, the Matlab function `KopEdw` calls the function `KopEdw3D` of Matlab Listing I.115.

After solution for the interface-contrast sources, we compute the 3D scattered field at the receivers with the help of Eq. (3.194), while for the 2D case we use

$$\begin{aligned}
\langle\hat{E}_1^{sct}\rangle(\boldsymbol{x}_T^R) &= 2\sum_{m=1}^{N}\,[\partial_2\langle\hat{G}\rangle(\boldsymbol{x}_T^R - \boldsymbol{x}_{T,m}^{(2)})\,\partial\hat{w}_{1,m}^{(2)} - \partial_2\langle\hat{G}\rangle(\boldsymbol{x}_T^R - \boldsymbol{x}_{T,m}^{(1)})\,\partial\hat{w}_{2,m}^{(1)}]\,,\\
\langle\hat{E}_2^{sct}\rangle(\boldsymbol{x}_T^R) &= 2\sum_{m=1}^{N}\,[\partial_1\langle\hat{G}\rangle(\boldsymbol{x}_T^R - \boldsymbol{x}_{T,m}^{(1)})\,\partial\hat{w}_{2,m}^{(1)} - \partial_1\langle\hat{G}\rangle(\boldsymbol{x}_T^R - \boldsymbol{x}_{T,m}^{(2)})\,\partial\hat{w}_{1,m}^{(2)}]\,.
\end{aligned} \tag{3.202}$$

In Section 3.5.2, we present the Matlab script and the Matlab functions to compute the final results for the scattered field at the receivers.

We use Matlab's built-in function BiCSTAB as the iterative solution method, see the function `ITERBiCGSTABEdw` of Matlab Listing I.117 and the associated function `Edwvector2matrix` of Matlab Listing I.118.

3.5.2 Matlab Codes Integral Equations for Zero Wave-Speed Contrast

To initialize the electromagnetic wave-field parameters, we start again with the function `initEM` of Matlab Listing I.82. To compare the results with the ones of the previous section (Section 3.5), we keep the relative electric permittivity of the scatterer unchanged, viz., $\varepsilon_{sct} = 1.75$ and take the relative magnetic permeability of the scatterer as $\mu_{sct} = 1/\varepsilon_{sct} = 1/1.75$. For these parameters, there is no contrast in wave speed. We use the same grid size with $N1 = 240$, $N2 = 200$, and $N3 = 161$, with a mesh width $\mathrm{d}x = 1$.

Matlab Listing I.115 The function `KopEdw3D` to carry out the $\mathcal{K}^E$ operators.

```
function [KE] = KopEdw3D(dw, input)

factor = 2 / (input.gamma_0^2 * input.dx); % for num dif Green function

Kdw{2,1}  = factor * Kop(dw{2,1},input.FFTG);        % input dw_2 on grid(1)
Kdw{3,1}  = factor * Kop(dw{3,1},input.FFTG);        % input dw_3 on grid(1)

Kdw{1,2}  = factor * Kop(dw{1,2},input.FFTG);        % input dw_1 on grid(2)
Kdw{3,2}  = factor * Kop(dw{3,2},input.FFTG);        % input dw_3 on grid(2)

Kdw{1,3}  = factor * Kop(dw{1,3},input.FFTG);        % input dw_1 on grid(3)
Kdw{2,3}  = factor * Kop(dw{2,3},input.FFTG);        % input dw_2 on grid(3)

% Compute tangential field component 1 on grid(2) ------------------------
  temp = Kdw{1,2} - interpolate(Kdw{2,1},2,1,input);
            KE{1,2} = gradj(temp,2,input);
  temp = interpolate(Kdw{1,3},2,3,input) - interpolate(Kdw{3,1},2,1,input);
  KE{1,2} = KE{1,2} + gradj(temp,3,input);

% Compute tangential field component 1 on grid(3) ------------------------
  temp = Kdw{1,3} - interpolate(Kdw{3,1},3,1,input);
            KE{1,3} = gradj(temp,3,input);
  temp = interpolate(Kdw{1,2},3,2,input) - interpolate(Kdw{2,1},3,1,input);
  KE{1,3} = KE{1,3} + gradj(temp,2,input);


% Compute tangential field component 2 on grid(3) ------------------------
  temp = Kdw{2,3} - interpolate(Kdw{3,2},3,2,input);
            KE{2,3} = gradj(temp,3,input);
  temp = interpolate(Kdw{2,1},3,1,input) - interpolate(Kdw{1,2},3,2,input);
  KE{2,3} = KE{2,3} + gradj(temp,1,input);

% Compute tangential field component 2 on grid(1) ------------------------
  temp = Kdw{2,1} - interpolate(Kdw{1,2},1,2,input);
            KE{2,1} = gradj(temp,1,input);
  temp = interpolate(Kdw{2,3},1,3,input) - interpolate(Kdw{3,2},1,2,input);
  KE{2,1} = KE{2,1} + gradj(temp,3,input);


% Compute tangential field component 3 on grid(1) ------------------------
  temp = Kdw{3,1} - interpolate(Kdw{1,3},1,3,input);
            KE{3,1} = gradj(temp,1,input);
  temp = interpolate(Kdw{3,2},1,2,input) - interpolate(Kdw{2,3},1,3,input);
  KE{3,1} = KE{3,1} + gradj(temp,2,input);

% Compute tangential field component 3 on grid(2) ------------------------
  temp = Kdw{3,2} - interpolate(Kdw{2,3},2,3,input);
            KE{3,2} = gradj(temp,2,input);
  temp = interpolate(Kdw{3,1},2,1,input) - interpolate(Kdw{1,3},2,3,input);
  KE{3,2} = KE{3,2} + gradj(temp,1,input);
```

Matlab Listing I.116 The function `KopEdw` to carry out the $\mathcal{K}^E$ operators.

```
function [KE] = KopEdw(dw, input)
global nDIM;

if nDIM == 2;
factor = 2 / (input.gamma_0^2 * input.dx);  % for num dif Green function

Kdw{2,1}  = factor * Kop(dw{2,1},input.FFTG);        % input dw_2 on grid(1)
Kdw{1,2}  = factor * Kop(dw{1,2},input.FFTG);        % input dw_1 on grid(2)

% Compute tangential component (1) on grid(2)
  KE{1,2} = gradj( Kdw{1,2}-interpolate(Kdw{2,1}, 2,1,input), 2,input);

% Compute tangential component (2) on grid(1)
  KE{2,1} = gradj( Kdw{2,1}-interpolate(Kdw{1,2}, 1,2,input), 1,input);

elseif nDIM == 3;

  [KE] = KopEdw3D(dw,input);

end % if
```

After the initialization of the electromagnetic parameters, we start the main Matlab script `ForwardBiCGSTABFFTEdw` of Matlab Listing I.119 with the four steps: (1) Computation of scattering by a canonical object; (2) Computation of the incident wave fields; (3) Solution of the set of integral equations; and (4) Computation of scattering data, based on the solution of the integral equations.

In step (1) of `ForwardBiCGSTABFFTEdw`, script `EMForwardCanonicalObjects` of Matlab Listing I.139 of Appendix 3.A.2 is called. It computes both the electric and magnetic wave fields at a number of receivers, for either the 2D circular cylinder or the 3D sphere. These data are used to benchmark the scattered field based on the numerical solution of our present integral equation. Next, the Matlab function `plotEMcontrast` of Matlab Listing I.86 plots the electric-permittivity and magnetic-permeability distributions, which are the electromagnetic input parameters. Both for the 2D case and the 3D case, the results in the cross-sectional plane $x_3 = 0$ are already shown in Figure 3.18.

In step (2) of `ForwardBiCGSTABFFTEdw`, we compute the tangential components of the electric incident wave field along the grid coordinates, either in 2D and 3D. The function `IncEtaufield` is listed as Matlab Listing I.120. It starts with the 2D case and it calls `IncEtaufield3D` to compute the 3D wave fields, see Matlab Listing I.121. Note again that, e.g. the cell array Einc_tau{2, 1} stands for the component $\hat{E}_2^{inc}$ on the grid (2).

In step (3) of `ForwardBiCGSTABFFTEdw`, we solve the integral equations for the interface-contrast sources dw{j,k} $= \partial\hat{w}_j^{(k)}$, $j \neq k$, by calling the function `ITERBiCGSTABEdw` of Matlab Listing I.117. Subsequently, we use the function `plotEMInterfaceSourceEdw` of Matlab Listing I.122. In the cross-sectional plane $x_3 = 0$, it plots the reflectivity coefficients and the interface-contrast sources.

For the 2D case, it plots the absolute value of reflection coefficient $R^{(1)}$ together with the absolute value of the interface-contrast source $\partial\hat{w}_2^{(1)}$ on grid (1), and the absolute

Matlab Listing I.117 The function `ITERBiCGSTABEdw` to use Matlab's built-in function BiCGSTAB.

```
function [dw] = ITERBiCGSTABEdw(E_inc, input)
% BiCGSTAB scheme for contrast source integral equation Aw = b
  global nDIM;
  itmax = 1000; Errcri = input.Errcri; [N,~] = size(input.Rfl{1}(:));

  if nDIM == 2;
                    b(      1:  N,1) = input.Rfl{1}(:) .* E_inc{2,1}(:);
                    b(  N+1:2*N,1) = input.Rfl{2}(:) .* E_inc{1,2}(:);

  elseif nDIM == 3;
                    b(      1:  N,1) = input.Rfl{1}(:) .* E_inc{2,1}(:);
                    b(  N+1:2*N,1) = input.Rfl{1}(:) .* E_inc{3,1}(:);
                    b(2*N+1:3*N,1) = input.Rfl{2}(:) .* E_inc{1,2}(:);
                    b(3*N+1:4*N,1) = input.Rfl{2}(:) .* E_inc{3,2}(:);
                    b(4*N+1:5*N,1) = input.Rfl{3}(:) .* E_inc{1,3}(:);
                    b(5*N+1:6*N,1) = input.Rfl{3}(:) .* E_inc{2,3}(:);
  end; % if

   w  = bicgstab(@(w) Aw(w,input), b, Errcri, itmax);      % call BICGSTAB

  [dw] = Edwvector2matrix(w,input);

end %------------------------------------------------------------------------

function y = Aw(w,input)
  global nDIM;        [N,~] = size(input.CHI_eps(:));

   [dw] = Edwvector2matrix(w,input);

  [Kdw] = KopEdw(dw,input);

  if nDIM == 2;
             y(      1:  N,1) = dw{2,1}(:) - input.Rfl{1}(:) .* Kdw{2,1}(:);
             y(  N+1:2*N,1) = dw{1,2}(:) - input.Rfl{2}(:) .* Kdw{1,2}(:);

  elseif nDIM == 3;
             y(      1:  N,1) = dw{2,1}(:) - input.Rfl{1}(:) .* Kdw{2,1}(:);
             y(  N+1:2*N,1) = dw{3,1}(:) - input.Rfl{1}(:) .* Kdw{3,1}(:);
             y(2*N+1:3*N,1) = dw{1,2}(:) - input.Rfl{2}(:) .* Kdw{1,2}(:);
             y(3*N+1:4*N,1) = dw{3,2}(:) - input.Rfl{2}(:) .* Kdw{3,2}(:);
             y(4*N+1:5*N,1) = dw{1,3}(:) - input.Rfl{3}(:) .* Kdw{1,3}(:);
             y(5*N+1:6*N,1) = dw{2,3}(:) - input.Rfl{3}(:) .* Kdw{2,3}(:);
  end; %if
end
```

Matlab Listing I.118 The function `Edwvector2matrix` required by the function `ITERBiCGSTABEdw`.

```
function [dw] = Edwvector2matrix(w, input)
% Modify vector output from 'bicgstab' to matrices for further computation
  global nDIM;

  [N,~] = size(input.Rfl{1}(:));
     dw = cell(nDIM,nDIM);

  if nDIM == 2;                     DIM = [input.N1,input.N2];

   dw{2,1} = reshape(w(     1:  N,1),DIM);
   dw{1,2} = reshape(w(  N+1:2*N,1),DIM);

  elseif  nDIM == 3;                DIM = [input.N1,input.N2,input.N3];

   dw{2,1} = reshape(w(     1:  N,1),DIM);
   dw{3,1} = reshape(w(  N+1:2*N,1),DIM);
   dw{1,2} = reshape(w(2*N+1:3*N,1),DIM);
   dw{3,2} = reshape(w(3*N+1:4*N,1),DIM);
   dw{1,3} = reshape(w(4*N+1:5*N,1),DIM);
   dw{2,3} = reshape(w(5*N+1:6*N,1),DIM);

  end;  %if
end
```

Matlab Listing I.119 The `ForwardBiCGSTABFFTEdw` script to solve the interface-contrast sources $\partial\hat{w}$.

```
clear all; clc; close all; clear workspace;
input = initEM();

% (1) Compute alanalytically scattered field data
      EMForwardCanonicalObjects

%     Input interface contrast
      [input] = initEMcontrastIntf(input);

% (2) Compute incident field ---------------------------------------------
      [Einc_tau] = IncEtaufield(input);

% (3) Solve integral equation for contrast source with FFT ---------------
tic;  [dw] = ITERBiCGSTABEdw(Einc_tau,input);
toc
% Plot interface contrast and interface sources
      plotEMInterfaceSourceEdw(dw,input);

% (4) Compute synthetic data and plot fields and data --------------------
      [Edata] = DOPmudw(dw,input);
      displayEdata(Edata,input);
```

Matlab Listing I.120 The function `IncEtaufield` to compute the tangential components of the incident electric wave field from a vertically oriented dipole.

```
function [Einc_tau] = IncEtaufield(input)
global nDIM;

gam0 = input.gamma_0;  xS = input.xS;    dx = input.dx;

if nDIM == 2;

  delta  = (pi)^(-1/2) * dx;          % radius circle with area of dx^2
  factor = 2 * besseli(1,gam0*delta) / (gam0*delta);

  % Shifted grid (1) ------------------------------------------------

  X1   = input.X1 + dx/2 - xS(1);
  X2   = input.X2 - xS(2);
  DIS  = sqrt(X1.^2 + X2.^2);
  X1   = X1./DIS;
  X2   = X2./DIS;
    G  =   factor *          1/(2*pi).* besselk(0,gam0*DIS);
    dG = - factor * gam0 .* 1/(2*pi).* besselk(1,gam0*DIS);
  dG21 =  2*X2.*X1 .* (-dG./DIS) + gam0^2 * X2.*X1 .* G;

  Einc_tau{2,1} = - dG21;

  % Shifted grid (2) ------------------------------------------------

  X1   = input.X1 - xS(1);
  X2   = input.X2 + dx/2 - xS(2);
  DIS  = sqrt(X1.^2 + X2.^2);
  X1   = X1./DIS;
  X2   = X2./DIS;
   G   =   factor *          1/(2*pi).* besselk(0,gam0*DIS);
  dG   = - factor * gam0 .* 1/(2*pi).* besselk(1,gam0*DIS);
  dG11 = (2*X1.*X1 - 1)   .* (-dG./DIS) + gam0^2 * X1.*X1 .* G;

  Einc_tau{1,2} = - (-gam0^2 * G + dG11);

elseif nDIM == 3 ;

 [Einc_tau] = IncEtaufield3D(input);

end % if
```

value of reflection coefficient $R^{(2)}$ together with the absolute values of the interface-contrast source $\partial\hat{w}_1^{(2)}$ on grid (2). The results are presented in Figure 3.22. Note that for $x_1 = 0$, both the reflection factor $\hat{R}^{(1)}$ and the interface-contrast source $\partial\hat{w}_2^{(1)}$ vanish at the circular boundary. For $x_2 = 0$, both the reflection factor $\hat{R}^{(2)}$ and the interface-contrast source $\partial\hat{w}_1^{(2)}$ vanish at the circular boundary.

For the 3D case, the function `plotEMInterfaceSourceEdw` of Matlab Listing I.122 plots the absolute values of the reflection coefficient $R^{(1)}$ with the interface-contrast source $\partial\hat{w}_2^{(1)}$ and $\hat{w}_3^{(1)}$ on

Matlab Listing I.121 The function `IncEtaufield3D` to compute the tangential components of the 3D incident electric wave field from a vertically oriented dipole.

```
function [Einc_tau] = IncEtaufield3D(input)

gam0 = input.gamma_0;  xS = input.xS;     dx = input.dx;
delta  = (4*pi/3)^(-1/3) * dx;      % radius sphere with area of dx^3
arg    = gam0*delta; factor = 3 * (cosh(arg) - sinh(arg)/arg) / arg^2;

% Shifted grid (1) ---------------------------------------------------------
   X1    = input.X1+dx/2-xS(1);
   X2    = input.X2-xS(2);
   X3    = input.X3-xS(3);
   DIS   = sqrt(X1.^2 + X2.^2 + X3.^2);
   X1    = X1./DIS;
   X2    = X2./DIS;
   X3    = X3./DIS;
    G    = factor * exp(-gam0*DIS) ./ (4*pi*DIS);
   dG21 = ( 3*X2.*X1      .* (gam0./DIS + 1./DIS.^2) + gam0^2*X2.*X1 ) .* G;
   dG31 = ( 3*X3.*X1      .* (gam0./DIS + 1./DIS.^2) + gam0^2*X3.*X1 ) .* G;

   Einc_tau{2,1} = - dG21;
   Einc_tau{3,1} = - dG31;

% Shifted grid (2) ---------------------------------------------------------
   X1    = input.X1-xS(1);
   X2    = input.X2+dx/2-xS(2);
   X3    = input.X3-xS(3);
   DIS   = sqrt(X1.^2 + X2.^2 + X3.^2);
   X1    = X1./DIS;
   X2    = X2./DIS;
   X3    = X3./DIS;
     G   = factor * exp(-gam0*DIS) ./ (4*pi*DIS);
   dG11 = ((3*X1.*X1 -1).* (gam0./DIS + 1./DIS.^2) + gam0^2*X1.*X1 ) .* G;
   dG31 = ( 3*X3.*X1      .* (gam0./DIS + 1./DIS.^2) + gam0^2*X3.*X1 ) .* G;

   Einc_tau{1,2} =  - (-gam0^2 * G + dG11);
   Einc_tau{3,2} =  - dG31;

% Shifted grid (3) ---------------------------------------------------------
   X1     = input.X1-xS(1);
   X2     = input.X2-xS(2);
   X3     = input.X3+dx/2-xS(3);
   DIS    = sqrt(X1.^2 + X2.^2 + X3.^2);
   X1     = X1./DIS;
   X2     = X2./DIS;
   X3     = X3./DIS;
    G     = factor * exp(-gam0*DIS) ./ (4*pi*DIS);
   dG11 = ((3*X1.*X1 -1).* (gam0./DIS + 1./DIS.^2) + gam0^2*X1.*X1 ) .* G;
   dG21 = ( 3*X2.*X1      .* (gam0./DIS + 1./DIS.^2) + gam0^2*X2.*X1 ) .* G;

   Einc_tau{1,3} = - (-gam0^2 * G + dG11);
   Einc_tau{2,3} = - dG21;
```

Matlab Listing I.122 The function `plotEMInterfaceSourceEdw` to plot the distributions of the reflection coefficients and the tangential interface-contrast sources.

```
function plotEMInterfaceSourceEdw(dw, input)
global nDIM;
dx =input.dx;  Rfl = input.Rfl;

if  nDIM == 2;       % Plot 2D contrast/source distribution ------------------

    x1 = input.X1(:,1);    x2 = input.X2(1,:);
    set(figure,'Units','centimeters','Position',[5 5 18 12]);
    subplot(1,2,1);
       IMAGESC(x1+dx/2,x2,Rfl{1});
       title('\fontsize{13} R^{(1)}');
    subplot(1,2,2);
     IMAGESC(x1+dx/2,x2,abs(dw{2,1}));
       title('\fontsize{12} |dw_2| on grid(1)');
    set(figure,'Units','centimeters','Position',[5 5 18 12]);
    subplot(1,2,1);
       IMAGESC(x1,x2+dx/2,Rfl{2});
       title('\fontsize{13} R^{(2)}');
    subplot(1,2,2);
       IMAGESC(x1,x2+dx/2,abs(dw{1,2}));
       title('\fontsize{13} |dw_1| on grid(2)');

elseif nDIM == 3;      % Plot 3D contrast/source distribution --------------
                       % at x3 = 0 or x3 = dx/2

    x1 = input.X1(:,1,1);   x2 = input.X2(1,:,1);
    N3cross = floor(input.N3/2+1);
    set(figure,'Units','centimeters','Position',[5 5 18 12]);
    subplot(1,3,1);
       IMAGESC(x1+dx/2,x2,  Rfl{1}(:,:,N3cross));
       title('\fontsize{13} R^{(1)}');
    subplot(1,3,2);
       IMAGESC(x1+dx/2,x2,abs(dw{2,1}(:,:,N3cross)));
       title('\fontsize{12} |dw_2| on grid(1)');
    subplot(1,3,3);
       IMAGESC(x1+dx/2,x2,abs(dw{3,1}(:,:,N3cross)));
       title('\fontsize{12} |dw_3| on grid(1)');

    set(figure,'Units','centimeters','Position',[5 5 18 12]);
    subplot(1,3,1);
       IMAGESC(x1+dx/2,x2,  Rfl{2}(:,:,N3cross));
       title('\fontsize{12} R^{(2)}');
    subplot(1,3,2);
       IMAGESC(x1+dx/2,x2,abs(dw{1,2}(:,:,N3cross)));
       title('\fontsize{12} |dw_1| on grid(2)');
    subplot(1,3,3);
       IMAGESC(x1+dx/2,x2,abs(dw{3,2}(:,:,N3cross)));
       title('\fontsize{12} |dw_3| on grid(2)');

end % if
```

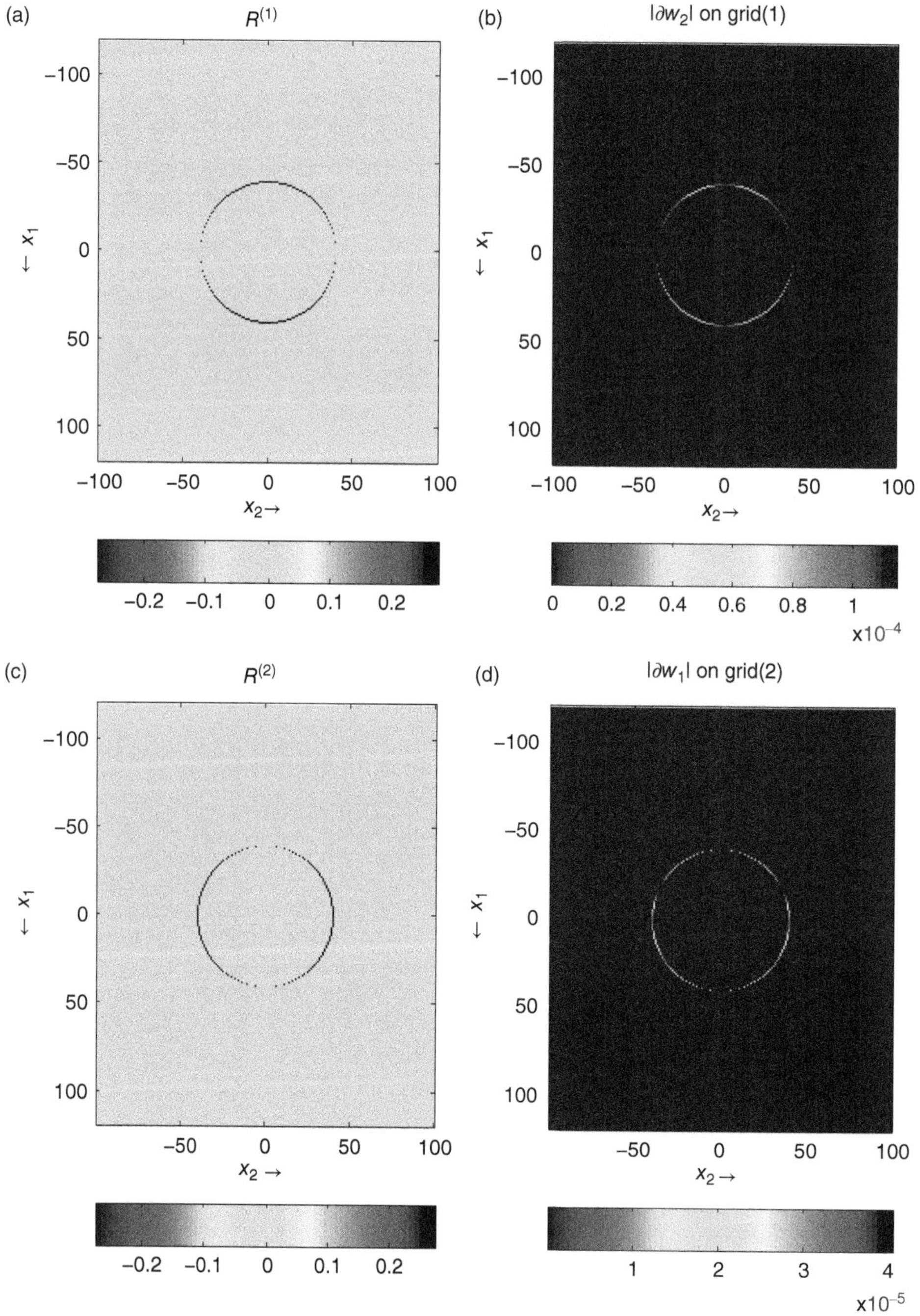

Figure 3.22 Plots of the reflection coefficient (a) $R^{(1)}$ with the interface-contrast source (b) ∂w_2 on grid (1), and the reflection coefficient (c) $R^{(2)}$ with the interface-contrast sources (d) ∂w_1 on grid (2), produced by `ITERBiCGSTABFFTEdw`, for 2D.

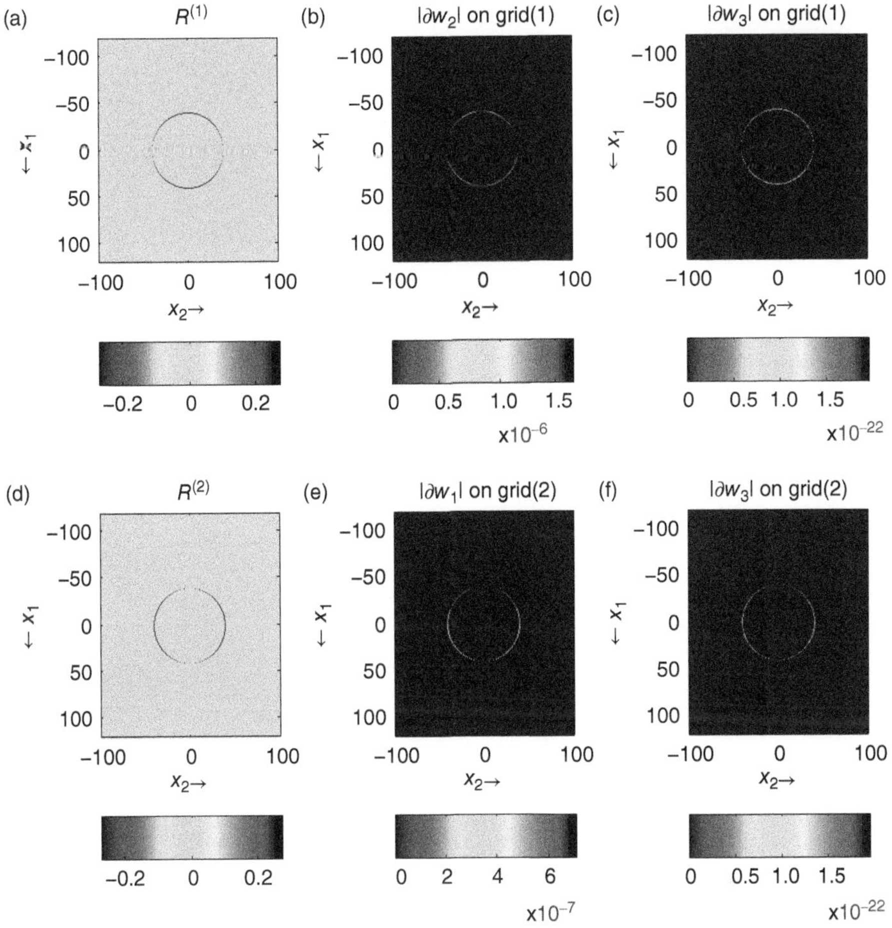

Figure 3.23 Plots of the reflection coefficient (a) $R^{(1)}$ with the interface-contrast source (b,c) ∂w_2 and ∂w_3 on grid (1), and the reflection coefficient (d) $R^{(2)}$ with the interface-contrast sources (e,f) ∂w_1 and ∂w_3 on grid (2), produced by `ITERBiCGSTABFFTEdw`, for 3D.

grid (1), and the reflection coefficient $R^{(2)}$ with the interface-contrast sources $\partial\hat{w}_1^{(2)}$ and $\hat{w}_3^{(2)}$ on grid (2), respectively. The results are presented in Figure 3.23. Note that in view of the 3D symmetry, the electric field is even with respect to the plane $x_3 = 0$, so that at this plane $\partial\hat{w}_3$ vanishes, both for grid (1) and grid (2). In the present figures, the interface-contrast sources are slightly visible, because of the fine grid size.

In step (iv) of the Matlab script `ForwardBiCGSTABFFTFFTwEH`, the scattered field at the selected receiver points are computed. The interface-contrast sources $\partial\hat{w}_j^{(k)}$ are input for the function `DOPmudw` of Matlab Listing I.123. This Matlab function computes the 2D electric-field strengths and calls the function `DOP3Dmudw` of Matlab Listing I.124 to compute the 3D electric-field components. The programs follow from the weak form of the scattered electromagnetic wave fields of Eqs. (3.194) and (3.202). The resulting data are compared to the analytical Bessel-function data stored in MAT-files. The function `displayEdata.m` of Matlab Listing I.93 loads the particular file and plot both the exact data and the numerical data. These data are presented in Figures 3.20 and 3.21. We use the error norms of Eqs. (3.107) and (3.108).

Matlab Listing I.123 The function `DOPmudw` to compute the 2D scattered electromagnetic field strengths at a number of receivers.

```
function [Edata] = DOPmudw(dw, input)
global nDIM;

dx =input.dx; gam0=input.gamma_0; xR = input.xR;

Edata = zeros(1,input.NR);

if nDIM == 2
  delta  = (pi)^(-1/2) * dx;           % radius circle with area of dx^2
  factor = 2 * besseli(1,gam0*delta) / (gam0*delta);

   for p = 1 : input.NR
   % Shifted grid (1) ---------------------------------------------
     X1   = xR(1,p)-input.X1-dx/2;
     X2   = xR(2,p)-input.X2;
     DIS  = sqrt(X1.^2 + X2.^2);
     X1   = X1./DIS;       X2 = X2./DIS;
     d_G  = - factor * gam0 * (1/(2*pi)) .* besselk(1,gam0*DIS);
    d1_G  = X1 .* d_G;  d2_G  = X2 .* d_G;

   E1rfl  = - 2 * dx * sum(d2_G(:) .* dw{2,1}(:));
   E2rfl  =   2 * dx * sum(d1_G(:) .* dw{2,1}(:));

   % Shifted grid (2) ---------------------------------------------
     X1   = xR(1,p)-input.X1;
     X2   = xR(2,p)-input.X2-dx/2;
     DIS  = sqrt(X1.^2 + X2.^2);
     X1   = X1./DIS;       X2 = X2./DIS;
     d_G  = - factor * gam0 * (1/(2*pi)) .* besselk(1,gam0*DIS);
    d1_G  = X1 .* d_G;  d2_G  = X2 .* d_G;

   E1rfl = E1rfl + 2 * dx * sum(d2_G(:) .* dw{1,2}(:));
   E2rfl = E2rfl - 2 * dx * sum(d1_G(:) .* dw{1,2}(:));

   Edata(1,p) = sqrt(abs(E1rfl)^2 + abs(E2rfl)^2);
   end % p_loop

elseif nDIM == 3
 [Edata] = DOP3Dmudw(dw, input);
end % if
```

Comparing the current results of Figures 3.24 and 3.25 with the results of Figures 3.20 and 3.21, we immediately see the significant reduction of the error norms. In 2D, the error norm decreases from 8.8% to 1.3%, while in 3D, the error norm reduces from 9.1% to 0.7%.

Both in 2D and 3D, and using the same error criterion of 0.001, the number of iterations decreases from iterations 13, for the solution of the volume-integral equations, to 3 iterations for the solution of the interface-integral equations. Then, in 3D, the computation time is 1115 seconds for the volume-integral equations and 299 seconds for the interface-integral equations.

Matlab Listing I.124 The function `DOP3Dmudw` to compute the 3D scattered electromagnetic field strengths at a number of receivers.

```
function [Edata] = DOP3Dmudw(dw, input)

dx =input.dx; gam0=input.gamma_0; xR = input.xR;

delta  = (4*pi/3)^(-1/3) * dx;      % radius sphere with area of dx^3
arg    = gam0*delta;
factor = 3 * (cosh(arg) - sinh(arg)/arg) / arg^2;

Edata = zeros(1,input.NR);

for p = 1 : input.NR
  % Shifted grid (1) ---------------------------------------------------------
    X1   = xR(1,p)-input.X1-dx/2;
    X2   = xR(2,p)-input.X2;
    X3   = xR(3,p)-input.X3;
    DIS  = sqrt(X1.^2+X2.^2+X3.^2);
    X1   = X1./DIS;  X2=X2./DIS;  X3= X3./DIS;
      G = factor * exp(-gam0*DIS)./(4*pi*DIS);  dG = -(gam0+1./DIS) .* G;
   d1_G = X1.*dG;      d2_G = X2.*dG;     d3_G = X3.*dG;
  E1rfl = -2*dx^2 * sum( d2_G(:).*dw{2,1}(:) + d3_G(:).*dw{3,1}(:));
  E2rfl = -2*dx^2 * sum(-d1_G(:).*dw{2,1}(:)                       );
  E3rfl = -2*dx^2 * sum(                     - d1_G(:).*dw{3,1}(:));

  % Shifted grid (2) ---------------------------------------------------------
    X1   = xR(1,p)-input.X1;
    X2   = xR(2,p)-input.X2-dx/2;
    X3   = xR(3,p)-input.X3;
    DIS  = sqrt(X1.^2+X2.^2+X3.^2);
    X1   = X1./DIS;  X2=X2./DIS;  X3= X3./DIS;
      G = factor * exp(-gam0*DIS)./(4*pi*DIS);  dG = -(gam0+1./DIS) .* G;
   d1_G = X1.*dG;      d2_G = X2.*dG;     d3_G = X3.*dG;
  E1rfl = E1rfl -2*dx^2 * sum(-d2_G(:).*dw{1,2}(:)                       );
  E2rfl = E2rfl -2*dx^2 * sum( d1_G(:).*dw{1,2}(:) + d3_G(:).*dw{3,2}(:));
  E3rfl = E3rfl -2*dx^2 * sum(                     - d2_G(:).*dw{3,2}(:));

  % Shifted grid(3) ----------------------------------------------------------
     X1 = xR(1,p)-input.X1;
     X2 = xR(2,p)-input.X2;
     X3 = xR(3,p)-input.X3-dx/2;
    DIS = sqrt(X1.^2+X2.^2+X3.^2);
    X1 = X1./DIS;  X2=X2./DIS;  X3= X3./DIS;
      G = factor * exp(-gam0*DIS)./(4*pi*DIS);  dG = -(gam0+1./DIS) .* G;
   d1_G = X1.*dG;      d2_G = X2.*dG;     d3_G = X3.*dG;
  E1rfl = E1rfl -2*dx^2 * sum(-d3_G(:).*dw{1,3}(:)                       );
  E2rfl = E2rfl -2*dx^2 * sum(                     - d3_G(:).*dw{2,3}(:));
  E3rfl = E3rfl -2*dx^2 * sum( d1_G(:).*dw{1,3}(:) + d2_G(:).*dw{2,3}(:));

  Edata(1,p) = sqrt(abs(E1rfl)^2  + abs(E2rfl)^2  + abs(E3rfl^2));

end % p_loop
```

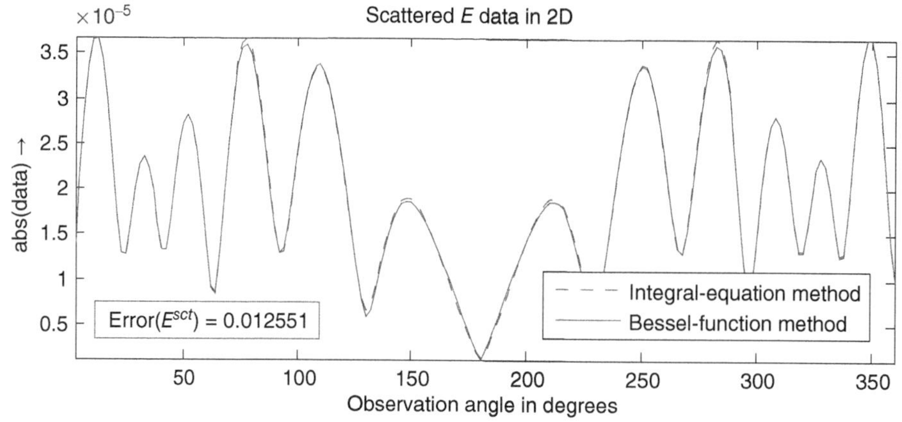

Figure 3.24 The 2D scattered electric-field amplitudes at the receivers.

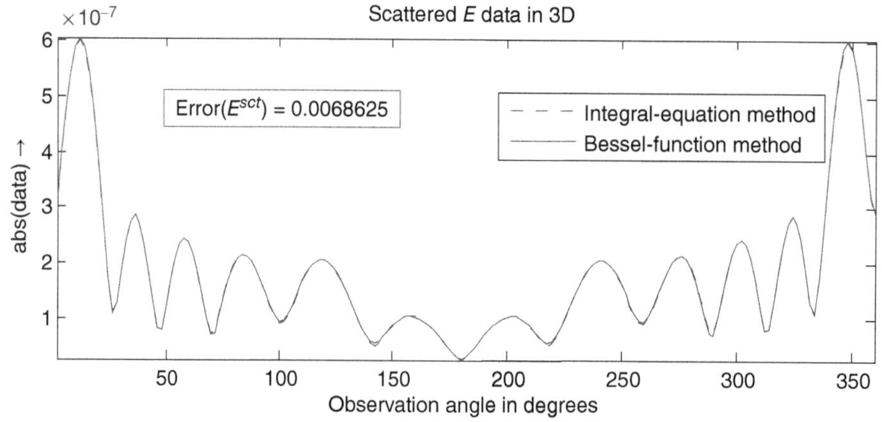

Figure 3.25 The 3D scattered electric-field amplitudes at the receivers.

In our treatise, addressing the contrast-source integral equations as a computational tool to simulate the scattering of acoustic and electromagnetic waves, we conclude the first part by suggesting that it may be advantageous to replace volume integrals by interface integrals, when spatial derivatives, via the Green's function operator, act on discontinuities in the medium parameters of the scattering object.

3.6 Time-Domain Solution of Contrast Source Integral Equation

So far we have considered the solution of the contrast-source equation for a certain value of $s = \epsilon - \mathrm{i}2\pi f \approx -\mathrm{i}2\pi f$, where ϵ is a vanishing positive constant. In our computer codes, we have taken $\epsilon = 10^{-16}$. The time-domain solution for the electric field vector $\boldsymbol{E} = \boldsymbol{E}(\boldsymbol{x}, t)$ is obtained from the frequency-domain solutions $\hat{\boldsymbol{E}}(\boldsymbol{x}, s)$ with the help of the Bromwich integral,

$$\boldsymbol{E}(\boldsymbol{x}, t) = \frac{1}{2\pi\mathrm{i}} \int_{\epsilon-\mathrm{i}\infty}^{\epsilon+\mathrm{i}\infty} \exp(st)\, \hat{\boldsymbol{E}}(\boldsymbol{x}, s)\, \mathrm{d}s. \tag{3.203}$$

In the limiting case that $\epsilon \downarrow 0$, we arrive at the inverse Fourier transform, i.e.

$$\boldsymbol{E}(\boldsymbol{x}, t) = \int_{-\infty}^{\infty} \exp(-\mathrm{i}2\pi f t)\, \hat{\boldsymbol{E}}(\boldsymbol{x}, s)\, \mathrm{d}f. \tag{3.204}$$

Discretization of this transform leads us to the inverse discrete Fourier transform,

$$\boldsymbol{E}(\boldsymbol{x}, t_m) = \Delta f \sum_{n=-\frac{1}{2}N+1}^{\frac{1}{2}N} \exp\left(-\mathrm{i}2\pi \frac{m\, n}{N}\right) \hat{\boldsymbol{E}}(\boldsymbol{x}, s_n), \tag{3.205}$$

where

$$t_m = m\, \Delta t, \quad f_n = n\, \Delta f, \quad s_n = \epsilon - \mathrm{i}\, 2\pi f_n, \qquad \text{and} \;\; \Delta t\, \Delta f = \frac{1}{N}, \tag{3.206}$$

with $m = -\frac{1}{2}N + 1 : \frac{1}{2}N$. The forward discrete Fourier transform is given by

$$\hat{\boldsymbol{E}}(\boldsymbol{x}, s_n) = \Delta t \sum_{m=-\frac{1}{2}N+1}^{\frac{1}{2}N} \exp\left(\mathrm{i}2\pi \frac{m\, n}{N}\right) \boldsymbol{E}(\boldsymbol{x}, t_m)\,, \tag{3.207}$$

with $n = -\frac{1}{2}N + 1 : \frac{1}{2}N$. Note that our definitions of the discrete Fourier transforms are different from the standard ones, as far as normalization and numbering are concerned [37]. However, we prefer the situation where the normalization and numbering of the discrete transform is closely related to the definition and symmetry properties of the continuous form.

In view of causality, we can restrict our summation in the forward discrete Fourier transform to positive values of n, viz.,

$$\boxed{\hat{\boldsymbol{E}}(\boldsymbol{x}, s_n) = \Delta t \sum_{m=0}^{\frac{1}{2}N} \exp\left(\mathrm{i}2\pi \frac{m\, n}{N}\right) \boldsymbol{E}(\boldsymbol{x}, t_m).} \tag{3.208}$$

Further, $\boldsymbol{E}(\boldsymbol{x}, t_m)$ is a real function in the time domain and the imaginary part of $\hat{\boldsymbol{E}}(\boldsymbol{x}, s_n)$ is an odd function of n. Then, the inverse discrete Fourier transform is obtained as

$$\boxed{\boldsymbol{E}(\boldsymbol{x}, t_m) = \Delta f \sum_{n=0}^{\frac{1}{2}N} \exp\left(-\mathrm{i}2\pi \frac{m\, n}{N}\right) \hat{\boldsymbol{E}}(\boldsymbol{x}, s_n).} \tag{3.209}$$

We note that in the last two expressions of our discrete transforms, symmetrical domains around $m = 0$ and $n = 0$ have to be chosen. Using FFT routines, space allocation of zero values (zero padding) for the negative values of n and m are required.

At this point, we choose the wavelet $Q^E(t)$. The frequency-domain results will be calculated for $\hat{Q}^E(s_n) = Z_0 \hat{M}(s_n)/\hat{\gamma}_0 = 1$. Before computing the time-domain results, we have to define $Q^E(t)$ and multiply the frequency domain results by $\hat{Q}^E(s_n)$. We choose the first derivative of the Gaussian as our frequency-domain wavelet. It is computed with the help of the Matlab function `WaveletE` of Matlab Listing I.125. The resulting frequency- and time-domain functions are presented in Figure 3.26.

Matlab Listing I.125 The function WaveletE to compute the first derivative of a Gaussian wavelet using Matlab's function gauswavf, both in the frequency domain and in the time domain.

```
function [input] = WaveletE(input)
Nfft = 512;                  fsamples = Nfft/2;
df = 60e6 / fsamples;        dt = 1 / (df*Nfft);

% First derivative of Gaussian Wavelet using Matlab function gauswavf
  Nwavel = 21;
  [Waveltime,~] = gauswavf(-5,5,Nwavel,1);

% Transform real function to frequency domain
  Wavelfrq = dt * conj(fft(Waveltime,Nfft));
  Wavelfrq(Nfft/2+1:Nfft) = 0;          % restrict to positive frequencies

% Transform to real function in time domain
  Waveltime = 2 * df * real(fft(Wavelfrq,Nfft));

figure;
subplot(2,1,1);
plot((0:fsamples-1)*df/1e6,abs(Wavelfrq(1:fsamples)),'b','LineWidth',1.5);
  axis('tight');
  xlabel(' frequency [MHz] \rightarrow ')
  ylabel(' |W(f)| \rightarrow ');
  legend('frequency amplitude spectrum');
subplot(2,1,2);
plot((0:fsamples-1)*dt*1e6,Waveltime(1:fsamples),'r','LineWidth',1.5);
  axis('tight');
  xlabel(' time [{\mu}s] \rightarrow ')
  ylabel(' W(t) \rightarrow ');
  legend('Wavelet in time domain')

input.Nfft      = Nfft;
input.fsamples  = fsamples;
input.Wavelfrq  = Wavelfrq(1:fsamples);
input.df        = df;
input.dt        = dt
end
```

We start with the single-frequency script ForwardCGFFTwE of Matlab Listing I.85, and replace ITERCGwE by ITERBiCGSTABwE. To deal with the complete frequency spectrum of the wavelet, the script TimedomainBiCGSTABFFTwE is used; it is listed as Matlab Listing I.126. Note that the frequency band ranges from 0 to 50 MHz. To accommodate the high-frequency range, we reduce our grid size from 2 to 1 m. To get a better picture of the wave propagation in the time domain, we take a considerably larger observation domain by choosing $N_1 = N_2 = 600$ in the 2D cross-section. In view of this larger domain and the number of computations for

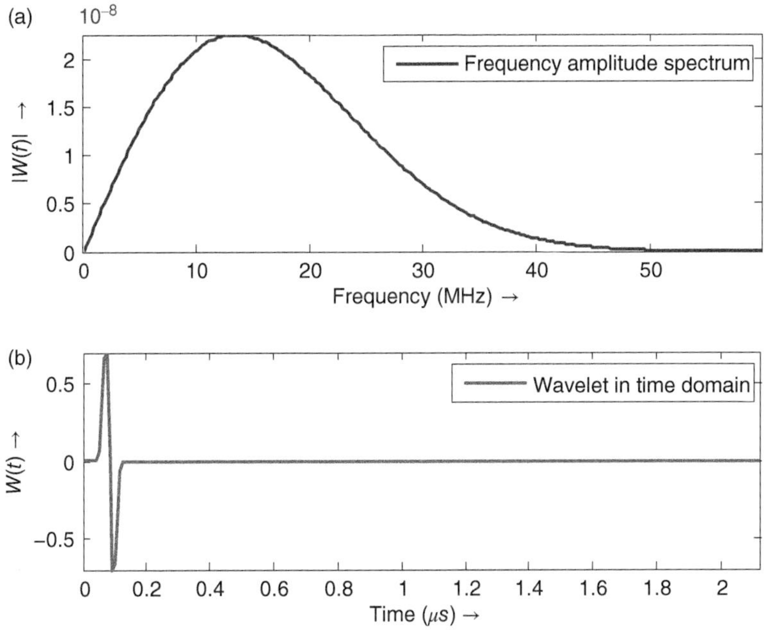

Figure 3.26 The wavelet $\hat{Q}^E(s)$ in the frequency domain (a) and $Q^E(t)$ in the time domain (b).

many frequencies, we only consider the 2D case. We put the source position at $(0, -170)$ m. Then, it is located on the left-hand side of the cylinder, and the wave propagates from the source in the right-hand direction toward the circular cylinder. Note that for the latter source position, the vertical dipole is not oriented in the radial direction, and the analytical results of Appendix 3.A.1 cannot be used for bench marking the results in the frequency domain. We keep the relative permittivity equal to 1.75 and set the relative permeability equal to one.

In step (0) of the frequency loop in `TimedomainBiCGSTABFFTwE`, we increase the error criterion by a factor which is inversely proportional to the amplitude of the normalized spectrum of the wavelet. Subsequently, for each frequency, we have to recall the Matlab functions to recompute (1) the Green function, (2) the incident wave field, and (3) the solution of the integral equation. After solution for the contrast sources, in step (4), the total wave field on the grid is computed. Finally, the total wave fields $\hat{E}(\boldsymbol{x}_T, s_n)$ are transformed to the time domain and only the values for positive times are retained.

For some specific values of time instants t_m, we call the script `SnapshotE` of Matlab Listing I.127. In fact, we compute the power in dB of the normalized total electric-field vector. For each snapshot, the maximum limit of `caxis` is chosen to be 60 dB.

The results are presented in Figure 3.27. In Figure 3.28, we zoomed in to see the wave propagation in detail within the circle cylinder. In the first few pictures of both figures, it is obvious that the wave propagates slower through the object than outside the object. For larger time instants,

Matlab Listing I.126 The script `TimedomainBiCGSTABFFTwE` to compute the time-domain total wave field $\mathbf{E}(\mathbf{x}_T, t)$ on a 2D grid; only permittivity contrast

```
clear all; clc; close all; clear workspace;
   input = initEM();
   input = WaveletE(input);
Wavelfrq = input.Wavelfrq;
Wavelmax = max(abs(Wavelfrq(:)));

% Redefine frequency-independent parameters
  input.N1 = 600;                          % number of samples in x_1
  input.N2 = 600;                          % number of samples in x_2
  input.dx = 1;                            % grid size
    x1 = -(input.N1+1)*input.dx/2 + (1:input.N1)*input.dx;
    x2 = -(input.N2+1)*input.dx/2 + (1:input.N2)*input.dx;
  [input.X1,input.X2] = ndgrid(x1,x2);
   input.xS = [0 ,-170];                   % source position
  [input]   = initEMcontrast(input);       % input contrast

Errcrr_0 = input.Errcri;
Ef     = cell(1,2);
Ef{1} = zeros(input.N1,input.N2,input.Nfft);
Ef{2} = zeros(input.N1,input.N2,input.Nfft);
for f = 2 : input.fsamples
% (0) Make error criterion frequency dependent
      factor = abs(Wavelfrq(f))/Wavelmax;
      Errcri = min([Errcrr_0 / factor, 0.999]);
      input.Errcri = Errcri;

% (1) Redefine frequency parameters
      freq = (f-1) * input.df;          disp(['freq sample: ',num2str(f)]);
      s = 1e-16 - 1i*2*pi*freq;           % LaPlace parameter
      input.gamma_0 = s/input.c_0;        % propagation coefficient
      input = initFFTGreen(input);        % compute FFT of Green function

% (2) Compute incident field ------------------------------------------------
      [E_inc,~] = IncEMwave(input);

% (3) Solve integral equation for contrast source with FFT ------------------
      [w_E] = ITERBiCGSTABwE(E_inc,input);

% (4) Compute total wave field on grid and add to frequency components
      E_sct = KopE(w_E,input);
      Ef{1}(:,:,f) = Wavelfrq(f) .* (E_inc{1}(:,:) + E_sct{1}(:,:));
      Ef{2}(:,:,f) = Wavelfrq(f) .* (E_inc{2}(:,:) + E_sct{2}(:,:));
end % frequency loop
Et{1} = 2 * input.df * real(fft(Ef{1},[],3));
Et{2} = 2 * input.df * real(fft(Ef{2},[],3));                  clear Ef;
Etot = sqrt(abs(Et{1}).^2+ abs(Et{2}).^2 );                     clear Et;
Etime(:,:,1:input.fsamples) = Etot(:,:,1:input.fsamples);      clear Etot;

save Etime;
SnapshotE;                        % Make snapshots for a few time instants
```

Matlab Listing I.127 The function `SnapshotE` to present snapshots of the time-domain total wave field $E(x_T, t)$ on a 2D grid.

```
% This script is a continuation of TimedomainBiCGSTABFFTEdw
clc;

xS(1) = input.xS(1);
xS(2) = input.xS(2);
set(figure,'Units','centimeters','Position',[0 -1 18 32]);

for n = 1 : 12;

   subplot(4,3,n)
   n_t = (10 + n * 20);

% Compute the time domain power in dB
EdB = -20 * log10(Etime/max(max(Etime(:,:,n_t))));

% Make snapshot
   imagesc(input.X2(1,:),input.X1(:,1),EdB(:,:,n_t)); grid on;
   t = n_t * input.dt *1e6;          % in microseconds
   title(['t = ',num2str(t),' {\mu}s']);
   hold on; colormap(hot);   axis('square');   caxis([0, 60]);
   plot(xS(2),xS(1),'ko','MarkerFaceColor',[0 0 0],'MarkerSize',3);
   phi =0:0.001:2*pi;
   plot(input.a*cos(phi),input.a*sin(phi),'b','LineWidth',1.2);
   pause(0.1)
   hold off;

end % if
```

we observe that a secondary wave is generated that propagates along both sides of the circular interface of cylinder. When the two wave fronts meet, they form a cardioid wave front together. After further propagation from this point, a new cardioid wave front is formed and is completed at the opposite point of the circular interface. We observe that the circular object acts as a secondary interface-contrast source.

Next, we study the time-domain results for the case that the wave-speed contrast vanishes. Then, the relative permeability is equal to the inverse of the relative permittivity. We now use the function `TimedomainBiCGSTABFFTEdw` of Matlab Listing I.128 instead of the function `Timedomain-BiCGSTABFFTwE` of Matlab Listing I.126. In principle, it uses a slightly modified version of the frequency-domain script `ForwardBiCGSTABFFTEdw` of Matlab Listing I.119. The results are presented in Figures 3.29 and 3.30. In the first few pictures of these figures, we see that, inside and outside the object, the wave propagates at the same speed. There is no refraction along the interface. In the fourth picture of Figure 3.30, we see scattering waves with wave fronts perpendicular to the interface. In the overall picture, we observe that the scattering is smaller than in the case of only permittivity contrast.

Matlab Listing I.128 The script `TimedomainBiCGSTABFFTEdw` to compute the time-domain total wave field $E(x_T, t)$ on a 2D grid; zero wave-speed contrast.

```
clear all; clc; close all; clear workspace;
input = initEM();

input = WaveletE(input);
Wavelfrq = input.Wavelfrq;
Wavelmax = max(abs(Wavelfrq(:)));

% Redefine frequency-independent parameters
  input.N1 = 600;     N1 = input.N1;              % number of samples in x_1
  input.N2 = 600;     N2 = input.N2;              % number of samples in x_2
  input.dx = 1;                                 % grid size
    x1 = -(input.N1+1)input.dx/2 + (1:input.N1)*input.dx;
    x2 = -(input.N2+1)*input.dx/2 + (1:input.N2)*input.dx;
  [input.X1,input.X2] = ndgrid(x1,x2);
   input.xS  = [0 ,-170];                        % source position
  [input] = initEMcontrast(input);               % Input contrast
  [input] = initEMcontrastIntf(input);           % Input interface contrast

Errcrr_0 = input.Errcri;            Ef = cell(2,2);
                                    Ef{2,1} = zeros(N1,N2,input.Nfft);
                                    Ef{1,2} = zeros(N1,N2,input.Nfft);
for f = 2 : input.fsamples
% (0) Make error criterion frequency dependent
       factor = abs(Wavelfrq(f))/Wavelmax;
       Errcri = min([Errcrr_0 / factor, 0.999]);
       input.Errcri = Errcri;
% (1) Redefine frequency parameters
       freq = (f-1) * input.df;            disp(['freq sample: ',num2str(f)]);
       s = 1e-16 - 1i*2*pi*freq;            % LaPlace parameter
       input.gamma_0 = s/input.c_0;         % propagation coefficient
       input = initFFTGreen(input);         % compute FFT of Green function
% (2) Compute incident field ------------------------------------------
       [Einc_tau] = IncEtaufield(input);
% (3) Solve integral equation for contrast source with FFT --------------
       [dw] = ITERBiCGSTABEdw(Einc_tau,input);
% (4) Compute total wave field on grid and add to frequency components
       E_sct = KopEdw(dw,input);
       % take average value at midpoints between interfaces
       E_sct{2,1}(2:N1,:) = (E_sct{2,1}(1:N1-1,:) + E_sct{2,1}(2:N1,:))/2;
       E_sct{1,2}(:,2:N2) = (E_sct{1,2}(:,1:N2-1) + E_sct{1,2}(:,2:N2))/2;
       Ef{2,1}(:,:,f) = Wavelfrq(f).*(Einc_tau{2,1}(:,:)+E_sct{2,1}(:,:));
       Ef{1,2}(:,:,f) = Wavelfrq(f).*(Einc_tau{1,2}(:,:)+E_sct{1,2}(:,:));
end; % frequency loop

Et{2,1} = 2 * input.df * real(fft(Ef{2,1},[],3));
Et{1,2} = 2 * input.df * real(fft(Ef{1,2},[],3));          clear Ef;
Etot = sqrt(abs(Et{2,1}).^2+ abs(Et{1,2}).^2 );            clear Et;
Etime(:,:,1:input.fsamples) = Etot(:,:,1:input.fsamples);  clear Etot;

SnapshotE;                           % Make snapshots for a few time instants
```

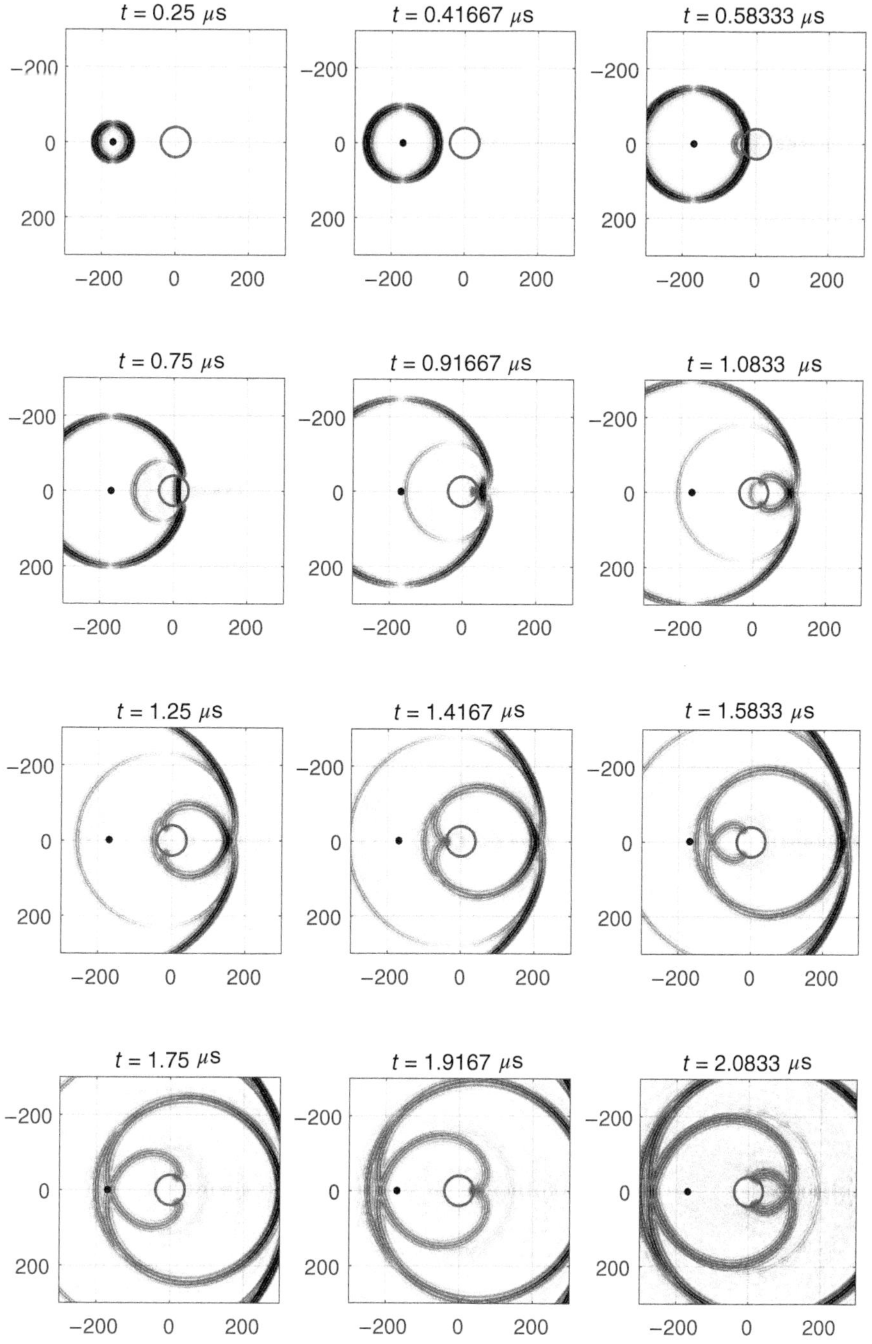

Figure 3.27 Snapshots of the 2D electromagnetic wave propagation in a model of a circular cylinder. The source is located at $x_1^S = 0$ of the vertical axis and $x_2^S = -170$ m of the horizontal axis. There is only permittivity contrast.

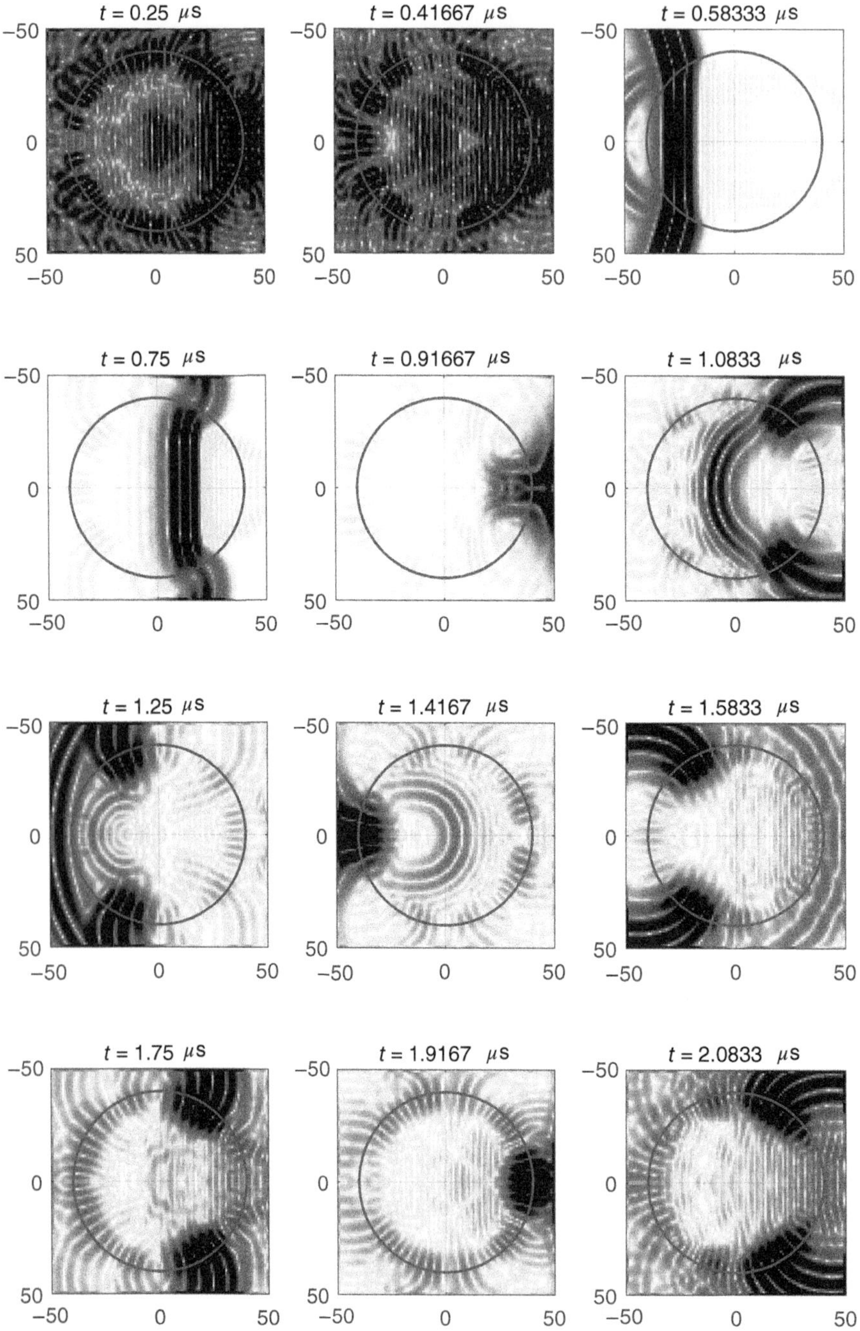

Figure 3.28 Similar to Figure 3.27, but the grid is reduced to 100 m by 100 m. Each image is normalized to the maximum value of the wave-field power (in dB). The wave-field values of the first picture only show the numerical noise.

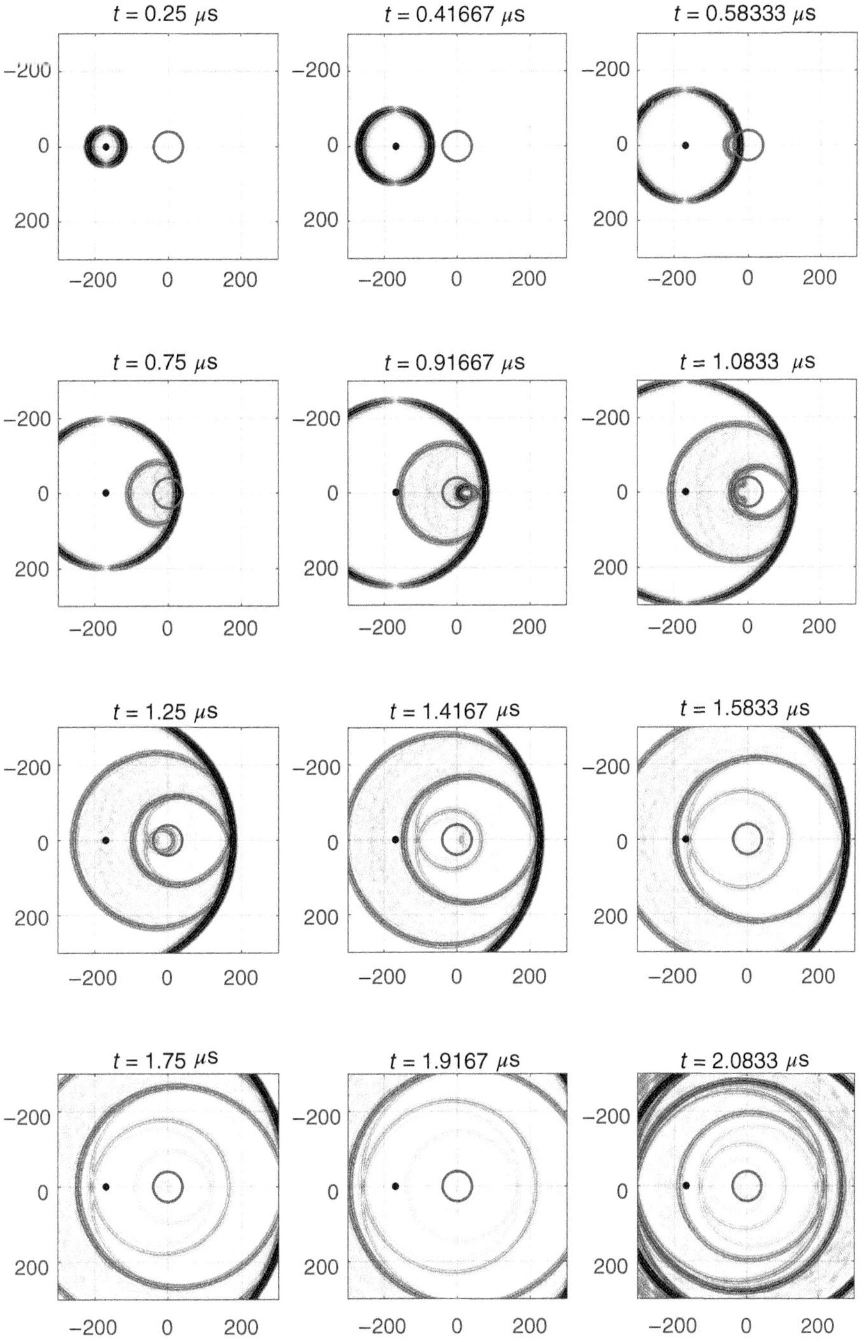

Figure 3.29 Snapshots of the 2D electromagnetic wave propagation in a model of a circular cylinder. The source is located at $x_1^S = 0$ of the vertical axis and $x_2^S = -170$ m of the horizontal axis. There is zero wave-speed contrast.

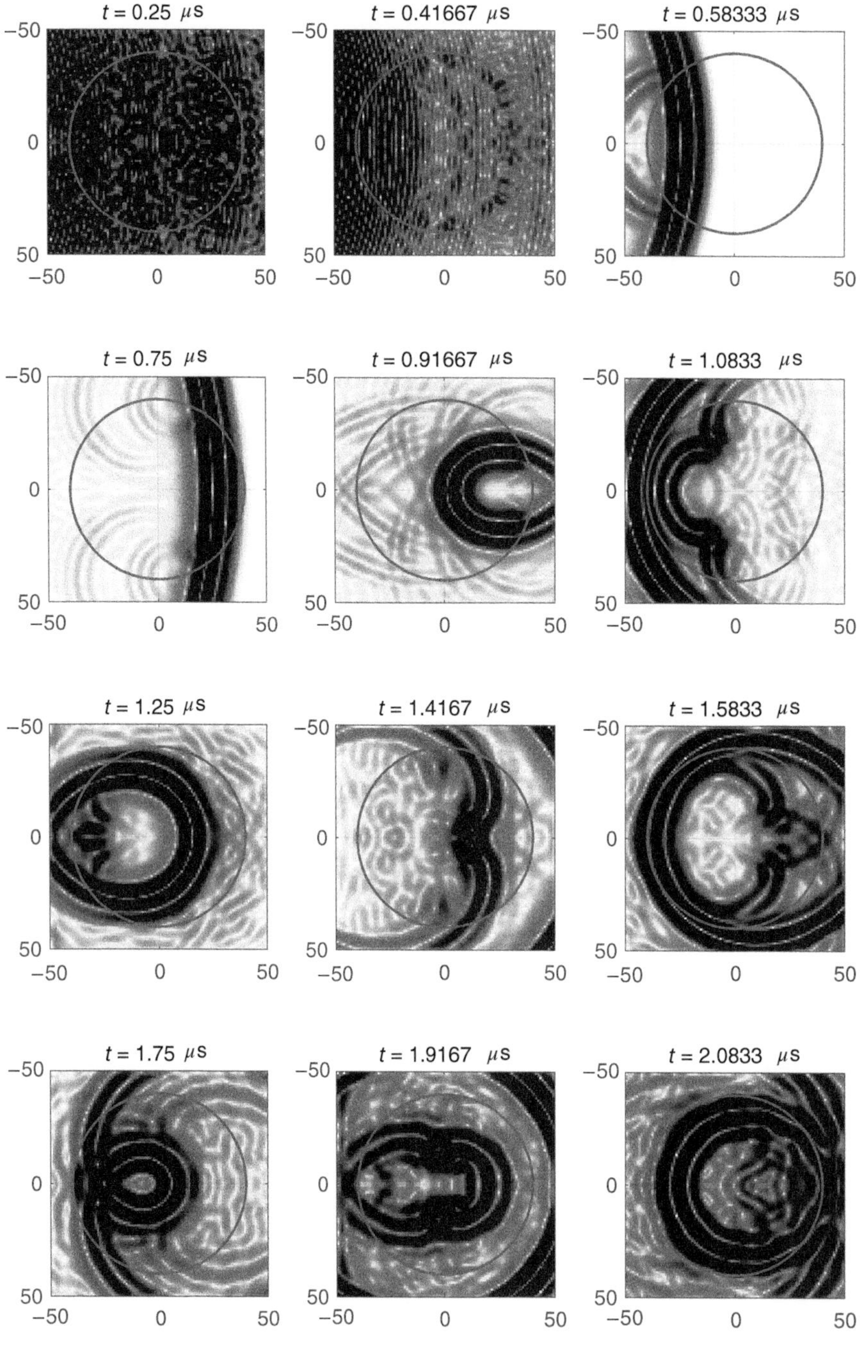

Figure 3.30 Similar to Figure 3.29, but the grid is reduced to 100 m by 100 m. Each image is normalized to the maximum value of the wave-field power (in dB). The wave-field values of the first picture only show the numerical noise.

3.A Scattering by a Simple Canonical Configuration

In order to have a necessary check on the correctness of the coding of the domain integral equations at hand, we present the analytical expressions for some configurations. We do not consider the 1D case because the codes are very similar to the ones of the 1D scalar case treated in Chapter 1. Further, the 1D electromagnetic case does not provide *a priori* insight into the scattering properties of the 2D and 3D configurations.

The 3D problem of the electromagnetic scattering by a homogeneous sphere is well-documented in the literature. The solution in spherical coordinates is known as the Mie series, and the development of a general computer code is not trivial. In the present book, one of the objectives is to use the computational solution as a code validation and benchmark for the solution of the pertaining contrast-source type integral equations. Therefore, we suffice to consider an electromagnetic source that yields the least complicated code, viz. either the electric dipole directed in the radial direction or the magnetic dipole oriented in the radial direction. We consider the radially oriented electric dipole. The changes for the magnetic dipole source is to interchange $\boldsymbol{E}$ and $\boldsymbol{H}$, and to interchange $\hat{\varepsilon}$ and $-\hat{\mu}$. We will use the scalar Debye potential to formulate the 3D electromagnetic scattering problem at hand.

In a homogeneous subdomain, the electromagnetic field can be decomposed into TE fields with electric-field vector transversal to the radial direction and TM fields with magnetic-field vector transversal to the radial direction. These two electromagnetic field constituents are characterized by two scalar Debye potentials, which are only valid in a homogeneous source-free region. For instance, the field of an electric dipole, that is not located at the origin of our coordinate system, cannot be expressed in terms of a TM type Debye potential only, for any region which includes the dipole, unless the dipole is directed in the radial direction, see page 487 of [26]. This dipole source produces only a TM field, where $E_r \neq 0$ and $H_r = 0$. This is agreement with the vertical dipole in the Cartesian x_1-direction, where Eq. (3.44) shows that the magnetic-field component $H_1 = 0$. For $\theta = \frac{1}{2}\pi$ and $\phi = 0$, the radial direction coincides with the vertical direction.

Similar as in Chapters 1 and 2, we will present the Matlab code for computing the scattered field outside the scattering object at hand. Therefore, we again need the Matlab scripts to initialize the various parameters. The source wavelet is chosen to be $\hat{M}Z_0/\hat{\gamma}_0 = 1$. The pertaining `initEM` function is given in Matlab Listing I.82. The global parameter `nDIM` assigns the dimension of the space at hand. In our case either two or three. The electromagnetic wave speed in the embedding is set to c_0=3e8, while in the scattering domain, we have set the relative permittivity to $\varepsilon_{sct} = 1.75$ and the relative permeability to $\mu_{sct} = 1$. The frequency of operation is f = 10e6 Hz. Note that the wave speed and the frequency are chosen in such a way that the wavelength has the same value as the one chosen in Chapters 1 and 2.

The `initEM` function calls the function `initSourceReceiver`, which does not need to be changed. It initializes the locations of the source and the receivers, and it is listed in Chapter 1. The `initEM` function further calls the function `initEMgrid` of Matlab Listing I.83, which defines the numerical grid in 2D and 3D. In view of the properties of the electric- and magnetic-field components, we prefer an odd number of discretization points along the Cartesian axes. It facilitates to show in the pictures that the odd components vanish at a vertical plane through the x_1-axis.

3.A.1 2D Scattering by a Circular Cylinder

In order to generate synthetic data for a two-dimensional configuration, we consider the scattering problem by a homogeneous, circular cylinder (Figure 3.A.1). In the literature, the 2D problem of the electromagnetic scattering by a homogeneous circular cylinder is almost exclusively presented for a monopole line source, which is difficult to compare with that of the dipole source in the 3D problem. We therefore consider the 2D electric dipole line source. We use a 2D scalar Debye potential to formulate the latter scattering problem. We define the spatial position by $\boldsymbol{x}_T = (x_1, x_2)$. The cylinder has a radius a. The medium in the interior of the cylinder is characterized by the constant permittivity $\hat{\varepsilon}_{sct}$, the constant permeability $\hat{\mu}_{sct}$, and the constant wave speed $\hat{c}_{sct}$. The center of the cylinder is at $x_1 = 0,\ x_2 = 0$.

In order to solve our scattering problem at hand, we introduce polar coordinates adapted to the geometry of the circular cylinder,

$$x_1 = r\ \cos(\phi), \quad x_2 = r\ \sin(\phi), \quad 0 \leq \phi < 2\pi\,. \tag{3.A.1}$$

Similarly for the source and receiver coordinates, we introduce

$$x_1^S = r^S\ \cos(\phi^S), \quad x_2^S = r^S\ \sin(\phi^S), \quad 0 \leq \phi^S < 2\pi\,, \tag{3.A.2}$$

$$x_1^R = r^R\ \cos(\phi^R), \quad x_2^R = r^R\ \sin(\phi^R), \quad 0 \leq \phi^R < 2\pi\,. \tag{3.A.3}$$

We now consider a radially oriented electric line dipole. This dipole is located at $\boldsymbol{x}_T^S$ and is infinitely long in the x_3-direction parallel to the circular cylinder. Therefore, we will use the 2D variant of the TM type of Debye potential.

3.A.1.1 TM Green Function of the Circular Cylinder

Before discussing the electromagnetic field solution of a radially oriented electric dipole, we first define the 2D TM Green function of the circular cylinder. Let this function satisfy the 2D modified Helmholtz equations outside and inside the sphere:

$$\begin{aligned} [\,\boldsymbol{\nabla}_T \cdot \boldsymbol{\nabla}_T - \hat{\gamma}_0^2\,]\hat{G}^{\mathrm{TM}}(\boldsymbol{x}_T|\boldsymbol{x}_T^S) &= -\delta(\boldsymbol{x}_T - \boldsymbol{x}_T^S), \quad && r > a, \\ [\boldsymbol{\nabla}_T \cdot \boldsymbol{\nabla}_T - \hat{\gamma}_{\mathrm{sct}}^2]\hat{G}^{\mathrm{TM}}(\boldsymbol{x}_T|\boldsymbol{x}_T^S) &= 0, && r < a, \end{aligned} \tag{3.A.4}$$

where $\hat{\gamma}_0 = s/c_0 = s(\varepsilon_0\mu_0)^{\frac{1}{2}}$ and $\hat{\gamma}_{sct} = s(\hat{\varepsilon}_{sct}\hat{\mu}_{sct})^{\frac{1}{2}}$. We assume that the source is located outside the sphere ($r^S > a$). For convenience, let this TM Green function satisfy the interface conditions:

$$\lim_{r\downarrow a} \hat{G}^{\mathrm{TM}}(\boldsymbol{x}_T|\boldsymbol{x}_T^S) = \lim_{r\uparrow a} \hat{G}^{\mathrm{TM}}(\boldsymbol{x}_T|\boldsymbol{x}_T^S), \tag{3.A.5}$$

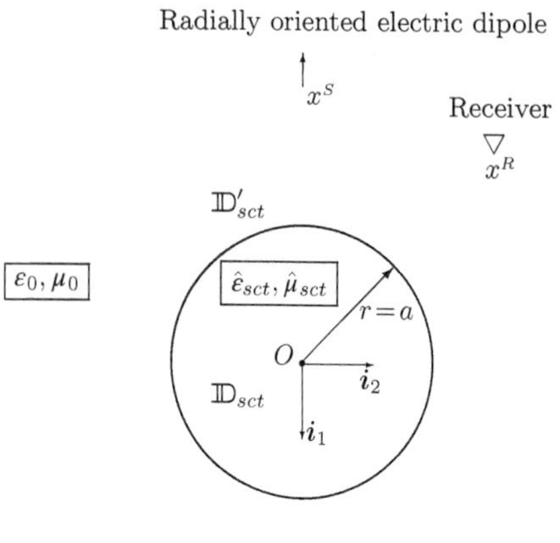

Figure 3.A.1 The 2D configuration with homogeneous circular cylinder.

$$\frac{1}{\varepsilon_0}\lim_{r\downarrow a} \partial_r \hat{G}^{\mathrm{TM}}(\boldsymbol{x}_T|\boldsymbol{x}_T^S) = \frac{1}{\hat{\varepsilon}_{sct}}\lim_{r\uparrow a} \partial_r \hat{G}^{\mathrm{TM}}(\boldsymbol{x}_T|\boldsymbol{x}_T^S). \tag{3.A.6}$$

The solution that satisfies the modified Helmholtz equations of (3.A.4) is written as

$$\begin{aligned} \hat{G}^{\mathrm{TM}}(\boldsymbol{x}_T|\boldsymbol{x}_T^S) &= \hat{G}^{inc}(\boldsymbol{x}_T - \boldsymbol{x}_T^S) + \hat{G}^{rfl}(\boldsymbol{x}_T|\boldsymbol{x}_T^S), \quad && r > a, \\ \hat{G}^{\mathrm{TM}}(\boldsymbol{x}_T|\boldsymbol{x}_T^S) &= \hat{G}^{int}(\boldsymbol{x}_T|\boldsymbol{x}_T^S), && r < a. \end{aligned} \tag{3.A.7}$$

where

$$\hat{G}^{inc}(\boldsymbol{x}_T - \boldsymbol{x}_T^S) = \frac{1}{2\pi} K_0(\hat{\gamma}_0 |\boldsymbol{x}_T - \boldsymbol{x}_T^S|). \tag{3.A.8}$$

In a spherical coordinate system, it is given by Eq. (1.A.27). For $r \le r^S$, we have

$$\hat{G}^{inc}(\boldsymbol{x}_T - \boldsymbol{x}_T^S) = \frac{1}{2\pi} \sum_{m=0}^{\infty} \epsilon_m I_m(\hat{\gamma}_0 r)\, K_m(\hat{\gamma}_0 r^S)\, \cos[m(\phi - \phi^S)]. \tag{3.A.9}$$

Taking into account the causality condition for the reflected constituent, we write the reflected wave field outside the scattering cylinder as

$$\hat{G}^{rfl}(\boldsymbol{x}_T|\boldsymbol{x}_T^S) = \frac{1}{2\pi} \sum_{m=0}^{\infty} \epsilon_m \hat{A}_m\, K_m(\hat{\gamma}_0 r)\, K_m(\hat{\gamma}_0 r^S)\, \cos[m(\phi - \phi^S)]. \tag{3.A.10}$$

Inside the cylinder, the Green function has to be bounded at $r = 0$, hence, we write this interior constituent as

$$\hat{G}^{int}(\boldsymbol{x}_T|\boldsymbol{x}_T^S) = \frac{1}{2\pi} \sum_{m=0}^{\infty} \epsilon_m \hat{B}_m\, I_m(\hat{\gamma}_{sct} r)\, K_m(\hat{\gamma}_0 r^S)\, \cos[m(\phi - \phi^S)]. \tag{3.A.11}$$

The unknown expansion factors $\hat{A}_m$ and $\hat{B}_m$ follow from the interface conditions at $r = a$ of Eqs. (3.A.5) and (3.A.6). Substitution of the expressions for the Green function into these interface conditions leads, for each m, to the following set of equations:

$$I_m(\hat{\gamma}_0 a) + \hat{A}_m\, K_m(\hat{\gamma}_0 a) = \hat{B}_m\, I_m(\hat{\gamma}_{sct} a), \tag{3.A.12}$$

$$\frac{1}{s\varepsilon_0}[\partial_r I_m(\hat{\gamma}_0 r) + \hat{A}_m \partial_r K_m(\hat{\gamma}_0 r)]_{r=a} = \frac{1}{s\hat{\varepsilon}_{sct}} \hat{B}_m\, \partial_r I_m(\hat{\gamma}_{sct} r)\Big|_{r=a}. \tag{3.A.13}$$

Introducing the impedance

$$\hat{Z}_{sct} = \frac{\hat{\gamma}_{sct}}{s\hat{\varepsilon}_{sct}} = \left(\frac{\hat{\mu}_{sct}}{\hat{\varepsilon}_{sct}}\right)^{\frac{1}{2}}, \tag{3.A.14}$$

and solving the equations for the coefficients A_m and B_m leads to

$$\hat{A}_m = -\frac{(\hat{Z}_{sct}/Z_0)\, \partial I_m(\hat{\gamma}_{sct} a)\, I_m(\hat{\gamma}_0 a) - \partial I_m(\hat{\gamma}_0 a)\, I_m(\hat{\gamma}_{sct} a)}{(\hat{Z}_{sct}/Z_0)\, \partial I_m(\hat{\gamma}_{sct} a)\, K_m(\hat{\gamma}_0 a) - \partial K_m(\hat{\gamma}_0 a)\, I_m(\hat{\gamma}_{sct} a)} \tag{3.A.15}$$

and

$$\hat{B}_m = (I_m(\hat{\gamma}_0 a) + \hat{A}_m\, K_m(\hat{\gamma}_0 a))/I_m(\hat{\gamma}_{sct} a)\,, \tag{3.A.16}$$

where $\partial I_m(\cdot)$ and $\partial K_m(\cdot)$ denote the derivatives of the functions I_m and K_m with respect to their arguments. Their expressions are given in Eq. (1.B.33). Substituting the factors $\hat{A}_m$ in the expression for the reflected wave field of Eq. (3.A.10) and taking $\boldsymbol{x}_T = \boldsymbol{x}_T^R$, we obtain

$$\hat{G}^{rfl}(\boldsymbol{x}_T^R|\boldsymbol{x}_T^S) = \frac{1}{2\pi} \sum_{m=0}^{\infty} \epsilon_m \hat{A}_m\, K_m(\hat{\gamma}_0 r^R)\, K_m(\hat{\gamma}_0 r^S)\, \cos[m(\phi^R - \phi^S)]. \tag{3.A.17}$$

3.A.1.2 Electromagnetic Field Strengths

Let us now consider a radially oriented dipole with electric current density $\boldsymbol{J}_T = \hat{M}\ \delta(\boldsymbol{x}_T - \boldsymbol{x}_T^S)\ \boldsymbol{i}_r$, where $r^S > a$ and $\boldsymbol{i}_r$ is the unit vector in the radial direction. Note that the TM electric vector potential, defined as

$$\begin{aligned}\hat{\boldsymbol{A}}^{\mathrm{TM}}(\boldsymbol{x}_T|\boldsymbol{x}_T^S) &= \hat{\boldsymbol{A}}^{inc}(\boldsymbol{x}_T - \boldsymbol{x}_T^S) + \hat{\boldsymbol{A}}^{rfl}(\boldsymbol{x}_T|\boldsymbol{x}_T^S), \quad r > a,\\ \hat{\boldsymbol{A}}^{\mathrm{TM}}(\boldsymbol{x}_T|\boldsymbol{x}_T^S) &= \hat{\boldsymbol{A}}^{int}(\boldsymbol{x}_T|\boldsymbol{x}_T^S), \quad r < a,\end{aligned} \tag{3.A.18}$$

is oriented in the radial direction, where the different constituents are given by

$$\hat{\boldsymbol{A}}^{inc}(\boldsymbol{x}_T|\boldsymbol{x}_T^S) = \hat{M}\ \hat{G}^{inc}(\boldsymbol{x}_T|\boldsymbol{x}_T^S)\ \frac{r}{r^S}\boldsymbol{i}_r, \quad r < r^S, \tag{3.A.19}$$

while

$$\hat{\boldsymbol{A}}^{rfl}(\boldsymbol{x}_T|\boldsymbol{x}_T^S) = \hat{M}\ \hat{G}^{rfl}(\boldsymbol{x}_T|\boldsymbol{x}_T^S)\ \frac{r}{r^S}\boldsymbol{i}_r, \quad r > a, \tag{3.A.20}$$

and

$$\hat{\boldsymbol{A}}^{int}(\boldsymbol{x}_T|\boldsymbol{x}_T^S) = \hat{M}\ \hat{G}^{int}(\boldsymbol{x}_T|\boldsymbol{x}_T^S)\ \frac{r}{r^S}\boldsymbol{i}_r, \quad r < a\,. \tag{3.A.21}$$

In the literature, the 3D form of the Green function $r\ \hat{G}^{\mathrm{TM}}$, for $r \neq r^S$, is denoted as the electric Debye potential. We have added the normalization factor r^S to ensure that at the source position the factor r/r^S is equal to one and the source amplitude has not been changed.

In the domain $r < r^S$, we require that the pertaining electromagnetic field strengths satisfy the Maxwell equations, the compatibility relations, and the continuity conditions at the cylinder surface. Let the magnetic-field vector be obtained from the vector potential $\hat{\boldsymbol{A}}$ as

$$\hat{\boldsymbol{H}} = \boldsymbol{\nabla}_T \times \hat{\boldsymbol{A}}_T^{\mathrm{TM}} = \hat{M}\ \boldsymbol{\nabla}_T \times \hat{G}^{\mathrm{TM}}\frac{r}{r^S}\boldsymbol{i}_r, \tag{3.A.22}$$

with the property that the magnetic field is divergence-free. Then, the magnetic-field components are

$$\hat{H}_1 = 0, \quad \hat{H}_\phi = 0, \quad \hat{H}_3 = -\hat{M}\frac{1}{r^S}\partial_\phi \hat{G}^{\mathrm{TM}}. \tag{3.A.23}$$

Since we constructed $\hat{G}^{\mathrm{TM}}$ such that it is continuous at $r = a$, we observe that the tangential components of the magnetic field satisfy the continuity condition through the interface at $r = a$. The Maxwell equations in the homogeneous domain $r > a$ and $r \neq r^S$, where $\hat{\varepsilon} = \varepsilon_0$, and in the homogeneous domain $r < a$, where $\hat{\varepsilon} = \hat{\varepsilon}_{sct}$, require that the electric-field vector satisfies

$$\hat{\boldsymbol{E}} = \frac{1}{s\hat{\varepsilon}}\boldsymbol{\nabla}_T \times \hat{H}_3\boldsymbol{i}_3, \tag{3.A.24}$$

with the property that, in the two homogeneous and source-free regions, the electric-field strength $\boldsymbol{E}$ is divergence-free. The radial component of the electric-field strength is obtained as

$$\hat{E}_r = \frac{1}{s\hat{\varepsilon}}\frac{1}{r}\partial_\phi H_3, \tag{3.A.25}$$

while the tangential components are given by

$$\hat{E}_\phi = -\frac{1}{s\hat{\varepsilon}}\partial_r \hat{H}_3, \quad \hat{E}_3 = 0. \tag{3.A.26}$$

Note that, since $E_r \neq 0$ and $H_r = 0$, we deal with a TM field with respect to the radial direction, but, since $\boldsymbol{H}_T = 0$ and $H_3 \neq 0$, we deal with a 2D H-polarized field in the x_3-direction.

Substituting Eq. (3.A.23) into Eq. (3.A.25) and (3.A.26), the radial component of the electric-field strength is obtained from

$$\hat{E}_r = -\frac{\hat{M}}{s\hat{\varepsilon}}\frac{1}{rr^S}\partial_\phi^2\hat{G}^{\mathrm{TM}}, \tag{3.A.27}$$

while the tangential components are given by

$$\hat{E}_\phi = \frac{\hat{M}}{s\hat{\varepsilon}}\frac{1}{r^S}\partial_r\partial_\phi\hat{G}^{\mathrm{TM}}, \quad \hat{E}_3 = 0. \tag{3.A.28}$$

In view of Eq. (3.A.6), we observe that the continuity of the tangential components of the electric-field strength is satisfied as well.

As next step, we substitute the Bessel-function expansions into the electromagnetic wave field expressions. The expression for $\hat{E}_r$ can be simplified. Using the expression for $r^2\nabla_T \cdot \nabla_T$ in cylindrical coordinates,

$$r^2\nabla_T \cdot \nabla_T\hat{G}^{\mathrm{TM}} = r\partial_r(r\partial_r\hat{G}^{\mathrm{TM}}) + \partial_\phi^2\hat{G}^{\mathrm{TM}}, \tag{3.A.29}$$

and $\nabla \cdot \nabla\hat{G}^{\mathrm{TM}} = \hat{\gamma}^2\hat{G}^{\mathrm{TM}}$, for $r \neq r^S$, we obtain

$$E_r = \frac{\hat{M}}{s\hat{\varepsilon}}\frac{1}{rr^S}[r\partial_r(r\partial_r\hat{G}^{\mathrm{TM}}) - (\hat{\gamma}r)^2\hat{G}^{\mathrm{TM}}], \tag{3.A.30}$$

where $\hat{\gamma} = \hat{\gamma}_0$ for $r > a$, and $\hat{\gamma} = \hat{\gamma}_{sct}$ for $r < a$. Before we substitute the expressions for G^{TM} into this expression, we note that the differential equation for modified Bessel functions reads:

$$r\partial_r\,\{r\partial_r f_m(\hat{\gamma}r)\} - (\hat{\gamma}r)^2 f_m(\hat{\gamma}r) = m^2 f_m(\hat{\gamma}r), \tag{3.A.31}$$

where $f_m(z)$ is either $I_m(z)$ or $K_m(z)$.

With this equation, the radial component of the incident electric-field strength in the domain $r < r^S$ follows from Eq. (3.A.9) as

$$\hat{E}_r^{inc} = \frac{Z_0\hat{M}}{\hat{\gamma}_0}\frac{1}{2\pi rr^S}\sum_{m=1}^{\infty}2m^2 I_m(\hat{\gamma}_0 r)\,K_m(\hat{\gamma}_0 r^S)\,\cos[m(\phi - \phi^S)], \tag{3.A.32}$$

where we have also used that $1/s\varepsilon_0 = Z_0/\hat{\gamma}_0$. We remark that this result could directly be obtained from Eq. (3.A.27) by using

$$\partial_\phi^2\cos[m(\phi - \phi^S)] = -m^2\cos[m(\phi - \phi^S)]. \tag{3.A.33}$$

The tangential component of the incident electric-field follows from Eqs. (3.A.28) and (3.A.9) as

$$\hat{E}_\phi^{inc} = \frac{Z_0\hat{M}}{\hat{\gamma}_0}\frac{\hat{\gamma}_0}{2\pi r^S}\sum_{m=1}^{\infty} -2m\,\partial I_m(\hat{\gamma}_0 r)\,K_m(\hat{\gamma}_0 r^S)\,\sin[m(\phi - \phi^S)], \tag{3.A.34}$$

and the nonzero component of the magnetic field follows from Eq. (3.A.23) as

$$Z_0\hat{H}_3^{inc} = \frac{Z_0\hat{M}}{\hat{\gamma}_0}\frac{-\hat{\gamma}_0}{2\pi r^S}\sum_{m=1}^{\infty} -2m\,I_m(\hat{\gamma}_0 r)\,K_m(\hat{\gamma}_0 r^S)\,\sin[m(\phi - \phi^S)]. \tag{3.A.35}$$

Similarly, the reflected field strengths are given by

$$
\begin{aligned}
\hat{E}_r^{rfl} &= \frac{Z_0\hat{M}}{\hat{\gamma}_0}\frac{1}{2\pi r r^S}\sum_{m=1}^{\infty} 2m^2\,\hat{A}_m\,K_m(\hat{\gamma}_0 r)\,K_m(\hat{\gamma}_0 r^S)\,\cos[m(\phi-\phi^S)],\\
\hat{E}_\phi^{rfl} &= \frac{Z_0\hat{M}}{\hat{\gamma}_0}\frac{\hat{\gamma}_0}{2\pi r^S}\sum_{m=1}^{\infty} -2m\,\hat{A}_m\,\partial K_m(\hat{\gamma}_0 r)\,K_m(\hat{\gamma}_0 r^S)\,\sin[m(\phi-\phi^S)],\\
Z_0\hat{H}_3^{rfl} &= \frac{Z_0\hat{M}}{\hat{\gamma}_0}\frac{-\hat{\gamma}_0}{2\pi r^S}\sum_{m=1}^{\infty} -2m\,\hat{A}_m\,K_m(\hat{\gamma}_0 r)\,K_m(\hat{\gamma}_0 r^S)\,\sin[m(\phi-\phi^S)],
\end{aligned}
\tag{3.A.36}
$$

and the interior field strengths as

$$
\begin{aligned}
\hat{E}_r^{int} &= \frac{Z_0\hat{M}}{\hat{\gamma}_0}\frac{\varepsilon_0}{\hat{\varepsilon}_{sct}}\frac{1}{2\pi r r^S}\sum_{m=1}^{\infty} 2m^2\,\hat{B}_m\,I_m(\hat{\gamma}_{sct} r)\,K_m(\hat{\gamma}_0 r^S)\,\cos[m(\phi-\phi^S)],\\
\hat{E}_\phi^{int} &= \frac{Z_0\hat{M}}{\hat{\gamma}_0}\frac{\varepsilon_0}{\hat{\varepsilon}_{sct}}\frac{\hat{\gamma}_{sct}}{2\pi r^S}\sum_{m=1}^{\infty} -2m\,\hat{B}_m\,\partial I_m(\hat{\gamma}_{sct} r) K_m(\hat{\gamma}_0 r^S)\sin[m(\phi-\phi^S)],\\
Z_0\hat{H}_3^{int} &= \frac{Z_0\hat{M}}{\hat{\gamma}_0}\frac{-\hat{\gamma}_0}{2\pi r^S}\sum_{m=1}^{\infty} -2m\,\hat{B}_m\,I_m(\hat{\gamma}_{sct} r) K_m(\hat{\gamma}_0 r^S)\sin[m(\phi-\phi^S)],
\end{aligned}
\tag{3.A.37}
$$

To transform these polar electric-field components to Cartesian vectors, we use the transformation:

$$
\begin{aligned}
\hat{E}_1 &= \cos(\phi)\,\hat{E}_r - \sin(\phi)\,\hat{E}_\phi,\\
\hat{E}_2 &= \sin(\phi)\,\hat{E}_r + \cos(\phi)\,\hat{E}_\phi.
\end{aligned}
\tag{3.A.38}
$$

The reflected electric-field strength at the receiver points $\boldsymbol{x}_T^R$, given by

$$
\begin{aligned}
\hat{E}_r^{rfl}(\boldsymbol{x}_T^R|\boldsymbol{x}_T^S) &= \frac{Z_0\hat{M}}{\hat{\gamma}_0}\frac{1}{2\pi r^R r^S}\sum_{m=1}^{\infty} 2m^2\,\hat{A}_m\,K_m(\hat{\gamma}_0 r^R)\,K_m(\hat{\gamma}_0 r^S)\,\cos[m(\phi^R-\phi^S)],\\
\hat{E}_\phi^{rfl}(\boldsymbol{x}_T^R|\boldsymbol{x}_T^S) &= \frac{Z_0\hat{M}}{\hat{\gamma}_0}\frac{\hat{\gamma}_0}{2\pi r^S}\sum_{m=1}^{\infty} -2m\,\hat{A}_m\,\partial K_m(\hat{\gamma}_0 r^R)\,K_m(\hat{\gamma}_0 r^S)\,\sin[m(\phi^R-\phi^S)],\\
Z_0\hat{H}_3^{rfl}(\boldsymbol{x}_T^R|\boldsymbol{x}_T^S) &= \frac{Z_0\hat{M}}{\hat{\gamma}_0}\frac{-\hat{\gamma}_0}{2\pi r^S}\sum_{m=1}^{\infty} -2m\,\hat{A}_m\,K_m(\hat{\gamma}_0 r^R)\,K_m(\hat{\gamma}_0 r^S)\,\sin[m(\phi^R-\phi^S)],
\end{aligned}
\tag{3.A.39}
$$

is not only valid in the domain $a < r^R < r^S$ but by analytic continuation also in the extended domain $a < r^R < \infty$.

3.A.1.3 Matlab Codes for Circular Cylinder (2D)

In the analysis of scattering by a circular cylinder, we have introduced polar coordinates and have expanded the wave fields in a series of modified Bessel functions. To determine the number of terms to take into account in this series, we compare the analytic solution of the incident field, Eqs. (3.49)–(3.54), with the results of the series solution, Eqs. (3.A.32)–(3.A.35).

We use the function `plotFieldError` of Matlab Listing I.129 to plot the absolute value of the exact field component and its pertaining series solution. The script for this comparison is given in `EMincidentTest2D` of Matlab Listing I.130.

The following steps are carried out.

(1) We compute the incident field analytically, denoted as $\hat{E}_{1,exact}^{inc}$, $\hat{E}_{2,exact}^{inc}$, and $Z_0\hat{H}_{3,exact}^{inc}$.
(2) We compute the incident field using the Bessel-series solution with 50 terms.
(3) We transform the polar field vectors to Cartesian vectors. The resulting field is denoted as $\hat{E}_1^{inc}$, $\hat{E}_2^{inc}$, and $Z_0\hat{H}_3^{inc}$. Using the Matlab function `plotFieldError`, these field components are

Matlab Listing I.129 The function `plotFieldError` to plot a field component in either the 2D or the 3D cross-section and compute the absolute difference between the exact component and its component using the Bessel-series expansion.

```
function plotFieldError(Title,Field_exact,Field,x1,x2,input)
global nDIM;  a = input.a;    PHI = 0:.01:2*pi;


if nDIM == 3;  N3cross =floor(input.N3/2+1);
  Field_exact = Field_exact(:,:,N3cross);
  Field       = Field(:,:,N3cross);
end

set(figure,'Units','centimeters','Position',[5 5 18 10]);
subplot(1,3,1)
IMAGESC(x1,x2,abs(Field_exact))
    title(['\fontsize{10}', Title,'_{exact}'])
    hold on;  plot(a*cos(PHI),a*sin(PHI),'w');
subplot(1,3,2)
IMAGESC(x1,x2,abs(Field))
    title(['\fontsize{10}', Title])
    hold on;  plot(a*cos(PHI),a*sin(PHI),'w');
subplot(1,3,3)
IMAGESC(x1,x2,abs(Field - Field_exact))
    title('\fontsize{10} abs(Error)');
    hold on; plot(a*cos(PHI),a*sin(PHI),'w');
```

plotted in the left and middle pictures of Figure 3.2. From these plots of the incident fields, it is difficult to guess the accuracy achieved. Therefore, for each field component f, we plot the absolute difference, Error $= \hat{f}^{inc} - \hat{f}^{inc}_{exact}$, as well (see right-hand side pictures). From these error plots, if r tends to r^S we observe that the error increases very strongly. This is visible at the corners of the error pictures.

However, the error in the transmitted and the reflected field components are solely determined by the incident field at the boundary of the circular cylinder. Therefore, a more realistic error of the scattering problem at hand is the mean error of the incident-field approximation over the circle at $r = a$, viz.,

$$\langle \text{Error} \rangle = \frac{\sum_{\phi_n \in [0,2\pi]} |\hat{f}^{inc}(r = a, \phi_n) - \hat{f}^{inc}_{exact}(r = a, \phi_n)|}{\sum_{\phi_n \in [0,2\pi]} |\hat{f}^{inc}_{exact}(r = a, \phi_n)|}. \quad (3.A.40)$$

This error is calculated with the help of the script `EMincidentTest2Derror` of Matlab Listing I.131. For $M = 20$, the mean error for the field components $\hat{E}_1$, $\hat{E}_2$, and $Z_0\hat{H}_3$ amounts to 4.9×10^{-6}, 1.4×10^{-6}, and 0.6×10^{-6}, respectively. For our scattering problem at hand, this is sufficiently small to consider that only 20 terms in the series of Bessel functions are enough to benchmark the numerical solution of the integral equation.

Since we are particularly interested in the contrast sources as fundamental unknowns to be solved from the corresponding integral equations, we next use the script `EMintCircle` of Matlab Listing I.132, in which the interior wave field is computed. The following steps are taken:

Matlab Listing I.130 The `EMincidentTest2D` script to compute the incident wave field.

```
clear all; clc; close all; clear workspace
input = initEM(); gam0 = input.gamma_0;

% (1) Cartesian coordinates: electric dipole in negative x_1 -----------

  xS   = input.xS;
  X1   = input.X1-xS(1);
  X2   = input.X2-xS(2);
  DIS  = sqrt(X1.^2 + X2.^2); X1 = X1./DIS;  X2 = X2./DIS;
   G   = 1/(2*pi).* besselk(0,gam0*DIS);
  dG   = - gam0 .* 1/(2*pi).* besselk(1,gam0*DIS);
  dG11 = (2*X1.*X1 - 1) .* (-dG./DIS) + gam0^2 * X1.*X1 .* G;
  dG21 =  2*X2.*X1      .* (-dG./DIS) + gam0^2 * X2.*X1 .* G;

  E{1} = - (-gam0^2 * G + dG11);
  E{2} = - dG21;
 ZH{3} = gam0 * X2 .* dG;

% (2) Transform Cartesian coordinates to polar ccordinates -------------

  rS = sqrt(xS(1)^2+xS(2)^2);        phiS = atan2(xS(2),xS(1));
  X1 = input.X1;
  X2 = input.X2;
  R  = sqrt(X1.^2+X2.^2+1e-16);      PHI  = atan2(X2,X1);

  % and compute Bessel series with M terms ----------------------------
  M = 50;                                  % increase M for more accuracy
  Er = zeros(size(R));  Ephi = zeros(size(R));  ZH3 = zeros(size(R));
  for m = 1 : M
    arg0 = gam0*R;          Ib0 = besseli(m,arg0);
                           dIb0 = besseli(m+1,arg0) + m./arg0 .* Ib0;
                            KbS = besselk(m,gam0*rS);
     Er   = Er   + 2*m^2 .* Ib0 .* KbS .* cos(m*(PHI-phiS));
    Ephi  = Ephi - 2*m   .*dIb0 .* KbS .* sin(m*(PHI-phiS));
    ZH3   = ZH3  - 2*m   .* Ib0 .* KbS .* sin(m*(PHI-phiS));
  end % m_loop

  Er   =          1/(2*pi) * Er./R ./rS;
  Ephi =  gam0 *  1/(2*pi) * Ephi  ./rS;
  ZH3  = -gam0 *  1/(2*pi) * ZH3   ./rS;

% (3) Determine mean error and plot error in Cartesian vectors ---------

  E1 = cos(PHI) .* Er - sin(PHI) .* Ephi;
  E2 = sin(PHI) .* Er + cos(PHI) .* Ephi;

  x1 = input.X1(:,1);   x2 = input.X2(1,:);
  plotFieldError('2D: abs(E_1)',   E{1},  E1,  x1,x2,input);
  plotFieldError('2D: abs(E_2)',   E{2},  E2,  x1,x2,input);
  plotFieldError('2D: abs(ZH_3)', ZH{3}, ZH3,  x1,x2,input);
```

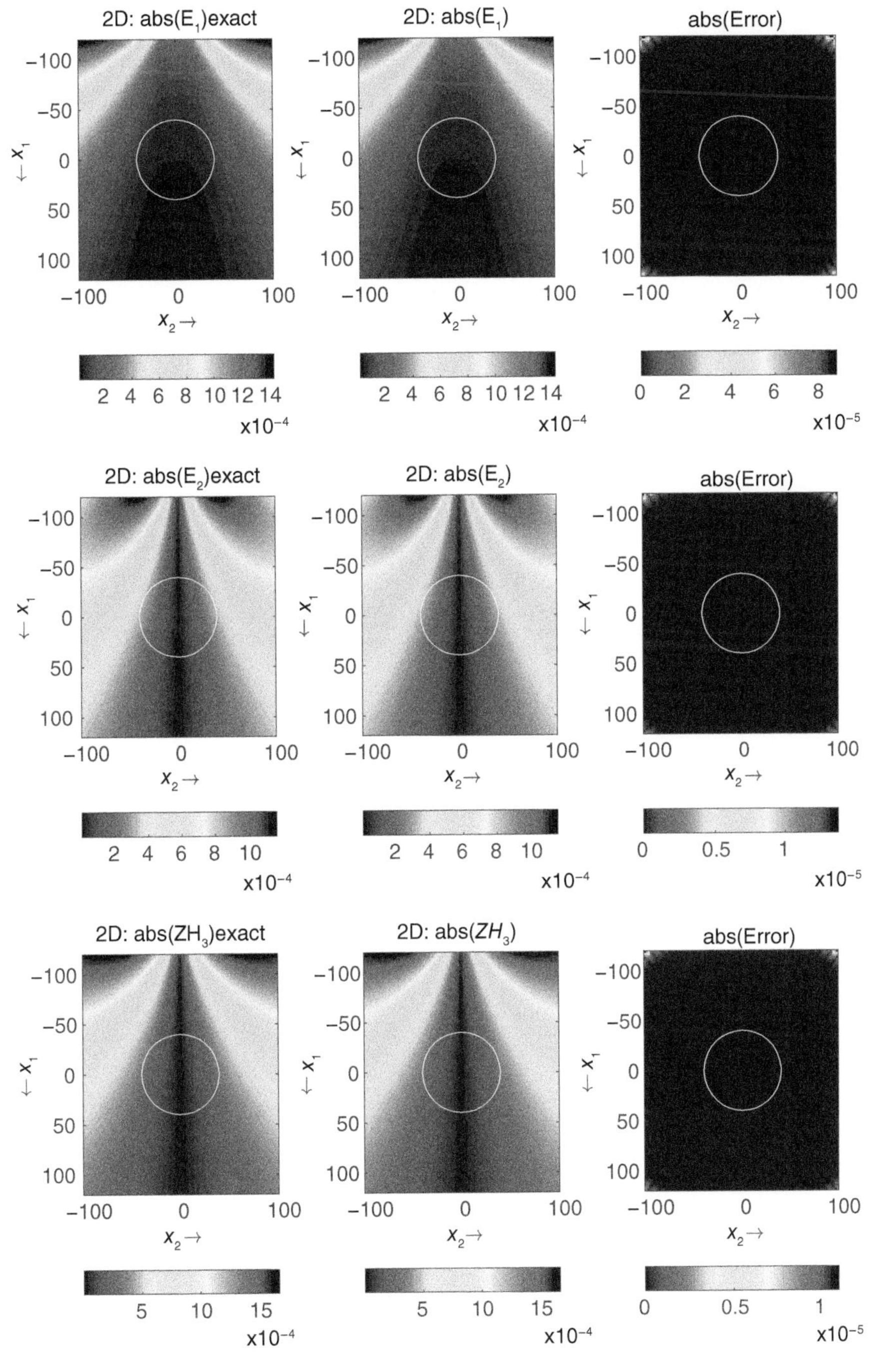

Figure 3.A.2 Comparison of the 2D incident-wave-field results.

Matlab Listing I.131 The `EMincidentTest2Derror` script to compute the error in the incident wave field.

```
clear all; clc; close all; clear workspace
input = initEM();
gam0  = input.gamma_0;
   a  = input.a;
  PHI = 0:.01:2*pi;

% (1) Cartesian coordinates: electric dipole in negative x_1 ------------

   xS  = input.xS;
   x1  = a*cos(PHI)-xS(1);        x2 = a*sin(PHI)-xS(2);
  DIS  = sqrt(x1.^2 + x2.^2);     x1 = x1./DIS;   x2 = x2./DIS;
   G   = 1/(2*pi).* besselk(0,gam0*DIS);
  dG   = - gam0 .* 1/(2*pi).* besselk(1,gam0*DIS);
  dG11 = (2*x1.*x1 - 1) .* (-dG./DIS) + gam0^2 * x1.*x1 .* G;
  dG21 =  2*x2.*x1      .* (-dG./DIS) + gam0^2 * x2.*x1 .* G;
  E{1} = - (-gam0^2 * G + dG11);
  E{2} = - dG21;
 ZH{3} = gam0 * x2 .* dG;

% (2) Polar coordinates: expansion in bessel functions ----------------

  rS   = sqrt(xS(1)^2+xS(2)^2);
  phiS = atan2(xS(2),xS(1));
  Er   = zeros(size(PHI));
  Ephi = zeros(size(PHI));
  ZH3  = zeros(size(PHI));

  M = 20;                                   % increase M for more accuracy
  for m = 1 : M
    Ib0  = besseli(m,gam0*a);  dIb0 = besseli(m+1,gam0*a)+m*Ib0/(gam0*a);
    KbS  = besselk(m,gam0*rS);
    Er   = Er   + 2*m^2 * Ib0 *  KbS .* cos(m*(PHI-phiS));
    Ephi = Ephi - 2*m   *dIb0 *  KbS .* sin(m*(PHI-phiS));
    ZH3  = ZH3  - 2*m   * Ib0 *  KbS .* sin(m*(PHI-phiS));
  end % m_loop

  Er   =          1/(2*pi) * Er/a /rS;
  Ephi =   gam0 * 1/(2*pi) * Ephi /rS;
  ZH3  =  -gam0 * 1/(2*pi) * ZH3  /rS;

  E1 = cos(PHI) .* Er - sin(PHI) .* Ephi;
  E2 = sin(PHI) .* Er + cos(PHI) .* Ephi;

% Print the normalized error on the circular boundary at r = a ----------
  norm_Error = norm(E{1}(:)-E1(:),1) / norm(E{1}(:),1);
               disp(['Error ',' E_1',' = ',num2str(norm_Error)]);
  norm_Error = norm(E{2}(:)-E2(:),1) / norm(E{2}(:),1);
               disp(['Error ',' E_2',' = ',num2str(norm_Error)]);
  norm_Error = norm(ZH{3}(:)-ZH3(:),1) / norm(ZH{3}(:),1);
               disp(['Error ','ZH_3',' = ',num2str(norm_Error)]);
```

Matlab Listing I.132 The script `EMintCircle` to compute the contrast sources.

```
clear all; clc; close all; clear workspace
input = initEM();
      a = input.a;
   gam0 = input.gamma_0;
eps_sct = input.eps_sct;
 mu_sct = input.mu_sct;
gam_sct = gam0 * sqrt(eps_sct*mu_sct);
 Z_sct  = sqrt(mu_sct/eps_sct);

% (1) Transform Cartesian coordinates to polar cOordinates ----------------
  xS = input.xS;
  X1 = input.X1;  X2 = input.X2;
  rS = sqrt(xS(1)^2+xS(2)^2);      phiS = atan2(xS(2),xS(1));
   R = sqrt(X1.^2+X2.^2+1e-16);  PHI  = atan2(X2,X1);

% (2) Compute coefficients of Bessel series expansion ---------------------
  arg0 = gam0 * input.a;
  args = gam_sct*input.a;
  M = 20;                                    % increase M for more accuracy
  A = zeros(1,M); B = zeros(1,M);
  for m = 1 : M
    Ib0 = besseli(m,arg0);        dIb0 =  besseli(m+1,arg0) + m/arg0 * Ib0;
    Ibs = besseli(m,args);        dIbs =  besseli(m+1,args) + m/args * Ibs;
    Kb0 = besselk(m,arg0);        dKb0 = -besselk(m+1,arg0) + m/arg0 * Kb0;
    denominator = Z_sct * dIbs*Kb0 - dKb0*Ibs;
    A(m)    =     -(Z_sct * dIbs*Ib0 - dIb0*Ibs) / denominator;
    B(m)    = (Ib0 + A(m) * Kb0) / Ibs;
  end

% (3) Compute interior wave field using M terms of Bessel series expansion
    Er = zeros(size(R));   Ephi = zeros(size(R));  ZH3 = zeros(size(R));
  for m = 1 : M
    args = gam_sct*R;     Ibs = besseli(m,args);
                         dIbs = besseli(m+1,args) + m./args .* Ibs;
                          KbS = besselk(m,gam0*rS);
    Er    = Er    + B(m)*2*m^2.* Ibs  .* KbS .* cos(m*(PHI-phiS));
    Ephi = Ephi - B(m)*2*m   .*dIbs .* KbS .* sin(m*(PHI-phiS));
  end % m_loop
  Er    = (          1/eps_sct) * 1/(2*pi) * Er./R ./rS;
  Ephi = (gam_sct/eps_sct) * 1/(2*pi) * Ephi   ./rS;

% (4) Transform to Cartesian vectors ----------------------------------------
  E1 = cos(PHI) .* Er  - sin(PHI) .* Ephi;
  E2 = sin(PHI) .* Er  + cos(PHI) .* Ephi;

% (5) Plot contrast sources
  w_E{1} = input.CHI_eps .* E1;
  w_E{2} = input.CHI_eps .* E2;

  plotContrastSourcewE(w_E,input);                  save CONTRASTSOURCE w_E;
```

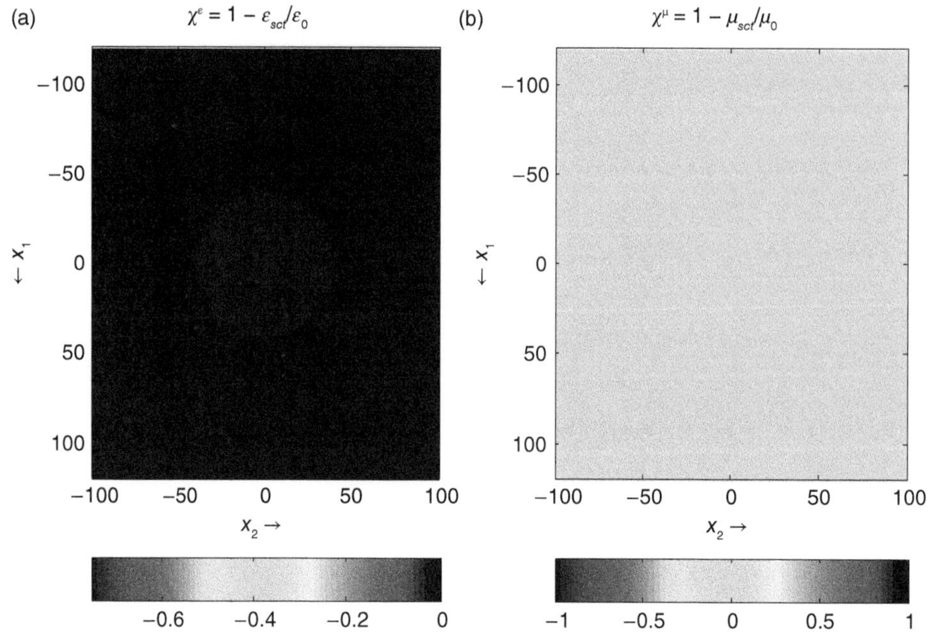

Figure 3.A.3 Permittivity contrast and zero permeability contrast.

(1) We transform the Cartesian coordinates to the polar coordinates.
(2) We compute the coefficients A and B.
(3) We compute the interior electric field using Eq. (3.A.37).
(4) We transform the polar field vectors to Cartesian vectors, cf. Eq. (3.A.38).
(5) We compute the contrast sources $\hat{w}_1^E = \hat{\chi}^\varepsilon \hat{E}_1$ and $\hat{w}_2^E = \hat{\chi}^\varepsilon \hat{E}_2$.

The input permittivity- and permeability-contrast distributions $\hat{\chi}^\varepsilon$ are obtained from the χ^μ are presented in Figure 3.A.3. The contrast sources pictures are generated by Matlab function `plot-ContrastSourcewE` of Matlab Listing I.90 and presented in Figure 3.A.4. The results serve as a benchmark of the integral equation method.

For benchmarking purposes of the final scattered field computed with the help of the integral representations, we employ the function `EmsctCircle` of Matlab Listing I.133. The following steps are taken:

(1) We transform the Cartesian receiver coordinates to the polar receiver coordinates.
(2) We compute the coefficients A and the reflected electric field at the receivers using Eq. (3.A.36).
(3) We transform the polar field vectors to Cartesian vectors, cf. Eq. (3.A.38). and display the result with the help of the function `displayEdata` of Matlab Listing I.93 and `displayHdata` of Matlab Listing I.94. The amplitudes of the scattered electric and magnetic wave-fields are shown in Figures 3.A.5 and 3.A.6. For benchmarking needs, the scattered field data are stored in the MAT-files EDATA2D.mat and HDATA2D.mat (Figures 3.A.5 and 3.A.6)

Matlab Listing I.133 The `EmsctCircle` script to compute the scattered wave field.

```
clear all; clc; close all;
input = initEM();
      a = input.a;
   gam0 = input.gamma_0;
eps_sct = input.eps_sct;
 mu_sct = input.mu_sct;
gam_sct = gam0 * sqrt(eps_sct*mu_sct);
 Z_sct  = sqrt(mu_sct/eps_sct);

% (1) Transform Cartesian coordinates to polar coordinates ----------------
  xR = input.xR;
  xS = input.xS;
  rR = sqrt(xR(1,:).^2 + xR(2,:).^2);  phiR = atan2(xR(2,:),xR(1,:));
  rS = sqrt(xS(1)^2+xS(2)^2);          phiS = atan2(xS(2),xS(1));

% (2) Compute coefficients of Bessel series expansion ----------------------
  arg0 = gam0 * input.a;  args = gam_sct * input.a;
  M = 20;     A = zeros(1,M);                % increase M for more accuracy
  for m = 1 : M
    Ib0 = besseli(m,arg0);         dIb0 =  besseli(m+1,arg0) + m/arg0 * Ib0;
    Ibs = besseli(m,args);         dIbs =  besseli(m+1,args) + m/args * Ibs;
    Kb0 = besselk(m,arg0);         dKb0 = -besselk(m+1,arg0) + m/arg0 * Kb0;
    denominator = Z_sct * dIbs*Kb0 - dKb0*Ibs;
    A(m)  =     -(Z_sct * dIbs*Ib0 - dIb0*Ibs) / denominator;
  end

% (3) Compute reflected Er field at receivers (data) -----------------------
  Er = zeros(size(rR));  Ephi = zeros(size(rR));  ZH3 = zeros(size(rR));
  for m = 1 : M
    arg0 = gam0*rR;  Kb0 =  besselk(m,arg0);
                    dKb0 = -besselk(m+1,arg0) + m./arg0 .* Kb0;
                     KbS = besselk(m,gam0*rS);
    Er   = Er   + A(m)*2*m^2.* Kb0.* KbS .*cos(m*(phiR-phiS));
    Ephi = Ephi - A(m)*2*m  .*dKb0.* KbS .*sin(m*(phiR-phiS));
    ZH3  = ZH3  - A(m)*2*m  .* Kb0.* KbS .*sin(m*(phiR-phiS));
  end % m_loop

  Er   =          1/(2*pi) * Er./rR ./rS;
  Ephi =  gam0 * 1/(2*pi) * Ephi   ./rS;
  ZH3  = -gam0 * 1/(2*pi) * ZH3    ./rS;
  E{1} =   cos(phiR) .* Er  - sin(phiR) .* Ephi;
  E{2} =   sin(phiR) .* Er  + cos(phiR) .* Ephi;
  Edata2D = sqrt(abs(E{1}).^2 + abs(E{2}).^2);
  Hdata2D = abs(ZH3);

  if exist(fullfile(cd, 'EDATA2D.mat'), 'file'); delete EDATA2D.mat; end
  if exist(fullfile(cd, 'HDATA2D.mat'), 'file'); delete HDATA2D.mat; end

  displayEdata(Edata2D,input);                save EDATA2D Edata2D;
  displayHdata(Hdata2D,input);                save HDATA2D Hdata2D;
```

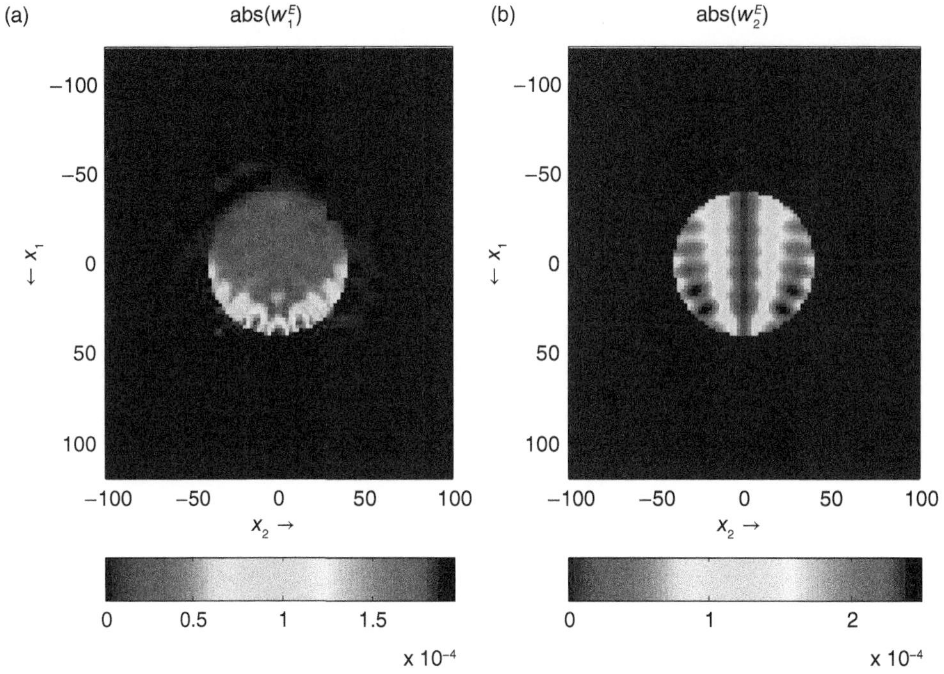

Figure 3.A.4 The two components of the contrast sources in 2D.

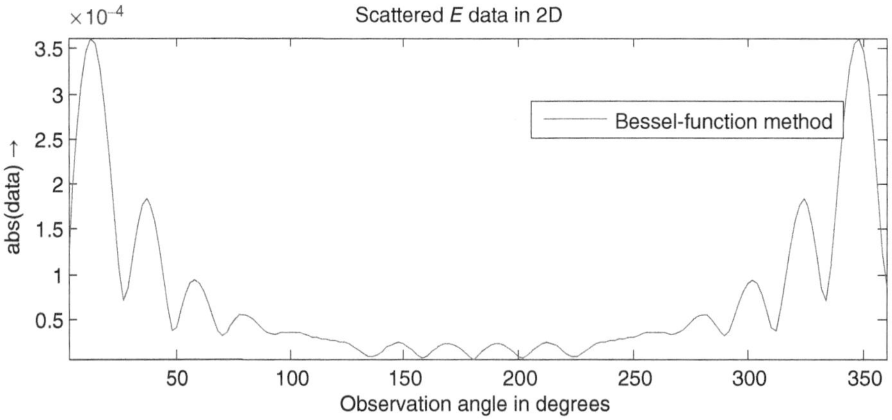

Figure 3.A.5 The scattered electric wave-field results at the receivers.

3.A.2 3D Scattering by a Sphere

In order to generate synthetic data for a 3D scatterer, we consider the scattering problem by a homogeneous, sphere (Figure 3.A.7). The sphere has a radius a. The medium in the interior of the cylinder is characterized by the constant permittivity $\hat{\varepsilon}_{sct}$, the constant permeability $\hat{\mu}_{sct}$ and the constant wave speed $\hat{c}_{sct}$. The center of the sphere is at $\boldsymbol{x} = (0, 0, 0)$.

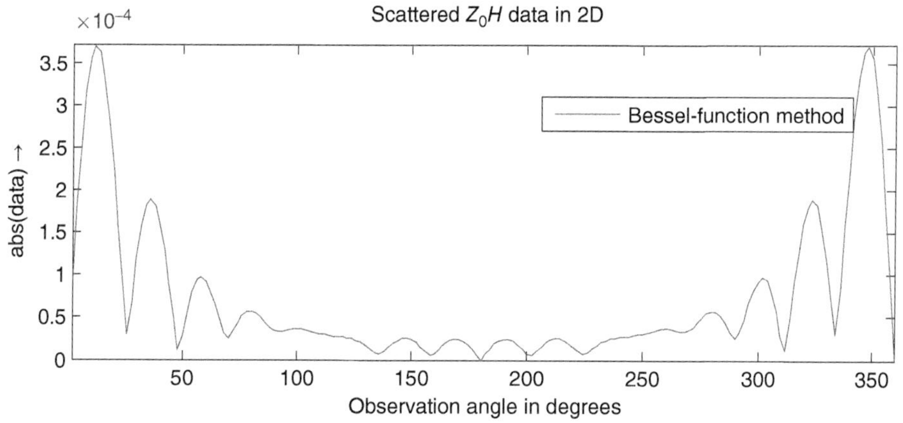

Figure 3.A.6 The scattered magnetic wave-field results at the receivers.

In order to solve our scattering problem at hand, we introduce spherical coordinates adapted to the geometry of the sphere,

$$x_1 = r\ \sin(\theta)\ \cos(\phi), \quad x_2 = r\ \sin(\theta)\ \sin(\phi), \quad x_3 = r\ \cos(\theta)\,, \tag{3.A.41}$$

with $0 \leq \phi < 2\pi$ and $0 \leq \theta < \pi$. Similarly for the source and receiver coordinates, we introduce

$$x_1^S = r^S\ \sin(\theta^S)\ \cos(\phi^S), \quad x_2^S = r^S\ \sin(\theta^S)\ \sin(\phi^S), \quad x_3^S = r^S\ \cos(\theta^S)\,, \tag{3.A.42}$$

$$x_1^R = r^R\ \sin(\theta^R)\ \cos(\phi^R), \quad x_2^R = r^R\ \sin(\theta^R)\ \sin(\phi^R), \quad x_3^R = r^R\ \cos(\theta^R)\,. \tag{3.A.43}$$

3.A.2.1 TM Green Function of the Sphere

Before discussing the electromagnetic field solution of a radially oriented electric dipole, we first define the TM Green function of the sphere. Let this function satisfy the modified Helmholtz equations outside and inside the sphere:

$$\begin{aligned} [\,\nabla\cdot\nabla - \hat{\gamma}_0^2\,]\hat{G}^{\mathrm{TM}}(\boldsymbol{x}|\boldsymbol{x}^S) &= -\delta(\boldsymbol{x}-\boldsymbol{x}^S), \quad && r > a, \\ [\nabla\cdot\nabla - \hat{\gamma}_{sct}^2]\hat{G}^{\mathrm{TM}}(\boldsymbol{x}|\boldsymbol{x}^S) &= 0, && r < a, \end{aligned} \tag{3.A.44}$$

where $\hat{\gamma}_0 = s/c_0 = s(\varepsilon_0\mu_0)^{\frac{1}{2}}$ and $\hat{\gamma}_{sct} = s(\hat{\varepsilon}_{sct}\hat{\mu}_{sct})^{\frac{1}{2}}$. We assume that the source is located outside the sphere ($r^S > a$). For convenience, let this TM Green function satisfy the interface conditions:

$$\lim_{r\downarrow a} \hat{G}^{\mathrm{TM}}(\boldsymbol{x}|\boldsymbol{x}^S) = \lim_{r\uparrow a} \hat{G}^{\mathrm{TM}}(\boldsymbol{x}|\boldsymbol{x}^S), \tag{3.A.45}$$

$$\frac{1}{s\varepsilon_0}\lim_{r\downarrow a}\frac{1}{r}\partial_r\{r\hat{G}^{\mathrm{TM}}(\boldsymbol{x}_T|\boldsymbol{x}_T^S)\} = \frac{1}{s\hat{\varepsilon}_{sct}}\lim_{r\uparrow a}\frac{1}{r}\partial_r\{r\hat{G}^{\mathrm{TM}}(\boldsymbol{x}_T|\boldsymbol{x}_T^S)\}. \tag{3.A.46}$$

Note that we use here a more complicated interface condition in spherical coordinates. The solution that satisfies the modified Helmholtz equations of (3.A.44) is written as

$$\begin{aligned} &\hat{G}^{\mathrm{TM}}(\boldsymbol{x}|\boldsymbol{x}^S) = \hat{G}^{inc}(\boldsymbol{x}-\boldsymbol{x}^S) + \hat{G}^{rfl}(\boldsymbol{x}|\boldsymbol{x}^S), \quad r > a, \\ &\hat{G}^{\mathrm{TM}}(\boldsymbol{x}|\boldsymbol{x}^S) = \hat{G}^{int}(\boldsymbol{x}|\boldsymbol{x}^S), \quad r < a. \end{aligned} \tag{3.A.47}$$

where

$$\hat{G}^{inc}(\boldsymbol{x}-\boldsymbol{x}^S) = \frac{\exp(-\hat{\gamma}_0|\boldsymbol{x}-\boldsymbol{x}^S|)}{4\pi|\boldsymbol{x}-\boldsymbol{x}^S|}. \tag{3.A.48}$$

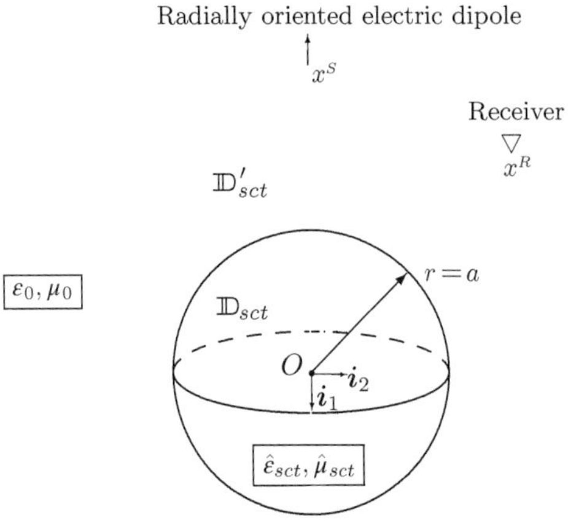

Figure 3.A.7 The 3D configuration with homogeneous sphere.

In a spherical coordinate system, it is given by Eq. (1.A.44). For $r \leq r^S$, we have

$$\hat{G}^{inc}(\boldsymbol{x}-\boldsymbol{x}^S) = \frac{\hat{\gamma}_0}{2\pi^2}\sum_{n=0}^{\infty}(2n+1)\, i_n^{(1)}(\hat{\gamma}_0 r)\, k_n(\hat{\gamma}_0 r^S) P_n[\cos(\boldsymbol{x},\boldsymbol{x}^S)]. \tag{3.A.49}$$

Taking into account the causality condition for the reflected constituent, we write, for $r > a$,

$$\hat{G}^{rfl}(\boldsymbol{x}|\boldsymbol{x}^S) = \frac{\hat{\gamma}_0}{2\pi^2}\sum_{n=0}^{\infty}(2n+1)\, \hat{A}_n\, k_n(\hat{\gamma}_0 r)\, k_n(\hat{\gamma}_0 r^S)\, P_n[\cos(\boldsymbol{x},\boldsymbol{x}^S)]. \tag{3.A.50}$$

Inside the sphere, the Green function has to be bounded at $r = 0$; hence, we write this interior constituent as

$$\hat{G}^{int}(\boldsymbol{x}|\boldsymbol{x}^S) = \frac{\hat{\gamma}_0}{2\pi^2}\sum_{n=0}^{\infty}(2n+1)\, \hat{B}_n\, i_n^{(1)}(\hat{\gamma}_{sct} r)\, k_n(\hat{\gamma}_0 r^S)\, P_n[\cos(\boldsymbol{x},\boldsymbol{x}^S)]. \tag{3.A.51}$$

The unknown expansion factors $\hat{A}_n$ and $\hat{B}_n$ follow from the interface conditions at $r = a$ of Eqs. (3.A.45) and (3.A.46). Substitution of the expressions for the Green function into these interface conditions leads, for each n, to the following set of equations,

$$i_n^{(1)}(\hat{\gamma}_0 r) + \hat{A}_n\, k_n(\hat{\gamma}_0 r) = \hat{B}_n\, i_n^{(1)}(\hat{\gamma}_{sct} r), \tag{3.A.52}$$

$$\frac{1}{s\varepsilon_0}\left[\frac{1}{r}\partial_r\{r\, i_n^{(1)}(\hat{\gamma}_0 r)\} + \hat{A}_n \frac{1}{r}\partial_r\{r\, k_n(\hat{\gamma}_0 r)\}\right] = \frac{1}{s\hat{\varepsilon}_{sct}}\left[\hat{B}_n\, \frac{1}{r}\partial_r\{r\, i_n^{(1)}(\hat{\gamma}_{sct} r)\}\right], \tag{3.A.53}$$

when $r = a$. Solving these equations for the coefficients $\hat{A}_n$ and $\hat{B}_n$ leads to

$$\hat{A}_n = -\frac{\hat{Z}_{sct}\; \tilde{\partial} i_n^{(1)}(\hat{\gamma}_{sct} a)\; i_n^{(1)}(\hat{\gamma}_0 a) - Z_0\; \tilde{\partial} i_n^{(1)}(\hat{\gamma}_0 a)\; i_n^{(1)}(\hat{\gamma}_{sct} a)}{\hat{Z}_{sct}\; \tilde{\partial} i_n^{(1)}(\hat{\gamma}_{sct} a)\; k_n(\hat{\gamma}_0 a) - Z_0\; \tilde{\partial} k_n(\hat{\gamma}_0 a)\; i_n^{(1)}(\hat{\gamma}_{sct} a)} \tag{3.A.54}$$

and

$$\hat{B}_m = \frac{i_n^{(1)}(\hat{\gamma}_0 a) + \hat{A}_m\, k_n(\hat{\gamma}_0 a)}{i_n^{(1)}(\hat{\gamma}_{sct} a)}, \tag{3.A.55}$$

where $\tilde{\partial} i_n^{(1)}(z)$ and $\tilde{\partial} k_n(z)$ are not given by Eq. (1.B.49). In fact we redefine them as

$$\tilde{\partial} i_n^{(1)}(z) := \frac{1}{z}\partial_z\{z\, i_n^{(1)}(z)\} \quad \text{and} \quad \tilde{\partial} k_n(z) := \frac{1}{z}\partial_z\{z\, k_n(z)\}. \tag{3.A.56}$$

Carrying out the differentiations with respect to z and using the right-hand equalities of Eq. (1.B.49), we obtain the modified definitions:

$$\begin{aligned} \tilde{\partial} i_n^{(1)}(z) &:= i_{n+1}^{(1)} + \frac{n+1}{z} i_n^{(1)}, \\ \tilde{\partial} k_n(z) &:= -k_{n+1} + \frac{n+1}{z} k_n. \end{aligned} \tag{3.A.57}$$

In numerical computations using Matlab coding, the functions $i_n^{(1)}$ and k_n are not available. We therefore use their expressions in terms of modified Bessel functions of fractional order, viz.,

$$i_n^{(1)}(z) = \sqrt{\frac{1}{2}\pi/z}\, I_{n+\frac{1}{2}}(z) \quad \text{and} \quad k_n(z) = \sqrt{\frac{1}{2}\,\pi/z}\, K_{n+\frac{1}{2}}(z). \tag{3.A.58}$$

3.A.2.2 Electromagnetic Field Strengths

Let us now consider a radially oriented dipole with electric current density $\boldsymbol{J} = \hat{M}\ \delta(\boldsymbol{x} - \boldsymbol{x}^S)\boldsymbol{i}_r$, where $r^S > a$ and $\boldsymbol{i}_r$ is the unit vector in the radial direction. Then the electric vector potential, defined as

$$\begin{aligned} &\hat{\boldsymbol{A}}^{\rm TM}(\boldsymbol{x}|\boldsymbol{x}^S) = \hat{\boldsymbol{A}}^{inc}(\boldsymbol{x} - \boldsymbol{x}^S) + \hat{\boldsymbol{A}}^{rfl}(\boldsymbol{x}|\boldsymbol{x}^S), \quad r > a, \\ &\hat{\boldsymbol{A}}^{\rm TM}(\boldsymbol{x}|\boldsymbol{x}^S) = \hat{\boldsymbol{A}}^{int}(\boldsymbol{x}|\boldsymbol{x}^S), \quad r < a, \end{aligned} \tag{3.A.59}$$

is oriented in the radial direction, where is different constituents are given by

$$\hat{\boldsymbol{A}}^{inc}(\boldsymbol{x}|\boldsymbol{x}^S) = \hat{M}\ \hat{G}^{inc}(\boldsymbol{x}|\boldsymbol{x}^S)\ \frac{r}{r^S}\boldsymbol{i}_r, \quad r < r^S, \tag{3.A.60}$$

while

$$\hat{\boldsymbol{A}}^{rfl}(\boldsymbol{x}|\boldsymbol{x}^S) = \hat{M}\ \hat{G}^{rfl}(\boldsymbol{x}|\boldsymbol{x}^S)\ \frac{r}{r^S}\boldsymbol{i}_r, \quad r > a, \tag{3.A.61}$$

and

$$\hat{\boldsymbol{A}}^{int}(\boldsymbol{x}|\boldsymbol{x}^S) = \hat{M}\ \hat{G}^{int}(\boldsymbol{x}|\boldsymbol{x}^S)\ \frac{r}{r^S}\boldsymbol{i}_r, \quad r < a\,. \tag{3.A.62}$$

In the literature, the Green function $r\hat{G}^{\rm TM}$, for $r \neq r^S$, is denoted as the electric Debye potential. We have added the normalization factor r^S to ensure that at the source position the factor r/r^S is equal to one and the source amplitude has not been changed.

We first show that the pertaining electromagnetic field strengths satisfy the Maxwell equations, the compatibility relations, and the continuity condition at the sphere surface. The magnetic-field vector is obtained from the vector potential $\hat{\boldsymbol{A}}$ as

$$\hat{\boldsymbol{H}} = \boldsymbol{\nabla} \times \hat{\boldsymbol{A}}^{\rm TM} = \hat{M}\boldsymbol{\nabla} \times \hat{G}^{\rm TM}\frac{r}{r^S}\boldsymbol{i}_r, \tag{3.A.63}$$

with the property that the magnetic fields in both homogeneous domains are divergence-free. Then, the magnetic-field components in a spherical coordinate system are

$$\hat{H}_r = 0, \quad \hat{H}_\theta = \hat{M}\frac{1}{r^S\sin(\theta)}\partial_\phi\hat{G}^{\rm TM}, \quad \hat{H}_\phi = -\hat{M}\frac{1}{r^S}\partial_\theta\hat{G}^{\rm TM}. \tag{3.A.64}$$

Note that, at the spherical surface, the discontinuous component of the magnetic-field strength is zero. Since we constructed $\hat{G}^{\rm TM}$ such that it is continuous at $r = a$, we observe that the tangential components of the magnetic field satisfy the continuity condition through the interface at $r = a$. The Maxwell equations in the homogeneous domain $r > a$ and $r \neq r^S$, where $\hat{\varepsilon} = \varepsilon_0$, and in the homogeneous domain $r < a$, where $\hat{\varepsilon} = \hat{\varepsilon}_{sct}$, require that the electric-field vector satisfies

$$\hat{\boldsymbol{E}} = \frac{1}{s\hat{\varepsilon}}\boldsymbol{\nabla} \times \hat{\boldsymbol{H}}, \tag{3.A.65}$$

with the property that, in the two homogeneous and source-free regions, the electric-field strength $\boldsymbol{E}$ is divergence free.

The radial component of the electric-field strength is obtained as

$$\hat{E}_r = \frac{1}{s\hat{\varepsilon}}\left[\frac{1}{r\sin(\theta)}\partial_\theta\{\sin(\theta)H_\phi\} - \frac{1}{r\sin(\theta)}\partial_\phi H_\theta\right], \tag{3.A.66}$$

while the tangential components are given by

$$\hat{E}_\theta = -\frac{1}{s\hat{\varepsilon}}\frac{1}{r}\partial_r(r\hat{H}_\phi), \quad \hat{E}_\phi = \frac{1}{s\hat{\varepsilon}}\frac{1}{r}\partial_r(r\hat{H}_\theta). \tag{3.A.67}$$

Substituting Eq. (3.A.64) into Eq. (3.A.66) and (3.A.67), the radial component of the electric-field strength is obtained from

$$\hat{E}_r = \frac{\hat{M}}{s\hat{\varepsilon}}\frac{1}{rr^S}\left[-\frac{1}{\sin(\theta)}\partial_\theta\{(\sin(\theta)\partial_\theta\hat{G}^{\mathrm{TM}}\} - \frac{1}{\sin^2(\theta)}\partial_\phi^2\hat{G}^{\mathrm{TM}}\right], \tag{3.A.68}$$

while the tangential components are given by

$$\hat{E}_\theta = \frac{\hat{M}}{s\hat{\varepsilon}}\frac{1}{r^S}\frac{1}{r}\partial_r(r\partial_\theta\hat{G}^{\mathrm{TM}}), \quad \hat{E}_\phi = \frac{\hat{M}}{s\hat{\varepsilon}}\frac{1}{r^S\sin(\theta)}\frac{1}{r}\partial_r(r\partial_\phi\hat{G}^{\mathrm{TM}}). \tag{3.A.69}$$

In view of Eq. (3.A.46), we observe that the continuity of the tangential components of the electric-field strength is satisfied as well.

As next step, we substitute the Bessel-function expansions into the electromagnetic wave-field expressions. The expression for $\hat{E}_r$ can be simplified. Using the expression for $r^2\boldsymbol{\nabla}\cdot\boldsymbol{\nabla}$ in spherical coordinates,

$$r^2\boldsymbol{\nabla}\cdot\boldsymbol{\nabla}\hat{G}^{\mathrm{TM}} = \partial_r(r^2\partial_r\hat{G}^{\mathrm{TM}}) + \frac{1}{\sin(\theta)}\partial_\theta\{\sin(\theta)\partial_\theta\hat{G}^{\mathrm{TM}}\} + \frac{1}{\sin^2(\theta)}\partial_\phi^2\hat{G}^{\mathrm{TM}}, \tag{3.A.70}$$

and $\boldsymbol{\nabla}\cdot\boldsymbol{\nabla}\hat{G}^{\mathrm{TM}} = \hat{\gamma}^2\hat{G}^{\mathrm{TM}}$, for $r \neq r^S$, we obtain

$$E_r = \frac{\hat{M}}{s\hat{\varepsilon}}\frac{1}{rr^S}[\partial_r(r^2\partial_r\hat{G}^{\mathrm{TM}}) - (\hat{\gamma}r)^2\hat{G}^{\mathrm{TM}}], \tag{3.A.71}$$

where $\hat{\gamma} = \hat{\gamma}_0$ for $r > a$, and $\hat{\gamma} = \hat{\gamma}_{sct}$ for $r < a$. Before we substitute the expressions for G^{TM} into this expression, we note that the differential equation for modified spherical Bessel functions reads:

$$\partial_r\,\{r^2\partial_r f_n(\hat{\gamma}r)\} - (\hat{\gamma}r)^2 f_n(\hat{\gamma}r) = n(n+1)f_n(\hat{\gamma}r), \tag{3.A.72}$$

where $f_n(z)$ is either $i_n^{(1)}(z)$ or $k_n(z)$.

With this equation, the radial component of the incident electric-field strength in the domain $r < r^S$ follows from Eq. (3.A.49) as

$$\hat{E}_r^{inc} = \frac{Z_0\hat{M}}{\hat{\gamma}_0}\frac{\hat{\gamma}_0}{2\pi^2 rr^S}\sum_{n=1}^{\infty}(2n+1)n(n+1)\,i_n^{(1)}(\hat{\gamma}_0 r)\,k_n(\hat{\gamma}_0 r^S)\,P_n[\cos(\boldsymbol{x},\boldsymbol{x}^S)], \tag{3.A.73}$$

where we have used that $1/s\varepsilon_0 = Z_0/\hat{\gamma}_0$. The tangential components of the incident electric-field strength follow from Eq. (3.A.69) as

$$\hat{E}_\theta^{inc} = \frac{Z_0\hat{M}}{\hat{\gamma}_0}\frac{\hat{\gamma}_0^2}{2\pi^2 r^S}\sum_{n=1}^{\infty}(2n+1)\,\tilde{\partial}i_n^{(1)}(\hat{\gamma}_0 r)\,k_n(\hat{\gamma}_0 r^S)\,\partial_\theta P_n[\cos(\boldsymbol{x},\boldsymbol{x}^S)], \tag{3.A.74}$$

and

$$\hat{E}_\phi^{inc} = \frac{Z_0\hat{M}}{\hat{\gamma}_0}\frac{\hat{\gamma}_0^2}{2\pi^2 r^S}\sum_{n=1}^{\infty}(2n+1)\,\tilde{\partial}i_n^{(1)}(\hat{\gamma}_0 r)\,k_n(\hat{\gamma}_0 r^S)\,\frac{\partial_\phi P_n[\cos(\boldsymbol{x},\boldsymbol{x}^S)]}{\sin(\theta)}, \tag{3.A.75}$$

where $\tilde{\partial} i_n^{(1)}$ is defined in the first relation of Eq. (3.A.57). The nonzero magnetic-field components follow from Eqs. (3.A.63) and (3.A.69) as

$$\hat{H}_\theta^{inc} = \frac{Z_0\hat{M}}{\hat{\gamma}_0}\frac{\hat{\gamma}_0^2}{2\pi^2 r^S}\sum_{n=1}^{\infty}(2n+1)\, i_n^{(1)}(\hat{\gamma}_0 r)\, k_n(\hat{\gamma}_0 r^S)\, \frac{\partial_\phi P_n[\cos(\boldsymbol{x},\boldsymbol{x}^S)]}{\sin(\theta)}, \tag{3.A.76}$$

and

$$\hat{H}_\phi^{inc} = -\frac{Z_0\hat{M}}{\hat{\gamma}_0}\frac{\hat{\gamma}_0^2}{2\pi^2 r^S}\sum_{n=1}^{\infty}(2n+1)\, i_n^{(1)}(\hat{\gamma}_0 r)\, k_n(\hat{\gamma}_0 r^S)\, \partial_\theta P_n[\cos(\boldsymbol{x},\boldsymbol{x}^S)], \tag{3.A.77}$$

Similarly, the reflected electric- and magnetic-field strengths are given by

$$\begin{aligned}
\hat{E}_r^{rfl} &= \frac{Z_0\hat{M}}{\hat{\gamma}_0}\frac{\hat{\gamma}_0}{2\pi r r^S}\sum_{n=1}^{\infty}(2n+1)n(n+1)\, \hat{A}_n\, k_n(\hat{\gamma}_0 r)\, k_n(\hat{\gamma}_0 r^S)\, P_n[\cos(\boldsymbol{x},\boldsymbol{x}^S)],\\
\hat{E}_\theta^{rfl} &= \frac{Z_0\hat{M}}{\hat{\gamma}_0}\frac{\hat{\gamma}_0^2}{2\pi^2 r^S}\sum_{n=1}^{\infty}(2n+1)\, \hat{A}_n\, \tilde{\partial} k_n(\hat{\gamma}_0 r)\, k_n(\hat{\gamma}_0 r^S)\, \partial_\theta P_n[\cos(\boldsymbol{x},\boldsymbol{x}^S)],\\
\hat{E}_\phi^{rfl} &= \frac{Z_0\hat{M}}{\hat{\gamma}_0}\frac{\hat{\gamma}_0^2}{2\pi^2 r^S}\sum_{n=1}^{\infty}(2n+1)\, \hat{A}_n\, \tilde{\partial} k_n(\hat{\gamma}_0 r)\, k_n(\hat{\gamma}_0 r^S)\, \frac{\partial_\phi P_n[\cos(\boldsymbol{x},\boldsymbol{x}^S)]}{\sin(\theta)},\\
\hat{H}_\theta^{rfl} &= \frac{Z_0\hat{M}}{\hat{\gamma}_0}\frac{\hat{\gamma}_0^2}{2\pi^2 r^S}\sum_{n=1}^{\infty}(2n+1)\, \hat{A}_n\, k_n(\hat{\gamma}_0 r)\, k_n(\hat{\gamma}_0 r^S)\, \frac{\partial_\phi P_n[\cos(\boldsymbol{x},\boldsymbol{x}^S)]}{\sin(\theta)},\\
\hat{H}_\phi^{rfl} &= -\frac{Z_0\hat{M}}{\hat{\gamma}_0}\frac{\hat{\gamma}_0^2}{2\pi^2 r^S}\sum_{n=1}^{\infty}(2n+1)\, \hat{A}_n\, k_n(\hat{\gamma}_0 r)\, k_n(\hat{\gamma}_0 r^S)\, \partial_\theta P_n[\cos(\boldsymbol{x},\boldsymbol{x}^S)],
\end{aligned} \tag{3.A.78}$$

where $\tilde{\partial} k_n$ is defined in the second relation of Eq. (3.A.57). The interior field strengths are given by

$$\begin{aligned}
\hat{E}_r^{int} &= \frac{Z_0\hat{M}}{\hat{\gamma}_0}\frac{\hat{\gamma}_0}{2\pi^2 r r^S}\frac{\hat{\varepsilon}_0}{\hat{\varepsilon}_{sct}}\sum_{n=1}^{\infty}(2n+1)n(n+1)\, \hat{B}_n\, i_n^{(1)}(\hat{\gamma}_{sct} r)\, k_n(\hat{\gamma}_0 r^S)\, P_n[\cos(\boldsymbol{x},\boldsymbol{x}^S)],\\
\hat{E}_\theta^{int} &= \frac{Z_0\hat{M}}{\hat{\gamma}_0}\frac{\hat{\gamma}_0\hat{\gamma}_{sct}}{2\pi^2 r^S}\frac{\hat{\varepsilon}_0}{\hat{\varepsilon}_{sct}}\sum_{n=1}^{\infty}(2n+1)\, \hat{B}_n\, \tilde{\partial} i_n^{(1)}(\hat{\gamma}_{sct} r)\, k_n(\hat{\gamma}_0 r^S)\, \partial_\theta P_n[\cos(\boldsymbol{x},\boldsymbol{x}^S)],\\
\hat{E}_\phi^{int} &= \frac{Z_0\hat{M}}{\hat{\gamma}_0}\frac{\hat{\gamma}_0\hat{\gamma}_{sct}}{2\pi^2 r^S}\frac{\hat{\varepsilon}_0}{\hat{\varepsilon}_{sct}}\sum_{n=1}^{\infty}(2n+1)\, \hat{B}_n\, \tilde{\partial} i_n^{(1)}(\hat{\gamma}_{sct} r)\, k_n(\hat{\gamma}_0 r^S)\, \frac{\partial_\phi P_n[\cos(\boldsymbol{x},\boldsymbol{x}^S)]}{\sin(\theta)},\\
\hat{H}_\theta^{int} &= \frac{Z_0\hat{M}}{\hat{\gamma}_0}\frac{\hat{\gamma}_0\hat{\gamma}_{sct}}{2\pi^2 r^S}\frac{\hat{\varepsilon}_0}{\hat{\varepsilon}_{sct}}\sum_{n=1}^{\infty}(2n+1)\, \hat{B}_n\, \tilde{\partial} i_n^{(1)}(\hat{\gamma}_{sct} r)\, k_n(\hat{\gamma}_0 r^S)\, \frac{\partial_\phi P_n[\cos(\boldsymbol{x},\boldsymbol{x}^S)]}{\sin(\theta)},\\
\hat{H}_\phi^{int} &= -\frac{Z_0\hat{M}}{\hat{\gamma}_0}\frac{\hat{\gamma}_0\hat{\gamma}_{sct}}{2\pi^2 r^S}\frac{\hat{\varepsilon}_0}{\hat{\varepsilon}_{sct}}\sum_{n=1}^{\infty}(2n+1)\, \hat{B}_n\, \tilde{\partial} i_n^{(1)}(\hat{\gamma}_{sct} r)\, k_n(\hat{\gamma}_0 r^S)\, \partial_\theta P_n[\cos(\boldsymbol{x},\boldsymbol{x}^S)].
\end{aligned} \tag{3.A.79}$$

The tangential derivatives $\partial_\theta P_n[\cos(\boldsymbol{x},\boldsymbol{x}^S)]$ and $\partial_\phi P_n[\cos(\boldsymbol{x},\boldsymbol{x}^S)]/\sin(\theta)$ are calculated as

$$\begin{aligned}
\partial_\theta P_n[\cos(\boldsymbol{x},\boldsymbol{x}^S)] &= [\partial_\theta \cos(\boldsymbol{x},\boldsymbol{x}^S)]\, \partial P_n[\cos(\boldsymbol{x},\boldsymbol{x}^S)],\\
\partial_\phi P_n[\cos(\boldsymbol{x},\boldsymbol{x}^S)] &= [\partial_\phi \cos(\boldsymbol{x},\boldsymbol{x}^S)]\, \partial P_n[\cos(\boldsymbol{x},\boldsymbol{x}^S)].
\end{aligned} \tag{3.A.80}$$

At this point, it is convenient to use the relation, cf. Eq. (1.A.42),

$$\cos(\boldsymbol{x},\boldsymbol{x}^S) = \cos(\theta)\cos(\theta^S) + \sin(\theta)\sin(\theta^S)\cos(\phi-\phi^S), \tag{3.A.81}$$

and its tangential derivatives given by

$$\partial_\theta \cos(\boldsymbol{x}, \boldsymbol{x}^S) = -\sin(\theta)\cos(\theta^S) + \cos(\theta)\sin(\theta^S)\cos(\phi - \phi^S),$$
$$\frac{\partial_\phi \cos(\boldsymbol{x}, \boldsymbol{x}^S)}{\sin(\theta)} = -\sin(\theta^S)\sin(\phi - \phi^S). \tag{3.A.82}$$

The derivative of the Legendre polynomial with respect to its argument, $\partial P_n(z) = \partial_z P_n(z)$, is obtained from its recurrence relation

$$\partial P_0(z) = 0, \qquad \partial P_n(z) = n\, P_n(z) + z\, \partial P_{n-1}(z), \quad n = 1,\ 2,\ \dots . \tag{3.A.83}$$

To transform the polar electric-field components to Cartesian vectors, we use the transformation:

$$\hat{E}_1 = \sin(\theta)\cos(\phi)\ \hat{E}_r + \cos(\theta)\cos(\phi)\ \hat{E}_\theta - \sin(\phi)\ \hat{E}_\phi,$$
$$\hat{E}_2 = \sin(\theta)\sin(\phi)\ \hat{E}_r + \cos(\theta)\sin(\phi)\ \hat{E}_\theta + \cos(\phi)\ \hat{E}_\phi,$$
$$\hat{E}_3 = \cos(\theta)\hat{E}_r - \sin(\theta)\ \hat{E}_\theta\ , \tag{3.A.84}$$

with similar transformation for the magnetic-field components.

Apart from the amplitude factor $Z_0\hat{M}/\hat{\gamma}_0$, the nonzero reflected electric field and magnetic field at the receiver points $\boldsymbol{x}^R$ are given by

$$\hat{E}_r^{rfl}(\boldsymbol{x}^R|\boldsymbol{x}^S) = \frac{\hat{\gamma}_0}{2\pi^2 r r^S}\sum_{n=1}^{\infty}(2n+1)n(n+1)\,\hat{A}_n\ k_n(\hat{\gamma}_0 r^R)\ k_n(\hat{\gamma}_0 r^S)\ P_n[\cos(\boldsymbol{x}^R, \boldsymbol{x}^S)],$$
$$\hat{E}_\theta^{rfl}(\boldsymbol{x}^R|\boldsymbol{x}^S) = \frac{\hat{\gamma}_0^2}{2\pi^2 r^S}\sum_{n=1}^{\infty}(2n+1)\,\hat{A}_n\ \tilde{\partial k}_n(\hat{\gamma}_0 r^R)\ k_n(\hat{\gamma}_0 r^S)\ \partial_\theta P_n[\cos(\boldsymbol{x}^R, \boldsymbol{x}^S)],$$
$$\hat{E}_\phi^{rfl}(\boldsymbol{x}^R|\boldsymbol{x}^S) = \frac{\hat{\gamma}_0^2}{2\pi^2 r^S}\sum_{n=1}^{\infty}(2n+1)\,\hat{A}_n\ \tilde{\partial k}_n(\hat{\gamma}_0 r^R)\ k_n(\hat{\gamma}_0 r^S)\ \frac{\partial_\phi P_n[\cos(\boldsymbol{x}^R, \boldsymbol{x}^S)]}{\sin(\theta^R)},$$
$$\hat{H}_\theta^{rfl}(\boldsymbol{x}^R|\boldsymbol{x}^S) = \frac{\hat{\gamma}_0^2}{2\pi^2 r^S}\sum_{n=1}^{\infty}(2n+1)\,\hat{A}_n\ k_n(\hat{\gamma}_0 r^R)\ k_n(\hat{\gamma}_0 r^S)\ \frac{\partial_\phi P_n[\cos(\boldsymbol{x}^R, \boldsymbol{x}^S)]}{\sin(\theta^R)},$$
$$\hat{H}_\phi^{rfl}(\boldsymbol{x}^R|\boldsymbol{x}^S) = -\frac{\hat{\gamma}_0^2}{2\pi^2 r^S}\sum_{n=1}^{\infty}(2n+1)\,\hat{A}_n\ k_n(\hat{\gamma}_0 r^R)\ k_n(\hat{\gamma}_0 r^S)\ \partial_\theta P_n[\cos(\boldsymbol{x}^R, \boldsymbol{x}^S)], \tag{3.A.85}$$

is not only valid in the domain $a < r^R < r^S$ but by analytic continuation also in the extended domain $a < r^R < \infty$.

Note that in the polar coordinate system the simple source/receiver reciprocity only holds for the radial field vector components.

3.A.2.3 Matlab Codes for Sphere (3D)

In the analysis of scattering by a sphere, we have introduced spherical coordinates and have expanded the wave fields in a series of modified spherical Bessel functions. To determine the number of terms to take into account in this series, we compare the analytic solution of the incident field, Eqs. (3.43))–(3.45)), with the results of the series solution, Eqs. (3.A.73)–(3.A.77).

The script `EMincidentTest3D` for this comparison is given in Matlab Listing I.134 as follows:

(1) We compute the incident field analytically and we denote it here a as $\hat{E}^{inc}_{1,exact}$, $\hat{E}^{inc}_{2,exact}$, $\hat{E}^{inc}_{3,exact}$, $Z_0\hat{H}^{inc}_{2,exact}$, and $Z_0\hat{H}^{inc}_{3,exact}$. The vertical component of the incident magnetic field is identically zero.

(2) We call `IncEMsphere` to compute the incident field using the Bessel-series solution with 50 terms. Here, we denote the resulting Cartesian field components as $\hat{E}_1^{inc}$, $\hat{E}_2^{inc}$, $\hat{E}_3^{inc}$, $Z_0\hat{H}_2^{inc}$, and $Z_0\hat{H}_3^{inc}$.

(3) Using the Matlab function `plotFieldError` of Matlab Listing I.129, in the cross-section $x_3 = 0$, we plot those field components which are symmetrical with respect to the plane x = 0, viz. $\hat{E}_1^{inc}$, $\hat{E}_2^{inc}$, and $Z_0\hat{H}_3^{inc}$. This is achieved by considering an odd number of discretization points in the x_3 direction. These field components are plotted in the left and middle pictures of Figure 3.8. From the plots of the incident fields, it is difficult to guess the accuracy achieved. Therefore, for each field component f, we compute the absolute difference, Error = $\hat{f}^{inc} - \hat{f}^{inc}_{exact}$, as well (see right-hand side pictures).

Matlab Listing I.134 The `EMincidentTest3D` script to compute the incident wave field.

```
clear all; clc; close all; clear workspace
input = initEM(); gam0 = input.gamma_0; xS = input.xS;

% (1) Cartesian coordinates: field from electric dipole in negative x_1 ---
  X1   = input.X1-xS(1);     X2 = input.X2-xS(2);      X3 = input.X3-xS(3);
  DIS  = sqrt(X1.^2 + X2.^2 + X3.^2);
  X1   = X1 ./ DIS;          X2 = X2 ./ DIS;           X3 = X3 ./ DIS;
   G   = exp(-gam0*DIS) ./ (4*pi*DIS);
   dG  = -(gam0 + 1./DIS) .* G;
  dG11 = ((3*X1.*X1 -1).* (gam0./DIS + 1./DIS.^2) + gam0^2 *X1.*X1) .* G;
  dG21 = ( 3*X2.*X1     .* (gam0./DIS + 1./DIS.^2) + gam0^2 *X2.*X1) .* G;
  dG31 = ( 3*X3.*X1     .* (gam0./DIS + 1./DIS.^2) + gam0^2 *X3.*X1) .* G;

  E{1}  = - (-gam0^2 * G + dG11);
  E{2}  = - dG21;
  E{3}  = - dG31;

  % H1 is zero !
  ZH{2} = -gam0 * X3 .* dG;
  ZH{3} =  gam0 * X2 .* dG;

% (2) Compute the field in terms  of Bessel series ----------------------
  [E1,E2,E3,ZH2,ZH3] = IncEMsphere(input);

% (3) Plot symmetric fields with respect to x3=0 at (x3=0 or x3=dx/2) -----
  N3cross = floor(input.N3/2+1);
  x1 = input.X1(:,1,N3cross);   x2 = input.X2(1,:,N3cross);

  plotFieldError('3D: abs(E_1)', E{1}, E1, x1,x2,input);
  plotFieldError('3D: abs(E_2)', E{2}, E2, x1,x2,input);
  plotFieldError('3D: abs(Z_0H_3)', ZH{3}, ZH3, x1,x2,input);
```

From these error plots, if r tends to r^S we observe that the error increases very strongly. This is visible at the corners of the error pictures. However, the error in the transmitted and the reflected field components are solely determined by the incident field at the boundary of the circular cylinder. Therefore, a more realistic error of scattering at hand is the mean error of the incident-field

Matlab Listing I.135 The `IncEMsphere` script to compute the incident wave field.

```
function [E1,E2,E3,ZH2,ZH3] = IncEMsphere(input)
gam0 = input.gamma_0;

% Transform Cartesian coordinates to spherical coordinates ---------------
  xS   = input.xS;
  rS   = sqrt(xS(1)^2+xS(2)^2+xS(3)^2);
  phiS = atan2(xS(2),xS(1));               thetaS = acos(xS(3)/rS);
  X1   = input.X1;        X2 = input.X2;       X3 = input.X3;
  R    = sqrt(X1.^2+X2.^2+X3.^2+1e-16);
  PHI  = atan2(X2,X1);                      THETA  = acos(X3./R);

% and compute incident wave as Bessel series with 0:N terms ---------------
  N = 50;                                 % increase N for more accuracy
  COS     =  cos(THETA)*cos(thetaS)+sin(THETA)*sin(thetaS).*cos(PHI-phiS);
  dTHETA = -sin(THETA)*cos(thetaS)+cos(THETA)*sin(thetaS).*cos(PHI-phiS);
  dPHI_sin=                                 - sin(thetaS).*sin(PHI-phiS);
  Pn   = zeros(size(R));   Pn_1 = Pn;  Pn_2 = Pn;   dPn = Pn;
  Er   = zeros(size(R));   Etheta = zeros(size(R));  Ephi = zeros(size(R));
                           ZHtheta = zeros(size(R)); ZHphi = zeros(size(R));
  [Pn,Pn_1,Pn_2] = Legendre(0,COS,Pn,Pn_1,Pn_2);
  for n = 1 : N
    dPn = n*Pn + COS.*dPn; [Pn,Pn_1,Pn_2] = Legendre(n,COS,Pn,Pn_1,Pn_2);
    arg = gam0*R;  fctr=sqrt(pi/2./arg); Ib0 = fctr.* besseli(n+1/2,arg);
          dIb0 = fctr.*besseli(n+3/2,arg) + (n+1)./arg.*Ib0;
    arg = gam0*rS; fctr=sqrt(pi/2./arg); KbS = fctr * besselk(n+1/2,arg);
    Er      = Er + n*(n+1)*(2*n+1)* Ib0.* KbS              .*  Pn;
    Etheta  = Etheta      +(2*n+1)*dIb0.* KbS .* dTHETA    .* dPn;
    Ephi    = Ephi        +(2*n+1)*dIb0.* KbS .* dPHI_sin  .* dPn;
   ZHtheta  = ZHtheta     +(2*n+1)* Ib0.* KbS .* dPHI_sin  .* dPn;
   ZHphi    = ZHphi       -(2*n+1)* Ib0.* KbS .* dTHETA    .* dPn;
  end % n_loop
   Er      = gam0   * Er      ./ (2*pi^2 .*R*rS);
   Etheta  = gam0^2 * Etheta  ./ (2*pi^2     *rS);
   Ephi    = gam0^2 * Ephi    ./ (2*pi^2     *rS);
% ZHr      = 0    for TM case
  ZHtheta  = gam0^2 * ZHtheta ./ (2*pi^2     *rS);
  ZHphi    = gam0^2 * ZHphi   ./ (2*pi^2     *rS);

% Transform to Cartesian vectors
 E1 = sin(THETA).*cos(PHI).*Er+cos(THETA).*cos(PHI).*Etheta-sin(PHI).*Ephi;
 E2 = sin(THETA).*sin(PHI).*Er+cos(THETA).*sin(PHI).*Etheta+cos(PHI).*Ephi;
 E3 = cos(THETA)            .*Er-sin(THETA)            .*Etheta;
 % ZH1= 0;
 ZH2 = cos(THETA).*sin(PHI).*ZHtheta + cos(PHI).*ZHphi;
 ZH3 = -sin(THETA)           .*ZHtheta;
```

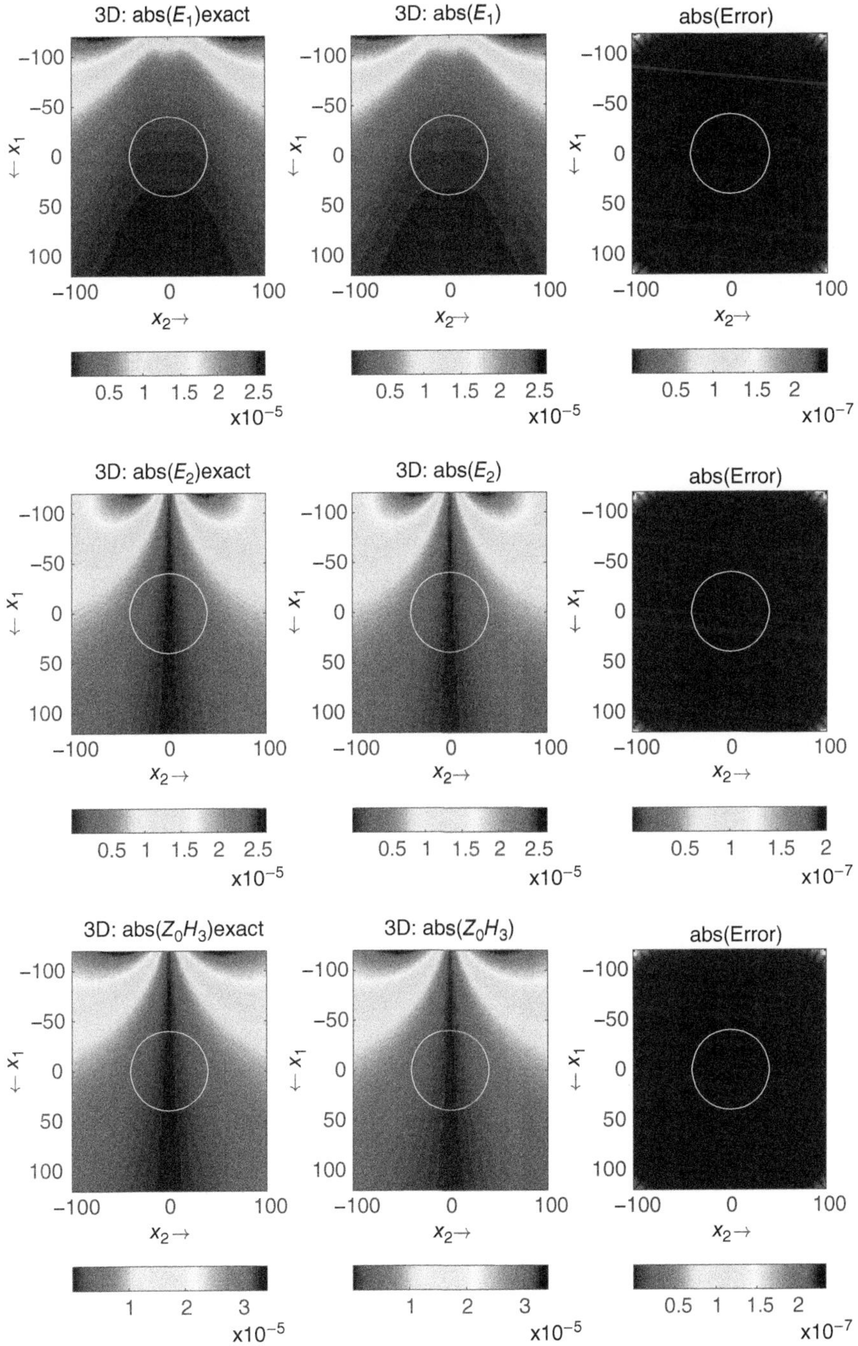

Figure 3.A.8 Comparison of the 3D incident-wave-field results.

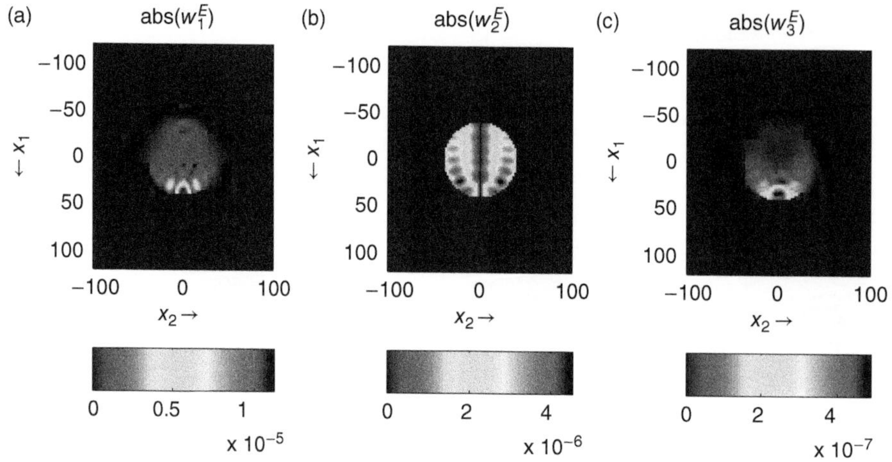

Figure 3.A.9 The three components of the contrast sources at the plane $x_3 = 0$.

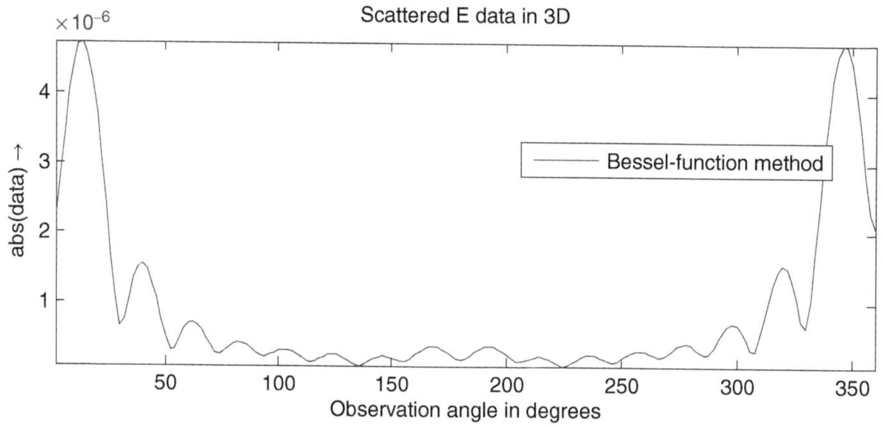

Figure 3.A.10 The scattered electric wave-field results at the receivers.

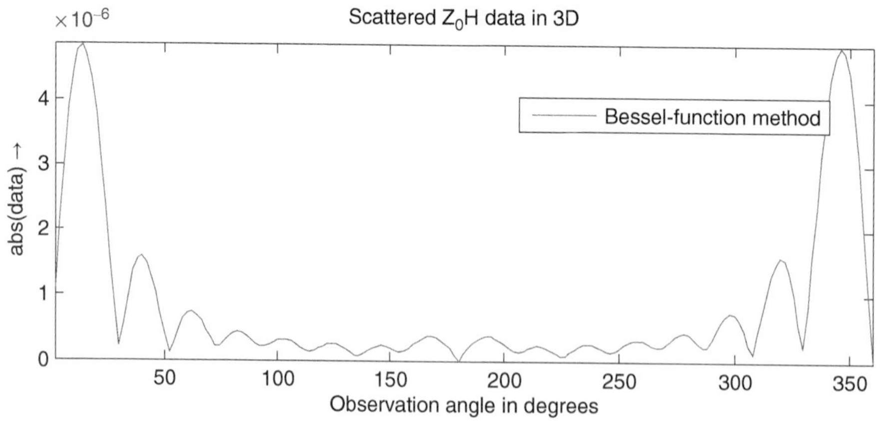

Figure 3.A.11 The scattered magnetic wave-field results at the receivers.

Matlab Listing I.136 The `EMincidentTest3Derror` script to compute the incident wave field.

```
clear all; clc; close all; clear workspace
input = initEM(); gam0 = input.gamma_0; xS = input.xS;
   a = input.a; theta = 0:.01:pi;      phi = 0:.01:2*pi;
   [THETA,PHI] = ndgrid(theta,phi);
% (1) Cartesian coordinates: field from electric dipole in negative X_1 --
   X1   = a*sin(THETA).*cos(PHI)-xS(1); X2 = a*sin(THETA).*sin(PHI)-xS(2);
   X3   = a*cos(THETA)-xS(3);
  DIS   = sqrt(X1.^2 + X2.^2 + X3.^2);
   X1   = X1 ./ DIS;         X2 = X2 ./ DIS;          X3 = X3 ./ DIS;
    G   = exp(-gam0*DIS) ./ (4*pi*DIS);
   dG   = -(gam0 + 1./DIS) .* G;
  dG11 = ((3*X1.*X1 -1).* (gam0./DIS + 1./DIS.^2) + gam0^2 *X1.*X1) .* G;
  dG21 = ( 3*X2.*X1     .* (gam0./DIS + 1./DIS.^2) + gam0^2 *X2.*X1) .* G;
  dG31 = ( 3*X3.*X1     .* (gam0./DIS + 1./DIS.^2) + gam0^2 *X3.*X1) .* G;
 E{1}  = - (-gam0^2 * G + dG11);
 E{2}  = - dG21;
 E{3}  = - dG31;
 ZH{2} = -gam0 * X3 .* dG;
 ZH{3} =  gam0 * X2 .* dG;

[E1,E2,E3,ZH2,ZH3] = IncEMsphereError(input,THETA,PHI);

 % Print the normalized error on the sphere boundary at r = a
 norm_Error = norm(E{1}(:)-E1(:),1) / norm(E{1}(:),1);
              disp(['Error ',' E_1',' = ',num2str(norm_Error)]);
 norm_Error = norm(E{2}(:)-E2(:),1) / norm(E{2}(:),1);
              disp(['Error ',' E_2',' = ',num2str(norm_Error)]);
 norm_Error = norm(E{3}(:)-E3(:),1) / norm(E{3}(:),1);
              disp(['Error ',' E_3',' = ',num2str(norm_Error)]);

 norm_Error = norm(ZH{2}(:)-ZH2(:),1) / norm(ZH{2}(:),1);
              disp(['Error ','ZH_2',' = ',num2str(norm_Error)]);
 norm_Error = norm(ZH{3}(:)-ZH3(:),1) / norm(ZH{3}(:),1);
              disp(['Error ','ZH_3',' = ',num2str(norm_Error)]);
```

approximation over the circle at $r = a$, viz.,

$$\langle \text{Error} \rangle = \frac{\sum_{\theta_n \in [0,pi];\phi_n \in [0,2\pi]} |\hat{f}^{inc}(r = a, \theta_n, \phi_n) - \hat{f}^{inc}_{exact}(r = a, \theta_n, \phi_n)|}{\sum_{\theta_n \in [0,pi];\phi_n \in [0,2\pi]} |\hat{f}^{inc}_{exact}(r = a, \theta_n, \phi_n)|}. \tag{3.A.86}$$

This error is calculated with the help of the script `EMincidentTest3Derror` of Matlab Listing I.136. For $N = 20$, the mean error at $r = a$ for the symmetrical field components $\hat{E}_1$, $\hat{E}_2$, and $Z_0\hat{H}_3$ amounts to 2.7×10^{-6}, 1.0×10^{-6}, and 0.4×10^{-6}, respectively. For our scattering problem at hand, this is sufficiently small to consider that only 20 terms in the series of Bessel functions are enough to benchmark the numerical solution of the integral equation.

Matlab Listing I.137 The `EmintSphere` script to compute the contrast sources in the interior of the sphere.

```
clear all; clc; close all; clear workspace
input = initEM();    a = input.a;    gam0 = input.gamma_0;  xS = input.xS;
eps_sct = input.eps_sct;                   mu_sct = input.mu_sct;
gam_sct = gam0 * sqrt(eps_sct*mu_sct); Z_sct  = sqrt(mu_sct/eps_sct);

% (1) Transform Cartesian coordinates to polar ccordinates ----------------
  rS   = sqrt(xS(1)^2+xS(2)^2+xS(3)^2);
  phiS = atan2(xS(2),xS(1));               thetaS  = acos(xS(3)/rS);
  X1   = input.X1;        X2 = input.X2;       X3 = input.X3;
  R    = sqrt(X1.^2+X2.^2+X3.^2+1e-16);
  PHI  = atan2(X2,X1);                    THETA   = acos(X3./R);
% (2) Compute coefficients of series expansion ---------------------------
  N = 20;                                  % increase N for more accuracy
  A = zeros(1,N);  B = zeros(1,N);
  arg0 = gam0 * input.a;  args = gam_sct * input.a;
  for n = 1 : N
    Ib0 = besseli(n+1/2,arg0);  dIb0 = besseli(n+3/2,arg0)+(n+1)/arg0*Ib0;
    Ibs = besseli(n+1/2,args);  dIbs = besseli(n+3/2,args)+(n+1)/args*Ibs;
    Kb0 = besselk(n+1/2,arg0);  dKb0 =-besselk(n+3/2,arg0)+(n+1)/arg0*Kb0;
    A(n) = -(Z_sct *dIbs*Ib0 - dIb0*Ibs) / (Z_sct *dIbs*Kb0 - dKb0*Ibs);
  % factor sqrt(pi^2/4/argo/args) in numerator and denominator is omitted
    B(n) = sqrt(args/arg0) * (Ib0 + A(n)*Kb0) / Ibs;
  end
% (3) Compute interior wave field using N terms of Bessel series expansion
  COS      =  cos(THETA)*cos(thetaS)+sin(THETA)*sin(thetaS).*cos(PHI-phiS);
  dTHETA   = -sin(THETA)*cos(thetaS)+cos(THETA)*sin(thetaS).*cos(PHI-phiS);
  dPHI_sin=                                  - sin(thetaS).*sin(PHI-phiS);
  Pn  = zeros(size(R));  Pn_1 = Pn;  Pn_2 = Pn;    dPn  = Pn;
  Er  = zeros(size(R));  Etheta = zeros(size(R));  Ephi = zeros(size(R));
  [Pn,Pn_1,Pn_2] = Legendre(0,COS,Pn,Pn_1,Pn_2);
  for n = 1 : N
    dPn = n*Pn + COS.*dPn; [Pn,Pn_1,Pn_2] = Legendre(n,COS,Pn,Pn_1,Pn_2);
    arg = gam_sct*R; fctr =sqrt(pi/2./arg); Ibs = fctr.*besseli(n+1/2,arg);
                     dIbs = fctr .* besseli(n+3/2,arg) + (n+1)./arg.*Ibs;
    arg = gam0*rS;   fctr =sqrt(pi/2 /arg); KbS = fctr *besselk(n+1/2,arg);
    Er      = Er     + B(n) *n*(n+1)*(2*n+1).* Ibs .* KbS           .* Pn;
    Etheta = Etheta + B(n) *          (2*n+1).*dIbs .* KbS .* dTHETA   .*dPn;
    Ephi    = Ephi   + B(n) *          (2*n+1).*dIbs .* KbS .* dPHI_sin.*dPn;
  end % n_loop
  Er      = (gam0        /eps_sct) * Er     ./ (2*pi^2 .*R*rS);
  Etheta = (gam0*gam_sct/eps_sct) * Etheta ./ (2*pi^2    *rS);
  Ephi    = (gam0*gam_sct/eps_sct) * Ephi   ./ (2*pi^2    *rS);
% (4) Transform to Cartesian vectors ---------------------------------------
E1 = sin(THETA).*cos(PHI).*Er+cos(THETA).*cos(PHI).*Etheta-sin(PHI).*Ephi;
E2 = sin(THETA).*sin(PHI).*Er+cos(THETA).*sin(PHI).*Etheta+cos(PHI).*Ephi;
E3 = cos(THETA)          .*Er-sin(THETA)          .*Etheta                ;
w_E{1} = input.CHI_eps .* E1;
w_E{2} = input.CHI_eps .* E2;
w_E{3} = input.CHI_eps .* E3;
plotContrastSourcewE(w_E,input);  save CONTRASTSOURCE w_E;
```

Matlab Listing I.138 The `EmsctSphere` script to compute the scattered field.

```
clear all; clc; close all;  input = initEM();  a = input.a;
   gam0 = input.gamma_0;  eps_sct = input.eps_sct;  mu_sct = input.mu_sct;
gam_sct = gam0 * sqrt(eps_sct*mu_sct);        Z_sct  = sqrt(mu_sct/eps_sct);
if exist(fullfile(cd, 'EDATA3D.mat'), 'file');    delete EDATA3D.mat;  end
if exist(fullfile(cd, 'HDATA3D.mat'), 'file');    delete HDATA3D.mat;  end
% (1) Transform Cartesian coordinates to polar coordinates ----------------
  xS   = input.xS;               rS = sqrt(xS(1)^2+xS(2)^2+xS(3)^2);
  phiS = atan2(xS(2),xS(1));          thtS = acos(xS(3)/rS);
  xR   = input.xR;               rR = sqrt(xR(1,:).^2+xR(2,:).^2+xR(3,:).^2);
  phiR = atan2(xR(2,:),xR(1,:));   thtR = acos(xR(3,:)./rR(1,:));
% (2) Compute Bessel series expansion at the receiver points ---------------
  N = 20;                                   % increase N for more accuracy
  A = zeros(1,N);  arg0 = gam0 * input.a;  args = gam_sct * input.a;
  for n = 1 : N
    Ib0 = besseli(n+1/2,arg0);  dIb0 = besseli(n+3/2,arg0)+(n+1)/arg0*Ib0;
    Ibs = besseli(n+1/2,args);  dIbs = besseli(n+3/2,args)+(n+1)/args*Ibs;
    Kb0 = besselk(n+1/2,arg0);  dKb0 =-besselk(n+3/2,arg0)+(n+1)/arg0*Kb0;
    A(n) = -(Z_sct *dIbs*Ib0 - dIb0*Ibs) / (Z_sct *dIbs*Kb0 - dKb0*Ibs);
  end % factor sqrt(pi^2/4/argo/args) in numerator/denominator is omitted
  COS   = cos(thtR)*cos(thtS)+sin(thtR)*sin(thtS).*cos(phiR-phiS);
 dthtR =-sin(thtR)*cos(thtS)+cos(thtR)*sin(thtS).*cos(phiR-phiS);
 dphiR_sin=                          - sin(thtS).*sin(phiR-phiS);
  Pn   = zeros(size(rR));  Pn_1 = Pn;   Pn_2 = Pn;    dPn = Pn;
  Er   = zeros(size(rR));  Etht = Er;   Ephi = Er;  ZHtht = Er; ZHphi = Er;
  [Pn,Pn_1,Pn_2] = Legendre(0,COS,Pn,Pn_1,Pn_2);
  for n = 1 : N
    dPn = n*Pn + COS.*dPn; [Pn,Pn_1,Pn_2] = Legendre(n,COS,Pn,Pn_1,Pn_2);
    arg = gam0*rR; fctr = sqrt(pi/2./arg); Kb0 = fctr.*besselk(n+1/2,arg);
                     dKb0 = - fctr .* besselk(n+3/2,arg) + (n+1)./arg.*Kb0;
    arg = gam0*rS; fctr = sqrt(pi/2 /arg); KbS = fctr *besselk(n+1/2,arg);
    Er    = Er    + A(n)*n*(n+1)*(2*n+1) * Kb0 .* KbS             .* Pn;
    Etht  = Etht  + A(n)         *(2*n+1) *dKb0 .* KbS .* dthtR     .*dPn;
    Ephi  = Ephi  + A(n)         *(2*n+1) *dKb0 .* KbS .* dphiR_sin.*dPn;
   ZHtht = ZHtht + A(n)         *(2*n+1) * Kb0 .* KbS .* dphiR_sin.*dPn;
   ZHphi = ZHphi - A(n)         *(2*n+1) * Ib0 .* KbS .* dthtR     .*dPn;
  end % n_loop
   Er   = gam0   * Er    ./ (2*pi^2 .*rR*rS);
   Etht = gam0^2 * Etht  ./ (2*pi^2       *rS);
   Ephi = gam0^2 * Ephi  ./ (2*pi^2       *rS);
  ZHtht = gam0^2 * ZHtht ./ (2*pi^2       *rS);
  ZHphi = gam0^2 * ZHtht ./ (2*pi^2       *rS);
 E1 = sin(thtR).*cos(phiR).*Er+cos(thtR).*cos(phiR).*Etht-sin(phiR).*Ephi;
 E2 = sin(thtR).*sin(phiR).*Er+cos(thtR).*sin(phiR).*Etht+cos(phiR).*Ephi;
 E3 =              cos(thtR).*Er            - sin(thtR).*Etht;
ZH1 = cos(thtR).*cos(phiR).*ZHtht - sin(phiR).*ZHphi;
ZH2 = cos(thtR).*sin(phiR).*ZHtht + cos(phiR).*ZHphi;
ZH3 =           -sin(thtR).*ZHtht;
  Edata3D = sqrt(abs(E1).^2 + abs(E2).^2 + abs(E3).^2);
  Hdata3D = sqrt(abs(ZH1).^2 + abs(ZH2).^2 + abs(ZH3).^2);
  displayEdata(Edata3D,input);             save EDATA3D Edata3D;
  displayHdata(Hdata3D,input);             save HDATA3D Hdata3D;
```

In the next script of Matlab Listing I.137, the interior wave field is computed. The following steps are taken:

(1) We transform the Cartesian coordinates to the polar coordinates.
(2) We compute the coefficients A and B.
(3) We compute the interior electric field using Eq. (3.A.79).
(4) We transform this polar electric-field vectors to Cartesian vectors using Eq. (3.A.84).
(5) We compute the contrast sources $\hat{w}_1^E = \hat{\chi}^\varepsilon \hat{E}_1$ and $\hat{w}_2^E = \hat{\chi}^\varepsilon \hat{E}_2$ and plot these contrast sources with the Matlab function `plotContrastSourcewE` of Matlab Listing I.90

Finally, using the Matlab function `EMsctSphere` of Matlab Listing I.138, the scattered electric- and magnetic-field components at the receivers are computed. Their absolute values are plotted using the Matlab functions `displayEdata` and `displayHdata` of Matlab Listing I.93 and I.94, respectively. The resulting contrast sources at the cross-sectional plane $x_3 = 0$ are presented in Figure 3.A.9. Note that at this plane the component $\hat{w}_3^E$ vanishes.

The amplitudes of the scattered electric and magnetic wave-fields are shown in Figures 3.A.8. and 3.A.9. For benchmarking needs, the scattered field data are stored in the MAT-files EDATA3D.mat and HDATA3D.mat (Figures 3.A.10 and 3.A.11).

3.A.3 Scattered-Field Computations Canonical Objects

After completion of the description of the various Matlab scripts to compute the detailed behavior of the scattering by the canonical objects, we finally present a combined script to compute the scattered field at a number of receivers points, see Matlab Listing I.139. This latter script should be used, before we run a particular Matlab script for solving an integral equation. The resulting field values are used as a benchmark for the wave fields computed with the integral equation methods.

Matlab Listing I.139 The `EMForwardCanonicalObjects` script to compute the scattered wave fields for canonical objects.

```
clear all; clc; close all; clear workspace
input = initEM();
global nDIM;

if nDIM == 2
   % Compute scattered acoustic field at receivers around circular cylinder
   disp('Running EMsctCircle'); EMsctCircle;

elseif nDIM == 3
   % Compute scattered acoustic field at receivers around sphere
   disp('Running EMsctSphere'); EMsctSphere;

end % if
```

Part II

Inverse Scattering Problem

The inverse scattering problem consists of determination of the material contrast χ from a knowledge of the incident fields in the object domain $\mathbb{D}$ and the scattered fields in the measurements domain $\mathbb{M}$. In general, this problem is both nonlinear and ill-posed. If the scattered field data are the true scattered field values for an object, then there is a solution to the inverse problem. Whether there is only one solution is a subject of ongoing research. In $\mathbb{R}^3$, it is known that if the scattered field is measured exactly on $\mathbb{M}$ for every incident plane wave, then there is only one χ that will generate these scattered fields; that is, there is a unique solution of the inverse problem (Colton and Kress [12]). No results are available if the scattered fields are known on $\mathbb{M}$ for only a finite number of incident directions. Moreover, in $\mathbb{R}^2$, the uniqueness property has not been established even if the scattered field is known for all incident directions.

Despite the lack of rigorous uniqueness results, the inverse problem has been cast into an optimization problem with some success. Specifically, we recast the inverse scattering problem into an optimization problem of finding χ to minimize the error in the data on the measurements domain $\mathbb{M}$, subject to the constraint that the object equation in $\mathbb{D}$ is satisfied in some sense. The existence of a minimizer can be guaranteed by a suitable choice of the class of admissible values of χ. However, whether this minimizer is related to a local or global minimum remains an ongoing concern.

Tests of the inversion procedure are often made with synthetic data using the same model for computing both predicted and estimated measurements. These tests lead to unreal and too optimistic results. Colton and Kress [12, pp. 121 and 289] coined it as "*the inverse crime*". In addition to warning against committing an inverse crime, they emphasize that it is crucial that the synthetic data are obtained by a forward solver that is unrelated to the inverse solver. In practice, the test of an inversion process avoiding the inverse crime could be done using a model for the numerically produced data and a different one to invert the data. In this book, we are in the ideal situation that our canonical objects are modeled both analytically and numerically, but in a completely different way. In addition, the configurations are different, viz., an object with smooth boundaries in the analytical model and staircase-shaped boundaries in the numerical world.

In this book, we restrict ourselves to the reconstruction of a single contrast function. For scalar waves, it is the wave-speed contrast; for acoustic waves, it is the mass-density contrast; for electromagnetic waves, it is the electrical permittivity.

Forward and Inverse Scattering Algorithms based on Contrast Source Integral Equations, First Edition. Peter M. van den Berg.

Companion website: www.wiley.com/go/vandenBerg/ScatteringAlgorithms

4

Scalar Wave Inversion

The problem of reconstructing the wave-speed contrast of a bounded object immersed in a known background medium, from a knowledge of how the object scatters known incident radiation, has received a tremendous amount of attention in the past decades, see e.g. Pike and Sabatier [38]. Almost all reconstruction algorithms rely in some way upon the domain integral equation for the wave field (Lippmann–Schwinger equation) inside the scattering object as well as the related integral equation for the wave field outside the object. In Part I of this book, we have shown that the contrast-source integral equations have many advantages for computation of the scattered wave fields for known wave-speed contrast. Therefore, we discuss the use of these types of integral equation methods as an efficient technique to solve the nonlinear inverse scattering problem.

4.1 Notation

Let $\mathbb{D}_{sct}$ denote a bounded, not necessarily connected, scattering object whose location and wave-speed contrast $\hat{\chi}^c(\boldsymbol{x}) = 1 - (c_0/\hat{c}_{sct}(\boldsymbol{x}))^2$, see Eq. (1.29), are unknown, but are known to lie within another, larger, bounded simply connected domain $\mathbb{D}$. Observe that if $\boldsymbol{x}$ is not in $\mathbb{D}_{sct}$ then $\hat{\chi}^c$ vanishes, but if the location of $\mathbb{D}_{sct}$ is unknown then it is not known *a priori* where $\hat{\chi}^c$ vanishes. However, with the assumption that $\mathbb{D}_{sct} \subset \mathbb{D}$, it is known that $\hat{\chi}^c$ vanishes for $\boldsymbol{x}$ outside $\mathbb{D}$. Let $\mathbb{M}$ denote the domain with a discrete collection of receiver points outside of $\mathbb{D}$, then the scattered wave field is defined as $\hat{u}_{\mathbb{M}}^{sct}$, see Figure 4.1. As model for this field, we use the contrast-source integral representation, cf. Eq. (1.36),

$$\hat{u}_{\mathbb{M}}^{sct}(\boldsymbol{x}_p^R, \boldsymbol{x}_q^S) := \hat{u}^{sct}(\boldsymbol{x}_p^R, \boldsymbol{x}_q^S) = \int_{\boldsymbol{x}' \in \mathbb{D}} \hat{\gamma}_0^2 \, \hat{G}(\boldsymbol{x}_p^R - \boldsymbol{x}') \, \hat{w}(\boldsymbol{x}', \boldsymbol{x}_q^S) \, \mathrm{d}V, \quad (\boldsymbol{x}_p^R, \boldsymbol{x}_q^S) \in \mathbb{M} . \tag{4.1}$$

Here, the set of $\boldsymbol{x}_p^R$, $p = 1, 2, \ldots$, denotes the collection of receiver points in the domain $\mathbb{M}$ and the set of $\boldsymbol{x}_q^S$, $q = 1, 2, \ldots$, denotes the collection of source points in the domain $\mathbb{M}$. The contrast source is given by

$$\hat{w}(\boldsymbol{x}, \boldsymbol{x}_q^S) = \hat{\chi}^c(\boldsymbol{x}) \, \hat{u}(\boldsymbol{x}, \boldsymbol{x}_q^S), \quad \boldsymbol{x} \in \mathbb{D}, \quad \boldsymbol{x}_q^S \in \mathbb{M}. \tag{4.2}$$

Forward and Inverse Scattering Algorithms based on Contrast Source Integral Equations, First Edition. Peter M. van den Berg.
© 2021 John Wiley & Sons, Inc. Published 2021 by John Wiley & Sons, Inc.
Companion website: www.wiley.com/go/vandenBerg/ScatteringAlgorithms

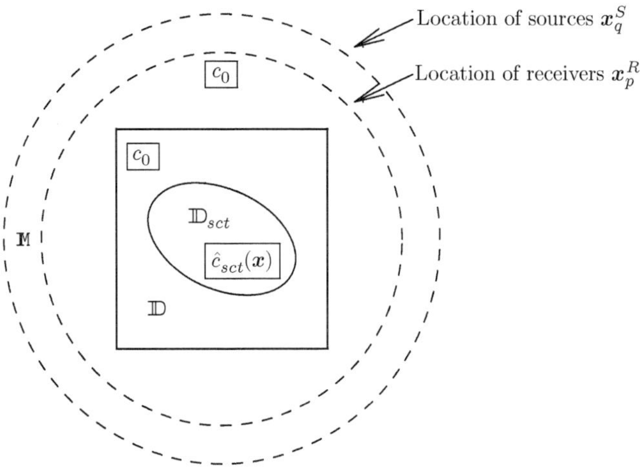

Figure 4.1 The inverse scattering configuration with the measurement domain $\mathbb{M}$ and the object domain $\mathbb{D}$. The measurement domain $\mathbb{M} \not\subset \mathbb{D}$. The scattering domain $\mathbb{D}_{sct} \subset \mathbb{D}$. In our numerical examples, the sources are distributed over the dashed outer circle and the receivers are distributed over the dashed inner circle.

Eq. (4.1) holds only if there is no noise and no error in the measurements. However, error-free data is extremely unlikely, and we do not assume that this equation holds exactly.

If $\hat{u}^{inc}(\boldsymbol{x}, \boldsymbol{x}_q^S)$ denotes an incident wave with propagation coefficient $\hat{\gamma}_0$ and source point $\boldsymbol{x}_q^S$, then the total wave field in $\mathbb{D}$ satisfies the contrast-source integral equation, cf. Eq. (1.37),

$$\hat{\chi}^c(\boldsymbol{x})\ \hat{u}^{inc}(\boldsymbol{x}, \boldsymbol{x}_q^S) = \hat{w}(\boldsymbol{x}, \boldsymbol{x}^S) - \hat{\chi}^c(\boldsymbol{x}) \int_{\boldsymbol{x}' \in \mathbb{D}} \hat{\gamma}_0^2\ \hat{G}(\boldsymbol{x} - \boldsymbol{x}')\ \hat{w}(\boldsymbol{x}', \boldsymbol{x}_q^S) \mathrm{d}V\ , \tag{4.3}$$

for $\boldsymbol{x} \in \mathbb{D}$ and $\boldsymbol{x}_q^S \in \mathbb{M}$. Rewriting Eqs. (4.1) and (4.3) in symbolic form, we have the data equations

$$\boxed{\hat{u}_{\mathbb{M}}^{sct}(\boldsymbol{x}_p^R, \boldsymbol{x}_q^S) = \hat{G}_{\mathbb{M}} \hat{w}\ , \qquad (\boldsymbol{x}_p^R, \boldsymbol{x}_q^S) \in \mathbb{M}\ ,} \tag{4.4}$$

with

$$\boxed{\hat{G}_{\mathbb{M}} \hat{w} = \int_{\boldsymbol{x}' \in \mathbb{D}} \hat{\gamma}_0^2\ \hat{G}(\boldsymbol{x}_p^R - \boldsymbol{x}')\ \hat{w}(\boldsymbol{x}', \boldsymbol{x}_q^S)\ \mathrm{d}V\ ,} \tag{4.5}$$

and the object equations (or state equations)

$$\boxed{\hat{\chi}^c(\boldsymbol{x})\ \hat{u}^{inc}(\boldsymbol{x}, \boldsymbol{x}_q^S) = \hat{w}(\boldsymbol{x}, \boldsymbol{x}_q^S) - \hat{\chi}^c(\boldsymbol{x})\ \hat{G}_{\mathbb{D}} \hat{w}\ , \quad \boldsymbol{x} \in \mathbb{D}\ , \quad \boldsymbol{x}_q^S \in \mathbb{M}\ ,} \tag{4.6}$$

with

$$\boxed{\hat{G}_{\mathbb{D}} \hat{w} = \int_{\boldsymbol{x}' \in \mathbb{D}} \hat{\gamma}_0^2\ \hat{G}(\boldsymbol{x} - \boldsymbol{x}')\ \hat{w}(\boldsymbol{x}', \boldsymbol{x}_q^S)\ \mathrm{d}V\ .} \tag{4.7}$$

The subscripts $\mathbb{M}$ and $\mathbb{D}$ on the operators are added to accentuate the location of the point $\boldsymbol{x}$, either in the measurement domain $\mathbb{M}$ or in the object domain $\mathbb{D}$, because the operators are identical in all other respects.

4.2 Synthetic Data

In this section, we discuss the computation of synthetic data, which will replace the measured data for our inversion examples. In this part of the book, we only present 2D computer algorithms. Although our theory and numerical codes apply to any 2D object, we only consider the circular cylinder. The forward model has been discussed in Chapter 1. Since we do not want to take any advantage of the symmetry of this circular cylinder, we consider the scattering by an off-centered circular cylinder. We first compare the two scattered-field data sets, obtained either analytically or numerically.

The wave-field parameters are initialized by the `init` function (Matlab Listing II.1). The global parameter `nDIM` assigns the 2D space to be used in our inversion codes. Subsequently, the wave

Matlab Listing II.1 The function `init` to initialize the 2D wave-field parameters.

```
function input = init()
% Time factor = exp(-iwt); Spatial units is in m; Source wavelet  Q = 1
  global nDIM;  nDIM = 2;                          % dimension of space

  input.c_0      = 1500;           % wave speed in embedding

  input.c_sct    = 2000;           % wave speed in scatterer (CHI =   0.4375)
% input.c_sct    = 1200;           % wave speed in scatterer (CHI = - 0.5625)

  f              = 50;                             % temporal frequency
  wavelength     = input.c_0 / f;                  % wavelength
  s              = 1e-16 - 1i*2*pi*f;              % LaPlace parameter
  input.gamma_0 = s/input.c_0;                     % propagation coefficient

  disp(['wavelength = ' num2str(wavelength)]);

% add input data to structure array 'input'
  input = initSourceReceiver(input);   % add location of source/receiver

  input = initGrid(input);             % add grid in 2D

  input = initFFTGreen(input);         % compute FFT of Green function

  input = initContrast(input);         % add contrast distribution

  input.Errcri = 1e-5;

  input.Noise = 0;
  if input.Noise == 1  % generate random numbers that are repeatable
    sigma  = 0.2;          % sigma = standard deviation and mu =0
    randn('state',1);    ReRand = randn(input.NR,input.NS) * sigma;
    randn('state',2);    ImRand = randn(input.NR,input.NS) * sigma;
    input.Rand = ReRand + 1i*ImRand;
  end
```

speed c_0 of the embedding is 1500 m/s and the constant wave speed of the scattering object is taken either $\hat{c}_{sct} = 2000$ m/s or $\hat{c}_{sct} = 1200$ m/s. The frequency of operation is 50 Hz. At this frequency, we have taken these specific values of $\hat{c}_{sct}$, because the associated contrast values $\hat{\chi}^c = 0.4375$ and $\hat{\chi}^c = -0.5625$ are about the maximum and minimum values for which our inversion method results in a convergent process. For larger contrast amplitudes and/or larger frequencies, one should use multifrequency data, see e.g. Bloemenkamp et al. [6]. But here we focus on the single-frequency analysis.

As output, the `init` function displays the operating wavelength. This facilitates the user to check the value $\lambda/\Delta x$. Although in the forward case often a discretization of 15 samples per wavelength is used, leading to an error of about 1 %, in the inverse case we choose a discretization of 10 samples of wavelength. In this way, the numerical data differs substantially from the analytical data. Hence, this difference may be seen as model noise. In any case, this error is within the range of 5–10%. The `init` function also calls the functions `initSourceReceiver`, `initGrid`, `initFFTGreen`, and `initContrast`, but now only in 2D. They initialize the locations of the sources and the receivers, the numerical grid and the contrast distribution $\hat{\chi}^c$, respectively. Next, the error criterion is given. We use a stronger one than the error criterion used in the first part of this book because we want to be sure that the difference between the analytical data and the numerical data is not caused by a convergence problem of the integral equation. Finally, the `init` function offers the option to add noise to the analytical data to increase the noise level between the analytical and the numerical data. We use the Matlab random number generator `randn` to generate two repeatable arrays with a size of input.NR by input.NS, consisting of normally distributed random numbers with standard deviation $\sigma = 0.2$ and zero mean. These two different arrays are used to construct a single complex-valued array input.Rand. In the function `initSourceReceiver` (Matlab Listing II.2), we have 50 sources on a circle with a radius of 170 m around the scattering object, while 50 receivers are located on a circle with a radius of 150 m around the scattering object.

The function `initGrid` (Matlab Listing II.3) defines a rectangular grid of N_1 by N_2 square subdomains with equal area of $(\Delta x)^2$.

The function `initFFTGreen` of Matlab Listing II.4 generates a 2D grid with circulant properties. The $N_{2\text{fft}}$ columns of array `temp.X1fft` are copies of the vector x_1, and the $N_{1\text{fft}}$ rows of array `temp.X2fft` are copies of the vector x_2. For efficient computations, the FFT numbers in each direction must have a power of two. These numbers are a factor of two greater than the grid

Matlab Listing II.2 The function `initSourceReceiver` to initialize the 2D source/receivers locations.

```
function input = initSourceReceiver(input)

                input.NS  = 50;                          % source positions
 input.src_phi(1:input.NS) = (1:input.NS) * 2*pi/input.NS;
   input.xS(1,1:input.NS) = 170 * cos(input.src_phi);
   input.xS(2,1:input.NS) = 170 * sin(input.src_phi);

                input.NR  = 50;                          % receiver positions
input.rcvr_phi(1:input.NR) = (1:input.NR) * 2*pi/input.NR;
   input.xR(1,1:input.NR) = 150 * cos(input.rcvr_phi);
   input.xR(2,1:input.NR) = 150 * sin(input.rcvr_phi);
```

Matlab Listing II.3 The function `initGrid` to initialize the 2D grid parameters.

```
function input = initGrid(input)

input.N1 = 50;                          % number of samples in x_1
input.N2 = 50;                          % number of samples in x_2
input.dx = 3;                           % with meshsize dx

x1 = -(input.N1+1)*input.dx/2 + (1:input.N1)*input.dx;
x2 = -(input.N2+1)*input.dx/2 + (1:input.N2)*input.dx;
[input.X1,input.X2] = ndgrid(x1,x2);
```

Matlab Listing II.4 The function `initFFTGreen` computes the Discrete Fourier transform of the repeated weak form of the 2D Green function.

```
function input = initFFTGreen(input)
  % make two-dimensional FFT grid
    N1fft        = 2^ceil(log2(2*input.N1));
    N2fft        = 2^ceil(log2(2*input.N2));
    x1(1:N1fft) = [0 : N1fft/2-1   N1fft/2 : -1 : 1] * input.dx;
    x2(1:N2fft) = [0 : N2fft/2-1   N2fft/2 : -1 : 1] * input.dx;
    [temp.X1fft,temp.X2fft] = ndgrid(x1,x2);

  % compute gam_0^2 * subdomain integrals  of Green function
    [IntG] = Green(temp,input);

  % apply n-dimensional Fast Fourier transform
    input.FFTG = fftn(IntG);
```

numbers N_1 and N_2. At the end of the function `initFFTGreen`, there is a call to Matlab function `Green` for computation of the repeated weak form of the Green functions, see Matlab Listing II.4. The function `initContrast` (Matlab Listing II.6) initializes the nominal values of the contrast distribution $\hat{\chi}^c(x_1, x_2)$. Inside $\mathbb{D}_{sct}$, the contrast is equal to the constant value given by $1 - (\hat{c}_0/\hat{c}_{sct})^2$. If this value is positive, we set input.signCHI = 1. If this value is negative, we set input.signCHI = −1. We may use this *a priori* information in our inversion routines. To create an off-centered cylinder, we define a new origin for both the computation of the contrast distribution and the analytical computations of the wave fields. The contrast distributions are presented either as an image plot of its real part or as a mesh plot of the absolute value. For a positive contrast distribution, the contrast values are presented in Figure 4.2, while the results for a negative contrast distribution are shown in Figure 4.3.

The Matlab script to compute the 2D scattered field both analytically and numerically is given in Matlab Listing II.7. The following steps are carried out:

(1) For the analytical computation of the wave field scattered from a circular object at a number of receivers, we call the Matlab script `DataSctCircle` of Matlab Listing II.8. The difference

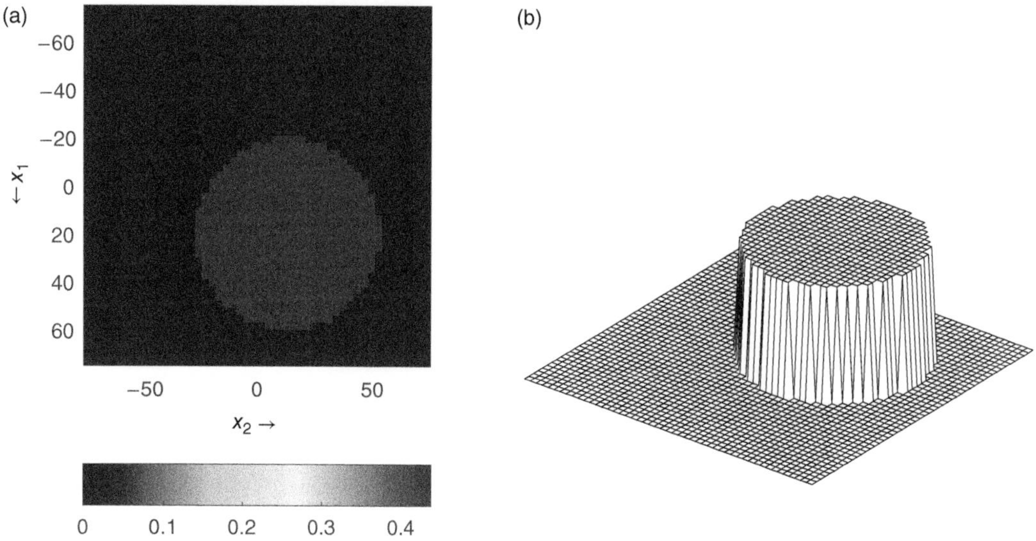

Figure 4.2 Plots of the nominal values of the contrast, for the case $c_0 = 1500$ m/s and $\hat{c}_{sct} = 2000$ m/s ($0 < \hat{\chi}^c < 0.4375$): (*left*) real part of contrast using Matlab's `imagesc` script; (*right*) absolute value of contrast using Matlab's `mesh` script.

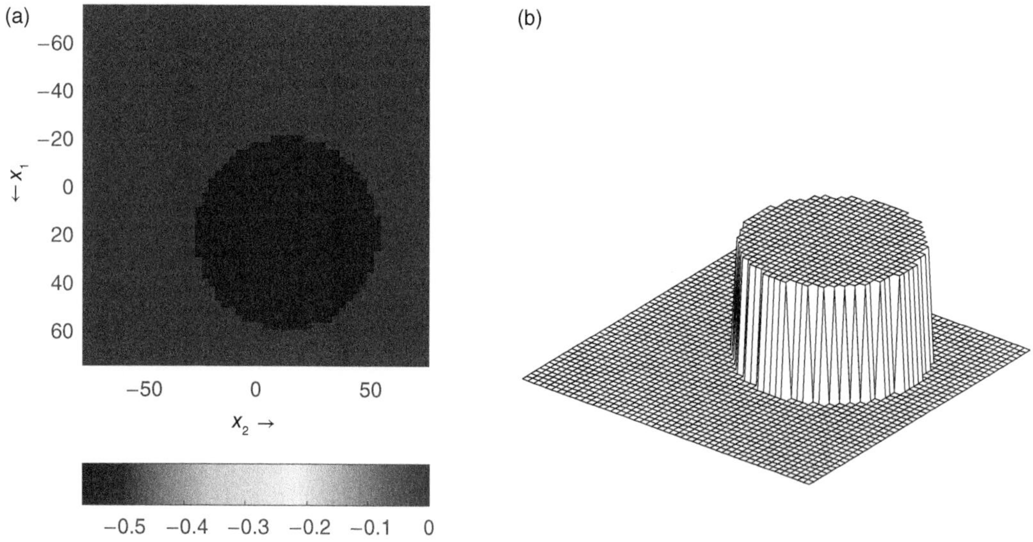

Figure 4.3 Plots of the nominal values of the contrast, for the case $c_0 = 1500$ m/s and $\hat{c}_{sct} = 1200$ m/s ($-0.5625 < \hat{\chi}^c < 0$): (*left*) real part of contrast using Matlab's `imagesc` script; (*right*) absolute value of contrast using Matlab's `mesh` script.

Matlab Listing II.5 The function `Green` to compute the repeated weak form of the 2D Green function.

```
function [IntG] = Green(temp,input)
  dx = input.dx;
gam0 = input.gamma_0;

 X1        = temp.X1fft;
 X2        = temp.X2fft;
DIS        = sqrt(X1.^2 + X2.^2);
DIS(1,1)   = 1;                    % avoid Green's singularity for DIS = 0
 G         = 1/(2*pi).* besselk(0,gam0*DIS);
 delta     = (pi)^(-1/2) * dx;     % radius circle with area of dx^2
 factor    = 2 * besseli(1,gam0*delta) / (gam0*delta);

% Integral of Green function including gam0^2
IntG       = (gam0^2 * dx^2) * factor^2 * G;
IntG(1,1)  = 1 - gam0*delta * besselk(1,gam0*delta) * factor;
```

Matlab Listing II.6 The function `initContrast` to initialize the 2D grid parameters.

```
function [input] = initContrast(input)
input.a  = 40;                 % radius circle cylinder
contrast = 1 - input.c_0^2/input.c_sct^2;

if contrast > 0;  input.signCHI =  1; end
if contrast < 0;  input.signCHI = -1; end
input.contrast = contrast;

input.xO(1) = input.a / 2;   % center coordinates of circle
input.xO(2) = input.a / 3;
          R = sqrt((input.X1-input.xO(1)).^2 + (input.X2-input.xO(2)).^2);
  input.CHI = contrast .* (R < input.a);

% Figures of contrast
x1 = input.X1(:,1);     x2 = input.X2(1,:);
figure(1);  IMAGESC(x1,x2,real(input.CHI));
figure(2);  mesh(abs(input.CHI),'edgecolor', 'k');
            view(37.5,45); axis xy; axis('off'); axis('tight')
            axis([1 input.N1 1 input.N2  0 1.5*max(abs(input.CHI(:)))])
```

with the code described in Matlab Listing I.28 of Chapter 1 is that we now have an off-centered cylinder and that the data have to be computed for a number of sources. These data are saved as synthetic data for our inversion methods.

(2) To facilitate the numerical computation of the incident fields, we compute the wave fields from a number of line sources. We compute their weak forms. They are denoted as G_S. Similarly, we

Matlab Listing II.7 The `Forwardw` script to compute synthetic 2D data.

```
clear all; clc; close all; clear workspace
input = init();

% (1) Compute analytically scattered field data ----------------------------
    disp('Running DataSctCircle');  DataSctCircle;

% (2) Compute Green function for sources and receivers ---------------------
    [G_S,G_R] = GreenSourcesReceivers(input);

% (3) Solve contrast source integral equation and compute synthetic data
    data = zeros(input.NR,input.NS);
       w = cell(1,input.NS);
    for q = 1 : input.NS
%       w{q} = ITERCGw(G_S{q},input);   % Note u_inc(:,:) = G_S{q}(:,:)
        w{q} = BICGSTABw(G_S{q},input);
      data(:,q) = DopM(G_R,w{q},input);
    end % q_loop
    save DataIntEq data;

% (4) Plot data at a number of receivers ------------------------------------
    figure(3);
    subplot(1,2,1)
       PlotData(dataCircle,input);
       title('|u^{sct}(x_p^R,x_q^S)|=|data_{Bessel}|')
    subplot(1,2,2)
       PlotData(data-dataCircle,input);
       title('|data_{Int.Eq.} - data_{Bessel}|')

  % Compute normalized difference of norms
  error = num2str(norm(data(:)-dataCircle(:),1)/norm(dataCircle(:),1));
  disp(['error=' error]);
```

compute the weak forms of the Green functions G_R associated with the wave fields measured at the receivers. Both types of Green functions are computed with the help of the Matlab function `GreenSourcesReceivers` of Matlab Listing II.11.

(3) We solve the integral equation for the contrast sources using either the CGFFT method or the BICGSTAB method. Since the latter converges much faster, we prefer the latter one. The integral equation is solved for each source location, indicated by the value q. For convenience, their 2D versions are given by the functions `ITERCGw` of Matlab Listing II.9 and `BICGSTABw` of Matlab Listing II.10. Using the contrast sources, we compute the scattered field at the receiver locations, see function `DopM` of Matlab Listing II.12.

(4) We plot the data using function `PlotData` of Matlab Listing II.13. To measure the quality of these data, we benchmark it against the analytical data of step (1), based on a Bessel function series.

Matlab Listing II.8 The `DataSctCircle` script to compute synthetic 2D data.

```
input  = init();
c_0      = input.c_0;        c_sct    = input.c_sct;
gam0     = input.gamma_0;  gam_sct  = input.gamma_0 * c_0/c_sct;

% (1) Compute coefficients of series expansion -------------------------------
    arg0 = gam0 * input.a;   args = gam_sct*input.a;
    M = 50;                             % increase M for more accuracy
    A = zeros(1,M+1);
    for m = 0 : M
     Ib0 = besseli(m,arg0);      dIb0 =   besseli(m+1,arg0) + m/arg0 * Ib0;
     Ibs = besseli(m,args);      dIbs =   besseli(m+1,args) + m/args * Ibs;
     Kb0 = besselk(m,arg0);      dKb0 = -besselk(m+1,arg0) + m/arg0 * Kb0;
     A(m+1) = - (gam_sct * dIbs*Ib0 - gam0 * dIb0*Ibs) ...
                /(gam_sct * dIbs*Kb0 - gam0 * dKb0*Ibs);
    end

% (2) Compute reflected field at the receivers for all sources (data) -----
    xR(1,:) = input.xR(1,:) - input.xO(1);  % shifted origin
    xR(2,:) = input.xR(2,:) - input.xO(2);  % shifted origin
    xS(1,:) = input.xS(1,:) - input.xO(1);  % shifted origin
    xS(2,:) = input.xS(2,:) - input.xO(2);  % shifted origin
    rR = sqrt(xR(1,:).^2 + xR(2,:).^2);   phiR = atan2(xR(2,:),xR(1,:));
    rS = sqrt(xS(1,:).^2 + xS(2,:).^2);   phiS = atan2(xS(2,:),xS(1,:));

    data2D = zeros(input.NR,input.NS);
    for p = 1 : input.NR
     for q = 1 : input.NS
      data2D(p,q) = A(1) * besselk(0,gam0*rR(p))*besselk(0,gam0*rS(q));
      for m = 1 : M
       data2D(p,q) = data2D(p,q) + 2*A(m+1)*cos(m*(phiR(p)-phiS(q))) ...
                     * besselk(m,gam0*rR(p))*besselk(m,gam0*rS(q));
      end % m_loop
     end % q_loop
    end % p_loop

    dataCircle = 1/(2*pi) * data2D;                         clear data2D;
    save DataAnalytic dataCircle;
```

For our two chosen values of $\hat{c}_{sct}$, in Figures 4.4 and 4.5, we show the amplitudes of the analytical data of the circular object (a) and the difference between the numerical data obtained from the integral equation method (b). Note that for $\hat{c}_{sct} > c_0$, the data vary slower than the data for $\hat{c}_{sct} < c_0$. In the first case, the numerical data approximate the analytical data better (Error = 5.4 %), while in the second case, the sharp variation of the data is more difficult to approximate with a rather crude grid (Error = 9.6 %).

With these results, we now discuss the nonlinear inverse scattering problem.

Matlab Listing II.9 The function `ITERCGw` to compute the contrast sources.

```
function [w] = ITERCGw(u_inc, input)
% CG_FFT scheme for contrast source integral equation
CHI = input.CHI;   FFTG = input.FFTG;

itmax    = 200;
Errcri   = input.Errcri;
it       = 0;                          % initialization of iteration
w        = zeros(size(u_inc));         % first guess for contrast source
r_D      = CHI.*u_inc;                 % first residual vector
Norm_D   = norm(r_D(:))^2;
eta_D    = 1 / Norm_D;                 % normalization factor
Error    = 1;                          % error norm initial error
fprintf('Error =          %g', Error);

% check adjoint operator [I-K*conj(CHI)] via inner product ----------------
%          dummy = r_D - AdjKop(conj(CHI).*r_D,FFTG);
%        Result1 = sum(r_D(:).*conj(dummy(:)));
%          dummy = r_D - CHI.* Kop(r_D,FFTG);
%        Result2 = sum(dummy(:).*conj(r_D(:)));
%        fprintf('Check adjoint: %e\n',abs(Result1-Result2));
% --------------------------------------------------------------------------

while (it < itmax) && (Error > Errcri)
    % determine conjugate gradient direction v
        g = r_D - AdjKop(conj(CHI).*r_D,FFTG);
        g = (abs(CHI) >= Errcri) .* g; % window for negligible contrast!!
      AN  = norm(g(:))^2;
      if it == 0
          v = g;
      else
          v = g + (AN/AN_1) * v;
      end
    % determine step length alpha
      Kv    = v - CHI.* Kop(v,FFTG);
      BN    = norm(Kv(:))^2;
      alpha = AN / BN;
    % update contrast source w and AN
      w     = w + alpha * v;
      AN_1  = AN;
    % update the residual error r
      r_D    = r_D - alpha * Kv;
      Norm_D = norm(g(:))^2;
      Error  = sqrt(eta_D * Norm_D);
      fprintf('\b\b\b\b\b\b\b\b%6f', Error);              it = it+1;
end % while

fprintf('\b\b\b\b\b\b\b\b%6f\n', Error);
disp(['Number of iterations is ' num2str(it)]);
if it == itmax
   disp(['itmax was reached:    err/norm = ' num2str(Error)]);
end
```

Matlab Listing II.10 The function `BICGSTABw` to compute the contrast sources.

```
function [w] = BICGSTABw(u_inc, input)
CHI = input.CHI;     Data_D = CHI.*u_inc;     FFTG = input.FFTG;

itmax  = 200;
Errcri = input.Errcri;
it     = 0;                          % initialization of iteration
w      = zeros(size(u_inc));         % first guess for contrast source
r_D    = Data_D;                     % first residual vector
Norm_D = norm(r_D(:))^2;             % normalization factor
eta_D  = 1 / Norm_D;
Error  = 1;                          % error norm initial error

fprintf('Error =         %g', Error);

while (it < itmax) && (Error > Errcri)

  % Determine gradient directions
    AN  = sum(r_D(:).*conj(Data_D(:)));
    if it == 0
        v = r_D;
    else
        v = r_D +(AN/AN_1) * v;
    end
  % determine step length alpha and update residual error r_D
    Kv    = v - CHI.* Kop(v,FFTG);
    BN    = sum(Kv(:).*conj(Data_D(:)));
    alpha = AN / BN;
    r_D   = r_D - alpha * Kv;
  % + successive overrelaxation (first step of GMRES)
    Kr    = r_D - CHI.* Kop(r_D,FFTG);
    beta  = sum(r_D(:).*conj(Kr(:))) / norm(Kr(:))^2;
  % update contrast source w ,v and AN
    w     = w + alpha * v   + beta * r_D;
    v     = (alpha / beta) * (v - beta * Kv);
    AN_1  = AN;
  % update the residual error r_D
    r_D   = r_D - beta * Kr;

    Norm_D = norm(r_D(:))^2;
    Error  = sqrt(eta_D * Norm_D);
    fprintf('\b\b\b\b\b\b\b\b\b%6f', Error);        it = it+1;

end % while

fprintf('\b\b\b\b\b\b\b\b\b%6f\n', Error);
disp(['Number of iterations is ' num2str(it)]);

if it == itmax
   disp(['itmax was reached:   err/norm = ' num2str(Error)]);
end
```

Matlab Listing II.11 The function `GreenSourcesReceivers` to compute the 2D Green functions for all source and receiver locations.

```
function [G_S,G_R] = GreenSourcesReceivers(input)

gam0 = input.gamma_0;
X1   = input.X1;
X2   = input.X2;
dx   = input.dx;

delta  = (pi)^(-1/2) * dx;            % radius circle with area of dx^2
factor = 2 * besseli(1,gam0*delta) / (gam0*delta);
                                      % factor for weak form if DIS > delta

 xS  = input.xS;
 G_S = cell(1,input.NS);
for q = 1 : input.NS
      DIS  = sqrt( (X1-xS(1,q)).^2 + (X2-xS(2,q)).^2 );
    G_S{q} = factor * 1/(2*pi) .* besselk(0,gam0*DIS);
end % q_loop

  xR  = input.xR;
  G_R = cell(1,input.NR);
for p = 1 : input.NR
      DIS  = sqrt((xR(1,p)-X1).^2 +(xR(2,p)-X2).^2);
   G_R{p}  = factor * 1/(2*pi) .* besselk(0,gam0*DIS);
end % p_loop
```

Matlab Listing II.12 The function `DopM` to compute synthetic 2D data, for each source location q.

```
function [GRw] = DopM(G_R,w,input)

gam0 = input.gamma_0;
dx   = input.dx;

GRw = zeros(input.NR,1);
for p = 1: input.NR
    GRw(p,1) = (gam0^2 * dx^2) * sum(G_R{p}(:).*w(:));
end  % p_loop
```

4.3 Nonlinear Inverse Scattering Problem

The inverse scattering problem consists of determining $\hat{\chi}^c$ from a knowledge of the measured data in the domain $\mathbb{M}$ and the incident fields in the domain $\mathbb{D}$. In the absence of other *a priori* information, Eqs. (4.6) and (4.7) are the only equations we have relating the unknown contrast sources $\hat{w}$ and the unknown contrast $\hat{\chi}^c$ in $\mathbb{D}$. The known quantities consist of the measured data in $\mathbb{M}$ and the incident wave fields in $\mathbb{D}$. Equations (4.6) and (4.7) are linear in each of the unknowns $\hat{\chi}^c$

Matlab Listing II.13 The function `PlotData` to compute synthetic 2D data, for each source location q.

```
function PlotData(data, input)

imagesc(1:input.NR,1:input.NS, abs(data));
    xlabel('Receiver number p \rightarrow');
    ylabel('Source number q \rightarrow');
    axis('equal','tight'); axis xy;
    colorbar('hor'); colormap jet;
```

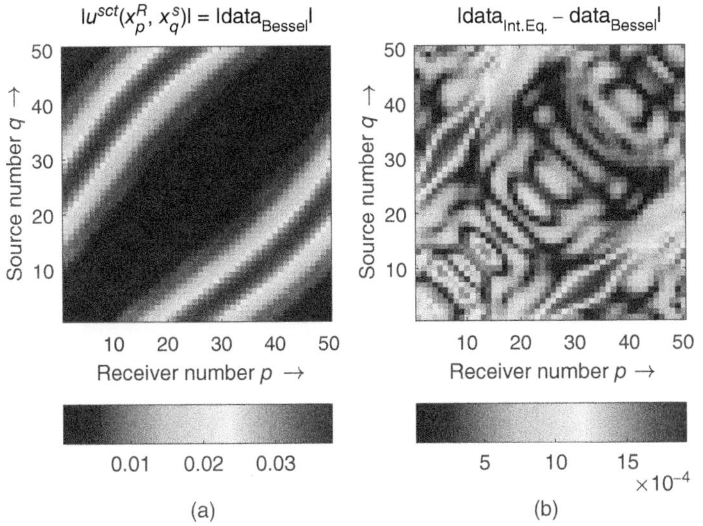

Figure 4.4 Plots of the data amplitudes as function of the source and receiver numbers, for the nominal values $c_0 = 1500$ m/s and $\hat{c}_{sct} = 2000$ m/s. The normalized global error in the integral equation data amounts to 5.4%.

and $\hat{w}$, but since both are unknown the problem is in fact mildly linear. Of course the dependence of the contrast sources $\hat{w}$ on the contrast $\hat{\chi}^c$ is highly nonlinear. This may be seen by writing the formal inverse of Eq. (4.6) as

$$\hat{w} = (I - \hat{\chi}^c \, \hat{G}_{\rm D})^{-1} \, (\hat{\chi}^c \, \hat{u}^{inc}), \tag{4.8}$$

where I is the identity operator. Substituting Eq. (4.8) into the data equation, Eq. (4.4), we obtain

$$\hat{u}^{sct}_{\rm M} = \hat{G}_{\rm M}[(I - \hat{\chi}^c \, \hat{G}_{\rm D})^{-1} \, (\hat{\chi}^c \, \hat{u}^{inc})], \tag{4.9}$$

wherein the nonlinearity of the inverse problem is clearly exposed. Approximating the inverse operator by

$$(I - \hat{\chi}^c \, \hat{G}_{\rm D})^{-1} \approx I \, , \tag{4.10}$$

leads to the Born approximation, while in iterative methods, where a sequence $\{\hat{\chi}^c_n\}$ is constructed, the approximation

$$(I - \hat{\chi}^c_n \, \hat{G}_{\rm D})^{-1} \approx (I - \hat{\chi}^c_{n-1} \, \hat{G}_{\rm D})^{-1} \tag{4.11}$$

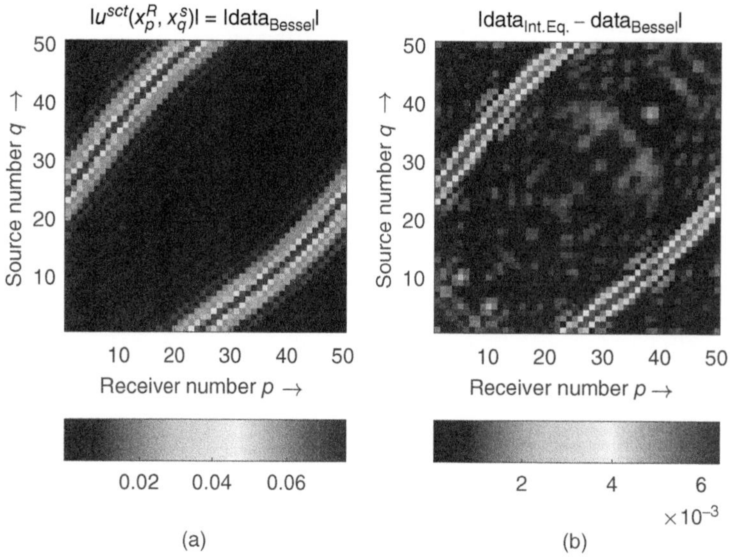

Figure 4.5 Plots of the data amplitudes as function of the source and receiver numbers for the nominal values $c_0 = 1500$ m/s and $\hat{c}_{sct} = 1200$ m/s. The normalized global error in the integral equation data amounts to 9.6%.

is similar to the iterative Born approach [49], where instead of the contrast-source type object equations, the field equations are used. Using the field equations, the method has been improved, for example, the distorted Born approach [10] and the Newton–Kantorovich method [42], which are equivalent (see [41]), and the Gauss–Newton method, for example [3]. Using these types of iterative methods to solve for the wave-speed contrast, we observe that in each iteration, it is necessary to compute the action of the operator $(I - \hat{\chi}^c\, \hat{G}_{\mathrm{D}})^{-1}$ for known $\hat{\chi}^c$. This means that a full forward scattering problem must be solved at each iterative step.

4.4 Inverse Contrast Source Problem

In order to avoid the necessity of solving forward problems completely, many workers started with the data equation,

$$\hat{u}_{\mathrm{M}}^{sct} = \hat{G}_{\mathrm{M}} \hat{w}\,. \tag{4.12}$$

After solving this linear equation for the contrast sources $\hat{w}$, the total wave field follows directly from the object equation,

$$\hat{u} = \hat{u}^{inc} + \hat{G}_{\mathrm{D}} \hat{w}\,, \quad \text{on } \mathbb{D}\,, \tag{4.13}$$

after which the unknown contrast follows from the constitutive relation

$$\hat{w} = \hat{\chi}^c\, \hat{u}\,. \tag{4.14}$$

Equation (4.12) is called by some a source-type integral equation and it has a long history. It is a classic ill-posed equation and for a time there was considerably attention paid to the question of uniqueness. It was shown that there exists nontrivial solutions of the homogeneous form of Eq. (4.12), although it was argued by some that uniqueness could be restored from physical considerations. A good summary of the debate is given by Daveney and Sherman [15] and the responses by Bojarski [7] and Stone [46]. It is not our intent to renew this controversy, since it is now well accepted that nontrivial solutions of Eq. (4.12) exist. Moreover, it has also been shown that the minimum norm solution of Eq. (4.12), the solution produced, by the conjugate gradient method, is not the appropriate physical solution. Nonetheless, this source-type equation has served as an essential ingredient in many inversion procedures, for example by Chew et al. [11]. Habashy et al. [19] presented an inversion method wherein the minimum norm solution of Eq. (4.12) is found first and then a basis for the orthogonal complement of this solution is constructed in terms of which the physical solution is sought to satisfy Eq. (4.13). van den Berg and Haak [53] proposed a variant of this technique wherein the full minimum norm solution is not found but rather it is sought iteratively, using conjugate gradient steps, with the contrast updated at each step to satisfy Eqs. (4.13) and (4.14). With this contrast, new contrast-source updates are determined. This approach yielded promising numerical results; however, the error did not decrease monotonically. In order to obtain an error reducing method, van den Berg and Kleinman [55] defined a cost functional that consists of two terms. Formally, we write it as

$$F = \frac{\sum_{\text{all } \boldsymbol{x}_q^S} \| \hat{u}_{\mathbb{M}}^{sct} - \hat{G}_{\mathbb{M}} \hat{w} \|_{\mathbb{M}}^2}{\sum_{\text{all } \boldsymbol{x}_q^S} \| \hat{u}_{\mathbb{M}}^{sct} \|_{\mathbb{M}}^2} + \frac{\sum_{\text{all } \boldsymbol{x}_q^S} \| \hat{\chi}^c \hat{u}^{inc} - \hat{w} + \hat{\chi}^c \, \hat{G}_{\mathbb{D}} \hat{w} \|_{\mathbb{D}}^2}{\sum_{\text{all } \boldsymbol{x}_q^S} \| \hat{\chi}^c \hat{u}^{inc} \|_{\mathbb{D}}^2} . \tag{4.15}$$

The norms in the first term imply summations for all receiver positions $\boldsymbol{x}_p^R \in \mathbb{M}$. Although in the literature we have formulated the norms in the second term as integrals, in this book we define these norms as summations over a discrete set of points $\boldsymbol{x}_m \in \mathbb{D}$. The precise formulation is given in Section 4.5.

By alternate updating the contrast sources and the contrast, while minimizing the combined functional, the solution of a full forward problem per iteration is not required. We shall outline this optimization scheme and demonstrate its implementation in Matlab.

4.5 Contrast Source Inversion

The contrast source inversion (CSI) method recasts the inversion problem as a minimization of the cost functional of Eq. (4.15), from which the contrast sources $\hat{w}$ and the material contrast $\hat{\chi}$ are reconstructed. We first introduce the short-hand notations for the unknown wave-field functions, $\hat{u}_q = \hat{u}(\boldsymbol{x}_m, \boldsymbol{x}_q^S)$, and the known wave-field functions,

$$\begin{aligned} \hat{u}_{\mathbb{M},q}^{sct} &= \hat{u}_{\mathbb{M}}^{sct}(\boldsymbol{x}_p^R, \boldsymbol{x}_q^S) , \\ \hat{u}_q^{inc} &= \hat{u}^{inc}(\boldsymbol{x}_m, \boldsymbol{x}_q^S) , \end{aligned} \tag{4.16}$$

for $p = 1, 2, \ldots, N^R$ and $q = 1, 2, \ldots, N^S$, where N^R is the total number of receivers and N^S is the total number of sources.

Then, in the CSI method, the sequences of the contrast sources $\hat{w}_q^{(n)} = \hat{w}^{(n)}(\boldsymbol{x}, \boldsymbol{x}_q^S)$ and the contrast $\hat{\chi}^{c(n)} = \hat{\chi}^{c(n)}(\boldsymbol{x})$, for $n = 1, 2, \ldots$, are iteratively found to minimize the cost functional

$$F^{(n)}(\hat{w}_q^{(n)}, \hat{\chi}^{c(n)}) = \hat{\eta}_{\mathbb{M}} F_{\mathbb{M}}(\hat{w}_q^{(n)}) + \hat{\eta}_{\mathbb{D}}^{(n)} F_{\mathbb{D}}(\hat{w}_q^{(n)}, \hat{\chi}^{c(n)}) , \tag{4.17}$$

where

$$F_{\mathrm{M}}(\hat{w}_q^{(n)}) = \sum_{q=1}^{N^S} \|\hat{r}_{\mathrm{M},q}^{(n)}\|_{\mathbb{M}}^2 \quad \text{with} \quad \hat{r}_{\mathrm{M},q}^{(n)} = \hat{u}_{\mathrm{M},q}^{sct} - \hat{G}_{\mathrm{M}}\hat{w}_q^{(n)}, \tag{4.18}$$

$$F_{\mathrm{D}}(\hat{w}_q^{(n)}, \hat{\chi}^{c(n)}) = \sum_{q=1}^{N^S} \|\hat{r}_{\mathrm{D},q}^{(n)}\|_{\mathbb{D}}^2 \quad \text{with} \quad \hat{r}_{\mathrm{D},q}^{(n)} = \hat{\chi}^{c(n)}\hat{u}_q^{inc} - \hat{w}_q^{(n)} + \hat{\chi}^{c(n)}\,\hat{G}_{\mathrm{D}}\hat{w}_q^{(n)}, \tag{4.19}$$

and normalization factors

$$\hat{\eta}_{\mathrm{M}} = \left[\sum_{q=1}^{N^S} \|\hat{u}_{\mathrm{M},q}^{sct}\|_{\mathbb{M}}^2\right]^{-1} \quad \text{and} \quad \hat{\eta}_{\mathrm{D}}^{(n)} = \left[\sum_{q=1}^{N^S} \|\hat{\chi}^{c(n)}\hat{u}_q^{inc}\|_{\mathbb{D}}^2\right]^{-1}. \tag{4.20}$$

Here, $\|\cdot\|_{\mathbb{M},\mathbb{D}}^2 = \langle\cdot,\cdot\rangle_{\mathbb{M},\mathbb{D}}$ denotes the squared norms for $\boldsymbol{x}_p^R \in \mathbb{M}$ and $\boldsymbol{x}_m \in \mathbb{D}$, respectively. The first term of Eq. (4.17) measures the residual error $\hat{r}_{\mathrm{M}}$ in the measurement domain $\mathbb{M}$ and the second term measures the residual error $\hat{r}_{\mathrm{D}}$ in the object domain $\mathbb{D}$. The normalization is chosen so that both terms are equal to one if $\hat{w}_q^{(n)} = 0$. The first term is a quadratic functional in $\hat{w}_q^{(n)}$, but the second one is nonlinear in $\hat{\chi}^{c(n)}$ and $\hat{w}_q^{(n)}$. To circumvent this nonlinearity, the CSI method employs an alternate updating scheme for these quantities. The CSI method starts with initial estimates for the contrast sources and the contrast. We discuss this later.

4.5.1 Discretization of Green's Operators and Norms

To facilitate the implementation of the mathematical expressions in a computer code, we first discuss the numerical discretization. We omit the hat-symbols to indicate the s-dependency. We employ the spatial discretization as outlined in Chapter 1. After discretization, the data operator $G_{\mathrm{M}}w$ is written as, cf. Eq. (4.5),

$$\boxed{G_{\mathrm{M}}w_q = \Delta V \sum_{m'=1}^{N} \gamma_0^2 \langle G\rangle(\boldsymbol{x}_p^R - \boldsymbol{x}_{m'})\, w(\boldsymbol{x}_{m'}, \boldsymbol{x}_q^S)\,, \quad \text{for } p = 1, 2, \dots, N^R\,,} \tag{4.21}$$

where N^R are the total number of receivers. The object operator $G_{\mathrm{D}}w$ is written as, cf. Eq. (4.7),

$$\boxed{G_{\mathrm{D}}w_q = \Delta V \sum_{m'=1}^{N} \gamma_0^2 \langle\langle G\rangle\rangle(\boldsymbol{x}_m - \boldsymbol{x}_{m'})\, w(\boldsymbol{x}_{m'}, \boldsymbol{x}_q^S)\,, \quad m = 1, 2, \dots, N\,.} \tag{4.22}$$

In the discrete formulation, the square norms for the errors in $\mathbb{M}$ and $\mathbb{D}$ are defined as

$$\|r_{\mathrm{M},q}^{(n)}\|_{\mathbb{M}}^2 = \sum_{p=1}^{N^R} |r_{\mathrm{M}}^{(n)}(\boldsymbol{x}_p^R, \boldsymbol{x}_q^S)|^2 \quad \text{and} \quad \|r_{\mathrm{D},q}^{(n)}\|_{\mathbb{D}}^2 = \sum_{m=1}^{N} |r_{\mathrm{D}}^{(n)}(\boldsymbol{x}_m, \boldsymbol{x}_q^S)|^2\,, \tag{4.23}$$

where $r_{\mathrm{M},q}^{(n)} = r_{\mathrm{M}}^{(n)}(\boldsymbol{x}_p^R, \boldsymbol{x}_q^S)$ and $r_{\mathrm{D},q}^{(n)} = r_{\mathrm{D}}^{(n)}(\boldsymbol{x}_m, \boldsymbol{x}_q^S)$. Similarly, the normalization factors are defined as

$$\eta_{\mathrm{M}} = \left[\sum_{q=1}^{N^S}\sum_{p=1}^{N^R} \|u_{\mathrm{M}}^{sct}(\boldsymbol{x}_p^R, \boldsymbol{x}_q^S)\|_{\mathbb{M}}^2\right]^{-1} \quad \text{and} \quad \eta_{\mathrm{D}}^{(n)} = \left[\sum_{q=1}^{N^S}\sum_{m=1}^{N} \|\chi^{c(n)}(\boldsymbol{x}_m) u^{inc}(\boldsymbol{x}_m, \boldsymbol{x}_q^S)\|_{\mathbb{D}}^2\right]^{-1}. \tag{4.24}$$

The discrete inner products $\langle \cdot, \cdot \rangle_{\mathrm{M}}$ and $\langle \cdot, \cdot \rangle_{\mathrm{D}}$ are defined in a similar way. With these redefinitions, we discuss the CSI scheme.

4.5.2 Updating the Contrast Sources

In each iteration, we first update the contrast sources $w_q^{(n)} = w^{(n)}(\boldsymbol{x}_m, \boldsymbol{x}_q^S)$ using the conjugate gradient step

$$\boxed{w_q^{(n)} = w_q^{(n-1)} + \alpha_q^{(n)} v_q^{(n)}\,,} \tag{4.25}$$

where, for each q, the parameter $\alpha_q^{(n)}$ is the step length and the function $v_q^{(n)} = v^{(n)}\boldsymbol{x}_m, \boldsymbol{x}_q^S)$ is the Polak–Ribière conjugate gradient direction,

$$\boxed{v_q^{(1)} = g_q^{(1)}\,, \quad v_q^{(n)} = g_q^{(n)} + \gamma_q^{(n)}\, v_q^{(n-1)}\,, \qquad n = 2, 3, \dots\,,} \tag{4.26}$$

and $g_q^{(n)} = g_q^{(n)}(\boldsymbol{x}_m, \boldsymbol{x}_q^S)$ is the gradient of the cost functional $F^{(n)}$ with respect to changes of $w_q^{(n)}$ evaluated at $w_q^{(n-1)}$. The parameter $\gamma_q^{(n)}$ in the Polak–Ribière direction of Eq. (4.26) is defined as

$$\boxed{\gamma_q^{(n)} = \frac{\mathrm{Re}\,[\langle g_q^{(n)}, g_q^{(n)} - g_q^{(n-1)} \rangle_{\mathrm{D}}]}{\| g_q^{(n-1)} \|_{\mathrm{D}}^2}\,.} \tag{4.27}$$

4.5.2.1 Gradient Directions

Using the definition of the Fréchet derivative, this gradient for the updating of the contrast sources is obtained as the sum of the gradient of $\eta_{\mathrm{M}} F_{\mathrm{M}}$ and the gradient of $\eta_{\mathrm{D}}^{(n-1\,)} F_{\mathrm{D}}$. The gradient of the qth term of the cost functional F_{M} is calculated as

$$\begin{aligned}
&\lim_{\epsilon \to 0} \frac{1}{\epsilon} \left[\sum_{p=1}^{N^R} |u_{\mathrm{M},q}^{sct} - G_{\mathrm{M}}\{w_q^{(n-1)} + \epsilon g_q^{(n)}\}|^2 - \sum_{p=1}^{N^R} |u_{\mathrm{M},q}^{sct} - G_{\mathrm{M}} w_q^{(n-1)}|^2 \right] \\
&\qquad = -2\,\mathrm{Re} \left[\sum_{p=1}^{N^R} r_{\mathrm{M},q}^{(n-1)}\, \overline{G_{\mathrm{M}} g_q^{(n)}} \right] = -2\,\mathrm{Re} \left[\sum_{p=1}^{N^R} \overline{g}_q^{(n)}\, G_{\mathrm{M}}^{\star} r_{\mathrm{M},q}^{(n-1)} \right],
\end{aligned} \tag{4.28}$$

where we have used that

$$r_{\mathrm{M},q}^{(n-1)} = u_{\mathrm{M},q}^{sct} - G_{\mathrm{M}} w_q^{(n-1)} \tag{4.29}$$

and ϵ is real valued. The overbar indicates complex conjugate. Apart from a constant factor, the last expression of Eq. (4.28) obtains its maximum variation when

$$g_{\mathrm{M},q}^{(n)} = G_{\mathrm{M}}^{\star} r_{\mathrm{M},q}^{(n-1)}\,. \tag{4.30}$$

Here, $G_{\mathrm{M}}^{\star}$ is the adjoint of G^{M} mapping functions of $\boldsymbol{x}_p^R \in \mathbb{M}$ to $\boldsymbol{x}_m \in \mathbb{D}$, viz.,

$$\boxed{G_{\mathrm{M}}^{\star} f_q = \Delta V \sum_{p=1}^{N^R} \overline{\gamma}_0^{-2}\, \langle \overline{G} \rangle (\boldsymbol{x}_p^R - \boldsymbol{x}_m)\, f(\boldsymbol{x}_p^R, \boldsymbol{x}_q^S).} \tag{4.31}$$

Note that we may interchange the two spatial positions in the argument of the Green function, because only the distance between these point plays a role. Next, for $\chi = \chi^{c(n-1)}$, the gradient of the

qth term of the cost functional F_{D} is calculated as

$$\lim_{\epsilon\to 0}\frac{1}{\epsilon}\sum_{m=1}^{N}\Big[|\chi u_q^{inc} - (w_q^{(n-1)} + \epsilon\, g_q^{(n)}) + \chi G_{\mathrm{D}}\{w_q^{(n-1)} + \epsilon\, g_q^{(n)}\}|^2 \\ -|\chi u_q^{inc} - w_q^{(n-1)} + \chi G_{\mathrm{D}} w_q^{(n-1)}|^2\Big] \\ = -2\,\mathrm{Re}\left[\sum_{m=1}^{N} r_{\mathrm{D},q}^{(n-1)}\,\overline{(g_q^{(n)} - \chi G_{\mathrm{D}} g_q^{(n)})}\right] = -2\,\mathrm{Re}\left[\sum_{m=1}^{N}\overline{g}_q^{(n)}\,(r_{\mathrm{D},q}^{(n-1)} - G_{\mathrm{D}}^{\star}\{\overline{\chi} r_{\mathrm{D},q}^{(n-1)}\})\right], \tag{4.32}$$

where we have used that

$$r_{\mathrm{D},q}^{(n-1)} = \chi^{c(n-1)} u_q^{inc} - w_q^{(n-1)} + \chi^{c(n-1)} G_{\mathrm{D}} w_q^{(n-1)} . \tag{4.33}$$

Apart of a constant factor, the last expression of Eq. (4.32) obtains its maximum variation, when

$$g_{\mathrm{D},q}^{(n)} = r_{\mathrm{D},q}^{(n-1)} - G_{\mathrm{D}}^{\star}\{\overline{\chi}^{c(n-1)} r_{\mathrm{D},q}^{(n-1)}\} . \tag{4.34}$$

Here $G_{\mathrm{D}}^{\star}$ is the adjoint of G_{D} mapping functions of $\boldsymbol{x}_m \in \mathbb{D}$ to $\mathbb{D}$, viz.,

$$\boxed{G_{\mathrm{D}}^{\star} f_q = \Delta V \sum_{m'=1}^{N} \overline{\gamma}_0^{-2}\langle\langle\overline{G}\rangle\rangle(\boldsymbol{x}_{m'} - \boldsymbol{x}_m)\, f(\boldsymbol{x}_{m'}, \boldsymbol{x}_q^S) .} \tag{4.35}$$

A linear superposition of Eqs. (4.30) and (4.34), with the proper normalization factors, provides us the gradient of the total cost functional $F^{(n)}$ as $g_q^{(n)} = \eta_{\mathrm{M}}\, g_{\mathrm{M},q}^{(n)} + \eta_{\mathrm{D}}^{(n-1)} g_{\mathrm{D},q}^{(n)}$, viz.,

$$\boxed{g_q^{(n)} = \eta_{\mathrm{M}}\, G_{\mathrm{M}}^{\star} r_{\mathrm{M},q}^{(n-1)} + \eta_{\mathrm{D}}^{(n-1)}[r_{\mathrm{D},q}^{(n-1)} - G_{\mathrm{D}}^{\star}\{\overline{\chi}^{c(n-1)} r_{\mathrm{D},q}^{(n-1)}\}].} \tag{4.36}$$

We remark that after discretization of the Green's operators and the norms, the calculation of the gradient with the help of the Fréchet derivatives may be replaced by defining the gradient as the negative value of the partial derivative of the qth term, $F_q^{(n-1)}$, of the cost functional $F^{(n-1)} = \sum_q F_q^{(n-1)}$ with respect to the contrast source $\overline{w}_{m',q}^{(n-1)} = \overline{w}^{(n-1)}(\boldsymbol{x}_{m'}, \boldsymbol{x}_q^S)$ as

$$g^{(n)}(\boldsymbol{x}_{m'}, \boldsymbol{x}_q^S) = -\frac{\partial F_q^{(n-1)}}{\partial \overline{w}_{m',q}^{(n-1)}} , \tag{4.37}$$

where $F_q^{(n-1)}$ follows from Eqs. (4.17)–(1.19). Hence,

$$g^{(n)}(\boldsymbol{x}_{m'}, \boldsymbol{x}_q^S) = -\eta_{\mathrm{M}} \sum_{p=1}^{N^R} \frac{\partial \overline{r}_{\mathrm{M},q}^{(n-1)}}{\partial \overline{w}_{m',q}^{(n-1)}}\, r_{\mathrm{M},q}^{(n-1)} - \eta_{\mathrm{D}}^{(n-1)} \sum_{m=1}^{N} \frac{\partial \overline{r}_{\mathrm{D},q}^{(n-1)}}{\partial \overline{w}_{m',q}^{(n-1)}}\, r_{\mathrm{D},q}^{(n-1)} , \tag{4.38}$$

in which

$$\overline{r}_{\mathrm{M},q}^{(n-1)} = \overline{u}_{\mathrm{M},q}^{sct} - \Delta V \sum_{m'=1}^{N} \overline{\gamma}_0^{-2}\langle\overline{G}\rangle(\boldsymbol{x}_p^R - \boldsymbol{x}_{m'})\,\overline{w}^{(n-1)}(\boldsymbol{x}_{m'}, \boldsymbol{x}_q^S), \tag{4.39}$$

for each point $\boldsymbol{x}_p^R \in \mathbb{M}$, $p = 1, 2, \dots, N^R$, and

$$\overline{r}_{\mathrm{D},q}^{(n-1)} = \overline{\chi}^{c(n-1)}\overline{u}_q^{inc} - \overline{w}_{m,q}^{(n-1)} + \overline{\chi}^{c(n-1)}\Delta V \sum_{m'=1}^{N} \overline{\gamma}_0^{-2}\langle\langle\overline{G}\rangle\rangle(\boldsymbol{x}_m - \boldsymbol{x}_{m'})\,\overline{w}^{(n-1)}(\boldsymbol{x}_{m'}, \boldsymbol{x}_q^S), \tag{4.40}$$

for each point $\boldsymbol{x}_m \in \mathbb{D}$, $m = 1, 2, \dots, N$. Carrying out the partial differentiation of the residual errors with respect to $\overline{w}_{m',q}^{(n-1)}$, we arrive at

$$-\frac{\partial \overline{r}_{\mathrm{M},q}^{(n-1)}}{\partial \overline{w}_{m',q}^{(n-1)}} = \Delta V \overline{\gamma}_0^{-2}\langle\overline{G}\rangle(\boldsymbol{x}_p^R - \boldsymbol{x}_{m'}) \tag{4.41}$$

and

$$-\frac{\partial \overline{r}_{\mathrm{D},q}^{(n-1)}}{\partial \overline{w}_{m',q}^{(n-1)}} = \delta_{m,m'} - \overline{\chi}^{c(n-1)} \Delta V \overline{\gamma}_0^2 \langle\langle \overline{G} \rangle\rangle (\boldsymbol{x}_m - \boldsymbol{x}_{m'}). \tag{4.42}$$

Substituting these expressions into Eq. (4.38), interchanging m and m' while using the definitions of the adjoints, we observe that the resulting value for the gradient of Eq. (4.37) is identical to the one of Eq. (4.36).

4.5.2.2 Calculation of the Step Length

For a given update direction $v_q^{(n)}$, the step length $\alpha_q^{(n)}$ is found as minimizer of the function $F_q^{(n)}$, hence

$$\alpha_q^{(n)} = \underset{\text{real } \alpha}{\arg\min} \, \{ F_q^{(n)}(w_q^{(n-1)} + \alpha \, v_q^{(n)}) \}, \tag{4.43}$$

where $F_q^{(n)}$ as function of α is given as

$$F_q^{(n)} = \eta_{\mathrm{M}} \| r_{\mathrm{M},q}^{(n-1)} - \alpha \, G_{\mathrm{M}} v_q^{(n)} \|_{\mathrm{M}}^2 + \eta_{\mathrm{D}}^{(n-1)} \| r_{\mathrm{D},q}^{(n-1)} - \alpha \, (v_q^{(n)} - \chi^{c(n-1)} G_{\mathrm{D}} v_q^{(n)}) \|_{\mathrm{D}}^2 \, . \tag{4.44}$$

Note that the cost functional $F_q^{(n)}$ is a quadratic function of α, so that only one minimizer is found. We consider only the real part of this minimizer and we arrive at

$$\begin{aligned} \alpha_q^{(n)} &= \frac{\mathrm{Re}\,[\eta_{\mathrm{M}} \langle r_{\mathrm{M},q}^{(n-1)}, G_{\mathrm{M}} v_q^{(n)} \rangle_{\mathrm{M}} + \eta_{\mathrm{D}}^{(n-1)} \langle r_{\mathrm{D},q}^{(n-1)}, v_q^{(n)} - \chi^{c(n-1)} G_{\mathrm{D}} v_q^{(n)} \rangle_{\mathrm{D}}]}{\eta_{\mathrm{M}} \, \| G_{\mathrm{M}} v_q^{(n)} \|_{\mathrm{M}}^2 + \eta_{\mathrm{D}}^{(n-1)} \, \| v_q^{(n)} - \chi^{c(n-1)} G_{\mathrm{D}} v_q^{(n)} \|_{\mathrm{D}}^2} \\ &= \frac{\mathrm{Re}\,[\eta_{\mathrm{M}} \langle G_{\mathrm{M}}^{\star} r_{\mathrm{M},q}^{(n-1)}, v_q^{(n)} \rangle_{\mathrm{D}} + \eta_{\mathrm{D}}^{(n-1)} \langle r_{\mathrm{D},q}^{(n-1)} - G_{\mathrm{D}}^{\star} \overline{\chi}^{c(n-1)} r_{\mathrm{D},q}^{(n-1)}, v_q^{(n)} \rangle_{\mathrm{D}}]}{\eta_{\mathrm{M}} \, \| G_{\mathrm{M}} v_q^{(n)} \|_{\mathrm{M}}^2 + \eta_{\mathrm{D}}^{(n-1)} \, \| v_q^{(n)} - \chi^{c(n-1)} G_{\mathrm{D}} v_q^{(n)} \|_{\mathrm{D}}^2} \, , \end{aligned} \tag{4.45}$$

while, using the expression for the gradient of Eq. (4.36), it simplifies to

$$\boxed{\alpha_q^{(n)} = \frac{\mathrm{Re}\,[\langle g_q^{(n)}, v_q^{(n)} \rangle_{\mathrm{D}}]}{\eta_{\mathrm{M}} \, \| G_{\mathrm{M}} v_q^{(n)} \|_{\mathrm{M}}^2 + \eta_{\mathrm{D}}^{(n-1)} \, \| v_q^{(n)} - \chi^{c(n-1)} G_{\mathrm{D}} v_q^{(n)} \|_{\mathrm{D}}^2} \, .} \tag{4.46}$$

We further note that $\gamma_q^{(n)}$ and $\alpha_q^{(n)}$ are real valued in a conjugate gradient solution of a linear problem. If the contrast does not change, we also deal with a linear problem for the solution of the contrast sources. This motivates our choice of taking $\gamma_q^{(n)}$ and $\alpha_q^{(n)}$ as real valued.

From Eqs. (4.25), (4.29), and (4.33) we observe that, after each update of the contrast sources, the residual errors become

$$\boxed{\begin{aligned} r_{\mathrm{M},q}^{(n)} &= r_{\mathrm{M},q}^{(n-1)} - \alpha_q^{(n)} G_{\mathrm{M}} v_q^{(n)} \, , \\ r_{\mathrm{D},q}^{(n)} &= r_{\mathrm{D},q}^{(n-1)} - \alpha_q^{(n)} (v_q^{(n)} - \chi^{c(n-1)} \, G_{\mathrm{D}} v_q^{(n)}). \end{aligned}} \tag{4.47}$$

We remark that both operators G_{M} and G_{D} do not depend on the source positions. As a consequence, we may take the parameters $\gamma_q^{(n)}$ and $\alpha_q^{(n)}$ independent of q, see [55]. For a multifrequency inverse problem, these Green's operators do depend on the frequency and $\gamma_q^{(n)}$ and $\alpha_q^{(n)}$ become frequency dependent as well. The present analysis facilitates a simple extension for this problem.

4.5.3 Updating the Contrast

Using the update for the contrast sources, the fields $u_q^{(n)} = u^{(n)}(\boldsymbol{x}_m, \boldsymbol{x}_q^S)$ are directly obtained as

$$u_q^{(n)} = u_q^{inc} + G_{\mathrm{D}} w_q^{(n)} = u_q^{(n-1)} + \alpha_q^{(n)} G_{\mathrm{D}} v_q^{(n)} \, , \quad n = 1, 2, \dots \, , \tag{4.48}$$

after which an update for the contrast follows from

$$\chi^{c(n)} = \arg\min_{\chi} \left\{ \sum_{q=1}^{N^S} \| \chi \, u_q^{(n)} - w_q^{(n)} \|_{\mathrm{D}}^2 \right\}. \tag{4.49}$$

The minimizer is found explicitly as

$$\chi_{csi}^{c(n)}(\boldsymbol{x}) = \frac{\sum_{q=1}^{N^S} w^{(n)}(\boldsymbol{x}_m, \boldsymbol{x}_q^S)\, \overline{u}^{(n)}(\boldsymbol{x}_m, \boldsymbol{x}_q^S)}{\sum_{q=1}^{N^S} |u^{(n)}(\boldsymbol{x}_m, \boldsymbol{x}_q^S)|^2}. \tag{4.50}$$

In cases that *a priori* information with respect to the contrast distribution is present, we can apply some restrictions to the update of $\chi_{csi}^{c(n)}$, e.g. when the contrast is real valued, we replace the update by the contrast update

$$\boxed{\chi_{csi}^{c(n)} = \frac{\mathrm{Re}\left[\sum_{q=1}^{N^S} w_q^{(n)} \, \overline{u}_q^{(n)}\right]}{\sum_{q=1}^{N^S} |u_q^{(n)}|^2}.} \tag{4.51}$$

In addition, when we know that in a certain region the contrast vanishes, we may enforce these values of $\chi_{csi}^{c(n)}$ to zero.

After the update $\chi_{csi}^{c(n)}$, the residual error becomes

$$\boxed{r_{\mathrm{D},q}^{(n)} = \chi_{csi}^{c(n)} u_q^{(n)} - w_q^{(n)},} \tag{4.52}$$

which shows that the actual computation of this error from Eq. (4.47) is superfluous. With this result the description of the algorithm is completed, except for the starting values for the contrast sources $w_q^{(0)}$.

4.5.4 Initial Estimate

We cannot start with $w_q^{(0)} = 0$, since then $\chi^{c(0)} = 0$ and the normalization factor $\eta_{\mathrm{D}}^{(0)}$ is undefined. Therefore, we choose as starting values the contrast sources that only minimize the cost functional

$$F_{\mathrm{M}}(w_q^{(0)}) = \sum_{q=1}^{N^S} \| u_{\mathrm{M},q}^{sct} - G_{\mathrm{M}} w_q^{(0)} \|_{\mathrm{M}}^2. \tag{4.53}$$

Using the gradient method, we take

$$\boxed{w_q^{(0)} = \alpha_q^{(0)} \, g_q^{(0)}, \text{ and } g_q^{(0)} = G_{\mathrm{M}}^{\star} u_{\mathrm{M},q}^{sct}.} \tag{4.54}$$

The cost functional is minimized when each term for q is minimized, i.e.

$$\alpha_q^{(0)} = \arg\min_{\alpha} \{ \| u_{\mathrm{M},q}^{sct} - \alpha \, G_{\mathrm{M}} w_q^{(0)} \|_{\mathrm{M}}^2 \}. \tag{4.55}$$

This leads directly to

$$\alpha_q^{(0)} = \frac{\langle u_{\mathrm{M},q}^{sct}, G_{\mathrm{M}} g_q^{(0)} \rangle_{\mathrm{M}}}{\| G_{\mathrm{M}} g_q^{(0)} \|_{\mathrm{M}}^2}, \tag{4.56}$$

or in simplified form as

$$\boxed{\alpha_q^{(0)} = \frac{\| g_q^{(0)} \|_{\mathrm{D}}^2}{\| G_{\mathrm{M}} g_q^{(0)} \|_{\mathrm{M}}^2}.} \tag{4.57}$$

The gradient $g_q^{(0)} = G_{\mathbb{M}}^{\star} u_{\mathbb{M},q}^{sct}$ is the back projection of the data domain $\mathbb{M}$ into the object domain $\mathbb{D}$, and is often called a *back propagation* of the wave-field data.

With this initial estimate $w_q^{(0)} = w^{(0)}(\boldsymbol{x}, \boldsymbol{x}_q^S)$ and $u_q^{(0)} \approx u_q^{inc}$, the contrast estimate is obtained from

$$\boxed{\chi_{csi}^{c(0)} = \frac{\sum_{q=1}^{N^S} w_q^{(0)} \, \overline{u}_q^{inc}}{\sum_{q=1}^{N^S} |u_q^{inc}|^2}\,.} \tag{4.58}$$

After having discussed all the details of the CSI method, in the next subsection we present the Matlab implementations of the method.

4.5.5 Matlab Codes for the CSI Method

In this subsection, we present the Matlab codes for the 2D version Contrast Source Inversion method. The associated Matlab script `InverseCSI` is given in Matlab Listing II.14. The following steps are carried out:

(1) We start with the computation of the analytical data, denoted as dataCircle, by using `DataSctCircle` of Matlab Listing II.8. If these data have been computed with the forward code `Forwardw` of Matlab Listing II.7, we load these analytical data from the computer memory. When we would like to investigate what the result is of an "Inverse Crime", where both the forward and inverse codes are based on the same computational model, we have the option to load the data from the integral equation model. If in the Matlab function `init` the value of the `input.Noise` is equal to 1, then complex-valued noise is added with a standard deviation of $0.2 \times \text{mean}(|\text{dataCircle}|)$. Adding this so-called computer noise is certainly needed to prevent an inverse crime. However, we are now in the situation that our analytical data and numerical data of the integral equation method differ about 5–10%, and we do not need to add extra computer noise.

(2) Similar to the forward code `Forwardw` of Matlab Listing II.7, we compute the weak forms of the Green functions G_S and G_R with the Matlab function `GreenSourcesReceivers` of Matlab Listing II.11.

(3) The heart of the CSI method is found in the function `ITERCGMD` of the Matlab Listing II.15. In this function, the maximum number of iterations is set to it = 256. Both the initial contrast sources and the initial contrast are estimated with the help of the function `InitialEstimate` of Matlab Listing II.16. The contrast sources are determined by back projection. Besides the data operator `DopM` of Matlab Listing II.12, we also need its adjoint `AdjDopM` of Matlab Listing II.17. The contrast sources follow from minimization of the squared norm on $\mathbb{M}$, viz. ErrorM, after which the contrast follows from Eq. (4.58) using the function `UpdateContrast` of Matlab Listing II.18. Note that this function is called for $u = u^{inc}$. Further, we use the *a priori* information that the contrast function is either positive or negative. Further, the starting value of the squared norm on $\mathbb{D}$, viz. ErrorD, is computed. With these initial estimates, we employ the iterative CG method to determine the contrast sources that minimize ErrorM and ErrorD simultaneously. In each iteration, the function `UpdateContrast` of Matlab Listing II.18 determines the updates of the contrast. During the iterations, each update of the contrast is shown as an image plot on the computer screen. After finishing the iterative method, the final reconstructed contrast is plotted, both as image plot and mesh plot.

(4) With the help of the contrast sources solved from the present CSI method, we compute again the scattered field at the receiver locations, see function `DopM` of Matlab Listing II.12, and we

Matlab Listing II.14 The `InverseCSI` to reconstruct the contrast χ^c.

```
clear all; clc; close all; clear workspace
input = init();

% (1) Compute Exact data using analytic Bessel function expansion ---------
      disp('Running DataSctCircle');   DataSctCircle;

      load DataAnalytic;
      % load DataIntEq;   dataCircle = data;         % Inverse Crime

      % add complex-valued noise to data
      if input.Noise == 1
         dataCircle = dataCircle + mean(abs(dataCircle(:))) * input.Rand;
      end

% (2) Compute Green function for sources and receivers --------------------
      [G_S,G_R] = GreenSourcesReceivers(input);

% (3) Apply CSI method ------------------------------------------------------
      [w,CHI] = ITERCGMD(G_S,G_R, dataCircle, input);

   % Figure of contrast as image and surface plot
     x1 = input.X1(:,1);    x2 = input.X2(1,:);
     figure(5); IMAGESC(x1,x2,real(CHI));

     figure(6); mesh(abs(CHI),'edgecolor', 'k');
                view(37.5,45); axis xy; axis('off'); axis('tight')
                axis([1 input.N1 1 input.N2 0 1.5*max(abs(input.CHI(:)))])

% (4) Compute model data explained by reconstructed contrast souces ------
      data = zeros(input.NR, input.NS);
      for q = 1 : input.NS
         data(:,q) = DopM(G_R,w{q},input);
      end % q_loop

   % plot data and compare to synthetic data
     error = num2str(norm(data(:)-dataCircle(:),1)/norm(dataCircle(:),1));
     disp(['data error=' error]);
     figure(7);
     subplot(1,2,1)
       imagesc(1:input.NR,1:input.NS, abs(data));
       xlabel('x_p^R \rightarrow');
       ylabel('x_q^S \rightarrow');
       axis('equal','tight'); axis xy; colorbar('hor'); colormap jet;
       title('|data_{Exact}|')
     subplot(1,2,2)
       imagesc(1:input.NR,1:input.NS, abs(data-dataCircle));
       xlabel('x_p^R \rightarrow');
       ylabel('x_q^S \rightarrow');
       axis('equal','tight'); axis xy; colorbar('hor'); colormap jet;
       title('|data_{Inverted} - data_{Exact}|')
```

Matlab Listing II.15 The function `ITERCGMD` to update w_q and χ^c.

```
function [w,CHI] = ITERCGMD(u_inc ,G_R, data , input)
global it
FFTG = input.FFTG;  NS = input.NS;
v = cell(1,NS);     g_1 = cell(1,NS);     AN_1 = cell(1,NS);

itmax = 256;
eta_M = 1 / norm(data(:))^2;      % normalization factor eta_M

[w,u,CHI,r_M,r_D,eta_D,ErrorD,ErrorM] = ...
                       InitialEstimate(eta_M,u_inc,G_R,data,input);
it = 1;
while (it <= itmax)             % alternate updating of w and CHI
 Norm_M = 0;
 for q = 1 : NS
   g  = eta_M * AdjDopM(G_R,r_M{q},input)  ...
         + eta_D *(r_D{q}-AdjKop(conj(CHI).*r_D{q},FFTG));
   AN = norm(g(:))^2;
   if it > 1 ; BN = sum( g(:) .* conj(g_1{q}(:)) ); end
   if it == 1
      v{q} = g;
   else
      v{q} = g + real((AN-BN)/AN_1{q}) * v{q};
   end
    g_1{q} = g;
   AN_1{q} = AN;
   % determine step length alpha
   AN     = sum( g(:) .* conj(v{q}(:)) );
   GRv    = DopM(G_R,v{q},input);
   BN_M   = norm(GRv(:))^2;
   Kv     = Kop(v{q},FFTG);
   Lv     = v{q} - CHI.* Kv;
   BN_D   = norm(Lv(:))^2;
   alpha = real( AN / (eta_M * BN_M + eta_D * BN_D) );
   % update contrast source w and  residual  error  in M
    w{q} =  w{q} + alpha *   v{q};
    u{q} =  u{q} + alpha *  Kv;
   r_M{q} = r_M{q} - alpha * GRv(:);
   Norm_M = Norm_M + norm(r_M{q}(:))^2;
 end %q_loop
 ErrorM = eta_M * Norm_M;

 % Update contrast  by  minimization of  Norm_D = || CHI * u - w ||^2
  [CHI,r_D,eta_D,ErrorD] = UpdateContrast(w,u,u_inc,ErrorM,input);
  disp(['Iteration = ', num2str(it)]);
  disp('-------------------------------------------------------------------');
  it = it+1;
end % while

Error  = sqrt(ErrorM + ErrorD);
disp(['Number of iterations = ' num2str(it)]);
disp(['Total Error in M and D = ' num2str(Error)]);
```

Matlab Listing II.16 The `InitialEstimate` script to update w_q and χ^c.

```
function [w,u,CHI,r_M,r_D,eta_D,ErrorD,ErrorM] = ...
                          InitialEstimate(eta_M,u_inc,G_R,data,input)
global it
FFTG = input.FFTG;  NS = input.NS;
 w   = cell(1,NS);     u  = cell(1,NS);    r_M = cell(1,NS);

% Determine contrast sources by back-projection and minimization of Norm_M
  it = 0;
  Norm_M = 0;
  for q = 1 : NS
    r_M{q}  = data(:,q);
    g       = AdjDopM(G_R,r_M{q},input);
    A       = norm(g(:))^2;
    GRv     = DopM(G_R,g,input);
    B       = norm(GRv(:))^2;
    alpha   = real(A/B);
    r_M{q}  = r_M{q} - alpha * GRv(:);
    w{q}    = alpha * g;
    u{q}    = u_inc{q} + Kop(w{q},FFTG);
   Norm_M   = Norm_M + norm(r_M{q}(:))^2;
  end %q_loop
  ErrorM = eta_M * Norm_M;

% Update contrast  by  minimization of  Norm_D = || CHI * u_inc - w ||^2
  [CHI,r_D,eta_D,ErrorD] = UpdateContrast(w,u_inc,u_inc,ErrorM,input);

  disp([' ErrorM = ',num2str(ErrorM),'    ErrorD = ',num2str(ErrorD)]);
  disp(['Iteration = ', num2str(it)]);                     saveCHI(CHI);
  disp('----------------------------------------------------------------');

% Figures of contrast
  x1 = input.X1(:,1);    x2 = input.X2(1,:);
  figure(3);  IMAGESC(x1,x2,real(CHI));
  figure(4);  mesh(abs(CHI),'edgecolor', 'k');
              view(37.5,45); axis xy; axis('off'); axis('tight')
              axis([1 input.N1 1 input.N2  0 1.5*max(abs(input.CHI(:)))]);
end
```

compare these data with the analytical data used in step (1). By plotting the differences and computing the error being the norm of these differences, we obtain an indication how good the data are explained by the inversion scheme at hand.

In Figures 4.6 and 4.7, the reconstructed contrast is plotted after the initial estimate (it = 0), for a positive and a negative contrast profile, respectively. We observe that the reconstructed profiles are very flat. This is due to the choice of $u = u^{inc}$ made in Eq. (4.58). If we substitute the updated value $u^{(0)} = u^{inc} + Gw^{(0)}$, a more peaked value is arrived at, which may lead to a wrong local minimum.

In Figures 4.8 and 4.9, the reconstructed contrast is plotted after 256 iterations. More iterations did not improve the reconstructed contrast profiles substantially. In both cases, we observe that, in the region of the circular scatterer, the contrast profiles are band-limited. The profile for the

Matlab Listing II.17 The function `AdjDopM` for the adjoint operator G_M^*.

```
function [adjGR] = AdjDopM(G_R,f,input)

gam0 = input.gamma_0;
dx   = input.dx;

adjGR = zeros(input.N1,input.N2);
for p = 1: input.NR
    adjGR = adjGR + conj(gam0^2 * dx^2 *  G_R{p}) * f(p);
end % p_loop
```

Matlab Listing II.18 The function `UpdateContrast` to update χ^c.

```
function [CHI,r_D,eta_D,ErrorD] = UpdateContrast(w,u,u_inc,ErrorM,input)

% (1) Compute CHI_csi
      INuw = zeros(input.N1,input.N2);
      INuu = zeros(input.N1,input.N2);
      for q = 1 : input.NS
         INuw = INuw + conj(u{q}) .* w{q};
         INuu = INuu + conj(u{q}) .* u{q};
      end %q_loop

   % Without a priori information
      CHI = INuw./INuu;

   % With a priori information of the overall sign of the contrast
      CHI = input.signCHI * abs(CHI);

% (2) Update the object error in D
      Norm_D = 0;    Norm = 0;     r_D{q} = cell(1,input.NS);
      for q = 1 : input.NS
         r_D{q} = CHI .* u{q} - w{q};
         Norm_D = Norm_D + norm(r_D{q}(:))^2;
         Norm   = Norm   + norm(CHI(:).*u_inc{q}(:))^2;
      end %q_loop
      eta_D = 1 / Norm;
      ErrorD = eta_D * Norm_D;
      disp([' ErrorM = ',num2str(ErrorM),'    ErrorD = ',num2str(ErrorD)]);

% (4) Show intermediate pictures of reconstruction
      x1 = input.X1(:,1);    x2 = input.X2(1,:);
      figure(5); IMAGESC(x1,x2,real(CHI));
      title('\fontsize{13} Reconstructed Contrast Re[\chi^c] ');
      pause(0.1)
```

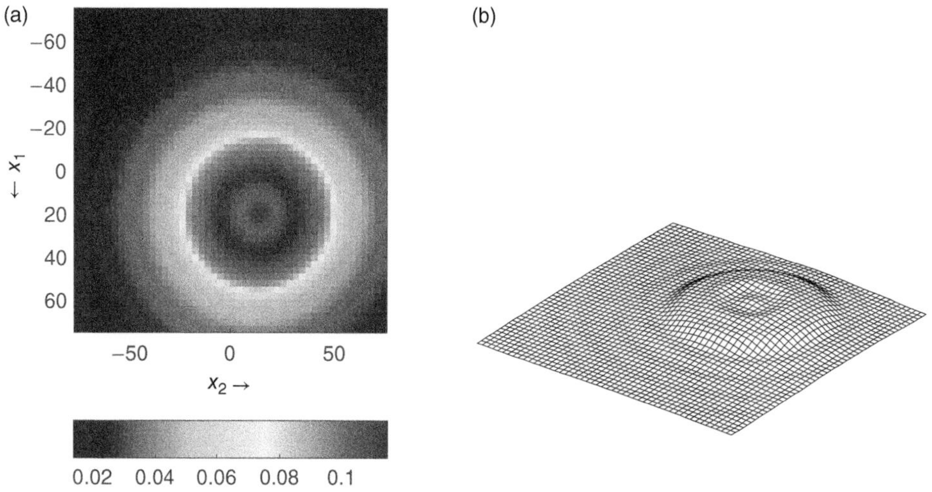

Figure 4.6 CSI method: plots of the contrast reconstructed after the initial estimate, for **noise-free** data and nominal values $c_0 = 1500$ m/s and $\hat{c}_{sct} = 2000$ m/s, with *a priori* information that the contrast is positive, using Matlab's `imagesc` script (a) and plotting the absolute value with Matlab's `mesh` script (b).

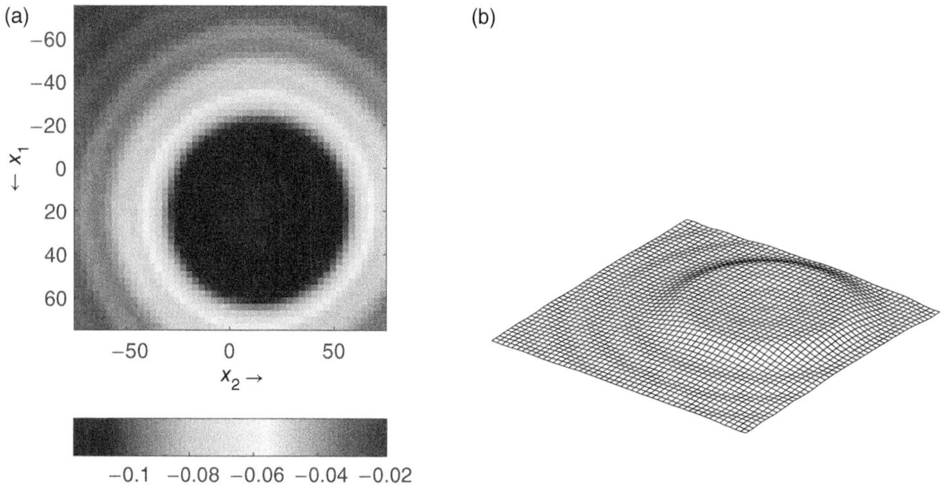

Figure 4.7 Similar to Figure 4.6 but now for nominal values $c_0 = 1500$ m/s and $\hat{c}_{sct} = 1200$ m/s, with *a priori* information that the contrast is negative.

case of positive contrast is a little more band-limited than the one for the case of negative contrast. Outside the area of the scatterer, the profiles are almost flat, because in Figure 4.8 the negative values of the contrast are enforced to zero, while in Figure 4.9 the positive values of the contrast are enforced to zero. This type of "regularization" enables excellent reconstructions. If we omit this *a priori information* for the present nominal parameters, useful contrast profiles are not arrived at. When we add our computer noise to the data, the reconstructed contrast after 256 iterations is

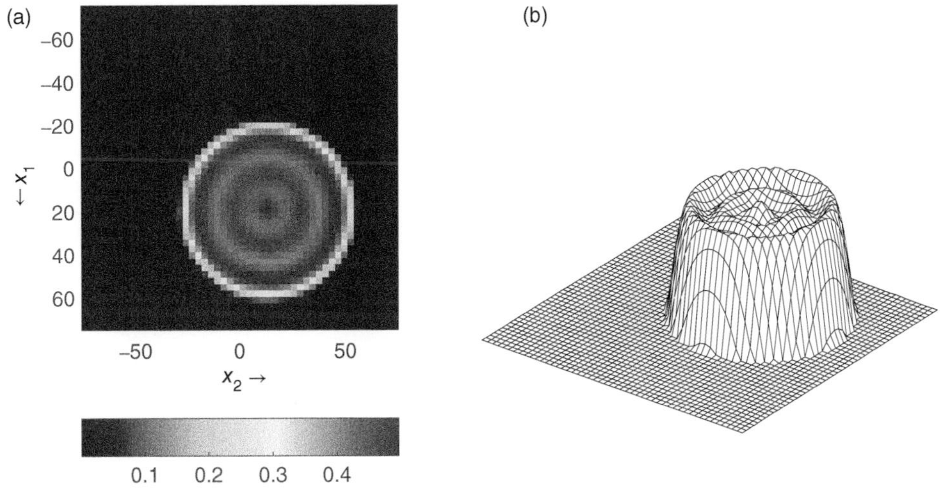

Figure 4.8 CSI method: plots of the contrast reconstructed after 256 iterations, for **noise-free** data and nominal values $c_0 = 1500$ m/s and $\hat{c}_{sct} = 2000$ m/s, with *a priori* information that the contrast is positive, using Matlab's `imagesc` script (a) and plotting the absolute value with Matlab's `mesh` script (b).

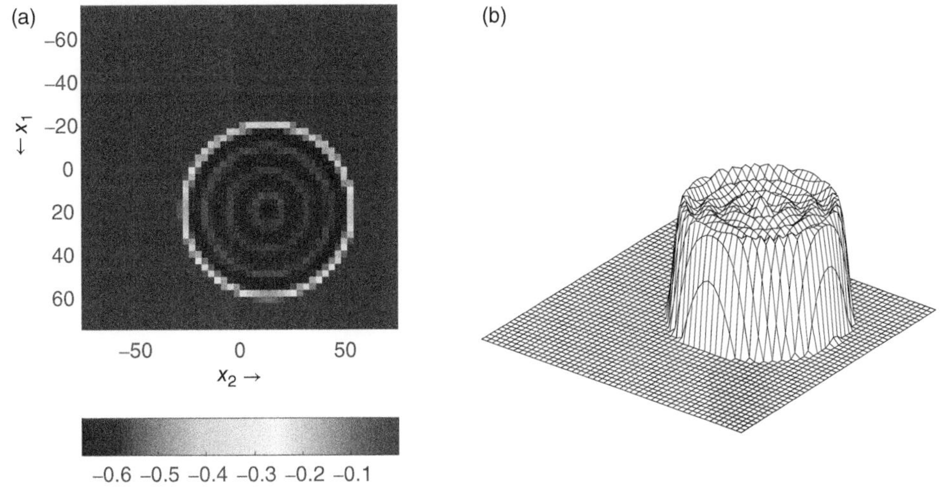

Figure 4.9 Similar to Figure 4.8, but now for nominal values $c_0 = 1500$ m/s and $\hat{c}_{sct} = 1200$ m/s ($-0.5625 < \hat{\chi}^c < 0$), with *a priori* information that the contrast is negative.

plotted in Figures 4.10 and 4.11. We observe that, in the that the contrast profile is not zero outside the scatterer. Positive peaks are present, which are compatible with the noise level in the data.

In Figures 4.12 and 4.13, on the left-hand sides we present the amplitude profile of the analytical data, while on the right-hand sides we plot the amplitude of the difference between the data modeled after 256 iterations and the analytic data. Obviously, the data is better explained when the differences are smaller. The global error in the inverted data amounts to 0.2% and 0.1%, for

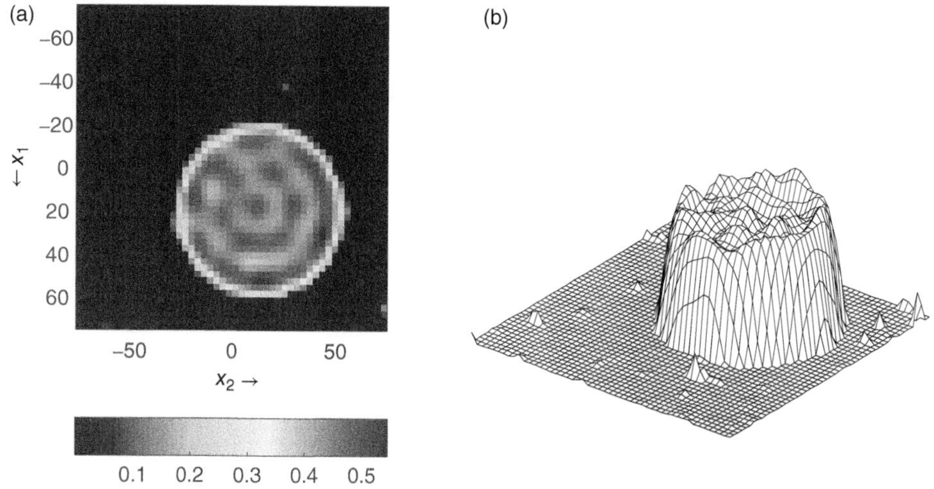

Figure 4.10 CSI method: plots of the contrast reconstructed after 256 iterations, for **noisy** data and nominal values $c_0 = 1500$ m/s and $\hat{c}_{sct} = 2000$ m/s, with *a priori* information that the contrast is positive, using Matlab's `imagesc` script (a) and plotting the absolute value with Matlab's `mesh` script (b).

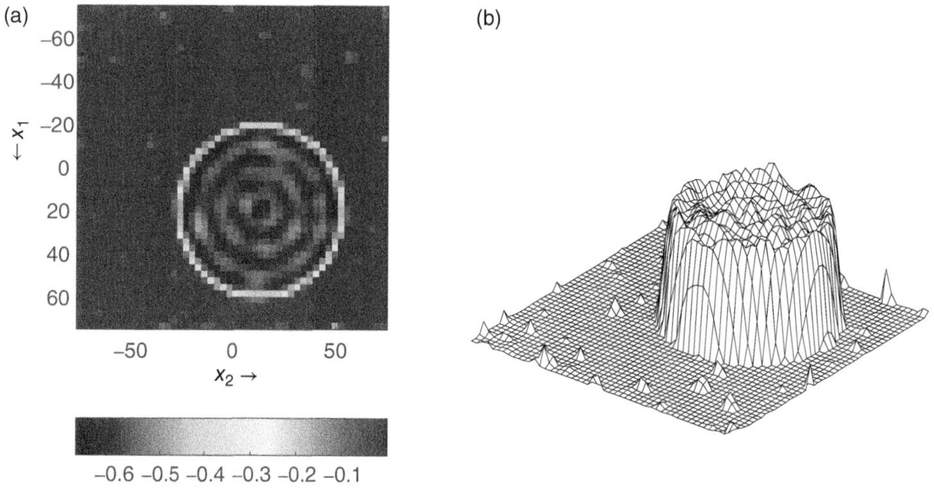

Figure 4.11 Similar to Figure 4.10, but now for nominal values $c_0 = 1500$ m/s and $\hat{c}_{sct} = 1200$ m/s, with *a priori* information that the contrast is negative.

$c_{sct} = 2000$ and 1200 m/s, respectively. However, for $c_{sct} = 2000$ m/s, the amplitudes are a factor of two larger than the ones for $c_{sct} = 1200$ m/s. Hence, it seems that the data of the latter case with the sharper data distribution are better explained. If we contaminate the analytical data with noise, we obtain the pictures of Figures 4.14 and 4.15. On Figures 4.14a and 4.15a, the plots of the analytical data include noise. From Figures 4.14b and 4.15b, we conclude that the noise level is more than 10% of the data amplitudes. This explains that the global error in the inverted data amounts now to 11% and 9%, for $c_{sct} = 2000$ and 1200 m/s, respectively.

In Section 4.2, we will discuss the addition of regularization to the CSI method, in particular the so-called Multiplicative Regularized Contrast Source Inversion (MRCSI) method.

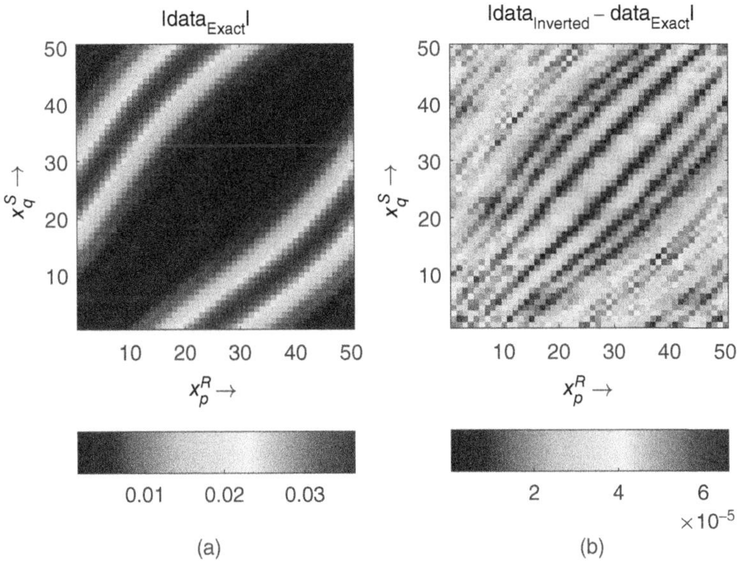

Figure 4.12 CSI method: plots of the **noise-free** data as function of the source and receiver numbers, nominal values $c_0 = 1500$ m/s and $\hat{c}_{sct} = 2000$ m/s, assuming that the contrast is positive. The global error in the inverted data amounts to 0.2%.

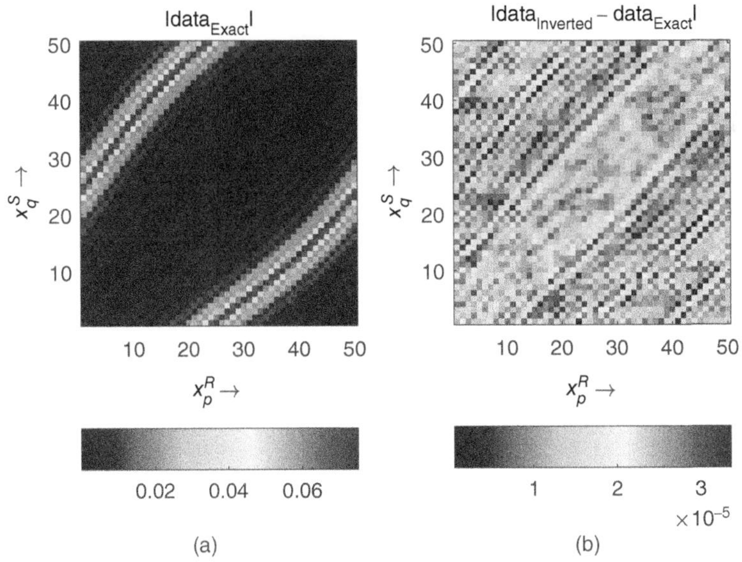

Figure 4.13 CSI method: plots of the **noise-free** data as function of the source and receiver numbers, for nominal values $c_0 = 1500$ m/s and $c_{sct} = 1200$ m/s, assuming that the contrast is negative. The global error in the inverted data amounts to 0.1%.

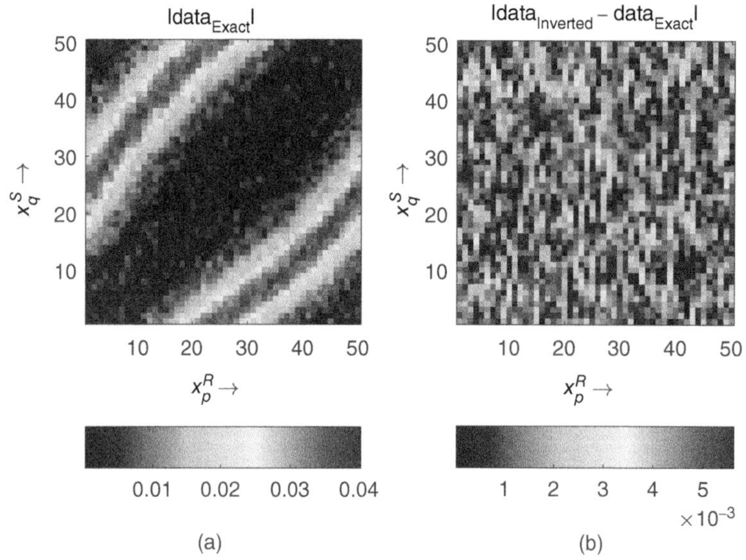

Figure 4.14 CSI method: plots of the **noisy** data as function of the source and receiver numbers, for nominal values $c_0 = 1500$ m/s and $c_{sct} = 2000$ m/s, assuming that contrast is positive. The global error in the inverted data amounts to 11%.

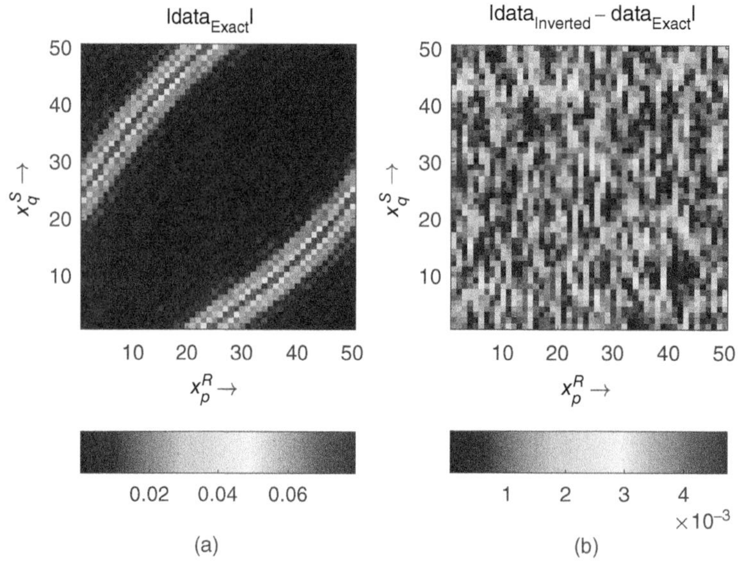

Figure 4.15 CSI method: Plots of the **noisy** data as function of the source and receiver numbers, for nominal values $c_0 = 1500$ m/s and $c_{sct} = 1200$ m/s, assuming that contrast is negative. The global error in the inverted data amounts to 9%.

4.6 Multiplicative Regularized Contrast Source Inversion

Although the inclusion of the object equation as in the second term of Eq. (4.15) can be considered as a physical regularization of the ill-posed data equation in the first term of Eq. (4.15), the inversion results may be improved by taking into account *a priori* information about the contrast profile. The standard way to include this information is to modify the functional of Eq. (4.17) by introducing an extra penalty function, viz.,

$$\begin{aligned}\mathcal{F}^{(n)}(\hat{w}_q^{(n)}, \hat{\chi}^{c(n)}) &= F^{(n)}(\hat{w}_q^{(n)}, \hat{\chi}^{c(n)}) + \hat{\eta}_R^{(n)} F_R^{(n)}(\hat{\chi}^{c(n)}) \,, \\ &= \hat{\eta}_{\mathrm{M}} F_{\mathrm{M}} \left(\hat{w}_q^{(n)} \right) + \hat{\eta}_{\mathrm{D}}^{(n)} F_{\mathrm{D}}(\hat{w}_q^{(n)}, \hat{\chi}^{c(n)}) + \hat{\eta}_R^{(n)} F_R^{(n)}(\hat{\chi}^{c(n)}) \,. \end{aligned} \tag{4.59}$$

The regularization function $F_R^{(n)}$ is a function of the contrast and $\hat{\eta}_R^{(n)}$ is a positive regularization parameter. As known in the literature, this additive regularization has a very positive effect on the quality of the reconstruction. The drawback is the presence of the regularization parameter in the cost functional, which with the present knowledge can only be determined through considerable numerical experimentation and *a priori* information of the desired reconstruction. The regularization parameter measures the trade-off between a good fit and an oscillatory solution. An important and still current problem is a proper choice of this parameter. There are several possible strategies that depend on the additional information concerning the problem and its solution. For instant, in order to apply the well-known discrepancy principle (see Morozov [34]), which is an *a posteriori* strategy for choosing $\hat{\eta}_R^{(n)}$ as a function of the noise level, the input level must be known. Other strategies that do not require knowledge of the input noise level are the generalized cross validation (see Golub et al. [18]) and the L-curve method (see Hansen [21]), but they require a large number of extra computations involving the wave-field operators both in the measurement domain and in the object domain.

Further, numerical experiments have shown that the results improve when we let the parameter $\hat{\eta}_R^{(n)}$ decrease with increasing number of iterations. In fact, a good choice seems to take this parameter proportionally to the value of the cost functional $\hat{\eta}_{\mathrm{D}}^{(n)} F_{\mathrm{D}}(\hat{w}_q^{(n)})$. This numerical experimentation has led to the idea of multiplicative regularization [54]. The cost functional is then written as the product of the physical cost functional and the regularization function, viz.,

$$\boxed{\mathcal{F}^{(n)}(\hat{w}_q^{(n)}, \hat{\chi}^{c(n)}) = F^{(n)}(\hat{w}_q^{(n)}, \hat{\chi}^{c(n)}) \, F_R^{(n)}(\hat{\chi}^{c(n)}) \,.} \tag{4.60}$$

Minimization of this functional with respect to changes in the contrast $\hat{\chi}^c$ will change the minimization procedure of the CSI reconstruction. Our aim is not to change the updating procedure of the contrast sources $\hat{w}_q$. At the beginning of each iteration, we have to replace the contrast update $\hat{\chi}^{c(n-1)}$ by its regularized counterpart, but the remainder of the contrast-source updating procedure is not changed, when we keep the regularization function $F_R^{(n)}$ to be equal to one during this part of the iteration. Then, only the updating of the contrast (for given $\hat{w}_q = \hat{w}_q^{(n)}$) has to be modified. We first specify the regularization function and subsequently we discuss the updating algorithm of the contrast.

In the remainder of this section, we again omit the hat-symbols to indicate the s-dependency.

4.6.1 Regularization Function for the Contrast Update

A common choice for the regularization term $F_R(\chi)$ is either the Tikhonov functional [48]

$$F_R(\chi) = \| \chi \|_{\mathrm{D}}^2 \,, \tag{4.61}$$

with $\chi = \chi^c$ or the total variation (TV) functional [43],

$$F_R^{(n)}(\chi) = \int_{x \in \mathbb{D}} (|\nabla \chi(x)|^2 + \delta^{(n)})^{\frac{1}{2}} \mathrm{d}V \,, \tag{4.62}$$

where $\nabla\chi = \nabla\chi^c$ is the gradient of the wave-speed contrast. The term $\delta^{(n)}$ is a real positive parameter originally introduced to restore differentiability of the TV. The minimization of the Tikhonov functional in Eq. (4.61) restrict the amplitudes of χ, but often it induces spurious oscillations, or ringing, when the image χ is discontinuous. Hence, this functional is unsuitable for applications where one wishes to recover sharp images modeled by discontinuities in χ. The minimization of the TV functional in Eq. (4.62) does not suffer from this problem, but it requires a nonlinear minimization due to its L_1-norm nature. In contrary to the L_1-norm, the L_2-norm of Eq. (4.61) favors a smooth profile, but edge-preserving effects of the regularization will be lost.

Inspired by the work on image restoration by Charbonnier et al. [9], we employ the TV-factor as a weighted edge-preserving regularization function with a L_2-norm nature. The weight function has an edge-preserving character. In each updating step of our contrast, we first calculate the contrast update $\chi_{csi}^{c(n)}$ of Eq. (4.50). This nonregularized contrast update is obtained analytically and it is denoted as $\chi_{csi}^{(n)}$. Subsequently, we define the regularization functional as [54]

$$\boxed{F_R^{(n)}(\chi) = \frac{1}{V} \int_{x \in \mathbb{D}} \frac{|\nabla \chi(x)|^2 + \delta^{(n)}}{|\nabla \chi_{csi}^{c(n)}(x)|^2 + \delta^{(n)}} \mathrm{d}V \,,} \tag{4.63}$$

where $V = \int_{x \in \mathbb{D}} \mathrm{d}V$. In the process of updating the contrast sources, this regularization functional has the required property that it is equal to one when $\chi = \chi_{csi}^{(n)}$.

Although the parameter $\delta^{(n)}$ is introduced for restoring differentiability of the TV regularizer, it also controls the influence of the regularization as a function of the number of iterations. For relatively large values of $\delta^{(n)}$, the regularization function is close to one and the regularization has no effect, while for relatively small values, the regularization function has a strong effect. We like to increase the influence of regularization as a function of the number of iterations by decreasing this control parameter $\delta^{(n)}$. Since the value of the object functional will decrease as a function of the number of iterations, in our discrete analysis we choose

$$\boxed{\delta^{(n)} = (1/\Delta x)^2 \, \eta_{\mathrm{D}}^{(n)} F_{\mathrm{D}}(w_q^{(n)}, \chi_{csi}^{(n)}) \,,} \tag{4.64}$$

where the mesh size Δx is included for dimensional purposes. This choice of the control operator is inspired by the idea that in the first few iterations, we do not need the minimization of the total variation and as the iterations proceed we want to increase the effect of regularization. If the iterative scheme has converged, the regularized contrast, χ, becomes equal to $\chi_{csi}^{c(n)}$ and the regularization function is equal to one. Then, the cost functional $\mathcal{F}_n$ is only determined by the original cost functional $F^{(n)} = \eta_{\mathrm{M}} F_{\mathrm{M}} + \eta_{\mathrm{D}}^{(n)} F_{\mathrm{D}}$, which measures the error in the satisfaction of wave field equations. Minimizing of the physical cost functional $F^{(n)}$ is the main objective of our wave-field inversion problem.

Another choice for the control parameter $\delta^{(n)}$ is the mean of the spatial variation, i.e.

$$\boxed{\delta^{(n)} = \frac{1}{V} \int_{x \in \mathbb{D}} |\nabla \chi_{csi}^{c(n)}(x)|^2 \, \mathrm{d}V \,.} \tag{4.65}$$

which has been employed in [52] for optical microscopy. In practice, at the end of the iterative procedure, the values of this choice of control parameter are larger than the ones using Eq. (4.64). This will provide a smoother profile in the reconstruction of the contrast.

4.6.2 Updating the Contrast with Multiplicative Regularization

As already said, we start with the observation that the contrast as minimizer of $F_{\mathbb{D}}$ with respect to changes of χ is given by

$$\chi_{csi}^{c(n)} = \arg\min_{\chi} \left\{ \sum_{q=1}^{N^S} \| \chi \, u_q^{(n)} - w_q^{(n)} \|_{\mathbb{D}}^2 \right\}, \tag{4.66}$$

or, see Eq. (4.50),

$$\chi_{csi}^{c(n)}(\boldsymbol{x}) = \frac{\sum_{q=1}^{N^S} w_q^{(n)}(\boldsymbol{x}) \, \overline{u}_q^{(n)}(\boldsymbol{x})}{\sum_{q=1}^{N^S} |u_q^{(n)}(\boldsymbol{x})|^2} . \tag{4.67}$$

We remark that the denominator $\sum_{q=1}^{N^S} |u_q^{(n)}|^2$ acts as a preconditioner for the minimization of Eq. (4.66). Therefore, our further analysis simplifies when we consider the minimization of Eq. (4.66) with respect to changes of χ as

$$\chi^{c(n)} = \arg\min_{\chi} \left\{ \int_{\boldsymbol{x}\in\mathbb{D}} |\chi(\boldsymbol{x}) - \chi_{csi}^{c(n)}(\boldsymbol{x})|^2 \, \mathrm{d}V \right\} . \tag{4.68}$$

Possible modifications of this contrast function, e.g. by including positivity constraints, are assumed to be included in the resulting $\chi_{csi}^{c(n)}$. At this point, we remark that this contrast is independent of the source positions, and the regularization is carried out in the spatial domain $\mathbb{D}$. In view of Eq. (4.68), we modify the multiplicative cost functional of Eq. (4.60) as

$$\mathcal{F}^{(n)}(w_q^{(n)}, \chi) := \left[\eta_{\mathrm{M}} F_{\mathrm{M}}(w_q^{(n)}) + \frac{\int_{\boldsymbol{x}\in\mathbb{D}} |\chi(\boldsymbol{x}) - \chi_{csi}^{c(n)}(\boldsymbol{x})|^2 \, \mathrm{d}V}{\int_{\boldsymbol{x}\in\mathbb{D}} |\chi_{csi}^{c(n)}(\boldsymbol{x})|^2 \, \mathrm{d}V} \right] F_R^{(n)}(\chi) \tag{4.69}$$

and replace Eq. (4.68) by

$$\chi^{c(n)} = \arg\min_{\chi} \{ \mathcal{F}^{(n)}(w_q^{(n)}, \chi) \} . \tag{4.70}$$

To minimize this relation, we opt for an algorithm that belongs to the class of an iterative regularization method. Convergence in one iteration is undesirable, because the contrast sources are not accurately enough determined and we can end up in a local minimum. In the literature, we employed a conjugate-gradient type of minimization. The structure of the minimization procedure is such that it will minimize the regularization functional with a large weighting factor in the beginning of the optimization process, when the factor in front of $F_R^{(n)}$ is large, and that it will gradually minimize more and more the error in the measurement and object domains, when the regularization factor $F_R^{(n)}$ remains a nearly constant value close to one. If noise is present in the image, the term F_{M} will remain at a large value during the whole optimization, and therefore the weight of the regularization factor will be more significant. Hence, the noise will, at all times, be suppressed in the reconstruction process and we automatically fulfill the need of a larger degree of regularization when the data set contains noise.

In this book, we discuss an alternative method to minimize Eq. (4.70). We first write the regularization function of Eq. (4.63) as

$$F_R^{(n)}(\chi) = \frac{(\Delta x)^2}{V} \int_{\boldsymbol{x}\in\mathbb{D}} [b^{(n)} \, |\boldsymbol{\nabla}\chi(\boldsymbol{x})|^2 + b^{(n)} \, \delta^{(n)}] \, \mathrm{d}V , \tag{4.71}$$

with weights $b^{(n)} = b^{(n)}(\boldsymbol{x})$ given by

$$b^{(n)}(\boldsymbol{x}) = \frac{1}{(\Delta x)^2} \frac{1}{|\boldsymbol{\nabla} \chi_{csi}^{c(n)}(\boldsymbol{x})|^2 + \delta^{(n)}} \,. \tag{4.72}$$

The factors $(\Delta x)^2$ and $1/(\Delta x)^2$ are inserted to simplify the numerical representations of the $\boldsymbol{\nabla}$ operators in terms of finite differences. Subsequently, we rewrite the first term on the right-hand side of Eq. (4.71). Using Gauss' integral theorem, we observe that

$$\int_{\mathbb{D}} \boldsymbol{\nabla} \cdot (\overline{\chi}(\boldsymbol{x})\; b^{(n)} \boldsymbol{\nabla} \chi)\, \mathrm{d}V = \int_{\partial \mathbb{D}} (\overline{\chi}(\boldsymbol{x})\; b^{(n)} \boldsymbol{\nabla} \chi) \cdot \boldsymbol{\nu}\, \mathrm{d}A = 0 \,, \tag{4.73}$$

because $\overline{\chi}$ vanishes on the boundary $\partial \mathbb{D}$ (with normal vector $\boldsymbol{\nu}$) of the domain $\mathbb{D}$. Therefore, integration by parts leads to

$$F_R^{(n)}(\chi) = \frac{(\Delta x)^2}{V} \int_{\boldsymbol{x} \in \mathbb{D}} [-\overline{\chi}(\boldsymbol{x})\; \boldsymbol{\nabla} \cdot (b^{(n)} \boldsymbol{\nabla} \chi) + b^{(n)} \delta^{(n)}]\, \mathrm{d}V \,. \tag{4.74}$$

To minimize Eq. (4.69), we calculate the first variation with respect to the function $\overline{\chi}(\boldsymbol{x})$, where $\boldsymbol{x}$ is a grid point of $\mathbb{D}$. With this calculus of variations, see Courant and Hilbert [13], Chapter IV, the associated Euler–Lagrange equation is arrived at,

$$\begin{aligned} &\left[\frac{\chi(\boldsymbol{x}) - \chi_{csi}^{c(n)}(\boldsymbol{x})}{\displaystyle\int_{\boldsymbol{x} \in \mathbb{D}} |\chi_{csi}^{c(n)}|^2\, \mathrm{d}V}\right] F_R^{(n)}(\chi) \\ &- \left[\eta_{\mathrm{M}} F_{\mathrm{M}}(w_q^{(n)}) + \frac{\displaystyle\int_{\boldsymbol{x} \in \mathbb{D}} |\chi(\boldsymbol{x}) - \chi_{csi}^{c(n)}(\boldsymbol{x})|^2\, \mathrm{d}V}{\displaystyle\int_{\boldsymbol{x} \in \mathbb{D}} |\chi_{csi}^{c(n)}|^2\, \mathrm{d}V}\right] \frac{(\Delta x)^2}{V} \boldsymbol{\nabla} \cdot [b^{(n)} \boldsymbol{\nabla} \chi(\boldsymbol{x})] = 0 \,. \end{aligned} \tag{4.75}$$

This is a nonlinear equation for $\chi(\boldsymbol{x})$. For our regularization method, a linear version of this equation suffices. After all, our only objective is to use this equation for a regularized version of the contrast. We write $\chi(\boldsymbol{x}) = \chi_{csi}^{c(n)}(\boldsymbol{x}) + \delta\chi(\boldsymbol{x})$ and we retain only the terms of first order in $\delta\chi$. Recognizing that $F_R^{(n)}(\chi_{csi}^{c(n)}) = 1$ and multiplying both sides of the resulting equation by $\int_{\boldsymbol{x}' \in \mathbb{D}} |\chi_{csi}^{c(n)}|^2\, \mathrm{d}V$, Eq. (4.75) becomes a linear equation for $\delta\chi(\boldsymbol{x})$. Adding the known term $\chi_{csi}^{c(n)}(\boldsymbol{x})$ to both sides of this equation, we obtain the linear equation for $\chi(\boldsymbol{x})$, viz.,

$$\boxed{\chi(\boldsymbol{x}) - \left[\eta_{\mathrm{M}} F_{\mathrm{M}} \frac{1}{V} \int_{\boldsymbol{x} \in \mathbb{D}} |\chi_{csi}^{c(n)}|^2\, \mathrm{d}V\right] (\Delta x)^2\; \boldsymbol{\nabla} \cdot [b^{(n)} \boldsymbol{\nabla} \chi(\boldsymbol{x})] = \chi_{csi}^{c(n)}(\boldsymbol{x}) \,,} \tag{4.76}$$

for $\boldsymbol{x} \in \mathbb{D}$. When we introduce a rectangular grid in the domain $\mathbb{D}$ and replace the gradient and divergence operators by a finite-difference approximation, we observe that Eq. (4.76) represents a linear system of equations for the unknown values $\chi(\boldsymbol{x})$. Its solution yields the regularized contrast value, viz. $\chi^{c(n)}$. In fact, this contrast quantity provides our regularized version of $\chi_{csi}^{c(n)}$. The strength of the regularization is determined by the mismatch in the measurement error $\eta_{\mathrm{M}} F_{\mathrm{M}}(w_q^{(n)})$ and the mean of $|\chi_{csi}^{c(n)}|^2$ over the domain $\mathbb{D}$.

4.6.3 Numerical Implementation of the Regularization

In this subsection, we discuss the numerical implementation of our regularization procedure. In 2D, we deal with $N_1 \times N_2$ subdomains, where the midpoints are denoted as $\{x_{i,j},\ i = 1 : N_1,\ j = 1 : N_2\}$. To compute the $\boldsymbol{\nabla}$ operators, central finite differences cannot be used because they disregard jumps in the contrast function at the central points. Therefore, we use the forward and backward finite-difference operators in the x_1- and x_2-directions, defined as

$$\begin{aligned}
\Delta_1\chi(x_{i,j}) &= \chi(x_{i+1,j}) - \chi(x_{i,j}),\\
\Delta_2\chi(x_{i,j}) &= \chi(x_{i,j+1}) - \chi(x_{i,j}),\\
\nabla_1\chi(x_{i,j}) &= \chi(x_{i,j}) - \chi(x_{i-1,j}),\\
\nabla_2\chi(x_{i,j}) &= \chi(x_{i,j}) - \chi(x_{i,j-1}),
\end{aligned} \tag{4.77}$$

respectively. To obtain a symmetric matrix in Eq. (4.76), we consider a superposition of forward and backward operations. To compute the weights $b^{(n)}$, we calculate $|\boldsymbol{\nabla}(\boldsymbol{x})|^2$ as

$$(\Delta x)^2|\boldsymbol{\nabla}\chi(x_{i,j})|^2 = \frac{1}{2}[|\Delta_1\chi(x_{i,j})|^2 + |\Delta_2\chi(x_{i,j})|^2 + |\nabla_1\chi(x_{i,j})|^2 + |\nabla_2\chi(x_{i,j})|^2]. \tag{4.78}$$

Replacing χ by $\chi_{csi}^{c(n)}$, we substitute this expression into the following relation for $b^{(n)}$:

$$b^{(n)}(x_{i,j}) = \frac{1}{(\Delta x)^2|\boldsymbol{\nabla}\chi_{csi}^{c(n)}(x_{i,j})|^2 + (\Delta x)^2\,\delta^{(n)}}\,. \tag{4.79}$$

Next we consider the calculation of $(\Delta x)^2\boldsymbol{\nabla}\cdot[b^{(n)}\boldsymbol{\nabla}\chi(x_{i,j})]$. We compute it as

$$\begin{aligned}
(\Delta x)^2\,\boldsymbol{\nabla}\cdot[b^{(n)}\boldsymbol{\nabla}\chi(x_{i,j})] = {}& \frac{1}{2}\{\Delta_1[b^{(n)}(x_{i,j})\nabla_1\chi(x_{i,j})] + \Delta_2[b^{(n)}(x_{i,j})\nabla_2\chi(x_{i,j})]\}\\
&+\frac{1}{2}\{\nabla_1[b^{(n)}(x_{i,j})\Delta_1\chi(x_{i,j})] + \nabla_2[b^{(n)}(x_{i,j})\Delta_2\chi(x_{i,j})]\}\,.
\end{aligned} \tag{4.80}$$

Using Eq. (4.77), we obtain

$$\begin{aligned}
(\Delta x)^2\,\boldsymbol{\nabla}\cdot[b^{(n)}\boldsymbol{\nabla}\chi(x_{i,j})] = {}& -\frac{1}{2}\,\big[\,b^{(n)}(x_{i+1,j}) + 2b^{(n)}(x_{i,j}) + b^{(n)}(x_{i,j+1})\\
&\quad+\,b^{(n)}(x_{i-1,j}) + 2b^{(n)}(x_{i,j}) + b^{(n)}(x_{i,j-1})\big]\,\chi(x_{i,j})\\
&+\frac{1}{2}[b^{(n)}(x_{i+1,j}) + b^{(n)}(x_{i,j})]\chi(x_{i+1,j})\\
&+\frac{1}{2}[b^{(n)}(x_{i-1,j}) + b^{(n)}(x_{i,j})]\chi(x_{i-1,j})\\
&+\frac{1}{2}[b^{(n)}(x_{i,j+1}) + b^{(n)}(x_{i,j})]\chi(x_{i,j+1})\\
&+\frac{1}{2}[b^{(n)}(x_{i,j-1}) + b^{(n)}(x_{i,j})]\chi(x_{i,j-1})\,.
\end{aligned} \tag{4.81}$$

In case $b^{(n)} = 1$ for all spatial points, we obtain the discretization rule for the 2D Laplacian $h^2\boldsymbol{\nabla}^2 u_{0,0} = (u_{1,0} + u_{0,1} + u_{-1,0} + u_{0,-1} - 4u_{0,0}) + O(h^4)$, with $h = \Delta x$, cf. equation 25.3.30 of Abramowitz and Stegun [1]. It shows that our TV regularization is a weighted form of the so-called Laplacian regularization. In our 2D discrete grid, the mean of $|\chi_{csi}^{(n)}|^2$ over the domain $\mathbb{D}$ is approximated as

$$\text{mean}(|\chi_{csi}^{c(n)}|^2) = \frac{1}{V}\int_{x\in\mathbb{D}} |\chi_{csi}^{c(n)}|^2\,\mathrm{d}V = \frac{1}{N_1\,N_2}\sum_{i=1}^{N_1}\sum_{j=1}^{N_2}|\chi_{csi}^{c(n)}(x_{i,j})|^2\,. \tag{4.82}$$

With these numerical implementations, the regularization equation of Eq. (4.76) has been fully specified.

4.6.4 Numerical Solution of Regularization Equation

To discuss the solution of Eq. (4.76), we write it as

$$A\chi = (D + R)\,\chi = \chi_{csi}^{c(n)}\,, \tag{4.83}$$

where the diagonal matrix D is related to the term with $\chi(x_{i,j})$ on the right-hand side of Eq. (4.81), while the R matrix is related to the other four terms. The diagonal constituent of $A\chi$ is given by

$$\begin{aligned} D\chi(x_{i,j}) = \; & 1 + \frac{1}{2}\,\eta_{\mathrm{M}}F_{\mathrm{M}}(w_{q,n})\ \mathrm{mean}(|\chi_{csi}^{c(n)}|^2) \\ & \times \left[\, b^{(n)}(x_{i+1,j}) + 2b^{(n)}(x_{i,j}) + b^{(n)}(x_{i,j+1}) \right. \\ & \left. + b^{(n)}(x_{i-1,j}) + 2b^{(n)}(x_{i,j}) + b^{(n)}(x_{i,j-1})\right]\, \chi(x_{i,j}) \end{aligned} \tag{4.84}$$

and the remainder $R\chi$ is obtained as

$$\begin{aligned} R\chi(x_{i,j}) = & -\frac{1}{2}\,\eta_{\mathrm{M}}F_{\mathrm{M}}(w_{q,n})\ \mathrm{mean}(|\chi_{csi}^{c(n)}|^2) \\ & \times \{[b^{(n)}(x_{i+1,j}) + b^{(n)}(x_{i,j})]\chi(x_{i+1,j}) \\ & + [b^{(n)}(x_{i-1,j}) + b^{(n)}(x_{i,j})]\chi(x_{i-1,j}) \\ & + [b^{(n)}(x_{i,j+1}) + b^{(n)}(x_{i,j})]\chi(x_{i,j+1}) \\ & + [b^{(n)}(x_{i,j-1}) + b^{(n)}(x_{i,j})]\chi(x_{i,j-1})\}\ . \end{aligned} \tag{4.85}$$

Since the matrix $(D + R)$ is a real and symmetric matrix with a positive dominant diagonal, we use a single Jacobi iteration, see e.g. [39, 44]. With initial estimate, $\chi = \chi_{csi}^{c(n)}$, the regularized contrast $\chi^{c(n)}$ is obtained as

$$\boxed{\chi^{c(n)} = D^{-1}(\chi_{csi}^{c(n)} - R\chi_{csi}^{c(n)})\,.} \tag{4.86}$$

Note that this Jacobi iteration is equivalent to

$$\chi^{c(n)} = \chi_{csi}^{c(n)} + D^{-1}(\chi_{csi}^{c(n)} - A\chi_{csi}^{c(n)})\,, \tag{4.87}$$

where $(\chi_{csi}^{c(n)} - A\chi_{csi}^{c(n)})$ is the residual error of Eq. (4.83) for the initial estimate $\chi = \chi_{csi}^{(n)}$. Although we do not use this relation, it exhibits some insight how the TV regularization is added to the initial values of $\chi_{csi}^{c(n)}$.

In Section 4.2, we will discuss the changes we have to make in the Matlab codes of the CSI method to arrive at the present MRCSI codes.

4.6.5 Matlab Codes for the MRCSI Method

The Matlab script `InverseMRCSI` is shown in Matlab Listing II.19. The only difference with the CSI version is that we have replaced the function `ITERCGMD` by the function `ITERCGMDreg` of Matlab Listing II.20. But this latter function also exhibits a single change, viz., the function `UpdateContrast` is replaced by the function `UpdateContrastReg` of Matlab Listing II.21, which encompasses the TV regularization.

The function `UpdateContrastReg` computes $\chi_{csi}^{(n)}$, using Eq. (4.67). Note that we do not use any *a priori* information, neither positivity or negativity of the contrast profile. From some numerical experiments, we have observed that the fastest convergence is achieved when we keep the value of $\chi_{csi}^{(n)}$ to be complex. The imaginary part will converge to small values. The reason is that in the first few iterations, the phase is incorrect and further iterations will correct this phase. By setting the imaginary part to zero, the amplitude of the contrast is reduced too much. In the various plots, we only plot the real part.

The function `TVregularizer` of Matlab Listing II.22 provides the TV regularization. We determine the weights $b^{(n)}$ and the choice of the control parameter $\delta^{(n)}$ of either Eq. (4.64) or Eq. (4.65). Further, we define a strength factor as the maximum of the data error, ErrorM, and the

Matlab Listing II.19 The script `InverseMRCSI` to reconstruct the contrast χ^c.

```
clear all; clc; close all; clear workspace
input = init();

% (1) Compute Exact data using analytic Bessel function expansion -------
    disp('Running DataSctCircle');   DataSctCircle;

    load DataAnalytic;
  % load DataIntEq;    dataCircle = data;        % Inverse Crime

    % add complex-valued noise to data
    if input.Noise == 1
      dataCircle = dataCircle + mean(abs(dataCircle(:))) * input.Rand;
    end

% (2) Compute Green function for sources and receivers ----------------
    [G_S,G_R] = GreenSourcesReceivers(input);

% (3) Apply MRCCSI method -------------------------------------------
    [w,CHI] = ITERCGMDreg(G_S,G_R,dataCircle,input);

% Figure of contrast
  x1 = input.X1(:,1);    x2 = input.X2(1,:);
  figure(5);  IMAGESC(x1,x2,real(CHI));

  figure(6);  mesh(abs(CHI),'edgecolor', 'k');
              view(37.5,45); axis xy; axis('off'); axis('tight')
              axis([1 input.N1 1 input.N2  0 1.5*max(abs(input.CHI(:)))])

% (4) Compute model data explained by reconstructed contrast souces -----
    data = zeros(input.NR,input.NS);
    for q = 1 : input.NS
      data(:,q) = DopM(G_R,w{q},input);
    end % q_loop

   % plot data and compare to synthetic data
   error = num2str(norm(data(:)-dataCircle(:),1)/norm(dataCircle(:),1));
   disp(['data error=' error]);
   figure(7);
   subplot(1,2,1)
     imagesc(1:input.NR,1:input.NS, abs(data));
     xlabel('x_p^R \rightarrow');
     ylabel('x_q^S \rightarrow');
     axis('equal','tight'); axis xy; colorbar('hor'); colormap jet;
     title('|data_{Exact}|')
   subplot(1,2,2)
     imagesc(1:input.NR,1:input.NS, abs(data-dataCircle));
     xlabel('x_p^R \rightarrow');
     ylabel('x_q^S \rightarrow');
     axis('equal','tight'); axis xy; colorbar('hor'); colormap jet;
     title('|data_{Inverted} - data_{Exact}|')
```

Matlab Listing II.20 The function `ITERCGMDreg` to update w_q and χ^c.

```
function [w,CHI] = ITERCGMDreg(u_inc ,G_R, data , input)
global it
FFTG = input.FFTG;  NS = input.NS;
v = cell(1,NS);    g_1 = cell(1,NS);    AN_1 = cell(1,NS);

itmax = 256;
eta_M = 1 / norm(data(:))^2;     % normalization factor eta_M

[w,u,CHI,r_M,r_D,eta_D,ErrorD ,ErrorM] = ...
                       InitialEstimate(eta_M,u_inc ,G_R,data ,input);
it = 1;
while (it <= itmax)          % alternate updating of w and CHI
 Norm_M = 0;
 for q = 1 : NS
   g  = eta_M * AdjDopM(G_R,r_M{q},input)  ...
        + eta_D *(r_D{q}-AdjKop(conj(CHI).*r_D{q},FFTG));
      g = (abs(CHI)>=0.001) .*  g;
   AN = norm(g(:))^2;
   if it > 1 ; BN = sum( g(:) .* conj(g_1{q}(:)) ); end
   if it == 1
      v{q} = g;
   else
      v{q} = g + real((AN-BN)/AN_1{q}) * v{q};
   end
     g_1{q} = g;
    AN_1{q} = AN;
  % determine step length alpha
  AN    = sum( g(:) .* conj(v{q}(:)) );
  GRv   = DopM(G_R,v{q},input);
  BN_M  = norm(GRv(:))^2;
  Kv    = Kop(v{q},FFTG);
  Lv    = v{q} - CHI.* Kv;
  BN_D  = norm(Lv(:))^2;
  alpha = real( AN / (eta_M * BN_M + eta_D * BN_D) );
  % update contrast source w and residual error in M
    w{q} =   w{q} + alpha *   v{q};
    u{q} =   u{q} + alpha *  Kv;
  r_M{q} = r_M{q} - alpha * GRv(:);
  Norm_M = Norm_M + norm(r_M{q}(:))^2;
 end %q_loop
 ErrorM = eta_M * Norm_M;

 % Update contrast by minimization of Norm_D = || CHI * u - w ||^2
  disp(['Iteration = ', num2str(it)]);                saveCHI(CHI);
  [CHI,r_D,eta_D,ErrorD]=UpdateContrastReg(w,u,u_inc ,ErrorM ,ErrorD ,input);
  disp('----------------------------------------------------------');
  it = it+1;
end % while

Error  = sqrt(ErrorM + ErrorD);
disp(['Number of iterations = ' num2str(it)]);
disp(['Total Error in M and D = ' num2str(Error)]);
```

Matlab Listing II.21 The function `UpdateContrastReg` to update χ_n^c using TV regularization.

```
function  [CHI, r_D, eta_D, ErrorD]  =  ...
                              UpdateContrastReg(w,u,u_inc ,ErrorM ,ErrorD ,input)

% (1) Compute CHI_csi
      INuw = zeros(input.N1,input.N2);       INuu = zeros(input.N1,input.N2);
      for q = 1 : input.NS
        INuw = INuw + conj(u{q}) .* w{q};
        INuu = INuu + conj(u{q}) .* u{q};
      end %q_loop
      CHI = INuw./INuu;

% (2) Add regularization by Jacobi iteration of Euler-Lagrange equation
      CHI = TVregularizer(CHI,ErrorM ,ErrorD ,input);

      % Enhance TV minimization by repeating the statements:
          % CHI = TVregularizer(CHI,ErrorM ,ErrorD ,input);
          % CHI = TVregularizer(CHI,ErrorM ,ErrorD ,input);

% (3) Update the object error in D
      Norm_D = 0;      Norm = 0;
      r_D = cell(1,input.NS);
      for q = 1 : input.NS
        r_D{q} = CHI .* u{q} - w{q};
         Norm_D = Norm_D + norm(r_D{q}(:))^2;
         Norm   = Norm + norm(CHI(:).*u_inc{q}(:))^2;
      end %q_loop
      eta_D = 1 / (Norm);
      ErrorD = eta_D * Norm_D;
      disp([' ErrorM = ',num2str(ErrorM),'     ErrorD = ',num2str(ErrorD)]);

% (4) Show intermediate pictures of reconstruction
      x1 = input.X1(:,1);
      x2 = input.X2(1,:);
      figure(5);
      IMAGESC(x1,x2,real(CHI));
      title('\fontsize{13} Reconstructed Contrast Re[\chi^c] ');
      pause(0.1)
```

value of 0.01. When the noise level of the data is very small, the data error can reach a very small quantity as well. To maintain some strength of the regularization, the strength factor is never smaller than 1%. Subsequently, the function `TVregularizer` calculates the diagonal matrix D, the remainder matrix R, and carry out the single Jacobi iteration of Eq. (4.86).

We further note that the TV minimization may be enhanced by repeating the function `TVregularizer` in step (2) of the function `UpdateContrastReg` a couple of times, e.g. three times. Then, 100 iterations of the MRCSI scheme are sufficient to achieve similar results.

In Figures 4.16 and 4.17, using the control parameter $\delta^{(n)}$ of Eq. (4.64), we present the contrast profile obtained after 256 iterations. For both cases of contrast profiles considered in this chapter, we observe excellent convergence. Comparing these results with the ones obtained from the CSI-method, we observe that the band-limitation in the area of the circular cylinder has vanished. In both cases, the edges of the reconstructed profile are exceptionally well preserved. For the case of positive contrast, in Figures 4.18 and 4.19, we also show the intermediate reconstructed contrast

Matlab Listing II.22 The function `TVregularizer` to determine the regularized version of $\chi_{csi}^{(n)}$.

```
function [CHI] = TVregularizer(CHI,ErrorM,ErrorD,input)
N1 = input.N1;  N2 = input.N2;

% (1) Determine weights b^2
    Ext_CHI                    = zeros(N1+2,N2+2);
    Ext_CHI(2:N1+1,2:N2+1) = CHI(1:N1,1:N2);

    Fwd = zeros(N1+2,N2+2);
    Bwd = zeros(N1+2,N2+2);

  % Symmetric forward and backward finite differences
    i = 2:N1+1;    %------------------------------------------------
      j = 1:N2+1; Fwd(i,j) = abs(Ext_CHI(i,j+1)-Ext_CHI(i,j)).^2;
      j = 2:N2+2; Bwd(i,j) = abs(Ext_CHI(i,j)-Ext_CHI(i,j-1)).^2;
    j = 2:N2+1;    %------------------------------------------------------------
      i = 1:N1+1; Fwd(i,j) = Fwd(i,j)+abs(Ext_CHI(i+1,j)-Ext_CHI(i,j)).^2;
      i = 2:N1+2; Bwd(i,j) = Bwd(i,j)+abs(Ext_CHI(i,j)-Ext_CHI(i-1,j)).^2;

  % Take delta^2 either as ErrorD or as mean value of profile gradient
      deltan = ErrorD;
    % deltan = mean(Fwd(:)+Bwd(:))/2;

    i = 2:N1+1; j = 2:N2+1; %------------------------------------------
      bn(i-1,j-1) = 1 ./ ((Bwd(i,j)+Fwd(i,j))/2 + deltan);

% (2) Determine regularized CHI by forward and backward differences
    Ext_bn                     = zeros(N1+2,N2+2);
    Ext_bn(2:N1+1,2:N2+1) = bn(1:N1,1:N2);
    Ext_CHI                    = zeros(N1+2,N2+2);
    Ext_CHI(2:N1+1,2:N2+1)= CHI(1:N1,1:N2);

    D      = zeros(N1+2,N2+2);
    R_CHI = zeros(N1+2,N2+2);

    factor   = max(ErrorM,0.01);  % strength of regularization
    MeanCHI = mean(abs(CHI(:)).^2);

% (3) UPDATE THE CONTRAST using a single Jacobi iteration
    i = 2:N1+1; j = 2:N2+1; %----------------------------------------------------
      D(i,j) = 1 + factor * MeanCHI /2                                    ...
                  .*( Ext_bn(i+1,j)+2*Ext_bn(i,j)+Ext_bn(i,j+1)         ...
                    + Ext_bn(i-1,j)+2*Ext_bn(i,j)+Ext_bn(i,j-1) );

      R_CHI(i,j) =  - factor * MeanCHI /2                                 ...
                  .*( (Ext_bn(i+1,j)+Ext_bn(i,j)) .* Ext_CHI(i+1,j)     ...
                     +(Ext_bn(i-1,j)+Ext_bn(i,j)) .* Ext_CHI(i-1,j)     ...
                     +(Ext_bn(i,j+1)+Ext_bn(i,j)) .* Ext_CHI(i,j+1)     ...
                     +(Ext_bn(i,j-1)+Ext_bn(i,j)) .* Ext_CHI(i,j-1) );

      CHI(i-1,j-1) = (1./D(i,j)).*(Ext_CHI(i,j)-R_CHI(i,j));
```

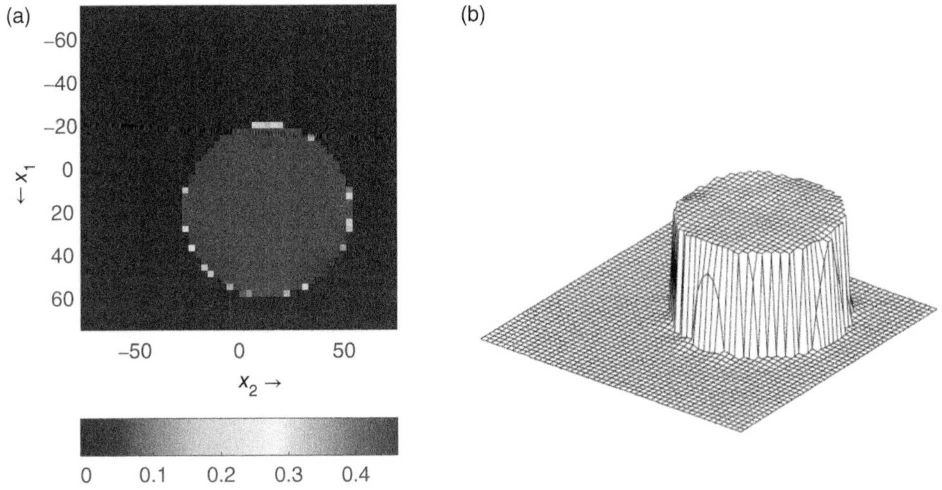

Figure 4.16 MRCSI method: plots of the contrast reconstructed after 256 iterations, for **noise-free** data and nominal values $c_0 = 1500$ m/s and $\hat{c}_{sct} = 2000$ m/s ($0 < \hat{\chi}^c < 0.4375$), using Matlab's `imagesc` script (a) and plotting the absolute value with Matlab's `mesh` script (b); $\delta^{(n)}$ of Eq. (4.64).

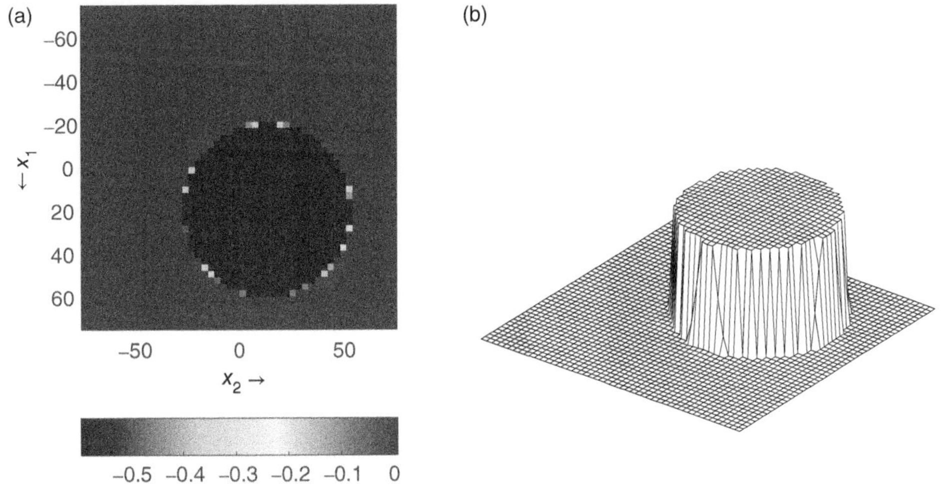

Figure 4.17 MRCSI method: plots of the contrast reconstructed after 256 iterations, for **noise-free** data and nominal values $c_0 = 1500$ m/s and $\hat{c}_{sct} = 1200$ m/s ($-0.5625 < \hat{\chi}^c < 0$), using Matlab's `imagesc` script (a) and plotting the absolute value with Matlab's `mesh` script (b); $\delta^{(n)}$ of Eq. (4.64).

values after a certain number of iterations. The pictures indicated by it = 0 represent the contrast values obtained from the initial estimate. In Figure 4.19, for increasing number of iterations, we see that the resolution of the circular interface of the scattering object increases. After 16 iterations, we see that the TV regularization becomes effective and the spatial variation within and outside the scattering object is kept to a minimum, while the location and jump of the interface becomes very precise. The edge-preserving property of our regularization is demonstrated in the final results.

If we contaminate the analytical data with noise, we obtain the pictures of Figures 4.20 and 4.21. From the image plots on the left, we observe sharper edges, because the data error ErrorM

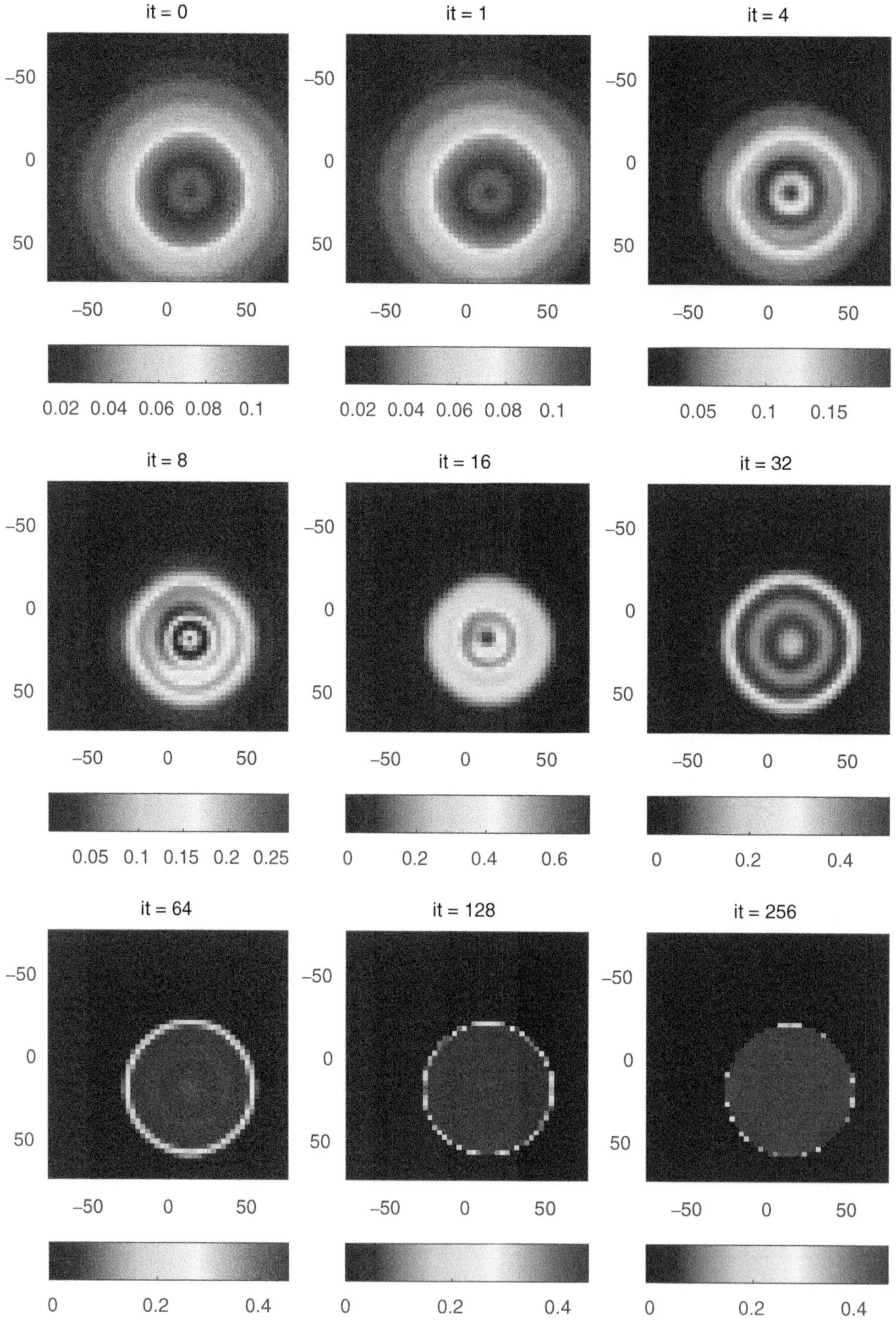

Figure 4.18 MRCSI method: image plots of the contrast reconstruction at different values of the iteration number (it), for **noise-free** data and nominal values $c_0 = 1500$ m/s and $\hat{c}_{sct} = 2000$ m/s ($0 < \hat{\chi}^c < 0.4375$).

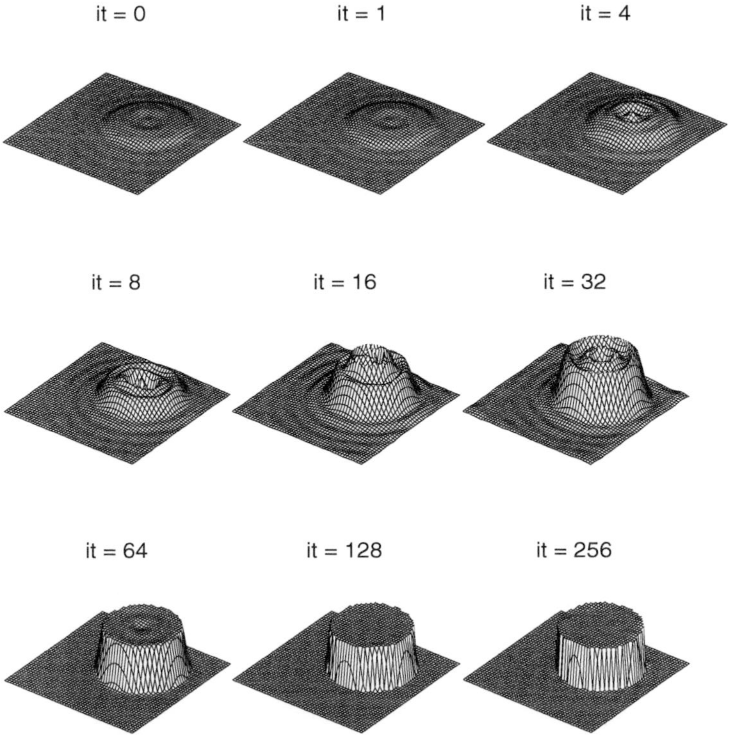

Figure 4.19 MRCSI method: mesh plots of the amplitude of the contrast reconstruction at different values of the iteration number (it), for **noise-free** data and nominal values $c_0 = 1500$ m/s and $\hat{c}_{sct} = 2000$ m/s ($0 < \hat{\chi}^c < 0.4375$). The contrast profile changes from a band-limited profile with low resolution (it = 0) to a discontinuous profile with high resolution (it = 256).

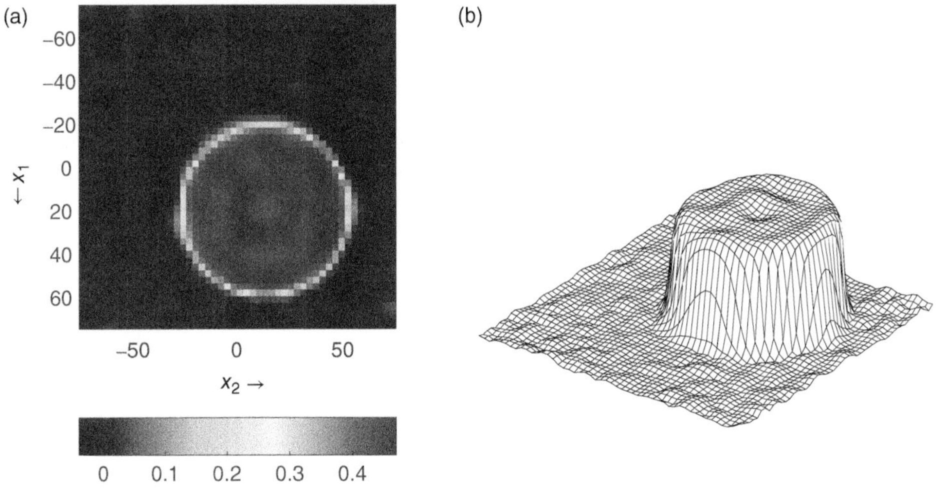

Figure 4.20 MRCSI method: plots of the contrast reconstructed after 256 iterations, for **noisy** data and nominal values $c_0 = 1500$ m/s and $\hat{c}_{sct} = 2000$ m/s ($0 < \hat{\chi}^c < 0.4375$), using Matlab's `imagesc` script (a) and plotting the absolute value with Matlab's `mesh` script (b).

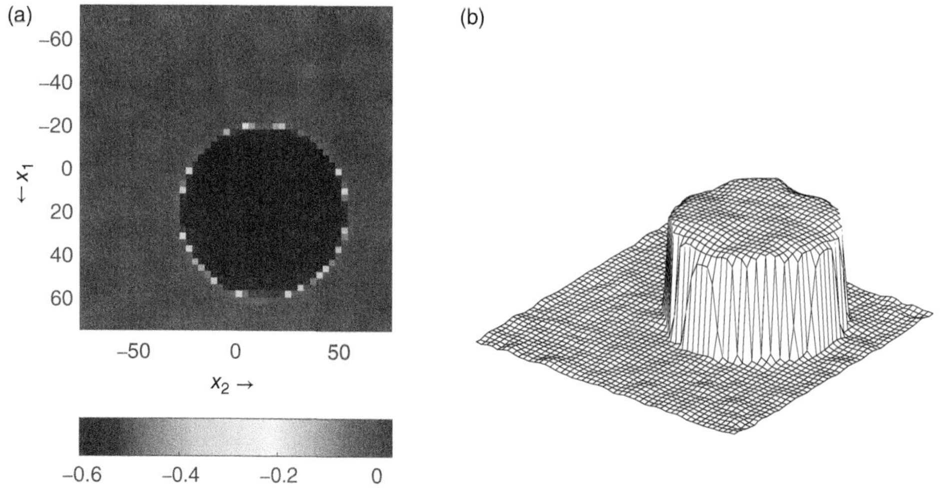

Figure 4.21 MRCSI method: plots of the contrast reconstructed after 256 iterations, for **noisy** data and nominal values $c_0 = 1500$ m/s and $\hat{c}_{sct} = 1200$ m/s ($-0.5625 < \hat{\chi}^c < 0$), using Matlab's `imagesc` script (a) and plotting the absolute value with Matlab's `mesh` script (b).

is an order larger, and therefore, the strength of the TV regularization is also larger. The smooth variations of the profile are caused by smoothed noise contributions.

On panel (a) of Figures 4.22 and 4.23, we present the amplitude profile of the analytical data. On panel (b), we plot the amplitude of the difference between the modeled data after 256 iterations and the analytical data. The global error in the inverted data amounts to 0.35% and 0.42%, for $c_{sct} = 2000$

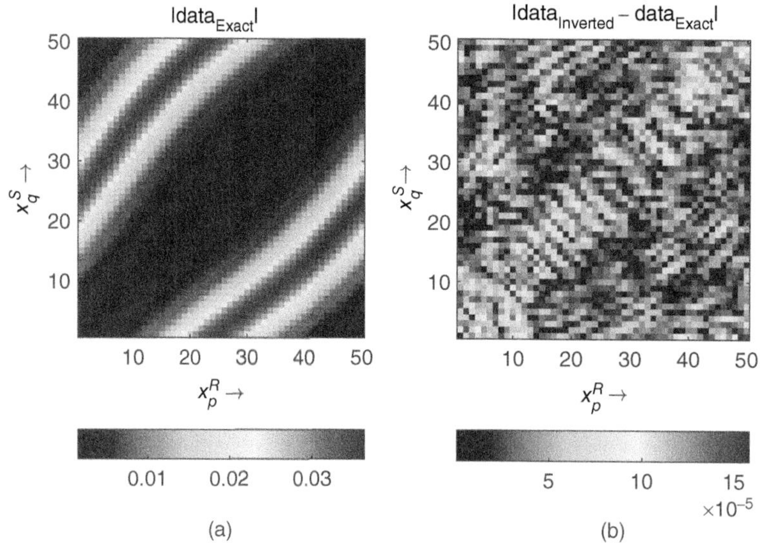

Figure 4.22 MRCSI method: plots of the inverted **noise-free** data as function of the source and receiver numbers, for nominal values $c_0 = 1500$ m/s and $c_{sct} = 2000$ m/s. The normalized global error in the inverted data amounts to 0.35%.

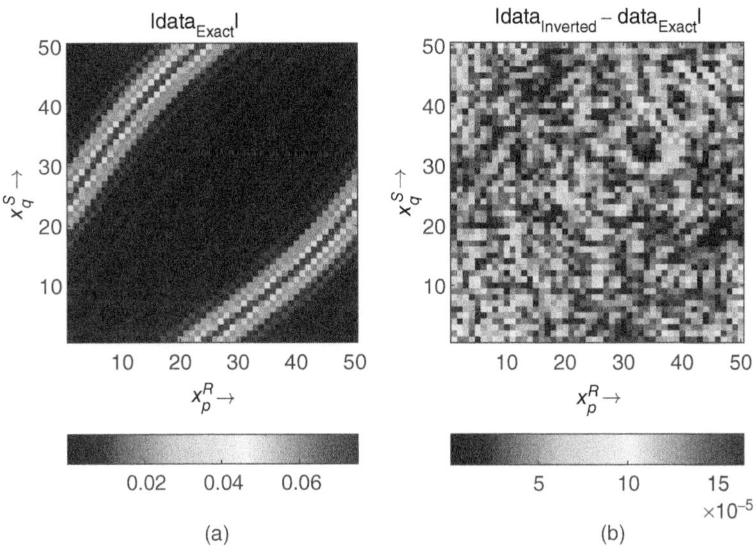

Figure 4.23 MRCSI method: plots of the inverted **noise-free** data as function of the source and receiver numbers, for nominal values $c_0 = 1500$ m/s and $\hat{c}_{sct} = 1200$ m/s. The normalized global error in the inverted data amounts to 0.42%.

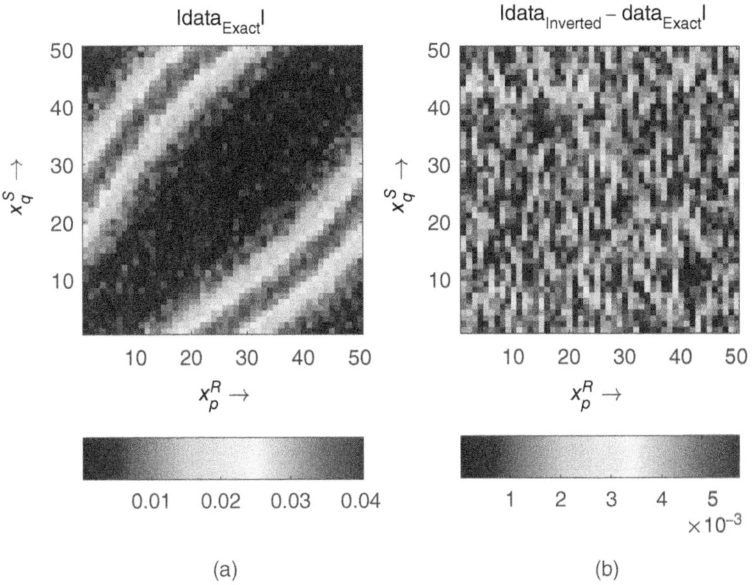

Figure 4.24 MRCSI method: plots of the inverted **noisy** data as function of the source and receiver numbers, for nominal values $c_0 = 1500$ m/s and $\hat{c}_{sct} = 2000$ m/s. The normalized global error in the inverted data amounts to 12%.

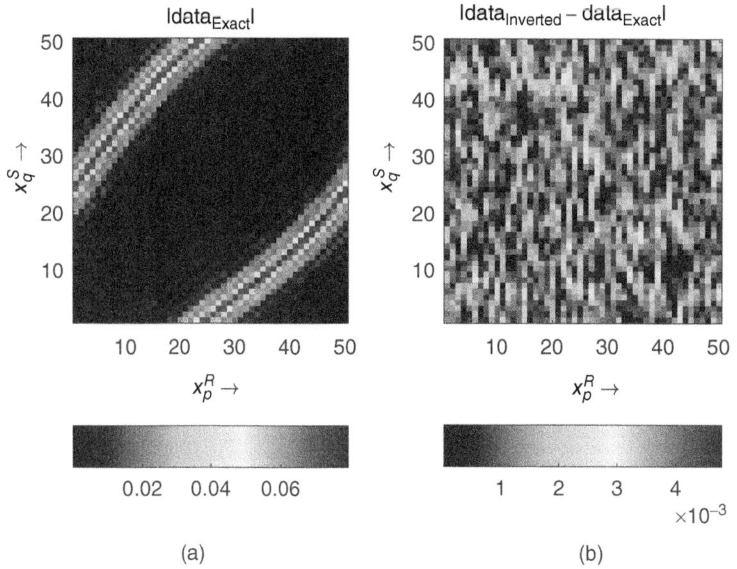

Figure 4.25 MRCSI method: plots of the inverted **noisy** data as function of the source and receiver numbers, for nominal values of $c_0 = 1500$ m/s and $c_{sct} = 1200$ m/s. The normalized global error in the inverted data amounts to 11%.

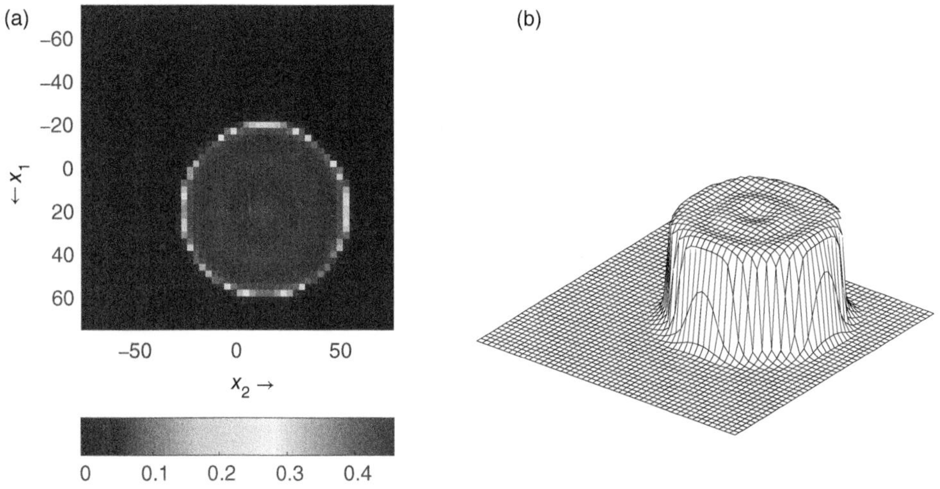

Figure 4.26 MRCSI method: plots of the contrast reconstructed after 256 iterations, for **noise-free** data and nominal contrast values between $\chi^c = 0$ and $\chi^c = 0.4375$, using Matlab's `imagesc` script (a) and plotting the absolute value with Matlab's `mesh` script (b); $\delta^{(n)}$ of Eq. (4.64).

and 1200 m/s, respectively. In case of noisy data, the data pictures are shown in Figures 4.24 and 4.25. Similar conclusions as in the CSI-method can be drawn.

Finally, in Figures 4.26 and 4.27, we show the reconstructed contrast profiles when we use the control parameter $\delta^{(n)}$ of Eq. (4.65) and noise-free data. We observe that now the reconstructed contrast profiles have been smoothed.

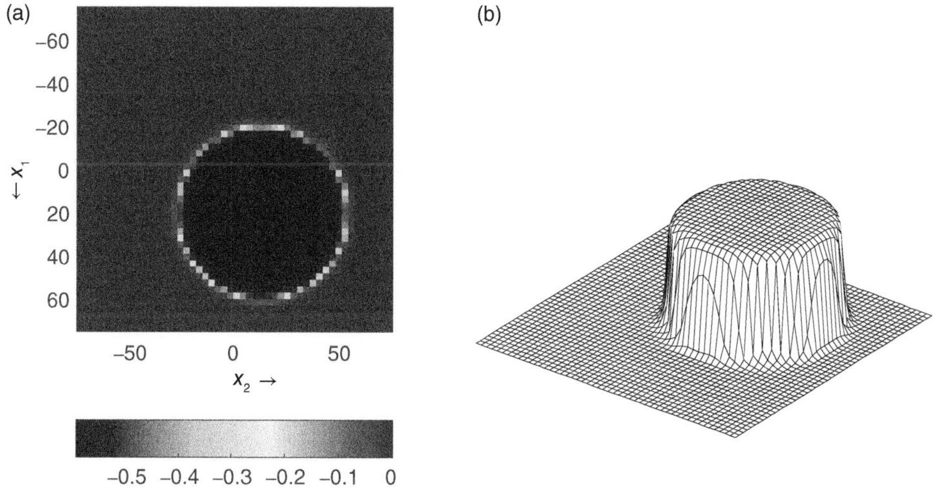

Figure 4.27 MRCSI method: plots of the contrast reconstructed after 256 iterations, for **noise-free** data and nominal contrast values between $\chi^c = -0.5625$ and $\chi^c = 0$, using Matlab's `imagesc` script (a) and plotting the absolute value with Matlab's `mesh` script (b); $\delta^{(n)}$ of Eq. (4.64).

4.7 CSI Method for Reconstruction of a Few Parameters

So far, we have considered the general case that the inversion methods determine the contrast values at each discretized point of the object domain $\mathbb{D}$. Both the CSI and MRCSI method determine in each iteration, n, the contrast function as

$$\chi_{csi}^{(n)}(\boldsymbol{x}) = \frac{\sum_{q=1}^{N^S} w_q^{(n)}(\boldsymbol{x})\, \overline{u}_q^{(n)}(\boldsymbol{x})}{\sum_{q=1}^{N^S} |u_q^{(n)}(\boldsymbol{x})|^2} . \tag{4.88}$$

After this step, the CSI method has assumed *a priori* information that the contrast values are either positive or negative. The MRCSI method has searched for minimum variation of the contrast profile. In principle, both methods are regularized solutions based on reduction of the total unknown parameters.

In case that the actual number of unknown parameters is limited, in the updating step of the contrast, we discuss an alternative inversion method. Our synthetic data are obtained for a circular cylinder. Its contrast function is defined by the coordinates x_1^O and x_2^O of the center point O, the real contrast value $\chi^O = 1 - (c_0/c_{sct})^2$ at the center point, and the radius a. Our objective is to determine these four unknown parameters iteratively, by matching a contrast function as function of these parameters. The most obvious choice is the Gauss–Newton method, see e.g. Fletcher [16]. This method needs the derivatives of the contrast function with respect to the unknown parameters. To obtain these derivatives, we replace the discontinuous contrast function by an analytic function that approximates the exact discontinuous contrast profile. We employ the function used for the Fermi–Dirac equilibrium distribution in solid state physics, e.g. [29], as

$$\chi_{anl}(\boldsymbol{x}) = \chi^O \frac{1}{1 + \exp[(R - a)/s]} , \quad \text{with } R = [(x_1 - x_1^O)^2 + (x_2 - x_2^O)^2]^{\frac{1}{2}} , \tag{4.89}$$

where x_1^O, x_2^O, χ^O, and a are the real profile parameters to be determined. Further, the parameter s determines the sharpness of the profile at the boundary of the circle; its value is of order Δx.

For sharp edges, however, is it difficult to match the derivatives of this analytical profile with the smoothed version of the contrast profile $\chi^{(n)}_{csi}(\boldsymbol{x})$. This is obtained from the weak form of the operators used in the CSI method. We therefore use a weak version of Eq. (4.89), in which we determine an average of the contrast values around the point $\boldsymbol{x}$. Since the profile varies in the R-direction only, we will suffice with an averaging in this direction, i.e.

$$\chi_{weak}(\boldsymbol{x}) = \chi^O \frac{1}{\Delta x} \int_{R-\frac{1}{2}\Delta x}^{R+\frac{1}{2}\Delta x} \frac{1}{1+\exp[(R-a)/s]} \, \mathrm{d}R \, , \tag{4.90}$$

In view that

$$-s\frac{\partial}{\partial R} \ln[1+\exp(-R/s)] = \frac{1}{1+\exp(R/s)} \, , \tag{4.91}$$

the integral of Eq. (4.90) can be calculated analytically. The result is

$$\chi_{weak}(\boldsymbol{x}) = -\chi^O \frac{\ln\left[1+\exp[-(R-a+\frac{1}{2}\Delta x)/s]\right] - \ln\left[1+\exp[-(R-a-\frac{1}{2}\Delta x)/s]\right]}{\Delta x/s} \, . \tag{4.92}$$

For convenience, we introduce the parameters P_k, $k = 1, 2, 3, 4$, as

$$P_1 = x_1^O \, , \quad P_2 = x_2^O \, , \quad P_3 = \chi^O = 1-(c_0/c_{sct})^2 \, , \quad P_4 = a \, . \tag{4.93}$$

The derivatives of χ_{weak} with respect to these parameters are

$$\begin{aligned} \frac{\partial \chi_{weak}}{\partial P_1} &= -\frac{x_1-P_1}{R}\frac{\partial \chi_{weak}}{\partial R} \, , & \frac{\partial \chi_{weak}}{\partial P_2} &= -\frac{x_2-P_2}{R}\frac{\partial \chi_{weak}}{\partial R} \, , \\ \frac{\partial \chi_{weak}}{\partial P_3} &= \frac{\chi_{weak}}{P_3} \, , & \frac{\partial \chi_{weak}}{\partial P_4} &= -\frac{\partial \chi_{weak}}{\partial R} \, , \end{aligned} \tag{4.94}$$

in which $\chi_{weak} = \chi_{weak}(\boldsymbol{x}, P_j)$ and $\partial\chi_{weak}/\partial R = \partial\chi_{weak}(\boldsymbol{x}, P_j)/\partial R$ are given by

$$\chi_{weak} = -P_3 \frac{\ln\left[1+\exp[-(R-P_4+\frac{1}{2}\Delta x)/s]\right] - \ln\left[1+\exp[-(R-P_4-\frac{1}{2}\Delta x)/s]\right]}{\Delta x/s} \tag{4.95}$$

and

$$\frac{\partial \chi_{weak}}{\partial R} = \frac{P_3}{\Delta x}\left[\frac{1}{1+\exp[(R-P_4+\frac{1}{2}\Delta x)/s]} - \frac{1}{1+\exp[(R-P_4-\frac{1}{2}\Delta x)/s]}\right] . \tag{4.96}$$

Note that the latter expression also follows as the central difference of Eq. (4.89). In the Gauss–Newton method, however, we prefer a function and derivatives with respect to its parameters, with consistently the same type of approximations being made. Hence, we consider the Eqs. (4.95) and (4.96) as a joint set of relations. But, after an update of the parameters P_k, we use the analytic form of Eq. (4.89), because there is another weakening procedure in the CSI step of updating the contrast sources.

The script `CheckAnlNum` of Matlab listing II.23 computes and plots the weak contrast function χ_{weak}, for a particular value of the sharpness parameter s. For our example of a positive contrast function discussed in the previous sections, Figure 4.28 shows χ_{weak} as function of $\boldsymbol{x} = x_{i,j}$, $i = 1, 2, \ldots, N_1, j = 1, 2, \ldots, N_2$, where we have chosen $s = 1.0\ \Delta x, s = 0.3\ \Delta x$ and $s = 0.1\ \Delta x$, respectively. The latter profile resembles very much the exact contrast χ^c of Figure 4.2. Further, in this Matlab script `CheckAnlNum`, the derivatives of the contrast function χ_{weak} with respect to the parameters P_k are calculated analytically. In general, analytic differentiation is not always possible.

Matlab Listing II.23 The function `CheckAnlNum` to plot the weak form χ_{weak} and to check the derivatives of χ_{weak} with respect to P_k.

```
clear all; clc; close all; clear workspace
input = init();     s = 1 * input.dx;

% (1) Determine weak profile
    P    = zeros(4);
    P(1) = input.xO(1);       P(2) = input.xO(2);
    P(3) = input.contrast;    P(4) = input.a;
    X1   = input.X1;          X2   = input.X2;
    R    = sqrt((X1-P(1)).^2 + (X2-P(2)).^2);  Rp = R - P(4) + input.dx/2;
                                                Rm = R - P(4) - input.dx/2;
    CHI_weak = -P(3) * (log(1+exp(-Rp/s))-log(1+exp(-Rm/s)))/(input.dx/s);
    if sum(isnan(CHI_weak(:))) > 0
       disp(['parameter s too small: increase value of ',num2str(s)]);
    end
    figure(13); IMAGESC(input.X1(:,1),input.X2(1,:),real(CHI_weak));
    figure(14); mesh(abs(CHI_weak),'edgecolor', 'k');
                view(37.5,45); axis xy; axis('off'); axis('tight');
                axis([1 input.N1 1 input.N2  0 1.5*max(abs(CHI_weak(:)))])

% (2) Determine analytical derivatives
    dCHI_R  =  P(3) * (1./(1+exp(Rp/s)) - 1./(1+exp(Rm/s))) / input.dx;
    dCHI_P1 = -(X1-P(1))./R .* dCHI_R;
    dCHI_P2 = -(X2-P(2))./R .* dCHI_R;
    dCHI_P3 = CHI_weak / P(3);
    dCHI_P4 = -dCHI_R;

% (3) Determine as check the numerical derivatives
    Norm = norm(CHI_weak(:));    eps = 1e-6;
    R = sqrt((X1-P(1)-eps).^2 + (X2-P(2)).^2); Rp = R - P(4) + input.dx/2;
                                                Rm = R - P(4) - input.dx/2;
    dCHI_P1num = (-P(3)*(log(1+exp(-Rp/s))-log(1+exp(-Rm/s))) ...
                 /(input.dx/s) - CHI_weak) / eps;
    disp(['Check dP1 = ',num2str( norm(dCHI_P1num(:)-dCHI_P1(:))/Norm)]);
    R = sqrt((X1-P(1)).^2 + (X2-P(2)-eps).^2); Rp = R - P(4) + input.dx/2;
                                                Rm = R - P(4) - input.dx/2;
    dCHI_P2num = (-P(3)*(log(1+exp(-Rp/s))-log(1+exp(-Rm/s))) ...
                 /(input.dx/s) - CHI_weak) / eps;
    disp(['Check dP2 = ',num2str( norm(dCHI_P2num(:)-dCHI_P2(:))/Norm)]);

    R = sqrt((X1-P(1)).^2 + (X2-P(2)).^2);     Rp = R - P(4) + input.dx/2;
                                                Rm = R - P(4) - input.dx/2;
    dCHI_P3num = (-(P(3)+eps)*(log(1+exp(-Rp/s))-log(1+exp(-Rm/s))) ...
                 /(input.dx/s) - CHI_weak) / eps;
    disp(['Check dP3 = ',num2str( norm(dCHI_P3num(:)-dCHI_P3(:))/Norm)]);

    R = sqrt((X1-P(1)).^2 + (X2-P(2)).^2); Rp = R - P(4)-eps + input.dx/2;
                                            Rm = R - P(4)-eps - input.dx/2;
    dCHI_P4num = (-P(3) * (log(1+exp(-Rp/s))-log(1+exp(-Rm/s))) ...
                 /(input.dx/s) - CHI_weak) / eps;
    disp(['Check dP4 = ',num2str( norm(dCHI_P4num(:)-dCHI_P4(:))/Norm)]);
```

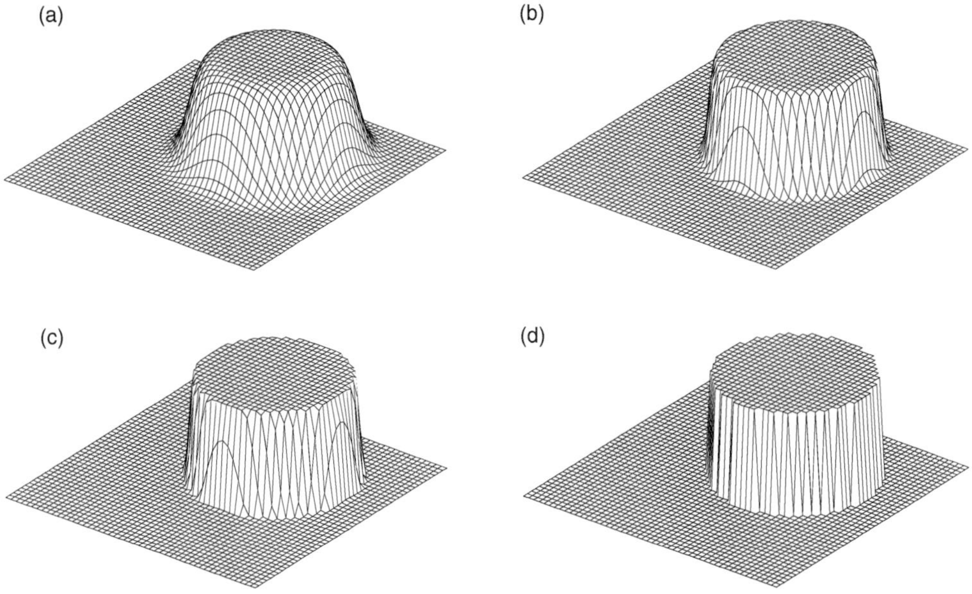

Figure 4.28 Plots of the weak forms χ_{weak} of the analytical contrast function; (a) $s = 1.0\ \Delta x$, (b) $s = 0.3\ \Delta x$, (c) $s = 0.1\ \Delta x$. For comparison, in (d) the exact function χ^c is presented as well.

Therefore, in this script, we also determine the profile derivatives numerically by defining these derivatives as

$$\frac{\partial \chi_{weak}}{\partial P_k} = \frac{\chi_{weak}(\boldsymbol{x}, P_{j\neq k}, P_k + \epsilon) - \chi_{weak}(\boldsymbol{x}, P_{j\neq k}, P_k)}{\epsilon}, \quad k = 1,\ 2,\ 3,\ 4\,, \tag{4.97}$$

in which ϵ should be small enough. For the current Matlab machine epsilon of 2.2204e-16, we take ϵ = 1e-6. Then, there are sufficient decimals to numerically compute the derivatives with an accuracy of 8–9 decimals. In our case, running the Matlab script `CheckAnlNum` confirms this observation.

4.7.1 Gauss–Newton Method for the Contrast Update

Let, in the CSI method, the initial estimates for the profile parameters be denoted as $P_k^{(0)}$, $k =$ 1, 2, 3, 4, then in every following CSI iteration, the updates for these parameters are defined as

$$P_k^{(n)} = P_k^{(n-1)} + \Delta P_k^{(n)}\,, \tag{4.98}$$

where $\Delta P_k^{(n)}$ is determined by carrying out a single Gauss–Newton iteration. Let, in each CSI iteration, the residual error be defined as

$$r^{(n)} = \chi_{csi}^{(n)}(\boldsymbol{x}) - \chi_{weak}(\boldsymbol{x}, P_j^{(n-1)})\,. \tag{4.99}$$

The Gauss–Newton algorithm [16] is based on a linear Taylor approximation of the residual $r^{(n)}$ as function of $\Delta P_k^{(n)}$, followed by a minimization of the norm $\| r^{(n)} \|_{\mathbb{D}}^2$. As result, the following set of linear equations for $\Delta P_k^{(n)}$ is arrived at:

$$\sum_{k=1}^{4} \left\langle \frac{\partial r^{(n)}}{\partial P_l^{(n-1)}}, \frac{\partial r^{(n)}}{\partial P_k^{(n-1)}} \right\rangle_{\mathbb{D}} \Delta P_k^{(n)} = -\left\langle \frac{\partial r^{(n)}}{\partial P_l^{(n-1)}}, r_n \right\rangle_{\mathbb{D}}, \quad l = 1,\ 2,\ 3\ ,4\,. \tag{4.100}$$

The derivatives of the residual error $r^{(n)}$ follow immediately from Eq. (4.99) as

$$\frac{\partial r^{(n)}}{\partial P_k^{(n-1)}} = -\frac{\partial \chi_{weak}}{\partial P_k^{(n-1)}}, \tag{4.101}$$

where the derivatives of χ_{weak} with respect to P_j follow from Eqs. (4.94) and (4.96).

After solving Eq. (4.100) for $\Delta P_k^{(n)}$, we use Eq. (4.98) and we subsequently update χ_{weak} with the help of Eq. (4.92). After carrying out the update step for the contrast sources, we return to the present Gauss–Newton algorithm explained in this section.

4.7.2 Matlab Codes for the Gauss–Newton Type Contrast Updating

The Matlab script `InverseCSIGaussNewton` is shown in Matlab Listing II.24. The difference with the CSI version is that we have replaced the function `ITERCGMD` by the function `ITERCGMDGaussNewton` of Matlab Listing II.25 and that we have added the function `DisplayParameters` to plot the change of the parameters $P_k^{(n)}$ as function of the iteration number n. We actually replaced the function `UpdateContrast` by the function `UpdateContrastGaussNewton` of Matlab Listing II.26, in which we include the function `GaussNewtonCSI` of Matlab Listing II.27. In this function, we start with the initial estimates for the parameters $P_{k,0}$. Our *a priori* knowledge is that the domain $\mathbb{D}$ encloses the circle. The largest

Matlab Listing II.24 The `InverseCSIGaussNewton` to reconstruct the contrast parameters P_k.

```
clear all; clc; close all; clear workspace
input = init();

% (1) Compute Exact data using analytic Bessel function expansion --------
      disp('Running DataSctCircle');   DataSctCircle;
      load DataAnalytic;
      %load DataIntEq;    dataCircle = data;
      % add complex-valued noise to data
       if input.Noise == 1
          dataCircle = dataCircle + mean(abs(dataCircle(:))) * input.Rand;
      end

% (2) Compute Green function for sources and receivers ------------------
      [G_S,G_R] = GreenSourcesReceivers(input);

% (3) Apply MRCCSI method -----------------------------------------------
      [w,CHI] = ITERCGMDGaussNewton(G_S,G_R, dataCircle , input);

% Figure of reconstructed contrast (analytic form)
  figure(5);  IMAGESC(input.X1(:,1),input.X2(1,:),real(CHI));
  figure(6);  mesh(abs(CHI),'edgecolor', 'k');
              view(37.5,45); axis xy; axis('off'); axis('tight')
              axis([1 input.N1 1 input.N2  0 1.5*max(abs(input.CHI(:)))]);

% Figure of reconstructed contrast parameters
  DisplayParameters;
```

Matlab Listing II.25 The function ITERCGMDGaussNewton to update the contrast sources w_q and the contrast profile χ_{weak}.

```
function [w,CHI] = ITERCGMDGaussNewton(u_inc ,G_R, data , input)
global it;
FFTG = input.FFTG;  NS = input.NS;
v = cell(1,NS);     g_1 = cell(1,NS);     AN_1 = cell(1,NS);

itmax = 250;
eta_M = 1 / norm(data(:))^2;      % normalization factor eta_M

[w,u,CHI,r_M,r_D,eta_D,ErrorD,ErrorM] = ...
                           InitialEstimate(eta_M,u_inc,G_R,data,input);
it = 1;
while (it < itmax)          % alternate updating of w and CHI
 Norm_M = 0;
 for q = 1 : NS
   g  = eta_M * AdjDopM(G_R,r_M{q},input)  ...
          + eta_D *(r_D{q}-AdjKop(conj(CHI).*r_D{q},FFTG));
   AN = norm(g(:))^2;
   if it > 1 ; BN = sum( g(:) .* conj(g_1{q}(:)) ); end
   if it == 1
      v{q} = g;
   else
      v{q} = g + real((AN-BN)/AN_1{q}) * v{q};
   end
    g_1{q} = g;
   AN_1{q} = AN;
   % determine step length alpha
   AN     = sum( g(:) .* conj(v{q}(:)) );
   GRv    = DopM(G_R,v{q},input);
   BN_M   = norm(GRv(:))^2;
   Kv     = Kop(v{q},FFTG);
   Lv     = v{q} - CHI.* Kv;
   BN_D   = norm(Lv(:))^2;
   alpha = real( AN / (eta_M * BN_M + eta_D * BN_D) );
   % update contrast source w and residual error in M
    w{q} =   w{q} + alpha *   v{q};
    u{q} =   u{q} + alpha *  Kv;
  r_M{q} = r_M{q} - alpha * GRv(:);
  Norm_M = Norm_M + norm(r_M{q}(:))^2;
 end %q_loop
 ErrorM = eta_M * Norm_M;
  % Update contrast by minimization of Norm_D = || CHI * u - w ||^2
  [CHI,r_D,eta_D,ErrorD]=UpdateContrastGaussNewton(w,u,u_inc,input);
  disp('------------------------------------------------------------');
  it = it+1;
end % while

Error  = sqrt(ErrorM + ErrorD);
disp(['Number of iterations = ' num2str(it)]);
disp(['Total Error in M and D = ' num2str(Error)]);
```

Matlab Listing II.26 The function `UpdateContrastGaussNewton` to update the contrast function χ_{weak}.

```
function [CHI,r_D,eta_D,ErrorD] = ...
                         UpdateContrastGaussNewton(w,u,u_inc,input)
global it; NS = input.NS;

% (1) Compute CHI_csi and determine ErrorD
      INuw = zeros(input.N1,input.N2);       INuu = zeros(input.N1,input.N2);
      for q = 1 : input.NS
        INuw = INuw + (conj(u{q}) .* w{q});
        INuu = INuu + real(conj(u{q}) .* u{q});
      end %q_loop
      CHI_csi = real(INuw./INuu);

% (2) Gauss Newton method for unknown location, contrast and radius
      [CHI,input] = GaussNewtonCSI(CHI_csi,input);
      ErrorCHI = norm(CHI(:)-input.CHI(:));
      txt = ['           ErrorCHI = ', num2str(ErrorCHI/norm(CHI(:)))];
      disp(['Iteration = ', num2str(it),txt]);

      Norm_D = 0;    Norm = 0;      r_D{q} = cell(1,input.NS);
      for q = 1 : NS
        r_D{q} = CHI .* u{q} - w{q};
        Norm_D = Norm_D + norm(r_D{q}(:))^2;
        Norm   = Norm + norm(CHI(:).*u_inc{q}(:))^2;
      end %q_loop
      eta_D = 1 / (Norm);
      ErrorD = eta_D * Norm_D;

% (3) Show intermediate pictures of reconstruction
      x1 = input.X1(:,1);    x2 = input.X2(1,:);
      figure(5); IMAGESC(x1,x2,real(CHI));
      title('\fontsize{13} Reconstructed Contrast Re[\chi^{anl}] ');
      pause(0.1)
```

circle that fits in this domain $\mathbb{D}$ must have its center in the origin of our coordinate system, hence $P_{1,0} = P_{2,0} = 0$.

The largest possible radius is half the length of the domain $\mathbb{D}$ of observation; hence, $P_{4,0} = 75$ m. The value of χ^O is estimated as twice the mean value of the contrast that we have obtained from back projection of the data, i.e. $P_{0,3} = 2\ \mathrm{mean}(\chi_0^{csi})$.

Furthermore, we performed some numerical experiments with the sharpness parameter. If the parameter is too small, the convergence is slow. The best strategy is to start with a relative large parameter and reduce it halfway through the iteration process. For all cases considered, we decided to use $s = 2\ \Delta x$ and after 100 iterations switch to $s = \Delta x$.

Subsequently, we compute the profile derivatives with respect to the profile parameters and construct the Jacobian matrix of the system of equations. After solving for ΔP_k, the profile parameters are updated.

Remark that for the determination of the parameters P_k, we use the weak form χ_{weak}, after which its analytic form χ_{anl}, for $s = 1$, is then passed on to the main CSI functions.

Matlab Listing II.27 The function `GaussNewtonCSI` to update the profile parameters P_k.

```
function [CHI,input] = GaussNewtonCSI(CHI_csi , input)
global it Pnom P_it
X1 = input.X1;  X2 = input.X2;

% Start with estimated contrast sources
if it == 1
   Pnom = [input.xO(1), input.xO(2), input.contrast,    input.a];
   P0   = [0, 0, 2*mean(CHI_csi(:)), 75];
   disp(['Nominal P: ', num2str(Pnom)]);
   disp(['Initial P: ', num2str(P0)]);
   P_it(1,:) = P0;   % save as global array for plotting
end

  P = P_it(it,:);
% Determine weak function and for a typical value of s
  s = 2 * input.dx;   if it > 100;  s =  1 * input.dx;  end

  R = sqrt((X1-P(1)).^2 + (X2-P(2)).^2);        Rp = R - P(4) + input.dx/2;
                                                  Rm = R - P(4) - input.dx/2;
  CHI_weak = -P(3) * (log(1+exp(-Rp/s))-log(1+exp(-Rm/s))) / (input.dx/s);
  dCHI_R   =  P(3) * (1./(1+exp(Rp/s)) - 1./(1+exp(Rm/s))) /  input.dx;

  if sum(isnan(CHI_weak(:))) > 0
       disp(['parameter s too small: increase value of ',num2str(s)]);
  end

 % Compute analytical derivatives with respect to P
  dCHI_P{1} = -(X1-P(1))./R .* dCHI_R;
  dCHI_P{2} = -(X2-P(2))./R .* dCHI_R;
  dCHI_P{3} = CHI_weak / P(3);
  dCHI_P{4} = -dCHI_R;

% Compute residual error, construct Jacobian matrix and update P
  Residual = CHI_csi - CHI_weak;
  GNerror  = norm(Residual(:))/norm(CHI_csi(:));
  disp(['              it= ',num2str(it),'  Residual= ',num2str(GNerror)]);
  b = zeros(1,4); A = zeros(4,4);
  for l = 1:4
      b(l) = sum(dCHI_P{l}(:).*Residual(:));
      for k = 1:4
          A(l,k) = sum(dCHI_P{l}(:).*dCHI_P{k}(:));
      end
  end
  DeltaP = real(b/A);  P = P + DeltaP; P(4) = abs(P(4)); % P(4)is positive
  disp(['          P: ', num2str(P)]);
  P_it(it+1,:) = P;  % save as global array for plotting

% Update contrast using analytic contrast profile
  R   = sqrt((X1-P(1)).^2 + (X2-P(2)).^2);
  CHI = P(3)./(1+exp((R-P(4))/s));
```

Matlab Listing II.28 The `DisplayParameters` to reconstruct the profile parameters P_k.

```
function DisplayParameters
global Pnom P_it
set(figure(9), 'Position', [0 0 800 500] );  N = length(P_it(:,1))-1;
subplot(1,4,1)
  plot(0:N,Pnom(1)*ones(N+1,1),':k','Linewidth',2);     hold on;
  plot(0:N,P_it(:,1),'-red','Linewidth',2);
  title('P_1 = x^O_1','Fontsize',11);
  xlabel('number of iterations','Fontsize',11);
  axis('tight'); axis([ 0 N 0 25]); grid on;
  legend('P nominal ', 'P iterated ','Location','SouthEast');
subplot(1,4,2)
  plot(0:N,Pnom(2)*ones(N+1,1),':k','Linewidth',2);     hold on;
  plot(0:N,P_it(:,2),'-red','Linewidth',2);
  title('P_2 = x^O_2','Fontsize',11);
  xlabel('number of iterations','Fontsize',11);
  axis('tight'); axis([ 0 N 0 25]);  grid on;
  legend('P nominal ', 'P iterated ','Location','SouthEast'); hold off;
subplot(1,4,3)
  plot(0:N,Pnom(3)*ones(N+1,1),':k','Linewidth',2);     hold on;
  plot(0:N,P_it(:,3),'-red','Linewidth',2);
  title('P_3 = 1 - c_0^2/c_{sct}^2','Fontsize',11);
  xlabel('number of iterations','Fontsize',11);
  axis('tight'); axis([ 0 N -0.8 0.8]);  grid on;
  legend('P nominal ', 'P iterated ','Location','East');
subplot(1,4,4)
  plot(0:N,Pnom(4)*ones(N+1,1),':k','Linewidth',2);     hold on;
  plot(0:N,P_it(:,4),'-red','Linewidth',2);
  title('P_4 = a','Fontsize',11);
  xlabel('number of iterations','Fontsize',11);
  axis('tight'); axis([ 0 N 20 80]);  grid on;
  legend('P nominal ', 'P iterated ','Location','NorthEast')
```

After 250 iterations, we finish the iterative procedure and call the function `DisplayParameters` of Matlab Listing II.28 to plot the profile parameters $P_k^{(n)}$ as function of the number of iterations. The results are presented as solid lines in Figure 4.29, both for positive contrast (a–d) and negative contrast (e–h), We also present the nominal values of the profile parameters, for which we compute the synthetic data. These data are the analytical data obtained from the Bessel function method. The nominal parameter values are shown as dotted lines. The switch of the sharpness parameter s at iteration number of 100 is clearly visible. The convergence is disturbed, but after a few iterations, the curves continue to follow their course, be it closer to the dotted line of the nominal value. We observe that the location of the center, described by the parameters P_1 and P_2, converges very fast to the nominal values. The values of the contrast parameter P_3 and the radius parameter P_4 converge to some error level, which depends on the discretization error. To investigate this, we reduced the grid size from $\Delta x = 3$ to $\Delta x = 1$. The results are presented in Figure 4.30, and they exhibit excellent convergence of all the parameters.

The final results after 250 iterations are listed in Tables 4.1 and 4.2, for positive contrast and negative contrast, respectively. We also show the influence of adding extra noise to the data, but

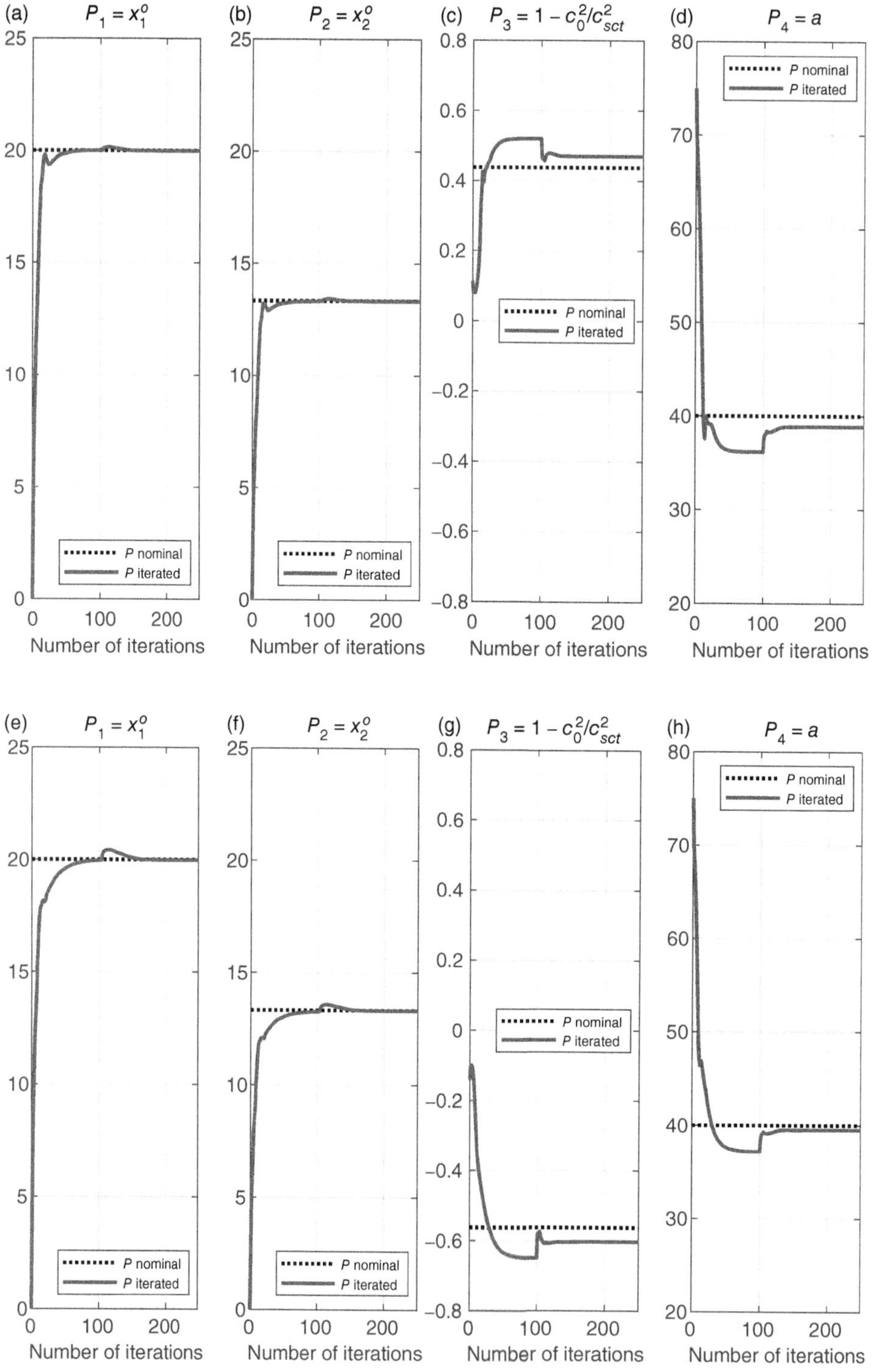

Figure 4.29 Plots of the profile parameters $P_k^{(n)}$ (solid lines) as function of the number of CSI iterations, for positive contrast profile (a–d) and negative contrast profile (e–h), respectively. For comparison, the nominal values of the circular cylinder are shown as dotted lines. The sharpness parameter is $s = 2.0\ \Delta x$, for iterations $n = 1, \ldots, 100$, and $s = \Delta x$, for iterations $n = 101, \ldots, 250$.

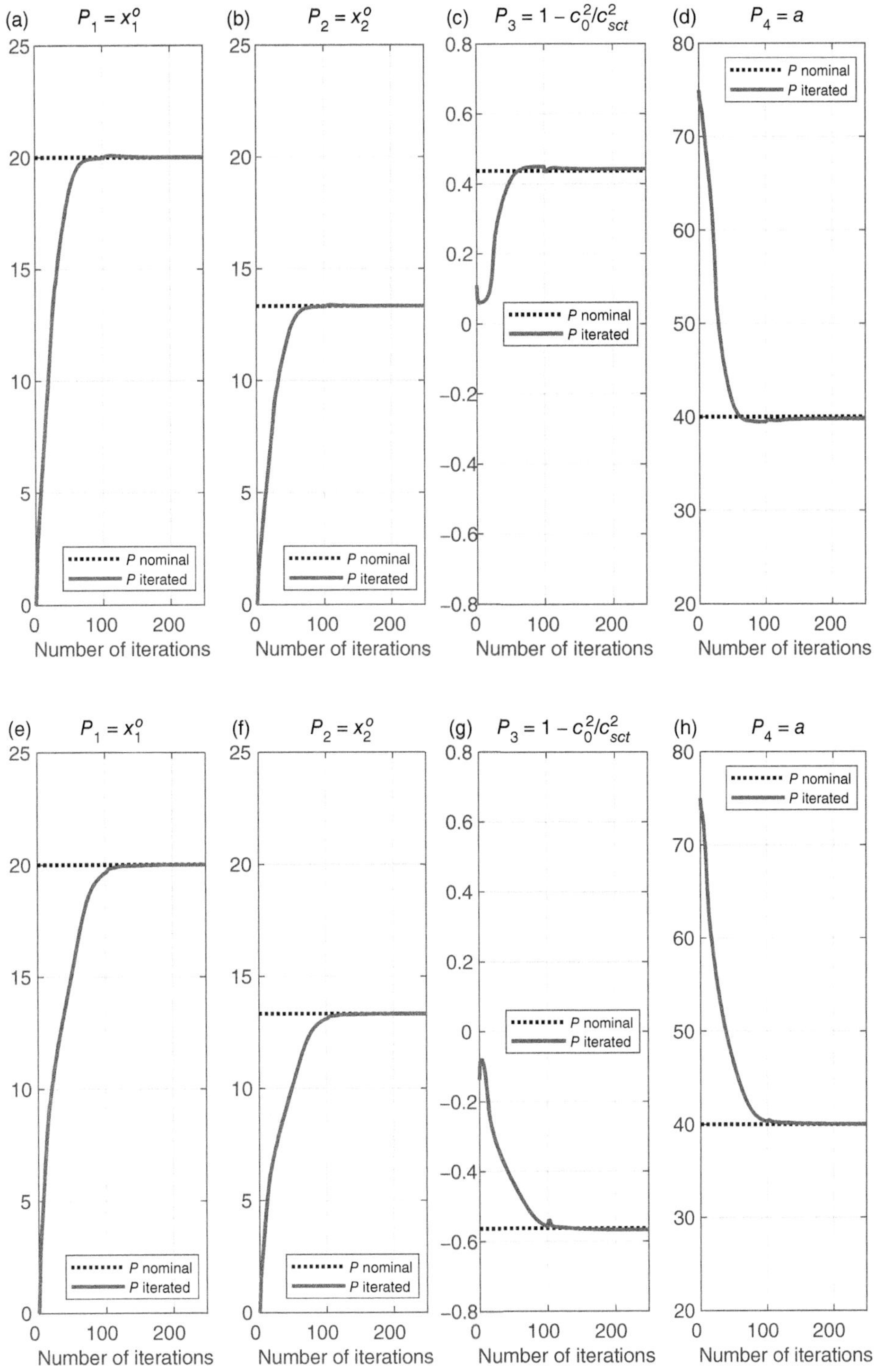

Figure 4.30 Similar to Figure 4.29, but now for a three times smaller grid size: $\Delta x = 1$; $N_1 = N_2 = 150$. Note that the transition from $s = 2.0\ \Delta x$ to $s = \Delta x$ is less visible. At this transition number of $n = 100$, convergence of all the parameters P_k is reached.

Table 4.1 Reconstructed values of P_k, for positive contrast with $c_{sct} = 2000$ m/s.

		$P_1 = x_1^O$	$P_2 = x_2^O$	$P_3 = 1 - c_0^2/c_{sct}^2$	$P_4 = a$	Error
	P_{nom}	**20.000**	**13.333**	**0.437**	**40.000**	
$\Delta x = 3$	Noise-free	19.997	13.330	0.470	38.838	0.071
$\Delta x = 3$	Noisy data	20.005	13.456	0.472	38.718	0.170
$\Delta x = 1$	Noise-free	19.999	13.332	0.442	39.808	0.007
$\Delta x = 1$	Noisy data	20.006	13.454	0.444	39.712	0.121

Table 4.2 Reconstructed values of P_k, for negative contrast with $c_{sct} = 1200$ m/s.

		$P_1 = x_1^O$	$P_2 = x_2^O$	$P_3 = 1 - c_0^2/c_{sct}^2$	$P_4 = a$	Error
	P_{nom}	**20.000**	**13.333**	**−0.561**	**40.000**	
$\Delta x = 3$	Noise-free	19.988	13.325	−0.607	39.376	0.067
$\Delta x = 3$	Noisy data	19.973	13.271	−0.607	39.401	0.134
$\Delta x = 1$	Noise-free	19.998	13.332	−0.567	40.006	0.008
$\Delta x = 1$	Noisy data	19.990	13.304	−0.566	40.044	0.091

from Tables 4.1 and 4.2 we do not observe major differences with the results of using noise-free data. The total Error $= [\text{Error}_\text{M} + \text{Error}_\text{D}]^{\frac{1}{2}}$ however, indicates that an extra noise error of 8–10% is present.

As far as the individual parameters are concerned, the differences in the contrast parameter P_3 are the largest. They are about 8% and 1%, for the cases of $\Delta x = 3$ and $\Delta x = 1$, respectively, while the parameter $P_4 = a$ shows discrepancies of 3% and 0.5%, for the positive case; for the negative case, these discrepancies are 1.5% and 0.1%. For all cases, the parameters P_1 and P_2 for the center location have been approximated very well.

We note that we also investigated the Gauss–Newton method when using the parameters of the analytical model based on the Bessel function series. In that case, we commit the inverse crime; but we could only reconstruct the profile parameters when their initial estimates are close to the nominal values, especially for the location parameters. However, in each iteration of our CSI method, the location follows from an approximated contrast profile $\chi_{csi}^{(n)}$, after updating the contrast sources. Moreover, we have no indication how many unknown parameters we can estimate in the current Gauss–Newton version of the CSI method.

4.8 CSI Methods for Phaseless Data

In many practical situations, it is not a trivial procedure to measure both amplitude and phase of wave-field data. In cases where phase information is available, the phase is often much more contaminated with noise than the amplitudes; reliable reconstruction is then difficult to realize. Therefore, there is increasing interest in the inversion of phaseless data. In principle, in a single-frequency inversion method, we lose half of the information. When using multiple frequencies, this problem becomes less serious because in multifrequency inversion the known behavior of matter as a function of frequency is taken into account. If the complete frequency spectrum of the amplitude data is known, we can use Kramers–Kronig relations [25] to determine the phase spectrum from the amplitude spectrum.

In this section, we discuss the CSI method for phaseless data. We again omit the hat-symbol denoting the frequency dependency. We assume that the incident field is known in whole space, both in amplitude and phase. So far, we have assumed that the data represent the scattered-field data and that they have determined from the actual field measurements by subtracting the incident-field values. In the present case of phaseless data, this subtraction is impossible, and therefore, we operate with the absolute value of the total field data $|u^{tot}_{\mathbb{M},q}|$.

In the measurements domain $\mathbb{M}$, the total field is modeled as

$$u^{mod}_{\mathbb{M},q} = u^{inc}_{\mathbb{M},q} + G_{\mathbb{M}} w_q \,, \tag{4.102}$$

where the incident field at the receiver is defined as

$$u^{inc}_{\mathbb{M},q} = u^{inc}(\boldsymbol{x}^R_p, \boldsymbol{x}^S_q) \,. \tag{4.103}$$

It is clear that only the definition of the first term of the CSI cost functional must be changed. There are two options: (i) Define the error norm $\text{Error}_{\mathbb{M}}$ of the residual error between the measured intensity and the modeled intensity; (ii) Define the error norm $\text{Error}_{\mathbb{M}}$ of the residual error between the measured amplitude and the modeled amplitude. We only discuss the modifications of this data norm. The changes to be made are similar for the CSI, MRCSI, and Gauss–Newton variants.

4.8.1 CSI Method for Measured Intensity Data

The CSI methods for intensity data has been discussed by Li et al. [32]. In our notation, for intensity data, the error norm in the measurements domain becomes, cf. Eq. (4.18),

$$F_{\mathbb{M}}(w^{(n)}_q) = \sum_{q=1}^{N^S} \| r^{(n)}_{\mathbb{M},q} \|^2_{\mathbb{M}} \,, \quad \text{with } r^{(n)}_{\mathbb{M},q} = |u^{tot}_{\mathbb{M},q}|^2 - |u^{mod(n)}_{\mathbb{M},q}|^2 \,, \tag{4.104}$$

and normalization factor

$$\eta_{\mathbb{M}} = \left[\sum_{q=1}^{N^S} \| \, |u^{tot}_{\mathbb{M},q}|^2 - |u^{inc}_{\mathbb{M},q}|^2 \, \|^2_{\mathbb{M}} \right]^{-1} . \tag{4.105}$$

As before, for zero contrast sources, the product of normalization factor and error norm is equal to one. With this new error norm, we have to determine the gradient direction. In [32], it is determined by writing the error norm in terms of real and imaginary parts. It is more elegant to use the definition via the Fréchet derivative, cf. Eq. (4.28). For each q, we now have

$$\begin{aligned}
&\lim_{\epsilon \to 0} \frac{1}{\epsilon} \left\{ \sum_{p=1}^{N^R} \left| |u^{tot}_{\mathbb{M},q}|^2 - |u^{mod(n-1)}_{\mathbb{M},q} + \epsilon\, G_{\mathbb{M}} g^{w(n)}_q|^2 \right|^2 - \sum_{p=1}^{N^R} \left| |u^{tot}_{\mathbb{M},q}|^2 - |u^{mod(n-1)}_{\mathbb{M},q}|^2 \right|^2 \right\} \\
&= \lim_{\epsilon \to 0} \frac{1}{\epsilon} \sum_{p=1}^{N^R} \left\{ |r^{(n-1)}_{\mathbb{M},q} - 2\epsilon \, \text{Re} \, [\overline{u}^{mod(n-1)}_{\mathbb{M},q} \; G_{\mathbb{M}} g^{w(n)}_q] + O(\epsilon^2)|^2 - |r^{(n-1)}_{\mathbb{M},q}|^2 \right\} \\
&= -4 \, \text{Re} \sum_{p=1}^{N^R} r^{(n-1)}_{\mathbb{M},q} \, \overline{u}^{mod(n-1)}_{\mathbb{M},q} \; G_{\mathbb{M}} g^{w(n)}_q = -4 \, \text{Re} \sum_{p=1}^{N^R} \overline{g}^{w(n)}_q \; G^{\star}_{\mathbb{M}} [u^{mod(n-1)}_{\mathbb{M},q} \, r^{(n-1)}_{\mathbb{M},q}] \,,
\end{aligned} \tag{4.106}$$

where we have used that

$$r_{\mathrm{M},q}^{(n-1)} = |u_{\mathrm{M},q}^{tot}|^2 - |u_{\mathrm{M},q}^{mod(n-1)}|^2 \,, \tag{4.107}$$

ϵ is real valued, and the expressions of the G_M operator, see Eq. (4.21), and its adjoint $G_\mathrm{M}^\star$, see Eq. (4.31). Apart from a constant factor, the last expression of Eq. (4.106) obtains its maximum variation when

$$g_{\mathrm{M},q}^{(n)} = G_\mathrm{M}^\star[2\, u_{\mathrm{M},q}^{mod(n-1)} r_{\mathrm{M},q}^{(n-1)}] \,. \tag{4.108}$$

If we compare this expression with the gradient of Eq. (4.30), we see that the residual error is multiplied by 2 $u_{\mathrm{M},q}^{mod(n-1)}$. This function is the approximation to the complex data (including the phase), made in the previous iteration. In this way, phase information is added to the residual error. Therefore, the back projection operator $G^\star$ is based on the phase information modeled in the previous iteration. If convergence is achieved, the approximation improves for an increasing number of iterations. At the end of the iterative procedure, we not only reconstructed the contrast profile but also reconstructed the missing phase in the measured data. We note that in this modified CSI method, the step length $\alpha_q^{(n)}$ is obtained as the root of a cubic equation [32].

The disadvantage of the method is that the function $u_{\mathrm{M},q}^{mod(n-1)}$ acts as weight for the residual error. If the error is relatively small, the modeled data will also be very small. This is a consequence of squaring the data in the error norm. It will reduce the resolution because a relative small data variation no longer contributes to the final contrast reconstructions

4.8.2 CSI Method for Measured Amplitude Data

For amplitude data, the error norm in the measurements domain becomes, cf. Eq. (4.18),

$$F_\mathrm{M}(w_q^{(n)}) = \sum_{q=1}^{N^S} \|r_{\mathrm{M},q}^{(n)}\|_\mathrm{M}^2 \,, \quad \text{with } r_{\mathrm{M},q}^{(n)} = |u_{\mathrm{M},q}^{tot}| - |u_{\mathrm{M},q}^{mod(n)}| \,, \tag{4.109}$$

and normalization factor

$$\eta_\mathrm{M} = \left[\sum_{q=1}^{N^S} \|\, |u_{\mathrm{M},q}| - |u_{\mathrm{M},q}^{inc}|\, \|_\mathrm{M}^2\right]^{-1} . \tag{4.110}$$

As before, for zero contrast sources, the product of normalization factor and error norm is equal to one. With this new error norm, we have to determine the gradient direction. For each q, the Fréchet derivative becomes

$$\begin{aligned}
&\lim_{\epsilon\to 0}\frac{1}{\epsilon}\left\{\sum_{p=1}^{N^R}\left||u_{\mathrm{M},q}^{tot}| - [|u_{\mathrm{M},q}^{mod(n-1)} + \epsilon\, G_\mathrm{M} g_q^{w(n)}|^2]^{\frac{1}{2}}\right|^2 - \sum_{p=1}^{N^R}\left||u_{\mathrm{M},q}^{tot}| - |u_{\mathrm{M},q}^{(n-1)}|\right|^2\right\}\\
&= \lim_{\epsilon\to 0}\frac{1}{\epsilon}\sum_{p=1}^{N^R}\left\{|r_{\mathrm{M},q}^{(n-1)} - \epsilon\,\mathrm{Re}\left[\frac{\overline{u}_{\mathrm{M},q}^{mod(n-1)}}{|u_{\mathrm{M},q}^{mod(n-1)}|}\, G_\mathrm{M} g_q^{w(n)}\right] + O(\epsilon^2)|^2 - |r_{\mathrm{M},q}^{(n-1)}|^2\right\}\\
&= -2\,\mathrm{Re}\sum_{p=1}^{N^R} r_{\mathrm{M},q}^{(n-1)}\,\frac{\overline{u}_{\mathrm{M},q}^{mod(n-1)}}{|u_{\mathrm{M},q}^{mod(n-1)}|}\, G_\mathrm{M} g_q^{w(n)} = -2\,\mathrm{Re}\sum_{p=1}^{N^R} \overline{g}_q^{w(n)}\, G_\mathrm{M}^\star\left[\frac{u_{\mathrm{M},q}^{mod(n-1)}}{|u_{\mathrm{M},q}^{mod(n-1)}|}\, r_{\mathrm{M},q}^{(n-1)}\right],
\end{aligned} \tag{4.111}$$

where we have used that

$$r_{\mathrm{M},q}^{(n-1)} = |u_{\mathrm{M},q}^{tot}| - |u_{\mathrm{M},q}^{mod(n-1)}| \tag{4.112}$$

and ϵ is real valued. Apart from a constant factor, the last expression of Eq. (4.111) obtains its maximum variation when

$$g^{(n)}_{\mathbb{M},q} = G^{\star}_{\mathbb{M}}[\varphi^{mod(n-1)}_{\mathbb{M},q}\ r^{(n-1)}_{\mathbb{M},q}] \,, \tag{4.113}$$

in which

$$\varphi^{mod(n-1)}_{\mathbb{M},q} = \frac{u^{mod(n-1)}_{\mathbb{M},q}}{|u^{mod(n-1)}_{\mathbb{M},q}|} \tag{4.114}$$

is the phase of $u^{mod(n-1)}_{\mathbb{M},q}$. This is the approximation to the missing phase of the data, made in the previous iteration. In this way, only the phase information is added to the residual error. Therefore, the back projection operator $G^{\star}$ is based on the phase information modeled in the previous iteration. However, the disadvantage of this method seems to be that the step length $\alpha^{(n)}_q$ should be found as the root of a transcendental equation.

4.9 Gauss–Newton Inversion

In Section 4.7, we have discussed a Gauss–Newton type of method to determine the finite set of parameters P_k, $k = 1, 2, \ldots$, representing a contrast profile $\chi^c(P_k)$, from a discrete set of contrast value $\chi_{csi}(\boldsymbol{x}_m)$, $m = 1, 2, \ldots, N$, computed in each iteration of the CSI method.

Originally, the Gauss–Newton method is an iterative procedure to determine the set of parameters P_k by the minimization of the the squared norm of the residual errors in $\mathbb{M}$. To discuss this method, we consider the Gauss–Newton method based on the contrast-source integral equations. In the nth iteration, we define the norm as

$$F_{\mathbb{M}}(P_k) = \| r^{(n)}_{\mathbb{M},q} \|^2_{\mathbb{M}} = \sum_{q=1}^{N^S}\sum_{p=1}^{N^R} | r^{(n)}_{\mathbb{M},q}(P_k)|^2 \,, \tag{4.115}$$

with

$$r^{(n)}_{\mathbb{M},q} = u^{sct}_{\mathbb{M}}(\boldsymbol{x}^R_p, \boldsymbol{x}^S_q) - G_{\mathbb{M}} w^{(n)}_q \,, \tag{4.116}$$

where the data operator $G_{\mathbb{M}} w^{(n)}_q = G_{\mathbb{M}} w^{(n)}(\boldsymbol{x}^R_p, \boldsymbol{x}^S_q)$ is written as, cf. Eq. (4.21),

$$G_{\mathbb{M}} w^{(n)}_q = \Delta V \sum_{m'=1}^{N} \gamma_0^2 \langle G\rangle(\boldsymbol{x}^R_p - \boldsymbol{x}_{m'})\ w^{(n)}(\boldsymbol{x}_{m'}, \boldsymbol{x}^S_q) \,. \tag{4.117}$$

The contrast source $w^{(n)}_q = w^{(n)}(\boldsymbol{x}_m, \boldsymbol{x}^S_q)$ is a solution of the contrast-source equation

$$w^{(n)}_q - \chi^{c(n-1)} G_{\mathbb{D}} w^{(n)}_q = \chi^{c(n-1)} u^{inc}_q, \quad q = 1, 2, \ldots, N^S \,, \tag{4.118}$$

where the object operator $G_{\mathbb{D}} w_q$ is written as, cf. Eq. (4.22),

$$G_{\mathbb{D}} w^{(n)}_q = \Delta V \sum_{m'=1}^{N} \gamma_0^2 \langle\langle G\rangle\rangle(\boldsymbol{x}_m - \boldsymbol{x}_{m'})\ w^{(n)}(\boldsymbol{x}_{m'}, \boldsymbol{x}^S_q) \,, \quad m = 1, 2, \ldots, N \,. \tag{4.119}$$

Let us assume that an approximation of the parameter set P_k is known from the previous iteration. We denote this set as $P^{(n-1)}_k$. Then, $\chi^{c(n-1)}(\boldsymbol{x}_m)$ is known as well and we solve Eq. (4.118) numerically. The resulting contrast source is substituted in Eq. (4.116). This procedure shows that the residual error is a function of the parameters P_k and can be computed for typical values of P_k.

An iterative minimization of the norm of Eq. (4.115) provides the appropriate tool to determine the set of parameters P_k.

Hence, starting with an initial guess $P_k^{(0)}$ for the minimum of the $F_{\rm M}(P_k)$, the Gauss–Newton method [16] continues with the iterations

$$P_k^{(n)} = P_k^{(n-1)} + \Delta P_k^{(n)} . \tag{4.120}$$

Carrying out a linear Taylor approximation of the residuals $r_{{\rm M},q}^{(n)}$ as function of $\Delta P_k^{(n)}$, followed by a minimization of the norm, we arrive at a system of linear equations for $\Delta P_k^{(n)}$, viz.,

$$\sum_{k=1,2,\dots} \left\langle \frac{\partial r_{{\rm M},q}^{(n)}}{\partial P_l}, \frac{\partial r_{{\rm M},q}^{(n)}}{\partial P_k} \right\rangle_{\rm M} \Delta P_k^{(n)} = -\left\langle \frac{\partial r_{{\rm M},q}^{(n)}}{\partial P_l}, r_{{\rm M},q}^{(n)} \right\rangle_{\rm M} , \quad l = 1,\ 2,\ \dots , \tag{4.121}$$

where the inner product follows from the definition of the norm of Eq. (4.115).

At this point, we remark that analytic derivatives of the residuals with respect to the parameters are not available, and we determine the profile derivatives numerically by defining these derivatives as

$$\frac{\partial r_{{\rm M},q}^{(n)}}{\partial P_k} = \frac{r_{{\rm M},q}^{(n)}(\boldsymbol{x}, P_{j\neq k}, P_k + \epsilon) - r_{{\rm M},q}^{(n)}(\boldsymbol{x}, P_{j\neq k}, P_k)}{\epsilon} , \quad k = 1,\ 2,\ \dots . \tag{4.122}$$

As an illustration, we consider the example of Section 4.7. Our synthetic data are obtained for a circular cylinder. Its contrast function is defined by the coordinates x_1^O and x_2^O of the center point O, the real contrast value $\chi^O = 1 - (c_0/c_{sct})^2$ at the center point, and the radius a. In our numerical procedure to model the data, we use the following parameter representation for the contrast profile:

$$\chi^c(\boldsymbol{x}) := P_3 \frac{1}{1 + \exp[(R - P_4)/s]} , \quad \text{with } R = [(x_1 - P_1)^2 + (x_2 - P_2)^2]^{\frac{1}{2}} , \tag{4.123}$$

where $P_1 = x_1^O$, $P_2 = x_2^O$, $P_3 = \chi^O$, and $P_4 = a$ are the real profile parameters to be determined. Further, the parameter s determines the sharpness of the profile at the boundary of the circular cylinder; we take $s = 0.3\ \Delta x$. A picture of this profile is shown in Figure 4.28(c).

Matlab Listing II.29 The function `FunDataSctIntEq`.

```
function [data\_est] = FunDataSctIntEq(P,G\_S,G\_R,input)
w = cell(1,input.NS); data\_est = zeros(input.NR,input.NS);

% Update contrast using analytic contrast profile
          s = 0.3 * input.dx;
          R = sqrt((input.X1-P(1)).^2 + (input.X2-P(2)).^2);
  input.CHI = P(3)./(1+exp((R-P(4))/s));

% Solve integral equation for contrast source with BICGSTAB method --------
  for q = 1 : input.NS
      w{q} = BICGSTABw(G\_S{q},input);
  end % q\_loop

% Compute synthetic data and plot fields and data -----------------------
  for q = 1 : input.NS
      data\_est(:,q) = DopM(G\_R,w{q},input);
  end % q\_loop
```

Matlab Listing II.30 The script `GaussNewtonNumeric` for a Gauss–Newton inversion using data computed with the integral equation method.

```
clear all; clc; close all; clear workspace
global Pnom P_it
input = init();    w = cell(1,input.NS); for q=1:input.NS, w{q} = 0; end
itmax = 15; eps = 1e-6;

% (1) Compute Exact data using analytic Bessel function expansion ---------
     disp('Running DataSctCircle');  DataSctCircle;
     load DataAnalytic;
     % Compute Green function for sources and receivers ------------------
      [G_S,G_R] = GreenSourcesReceivers(input);

% (2) Start with estimated contrast sources
   Pnom = [input.xO(1), input.xO(2), input.contrast,    input.a];
   P0   = [ 0, 0, input.contrast/10, 75];
   disp(['Nominal P : ', num2str(Pnom)]);
   disp(['Initial P0: ', num2str(P0)]);
   P_it(1,:) = P0;   % save as global array for plotting

 for it = 1: itmax
  % Compute residual error, construct Jacobian matrix and update P
    P = P_it(it,:);
    data_est = FunDataSctIntEq(P,G_S,G_R,input);
    Residual = dataCircle -  data_est;
    GNerror  = norm(Residual(:))/norm(dataCircle(:));
    disp(['            it= ',num2str(it),'  Residual= ',num2str(GNerror)]);

  % Compute numerical derivatives with respect to P
    P(1) = P(1) + eps;  data_plus = FunDataSctIntEq(P,G_S,G_R,input);
      dCHI_P{1} = (data_plus - data_est) /eps;       P(1) = P(1) - eps;
    P(2) = P(2) + eps;  data_plus = FunDataSctIntEq(P,G_S,G_R,input);
      dCHI_P{2} = (data_plus - data_est) /eps;       P(2) = P(2) - eps;
    P(3) = P(3) + eps;  data_plus = FunDataSctIntEq(P,G_S,G_R,input);
      dCHI_P{3} = (data_plus - data_est) /eps;       P(3) = P(3) - eps;
    P(4) = P(4) + eps;  data_plus = FunDataSctIntEq(P,G_S,G_R,input);
      dCHI_P{4} = (data_plus - data_est) /eps;       P(4) = P(4) - eps;

  % Compute residual error, construct Jacobian matrix and update P
    b = zeros(1,4); A = zeros(4,4);
    for l = 1:4
        b(l) = sum(conj(dCHI_P{l}(:)).*Residual(:));
        for k = 1:4
            A(l,k) = sum(conj(dCHI_P{l}(:)).*dCHI_P{k}(:));
        end
    end
    DeltaP = real(b/A);
    P = P + DeltaP;                  P(4) = abs(P(4)); % P(4) is positive
    disp(['          P: ', num2str(P)]);
    P_it(it+1,:) = P; % save as global array for plotting
 end % it_loop
 DisplayParameters;
```

4.9.1 Matlab Codes for Gauss–Newton Inversion

The Matlab function FunDataSctIntEq of Matlab Listing II.29 computes an update for the scattered field. It updates the contrast using Eq. (4.123), solves the integral equations using BICGSTABw.m of Matlab Listing II.10, and computes the scattered field using DopM of Matlab Listing II.12. This Matlab function FunDataSctIntEq is used in the Matlab script GaussNewtonNumeric of Matlab Listing II.30.

For comparison, we also present the Matlab codes for the Gauss–Newton inversion based on analytical data ("Inverse Crime"). The Matlab function FunDataSctCircle of Matlab Listing II.31 computes an update for the scattered field using the Bessel series expansion, and it is used in the Matlab script GaussNewtonAnalytic of Matlab Listing II.32.

Matlab Listing II.31 The function FunDataSctCircle.

```
function [data_est] = FunDataSctCircle(P,input)

c_0      = input.c_0;        c_sct   = c_0 / sqrt(1-P(3));
gam0     = input.gamma_0;  gam_sct = input.gamma_0 * c_0/c_sct;

% (1) Compute coefficients of series expansion ----------------------------
  arg0 = gam0 * P(4);   args = gam_sct * P(4);
  M = 100;                                    % increase M for more accuracy
  A = zeros(1,M+1);
  for m = 0 : M
    Ib0 = besseli(m,arg0);          dIb0 =  besseli(m+1,arg0) + m/arg0 * Ib0;
    Ibs = besseli(m,args);          dIbs =  besseli(m+1,args) + m/args * Ibs;
    Kb0 = besselk(m,arg0);          dKb0 = -besselk(m+1,arg0) + m/arg0 * Kb0;
    A(m+1) = - (gam_sct * dIbs*Ib0 - gam0 * dIb0*Ibs) ...
              /(gam_sct * dIbs*Kb0 - gam0 * dKb0*Ibs);
  end

% (2) Compute reflected field at the receivers for all sources (data) -----
  xR(1,:) = input.xR(1,:) - P(1);  % shifted origin
  xR(2,:) = input.xR(2,:) - P(2);  % shifted origin
  xS(1,:) = input.xS(1,:) - P(1);  % shifted origin
  xS(2,:) = input.xS(2,:) - P(2);  % shifted origin
  rR = sqrt(xR(1,:).^2 + xR(2,:).^2);  phiR = atan2(xR(2,:),xR(1,:));
  rS = sqrt(xS(1,:).^2 + xS(2,:).^2);  phiS = atan2(xS(2,:),xS(1,:));

  data2D = zeros(input.NR,input.NS);
  for p = 1 : input.NR
   for q = 1 : input.NS
    data2D(p,q) = A(1) * besselk(0,gam0*rR(p)) * besselk(0,gam0*rS(q));
    for m = 1 : M
     data2D(p,q) = data2D(p,q) + 2 * A(m+1) * cos(m*(phiR(p)-phiS(q))) ...
                        * besselk(m,gam0*rR(p)) * besselk(m,gam0*rS(q)) ;
    end % m_loop
   end % q_loop
  end % p_loop
  data_est= 1/(2*pi) * data2D;
```

Matlab Listing II.32 The function `GaussNewtonAnalytic` for a Gauss–Newton inversion using analytical data.

```
clear all; clc; close all; clear workspace
global Pnom P_it
input = init();
itmax = 15;
eps = 1e-6;

% (1) Compute Exact data using analytic Bessel function expansion --------
     disp('Running DataSctCircle');   DataSctCircle;
     load DataAnalytic;

% (2) Start with estimated contrast sources
   Pnom = [input.xO(1), input.xO(2), input.contrast,   input.a];
   P0   = [ 0, 0, input.contrast/10, 75];
   disp(['Nominal P : ', num2str(Pnom)]);
   disp(['Initial P0: ', num2str(P0)]);
   P_it(1,:) = P0;   % save as global array for plotting

 for it = 1: itmax
  % Compute residual error, construct Jacobian matrix and update P
    P = P_it(it,:);
    data_est = FunDataSctCircle(P,input);
    Residual = dataCircle -  data_est;
    GNerror  = norm(Residual(:))/norm(dataCircle(:));
    disp(['            it= ',num2str(it),'  Residual= ',num2str(GNerror)]);

  % Compute numerical derivatives with respect to P
    P(1) = P(1) + eps;  data_plus = FunDataSctCircle(P,input);
      dCHI_P{1} = (data_plus - data_est) /eps;        P(1) = P(1) - eps;
    P(2) = P(2) + eps;  data_plus = FunDataSctCircle(P,input);
      dCHI_P{2} = (data_plus - data_est) /eps;        P(2) = P(2) - eps;
    P(3) = P(3) + eps;  data_plus = FunDataSctCircle(P,input);
      dCHI_P{3} = (data_plus - data_est) /eps;        P(3) = P(3) - eps;
    P(4) = P(4) + eps;  data_plus = FunDataSctCircle(P,input);
      dCHI_P{4} = (data_plus - data_est) /eps;        P(4) = P(4) - eps;

  % Compute residual error, construct Jacobian matrix and update P
    b = zeros(1,4); A = zeros(4,4);
    for l = 1:4
        b(l) = sum(conj(dCHI_P{l}(:)).*Residual(:));
        for k = 1:4
            A(l,k) = sum(conj(dCHI_P{l}(:)).*dCHI_P{k}(:));
        end
    end
    DeltaP = real(b/A);
    P = P + DeltaP;                P(4) = abs(P(4)); % P(4) is positive
    disp(['         P: ', num2str(P)]);
    P_it(it+1,:) = P; % save as global array for plotting
 end % it_loop

 DisplayParameters;
```

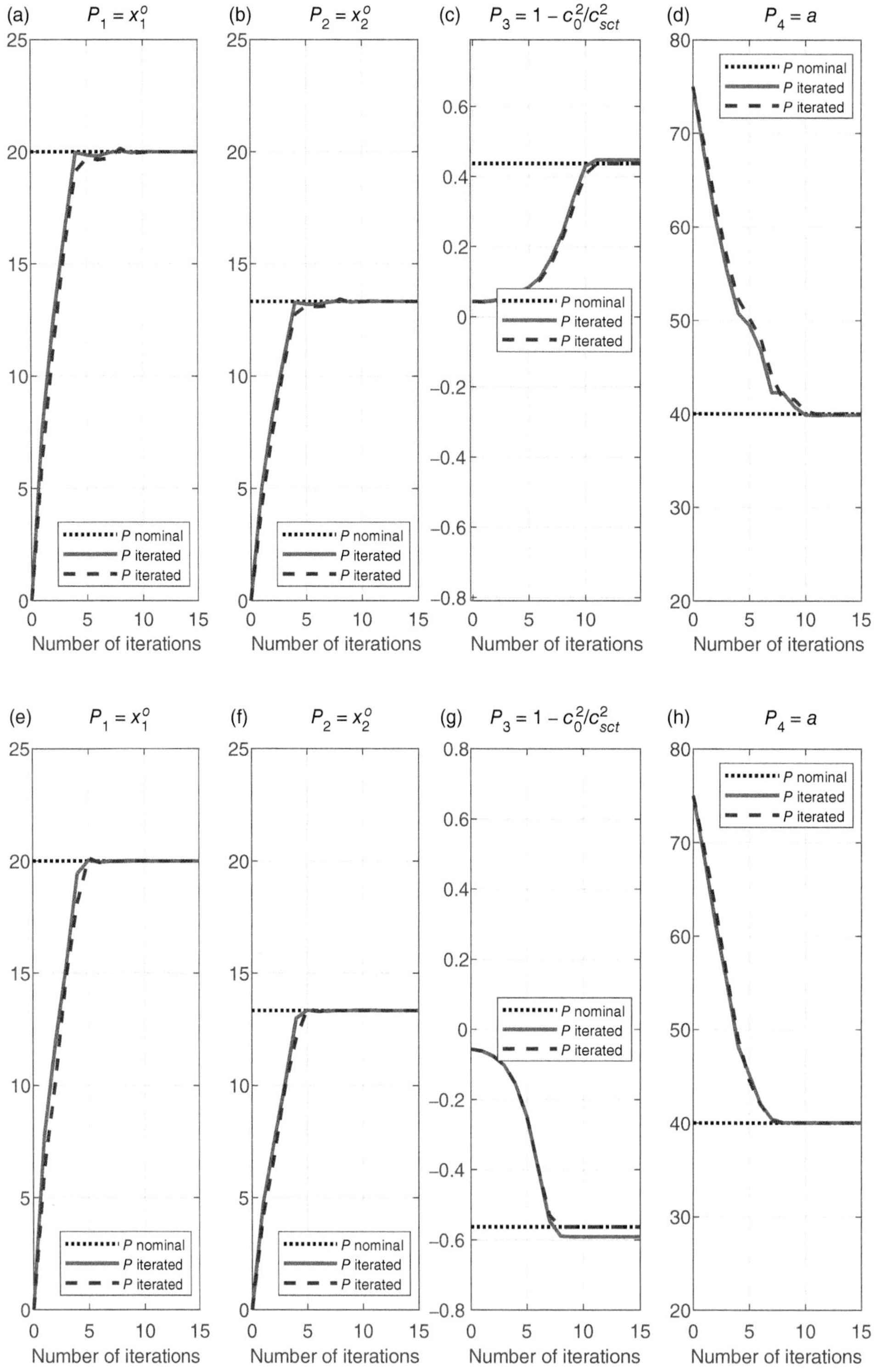

Figure 4.31 Plots of the parameters $P_k^{(n)}$ as function of the number of the Gauss–Newton iterations, for positive contrast profile (a–d) and negative contrast profile (e–h), respectively.

Both Matlab scripts call the function `DisplayParameters` of Matlab Listing II.28 to plot the profile parameters. In Figure 4.31, for both numerical and analytic data, we show the reconstructed parameters as a function of the number of iterations. Note that within about ten iterations, convergence has been achieved. There are no significant differences between the use of either numerical data (solid lines) or analytical data (dashed lines).

5

Acoustic Wave Inversion

In this chapter, we now consider the problem of reconstructing the acoustic parameters of a bounded object immersed in a known background medium, from a knowledge of how the object scatters known incident radiation.

5.1 Notation

Let $\mathbb{D}_{sct}$ denote a bounded, not necessarily connected, scattering object whose location, the compressibility contrast $\hat{\chi}^{\kappa} = 1 - (\hat{\kappa}_{sct}/\kappa_0)$ and the mass-density contrast $\hat{\chi}^{\rho} = 1 - (\hat{\rho}_{sct}/\rho_0)$, see Eqs. (2.53) and (2.54), are unknown, but are known to lie within another, larger, bounded simply connected domain $\mathbb{D}$. Observe that if $\boldsymbol{x}$ is not in $\mathbb{D}_{sct}$, then $\hat{\chi}^{\kappa}$ and $\hat{\chi}^{\rho}$ vanish, but if the location of $\mathbb{D}_{sct}$ is unknown, then it is not known *a priori*, where these contrast quantities vanish. However, with the assumption that $\mathbb{D}_{sct} \subset \mathbb{D}$, it is known that $\hat{\chi}^{\kappa}$ and $\hat{\chi}^{\rho}$ vanish for $\boldsymbol{x}$ outside $\mathbb{D}$. Let $\mathbb{M}$ denote a domain with a discrete collection of receiver points outside of $\mathbb{D}$, then the scattered pressure field is defined as $\hat{p}^{sct}_{\mathrm{M}}$. As a model for this field, we use the contrast-source integral representation, cf. Eq. (2.59),

$$\hat{p}^{sct}_{\mathrm{M}}(\boldsymbol{x}^R_p, \boldsymbol{x}^S_q) = \int_{\boldsymbol{x}' \in \mathbb{D}} [\hat{\gamma}_0^2 \hat{G}(\boldsymbol{x}^R_p - \boldsymbol{x}')\hat{w}^p(\boldsymbol{x}', \boldsymbol{x}^S_q) - \hat{\gamma}_0 \partial^R_k \hat{G}(\boldsymbol{x}^R - \boldsymbol{x}')\hat{w}^{Zv}_k(\boldsymbol{x}', \boldsymbol{x}^S_q)]\mathrm{d}V, \tag{5.1}$$

for $(\boldsymbol{x}^R_p, \boldsymbol{x}^S_q) \in \mathbb{M}$. Here, the set of $\boldsymbol{x}^R_p$, $p = 1, 2, \dots$ denotes the collection of receiver points in the domain $\mathbb{M}$, and the set of $\boldsymbol{x}^S_q$, $q = 1, 2, \dots$ denotes the collection of source points in the domain $\mathbb{M}$ (Figure 5.1). The scalar pressure contrast source $\hat{w}^p$ and the vectorial particle-velocity contrast source $\hat{w}^{Zv}_k$ are given by, cf. Eqs. (2.55) and (2.56),

$$\hat{w}^p(\boldsymbol{x}, \boldsymbol{x}^S_q) = \hat{\chi}^{\kappa}(\boldsymbol{x})\, \hat{p}(\boldsymbol{x}, \boldsymbol{x}^S_q)\,, \tag{5.2}$$

and

$$\hat{w}^{Zv}_j(\boldsymbol{x}, \boldsymbol{x}^S_q) = \hat{\chi}^{\rho}(\boldsymbol{x})\, Z_0 \hat{v}_j(\boldsymbol{x}, \boldsymbol{x}^S_q)\,. \tag{5.3}$$

Eq. (5.1) holds only if there is no noise and no error in the measurements. However, error-free data is extremely unlikely, and we do not assume that this equation holds exactly.

If $\hat{p}^{inc}(\boldsymbol{x}, \boldsymbol{x}^S_q)$ and $\hat{v}^{inc}_k(\boldsymbol{x}, \boldsymbol{x}^S_q)$ denote the pressure and the particle-velocity of an acoustic incident wave with propagation coefficient $\hat{\gamma}_0$ and source point $\boldsymbol{x}^S_q$, then the total acoustic field components

Forward and Inverse Scattering Algorithms based on Contrast Source Integral Equations, First Edition. Peter M. van den Berg.

Companion website: www.wiley.com/go/vandenBerg/ScatteringAlgorithms

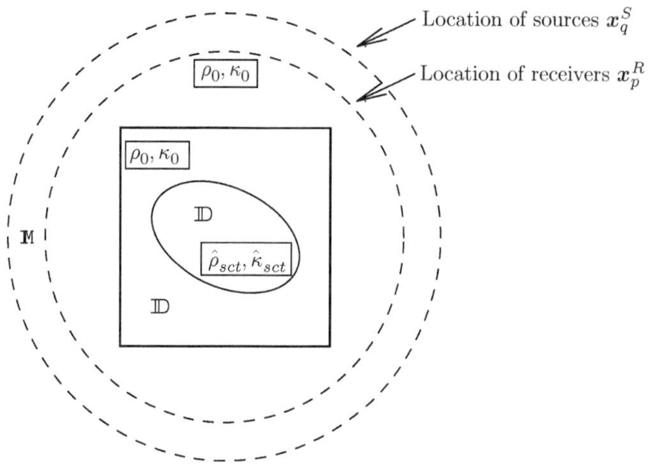

Figure 5.1 The inverse scattering configuration with the measurement domain 𝕄 and the object domain 𝔻. The measurement domain $\mathbb{M} \not\subset \mathbb{D}$. The scattering domain $\mathbb{D}_{sct} \subset \mathbb{D}$. In our numerical examples, the sources are distributed over the dashed outer circle, and the receivers are distributed over the dashed inner circle.

in 𝔻 satisfy the coupled set of contrast-source integral equations, cf. Eqs. (2.57) and (2.58),

$$\hat{\chi}^{\kappa}(\boldsymbol{x})\hat{p}^{inc}(\boldsymbol{x},\boldsymbol{x}_q^S) = \hat{w}^p(\boldsymbol{x},\boldsymbol{x}_q^S) - \hat{\chi}^{\kappa}(\boldsymbol{x})\left[\hat{\gamma}_0^2\int_{\boldsymbol{x}'\in\mathbb{D}} \hat{G}(\boldsymbol{x}-\boldsymbol{x}')\hat{w}^p(\boldsymbol{x}',\boldsymbol{x}_q^S)\,\mathrm{d}V - \hat{\gamma}_0\partial_k\int_{\boldsymbol{x}'\in\mathbb{D}} \hat{G}(\boldsymbol{x}-\boldsymbol{x}')\hat{w}_k^{Zv}(\boldsymbol{x}',\boldsymbol{x}_q^S)\,\mathrm{d}V\right] \tag{5.4}$$

and

$$\hat{\chi}^{\rho}(\boldsymbol{x})Z_0\hat{v}_j^{inc}(\boldsymbol{x},\boldsymbol{x}_q^S) = \hat{w}_j^{Zv}(\boldsymbol{x},\boldsymbol{x}_q^S) - \hat{\chi}^{\rho}(\boldsymbol{x})\left[-\hat{\gamma}_0\partial_j\int_{\boldsymbol{x}'\in\mathbb{D}} \hat{G}(\boldsymbol{x}-\boldsymbol{x}')\hat{w}^p(\boldsymbol{x}',\boldsymbol{x}_q^S)\,\mathrm{d}V + \partial_j\partial_k\int_{\boldsymbol{x}'\in\mathbb{D}} \hat{G}(\boldsymbol{x}-\boldsymbol{x}')\hat{w}_k^{Zv}(\boldsymbol{x}',\boldsymbol{x}_q^S)\,\mathrm{d}V + \hat{w}_j^{Zv}(\boldsymbol{x},\boldsymbol{x}_q^S)\right], \tag{5.5}$$

for $\boldsymbol{x} \in \mathbb{D}$ and $\boldsymbol{x}_q^S \in \mathbb{M}$. The simultaneous inversion of both the compressibility contrast and the mass-density contrast is discussed by van Dongen and Wright [59]. Here, we consider two special cases.

5.1.1 Compressibility Contrast Only

In the first case, we assume that the mass-density contrast vanishes. Then, the scattered field at the receiver points is given by

$$\hat{p}_{\mathbb{M}}^{sct}(\boldsymbol{x}_p^R,\boldsymbol{x}_q^S) = \int_{\boldsymbol{x}'\in\mathbb{D}} \hat{\gamma}_0^2\,\hat{G}(\boldsymbol{x}^R-\boldsymbol{x}')\hat{w}^p(\boldsymbol{x}',\boldsymbol{x}_q^S)\mathrm{d}V\,, \tag{5.6}$$

for $(\boldsymbol{x}_p^R,\boldsymbol{x}_q^S) \in \mathbb{M}$, while the contrast-source integral equation becomes

$$\hat{\chi}^{\kappa}(\boldsymbol{x})\,\hat{p}^{inc}(\boldsymbol{x},\boldsymbol{x}_q^S) = \hat{w}^p(\boldsymbol{x},\boldsymbol{x}_q^S) - \hat{\chi}^{\kappa}(\boldsymbol{x})\int_{\boldsymbol{x}'\in\mathbb{D}} \hat{\gamma}_0^2\,\hat{G}(\boldsymbol{x}-\boldsymbol{x}')\hat{w}^p(\boldsymbol{x}',\boldsymbol{x}_q^S)\,\mathrm{d}V\,, \tag{5.7}$$

for $\boldsymbol{x} \in \mathbb{D}$ and $\boldsymbol{x}_q^S \in \mathbb{M}$. In this case, we have $\hat{\chi}^{\kappa} := \hat{\chi}^c$, and we observe that the inversion problem of the compressibility contrast is identical to the inversion of the wave-speed contrast in Chapter 4. That is why this case is not dealt with further.

5.1.2 Mass-density Contrast Only

In the second case, we assume that the compressibility contrast vanishes. Then, the acoustic wave speed inside the scattering object is given by $\hat{c}_{sct} = c_0\sqrt{\rho_0/\hat{\rho}_{sct}}$. The scattered field at the receivers point is given by

$$\hat{p}^{sct}_{\mathbb{M}}(\boldsymbol{x}^R_p, \boldsymbol{x}^S_q) = -\int_{\boldsymbol{x}'\in\mathbb{D}} \hat{\gamma}_0\ \partial^R_k \hat{G}(\boldsymbol{x}^R_p - \boldsymbol{x}')\hat{w}^{Zv}_k(\boldsymbol{x}', \boldsymbol{x}^S_q)\ \mathrm{d}V\ , \tag{5.8}$$

for $(\boldsymbol{x}^R_p, \boldsymbol{x}^S_q) \in \mathbb{M}$, while the contrast-source integral equation becomes

$$\hat{\chi}^\rho(\boldsymbol{x})\ Z_0\hat{v}^{inc}_j(\boldsymbol{x}, \boldsymbol{x}^S_q) = \hat{w}^{Zv}_j(\boldsymbol{x}, \boldsymbol{x}^S_q) - \hat{\chi}^\rho(\boldsymbol{x})\ \left[\partial_j\partial_k \int_{\boldsymbol{x}'\in\mathbb{D}} \hat{G}(\boldsymbol{x} - \boldsymbol{x}')\hat{w}^{Zv}_k(\boldsymbol{x}', \boldsymbol{x}^S_q)\ \mathrm{d}V + \hat{w}^{Zv}_j(\boldsymbol{x}, \boldsymbol{x}^S_q)\right]\ , \tag{5.9}$$

for $\boldsymbol{x} \in \mathbb{D}$ and $\boldsymbol{x}^S_q \in \mathbb{M}$. In this case, we observe that there are two spatial differentiations in the operator of the integral equation, which need special attention. For this reason, we only discuss the inverse problem to find the mass-density.

5.2 Synthetic Data for Zero Compressibility Contrast

In this section, we discuss the computation of synthetic data, which will replace the measured data for our inversion examples. Since in this part of the book, we only present 2D computer algorithms, we consider the circular cylinder. The forward model has been discussed in Chapter 2. Since we do not want to take any advantage of the symmetry of this circular cylinder, we consider the scattering by an off-centered circular cylinder. We first compare the two scattered-field data sets, obtained either analytically using the Bessel-series solution or numerically using the integral-equation solution.

The wave-field parameters are initialized by the `initAC` function (Matlab Listing II.33). The global parameter `nDIM` assigns the 2D space to be used in our inversion codes. Subsequently, we consider only scattering objects with positive mass-density contrast. The mass density of the embedding is 2500 kg/m^3 and the constant mass density of the scattering object is 1500 kg/m^3. The wave speed c_0 of the embedding is 1500 m/s. For vanishing compressibility contrast, the constant wave speed of the scattering object follows from $\hat{c}_{sct} = c_0\sqrt{\rho_0/\hat{\rho}_{sct}} \approx 1936$ m/s. The frequency of operation is 50 Hz. At this frequency, our chosen mass-density contrast inside the scattering object, $\hat{\chi}^\rho = 0.4$, is about the maximum value for which our inversion method results in a convergent process.

As output, the `initAC` function displays the operating wavelength. This facilitates the user to check the value $\lambda/\Delta x$. In the forward case often a discretization of 15 samples per wavelength is chosen, leading to an error of about 1 %. In the inversion case, we choose a discretization of 10 samples of wavelength. In this way, the numerical data differs substantially from the analytical data. Hence, this difference may be seen as model noise. In our case, this error is within the range of 5–10%.

Further, `initAC` function calls the functions `initSourceReceiver`, `initGrid`, `initFFTGreenfun`, and `initAcousticContrast`, but now only in 2D. They initialize the locations of the sources and the receivers, the numerical grid and the contrast distribution, respectively. Next, the error criterion is given. We use a stronger one than the error criterion used in the first part of this book, because we want to be sure that the difference between the analytical data and the numerical data is not caused by a convergence problem of the integral equation.

Matlab Listing II.33 The function `initAC` to initialize the 2D acoustic wave-field parameters.

```
function input = initAC()
% Time factor = exp(-iwt)
% Spatial units is in m                         % Source wavelet  s rho_0 Q = 1

global nDIM;  nDIM = 2;                          % set dimension of space

input.c_0      = 1500;                           % wave speed in embedding
input.c_sct    = 1500;                           % wave speed in scatterer
input.rho_0    = 3000;                           % mass density in embedding
input.rho_sct  = 1500;                           % mass density in scatterer

f              = 50;                             % temporal frequency
wavelength     = input.c_0 / f;                  % wavelength
s              = 1e-16 - 1i*2*pi*f;              % LaPlace parameter
input.gamma_0 = s/input.c_0;                     % propagation coefficient
disp(['wavelength = ' num2str(wavelength)]);

% add input data to structure array 'input'
  input = initSourceReceiver(input);        % add location of source/receiver

  input = initGrid(input);                  % add grid in either 1D, 2D or 3D

  input = initFFTGreenfun(input);           % compute FFT of Green function

  input = initAcousticContrast(input);      % add contrast distribution

  input.Errcri = 1e-5;

  input.Noise = 0;

  if input.Noise == 1  % generate random numbers that are repeatable
    sigma  = 0.2;        % sigma = standard deviation and mu =0
    randn('state',1);    ReRand = randn(input.NR,input.NS) * sigma;
    randn('state',2);    ImRand = randn(input.NR,input.NS) * sigma;
    input.Rand = ReRand + 1i*ImRand;
  end

  % Kirchhoff type of approximation in object operators
  input.Kirchhoff = 0; % value = 0: no approximation
                       % value = 1: approximation in adjoint
                       % value = 2: plus approximation in forward operator
```

In the function `initSourceReceiver` (Matlab Listing II.2), we have 50 sources located on a circle with radius of 170 m around the scattering object, while 50 receivers are located on a circle with radius of 150 m around the scattering object. The function `initGrid` (Matlab Listing II.3) defines a rectangular grid of N_1 by N_2 square subdomains with equal area of $(\Delta x)^2$. The function `initFFTGreenfun` of Matlab Listing II.34 generates a 2D grid with circulant properties. At the end of this function, the Matlab function `Greenfun` is called to compute the repeated weak form of the Green functions, see Matlab Listing II.35. For later convenience, the spatial derivatives are computed as well. The function `initAcousticContrast` (Matlab Listing II.36) initializes the

Matlab Listing II.34 The function `initFFTGreenfun` computes the Discrete Fourier transform of the repeated weak form of the 2D Green function.

```
function [input] = initFFTGreenfun(input)

% make two-dimensional FFT grid
N1fft        = 2^ceil(log2(2*input.N1));
N2fft        = 2^ceil(log2(2*input.N2));
x1(1:N1fft) = [0 : N1fft/2-1   N1fft/2 : -1 : 1] * input.dx;
x2(1:N2fft) = [0 : N2fft/2-1   N2fft/2 : -1 : 1] * input.dx;
[temp.X1fft,temp.X2fft] = ndgrid(x1,x2);
sign_x1 = [0 ones (1,N1fft/2-1)  0  -ones(1,N1fft/2-1)];
sign_x2 = [0 ones (1,N2fft/2-1)  0  -ones(1,N2fft/2-1)];
[Sign.X1fft,Sign.X2fft] = ndgrid (sign_x1,sign_x2);

% compute gam_0^2 * subdomain integrals of Green function
% and interface integrals  of  derivatives of Green function
  [IntG,IntdG]  = Greenfun(temp,Sign,input);

% apply n-dimensional Fast Fourier transforms
  input.FFTG       = fftn(IntG);
  input.FFTdG{1} = fftn(IntdG {1});
  input.FFTdG{2} = fftn(IntdG {2});
```

Matlab Listing II.35 The function `Greenfun` to compute the repeated weak form of the 2D Green function.

```
function [IntG,IntdG] = Greenfun(temp,Sign,input)
gam0 = input.gamma_0;       dx   = input.dx;

  delta     = (pi)^(-1/2) * dx;              % radius circle with area of dx^2
  factor    = 2 * besseli(1,gam0*delta) / (gam0*delta);
  X1        = temp.X1fft;  X2 = temp.X2fft;
  DIS       = sqrt(X1.^2 + X2.^2);
DIS(1,1)    = 1;                          % avoid Green's singularity for DIS = 0
   G        = 1/(2*pi).* besselk(0,gam0*DIS) * factor^2;
   IntG     = (gam0^2 * dx^2) * G;                 % integral includes gam0^2
IntG(1,1)   = 1 - gam0*delta * besselk(1,gam0*delta) * factor; %----------
  d_G       = - gam0 * (1/(2*pi)) .* besselk(1,gam0*DIS) * factor^2;
  d1_G      = (Sign.X1fft .* X1./DIS) .* d_G;
  d2_G      = (Sign.X2fft .* X2./DIS) .* d_G;
IntdG{1}    = dx^2 * d1_G;      IntdG{1}(1,1) = 0;
IntdG{2}    = dx^2 * d2_G;      IntdG{2}(1,1) = 0;
```

contrast distribution $\hat{\chi}^{\rho}(x_1, x_2)$. To check whether the compressibility contrast vanishes, we also compute the compressibility from the mass densities and the wave speeds. To create an off-centered cylinder, we define a new origin for both the computation of the contrast distribution and the analytical computations of the wave fields. The contrast distributions are presented either as an image plot of its real part or as a surface plot of the absolute value. The distribution of the compressibility contrast and the mass-density contrast are shown in Figure 5.2.

Matlab Listing II.36 The function `initAcousticContrast` to initialize the acoustic parameters.

```
function [input] = initAcousticContrast(input)

input.a  = 40; % half width slab / radius circle cylinder / radius sphere
input.xO(1) = input.a / 2;   % center coordinates of circle
input.xO(2) = input.a / 3;
         R = sqrt((input.X1-input.xO(1)).^2 + (input.X2-input.xO(2)).^2);
shape = (R < input.a);

% (1) Compute compressibbily contrast and mass density contrast ----------
      kappa_0          = 1 / (input.rho_0   * input.c_0^2);
      kappa_sct        = 1 / (input.rho_sct * input.c_sct^2);
      contrast_kappa = 1 - kappa_sct /kappa_0;
      input.CHI_kap    = contrast_kappa .* shape;

   % Plot contrast values
     set(figure(1),'Units','centimeters','Position',[5 1 18 12]);
     subplot(1,2,1);
       IMAGESC(x1,x2, real(input.CHI_kap));
       title('\fontsize{13}\chi^{\kappa} = 1 - \kappa_{sct}/ \kappa_{0}');
     subplot(1,2,2);
       IMAGESC(x1,x2,input.CHI_rho);
       title('\fontsize{13}\chi^{\rho} = 1 - \rho_{sct}/ \rho_{0}');

% (2) Compute volume contrast and interface contrast --------------------
      c_contrast = 1 - input.c_0^2/input.xsxsc_sct^2;
      input.CHI  = c_contrast .* shape;
      rho        = input.rho_sct .* shape + (1-shape) .* input.rho_0;

      Rfl{1} = zeros(N1,N2);    Rfl{2} = zeros(N1,N2);
      Rfl{1}(1:N1-1,:) = (rho(2:N1,:) - rho(1:N1-1,:)) ...
                               ./(rho(2:N1,:) + rho(1:N1-1,:));
      Rfl{2}(:,1:N2-1) = (rho(:,2:N2) - rho(:,1:N2-1)) ...
                               ./(rho(:,2:N2) + rho(:,1:N2-1));
      input.Rfl = Rfl;

   % Plot contrast values
     set(figure(2),'Units','centimeters','Position',[5 1 18 12]);
     subplot(1,2,1);
       IMAGESC(x1+dx/2,x2,Rfl{1});
       axis('ij','equal','tight');  title('\fontsize{13} R^{(1)}');
       colorbar('hor')
    subplot(1,2,2);
      IMAGESC(x1,x2+dx/2,Rfl{2});
      axis('ij','equal','tight');  title('\fontsize{13} R^{(2)}');
      colorbar('hor');  colormap(jet);
```

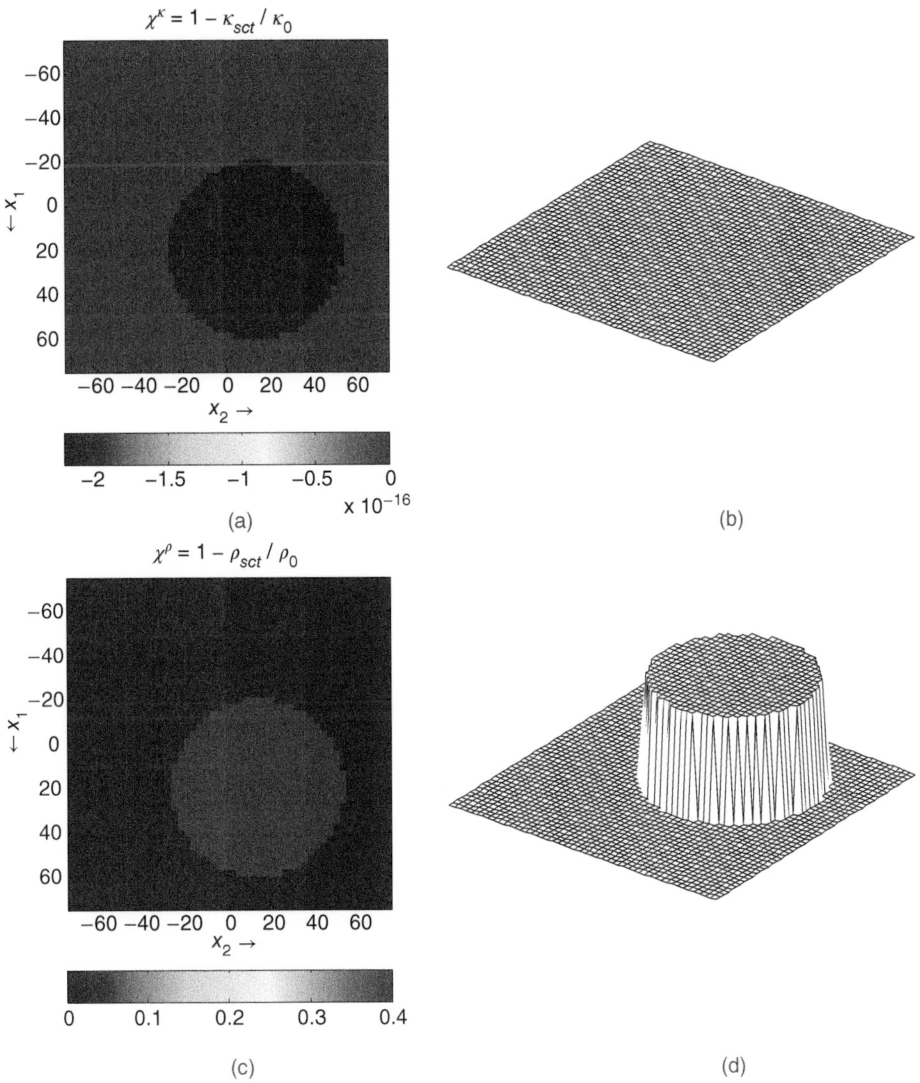

Figure 5.2 Plots of the compressibility contrast (a,b) and the mass-density contrast (c,d). These quantities are used to compute the synthetic data.

Finally, the `initAC` function offers the option to add noise to the analytical data to increase the noise level between the analytical and the numerical data. We use the Matlab random number generator `randn` to generate two repeatable arrays with a size of input.NR by input.NS, consisting of normally distributed random numbers with standard deviation $\sigma = 0.2$ and zero mean. These two different arrays are used to construct a single complex-valued array input.Rand.

The Matlab script `ForwardZv` to compute the 2D scattered field both analytically and numerically is given in Matlab Listing II.37. The following steps are carried out:

Matlab Listing II.37 The `ForwardwZv` script to compute synthetic 2D data.

```
clear all; clc; close all; clear workspace
input = initAC();

% (1) Compute analytically scattered field data -------------------------------
    disp('Running AcousticDataSctCircle');  AcousticDataSctCircle;

% (2) Compute Green functions for sources and receivers -----------------------
    [G_S,dG_S,G_R,dG_R] = dGreenSourcesReceivers(input);

% (3) Solve contrast source integral equations and compute sunthetic data
    data = zeros(input.NR,input.NS);
    w_Zv = cell(2,input.NS);
    for q = 1 : input.NS
        Zv_inc{1} = - (1/input.gamma_0) * dG_S{1,q};
        Zv_inc{2} = - (1/input.gamma_0) * dG_S{2,q};
      %[wZv]      = ITERCGwZv(Zv_inc,input);
        [wZv]     = BICGSTABwZv(Zv_inc,input);
        data(:,q) = DOPMwZv(dG_R,wZv,input);
    end % q_loop

    save DataIntEq data;

% (4) Plot data at a number of receivers ---------------------
    figure(5);
    subplot(1,2,1)
        PlotData(dataCircle,input);
        title('|p^{sct}(x_p^R,x_q^S)|=|data_{Bessel}|')
    subplot(1,2,2)
        PlotData(data-dataCircle,input);
        title('|data_{Int.Eq.} - data_{Bessel}|');
  % Compute normalized difference of norms
    error=num2str(norm(data(:)-dataCircle(:),1)/norm(dataCircle(:),1));
    disp(['error=' error]);
```

(1) For the analytical computation of the acoustic wave field scattered from a circular object at a number of receivers, the Matlab script `AcousticDataSctCircle` of Matlab Listing II.38 is called. The difference with the code described in Appendix 2.A.2 of Chapter 2 is that we now have an off-centered circle and that the data have to be computed for a number of sources as well. These data are saved as synthetic data for our inversion methods.

(2) To facilitate the numerical computation of the particle velocity of the incident fields, we compute the acoustic wave fields generated by a number of line sources. They are denoted as G_S. In the present chapter, we also need their spatial derivatives, which are denoted as dG_S. We compute their weak forms. Similarly, we compute the weak forms of the Green functions G_R associated with the acoustic wave fields measured at the receivers. We also need their spatial derivatives, which are denoted as dG_R. Both types of Green functions are computed with the help of the Matlab function `dGreenSourcesReceivers` of Matlab Listing II.39.

(3) We solve the integral equation for the contrast sources using either function `ITERCGwZv` of Matlab Listing II.40 or function BICGSTABwZv of Matlab Listing II.41. Since the latter

Matlab Listing II.38 The `AcousticDataSctCircle` script to compute synthetic 2D data.

```
clear all; clc; close all; clear workspace
input = initAC();

c_0   = input.c_0;         c_sct   = input.c_sct;
rho_0 = input.rho_0;       rho_sct = input.rho_sct;
gam0  = input.gamma_0;     gam_sct = input.gamma_0 * c_0/c_sct;
Z_0   = rho_0 * c_0;       Z_sct   = rho_sct * c_sct;

% (1) Compute coefficients of series expansion --------------------------
  arg0 = gam0 * input.a;   args = gam_sct*input.a;
  M = 50;                                  % increase M for more accuracy
  A = zeros(1,M+1);
  for m = 0 : M
    Ib0 = besseli(m,arg0);       dIb0 =  besseli(m+1,arg0) + m/arg0 * Ib0;
    Ibs = besseli(m,args);       dIbs =  besseli(m+1,args) + m/args * Ibs;
    Kb0 = besselk(m,arg0);       dKb0 = -besselk(m+1,arg0) + m/arg0 * Kb0;
    A(m+1) = - ((1/Z_sct) * dIbs*Ib0 - (1/Z_0) * dIb0*Ibs) ...
             /((1/Z_sct) * dIbs*Kb0 - (1/Z_0) * dKb0*Ibs);
  end

% (2) Compute reflected field at the receivers for all sources (data) -----
  xR(1,:) = input.xR(1,:) - input.xO(1);  % shifted origin
  xR(2,:) = input.xR(2,:) - input.xO(2);  % shifted origin
  xS(1,:) = input.xS(1,:) - input.xO(1);  % shifted origin
  xS(2,:) = input.xS(2,:) - input.xO(2);  % shifted origin
  rR = sqrt(xR(1,:).^2 + xR(2,:).^2);  phiR = atan2(xR(2,:),xR(1,:));
  rS = sqrt(xS(1,:).^2 + xS(2,:).^2);  phiS = atan2(xS(2,:),xS(1,:));

  data2D = zeros(input.NR,input.NS);
  for p = 1 : input.NR
   for q = 1 : input.NS
    data2D(p,q) = A(1) * besselk(0,gam0*rR(p)) * besselk(0,gam0*rS(q));
    for m = 1 : M
     data2D(p,q) = data2D(p,q) + 2 * A(m+1) * cos(m*(phiR(p)-phiS(q))) ...
                   * besselk(m,gam0*rR(p)) * besselk(m,gam0*rS(q)) ;
    end % m_loop
   end % q_loop
  end % p_loop

  dataCircle = 1/(2*pi) * data2D;                          clear data2D;
  save DataAnalytic dataCircle;
```

converges much faster, we prefer the latter one. The integral equation is solved for each source location, indicated by the value q. The forward operator and its adjoints are given by the functions `KOPwZv` of Matlab Listing II.42 and the function `adjKOPwZv` of Matlab Listing II.43. Note that the function BICGSTABwZv does not need the adjoints.

(4) Using the contrast sources, we compute the scattered field at the receiver locations, see function `DOPMwZv` of Matlab Listing II.44. For later convenience, these synthetic data are

Matlab Listing II.39 The function `dGreenSourcesReceivers` to compute the 2D Green functions for all source and receiver locations.

```
function [G_S,dG_S,G_R,dG_R] = dGreenSourcesReceivers (input)

gam0 = input.gamma_0;    X1 = input.X1;   X2 = input.X2;
dx   = input.dx;

delta  = (pi)^(-1/2) * dx;             % radius circle with area of dx^2
factor = 2 * besseli(1,gam0*delta) / (gam0*delta);
                                       % factor for weak form if DIS > delta
xS   = input.xS;
G_S = cell(1,input.NS); dG_S  = cell(2,input.NS);
for q = 1 : input.NS
      DIS   = sqrt((X1-xS(1,q)).^2 + (X2-xS(2,q)).^2);
    G_S{q}  = factor * 1/(2*pi) .* besselk(0,gam0*DIS);
       dG   = - gam0 * factor * 1/(2*pi).* besselk(1,gam0*DIS);
  dG_S{1,q} = (X1-xS(1,q))./DIS .* dG;
  dG_S{2,q} = (X2-xS(2,q))./DIS .* dG;
end %q_loop

xR   = input.xR;
G_R = cell(1,input.NR); dG_R = cell(2,input.NR);
for p = 1 : input.NR
       DIS  = sqrt((xR(1,p)-X1).^2 +(xR(2,p)-X2).^2);
    G_R{p}  = factor * 1/(2*pi) .* besselk(0,gam0*DIS);
       dG   = - gam0 * factor * 1/(2*pi).* besselk(1,gam0*DIS);
  dG_R{1,p} = - (xR(1,p)-X1)./DIS .* dG;
  dG_R{2,p} = - (xR(2,p)-X2)./DIS .* dG;
end % p_loop
```

saved as well. We plot the data using function `PlotData` of Matlab Listing II.13. To measure the quality of these data, we benchmark it against the analytical ones of step (1). For our acoustic example, in Figure 5.3, we present the amplitudes of the analytical data using the Bessel functions series (a,b) and the difference with regard to the numerical data obtained with the integral-equation method (c,d). The error is around 6.4%.

With these results, we now discuss the nonlinear inverse scattering problem.

5.3 Mass-density Contrast Source Inversion

The contrast-source inversion method recasts the acoustic inversion problem as a minimization of a cost functional, from which the contrast sources $\hat{w}_k^{Zv}$ and the mass-density contrast $\hat{\chi}^\rho$ are reconstructed. In our further analysis we omit the hat symbols. We first introduce the short-hand notations for the unknown wave field functions $Z_0 v_{j,q} = Z_0 v_j(\boldsymbol{x}_m, \boldsymbol{x}_q^S)$ and the known wave field functions,

$$\begin{aligned} p_{M,q}^{sct} &= p_M^{sct}(\boldsymbol{x}_p^R, \boldsymbol{x}_q^S)\,, \\ p_q^{inc} &= p^{inc}(\boldsymbol{x}_m, \boldsymbol{x}_q^S)\,, \\ Z_0 v_{j,q}^{inc} &= Z_0 v_j^{inc}(\boldsymbol{x}_m, \boldsymbol{x}_q^S)\,, \end{aligned} \tag{5.10}$$

Matlab Listing II.40 The function `ITERCGwZv` to compute the contrast sources.

```
function [w_Zv] = ITERCGwZv(Zv_inc,input)
global nDIM;   CHI_rho = input.CHI_rho;

w_Zv=cell(1,nDIM); g_Zv=cell(1,nDIM); r_Zv=cell(1,nDIM); v_Zv=cell(1,nDIM);
itmax  = 200; Errcri = input.Errcri;  it = 0; % initialization iteration
Norm_D = 0;
for n = 1:nDIM
   w_Zv{n} = zeros(size(Zv_inc{n}));
   r_Zv{n} = CHI_rho.*Zv_inc{n};     Norm_D = Norm_D + norm(r_Zv{n}(:))^2;
end
eta_D = 1 / Norm_D;                             % normalization factor
Error = 1;    fprintf('Error =         %g',Error);

while (it < itmax) && ( Error > Errcri)
% determine conjugate gradient direction v
  dummy=cell(1,nDIM); for n=1:nDIM; dummy{n}=conj(CHI_rho).*r_Zv{n}; end
  [KZv] = AdjKOPwZv(dummy,input);
  AN = 0;
  for n = 1:nDIM
   g_Zv{n} = r_Zv{n} - KZv{n};                % and window for small CHI_rho
   g_Zv{n} = (abs(CHI_rho) >= Errcri).*g_Zv{n};
   AN = AN + norm(g_Zv{n}(:))^2;
  end
  if it == 0
      for n=1:nDIM; v_Zv{n} = g_Zv{n};  end
  else
      for n=1:nDIM; v_Zv{n} = g_Zv{n} + AN/AN_1 * v_Zv{n}; end
  end
% determine step length alpha
  [KZv] = KOPwZv(v_Zv,input);
   BN = 0;
   for n = 1:nDIM
     KZv{n} = v_Zv{n} - CHI_rho .* KZv{n};    BN = BN + norm(KZv{n}(:))^2;
   end
   alpha = AN / BN;
 % update contrast sources and AN
   for n = 1:nDIM;  w_Zv{n} = w_Zv{n} + alpha * v_Zv{n}; end
   AN_1  = AN;
 % update residual errors
   Norm_D = 0;
   for n = 1:nDIM
     r_Zv{n} = r_Zv{n} - alpha * KZv{n};
     Norm_D  = Norm_D + norm(r_Zv{n}(:))^2;
   end
   Error = sqrt(eta_D * Norm_D); fprintf('\b\b\b\b\b\b\b\b%6f',Error);
   it = it+1;
 end % CG iterations

 fprintf('\b\b\b\b\b\b\b\b%6f\n',Error);
 disp(['Number of iterations is ' num2str(it)]);
 if it == itmax; disp(['itmax reached:  err/norm = ' num2str(Error)]); end
```

Matlab Listing II.41 The function `BICGSTABwZv` to compute the contrast sources.

```
function [w_Zv] = BICGSTABwZv(Zv_inc, input)
global nDIM;
w_Zv=cell(1,nDIM);   r_Zv=cell(1,nDIM);   v_Zv=cell(1,nDIM);
Data_Zv{1} = input.CHI_rho .* Zv_inc{1};
Data_Zv{2} = input.CHI_rho .* Zv_inc{2};
itmax   = 200;
it      = 0;                          % initialization of iteration
for n = 1:nDIM;   w_Zv{n} = zeros(size(Zv_inc{n}));  end
for n = 1:nDIM;   r_Zv{n} = Data_Zv{n};              end
Norm_D = norm(r_Zv{1}(:))^2 + norm(r_Zv{2}(:))^2;
eta_D   = 1 / Norm_D;                  % normalization factor
Error   = 1;                           % error norm initial error
fprintf('Error =          %g', Error);
while (it < itmax) && (Error > input.Errcri)
 % determine gradient directions
   AN = sum(r_Zv{1}(:).*conj(Data_Zv{1}(:))) + ...
        sum(r_Zv{2}(:).*conj(Data_Zv{2}(:)));
    if it == 0
       for n = 1:nDIM;  v_Zv{n} = r_Zv{n};  end
    else
       for n = 1:nDIM;  v_Zv{n} = r_Zv{n} + (AN/AN_1) * v_Zv{n}; end
    end
    AN_1 = AN;
 % determine step length alpha and update residual error r_D
   [Kv_Zv] = KOPwZv(v_Zv,input);
   for n = 1:nDIM; Kv_Zv{n} = v_Zv{n} - input.CHI_rho .* Kv_Zv{n}; end
         BN  = sum(Kv_Zv{1}(:).*conj(Data_Zv{1}(:))) + ...
               sum(Kv_Zv{2}(:).*conj(Data_Zv{2}(:)));
      alpha = AN / BN;
   for n = 1:nDIM;  r_Zv{n} = r_Zv{n} - alpha * Kv_Zv{n};  end
   % + successive overrelaxation (first step of GMRES)
   [Kr_Zv] = KOPwZv(r_Zv,input);
   for n=1:nDIM; Kr_Zv{n} = r_Zv{n} - input.CHI_rho .* Kr_Zv{n};  end
   beta = ( sum(r_Zv{1}(:).*conj(Kr_Zv{1}(:)))  + ...
            sum(r_Zv{2}(:).*conj(Kr_Zv{2}(:)))  ) ...
           /(norm(Kr_Zv{1}(:))^2 + norm(Kr_Zv{2}(:))^2);
   % update contrast source w ,v and the residual error r_Zv
   for n=1:nDIM
      w_Zv{n} = w_Zv{n} + *alpha * v_Zv{n} + beta * r_Zv{n};
      v_Zv{n} = (alpha / beta) * (v_Zv{n} - beta * Kv_Zv{n});
      r_Zv{n} = r_Zv{n} - beta * Kr_Zv{n};
   end
   Norm_D = norm(r_Zv{1}(:))^2 + norm(r_Zv{2}(:))^2 ;
   Error  = sqrt(eta_D * Norm_D);  fprintf('\b\b\b\b\b\b\b\b%6f',Error);
   it = it+1;
end % while
fprintf('\b\b\b\b\b\b\b\b%6f\n',Error);
disp(['Number of iterations is ' num2str(it)]);
if it == itmax
   disp(['itmax was reached:   err/norm = ' num2str (Error)]);
end
```

Matlab Listing II.42 The function `KOPwZv` to compute the operator $\mathcal{K}_{j,q}^{Zv(n)}$.

```
function [KZv] = KOPwZv(wZv, input)
global nDIM;
KZv  = cell(1,nDIM);
gam0 = input.gamma_0;
FFTG = input.FFTG;

% Acoustic operator with grad div differential operator
  KZv{1} = Kop(wZv{1},FFTG)/gam0^2;
  KZv{2} = Kop(wZv{2},FFTG)/gam0^2;

  KZv = graddiv(KZv,input);

  KZv{1} =  KZv{1} + wZv{1};
  KZv{2} =  KZv{2} + wZv{2};
```

Matlab Listing II.43 The function `AdjKOPwZv` to compute the adjoint of $\mathcal{K}_{j,q}^{Zv(n)}$.

```
function [KZv] = AdjKOPwZv(fZv, input)
global nDIM;
KZv = cell(1,nDIM);
gam0 = input.gamma_0;
FFTG = input.FFTG;

% Acoustic operator with grad div differential operator
  KZv{1} = AdjKop(fZv{1},FFTG)/conj(gam0^2);
  KZv{2} = AdjKop(fZv{2},FFTG)/conj(gam0^2);

  KZv = graddiv(KZv,input);

  KZv{1} = KZv{1} + fZv{1};
  KZv{2} = KZv{2} + fZv{2};
```

Matlab Listing II.44 The function `DOPMwZv` to compute synthetic 2D data, for each source location q.

```
function [GMw] = DOPMwZv(dG_R,w_Zv,input)

gam0 = input.gamma_0;
dx   = input.dx;

GMw = zeros(input.NR,1);
for p = 1 : input.NR
    GMw(p,1) =  gam0 * dx^2 * sum(dG_R{1,p}(:) .* w_Zv{1}(:)) ...
              + gam0 * dx^2 * sum(dG_R{2,p}(:) .* w_Zv{2}(:));
end % p_loop
```

Matlab Listing II.45 The function `AdjDOPMwZv` to compute the adjoint of the data operator.

```
function [adjGMZv] = AdjDOPMwZv(dG_R,f,input)

gam0 = input.gamma_0;
dx   = input.dx;

adjGMZv = cell(1,2);
adjGMZv{1} = zeros(input.N1,input.N2);
adjGMZv{2} = zeros(input.N1,input.N2);
for p = 1:input.NR
    adjGMZv{1} = adjGMZv{1} + conj(gam0 * dx^2 * dG_R{1,p}) * f(p);
    adjGMZv{2} = adjGMZv{2} + conj(gam0 * dx^2 * dG_R{2,p}) * f(p);
end % p_loop
```

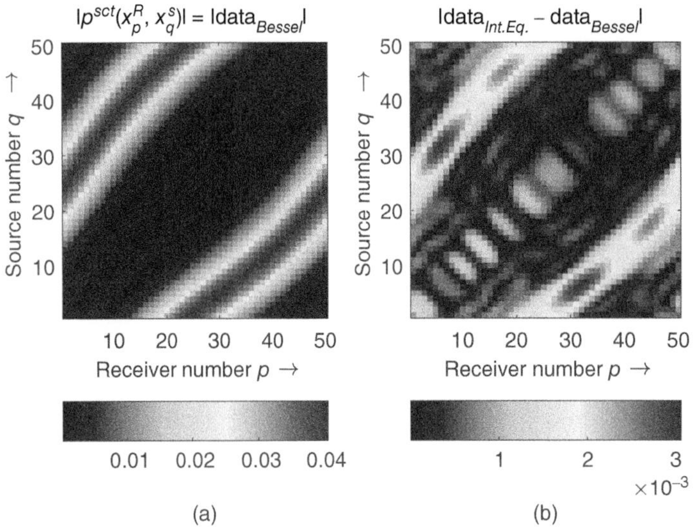

Figure 5.3 Plots of the data amplitudes as function of the source and receiver numbers, for $\rho_0 = 1500$ kg/m^3 and $\hat{\rho}_{sct} = 2500$ kg/m^3. The normalized global error in the integral-equation data amounts to 6.4%.

for $p = 1, 2, \ldots, N^R$ and $q = 1, 2, \ldots, N^S$, where N^R is the total number of receivers and N^S is the total number of sources.

Then, in the present acoustic MRCSI method, the sequences of the contrast sources $w^{Zv(n)}_{j,q} = w^{Zv(n)}_j(\boldsymbol{x}_m, \boldsymbol{x}^S_q)$, and the contrast $\chi^{\rho(n)} = \chi^{\rho(n)}(\boldsymbol{x}_m)$, for $n = 1, 2, \ldots$, are iteratively found to minimize the cost functional

$$F^{(n)} = \eta_{\mathrm{M}} F_{\mathrm{M}}(w^{Zv(n)}_{k,q}) + \eta^{(n)}_{\mathrm{D}} F^{Zv}_{\mathrm{D}}(w^{Zv(n)}_{k,q}, \chi^{\rho(n)}) , \tag{5.11}$$

where

$$F_{\mathrm{M}} = \sum_{q=1}^{N^S} \| r^{(n)}_{\mathrm{M},q} \|^2_{\mathrm{M}} , \quad \text{with} \quad \eta_{\mathrm{M}} = \left[\sum_{q=1}^{N^S} \| p^{sct}_{\mathrm{M},q} \|^2_{\mathrm{M}} \right]^{-1} , \tag{5.12}$$

$$F_{\mathrm{D}}^{Zv} = \sum_{q=1}^{N^S} \sum_k \| r_{\mathrm{D},k,q}^{Zv(n)} \|_{\mathrm{D}}^2 , \quad \text{with} \quad \eta_{\mathrm{D}}^{(n)} = \left[\sum_{q=1}^{N^S} \sum_k \| \chi^{\rho(n)} Z_0 v_{k,q}^{inc} \|_{\mathrm{D}}^2 \right]^{-1} . \tag{5.13}$$

In the present analysis, $\| \cdot \|^2 = \langle \cdot, \cdot \rangle_{\mathrm{M,D}}$ denotes the squared norm for $\boldsymbol{x}_p^R \in \mathbb{M}$ and $\boldsymbol{x}_m \in \mathbb{D}$, respectively. The residuals are given by

$$\begin{aligned} r_{\mathrm{M},q}^{(n)} &= p_{\mathrm{M},q}^{sct} + \gamma_0^{-1} \partial_k^R G_{\mathrm{M}} w_{k,q}^{Zv(n)} , \\ r_{\mathrm{D}j,q}^{Zv(n)} &= \chi^{\rho(n)} Z_0 v_{j,q}^{inc} - w_{j,q}^{Zv(n)} + \chi^{\rho(n)} \, [\gamma_0^{-2} \partial_j \partial_k G_{\mathrm{D}} w_{k,q}^{Zv(n)} + w_{j,q}^{Zv(n)}]. \end{aligned} \tag{5.14}$$

The operators G_{M} and G_{D} are defined in Eqs. (4.21) and (4.22). The first term of Eq. (5.11) measures the residual error $r_{\mathrm{M},q}^{(n)}$ in the measurement domain, while the second term measures the residual error $r_{\mathrm{D}j,q}^{Zv(n)}$ in the object domain. The normalization is chosen so that the two terms are equal to one if $w_{j,q}^{Zv(n)} = 0$. The residual in the first term is a linear function in $w_{j,q}^{Zv(n)}$, but the residual in the second term is nonlinear in $\chi^{\rho(n)}$ and $w_{j,q}^{Zv(n)}$. To deal with this nonlinearity, the CSI method employs an alternate updating scheme for these quantities.

5.3.1 Updating the Contrast Sources

In each iteration, we first update the contrast sources, $w_{j,q}^{Zv(n)} = w_j^{Zv(n)}(\boldsymbol{x}_m, \boldsymbol{x}_q^S)$, using the conjugate gradient step,

$$\boxed{w_{j,q}^{Zv(n)} = w_{j,q}^{Zv(n-1)} + \alpha_q^{(n)} v_{j,q}^{Zv(n)} ,} \tag{5.15}$$

where for each q and n, the parameters $\alpha_q^{(n)}$ are the step lengths and the functions $v_{j,q}^{Zv(n)}$ are the Polak-Ribière conjugate gradient directions,

$$\boxed{v_{j,q}^{Zv(1)} = g_{j,q}^{Zv(1)}, \quad v_{j,q}^{Zv(n)} = g_{j,q}^{Zv(n)} + \gamma_q^{(n)} \, v_{j,q}^{Zv(n-1)} \, , n = 2, 3, \dots ,} \tag{5.16}$$

where $g_{j,q}^{Zv(n)}$ are the gradients of the cost functional $F^{(n)}$ with respect to changes of $w_{j,q}^{Zv(n-1)}$. The parameter $\gamma_q^{(n)}$ in the Polak–Ribière direction of Eq. (5.16) is defined as

$$\boxed{\gamma_q^{(n)} = \frac{\mathrm{Re} \left[\sum_j \langle g_{j,q}^{Zv(n)}, g_{j,q}^{Zv(n)} - g_{j,q}^{Zv(n-1)} \rangle_{\mathrm{D}} \right]}{\sum_j \| g_{j,q}^{Zv(n-1)} \|_{\mathrm{D}}^2} .} \tag{5.17}$$

5.3.1.1 Gradient Directions

After discretization of the Green's operators and the norms, similar to the scalar gradient of Eq. (4.37), the gradient of the qth term of the cost functional $F^{(n-1)} = \sum_q F_q^{(n-1)}$ with respect to the contrast source, $\overline{w}_{j,m',q}^{Zv(n-1)} = \overline{w}_{j,q}^{Zv(n-1)}(\boldsymbol{x}_{m'}, \boldsymbol{x}_q^S)$, is defined as

$$g_j^{Zv(n)}(\boldsymbol{x}_{m'}, \boldsymbol{x}_q^S) = - \frac{\partial [F_q^{(n-1)}]}{\partial [\overline{w}_{j,m',q}^{Zv(n-1)}]} , \tag{5.18}$$

where the qth term of $F^{(n-1)}$ follows from Eqs. (5.11)–(5.13), and Eq. (5.14). Hence,

$$g_j^{Zv(n)}(\boldsymbol{x}_{m'},\boldsymbol{x}_q^S) = -\eta_{\mathrm{M}}\sum_{p=1}^{N^R}\frac{\partial[\overline{r}_{\mathrm{M},q}^{(n-1)}]}{\partial[\overline{w}_{j,m',q}^{Zv(n-1)}]}\, r_{\mathrm{M},q}^{(n-1)} - \eta_{\mathrm{D}}^{(n-1)}\sum_{m=1}^{N}\sum_{k}\frac{\partial[\overline{r}_{\mathrm{D},k,q}^{Zv(n-1)}]}{\partial[\overline{w}_{j,m',q}^{Zv(n-1)}]}\, r_{\mathrm{D},k,q}^{Zv(n-1)}. \tag{5.19}$$

Taking the partial differentiation with respect to $\overline{w}_{j,m',q}^{Zv(n-1)}$, and using Eqs. (5.14), (4.21), and (4.22), we arrive at

$$\begin{aligned}-\frac{\partial[\overline{r}_{\mathrm{M},q}^{(n-1)}]}{\partial[\overline{w}_{j,m',q}^{Zv(n-1)}]} &= -\overline{\gamma}_0^{-1}\Delta V\,\langle\partial_j^R\overline{G}\rangle(\boldsymbol{x}_p^R-\boldsymbol{x}_{m'})\,,\\ -\frac{\partial[\overline{r}_{\mathrm{D},k,q}^{Zv(n-1)}]}{\partial[\overline{w}_{j,m',q}^{Zv(n-1)}]} &= \left[\delta_{j,k}\delta_{m,m'} - \overline{\chi}^{\rho(n-1)}[\overline{\gamma}_0^{-2}\Delta V\langle\langle\partial_j\partial_k\overline{G}\rangle\rangle(\boldsymbol{x}_m-\boldsymbol{x}_{m'}) + \delta_{j,k}\delta_{m,m'}]\right].\end{aligned} \tag{5.20}$$

Substituting these expressions into Eq. (5.19), we obtain

$$\begin{aligned}g_j^{Zv(n)}(\boldsymbol{x}_{m'},\boldsymbol{x}_q^S) &= -\eta_{\mathrm{M}}\sum_{p=1}^{N^R}\Delta V\,\overline{\gamma}_0^{-1}\langle\partial_j^R\overline{G}\rangle(\boldsymbol{x}_p^R-\boldsymbol{x}_{m'})\,r_{\mathrm{M},q}^{(n-1)} + \eta_{\mathrm{D}}^{(n-1)}\\ &\times\left[r_{\mathrm{D},j,q}^{Zv(n-1)} - \sum_{m=1}^{N}\overline{\gamma}_0^{-2}\Delta V\langle\langle\partial_j\partial_k\overline{G}\rangle\rangle(\boldsymbol{x}_m-\boldsymbol{x}_{m'})\overline{\chi}^{\rho(n-1)}r_{\mathrm{D},k,q}^{Zv(n-1)} + \overline{\chi}^{\rho(n-1)}r_{\mathrm{D},j,q}^{Zv(n-1)}\right].\end{aligned} \tag{5.21}$$

Interchanging m' and m, we see that $g_{j,q}^{Zv(n)} = g_j^{Zv(n)}(\boldsymbol{x}_m,\boldsymbol{x}_q^S)$ may be written as

$$\boxed{\partial g_{j,q}^{Zv(n)} = \eta_{\mathrm{M}}[-\overline{\gamma}_0^{-1}\partial_j G_{\mathrm{M}}^{\star}\{r_{\mathrm{M},q}^{(n-1)}\}] + \eta_{\mathrm{D}}^{(n-1)}[r_{\mathrm{D},j,q}^{Zv(n-1)} - \overline{\gamma}_0^{-2}\partial_j\partial_k G_{\mathrm{D}}^{\star}\{\overline{\chi}^{\rho}r_{\mathrm{D},k,q}^{Zv(n-1)}\} - \overline{\chi}^{\rho}r_{\mathrm{D},j,q}^{Zv(n-1)}].} \tag{5.22}$$

The adjoints $G_{\mathrm{M}}^{\star}$ and $G_{\mathrm{D}}^{\star}$ are given by Eqs. (4.31) and (4.35).

5.3.1.2 Calculation of the Step Length

The step length $\alpha_q^{(n)}$ is found as minimizer of the cost functional $F^{(n)} = \sum_q F_q^{(n)}$, i.e.

$$\alpha_q^{(n)} = \underset{\text{real}\,\alpha}{\arg\min}\{F_q^{(n)}(w_{j,q}^{Zv(n-1)} + \alpha\, v_{j,q}^{Zv(n)})\}, \tag{5.23}$$

where $F_q^{(n)}$ as function of α is given by

$$F_q^{(n)} = \eta_{\mathrm{M}}\,\|\, r_{\mathrm{M},q}^{(n-1)} - \alpha\,\mathcal{K}_{\mathrm{M},q}^{Zv}\,\|_{\mathrm{M}}^2 + \eta_{\mathrm{D}}^{(n-1)}\sum_j\|\, r_{\mathrm{D},j,q}^{Zv(n-1)} - \alpha(v_{j,q}^{Zv(n)} - \chi^{\rho(n-1)}\mathcal{K}_{j,q}^{Zv(n)})\,\|_{\mathrm{D}}^2\,. \tag{5.24}$$

in which

$$\mathcal{K}_{\mathrm{M},q}^{Zv(n)} = \gamma_0^{-1}\partial_k^R G_{\mathrm{M}} v_{k,q}^{Zv(n)}, \tag{5.25}$$

and

$$\mathcal{K}_{j,q}^{Zv(n)} = \gamma_0^{-2}\partial_j\partial_k G_{\mathrm{D}} v_{k,q}^{Zv(n)} + v_{j,q}^{Zv(n)}\,. \tag{5.26}$$

The real part of the minimizer is found explicitly as

$$\alpha_q^{(n)} = \frac{\mathrm{Re}\left[\eta_{\mathrm{M}}\,\langle r_{\mathrm{M},q}^{(n-1)},\mathcal{K}_{\mathrm{M},q}^{Zv(n)}\rangle_{\mathrm{M}} + \eta_{\mathrm{D}}^{(n-1)}\sum_j\langle r_{\mathrm{D},j,q}^{Zv(n-1)}, v_{j,q}^{Zv(n)} - \chi^{\rho(n-1)}\mathcal{K}_{j,q}^{Zv(n)}\rangle_{\mathrm{D}}\right]}{\eta_{\mathrm{M}}\,\|\,\mathcal{K}_{\mathrm{M},q}^{Zv(n)}\,\|_{\mathrm{M}}^2 + \eta_{\mathrm{D}}^{(n-1)}\sum_j\|\, v_{j,q}^{Zv(n)} - \chi^{\rho(n-1)}\mathcal{K}_{j,q}^{Zv(n)}\,\|_{\mathrm{D}}^2}\,. \tag{5.27}$$

Using the expression for the gradient of Eq. (5.22), it simplifies to

$$\alpha_q^{(n)} = \frac{\mathrm{Re}\left[\sum_j \langle g_{j,q}^{Zv(n)}, v_{j,q}^{Zv(n)} \rangle_{\mathrm{D}}\right]}{\eta_{\mathrm{M}} \parallel \mathcal{K}_{\mathrm{M},q}^{Zv(n)} \parallel_{\mathrm{M}}^2 + \eta_{\mathrm{D}}^{(n-1)} \sum_j \parallel v_{j,q}^{Zv(n)} - \chi^{\rho(n-1)} \mathcal{K}_{j,q}^{Zv(n)} \parallel_{\mathrm{D}}^2} . \tag{5.28}$$

After each update of the contrast sources, the residual error in the measurement domain becomes

$$\boxed{r_{\mathrm{M},q}^{(n)} = r_{\mathrm{M},q}^{(n-1)} - \alpha_q^{(n)} \mathcal{K}_{\mathrm{M},q}^{Zv(n)} .} \tag{5.29}$$

5.3.2 Updating the Contrast

Using the update for the contrast sources, the fields $Z_0 v_{j,q}^{(n)} = Z_0 v_j^{(n)}(\boldsymbol{x}_m, \boldsymbol{x}_q^S)$ are directly obtained as

$$Z_0 v_{j,q}^{(n)} = Z_0 v_{j,q}^{Zv(n-1)} + \alpha_q^{(n)} \mathcal{K}_{j,q}^{Zv(n)} , \quad n = 1, 2, \dots , \tag{5.30}$$

where $\mathcal{K}_{j,q}^{Zv(n)}$ is defined by Eq. (5.26). Recall that these quantities depend on the position vector $\boldsymbol{x}_m$ and the source location $\boldsymbol{x}_q^S$. An update for the mass-density contrast is obtained from Eq. (5.3) as

$$\boxed{\chi_{csi}^{\rho(n)}(\boldsymbol{x}_m) = \frac{\sum_{q=1}^{N^S} \sum_k w_k^{Zv(n)}(\boldsymbol{x}_m, \boldsymbol{x}_q^S) \, \overline{Z_0 v_{k,q}^{(n)}}(\boldsymbol{x}_m, \boldsymbol{x}_q^S)}{\sum_{q=1}^{N^S} \sum_k |Z_0 v_{k,q}^{(n)}(\boldsymbol{x}_m, \boldsymbol{x}_q^S)|^2} .} \tag{5.31}$$

After computation of the contrast update $\chi_{csi}^{\rho(n)}$, the residual errors in the object domain become

$$\boxed{r_{\mathrm{D}j,q}^{Zv(n)} = \chi_{csi}^{\rho(n)} Z_0 v_{j,q}^{(n)} - w_{j,q}^{Zv(n)} .} \tag{5.32}$$

With this result the description of the algorithm is completed, except for the starting values for the contrast sources $w_{j,q}^{Zv(0)}$.

5.3.3 Initial Estimate

We cannot start with zero contrast sources $w_{j,q}^{Zv} = 0$, since then $\chi^{\rho(0)} = 0$ and the normalization factor $\eta_{\mathrm{D}}^{(0)}$ is undefined. Therefore, we choose as starting values the contrast sources that only minimize the cost functional

$$F_{\mathrm{M}}(w_{j,q}^{Zv(0)}) = \sum_{q=1}^{N^S} \parallel p_{\mathrm{M},q}^{sct} - \mathcal{K}_{\mathrm{M},q}^{Zv(0)} \parallel_{\mathrm{M}}^2 , \tag{5.33}$$

with

$$\mathcal{K}_{\mathrm{M},q}^{Zv(0)} = \gamma_0^{-1} \partial_k^R G_{\mathrm{M}} w_{k,q}^{Zv(0)} . \tag{5.34}$$

Using the gradient method, we take

$$\boxed{w_{j,q}^{p(0)} = \alpha_q^{(0)} g_{j,q}^{Zv(0)} , \qquad g_{j,q}^{Zv(0)} = \mathcal{K}_{\mathrm{M},j}^{\star} p_{\mathrm{M},q}^{sct} ,} \tag{5.35}$$

in which

$$\mathcal{K}_{\mathrm{M},j}^{\star} p_q^{sct} = -\overline{\gamma}_0^{-1} \partial_j^R \overline{G}_{\mathrm{M}} p_{\mathrm{M},q}^{sct} . \tag{5.36}$$

The cost functional is minimized when each term for q is minimized, i.e.

$$\alpha_q^{(0)} = \arg\min_{\alpha}\{\| p^{sct}_{\mathbb{M},q} - \alpha\, \mathcal{K}^{Zv(0)}_{\mathbb{M},q} \|^2_{\mathbb{M}}\}\,. \tag{5.37}$$

This leads directly to

$$\alpha_q^{(0)} = \frac{\langle p^{sct}_{\mathbb{M},q}, \mathcal{K}^{Zv(0)}_{\mathbb{M},q}\rangle_{\mathbb{M}}}{\| \mathcal{K}^{Zv(0)}_{\mathbb{M},q} \|^2_{\mathbb{M}}}\,, \tag{5.38}$$

or in simplified form as

$$\boxed{\alpha_q^{(0)} = \frac{\sum_j \| g^{Zv(0)}_{j,q} \|^2_{\mathbb{D}}}{\| \mathcal{K}^{Zv(0)}_{\mathbb{M},q} \|^2_{\mathbb{M}}}\,.} \tag{5.39}$$

The gradient $g^{Zv(0)}_{j,q} = \mathcal{K}^{\star}_{\mathbb{M},j} p^{sct}_{\mathbb{M},q}$ is the back projection of the data domain $\mathbb{M}$ into the object domain $\mathbb{D}$, and it is a *back propagation* of the pressure wave-field data $p^{sct}_{\mathbb{M},q}$ in the data domain and converted to particle-velocity field data with two components in the object domain.

With this initial estimate $w^{Zv(0)}_{j,q} = w^{Zv(0)}_j(\boldsymbol{x}_m, \boldsymbol{x}^S_q)$ and $Z_0 v^{(0)}_{k,q} \approx Z_0 v^{inc}_{k,q}$, the contrast estimate is obtained from

$$\boxed{\chi^{\rho(0)} = \frac{\sum_{q=1}^{N^S}\sum_k w^{Zv(0)}_{k,q}\, Z_0 \overline{v}^{inc}_{k,q}}{\sum_{q=1}^{N^S}\sum_k |Z_0 v^{inc}_{k,q}|^2}\,.} \tag{5.40}$$

After having discussed the details of the CSI method for the present acoustic case, in the next subsection, we discuss the regularization procedure.

5.3.4 Updating the Contrast with Multiplicative TV Regularization

To discuss the regularization of the mass-density contrast, we start with the observation that the mass-density contrast, as a minimizer of $F_{\mathbb{D}}$ with respect to changes of χ^{ρ}, is given by

$$\chi^{\rho(n)}_{csi} = \arg\min_{\chi}\left\{\sum_{q=1}^{N^S}\sum_k \| \chi\, Z_0 v^{(n)}_{k,q} - w^{\rho(n)}_{k,q} \|^2_{\mathbb{D}}\right\}, \tag{5.41}$$

or

$$\chi^{\rho(n)}_{csi}(\boldsymbol{x}) = \frac{\sum_{q=1}^{N^S}\sum_k w^{(n)}_{k,q}(\boldsymbol{x})\, \overline{Z_0 v^{(n)}_{k,q}}(\boldsymbol{x})}{\sum_{q=1}^{N^S}\sum_k |Z_0 v^{(n)}_{k,q}(\boldsymbol{x})|^2}\,. \tag{5.42}$$

Similar to the regularization procedure of the scalar case of Chapter 4, our further analysis simplifies when we consider the minimization of Eq. (5.41) with respect to changes of χ as

$$\chi^{\rho(n)} = \arg\min_{\chi}\left\{\int_{\boldsymbol{x}\in\mathbb{D}} |\chi(\boldsymbol{x}) - \chi^{\rho(n)}_{csi}(\boldsymbol{x})|^2 \mathrm{d}V\right\}. \tag{5.43}$$

Hence, the regularization procedure for the mass-density contrast is identical to the one for the wave-speed contrast of Eq. (4.68), we only have to replace the values of $\chi^{c(n)}_{csi}$ by $\chi^{\rho(n)}_{csi}$.

Matlab Listing II.46 The function `InitialEstimatewZv` to compute an initial estimate of the contrast sources $w^{Zv}_{j,q}$ and the acoustic particle velocity $Z_0 v_{k,q}$.

```
function [wZv,Zv,r_M,ErrorM] ...
                        = InitialEstimatewZv(eta_M,Zvinc,dG_R,data,input)
global nDIM
NS = input.NS;
wZv = cell(nDIM,NS); Zv = cell(nDIM,NS);  r_M = cell(1,NS);

Norm_M = 0;
dummyq = cell(1,nDIM);
for q = 1 : NS
    r_M{q}      = data(:,q);
    [gZv]       = AdjDOPMwZv(dG_R,r_M{q},input);
    A0          = norm(gZv{1}(:))^2 + norm(gZv{2}(:))^2 ;
    GMp         = DOPMwZv(dG_R,gZv,input);
    B0          = norm(GMp(:))^2;
    alpha       = real(A0/B0);
    r_M{q}      = r_M{q} - alpha * GMp(:);
    Norm_M      = Norm_M + norm(r_M{q})^2;
    wZv{1,q}    = alpha * gZv{1};
    wZv{2,q}    = alpha * gZv{2};
    dummyq{1}   = wZv{1,q};
    dummyq{2}   = wZv{2,q};
    [KZv]       = KOPwZv(dummyq,input);
    Zv{1,q}     = Zvinc{1,q} + KZv{1};
    Zv{2,q}     = Zvinc{2,q} + KZv{2};
end %qloop

ErrorM = sqrt(eta_M * Norm_M);
```

5.3.5 Matlab Codes for the Acoustic MRCSI Method

In this subsection, we present the Matlab codes for the 2D acoustic version of the present contrast inversion method. The associated Matlab script `AcMRCSI` is given in Matlab Listing II.47. The following steps are carried out:

(1) We start with the computation of analytical data for the acoustic case. We use `AcousticDataSctCircle` of Matlab Listing II.38. If these data have been computed with the forward code `ForwardwZv` of Matlab Listing II.37, we load these analytical data from the computer memory. When we would like to investigate what the result is of an "Inverse Crime," where both the forward and inverse codes are based on the same computational model, we then have the option to load the data from the integral-equation model. If in the Matlab function `initAC`, the value of the `input.Noise` is equal to 1, then there is complex-valued noise added with a standard deviation of $0.2 \times \text{mean}(|\text{dataCircle}|)$. Adding this so-called computer noise is certainly needed to prevent an inverse crime. However, we are now in the situation that our analytical data and numerical data of the integral-equation method differ about 5–10%, and that we do not need to add extra computer noise.

Matlab Listing II.47 The function AcMRCSI to reconstruct the mass-density contrast χ^ρ.

```
clear all; clc; close all; clear workspace
input = initAC();

% (1) Compute Exact data using analytic Bessel function expansion --------
    disp('Running AcousticDataSctCircle');    AcousticDataSctCircle;

    load DataAnalytic;

    % load DataIntEq;    dataCircle = data;          % Inverse Crime

    % add complex-valued noise to data
    if input.Noise == 1
       dataCircle = dataCircle + mean(abs(dataCircle(:))) * input.Rand;
    end

%  (2) Compute Green functions for sources and receivers ------------------
     [~,dG_S,~,dG_R] = dGreenSourcesReceivers(input);

%  (3) Apply MRCSI method ------------------------------------------------
     Zvinc = cell(2,input.NS);
     for q = 1 : input.NS
        Zvinc{1,q} = - (1/input.gamma_0) * dG_S{1,q};
        Zvinc{2,q} = - (1/input.gamma_0) * dG_S{2,q};
     end % q_loop
     clear G_S dG_S

     [wZv] = ITERCGMDwZv(Zvinc,dG_R,dataCircle,input);

%  (4) Compute model data explained by reconstructed contrast souces ------
     wZvq = cell(2,input.NS);
      data = zeros(input.NR,input.NS);
     for q = 1 : input.NS
         wZvq{1} = wZv{1,q};
         wZvq{2} = wZv{2,q};
       data(:,q) = DOPMwZv(dG_R,wZvq,input);
     end % q_loop

%  Plot data and compare to synthetic data
   figure(9);
   subplot(1,2,1);
      PlotData(dataCircle,input);
      title('|p^{sct}(x_p^R,x_q^S)|=|data_{Bessel}|')
   subplot(1,2,2);
      PlotData(data-dataCircle,input);
      title('|data_{Int.Eq.} - data_{Exact}|')

%  Compute normalized difference of norms
   error = num2str (norm(data (:)-dataCircle(:),1)/norm (dataCircle(:),1));
   disp(['error=' error]);
```

Matlab Listing II.48 The function `ITERCGMDwZv` for updating $w_{j,q}^{Zv}$ and χ^{ρ}.

```
function [wZv] = ITERCGMDwZv(Zvinc,dG_R,data,input)
global nDIM it
   NS = input.NS;
  AN1 = cell(1,NS);
 wZvq = cell(nDIM);        vZvq = cell(nDIM);        Zvq = cell(nDIM);
g1Zvq = cell(nDIM);       rDZvq = cell(nDIM);
g1Zv  = cell(nDIM,NS);     vZv  = cell(nDIM,NS);

itmax = 256;
eta_M = 1 / norm(data(:))^2;      % normalization factor eta_M

% (1) Initial estimate of contrast sources and mass-density contrast
   [wZv,Zv,r_M,ErrorM] = InitialEstimatewZv(eta_M,Zvinc,dG_R,data,input);
   it = 0;  ErrorD  = 1;
   disp(['Iteration = ', num2str(it)]);
   [CHI_rho,rZv,eta_D,ErrorD] ...
                     = UpdateContrastwZv(wZv,Zv,Zvinc,ErrorM,ErrorD,input);
   saveCHI(CHI_rho);
   disp('------------------------------------------------------------');

% (2) Iterative updating of contrast sources and mass-density contrast
   it = 1;
   while (it <= itmax)
     Norm_M = 0;
     for q = 1 : NS
       AN1q = AN1{q};
       for n = 1 : nDIM
        wZvq{n} =  wZv{n,q};    vZvq{n} = vZv{n,q};      Zvq{n} = Zv{n,q};
       g1Zvq{n} = g1Zv{n,q};     rMq     = r_M{q};     rDZvq{n} = rZv{n,q};
      end
      [wZvq,vZvq,Zvq,g1Zvq,AN1q,rMq] = InvITERwZv(CHI_rho,dG_R,wZvq, ...
                               vZvq,Zvq,g1Zvq,AN1q,rMq,rDZvq,eta_M,eta_D,input);
      AN1{q} = AN1q;
      for n = 1 : nDIM
        wZv{n,q} =  wZvq{n};  vZv{n,q} = vZvq{n};  Zv{n,q} = Zvq{n};
       g1Zv{n,q} = g1Zvq{n};     r_M{q} = rMq;
      end
      Norm_M = Norm_M + norm(r_M{q})^2;
    end % qloop
    ErrorM = eta_M * Norm_M;
    disp(['Iteration = ', num2str(it)]);
    [CHI_rho,rZv,eta_D,ErrorD] ...
                     = UpdateContrastwZv(wZv,Zv,Zvinc,ErrorM,ErrorD,input);
    saveCHI(CHI_rho);
     it = it+1;
   end % while
 disp('------------------------------------------------------------');
 Error  = sqrt(ErrorM + ErrorD);
 disp(['Number of iterations = ' num2str(it)]);
 disp(['Total Error in M and D = ' num2str(Error)]);
```

Matlab Listing II.49 The function `InvITERwZv` carrying out a single CG iteration.

```
function [wZv,vZv,ZvD,gZv,AN,r_M] ...
      = InvITERwZv(CHI_rho,dG_R,wZv,vZv,ZvD,g_1Zv,AN_1,r_M,r_D,     ...
                                                   eta_M,eta_D,input)
global nDIM it
LZv = cell(nDIM,input.NS);

% determine conjugate gradient directions
  dummy=cell(1,nDIM);
  for n=1:nDIM; dummy{n}=conj(CHI_rho).*r_D{n}; end
  [KZv]  = AdjKOPwZv(dummy,input);
  [gZv] = AdjDOPMwZv(dG_R,r_M,input);
      AN = 0;
  for n = 1:nDIM
     gZv{n} = eta_M * gZv{n} + eta_D * (r_D{n} - KZv{n});
         AN = AN + norm(gZv{n}(:))^2;
  end
  if it == 1
      for n = 1 : nDIM;   vZv{n} = gZv{n};   end
  else
      BN = 0;
      for n = 1:nDIM;    BN = BN + sum(gZv{n}(:).*conj(g_1Zv{n}(:))); end
      gamma = real((AN-BN)/AN_1);
      for n = 1 : nDIM; vZv{n} = gZv{n} + gamma * vZv{n};
      end
  end

% determine step length alpha
   GMp   = DOPMwZv(dG_R,vZv,input);
   BN_M  = norm(GMp(:))^2;
   [KZv] = KOPwZv(vZv,input);
   AN = 0; BN_D = 0;
   for n = 1:nDIM
     LZv{n} = vZv{n} - CHI_rho .* KZv{n};
       AN   = AN + sum(gZv{n}(:).*conj(vZv{n}(:)));
       BN_D = BN_D + norm(LZv{n}(:))^2;
   end
   alpha = real(AN) / (eta_M*BN_M + eta_D*BN_D);

 % update contrast sources
   for n = 1:nDIM
        wZv{n} = wZv{n} + alpha * vZv{n};
        ZvD{n} = ZvD{n} + alpha * KZv{n};
   end
%  update residual error in M
   r_M  = r_M - alpha * GMp(:);
```

Matlab Listing II.50 The function `UpdateContrastwZv` for updating the mass-density contrast χ^{ρ}.

```
function [CHI_rho,rZv,eta_D,ErrorD] ...
                    = UpdateContrastwZv(wZv,Zv,Zvinc,ErrorM,ErrorD,input)
global nDIM

rZv  = cell(nDIM,input.NS);
% Determine mass-density contrast
  INZv_wZv = zeros(input.N1,input.N2);
  INZv_Zv  = zeros(input.N1,input.N2);
  for q = 1 : input.NS
    INZv_wZv = INZv_wZv + conj(Zv{1,q}) .* wZv{1,q}   ...
                        + conj(Zv{2,q}) .* wZv{2,q};
    INZv_Zv  = INZv_Zv  + conj(Zv{1,q}) .*  Zv{1,q}   ...
                        + conj(Zv{2,q}) .*  Zv{2,q};
  end %q_loop
  CHI_rho  = INZv_wZv ./ INZv_Zv;

  if ErrorD < 1
     CHI_rho  = TVregularizer(CHI_rho,ErrorM,ErrorD,input);
  end

% Compute residual error in object domain
  Norm_D = 0;    Norm = 0;
  for q = 1 : input.NS
    rZv{1,q}  = CHI_rho .* Zv{1,q} - wZv{1,q};
    rZv{2,q}  = CHI_rho .* Zv{2,q} - wZv{2,q};
    Norm_D    = Norm_D + norm(rZv{1,q}(:))^2 + norm(rZv{2,q}(:))^2;
    Norm      = Norm   + norm(CHI_rho(:) .* Zvinc{1,q}(:))^2 ...
                       + norm(CHI_rho(:) .* Zvinc{2,q}(:))^2;
  end %q_loop
  eta_D = 1 / Norm;
  ErrorD = eta_D * Norm_D;
  disp([' ErrorM = ',num2str(ErrorM), '     ErrorD  = ',num2str(ErrorD)]);

% Show intermediate pictures of reconstruction of compressibility contrast
  x1 = input.X1(:,1);    x2 = input.X2(1,:);
  figure(5);
     IMAGESC(x1,x2,real(CHI_rho));
     title('\fontsize{13} Reconstructed Contrast \chi^{\rho}');
  figure(6);
     mesh(real(CHI_rho),'edgecolor', 'k');
     view(37.5,45);   axis('off');
     axis([1 input.N1 1 input.N2 [-1 1.01]*max(real(CHI_rho(:)))])
pause(0.1);
```

Matlab Listing II.51 The function `saveCHI` to store the contrast profiles.

```
function [] = saveCHI (CHI)
global it

if it ==   0; CHI_0   = real (CHI);  save Contrast0   CHI_0;    end
if it ==   2; CHI_2   = real (CHI);  save Contrast2   CHI_2;    end
if it ==   4; CHI_4   = real (CHI);  save Contrast4   CHI_4;    end
if it ==   8; CHI_8   = real (CHI);  save Contrast8   CHI_8;    end
if it ==  16; CHI_16  = real (CHI);  save Contrast16  CHI_16;   end
if it ==  32; CHI_32  = real (CHI);  save Contrast32  CHI_32;   end
if it ==  64; CHI_64  = real (CHI);  save Contrast64  CHI_64;   end
if it == 128; CHI_128 = real (CHI);  save Contrast128 CHI_128;  end
if it == 256; CHI_256 = real (CHI);  save Contrast256 CHI_256;  end
```

(2) Similar to the forward code `ForwardwZv` of Matlab Listing II.37, we compute the weak forms of the Green functions G_S and G_R and their spatial derivatives with the Matlab function `dGreenSourcesReceivers` of Matlab Listing II.39.
(3) The heart of the CSI method is found in the function `ITERCGMDwZv` of the Matlab Listing II.48. In this function, the maximum number of iterations is set to it = 256. Both the initial contrast sources and the initial contrast are estimated with the help of the function `InitialEstimatewZv` of Matlab Listing II.46. The contrast sources are determined by back projection. Besides the data operator `DOPMwZv` of Matlab Listing II.44, we also need its adjoint `AdjDOPMwZv` of Matlab Listing II.45. The contrast sources follow from minimization of the squared norm on $\mathbb{M}$, viz. ErrorM, after which the contrast follows from Eq. (5.42) with the help of function `UpdateContrastwZv` of Matlab Listing II.50. Further, the value of the squared norm on $\mathbb{D}$, viz. ErrorD, is computed. With these initial estimates, we employ the iterative CG method to determine the contrast sources that minimize ErrorM and ErrorD simultaneously. The function `InvITERwZv` of Matlab Listing II.49 carries out a single CG iteration, after which the function `UpdateContrast` of Matlab Listing II.18 determines an update of the contrast. After the contrast update of Eq. (5.42), the regularization is carried out using the function `TVregularizer` of Matlab Listing II.22. Note that regularization is not used in the initial estimate, because we start with an initial estimate of ErrorD = 1. During the iterations, each update of the contrast is shown as an image and a mesh plot on the computer screen. The contrast profiles are saved for a specific number of iterations, cf. the script `saveCHI` of Matlab Listing II.51.
(4) With the help of the contrast sources solved by the present inversion method, we compute again the scattered field at the receiver locations, see function `DOPMwZv` of Matlab Listing II.44, and we compare these data with the analytical data used in step (1). By plotting the differences and computing the error being the norm of these differences, we obtain an indication how good the data are explained by the inversion scheme at hand.

In Figures 5.4 and 5.5, after some number of iterations, we show the reconstructed contrast profiles. The pictures indicated by it = 0 represent the contrast profiles obtained from the initial estimates. From Figure 5.5, we see that the location of the circular boundary can be estimated based on the location of the maximum values of the estimated contrast. Within this range, the estimated contrast has significant negative values. This property is a consequence of the band limitation of the back-projection of the undersampled data. For an increasing number of iterations, these maximum contrast values increase as well. After 16 iterations, the TV regularization becomes effective

Matlab Listing II.52 The function `loadImages` to load the contrast profiles from storage and present them as image plots.

```
input = initAC;    x1 = input.X1(:,1);    x2 = input.X2(1,:);
close all

set(figure,'Units','centimeters','Position',[0 -1 20 30]);
sp = subplot(3,3,1);    load Contrast0 CHI_0;    IMAGESC(x1,x2,CHI_0);
     title('\fontsize{13} it = 0');    xlabel(''); ylabel('');
     pos=get(sp,'Position'); pos=pos+[0.05 0 0 0]; set(sp,'Position',pos);
sp = subplot(3,3,2);    load Contrast1    CHI_1;    IMAGESC(x1,x2,CHI_1);
     title('\fontsize{13} it = 1');    xlabel(''); ylabel('');
     pos=get(sp,'Position'); pos=pos+[0 0 0 0];    set(sp,'Position',pos);
sp = subplot(3,3,3);    load Contrast4    CHI_4;    IMAGESC(x1,x2,CHI_4);
     title('\fontsize{13} it = 4');    xlabel(''); ylabel('');
     pos=get(sp,'Position');pos=pos+[-0.05 0 0 0]; set(sp,'Position',pos);

set(figure,'Units','centimeters','Position',[0 -1 20 30]);
sp = subplot(3,3,1);    load Contrast8    CHI_8;    IMAGESC(x1,x2,CHI_8);
     title('\fontsize{13} it = 8');    xlabel(''); ylabel('');
     pos=get(sp,'Position'); pos=pos+[0.05 0 0 0]; set(sp,'Position',pos);
sp = subplot(3,3,2);    load Contrast16   CHI_16; IMAGESC(x1,x2,CHI_16);
     title('\fontsize{13} it = 16');    xlabel(''); ylabel('');
     pos=get(sp,'Position'); pos=pos+[0 0 0 0];    set(sp,'Position',pos);
sp = subplot(3,3,3);    load Contrast32   CHI_32; IMAGESC(x1,x2,CHI_32);
     title('\fontsize{13} it = 32');    xlabel(''); ylabel('');
     pos=get(sp,'Position');pos=pos+[-0.05 0 0 0]; set(sp,'Position',pos);

set(figure,'Units','centimeters','Position',[0 -1 20 30]);
sp = subplot(3,3,1);   load Contrast64   CHI_64;    IMAGESC(x1,x2,CHI_64);
   title('\fontsize{13} it = 64');   xlabel(''); ylabel('');
     pos=get(sp,'Position'); pos=pos+[0.05 0 0 0]; set(sp,'Position',pos);
sp = subplot(3,3,2);   load Contrast128 CHI_128; IMAGESC(x1,x2,CHI_128);
   title('\fontsize{13} it = 128'); xlabel(''); ylabel('');
     pos=get(sp,'Position'); pos=pos+[0 0 0 0];    set(sp,'Position',pos);
sp = subplot(3,3,3);   load Contrast256 CHI_256; IMAGESC(x1,x2,CHI_256);
   title('\fontsize{13} it = 256'); xlabel(''); ylabel('');
     pos=get(sp,'Position');pos=pos+[-0.05 0 0 0]; set(sp,'Position',pos);
```

and the spatial variation within and outside the region with the maximum contrast values is kept to a minimum. Note that the pictures of Figures 5.4 and 5.5 are obtained, by running the scripts `LoadImages` and `LoadMeshes` of Matlab Listings II.52 and II.53, respectively.

If we contaminate the analytical data with noise, we obtain the pictures of Figures 5.6 and 5.7 and if we compare the results of Figures 5.5 and 5.6 after 256 iterations, we see that the presence of noise in the data negatively influences the resolution of the reconstructed contrast profile.

Finally, in Figures 5.8 and 5.9, we show the differences between the analytical data amplitudes and the inverted data amplitudes obtained from the contrast sources reconstructed after 256 iterations. On the left-hand sides of these figures, we present the amplitudes of the "exact" data using the Bessel function series, while on the right-hand sides we show the differences between the inverted data and the exact data. Note that the global error between the exact data and the integral-equation

Matlab Listing II.53 The script `loadMeshes` to load the contrast profiles from storage and present them as mesh plots.

```
input = initAC;              x1 = input.X1(:,1);    x2 = input.X2(1,:);
close all

Axis = [1 input.N1 1 input.N2 [-1 1.01]*max(CHI_256(:))]; View=[37.5,45];

set(figure,'Units','centimeters','Position',[0 0 19 16]);
sp = subplot(3,3,1); mesh(CHI_0,'edgecolor', 'k');
    title('\fontsize{12} it = 0');   axis ('off'); axis(Axis); view(View);
    pt=get(sp,'Position'); pt=pt +[0.065 0 0 0]; set(sp,'Position',pt);
sp = subplot(3,3,2); mesh(CHI_1,'edgecolor', 'k');
    title('\fontsize{12} it = 1');   axis ('off'); axis (Axis); view(View);
    pt=get(sp,'Position'); pt=pt +[0.00 0 0 0]; set(sp,'Position',pt);
sp = subplot(3,3,3); mesh(CHI_4,'edgecolor', 'k');
    title('\fontsize{12} it = 4');   axis('off'); axis(Axis); view(View);
    pt=get(sp,'Position'); pt=pt+[-0.065 0 0 0]; set(sp,'Position',pt);

set(figure,'Units','centimeters','Position',[0 0 18 16]);
sp=subplot(3,3,1); mesh(CHI_8,'edgecolor', 'k');
    title('\fontsize{12} it = 8');   axis('off'); axis(Axis);  view(View);
    pt=get(sp,'Position'); pt= pt+[0.065 0 0 0]; set(sp,'Position',pt);
sp=subplot(3,3,2); mesh(CHI_16,'edgecolor', 'k');
    title('\fontsize{12} it = 16');  axis('off'); axis(Axis);  view(View);
    pt=get(sp,'Position'); pt=pt+[0.00 0 0 0]; set(sp,'Position',pt);
sp=subplot(3,3,3); mesh(CHI_32,'edgecolor', 'k');
    title('\fontsize{12} it = 32');  axis('off'); axis(Axis);  view(View);
  pt=get(sp,'Position'); pt= pt+[-0.065 0 0 0]; set(sp,'Position',pt);

set(figure,'Units','centimeters','Position',[0 0 18 16]);
sp=subplot(3,3,1); mesh(CHI_64,'edgecolor', 'k');
    title('\fontsize{12} it = 64');  axis('off'); axis(Axis);  view(View);
    pt=get(sp,'Position'); pt=pt+[0.065 0 0 0]; set(sp,'Position',pt);
sp=subplot(3,3,2); mesh(CHI_128,'edgecolor', 'k');
    title('\fontsize{12} it = 128'); axis('off'); axis(Axis);  view(View);
    pt=get(sp,'Position'); pt=pt+[0.00 0 0 0]; set(sp,'Position',pt);
sp=subplot(3,3,3); mesh(CHI_256,'edgecolor', 'k');
    title('\fontsize{12} it = 256'); axis('off'); axis(Axis);  view(View);
    pt=get(sp,'Position'); pt=pt+[-0.065 0 0 0]; set(sp,'Position',pt);
```

data amounts to 6.4%, while the global error between the analytical data and the inverted data amounts to 2.5% and 11% for noise-free data and noisy-data, respectively. Obviously, for noise-free data, the inverted model explains the analytical data better than our discretized model used in the forward calculations.

We note that the general case of the reconstruction of contrast in both compressibility and mass density for single-frequency pressure wave-field data does not lead to satisfactory inversion results. For this case, one should employ pressure wave-field data for more frequencies and/or particle-velocity wave-field data, see van Dongen and Wright [59].

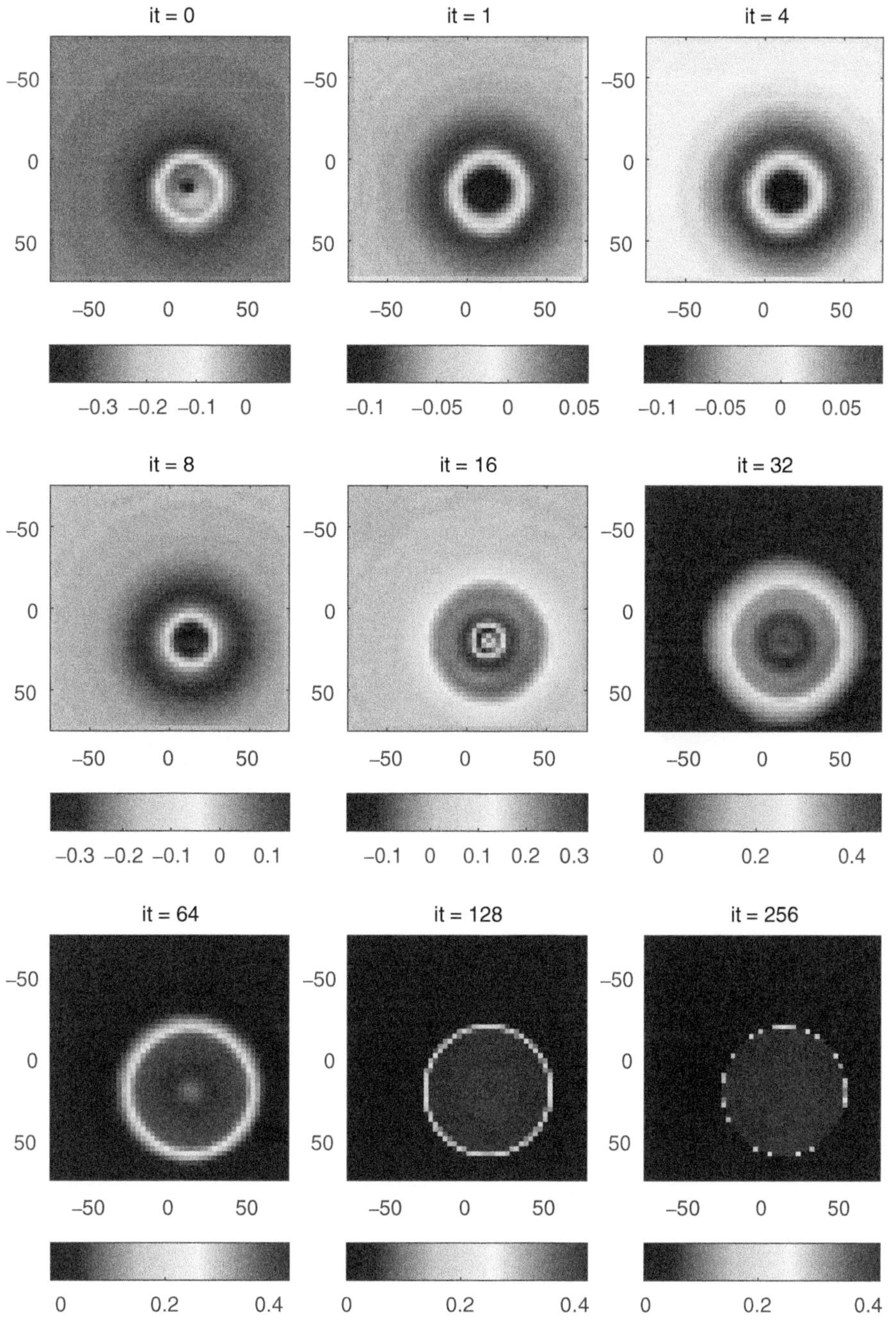

Figure 5.4 Acoustic MRCSI method: Image plots of the mass-density contrast reconstruction at different values of the iteration number (it), for **noise-free** data and nominal values $\rho_0 = 2500$ kg/m^3 and $\hat{\rho}_{sct} = 1500$ kg/m^3.

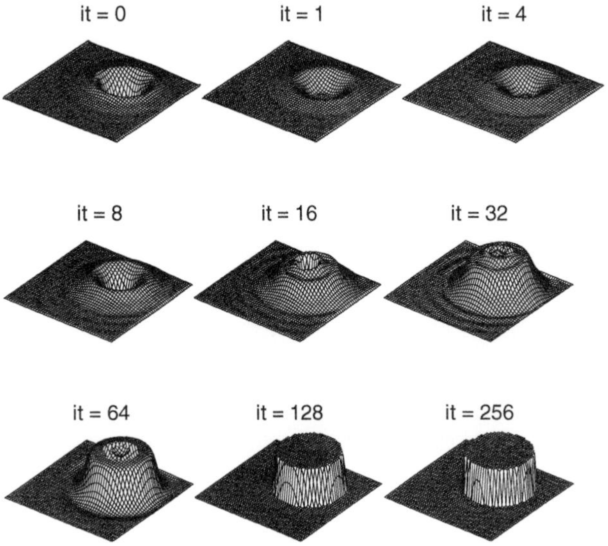

Figure 5.5 Acoustic MRCSI method: Mesh plots of the mass-density contrast reconstruction at different values of the iteration number (it), for **noise-free** data and nominal values $\rho_0 = 2500$ kg/m^3 and $\hat{\rho}_{sct} = 1500$ kg/m^3.

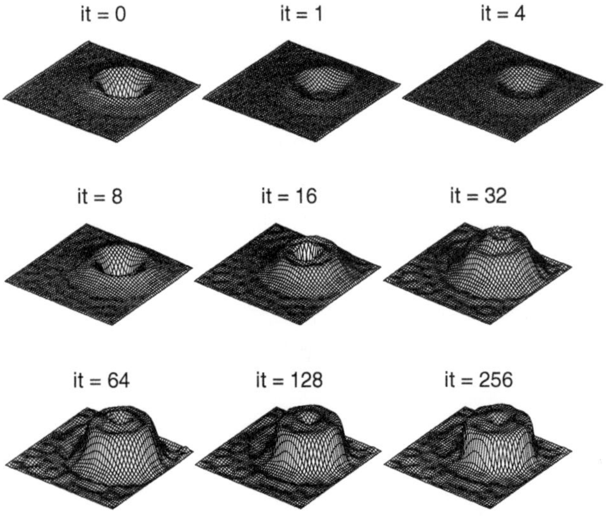

Figure 5.6 Acoustic MRCSI method: Mesh plots of the mass-density contrast reconstruction at different values of the iteration number (it), for **noisy** data and nominal values $\rho_0 = 2500$ kg/m^3 and $\hat{\rho}_{sct} = 1500$ kg/m^3.

5.4 Mass-density Interface Model for Zero Wave-Speed Contrast

Let us now consider the case that there is a vanishing wave-speed contrast, i.e. $\hat{\kappa}^{sct}/\kappa_0 = \rho_0/\hat{\rho}^{sct}$. When we discretize our scattering domain into a set of cubic subdomains with constant material properties, in Chapter 2, we have shown that we can model this case by rewriting the acoustic pressure as a contrast-source integral representation in terms of the interface sources.

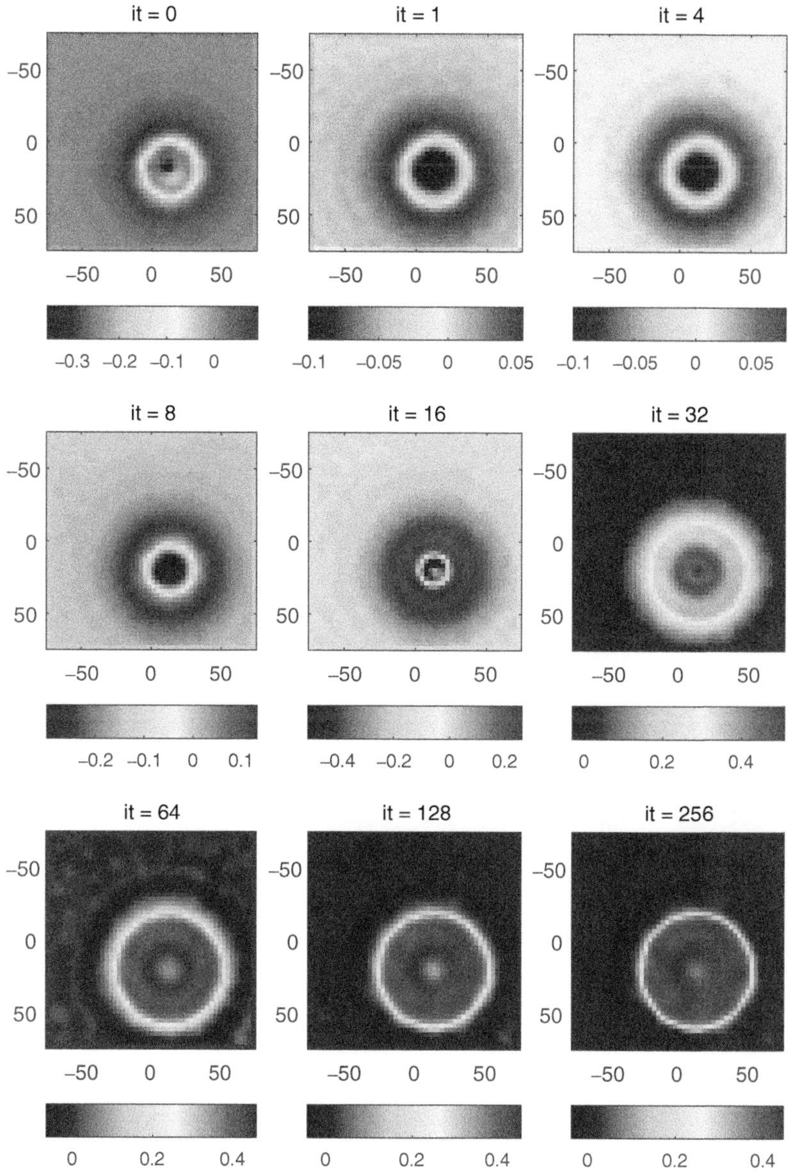

Figure 5.7 Acoustic MRCSI method: Image plots of the mass-density contrast reconstruction at different values of the iteration number (it), for **noisy** data and nominal values $\rho_0 = 2500$ kg/m^3 and $\hat{\rho}_{sct} = 1500$ kg/m^3.

From Eq. (2.126), it follows that

$$\hat{p}^{sct}_{\mathbb{M}}(\boldsymbol{x}^R_p,\boldsymbol{x}^S_q) = \sum_{k=1}^{3}\int_{\boldsymbol{x}'\in \mathrm{S}_k} 2\,\partial^R_k \hat{G}(\boldsymbol{x}^R_p-\boldsymbol{x}')\,\partial\hat{w}^{(k)}(\boldsymbol{x}',\boldsymbol{x}^S_q)\,\mathrm{d}A, \quad (\boldsymbol{x}^R_p,\boldsymbol{x}^S_q)\in\mathbb{M}\,, \tag{5.44}$$

where S_k are the interfaces in the Cartesian directions with normal vector ν_k. The interface contrast sources are denoted as $\partial\hat{w}^{(k)}(\boldsymbol{x}',\boldsymbol{x}^S_q) = \partial\hat{w}(\boldsymbol{x}',\boldsymbol{x}^S_q)\,\nu_k(\boldsymbol{x}')$, for $\boldsymbol{x}'\in \mathrm{S}_k$. The set of integral equations

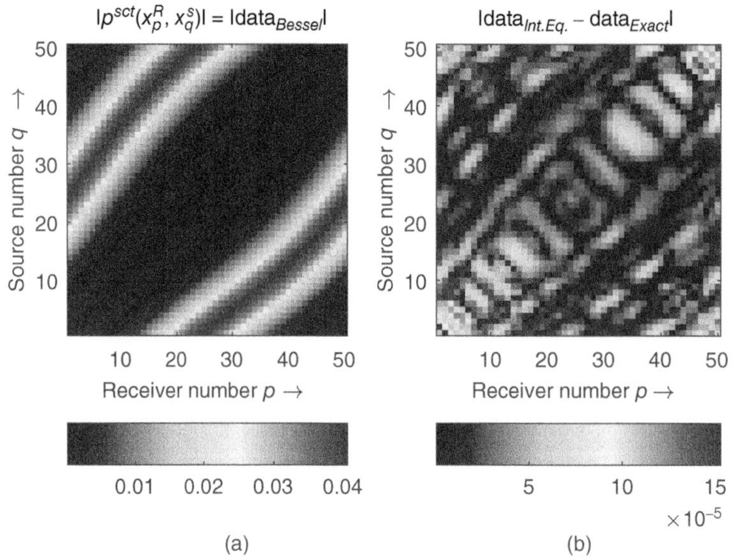

Figure 5.8 Acoustic MRCSI method: plots of the inverted **noise-free** data as function of the source and receiver numbers. The normalized global error in the inverted data amounts to 2.5%.

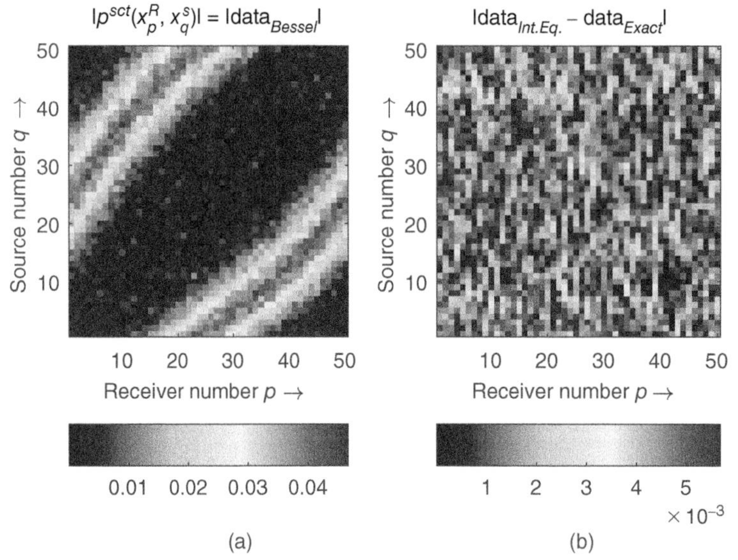

Figure 5.9 Acoustic MRCSI method: plots of the inverted **noisy** data as function of the source and receiver numbers The normalized global error in the inverted data amounts to 11%.

may then be written as, cf. Eq. (2.125),

$$\hat{R}^{(j)}\,\hat{p}^{inc}(\boldsymbol{x},\boldsymbol{x}_q^S) = \partial\hat{w}^{(j)}(\boldsymbol{x},\boldsymbol{x}_q^S) - \hat{R}^{(j)}\sum_{k=1}^{3}\int_{\boldsymbol{x}'\in S_{sct}} 2\,\partial_k\hat{G}(\boldsymbol{x}_p-\boldsymbol{x}')\,\partial\hat{w}^{(k)}(\boldsymbol{x}',\boldsymbol{x}_q^S)\,\mathrm{d}A\,, \tag{5.45}$$

Matlab Listing II.54 The function `initAC` to initialize the 2D acoustic wave-field parameters.

```
function input = initAC()
% Time factor = exp(-iwt)
% Spatial units is in m                          % Source wavelet  s rho_0 Q = 1

global nDIM;  nDIM = 2;                          % set dimension of space

input.c_0       = 1500;                          % wave speed in embedding
input.c_sct     = 1500;                          % wave speed in scatterer
input.rho_0     = 3000;                          % mass density in embedding
input.rho_sct   = 1500;                          % mass density in scatterer

f               = 50;                            % temporal frequency
wavelength      = input.c_0 / f;                 % wavelength
s               = 1e-16 - 1i*2*pi*f;             % LaPlace parameter
input.gamma_0 = s/input.c_0;                     % propagation coefficient
disp(['wavelength = ' num2str(wavelength)]);

% add input data to structure array 'input'
  input = initSourceReceiver(input);        % add location of source/receiver

  input = initGrid(input);                  % add grid in either 1D, 2D or 3D

  input = initFFTGreenfun(input);           % compute FFT of Green function

  input = initAcousticContrast(input);      % add contrast distribution

  input.Errcri = 1e-5;

  input.Noise = 0;

  if input.Noise == 1  % generate random numbers that are repeatable
    sigma  = 0.2;        % sigma = standard deviation and mu =0
    randn('state',1);   ReRand = randn(input.NR,input.NS) * sigma;
    randn('state',2);   ImRand = randn(input.NR,input.NS) * sigma;
    input.Rand = ReRand + 1i*ImRand;
  end

  % Kirchhoff type of approximation in object operators
  input.Kirchhoff = 0; % value = 0: no approximation
                       % value = 1: approximation in adjoint
                       % value = 2: plus approximation in forward operator
```

for $\boldsymbol{x} = \boldsymbol{x}^{(j)} \in \mathbb{S}_j \in \mathbb{D}, j = 1,\ 2,\ 3$. The interface contrast at the interfaces $\mathbb{S}_j$ in the $\boldsymbol{i}_j$-direction are denoted as $\hat{R}^{(j)} = \hat{R}(\boldsymbol{x}^{(j)})$, with

$$\hat{R}^{(j)} = \frac{\hat{\rho}^+ - \hat{\rho}^-}{\hat{\rho}^+ + \hat{\rho}^-}, \tag{5.46}$$

in which $\hat{\rho}^+$ and $\hat{\rho}^-$ are the values of the mass-density on both sides of the interface $\mathbb{S}_j$, normal to the positive and negative $\boldsymbol{i}_j$ direction, respectively. It is obvious that the inverse problem now consists

Matlab Listing II.55 The function `initAcousticContrast` to initialize the acoustic parameters.

```
function [input] = initAcousticContrast(input)

input.a   = 40; % half width slab / radius circle cylinder / radius sphere
input.xO(1) = input.a / 2;   % center coordinates of circle
input.xO(2) = input.a / 3;
          R = sqrt((input.X1-input.xO(1)).^2 + (input.X2-input.xO(2)).^2);
shape = (R < input.a);

% (1) Compute compressibbily contrast and mass density contrast ----------
      kappa_0         = 1 / (input.rho_0   * input.c_0^2);
      kappa_sct       = 1 / (input.rho_sct * input.c_sct^2);
      contrast_kappa  = 1 - kappa_sct /kappa_0;
      input.CHI_kap   = contrast_kappa .* shape;

   % Plot contrast values
     set(figure(1),'Units','centimeters','Position',[5 1 18 12]);
     subplot(1,2,1);
       IMAGESC(x1,x2, real(input.CHI_kap));
       title('\fontsize{13}\chi^{\kappa} = 1 - \kappa_{sct}/ \kappa_{0}');
     subplot(1,2,2);
       IMAGESC(x1,x2,input.CHI_rho);
       title('\fontsize{13}\chi^{\rho} = 1 - \rho_{sct}/ \rho_{0}');

% (2) Compute volume contrast and interface contrast ---------------------
      c_contrast = 1 - input.c_0^2/input.xsxsc_sct^2;
      input.CHI  = c_contrast .* shape;
      rho        = input.rho_sct .* shape + (1-shape) .* input.rho_0;

      Rfl{1} = zeros(N1,N2);    Rfl{2} = zeros(N1,N2);
      Rfl{1}(1:N1-1,:) = (rho(2:N1,:) - rho(1:N1-1,:)) ...
                                    ./(rho(2:N1,:) + rho(1:N1-1,:));
      Rfl{2}(:,1:N2-1) = (rho(:,2:N2) - rho(:,1:N2-1)) ...
                                    ./(rho(:,2:N2) + rho(:,1:N2-1));
      input.Rfl = Rfl;

   % Plot contrast values
     set(figure(2),'Units','centimeters','Position',[5 1 18 12]);
     subplot(1,2,1);
       IMAGESC(x1+dx/2,x2,Rfl{1});
       axis('ij','equal','tight');  title('\fontsize{13} R^{(1)}');
       colorbar('hor')
    subplot(1,2,2);
      IMAGESC(x1,x2+dx/2,Rfl{2});
      axis('ij','equal','tight');  title('\fontsize{13} R^{(2)}');
      colorbar('hor');  colormap(jet);
```

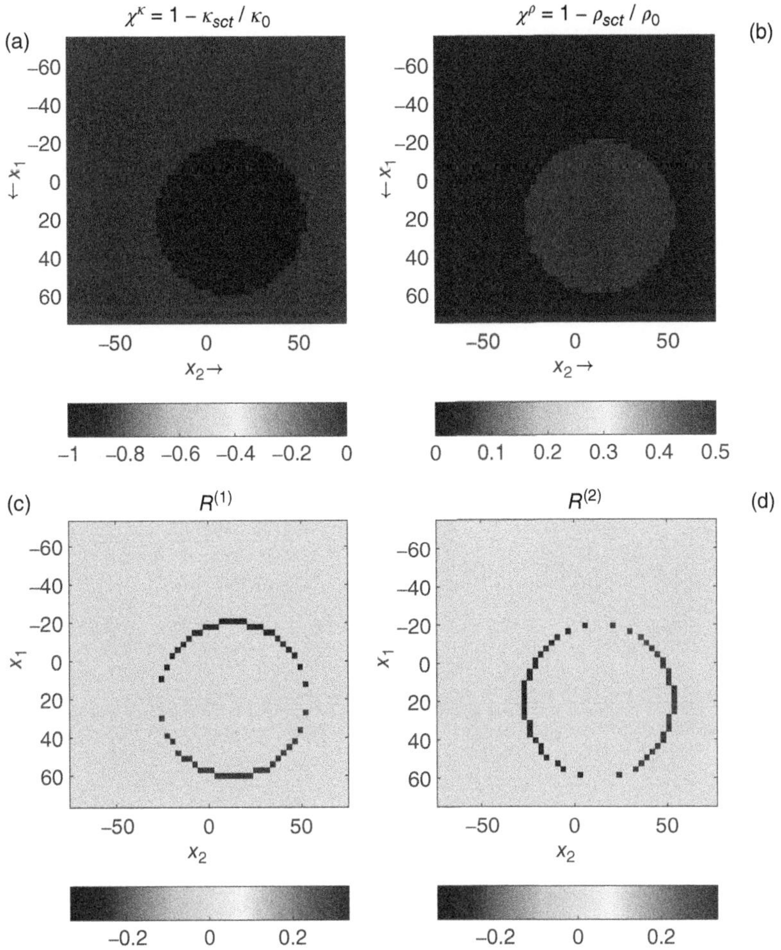

Figure 5.10 Plots of the real parts of the compressibility and the mass-density contrast (a,b), and the two components of the interface mass-density contrast (c,d).

of the determination of this interface contrast from the measurements, using the data equation of Eq. (5.44) and the object equation of Eq. (2.125).

5.4.1 Synthetic Data for Zero Wave-speed Contrast

We consider again the 2D case, and we first discuss the synthetic data for an off-centered circular cylinder. We compute two scattered-field data sets, obtained either analytically from the Bessel-series solution, or numerically from the integral-equation solution.

The wave-field parameters are initialized by the `initAC` function (Matlab Listing II.54). The global parameter `nDIM` assigns the 2D space to be used in our inversion codes. Subsequently, both the acoustic wave speed c_0 of the embedding and the acoustic wave speed $\hat{c}_{sct}$ of the scattering object are taken equal to each other, i.e. 1500 m/s. The mass density ρ_0 in the embedding is equal to 3000 kg/m^3, and the mass density in the scattering object is 1500 kg/m^3. The frequency of operation is 50 Hz. As output, the `initAC` function displays the operating wavelength. This facilitates

Matlab Listing II.56 The `Forwarddw` script to compute synthetic 2D data.

```
clear all; clc; close all; clear workspace
input = initAC();

% (1) Compute analytically scattered field data ---------------------------
     disp('Running AcousticDataSctCircle');  AcousticDataSctCircle;

     %load DataAnalytic;
     %load DataIntEq;    dataCircle = data;

data = zeros(input.NR,input.NS);
for q = 1 : input.NS
% (2) Compute incident field at different grids
     Pinc = IncPressureWave(q,input);

% (3) Solve contrast source integral equations with CGFFT method ---------
     % dw = ITERCGdw(Pinc,input);
       dw = BICGSTABdw(Pinc,input);

% (4) Compute synthetic data and plot fields and data ---------------------
     data(:,q) = DOPMdw(dw,input);
end % q_loop
  save DataIntEq data;

 % Plot data at a number of receivers -------------------
   figure(10);
   subplot(1,2,1)
     PlotData(dataCircle,input);
     title('|p^{sct}(x_p^R,x_q^S)|=|data_{Bessel}|')
   subplot(1,2,2)
     PlotData(data-dataCircle,input);

 % Compute normalized difference of norms
   error = num2str(norm(data(:)-dataCircle(:),1)/norm(dataCircle(:),1));
   disp(['error=' error]);        title('|data_{Int.Eq.} - data_{Bessel}|')
```

Matlab Listing II.57 The `IncPressureWave` script to compute incident pressure wavefield at the interfaces.

```
function [Pinc] = IncPressureWave(q,input)
gam0 = input.gamma_0;
xS   = input.xS;     dx = input.dx;
X1     = input.X1;    X2 = input.X2;

DIS     = sqrt( (X1+dx/2-xS(1,q)).^2 + (X2-xS(2,q)).^2 );
Pinc{1} = 1/(2*pi).* besselk(0,gam0*DIS);   % Grid (1) on shifted points

DIS     = sqrt( (X1-xS(1,q)).^2 + (X2+dx/2-xS(2,q)).^2 );
Pinc{2} = 1/(2*pi).* besselk(0,gam0*DIS);   % Grid (2) on shifted points
```

Matlab Listing II.58 The function `IterCGdw` to compute the contrast sources.

```
function [dw] = ITERCGdw(Pinc,input)
global nDIM;        Rfl = input.Rfl;

dw = cell(1,nDIM); dr = cell(1,nDIM); dg = cell(1,nDIM); dv = cell(1,nDIM);
itmax = 200;  Errcri = input.Errcri;  it = 0;  % initialization iteration
Norm_D = 0;
for n = 1:nDIM
   dw{n} = zeros(size(Pinc{n}));
   dr{n} = Rfl{n}.*Pinc{n};               Norm_D = Norm_D + norm(dr{n}(:))^2;
end
eta_D = 1 /Norm_D;
Error = 1;  fprintf('Error =          %g', Error);

while (it < itmax) && ( Error > Errcri)
 % determine conjugate gradient direction v
   dummy=cell(1,nDIM); for n=1:nDIM; dummy{n} = conj(Rfl{n}).*dr{n}; end
   [Kdf] = AdjKOPdw(dummy,input);
   AN = 0;
   for n = 1:nDIM
    dg{n} = dr{n} - Kdf{n}; % window for negligible Rfl !!
    dg{n} = (abs(Rfl{n}) >= Errcri) .* dg{n};
    AN = AN + norm(dg{n}(:))^2;
   end
   if it == 0
      for n=1:nDIM;  dv{n} = dg{n}; end
   else
      for n=1:nDIM;  dv{n} = dg{n} + AN/AN_1 * dv{n}; end
   end
% determine step length alpha
  [Kdv] = KOPdw(dv,input);
  BN = 0;
  for n = 1:nDIM
      Kdv{n} = dv{n} - Rfl{n} .* Kdv{n};
      BN = BN + norm(Kdv{n}(:))^2;
  end
   alpha = AN / BN;
 % update contrast sources, AN and update residual errors
   Norm_D = 0;
   for n = 1:nDIM
      dw{n} = dw{n} + alpha * dv{n};
      dr{n} = dr{n} - alpha * Kdv{n};
      Norm_D = Norm_D + norm(dr{n}(:))^2;
   end
   AN_1  = AN;
   Error = sqrt(eta_D * Norm_D); fprintf('\b\b\b\b\b\b\b\b%6f',Error);
   it = it+1;
end % CG iterations

fprintf('\b\b\b\b\b\b\b\b%6f\n',Error);
disp(['Number of iterations is ' num2str(it)]);
if it == itmax; disp(['itmax reached:  err/norm = ' num2str(Error)]); end
```

Matlab Listing II.59 The function `BICGSTABdw` to compute the contrast sources.

```
function [dw] = BICGSTABdw(Pinc,input)
global nDIM;
dw = cell(1,nDIM); dr = cell(1,nDIM); dv = cell(1,nDIM);
Data_dw{1} = input.Rfl{1} .* Pinc{1};
Data_dw{2} = input.Rfl{2} .* Pinc{2};
itmax    = 200;
it       = 0;                              % initialization of iteration
for n = 1:nDIM;  dw{n} = zeros(size(Pinc{n})); end
for n = 1:nDIM;  dr{n} = Data_dw{n};           end
Norm_D   = norm(dr{1}(:))^2 + norm(dr{2}(:))^2;
eta_D    = 1 / Norm_D;                     % normalization factor
Error    = 1;                              % error norm initial error
fprintf('Error =          %g',Error);
while (it <= itmax) && (Error > input.Errcri)
 % Determine gradient directions
   AN = sum(dr{1}(:).*conj(Data_dw{1}(:))) + ...
        sum(dr{2}(:).*conj(Data_dw{2}(:)));
      if it == 0
        for n = 1:nDIM; dv{n} = dr{n};  end
      else
        for n = 1:nDIM; dv{n} = dr{n} +  (AN/AN_1) * dv{n}; end
      end
   AN_1  = AN;
 % determine step length alpha and update residual error r_D
   [Kdv] = KOPdw(dv,input);
   for n = 1:nDIM; Kdv{n}   = dv{n} - input.Rfl{n} .* Kdv{n};  end
          BN   = sum(Kdv{1}(:).*conj(Data_dw{1}(:))) + ...
                 sum(Kdv{2}(:).*conj(Data_dw{2}(:)));
      alpha = AN / BN;
     for n = 1:nDIM; dr{n}   = dr{n} - alpha * Kdv{n};  end
    % + successive overrelaxation (first step of GMRES)
    [Kdr] = KOPdw(dr,input);
    for n = 1:nDIM; Kdr{n} = dr{n} - input.Rfl{n} .* Kdr{n};  end
    beta    = ( sum(dr{1}(:).*conj(Kdr{1}(:)))  + ...
                sum(dr{2}(:).*conj(Kdr{2}(:)))  ) ...
               /(norm(Kdr{1}(:))^2 + norm(Kdr{2}(:))^2);
    % update contrast source w ,v and the residual error r_D
     for n = 1:nDIM
        dw{n} = dw{n}  + alpha *  dv{n} + beta *  dr{n};
        dv{n} = (alpha / beta) * (dv{n} - beta * Kdv{n});
        dr{n}  = dr{n} - beta * Kdr{n};
     end
     Norm_D = norm(dr{1}(:))^2 + norm(dr{2}(:))^2 ;
     Error   = sqrt(eta_D * Norm_D);
     fprintf('\b\b\b\b\b\b\b\b%6f',Error);          it = it+1;
end % while
fprintf('\b\b\b\b\b\b\b\b%6f\n',Error);
disp(['Number of iterations is ' num2str(it)]);
if it == itmax
   disp(['itmax was reached:   err/norm = ' num2str(Error)]);
end
```

Matlab Listing II.60 The function `KOPdw` to compute the interface operators $\mathcal{K}$.

```
function [Kdw] = KOPdw(dw, input)

FFTdG{1} = input.FFTdG{1} * 2 / input.dx;
FFTdG{2} = input.FFTdG{2} * 2 / input.dx;

Kdiag  = Kop(dw{1},FFTdG{1}); % ------------------------------------------------
Kdw{1} = Kdiag;
Kdw{2} = interpolate(Kdiag,2,1,input);

Kdiag  = Kop(dw{2},FFTdG{2}); % ------------------------------------------------
Kdw{1} = Kdw{1} + interpolate(Kdiag,1,2,input);
Kdw{2} = Kdiag  + Kdw{2};
```

Matlab Listing II.61 The function `AdjKOPdw` to compute the adjoint of $\mathcal{K}$.

```
function [Kdf] = AdjKOPdw(df, input)
global nDIM;
FFTdG = cell(1,nDIM);
for n = 1:nDIM;   FFTdG{n} = input.FFTdG{n} * 2 / input.dx;   end

  arg  = df{1} + interpolate(df{2},1,2,input);
Kdf{1} = AdjKop(arg,FFTdG{1});

  arg  = df{2} + interpolate(df{1},2,1,input);
Kdf{2} = AdjKop(arg,FFTdG{2});
```

Matlab Listing II.62 The function `DOPMdw` to compute synthetic data2D data, for each source location q.

```
function [data] = DOPMdw(dw, input)
gam0 = input.gamma_0;     dx = input.dx;    xR = input.xR;
X1   = input.X1;          X2 = input.X2;

data = zeros(input.NR,1);
for p = 1 : input.NR
   DIS       = sqrt((xR(1,p)-(X1+dx/2)).^2 + (xR(2,p)-X2).^2);
   d_G       = - gam0 * (1/(2*pi)) .* besselk(1,gam0*DIS);
   d1_G      = ((xR(1,p)-(X1+dx/2))./DIS) .* d_G;
  data(p,1)  = data(p,1) + 2 * dx * sum(d1_G(:) .* dw{1}(:));

   DIS       = sqrt((xR(1,p)-X1).^2 + (xR(2,p)-(X2+dx/2)).^2);
   d_G       = - gam0 * (1/(2*pi)) .* besselk(1,gam0*DIS);
   d2_G      = ((xR(2,p)-(X2+dx/2))./DIS) .* d_G;
  data(p,1)  = data(p,1) + 2 * dx * sum(d2_G(:) .* dw{2}(:));
end % p_loop
```

Matlab Listing II.63 The function `AdjDOPMdw` to compute the adjoint of the data operator.

```
function [adjGMdw] = AdjDOPMdw(f,input)
gam0 = input.gamma_0;      dx = input.dx;      xR = input.xR
X1   = input.X1;           X2 = input.X2;

adjGMdw    = cell(1,2);
adjGMdw{1} = zeros(input.N1,input.N2);
adjGMdw{2} = zeros(input.N1,input.N2);
for p = 1: input.NR
    DIS       = sqrt((xR(1,p)-(X1+dx/2)).^2 + (xR(2,p)-X2).^2);
    d_G       = - gam0 * (1/(2*pi)) .* besselk(1,gam0*DIS);
    d1_G      = ((xR(1,p)-(X1+dx/2))./DIS) .* d_G;
  adjGMdw{1} = adjGMdw{1} + 2 * dx * conj(d1_G) * f(p);

    DIS       = sqrt((xR(1,p)-X1).^2 + (xR(2,p)-(X2+dx/2)).^2);
    d_G       = - gam0 * (1/(2*pi)) .* besselk(1,gam0*DIS);
    d2_G      = ((xR(2,p)-(X2+dx/2))./DIS) .* d_G;
  adjGMdw{2} = adjGMdw{2} + 2 * dx * conj(d2_G) * f(p);
end % p_loop
```

the user to check the value $\lambda/\Delta x$. In the forward case, often a discretization of 15 samples per wavelength is chosen. This leads to an error of about 1 %. In the inversion case, we choose a discretization of 10 samples per wavelength. In this way, the numerical data differs substantially from the analytical data. Hence, this difference may be seen as model noise. In our case, this error is within the range of 10–20%. This is a factor of two greater than the error of the previous section, because the interface-contrast errors are more sensitive to the data than the volume-contrast errors. Further, the `initAC` function calls the functions `initSourceReceiver`, `initGrid`, `initFFTGreen` and `initAcousticContrast`, but now only in 2D. They initialize the locations of the sources and the receivers, the numerical grid, and the contrast distribution, respectively. Next, the error criterion is given. We use a stronger one than the error criterion used in the first part of this book, because we want to be sure that the difference between the analytical data and the numerical data is not caused by a convergence problem of the integral equation. Again, the `initAC` function offers the option to add noise to the analytical data to increase the noise level between the analytical and the numerical data. Finally, for the purpose of the inverse problem, we apply some Kirchhoff type of approximations. This will be discussed later.

In the function `initSourceReceiver` of Matlab Listing II.2, we have 50 sources located on a circle with radius of 170 m around the scattering object, while 50 receivers are located on a circle with radius of 150 m around the scattering object. The function `initGrid` (Matlab Listing II.3) defines a rectangular grid of N_1 by N_2 square subdomains with equal area of $(\Delta x)^2$. The function `initFFTGreenfun` of Matlab Listing II.34 generates a 2D grid with circulant properties. For fast computations, the FFT numbers in each direction must have a power of two. These numbers are a factor of two greater than the grid numbers N_1 and N_2. At the end of function `initFFTGreenfun`, there is a call to the Matlab function `Greenfun` to compute the repeated weak form of the Green functions, see Matlab Listing II.35. The function `initAcousticContrast` of Matlab Listing II.55 creates an off-centered cylinder and initializes the compressibility contrast $\hat{\chi}^{\kappa}$ and the mass-density $\hat{\chi}^{\rho}$. Observe that $\hat{\kappa}_{sct}/\kappa_0 = \rho_0/\hat{\rho}_{sct} = 2$. Further, the two components of the interface mass-density contrast are computed, see Figure 5.10.

The Matlab script `Forwarddw` to compute the 2D scattered field both analytically and numerically is given in Matlab Listing II.56. The following steps are carried out:

1) For the analytical computation of the acoustic wave field at a number of receivers, the Matlab script `AcousticDataSctCircle` of Matlab Listing II.38 is called. These data are saved as synthetic data for our inversion methods. Note that in the analytical calculations, we need the values of the wave speed and the mass-density of the circular object, while in the numerical computations we only need its mass-density.
2) With respect to the incident fields, we compute the acoustic pressure from a number of line sources. Here, we have however to compute this wave field at the interfaces of our discrete grid. This is carried out by calling the function `IncPressureWave` of Matlab Listing II.57. Note that we do not calculate the weak form, but we prefer to use the strong form of this field at the interfaces. In the directions normal to these interfaces, we want to maintain the resolution.
3) We solve the integral equation for the contrast sources using either the CGFFT method of Matlab Listing II.58 or the BICGFFT method of Matlab Listing II.59. Since the latter converges much faster we prefer the latter one. The integral equation is solved for each source location indicated by the value q. The forward operator `KOPdw` and its adjoint `AdjKOPdw` are given by Matlab Listing II.60 and Matlab Listing II.61. Note that the BICGSTAB method does not need the adjoints.
4) Using the contrast sources, we compute the scattered field at the receiver locations, see function `DOPMdw` of Matlab Listing II.62. For later convenience, these synthetic data are saved as well. We plot the data using function `PlotData` of Matlab Listing II.13. To measure the quality of these data we benchmark it against the analytical ones of step (1). For our acoustic example, in Figure 5.11, we present the amplitudes of the analytical data of the circular cylinder (a) and the difference with regard to the numerical data obtained with the integral-equation method (b). The error is around 19%.

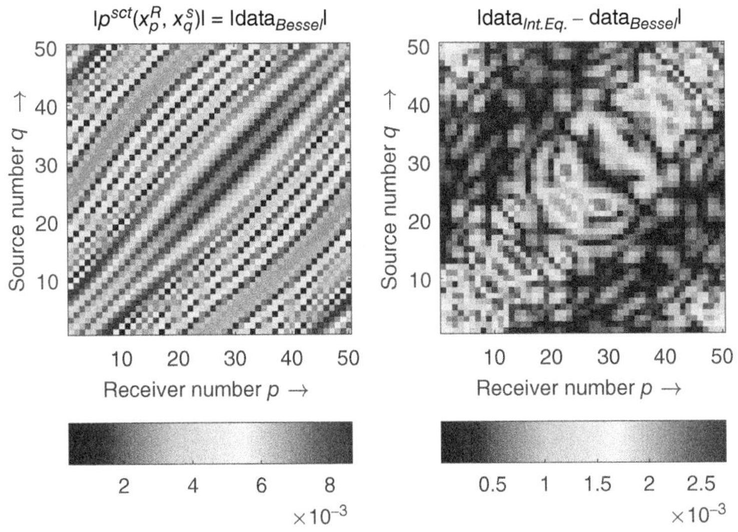

Figure 5.11 Plots of the data amplitudes as a function of the source and receiver numbers, for $\rho_0 = 1500$ kg/m^3 and $\hat{\rho}_{sct} = 3000$ kg/m^3. The normalized global error in the integral equation data amounts to 19%.

5.5 Mass-density Interface Contrast Source Inversion

For zero wave-speed contrast, we now discuss the nonlinear inverse scattering problem, using the Mass-density Interface Contrast Source Inversion (MICSI) model. This contrast source inversion method recasts the acoustic inversion problem as a minimization of a cost functional from which the interface contrast sources $\partial\hat{w}^{(k)}$ and the interface mass-density contrast $\hat{R}^{(j)}$ are reconstructed. In our further analysis, we omit the hat symbols. We first introduce the short-hand notations for the unknown wave field functions $p_q^{(j)} = p(\boldsymbol{x}_m^{(j)}, \boldsymbol{x}_q^S)$ and the known wave field-functions,

$$\begin{aligned} p_{\mathbb{M},q}^{sct} &= p_{\mathbb{M}}^{sct}(\boldsymbol{x}_p^R, \boldsymbol{x}_q^S) \,, \\ p_q^{inc(j)} &= p^{inc}(\boldsymbol{x}_m^{(j)}, \boldsymbol{x}_q^S) \,, \end{aligned} \tag{5.47}$$

for $p = 1, 2, \ldots, N^R$ and $q = 1, 2, \ldots, N^S$, where N^R is the total number of receivers and N^S is the total number of sources.

Then, in the present CSI method, the sequences of the contrast sources $\partial w_q^{(k,n)} = \partial w^{(k,n)}(\boldsymbol{x}, \boldsymbol{x}_q^S)$, and the contrast $\hat{R}^{(j,n)}$ for $n = 1, 2, \ldots,$ are iteratively found to minimize the cost functional

$$F^{(n)} = \eta_{\mathbb{M}} F_{\mathbb{M}}(w_q^{(k,n)}) + \eta_{\mathbb{D}}^{(n)} F_{\mathbb{D}}(w_q^{(k,n)}, R^{(k,n)}), \tag{5.48}$$

where

$$F_{\mathbb{M}} = \sum_{q=1}^{N^S} \| r_{\mathbb{M},q}^{(n)} \|_{\mathbb{M}}^2 \,, \quad \text{with} \quad \eta_{\mathbb{M}} = \left[\sum_{q=1}^{N^S} \| p_{\mathbb{M},q}^{sct} \|_{\mathbb{M}}^2 \right]^{-1} , \tag{5.49}$$

$$F_{\mathbb{D}} = \sum_{q=1}^{N^S} \sum_{k} \| \partial r_q^{(k,n)} \|_{\mathbb{D}}^2 \,, \quad \text{with} \quad \eta_{\mathbb{D}}^{(n)} = \left[\sum_{q=1}^{N^S} \sum_{k} \| R^{(k,n)} p_q^{inc(k)} \|_{\mathbb{D}}^2 \right]^{-1} . \tag{5.50}$$

In the present analysis, $\| \cdot \|^2 = \langle \cdot, \cdot \rangle_{\mathbb{M},\mathbb{D}}$ denotes the squared norms for $\boldsymbol{x}_p^R \in \mathbb{M}$ and $\boldsymbol{x}_m \in \mathbb{D}$, respectively. After discretization, the residuals are given by

$$\begin{aligned} r_{\mathbb{M},q}^{(n)} &= p_{\mathbb{M},q}^{sct} - 2\Delta x \sum_k \partial_k^R G_{\mathbb{M}} \, \partial w_q^{(k,n)} \,, \\ \partial r_q^{(j,n)} &= R^{(j,n)} p_q^{inc(j)} - \partial w_q^{(j,n)} + R^{(j,n)} 2\Delta x \sum_k \partial_k G_{\mathbb{D}}^{(j,k)} \, \partial w_q^{(k,n)} \,. \end{aligned} \tag{5.51}$$

The operators $G_{\mathbb{M}}$ and $G_{\mathbb{D}}^{(j,k)} = G_{\mathbb{D}}(\boldsymbol{x}_m^{(j)} - \boldsymbol{x}_m^{(k)})$ are defined in Eqs. (4.21) and (4.22). We remark that, for each k, the convolutional operators $G_{\mathbb{D}}$ operates on the interface sources defined on S_k only. The interaction between sources at different set ($j \neq k$) of interfaces is carried out by interpolation. The first term of Eq. (5.48) measures the residual error $r_{\mathbb{M},q}^{(n)}$ in the measurement domain, , while the second term measures the residual error $\partial r_q^{(j,n)}$ at the interfaces S_j in the object domain. The normalization is chosen so that the two terms are equal to one, for vanishing interface-contrast sources. The residual in the first term is a linear function of the interface-contrast source, but the residual in the second term is nonlinear in the interface contrast and the interface-contrast source. To deal with this nonlinearity, the MICSI method employs an alternate updating scheme for these quantities.

5.5.1 Updating the Interface-contrast Sources

In each iteration, we first update the interface-contrast sources, $\partial w_q^{(j,n)}$, using the conjugate gradient step,

$$\boxed{\partial w_q^{(j,n)} = \partial w_q^{(j,n-1)} + \alpha_q^{(n)} \partial v_q^{(j,n)}\,,} \tag{5.52}$$

where, for each q and n, the parameters $\alpha_q^{(n)}$ are the step lengths and the functions $\partial v_q^{(j,n)}$, are the Polak–Ribière conjugate gradient directions,

$$\boxed{\partial v_q^{(j,n)} = \partial g_q^{(j,n)},\ \partial v_q^{(j,n)} = \partial g_q^{(j,n)} + \gamma_q^{(n)}\,\partial v_q^{(j,n-1)}\,, n = 2, 3, \dots\,,} \tag{5.53}$$

where $\partial g_q^{(j,n)}$ are the gradients of the cost functional $F^{(n)}$ with respect to changes of $\partial w_q^{(j,n-1)}$. The parameter $\gamma_q^{(n)}$ in the Polak–Ribière direction of Eq. (5.53) is defined as

$$\boxed{\gamma_q^{(n)} = \frac{\mathrm{Re}\left[\sum_j \langle \partial g_q^{(j,n)}, \partial g_q^{(j,n)} - \partial g_q^{(j,n-1)}\rangle_{\mathrm{D}}\right]}{\sum_j \|\ \partial g_q^{(j,n-1)}\ \|_{\mathrm{D}}^2}\,.} \tag{5.54}$$

5.5.1.1 Gradient Directions

After discretization of the Green's operators and the norms, similar to the scalar gradient of Eq. (4.37), the gradient of the qth term of the cost functional $F^{(n-1)} = \sum_q F_q^{(n-1)}$ with respect to the interface-contrast source $\overline{\partial w}_{m',q}^{(j,n-1)} = \overline{\partial w}^{(j,n-1)}(\boldsymbol{x}_{m'}, \boldsymbol{x}_q^S)$ is defined as

$$\partial g^{(j,n)}(\boldsymbol{x}_{m'}, \boldsymbol{x}_q^S) = -\frac{\partial[F_q^{(j,n-1)}]}{\partial[\overline{\partial w}_{j,m',q}^{(n-1)}]}\,, \tag{5.55}$$

where the qth term of $F^{(n-1)}$ follow from Eqs. (5.48)–(5.50), and Eq. (5.51). Hence,

$$\partial g^{(j,n)}(\boldsymbol{x}_{m'}, \boldsymbol{x}_q^S) = -\eta_{\mathrm{M}} \sum_{p=1}^{N^R} \frac{\partial[\overline{r}_{\mathrm{M},q}^{(n-1)}]}{\partial[\overline{\partial w}_{m',q}^{(j,n-1)}]} \overline{r}_{\mathrm{M},q}^{(n-1)} - \eta_{\mathrm{D}}^{(n)} \sum_{m=1}^{N} \sum_k \frac{\partial[\overline{\partial r}_q^{(k,n-1)}]}{\partial[\overline{\partial w}_{m',q}^{(j,n-1)}]} \partial r_q^{(k,n-1)}. \tag{5.56}$$

Carrying out the partial differentiation of the complex conjugate of the residual errors with respect to $\overline{\partial w}_{m',q}^{(j,n-1)}$, and using Eqs. (5.51), (4.21), and (4.22), we arrive at

$$\begin{aligned}
-\frac{\partial[\overline{r}_{\mathrm{M},q}^{(n-1)}]}{\partial[\ \overline{\partial w}_{m',q}^{(j,n-1)}\ ]} &= 2\Delta x\ \partial_k^R \overline{G}(\boldsymbol{x}_p^R - \boldsymbol{x}_{m'})\,,\\
-\frac{\partial[\ \overline{\partial r}_{m',q}^{(j,n-1)}\ ]}{\partial[\ \overline{\partial w}_{m',q}^{(j,n-1)}\ ]} &= \delta_{j,k}\delta_{m,m'} - \overline{R}^{(j,n-1)} 2\Delta x \partial_k \overline{G}(\boldsymbol{x}_m - \boldsymbol{x}_{m'})\delta_{j,k}\,.
\end{aligned} \tag{5.57}$$

Substituting these expressions into Eq. (5.56), we obtain

$$\begin{aligned}
\partial g_q^{(j,n)} &= \eta_{\mathrm{M}} \sum_{p=1}^{N^R} 2\Delta x\ \partial_j^R \overline{G}(\boldsymbol{x}_p^R - \boldsymbol{x}_{m'})\ r_{\mathrm{M},q}^{(n-1)}\\
&\quad + \eta_{\mathrm{D}}^{(n-1)} \sum_{m=1}^{N} \left[\partial r_q^{(j,n-1)} - 2\Delta x \sum_k \partial_j \overline{G}(\boldsymbol{x}_m - \boldsymbol{x}_{m'}) \overline{R}^{(k,n-1)} \partial r_q^{(k,n-1)}\right].
\end{aligned} \tag{5.58}$$

Interchanging m' and m, we see that $\partial g_q^{(j,n)} = \partial g^{(j,n)}(\boldsymbol{x}_m, \boldsymbol{x}_q^S)$ may be written as

$$\boxed{\partial g_q^{(j,n)} = \eta_{\rm M}\ 2\Delta x \partial_j^R G_{\rm M}^\star \{r_{{\rm M},q}\} + \eta_{\rm D}^{(n-1)} \left[\partial r_q^{(j,n-1)} - 2\Delta x \sum_k \partial_j G_{\rm D}^\star \{\overline{R}^{(k,n-1)} \partial r_q^{(k,n-1)}\} \right].} \tag{5.59}$$

The adjoints $G_{\rm M}^\star$ and $G_{\rm D}^{\star(j,k)} = G_{\rm D}^\star(\boldsymbol{x}_m^{(j)} - \boldsymbol{x}_m^{(k)})$ are given by Eqs. (4.31) and (4.35).

5.5.1.2 Calculation of the Step Length

The step length $\alpha_q^{(n)}$ is found as minimizer of the cost functional $F^{(n)} = \sum_q F_q^{(n)}$, i.e.

$$\alpha_q^{(n)} = \underset{\text{real } \alpha}{\arg\min} \{F_q^{(n)}(\partial w_q^{(j,n-1)} + \alpha\ \partial v_q^{(j,n)})\}\ , \tag{5.60}$$

where $F_q^{(n)}$ as function of α is given by

$$F_q^{(n)} = \eta_{\rm M} \parallel r_{{\rm M},q}^{(n-1)} - \alpha\ \mathcal{K}_{{\rm M},q}^{(n)} \parallel_{\rm M}^2 + \eta_{\rm D}^{(n-1)} \sum_j \parallel r_{{\rm D},q}^{(j,n-1)} - \alpha\ (\partial v_{j,q}^{(j,n)} - R^{(j,n-1)} \mathcal{K}_q^{(j,n)}) \parallel_{\rm D}^2\ , \tag{5.61}$$

in which

$$\mathcal{K}_{{\rm M},q}^{(n)} = 2\Delta x \sum_k \partial_k^R G_{\rm M} \partial v_q^{(k,n)}\ , \tag{5.62}$$

and

$$\mathcal{K}_q^{(j,n)} = 2\Delta x \sum_k \partial_k G_{\rm D}^{(j,k)} \partial v_q^{(k,n)}\ . \tag{5.63}$$

The real part of the minimizer is found explicitly as

$$\alpha_q^{(n)} = \frac{\text{Re}\left[\eta_{\rm M}\ \langle r_{{\rm M},q}^{(n-1)}, \mathcal{K}_{{\rm M},q}^{(n)}\rangle + \eta_{\rm D}^{(n-1)} \sum_j \langle \partial r_{{\rm D},q}^{(j,n-1)}, \partial v_q^{(j,n)} - R^{(j,n-1)} \mathcal{K}_q^{(j,n)}\rangle_{\rm D}\right]}{\eta_{\rm M} \parallel \mathcal{K}_{{\rm M},q}^{(n)} \parallel_{\rm M}^2 + \eta_{\rm D}^{(n-1)} \sum_j \parallel \partial v_q^{(j,n)} - R^{(j,n-1)} \mathcal{K}_q^{(j,n)} \parallel_{\rm D}^2}\ . \tag{5.64}$$

Using the expression for the gradient of Eq. (5.22), it simplifies to

$$\alpha_q^{(n)} = \frac{\text{Re}\left[\sum_j \langle \partial g_q^{(j,n)}, \partial v_q^{(j,n)}\rangle_{\rm D}\right]}{\eta_{\rm M} \parallel \mathcal{K}_{{\rm M},q}^{(j)} \parallel_{\rm M}^2 + \eta_{\rm D}^{(n-1)} \sum_j \parallel \partial v_q^{(j,n)} - R^{(j,n-1)} \mathcal{K}_q^{(j,n)} \parallel_{\rm D}^2}\ . \tag{5.65}$$

After each update of the interface-contrast sources, the residual error in the measurement domain becomes

$$\boxed{r_{{\rm M},q}^{(n)} = r_{{\rm M},q}^{(n-1)} - \alpha_q^{(n)} \mathcal{K}_{{\rm M},q}^{(j,n)}\ .} \tag{5.66}$$

5.5.2 Updating the Interface Contrast

Using the update for the interface-contrast sources, the acoustic pressure fields $p_q^{(j,n)} = p^{(n)}(\boldsymbol{x}_m^{(j)}, \boldsymbol{x}_q^S)$ at the interfaces S_j are directly obtained as

$$p_q^{(j,n)} = p_q^{(j,n-1)} + \alpha_q^{(n)} \mathcal{K}_q^{(j,n)}\ , \quad n = 1, 2, \dots\ , \tag{5.67}$$

where $\mathcal{K}_q^{(j,n)}$ is defined by Eq. (5.63). Recall that these quantities depend on the position vector $\boldsymbol{x}_m$ and the source location $\boldsymbol{x}_q^S$. An update for the interface mass-density contrast $R^{(j,n)} = R^{(j,n)}(\boldsymbol{x}_m^{(j)})$ is obtained as

$$\boxed{R^{(j,n)} = \frac{\sum_{q=1}^{N^S} \partial w_q^{(j,n)} \, \overline{p}_q^{(j,n)}}{\sum_{q=1}^{N^S} |p_q^{(j,n)}|^2} .} \tag{5.68}$$

Using the interface contrast update, the residual errors at the interfaces become

$$\boxed{\partial r_q^{(j,n)} = R^{(j,n)} p_q^{(j,n)} - \partial w_q^{(j,n)} ,} \tag{5.69}$$

With this result the description of the algorithm is completed, except for the starting values for the interface-contrast sources $\partial w_q^{(j,0)}$.

5.5.3 Initial Estimate

We cannot start with zero interface contrast-sources $\partial w_q^{(j,0)} = 0$, since then $R^{(j,n)} = 0$ and the normalization factor $\eta_{\mathrm{D}}^{(0)}$ is undefined. Therefore, we choose as starting values the interface-contrast sources that only minimize the cost functional

$$F_{\mathrm{M}}(\partial w_q^{(j,0)}) = \sum_{q=1}^{N^S} \| \, p_{\mathrm{M},q}^{sct} - \mathcal{K}_{\mathrm{M},q}^{(0)} \, \|_{\mathrm{M}}^2 , \tag{5.70}$$

with

$$\mathcal{K}_{\mathrm{M},q}^{(0)} = 2\Delta x \sum_k \partial_k^R G_{\mathrm{M}} \partial v_q^{(k,0)} , \tag{5.71}$$

Using the gradient method, we take

$$\boxed{\partial w_q^{(j,0)} = \alpha_q^{(0)} \, g_q^{(j,0)} , \qquad \partial g_q^{(j,0)} = \mathcal{K}_{\mathrm{M},j}^{\star} p_{\mathrm{M},q}^{sct} ,} \tag{5.72}$$

in which

$$\mathcal{K}_{\mathrm{M},j}^{\star} p_{\mathrm{M},q}^{sct} = 2\Delta x \partial_j^R \overline{G}_{\mathrm{M}} p_{\mathrm{M},q}^{sct} \tag{5.73}$$

The cost functional is minimized when each term for q is minimized, i.e.

$$\alpha_q^{(0)} = \arg\min_{\alpha} \{ \| \, p_{\mathrm{M},q}^{sct} - \alpha \, \mathcal{K}_{\mathrm{M},q}^{(0)} \, \|_{\mathrm{M}}^2 \} . \tag{5.74}$$

This leads directly to

$$\alpha_q^{(0)} = \frac{\langle p_{\mathrm{M},q}^{sct}, \mathcal{K}_{\mathrm{M},q}^{(0)} \rangle_{\mathrm{M}}}{\| \, \mathcal{K}_{\mathrm{M},q}^{(0)} \, \|_{\mathrm{M}}^2} , \tag{5.75}$$

or in simplified form as

$$\boxed{\alpha_q^{(0)} = \frac{\sum_j \| \, \partial g_q^{(j,0)} \, \|_{\mathrm{D}}^2}{\| \, \mathcal{K}_{\mathrm{M},q}^{(0)} \, \|_{\mathrm{M}}^2} .} \tag{5.76}$$

The gradient $\partial g_q^{(j,0)} = \mathcal{K}^{\star}_{\mathbb{M},j} p^{sct}_{\mathbb{M},q}$ is the back projection of the data domain $\mathbb{M}$ into the object domain $\mathbb{D}$, and it is a *back propagation* of the pressure wave field data in the object domain and converted to interface data with two components $\partial w_j, j = 1, 2$ (in 2D).

With this initial estimate $\partial w_q^{(j,0)} = \partial w^{(j,0)}(\boldsymbol{x}, \boldsymbol{x}_q^S)$ and $p_q^{(j,0)} \approx p_q^{inc(j)}$, the interface contrast estimate on S_j is obtained from

$$\boxed{R^{(j,0)} = \frac{\sum_{q=1}^{N^S} \partial w_q^{(j,0)} \overline{p}_q^{inc(j)}}{\sum_{q=1}^{N^S} |p_q^{inc(j)}|^2}} \,. \tag{5.77}$$

5.5.4 Regularization by Resetting Small Interface-contrast Variation to Zero

Without taking any measures, the reconstructed interface-contrast values $R^{(j)} = R(\boldsymbol{x}_m^{(j)})$ exhibit ringing artifacts. These interface profiles oscillate around spatial sharp transitions. The main cause of these ringing artifacts is that the scattered field data is band-limited in space. Mathematically, it is called Gibbs phenomenon, and it is the way in which the Fourier series of a piecewise continuously differentiable function behaves at a jump discontinuity. The edge type of behavior of our interface jumps may be restored by setting the small variations to zero. In our interface-contrast model, the interfaces are characterized by positive and negative jumps. Between these jumps, the contrast variation should vanish. Comparison of our model data obtained from our integral model and the analytical synthetic data have shown discrepancies of the relative order of 10%, due to the stair-case approximations of the circular interface. Consistently, amplitude variations of the interface profile less than 10% of the maximum amplitude value of the interface profile may be neglected. Therefore, in each iteration, we reset the interface contrast $R^{(j)} = R(\boldsymbol{x}_m^{(j)})$ to its windowed version,

$$R^{(j)} := \chi^{\mathrm{S}_j} R^{(j)} \,, \tag{5.78}$$

in which the window is taken as

$$\chi^{\mathrm{S}_j} == \begin{cases} 0, & \text{if } R^{(j)} \leq 0.1 \max_{\boldsymbol{x}_m^{(j)} \in \mathrm{S}_j} |R^{(j)}| \,, \\ 1, & \text{if } R^{(j)} > 0.1 \max_{\boldsymbol{x}_m^{(j)} \in \mathrm{S}_j} |R^{(j)}| \,. \end{cases} \tag{5.79}$$

This procedure is denoted as regularization by windowing.

5.5.5 Matlab Codes for the MICSI Method

In this subsection, we present the Matlab codes for the 2D acoustic version of the present method. The associated Matlab script `MICSI` is given in Matlab Listing II.64. The following steps are carried out:

(1) We start with the computation of analytical data for the acoustic case. We use `AcousticDataSctCircle` of Matlab Listing II.38. If these data have been computed with the forward code `ForwardwZv` of Matlab Listing II.38. If these data have been computed with the forward code `InversionForwarddw` of Matlab Listing II.56, we load these analytical data from the computer memory. When we would like to investigate what the result is of an "Inverse Crime," where both the forward and inverse codes are based on the same computational model, we

Matlab Listing II.64 The function `MICSI` to reconstruct the interface contrast $R^{(1)}$ and $R^{(2)}$.

```
clear all; clc; close all; clear workspace
input = initAC();

% (1) Compute Exact data using analytic Bessel function expansion --------
      disp('Running AcousticDataSctCircle');   AcousticDataSctCircle;

      load DataAnalytic;
      % load DataIntEq;     dataCircle = data;          % Inverse Crime

      % add complex-valued noise to data
      if input.Noise == 1
         dataCircle = dataCircle + mean(abs(dataCircle(:))) * input.Rand;
      end

%   (2) Apply MRCCSI method -----------------------------------------------
        [dw] = ITERCGMDdw(dataCircle, input);

%   (3) Compute model data explained by reconstructed contrast sources ----
        data = zeros(input.NR, input.NS);
        for q = 1 : input.NS
          dwq{1} = dw{1,q};
          dwq{2} = dw{2,q};
          data(:,q) = DOPMdw(dwq, input);
        end % q_loop

%     plot data and compare to synthetic data
      figure(20);
      subplot(1,2,1)
        PlotData(dataCircle, input);
        title('|p^{sct}(x_p^R,x_q^S)|=|data_{Bessel}|')
      subplot(1,2,2)
        PlotData(data-dataCircle, input);
%     Compute normalized difference of norms
      error = num2str(norm(data(:)-dataCircle(:),1)/norm(dataCircle(:),1));
      disp(['error=' error]);        title('|data_{Int.Eq.} - data_{Exact}|')
```

have the option to load the data from the integral-equation model. If in the Matlab function `initAC`, the value of the `input.Noise` is equal to 1, then there is complex-valued noise added with a standard deviation of 0.2 × mean(|dataCircle|). Adding this so-called computer noise is certainly needed to prevent an inverse crime. However, we are now in the situation that our analytical data and numerical data of the integral-equation method differ more than 10/ In this book, we do not discuss approximations of our wave-field models. One of these is the Born approximation, which is a low-frequency approximation. In the range of frequencies we are interested in, this approximation is not very useful to approximate the integral operators based on volume-contrast sources. However, the diagonal part of the integral operator based on interface-contrast sources vanishes, for all frequencies and configurations, because the normal spatial derivative of the Green function in the normal direction to a plane interface vanishes. The off-diagonal contributions decay as function of the distance between observation point

Matlab Listing II.65 The function `InitialEstimatedw` to compute an initial estimate of the contrast sources $\partial w_q^{(j)}$ and the acoustic pressure p.

```
function [dw,P,P_inc,r_M,ErrorM] ...
                        = InitialEstimatedw(eta_M,data,input)
global nDIM

NS = input.NS;
dw = cell(nDIM,NS); P = cell(nDIM,NS);    P_inc = cell(nDIM,NS);
r_M = cell(1,NS);

Norm_M = 0;
dummyq = cell(1,nDIM);
for q = 1 : NS
    r_M{q}      = data(:,q);
    g_dw        = AdjDOPMdw(r_M{q},input);
    A0          = norm(g_dw{1}(:))^2 + norm(g_dw{2}(:))^2;
    dummyq{1}   = g_dw{1};
    dummyq{2}   = g_dw{2};
    GMp         = DOPMdw(dummyq,input);
    B0          = norm(GMp(:))^2;
    alpha       = real(A0/B0);
    r_M{q}      = r_M{q} - alpha * GMp(:);
    Norm_M      = Norm_M + norm(r_M{q})^2;
    dw{1,q}     = alpha * g_dw{1};
    dw{2,q}     = alpha * g_dw{2};
    dummyq{1}   = dw{1,q};
    dummyq{2}   = dw{2,q};
          Pinc  = IncPressureWave(q,input);
   P_inc{1,q}   = Pinc{1};
   P_inc{2,q}   = Pinc{2};

   if (input.Kirchhoff == 0)  || (input.Kirchhoff == 1)
      [Kdw]   = KOPdw(dummyq,input);
   elseif input.Kirchhoff == 2
      Kdw{1}  = 0 * dw{1,q};
      Kdw{2}  = 0 * dw{2,q};
    end
      P{1,q}  = P_inc{1,q} + Kdw{1};
      P{2,q}  = P_inc{2,q} + Kdw{2};
end %qloop

ErrorM = sqrt(eta_M * Norm_M);
```

$\boldsymbol{x}$ and integration points $\boldsymbol{x}'$. Neglecting the latter contributions, we arrive at an approximation very similar to the Kirchhoff approximation, that is used sometimes in seismic imaging, see Bleistein [5]. Basically, it is a high-frequency approximation. We use three variants in our inversion methods.

(a) If `input.Kirchhoff` $= 0$, no approximation is used.

(b) If `input.Kirchhoff` $= 1$, a *first approximation* is made in the adjoint operator. We neglect the scattering operator that represents the wave-field interactions between adjacent interfaces.

Matlab Listing II.66 The function `ITERCGMDdw` for updating $\partial w_q^{(j)}$, $R^{(1)}$ and $R^{(2)}$.

```
function [dw] = ITERCGMDdw(data, input)
global nDIM it
NS     = input.NS;
AN1    = cell(1,NS);
dwq    = cell(nDIM);      Pq = cell(nDIM);    vdwq = cell(nDIM);
g1dwq  = cell(nDIM);     drq = cell(nDIM);
g1dw   = cell(nDIM,NS);  vdw = cell(nDIM,NS);

itmax = 256;
eta_M = 1 / norm(data(:))^2;       % normalization factor eta_M

% (1) Initial estimate of interface sources and mass-density jumps
  [dw,P,P_inc,r_M,ErrorM] = InitialEstimatedw(eta_M,data,input);
   it = 0;
   disp(['Iteration = ', num2str(it)]);
  [Rfl,dr,eta_D,ErrorD] = UpdateContrastdw(dw,P,P_inc,ErrorM,input);
  saveRfl(Rfl);
  disp('-------------------------------------------------------------');

% (2) Iterative updates of interface sources and mass-density jumps
  it = 1;
  while (it <= itmax)
    Norm_M = 0;
    for q = 1 : NS
      AN1q = AN1{q};
      for n = 1:nDIM
         dwq{n} = dw{n,q};          Pq{n} = P{n,q};
       g1dwq{n} = g1dw{n,q};      vdwq{n} = vdw{n,q};        rMq = r_M{q};
         drq{n} = dr{n,q};
      end
      [dwq,vdwq,Pq,g1dwq,AN1q,rMq] ...
        = InvITERdw(Rfl,dwq,vdwq,Pq,g1dwq,AN1q,rMq,drq,eta_M,eta_D,input);
      AN1{q} = AN1q;
      for n=1:nDIM
         dw{n,q} = dwq{n};          P{n,q} = Pq{n};
       g1dw{n,q} = g1dwq{n};      vdw{n,q} = vdwq{n};        r_M{q} = rMq;
      end
      Norm_M = Norm_M + norm(r_M{q})^2;
   end % qloop
   ErrorM = eta_M *Norm_M;
   disp(['Iteration = ', num2str(it)]);
   [Rfl,dr,eta_D,ErrorD] = UpdateContrastdw(dw,P,P_inc,ErrorM,input);
   saveRfl(Rfl);
   it = it+1;
 end % while

 disp('-------------------------------------------------------------');
 Error = sqrt(ErrorM + ErrorD);
 disp(['Number of Iterations = ' num2str(it)]);
 disp(['Total Error in M and D = ' num2str(Error)]);
```

Matlab Listing II.67 The function `InvITERdw` carrying out a single CG iteration.

```
function [dw,vdw,P,g_dw,AN,r_M] ...
          = InvITERdw(Rfl,dw,vdw,P,g_1dw,AN_1,r_M,r_D,eta_M,eta_D,input)
global nDIM it
Ldv = cell(nDIM,input.NS); shape = cell(nDIM);
% determine conjugate gradient directions
 dummy = cell(1,nDIM);
 for n = 1:nDIM; dummy{n} = conj(Rfl{n}).*r_D{n}; end
 if input.Kirchhoff == 0; [Kdf] = AdjKOPdw(dummy,input); end
 if (input.Kirchhoff == 1) || (input.Kirchhoff == 2)
                   Kdf{1}= 0*dummy{1}; Kdf{2}= 0*dummy{2}; end
 [g_dw] = AdjDOPMdw(r_M,input);
     AN = 0;
 for n = 1:nDIM
   % set data gradients for small interface contrast to zero
   shape{n} = abs(Rfl{n}) > 0.10 * max(abs(Rfl{n}(:)));
   g_dw{n}  = shape{n} .* g_dw{n};
   g_dw{n}  = eta_M * g_dw{n} + eta_D * (r_D{n} - Kdf{n});
   AN = AN + norm(g_dw{n}(:))^2;
 end
 if it == 1
     for n = 1:nDIM; vdw{n} = g_dw{n};  end
 else
     BN = 0;
     for n = 1:nDIM; BN = BN + sum(g_dw{n}(:).* conj(g_1dw{n}(:))); end
     gamma = real((AN-BN)/AN_1);
     for n = 1:nDIM
       % set previous directions for small interface contrast to zero
       vdw{n} = g_dw{n} + gamma * shape{n} .* vdw{n};
     end
end
% determine step length alpha
  GMp   = DOPMdw(vdw,input);
  BN_M  = norm(GMp(:))^2;
  AN_D = 0;
  BN_D = 0;
  if (input.Kirchhoff == 0) || (input.Kirchhoff == 1)
     [Kdv] = KOPdw(vdw,input); end
  if input.Kirchhoff == 2;  Kdv{1}= 0*vdw{1};  Kdv{2}= 0*vdw{2}; end
  for n = 1:nDIM
    AN_D    = AN_D + sum(g_dw{n}(:).*conj(vdw{n}(:)));
   Ldv{n} = vdw{n} - Rfl{n} .* Kdv{n};
    BN_D  = BN_D + norm(Ldv{n}(:))^2;
  end
 alpha  = real(AN_D) / (eta_M*BN_M + eta_D*BN_D);
 % update contrast sources and AN
  for n = 1:nDIM
    dw{n} = dw{n} + alpha * vdw{n};
    P{n}  = P{n} + alpha * Kdv{n};
  end
 % update residual error in M
    r_M = r_M - alpha * GMp(:);
```

Matlab Listing II.68 The function `UpdateContrastdw` for updating the interface contrast $R^{(1)}$ and $R^{(2)}$.

```
function [Rfl,dr,eta_D,ErrorD] ...
                        = UpdateContrastdw(dw,P,P_inc,ErrorM,input)
global nDIM
dr  = cell(nDIM,input.NS);
% Determine mass density interface contrast
  INP1_dw1 = zeros(input.N1,input.N2);
  INP1_P1  = zeros(input.N1,input.N2);
  for q = 1 : input.NS
    INP1_dw1 = INP1_dw1 + conj(P{1,q}) .* dw{1,q};
    INP1_P1  = INP1_P1  + conj(P{1,q}) .*  P{1,q};
  end %q_loop
  Rfl{1} = INP1_dw1 ./ INP1_P1;
   shape = abs(Rfl{1}) > 0.10 * max(abs(Rfl{1}(:)));
  Rfl{1} = shape .* Rfl{1};                % Neglect small small variations

  INP2_dw2 = zeros(input.N1,input.N2);
  INP2_P2  = zeros(input.N1,input.N2);
  for q = 1 : input.NS
    INP2_dw2 = INP2_dw2 + conj(P{2,q}) .* dw{2,q};
    INP2_P2  = INP2_P2  + conj(P{2,q}) .* P{2,q};
  end %q_loop
  Rfl{2} = INP2_dw2 ./ INP2_P2;
   shape = abs(Rfl{2}) >  0.10 * max(abs(Rfl{2}(:)));
  Rfl{2} = shape .* Rfl{2};                % Neglect small small variations

% Compute residual error in object domain
  Norm_D = 0;    Norm = 0;
  for q = 1 : input.NS
    dr{1,q}  = Rfl{1} .*  P{1,q}   - dw{1,q};
    dr{2,q}  = Rfl{2} .*  P{2,q}   - dw{2,q};
    Norm_D = Norm_D + norm(dr{1,q}(:))^2 +  norm(dr{2,q}(:))^2 ;
    Norm   = Norm   + norm(Rfl{1}(:).*P_inc{1,q}(:))^2 ...
                    + norm(Rfl{2}(:).*P_inc{2,q}(:))^2;
  end %q_loop
  eta_D = 1 / Norm;
  ErrorD = eta_D * Norm_D;
  disp(['  ErrorM = ',num2str(ErrorM),'     ErrorD  = ',num2str(ErrorD)]);

% Compute the mass-density volume contrast from interface contrast
  CHI_rho = RfltoCHI(Rfl,input);

% Show intermediate pictures of reconstruction of compressibility contrast
  x1 = input.X1(:,1);    x2 = input.X2(1,:);
  set(figure(6),'Units','centimeters','Position',[5 1 27 12]);
  subplot(1,3,1); IMAGESC(x1+input.dx/2,x2,real(Rfl{1}));
                  title('\fontsize{13} R^{(1)}');
  subplot(1,3,2); IMAGESC(x1,x2+input.dx/2,real(Rfl{2}));
                  title('\fontsize{13} R^{(2)}');
  subplot(1,3,3); IMAGESC(x1,x2,real(CHI_rho));
                  title('\fontsize{13} 1-\rho_{sct}/ \rho_{0}'); pause(.1)
```

Matlab Listing II.69 The function `RfltoCHI` to convert the interface contrast to the mass-density distribution and to plot all contrast profiles.

```
function [CHI] = RfltoCHI(Rfl ,input)

N1 = input.N1 ;  N2 = input.N2 ;

% Determine the mass-density distribution
  factor1 = (1 + Rfl{1}) ./ (1 - Rfl{1});
% and integrate in  forward x_1 direction
  rhoF1 = ones(N1,N2);
  for i = 1 : N1-1
    rhoF1(i+1,:) = rhoF1(i,:) .* factor1(i,:);
  end
% and integrate in backward x_1 direction
  rhoB1 = ones(N1,N2);
  for i = 2 : N1
      I = N1-i+1;     rhoB1(I,:) = rhoB1(I+1,:) ./ factor1(I+1,:);
  end
  i = 2:N1-1; rhoB1(i,:) = rhoB1(i-1,:);              % with shift correction

% Determine the mass-density distribution
  factor2 = (1 + Rfl{2}) ./ (1 - Rfl{2});
% and integrate in forward x_2 direction
  rhoF2 = ones(N1,N2);
  for j = 1 : N2-1
    rhoF2(:,j+1) = rhoF2(:,j) .* factor2(:,j);
  end
% and integrate in backward x_2 direction
  rhoB2 = ones(N1,N2);
  for j = 2 : N2
      J = N2-j+1;     rhoB2(:,J) = rhoB2(:,J+1) ./ factor2(:,J+1);
  end
  j = 2:N2-1; rhoB2(:,j) = rhoB2(:,j-1);              % with shift correction

% determine mean value and mass-density contrast CHI
  CHI = 1- (rhoF1 + rhoB1 + rhoF2 + rhoB2)/4;
```

Then, the gradient $g_j^{(j,n)}$ of Eq. (5.59) becomes

$$g_j^{(j,n)} \approx \eta_{\mathrm{M}} \; 2\Delta x \partial_j^R G_{\mathrm{M}}^{\star}\{r_{\mathrm{M},q}\} + \eta_{\mathrm{D}}^{(n-1)} \partial r_q^{(j,n-1)} \; . \tag{5.80}$$

(c) If `input.Kirchhoff` = 2, a *a second approximation* is made in the forward operator by again neglecting the scattering operator that represents the wave-field interactions between adjacent interfaces. Because these interactions are often less than the noise and model errors in the measurement data, this type of approach is usually permitted. Then, the residuals in the measurement domain and the object domain become, cf. Eq. (5.51),

$$\begin{aligned} r_{\mathrm{M},q}^{(n)} &= p_{\mathrm{M},q}^{sct} - 2\Delta x \sum_k \partial_k G_{\mathrm{M}} \partial w_q^{(k,n)}, \\ \partial r_q^{(j,n)} &= R^{(j,n)} p_q^{inc(j)} - \partial w_q^{(j,n)} \; . \end{aligned} \tag{5.81}$$

Matlab Listing II.70 The function `saveRfl` to store the contrast profiles.

```
function [] = saveRfl(Rfl)
global it

if it == 0;   Rfl_0{1}   = real(Rfl{1});
              Rfl_0{2}   = real(Rfl{2});   save Contrast0   Rfl_0;   end

if it == 32;  Rfl_32{1}  = real(Rfl{1});
              Rfl_32{2}  = real(Rfl{2});   save Contrast32  Rfl_32;  end

if it == 64;  Rfl_64{1}  = real(Rfl{1});
              Rfl_64{2}  = real(Rfl{2});   save Contrast64  Rfl_64;  end

if it == 128; Rfl_128{1} = real(Rfl{1});
              Rfl_128{2} = real(Rfl{2});   save Contrast128 Rfl_128; end

if it == 256; Rfl_256{1} = real(Rfl{1});
              Rfl_256{2} = real(Rfl{2});   save Contrast256 Rfl_256; end
```

This greatly simplifies the inversion procedure. The computation of the object operator KOPdw is avoided and the updates for the total pressure wave fields are unnecessary. The total pressure wave fields do not change and remain the same, namely equal to the incident wave field.

(2) The heart of the CSI method is found in the function `ITERCGMDdw` of the Matlab Listing II.66. In this function, the maximum number of iterations is set to it = 256. Both the initial interface-contrast sources and the initial interface contrast are estimated with the help of the function `InitialEstimatedw` of Matlab Listing II.65. The interface-contrast sources are determined by back projection. Besides the data operator `DOPMdw` of Matlab Listing II.62, we also need its adjoint `AdjDOPMdw` of Matlab Listing II.63. The interface-contrast sources follow from minimization of the squared norm on $\mathbb{M}$, viz. ErrorM, after which the contrast follows from Eq. (5.68) with the help of function `UpdateContrastdw`. If `input.Kirchhoff` = 2, we neglect the interaction between the interfaces and exclude the operation Kdw = KOPdw. Further, the value of the squared norm on $\mathbb{D}$, viz. ErrorD, is computed. With these initial estimates, we employ the iterative CG method to determine the contrast sources that minimize ErrorM and ErrorD simultaneously.

The function `InvITERdw` of Matlab Listing II.67 carries out a single CG iteration. If `input.Kirchhoff` = 1 or `input.Kirchhoff` = 2, the adjoint operation Kdf = AdjKOPdw is excluded. If `input.Kirchhoff` = 2, the operation Kdv = KOPdw is excluded as well.

Next, the function `UpdateContrast` of Matlab Listing II.68 determines the updates of the interface contrast. After the contrast update of Eq. (5.68), the regularization is carried out by resetting small interface contrast variation to zero, cf. Eqs. (5.78) and (5.79). In Chapter 1, we have remarked that for the CG solution of the contrast source integral equation, it is not wise to update the contrast source at locations where the contrast is negligible small. Therefore, at the locations where the interface contrast $R^{(1)}$ and $R^{(2)}$ are less than 10% of their maximum value in the domain $\mathbb{D}$, we window the data gradient and the CG direction.

During each iteration, we also compute the mass-density distribution related to the inverted interface contrasts. Using Eq. (5.46), we can express the mass density ρ^{+} in terms of the mass density ρ^{-}. In the forward direction of $\boldsymbol{i}_j$, for $m = 1, 2, \dots, N_j - 1$, the mass density is obtained recursively as

$$\rho(\boldsymbol{x} + (m+1)\Delta x\, \boldsymbol{i}_j) = \frac{1 + R^{(j)}\left(\boldsymbol{x} + (m+\frac{1}{2})\Delta x\, \boldsymbol{i}_j\right)}{1 - R^{(j)}\left(\boldsymbol{x} + (m+\frac{1}{2})\Delta x \boldsymbol{i}_j\right)} \rho(\boldsymbol{x} + m\Delta x\, \boldsymbol{i}_j)\,. \tag{5.82}$$

We can also operate in the backward direction of $\boldsymbol{i}_j$, for $m = N_j - 1, N_j - 2, \ldots, 1$, viz.,

$$\rho(\boldsymbol{x} + m\Delta x\, \boldsymbol{i}_j) = \frac{1 - R^{(j)}\left(\boldsymbol{x} + (m+\frac{1}{2})\Delta x\, \boldsymbol{i}_j\right)}{1 + R^{(j)}\left(\boldsymbol{x} + (m+\frac{1}{2})\Delta x \boldsymbol{i}_j\right)} \rho(\boldsymbol{x} + (m+1)\Delta x\, \boldsymbol{i}_j)\,. \tag{5.83}$$

In principle, with $j = 1$ and 2, we get four different values for the mass density at each grid point. To reduce numerical errors, we use the mean value of these four results. This procedure is carried out by the function `RfltoCHI` of Matlab Listing II.69 and called by the Matlab function `UpdateContrast` of Matlab Listing II.68. During the iterations, each update of the interface contrast $R^{(1)}$ and $R^{(2)}$, and the mass-density contrast $\chi^\rho = 1 - \rho_{sct}/\rho_0$ are shown as an image plot on the computer screen. For later convenience, for specific numbers of iterations the interface contrast distributions are saved, cf. the script `saveRfl` of Matlab Listing II.70.

(3) With the help of the interface-contrast sources solved from the present inversion method, we compute again the scattered field at the receiver locations, see function `DOPMdw` of Matlab Listing II.62, and we compare these data with the analytical data used in step (1). By plotting the differences and computing the error being the norm of these differences, we obtain an indication how good the data are explained by the inversion scheme at hand.

In Figuress 5.12 and 5.13, we show the reconstructed interface-contrast distributions (*left and middle figures*) together with the mass-density contrast (*right figures*). These figures are made with the help of the Matlab script `loadRfl` of Matlab Listing II.71 and the function `Convert` of Matlab Listing II.72. We observe some blocky noise in the reconstructed interface contrast distributions. During the transition process from interface-contrast to volume contrast, these noisy spots are converted into straight lines in the Cartesian directions. To suppress this noise in the mass-density contrast a little bit, we use a simple two-points moving-average method. The filtered interface source distributions are then converted to a volume contrast, using the function `RfltoCHI`.

The pictures on the right-hand sides are determined from the reconstructed interface contrasts with the help of function `RfltoCHI` of Matlab Listing II.69, except for the the right-hand side of the pictures indicated by `Exact`. It presents the discrete mass-density distribution of the nominal mass-density contrast. The interface-contrast distributions $R^{(1)}$ and $R^{(2)}$ on the left and middle pictures are determined from this mass-density model. From the iteration results of Figure 5.12, we see very strong oscillation contrast profiles, due to band limitation. Beyond 64 iterations (see Figure 5.13) we observe that the regularization, by enforcing small variations in the interface contrast to zero, becomes effective. After 256 iterations, no further improvement of the resolution in the interface contrast is feasible. The width of the interfaces that have been reconstructed can be measured directly from the figures presented. In the picture labeled as it = 256, we observe a resolution of approximately two pixel dimensions, i.e. $2\Delta x = 6$ m. At the present operating wavelength $\lambda = 30$ m, we note that a resolution of approximately $1/5\ \lambda$ has been achieved.

To check whether this is the correct conclusion, we refine the grid of our discrete model by a factor of two. The inversion results for $\Delta x = 1.5$ m are presented in Figures 5.14 and 5.15. We immediately

Matlab Listing II.71 The script `loadRfl` to load the contrast sources from storage and present them as image plots.

```
close all
input = initAC;

x1 = input.X1(:,1);    x2 = input.X2(1,:);    dx = input.dx;

load Contrast0   Rfl_0;    Convert(Rfl_0,0,input) ;
load Contrast32  Rfl_32;   Convert(Rfl_32,32,input) ;
load Contrast64  Rfl_64;   Convert(Rfl_64,64,input) ;
load Contrast128 Rfl_128;  Convert(Rfl_128,128,input) ;
load Contrast256 Rfl_256;  Convert(Rfl_256,256,input) ;

% Plot exact case
set(figure,'Units','centimeters','Position',[0 2 17 9]); snapnow;

sp = subplot(1,3,1);
     IMAGESC(x1+dx/2,x2,real(input.Rfl{1}));   xlabel(''); ylabel('');
     title('\fontsize{12} R^{(1)}');                colorbar('hor');
     text(-100, -100,'Exact ',      'FontSize',15);
     pos = get(sp,'Position');    pos = pos + [0.02 0 0 0];
     set(sp,'Position',pos);

sp = subplot(1,3,2);
     IMAGESC(x1,x2+dx/2,real(input.Rfl{2}));   xlabel(''); ylabel('');
     title('\fontsize{12} R^{(2)}');                colorbar('hor');
     pos = get(sp,'Position');    pos = pos + [0 0 0 0];
     set(sp,'Position',pos);

sp = subplot(1,3,3);
     IMAGESC(x1,x2,real(input.CHI_rho));        xlabel(''); ylabel('');
     title('\fontsize{13} 1-\rho_{sct}/ \rho_{0}');    colorbar('hor');
     pos = get(sp,'Position');    pos = pos - [0.02 0 0 0];
     set(sp,'Position',pos);
```

see a clearer picture of the band limitation, and we observe that the final reconstructed interface contrast has a resolution between 3 and 4 pixels. This means that our interface imaging method has the potential to create images with a resolution less than $1/5\ \lambda$. We also increased the frequency f to 100 Hz. For $\Delta x = 1.5$ m, we observed a resolution of $0.1 - 0.2\lambda$.

Finally, in Figures 5.16 and 5.17, we show the differences between the analytical data amplitudes and the inverted data amplitudes obtained from the contrast sources reconstructed after 256 iterations. On the left-hand sides of these figures, we present the amplitudes of the "exact" data using the Bessel function series, while on the right-hand sides, we show the differences between the inverted data and the exact data. Note that the global error between the exact data and the integral-equation data amounts to 19%, while the global error between the analytical data and the inverted data amounts to 2.7% and 13% for noise-free data and noisy-data, respectively. Obviously, for noise-free data, the inverted model explains the analytical data better than our discretized model used in the forward calculations.

Matlab Listing II.72 The function `convert` to convert the interface contrast to thethe mass-density distribution and to plot all contrast profiles

```
function [] = Convert(Rfl, it, input)
x1 = input.X1(:,1);    x2 = input.X2(1,:);      dx = input.dx;

MA = ones(1,2)/2;                   % apply a two-points moving average to Rfl
for j = 1:input.N2
   Rfl{1}(:,j)  = conv( Rfl{1}(1:input.N1,j),MA,'same');
   Rfl{2}(:,j)  = conv( Rfl{2}(1:input.N1,j),MA,'same');
end
for i = 1:input.N1
    Rfl{1}(i,:)  = conv( Rfl{1}(i,1:input.N2),MA,'same');
    Rfl{2}(i,:)  = conv( Rfl{2}(i,1:input.N2),MA,'same');
end

CHI_rho = RfltoCHI(Rfl,input); % convert to mass-density volume contrast

set(figure,'Units','centimeters','Position',[0 2 17 9]); snapnow;

sp = subplot(1,3,1);
     IMAGESC(x1+dx/2,x2,Rfl{1}); xlabel(''), ylabel('');
     title('\fontsize{12} R^{(1)}'); colorbar('hor');
     text(-70, -100,num2str(it),'FontSize',15);
     text(-100, -100,'it = ',    'FontSize',15);
     pos = get(sp,'Position');    pos = pos + [0.02 0 0 0];
     set(sp,'Position',pos);

sp = subplot(1,3,2);
     IMAGESC(x1,x2+dx/2,Rfl{2});   xlabel(''), ylabel('');
     title('\fontsize{12} R^{(2)}');   colorbar('hor');
     pos = get(sp,'Position');    pos = pos + [0 0 0 0];
     set(sp,'Position',pos);

sp = subplot(1,3,3);
     IMAGESC(x1,x2,real(CHI_rho));          xlabel(''); ylabel('');
     title('\fontsize{13} 1-\rho_{sct}/ \rho_{0}');  colorbar('hor');
     pos = get(sp,'Position');    pos = pos - [0.02 0 0 0];
     set(sp,'Position',pos);
```

5.5.6 Kirchhoff Type of Approximations

We now examine the usefulness of the Kirchhoff approximation. After 256 iterations, we present the various reconstruction results in Figures 5.18 and 5.19.

- For $\Delta x = 3$ m, $f = 50$ Hz and **noise-free** data, the images are shown in Figure 5.18a. The normalized global error in the inverted data is 3.1%.
- When we ignore the interaction between the interfaces, by excluding the adjoint operation Kdf = AdjKOPdw (`input.Kirchhoff` = 1), we get the images of Figure 5.18b. The normalized global error in the inverted data is 2.0%.
- When we also ignore the interaction between the interfaces by excluding the operation Kdw = KOPdw (`input.Kirchhoff` = 2), we obtain the images of Figure 5.18c. Then the normalized global error in the inverted data is 1.7%.

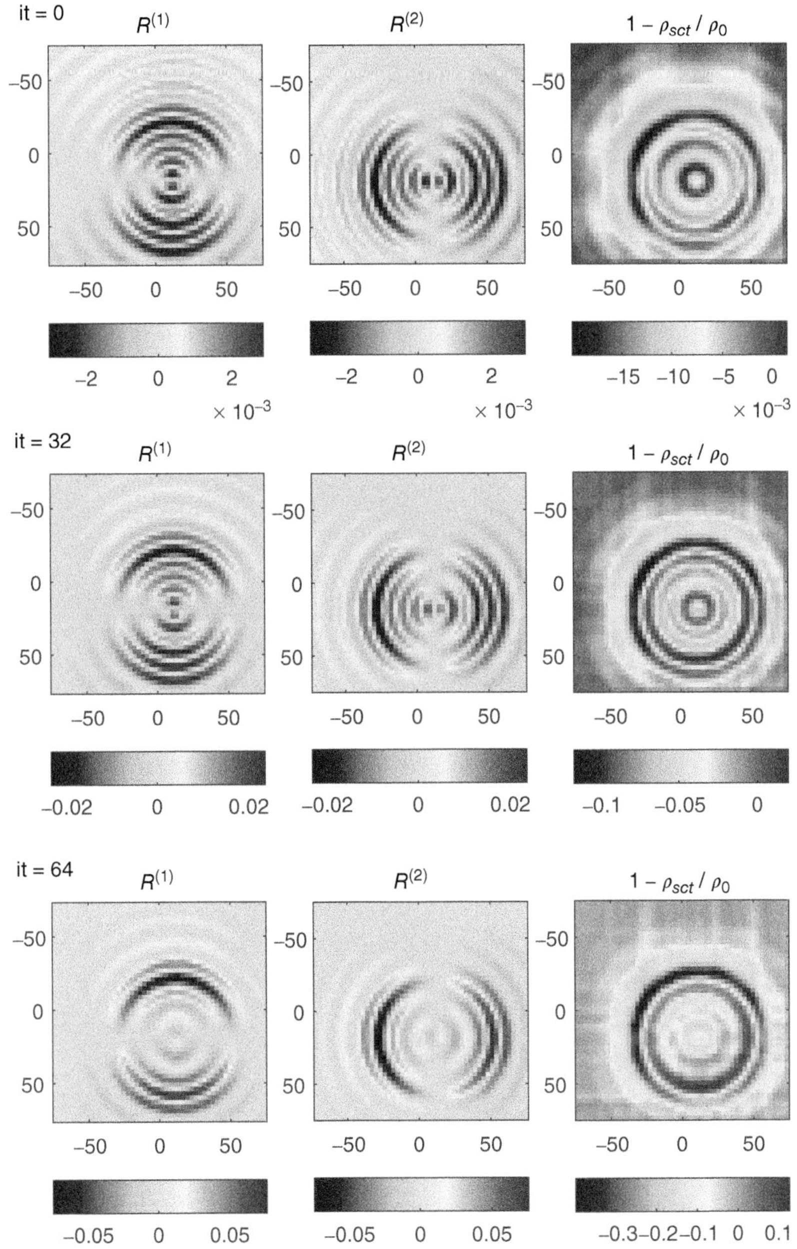

Figure 5.12 MICSI method: image plots of the reconstruction of the interface contrast for the initial estimate it = 0, iteration number it = 32, and it = 64, for **noise-free** data and nominal values $\rho_0 = 3000$ kg/m^3 and $\hat{\rho}_{sct} = 1500$ kg/m^3.

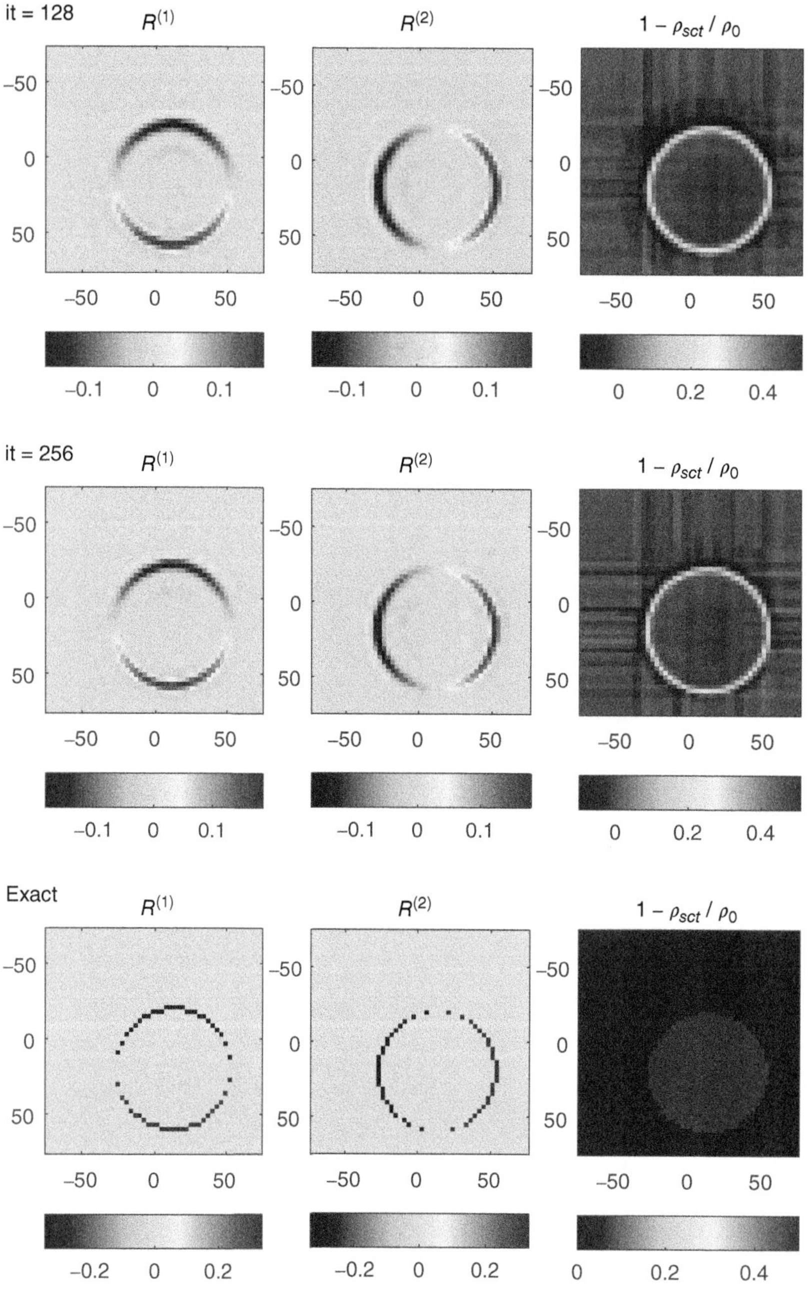

Figure 5.13 MICSI method: image plots of the reconstruction of the interface contrast at it = 128 and it = 256, for **noise-free** data and nominal values $\rho_0 = 3000$ kg/m^3 and $\hat{\rho}_{sct} = 1500$ kg/m^3. At the bottom pictures, we show the "exact" profiles.

- When we do not use the Kirchhoff type of approximations, but we use **noisy** data, the images are presented in Figure 5.19a. The normalized global error in the inverted data is 13%.
- When we return to **noise-free** data, we increase the frequency f from 50 to 100 Hz and ignore the interaction between the interfaces in both the adjoint and forward operator

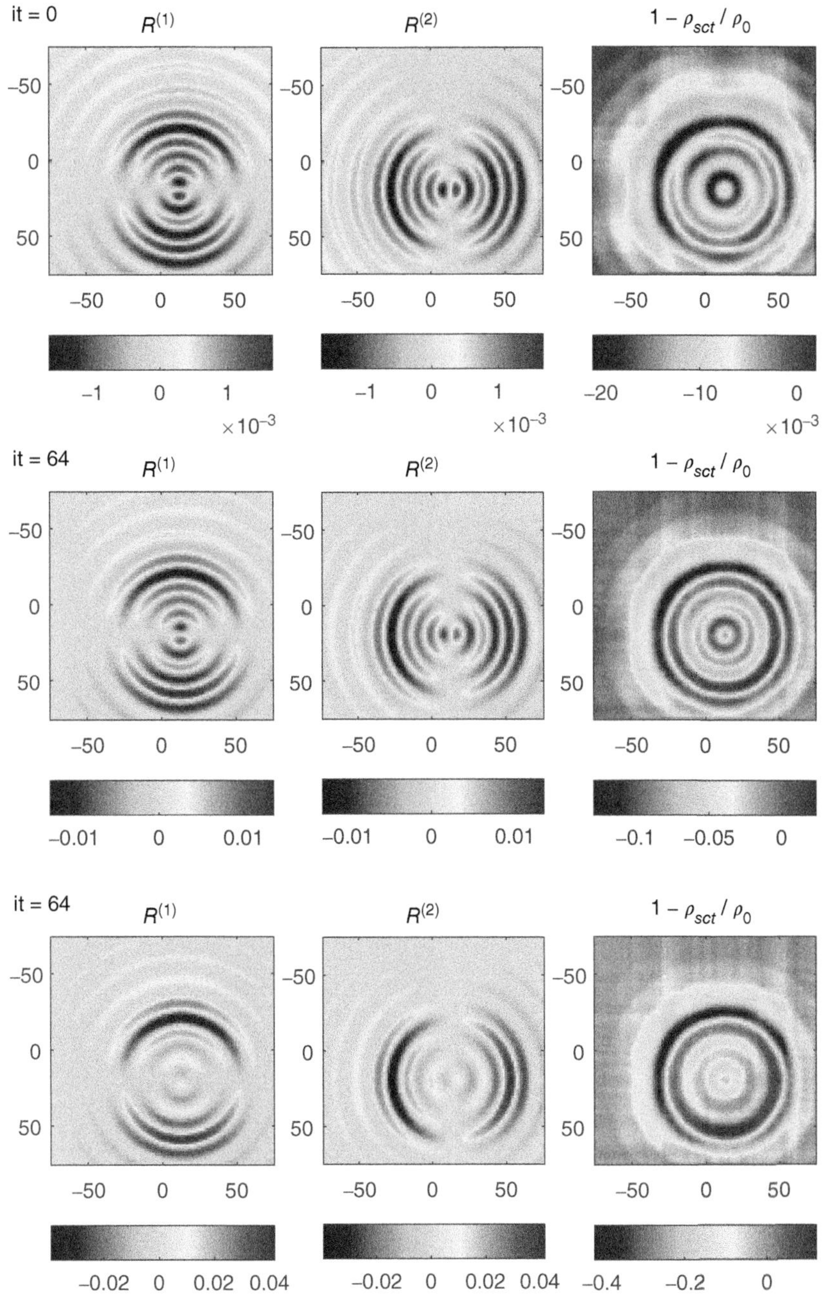

Figure 5.14 Similar to Figure 5.12, but the grid size is reduced from 3 to 1.5 m.

(`input.Kirchhoff` = 2), we obtain the images of Figure 5.19b. Then the normalized global error in the inverted data is 3.8%.

- When we subsequently reduce the grid size from 3 to 1.5 m and keep the same parameters of the previous case, we obtain the image of Figure 5.19c. Then the normalized global error in the inverted data is 2.1%.

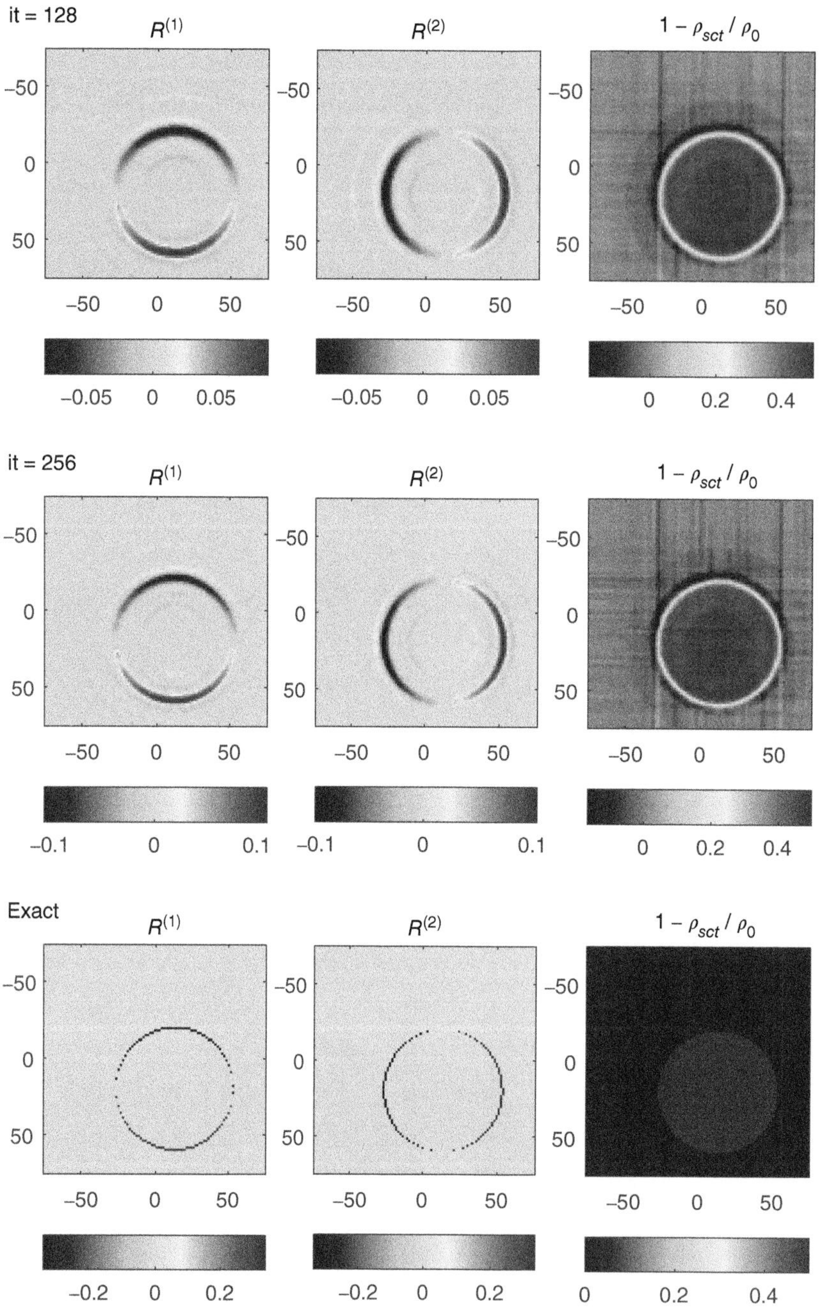

Figure 5.15 Similar to Figure 5.13, but the grid size is reduced from 3 to 1.5 m.

In summary, we first note that the data error is not very sensitive to our inversion model. Although the usage of noisy data results in a data error of 13%, our interface reconstruction appears to be very robust.

Second, as expected, neglecting multiple scattering between the interfaces in our adjoint operator AdjKOPdw is allowed. This simplifies the inversion method.

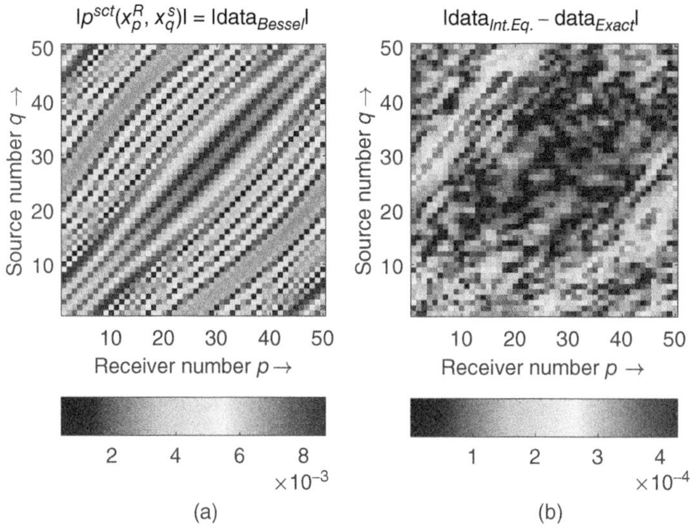

Figure 5.16 MICSI method: plots of the inverted **noise-free** data as function of the source and receiver numbers. The normalized global error in the inverted data amounts to 2.7%.

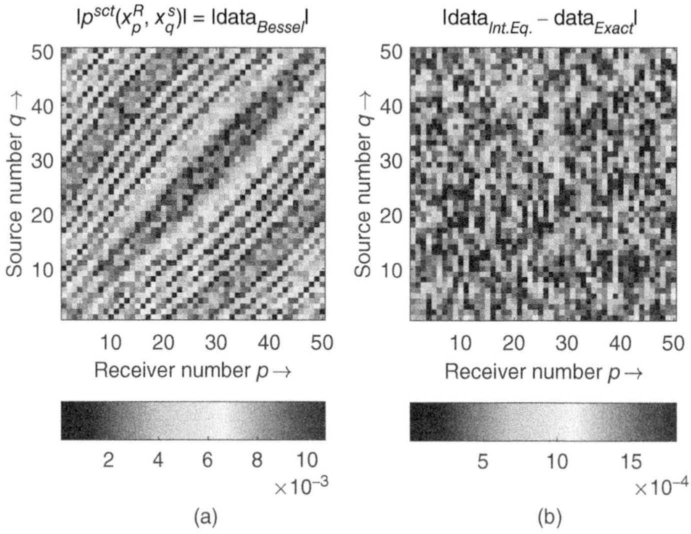

Figure 5.17 MICSI method: plots of the inverted **noisy** data as function of the source and receiver numbers. The normalized global error in the inverted data amounts to 13%.

Third, the neglection of multiple scattering between the interfaces in our forward operator KOPdw affects the reconstruction of the mass density in the interior of the object. Some incorrect interface residues remain visible in the images. As a result, the internal mass distribution shows some nonzero reconstruction results. However, the inversion procedure simplifies considerably. This observation led to the publication of van der Neut et al. [57], which describes a time-domain interface-contrast imaging procedure, assuming vanishing wave-speed contrast.

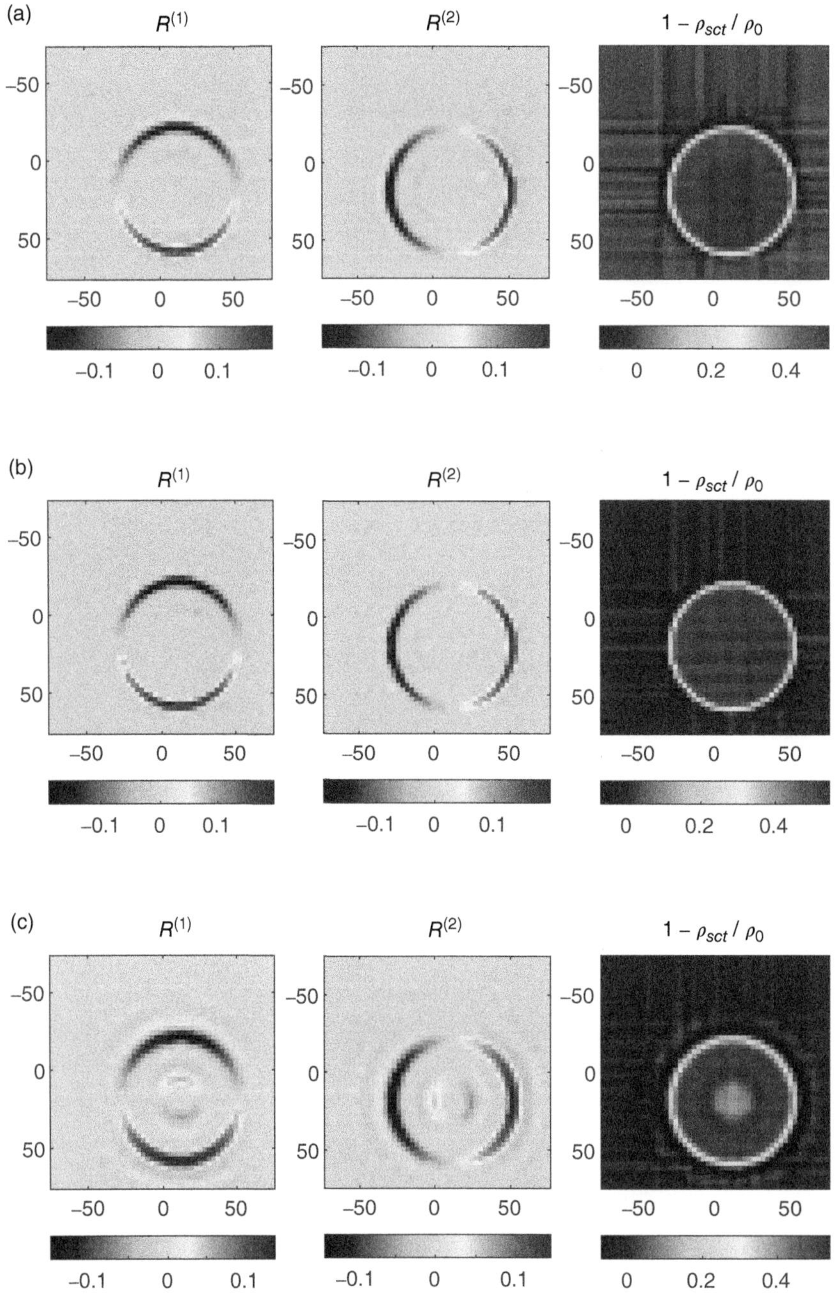

Figure 5.18 MICSI method: image plots of the reconstruction of the interface contrast after 256 iterations for **noise-free** data: (a) no approximations, cf. Figure 5.13; (b) without scattering operator in adjoint; (c) without scattering operator in both adjoint and forward operator.

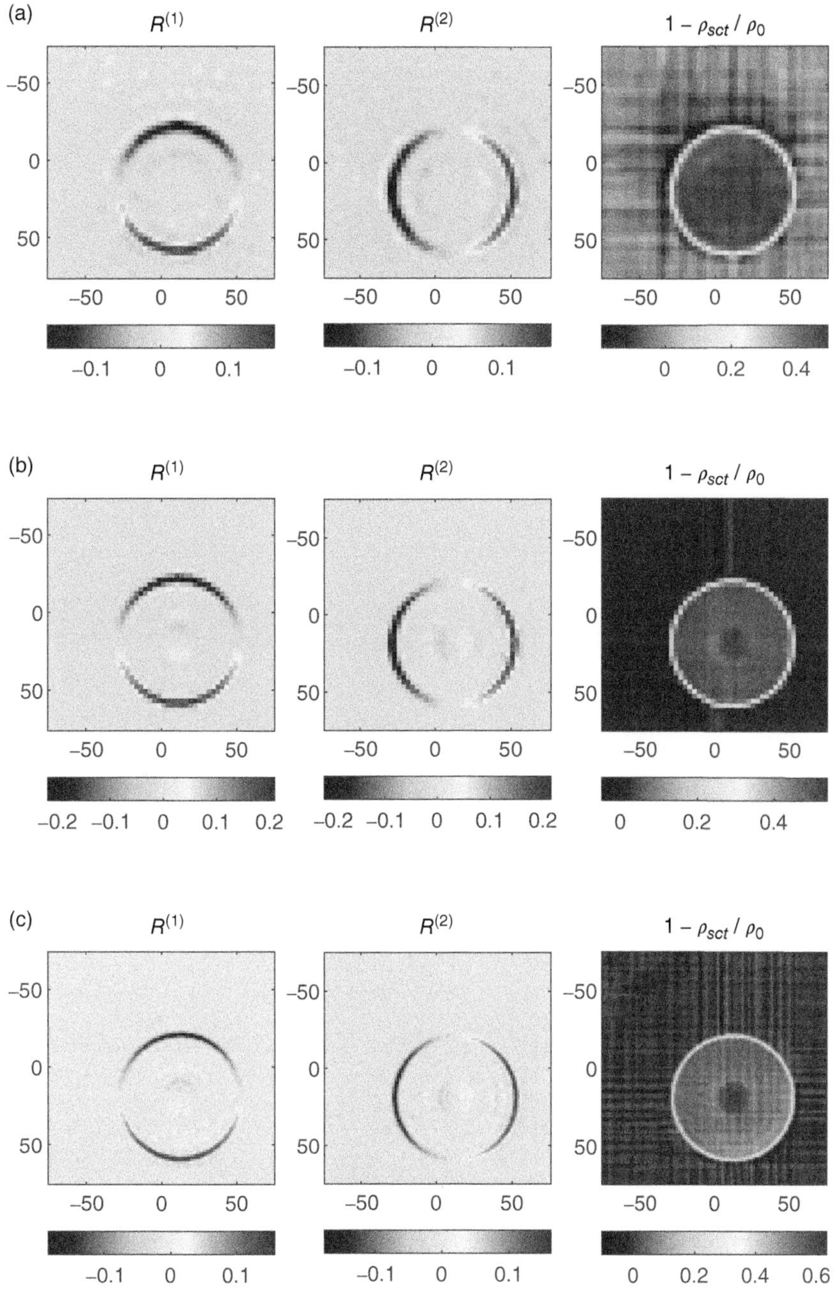

Figure 5.19 MICSI merhod: image plots of the reconstruction of the interface contrast after 256 iterations: (a) Similar to Figure 5.18a, but now for **noisy** data; (b) Similar to Figure 5.18c, but now for $f = 100$ Hz, (c) Similar to (b) of this figure, but now for $\Delta x = 1.5$ m.

We remark that the general case of the reconstruction of contrast in both wave speed and mass density does not lead to satisfactory inversion results, if we only use single-frequency pressure wave-field data In this case, one must use either pressure wave-field data for more frequencies or record particle-velocity wave-field data as well.

6

Electromagnetic Wave Inversion

In this chapter, we now consider the problem of reconstructing the electromagnetic parameters of a bounded object immersed in a known background medium from a knowledge of how the object scatters known incident radiation.

6.1 Notation

Let $\mathbb{D}_{sct}$ denote a bounded, not necessarily connected, scattering object whose location, the permittivity contrast $\hat{\chi}^{\varepsilon} = 1 - (\hat{\varepsilon}_{sct}/\varepsilon_0)$ and the permeability contrast $\hat{\chi}^{\mu} = 1 - (\hat{\mu}_{sct}/\mu_0)$, are unknown but are known to lie within another, larger, bounded simply connected domain $\mathbb{D}$. Observe that if $\boldsymbol{x}$ is not in $\mathbb{D}_{sct}$, then $\hat{\chi}^{\varepsilon}$ and $\hat{\chi}^{\mu}$ vanish, but if the location of $\mathbb{D}_{sct}$ is unknown, then it is not known *a priori*, where these contrast quantities vanish. However, with the assumption that $\mathbb{D}_{sct} \subset \mathbb{D}$, it is known that $\hat{\chi}^{\varepsilon}$ and $\hat{\chi}^{\mu}$ vanish for $\boldsymbol{x}$ outside $\mathbb{D}$. Let $\mathbb{M}$ denote a domain with a discrete collection of receiver points outside of $\mathbb{D}$, then the scattered electric and magnetic field vectors are defined as $\hat{\boldsymbol{E}}_{\mathrm{M}}^{sct}$ and $Z_0\hat{\boldsymbol{H}}_{\mathrm{M}}^{sct}$ (Figure 6.1). As a model for this field, we use the contrast-source integral representation, cf. Eqs. (3.72) and (3.73),

$$\hat{\boldsymbol{E}}_{\mathrm{M}}^{sct}(\boldsymbol{x}_p^R, \boldsymbol{x}_q^S) = \int_{\boldsymbol{x}' \in \mathbb{D}} \big[(-\hat{\gamma}_0^2 + \boldsymbol{\nabla}\boldsymbol{\nabla}\cdot)\hat{G}(\boldsymbol{x}_p^R - \boldsymbol{x}')\, \hat{\boldsymbol{w}}^E(\boldsymbol{x}', \boldsymbol{x}_q^S) \\ -\hat{\gamma}_0 \boldsymbol{\nabla} \times \hat{G}(\boldsymbol{x} - \boldsymbol{x}')\, \hat{\boldsymbol{w}}^H(\boldsymbol{x}', \boldsymbol{x}_q^S)\big]\, \mathrm{d}V, \tag{6.1}$$

and

$$Z_0\hat{\boldsymbol{H}}_{\mathrm{M}}^{sct}(\boldsymbol{x}_p^R, \boldsymbol{x}_q^S) = \int_{\boldsymbol{x}' \in \mathbb{D}} \big[\hat{\gamma}_0 \boldsymbol{\nabla} \times \hat{G}(\boldsymbol{x}_p^R - \boldsymbol{x}')\, \hat{\boldsymbol{w}}^E(\boldsymbol{x}', \boldsymbol{x}_q^S) \\ +(-\hat{\gamma}_0^2 + \boldsymbol{\nabla}\boldsymbol{\nabla}\cdot)\hat{G}(\boldsymbol{x} - \boldsymbol{x}')\, \hat{\boldsymbol{w}}^H(\boldsymbol{x}', \boldsymbol{x}_q^S)\big]\, \mathrm{d}V, \tag{6.2}$$

for $(\boldsymbol{x}_p^R, \boldsymbol{x}_q^S) \in \mathbb{M}$. Here, the set of $\boldsymbol{x}_p^R$, $p = 1, 2, \ldots$ denotes the collection of receiver points in the domain $\mathbb{M}$ and the set of $\boldsymbol{x}_q^S$, $q = 1, 2, \ldots$ denotes the collection of source points in the domain $\mathbb{M}$. The electromagnetic contrast sources $\boldsymbol{w}^E$ and $\boldsymbol{w}^H$ are given by

$$\hat{\boldsymbol{w}}^E(\boldsymbol{x}, \boldsymbol{x}_q^S) = \hat{\chi}^{\varepsilon}(\boldsymbol{x})\, \hat{\boldsymbol{H}}(\boldsymbol{x}, \boldsymbol{x}_q^S) \tag{6.3}$$

and

$$\hat{\boldsymbol{w}}^H(\boldsymbol{x}, \boldsymbol{x}_q^S) = \hat{\chi}^{\mu}(\boldsymbol{x})\, Z_0\hat{\boldsymbol{H}}(\boldsymbol{x}, \boldsymbol{x}_q^S). \tag{6.4}$$

Equations (6.1) and (6.2) hold only if there is no noise or error in the measurements. However, error-free data are extremely unlikely, and we do not assume that this equation holds exactly.

Forward and Inverse Scattering Algorithms based on Contrast Source Integral Equations, First Edition. Peter M. van den Berg.

Companion website: www.wiley.com/go/vandenBerg/ScatteringAlgorithms

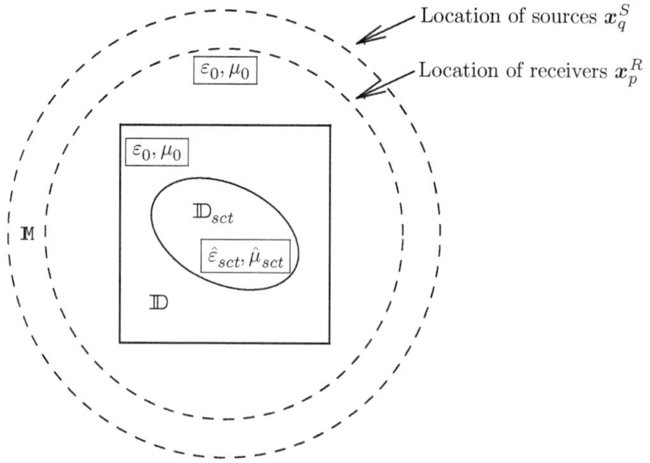

Figure 6.1 The inverse scattering configuration with the measurement domain $\mathbb{M}$ and the object domain $\mathbb{D}$. The measurement domain $\mathbb{M} \subset \mathbb{D}$. The scattering domain $\mathbb{D}_{sct} \subset \mathbb{D}$. In our numerical examples, the sources are distributed over the dashed outer circle and the receivers are distributed over the dashed inner circle.

If $\hat{\boldsymbol{E}}^{inc}(\boldsymbol{x}, \boldsymbol{x}_q^S)$ and $\hat{\boldsymbol{H}}^{inc}(\boldsymbol{x}, \boldsymbol{x}_q^S)$ denote the electromagnetic field strengths of an incident wave with propagation coefficient $\hat{\gamma}_0$ and source point $\boldsymbol{x}_q^S$, then the total electromagnetic field components in $\mathbb{D}$ satisfy the coupled set of contrast-source integral equations, cf. Eqs. (3.74) and (3.75),

$$\begin{aligned} \chi^{\varepsilon}(\boldsymbol{x})\hat{\boldsymbol{E}}^{inc}(\boldsymbol{x}, \boldsymbol{x}_q^S) &= \hat{\boldsymbol{w}}^E(\boldsymbol{x}, \boldsymbol{x}_q^S) \\ &\quad -\hat{\chi}^{\varepsilon}(\boldsymbol{x}) \Bigg[(-\hat{\gamma}_0^2 + \boldsymbol{\nabla}\boldsymbol{\nabla}\cdot) \int_{\boldsymbol{x}' \in \mathbb{D}} \hat{G}(\boldsymbol{x} - \boldsymbol{x}')\, \hat{\boldsymbol{w}}^E(\boldsymbol{x}', \boldsymbol{x}_q^S)\, \mathrm{d}V \\ &\quad - \hat{\gamma}_0 \boldsymbol{\nabla} \times \int_{\boldsymbol{x}' \in \mathbb{D}} \hat{G}(\boldsymbol{x} - \boldsymbol{x}')\, \hat{\boldsymbol{w}}^H(\boldsymbol{x}', \boldsymbol{x}_q^S)\, \mathrm{d}V \Bigg], \end{aligned} \tag{6.5}$$

$$\begin{aligned} \hat{\chi}^{\mu}(\boldsymbol{x})Z_0\hat{\boldsymbol{H}}^{inc}(\boldsymbol{x}, \boldsymbol{x}_q^S) &= \hat{\boldsymbol{w}}^H(\boldsymbol{x}, \boldsymbol{x}_q^S) \\ &\quad -\hat{\chi}^{\mu}(\boldsymbol{x}) \Bigg[\hat{\gamma}_0 \boldsymbol{\nabla} \times \int_{\boldsymbol{x}' \in \mathbb{D}} \hat{G}(\boldsymbol{x} - \boldsymbol{x}')\, \hat{\boldsymbol{w}}^E(\boldsymbol{x}', \boldsymbol{x}_q^S)\, \mathrm{d}V \\ &\quad +(-\hat{\gamma}_0^2 + \boldsymbol{\nabla}\boldsymbol{\nabla}\cdot) \int_{\boldsymbol{x}' \in \mathbb{D}} \hat{G}(\boldsymbol{x} - \boldsymbol{x}')\, \hat{\boldsymbol{w}}^H(\boldsymbol{x}', \boldsymbol{x}_q^S)\, \mathrm{d}V \Bigg], \end{aligned} \tag{6.6}$$

for $\boldsymbol{x} \in \mathbb{D}$ and $\boldsymbol{x}_q^S \in \mathbb{M}$.

Before we discuss the inversion of both the permittivity contrast, we consider the special case for zero permeability contrast.

6.1.1 Permittivity Contrast Only

In the case that the permeability contrast vanishes, the scattered field at the receiver points is given by

$$\hat{\boldsymbol{E}}_{\mathbb{M}}^{sct}(\boldsymbol{x}_p^R, \boldsymbol{x}_q^S) = \int_{\boldsymbol{x}' \in \mathbb{D}} (-\hat{\gamma}_0^2 + \boldsymbol{\nabla}\boldsymbol{\nabla}\cdot)\hat{G}(\boldsymbol{x}_p^R - \boldsymbol{x}')\, \hat{\boldsymbol{w}}^E(\boldsymbol{x}', \boldsymbol{x}_q^S)\, \mathrm{d}V \tag{6.7}$$

and

$$Z_0\hat{\boldsymbol{H}}_{\mathbb{M}}^{sct}(\boldsymbol{x}_p^R, \boldsymbol{x}_q^S) = \int_{\boldsymbol{x}' \in \mathbb{D}} \hat{\gamma}_0 \boldsymbol{\nabla} \times \hat{G}(\boldsymbol{x}_p^R - \boldsymbol{x}')\, \hat{\boldsymbol{w}}^E(\boldsymbol{x}', \boldsymbol{x}_q^S)\, \mathrm{d}V, \tag{6.8}$$

for $\boldsymbol{x}_p^R \in \mathbb{M}$, while the contrast-source integral equation becomes

$$\chi^{\varepsilon}(\boldsymbol{x})\hat{\boldsymbol{E}}^{inc}(\boldsymbol{x},\boldsymbol{x}_q^S) = \hat{\boldsymbol{w}}^E(\boldsymbol{x},\boldsymbol{x}_q^S) - \hat{\chi}^{\varepsilon}(\boldsymbol{x})(-\hat{\gamma}_0^2 + \boldsymbol{\nabla}\boldsymbol{\nabla}\cdot)\int_{\boldsymbol{x}'\in\mathbb{D}} \hat{G}(\boldsymbol{x}-\boldsymbol{x}')\,\hat{\boldsymbol{w}}^E(\boldsymbol{x}',\boldsymbol{x}_q^S)\,\mathrm{d}V, \tag{6.9}$$

for $\boldsymbol{x} \in \mathbb{D}$ and $\boldsymbol{x}_q^S \in \mathbb{M}$.

In the 2D case, Eq. (3.49) presents the incident wave field for a radially oriented electric dipole in the vertical direction. In general, for an arbitrary direction, the Cartesian components of the 2D electric-field strengths are

$$E_1^{inc}(\boldsymbol{x}_T|\boldsymbol{x}_{T,q}^S) = \frac{Z_0\,M}{\gamma_0}\left[(-\gamma_0^2+\partial_1\partial_1)G(\boldsymbol{x}_T-\boldsymbol{x}_{T,q}^S)\frac{x_{1,q}^S}{|\boldsymbol{x}_{T,q}^S|} + \partial_1\partial_2 G(\boldsymbol{x}_T-\boldsymbol{x}_{T,q}^S)\frac{x_{2,q}^S}{|\boldsymbol{x}_{T,q}^S|}\right],$$

$$E_2^{inc}(\boldsymbol{x}_T|\boldsymbol{x}_{T,q}^S) = \frac{Z_0\,M}{\gamma_0}\left[(-\gamma_0^2+\partial_2\partial_2)G(\boldsymbol{x}_T-\boldsymbol{x}_{T,q}^S)\frac{x_{2,q}^S}{|\boldsymbol{x}_{T,q}^S|} + \partial_2\partial_1 G(\boldsymbol{x}_T-\boldsymbol{x}_{T,q}^S)\frac{x_{1,q}^S}{|\boldsymbol{x}_{T,q}^S|}\right]. \tag{6.10}$$

This expression is used in the numerical example to be discussed.

6.2 Synthetic Data for Zero Permeability Contrast

In this section, we discuss the computation of synthetic data, which will replace the measured data for our inversion examples. Since in this part of the book we only present 2D computer algorithms, we consider the circular cylinder. The forward electromagnetic model has been discussed in Chapter 3.

Note that we do not consider the scattering by an off-centered circular cylinder. In Chapter 3, the analytic solution for the scattered electromagnetic field is only evaluated, when the dipole sources are directed in the radial direction of the circular cylinder. Therefore, we restrict our scattering object to a circular cylinder with center at the origin of our coordinate system. We first compare the two scattered-field data sets, obtained either analytically using the Bessel-series solution or numerically using the integral-equation solution.

The electromagnetic wave-field parameters are initialized by the `initEM` function (Matlab Listing II.73). The global parameter `nDIM` assigns the 2D space to be used in our inversion codes. Subsequently, the wave speed c_0 of the embedding is 3×10^8 m/s and the relative permittivity of the scattering object is taken as $\hat{\varepsilon}_{sct}/\varepsilon_0 = 1.55$. The value of the relative permeability is chosen to be one. The frequency of operation is 10 MHz. The wavelength in free space amounts to 30 m. Then, our chosen value for the relative permittivity is about the maximum value, for which our inversion method let us say results in a convergent process.

As output, the `initEM` function displays the operating wavelength. This facilitates the user to check the value $\lambda/\Delta x$. Although in the forward case, often a discretization of 15 samples per wavelength, leading to an error of the order of a few percents, in the inversion case we choose a discretization of 10 samples of wavelength. In this way, the numerical data differs substantially from the analytical data. Hence, this difference may be seen as model noise. In our case, this error is of the order of 10%. Further, the `init` function also calls the functions `initSourceReceiver`, `initEMGrid`, `initFFTGreen`, and `initEMContrast`, but now only in 2D. They initialize the locations of the sources and the receivers, the numerical grid, and the contrast distribution $\hat{\chi}^{\varepsilon}$, respectively. Next, the error criterion is given. We use a stronger one than the error criterion used in the first part of this book, because we want to be sure that the difference between the analytical data and the numerical data is not caused by a convergence problem of the integral equation.

Matlab Listing II.73 The function `initEM` to initialize the 2D electromagnetic wave-field parameters.

```
function input = initEM()
% Time factor = exp(-iwt)
% Spatial units is in m
% Source wavelet M Z_0 / gamma_0  = 1   (Z_0 M = gamma_0)

global nDIM;  nDIM = 2;              % set dimension of space
if nDIM ==1;  disp(('')nDIM should be either 2 or 3('')); return; end
input.c_0     = 3e8;                 % wave speed in embedding
input.eps_sct = 1.55;                % relative permittivity (eps_sct/eps_0)
input.mu_sct  = 1.000006;            % relative permeability (mu_sct/mu_0)

f             = 10e6;                 % temporal frequency
wavelength    = input.c_0 / f;        % wavelength
s             = 1e-16 - 1i*2*pi*f;    % LaPlace parameter
input.gamma_0 = s/input.c_0;          % propagation coefficient
disp([('')wavelength = ('') num2str(wavelength)]);

% add input data to structure array ('')input('')
  input = initSourceReceiver(input);  % add location of source/receiver

  input = initEMgrid(input);          % add grid in either 2D or 3D

  input = initFFTGreen(input);        % compute FFT of Green function

  input = initEMcontrast(input);      % add contrast distribution

  input.Errcri = 1e-5;

  input.Noise = 0;

  if input.Noise == 1 % generate random numbers that are repeatable
    sigma  = 0.2;        % sigma = standard deviation and mu =0
    randn(('')state(''),1);  ReRand = randn(input.NR,input.NS) * sigma;
    randn(('')state(''),2);  ImRand = randn(input.NR,input.NS) * sigma;
    input.Rand = ReRand + 1i*ImRand;
  end
```

Finally, the `initEM` function offers the option to add noise to the analytical data to increase the noise level between the analytical and the numerical data. We use the Matlab random number generator `randn` to generate two repeatable arrays with a size of input.NR by input.NS, consisting of normally distributed random numbers with standard deviation $\sigma = 0.2$ and zero mean. These two different arrays are used to construct a single complex-valued array input.Rand. In the function `initSourceReceiver` (Matlab Listing II.2), we have 50 sources located on a circle with radius of 170 m around the scattering object, while 50 receivers are located on a circle with radius of 150 m around the scattering object. The function `initEMGrid` is similar to the one of Matlab Listing II.3. It defines a rectangular grid of N_1 by N_2 square subdomains with equal area of $(\Delta x)^2$,

Matlab Listing II.74 The `initEMgrid` script to initialize the grid parameters.

```
function input = initEMgrid(input)

   input.N1 = 50;                        % number of samples in x_1
   input.N2 = 50;                        % number of samples in x_2
   input.dx = 3;                         % with meshsize dx

   x1 = -(input.N1+1)*input.dx/2 + (1:input.N1)*input.dx;
   x2 = -(input.N2+1)*input.dx/2 + (1:input.N2)*input.dx;
   [input.X1,input.X2] = ndgrid(x1,x2);
```

Matlab Listing II.75 The `initEMcontrast` function to initialize the electromagnetic parameters.

```
function [input] = initEMcontrast(input)

input.a    = 40;        % radius circle cylinder
         R = sqrt(input.X1.^2 + input.X2.^2);

input.CHI_eps = (1-input.eps_sct) * (R < input.a);
input.CHI_mu  = (1-input.mu_sct)  * (R < input.a);

% Figures of contrast
x1 = input.X1(:,1);    x2 = input.X2(1,:);
figure(1); IMAGESC(x1,x2,real(input.CHI_eps));
title(('')\fontsize{13} \chi^{\epsilon} = 1 - \epsilon_{sct}/ \epsilon_{0}(''));
figure(2); mesh(abs(input.CHI_eps),('')edgecolor(''), ('')k(''));
           view(37.5,45); axis xy; axis(('')off('')); axis(('')tight(''))
           axis([1 input.N1 1 input.N2 0 1.5*max(abs(input.CHI_eps(:)))])
figure(3); IMAGESC(x1,x2,real(input.CHI_mu));
title(('')\fontsize{13}\chi^{\mu} = 1 - \mu_{sct}/ \mu_{0}(''));
figure(4); mesh(abs(input.CHI_mu),('')edgecolor(''), ('')k(''));
           view(37.5,45); axis xy; axis(('')off('')); axis(('')tight(''))
           axis([1 input.N1 1 input.N2 0 10000*max(abs(input.CHI_mu(:)))])
```

where $\Delta x = 3$ m. The function `initFFTGreen` of Matlab Listing II.4 generates a 2D grid with circulant properties. The $N_{2\text{fft}}$ columns of the output array $X_{1\text{fft}}$ are copies of the vector x_1 and the $N_{1\text{fft}}$ rows are copies of the vector x_2. For fast computations, the FFT numbers in each direction must have a power of two. These numbers are a factor of two greater than the grid numbers N_1 and N_2. At the end of function `initFFTGreen`, there is a call for the Matlab function `Green` to compute the repeated weak form of the Green functions, see Matlab Listing II.5. The function `initEMcontrast` (Matlab Listing II.75) initializes the permittivity-contrast distribution $\hat{\chi}^{\varepsilon}(x_1,x_2)$ and the permeability-contrast distribution $\hat{\chi}^{\mu} = 0$. The contrast distributions are

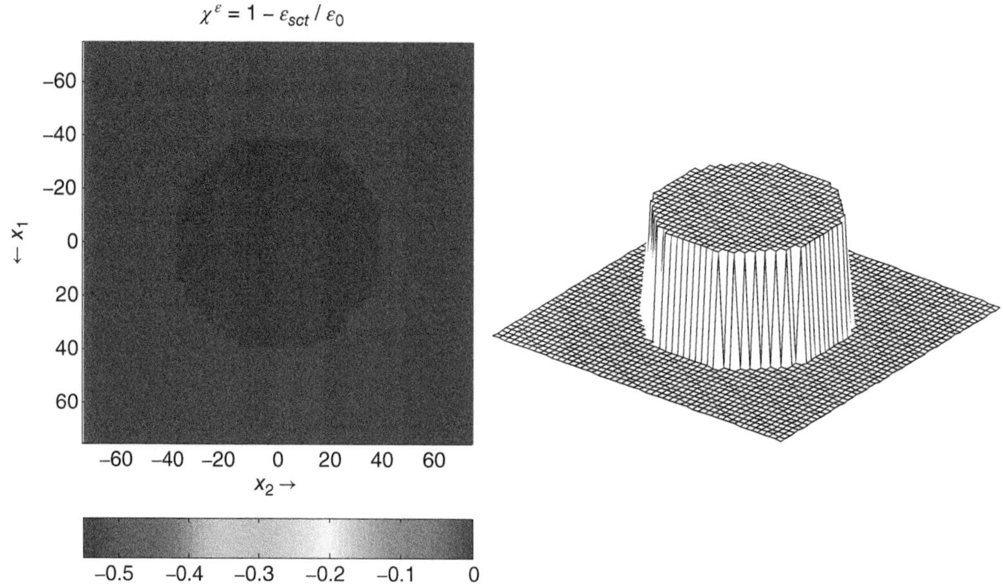

Figure 6.2 Plot of the permittivity contrast. This quantity is used to compute the synthetic data.

presented either as an image plot of its real part or as a surface plot of the absolute value. The distribution of the permittivity contrast is shown in Figure 6.2.

The Matlab script `ForwardwE` to compute the 2D scattered field is given in Matlab Listing II.76. The following steps are carried out:

(1) For the analytical computation of the electromagnetic wave field scattered from a circular object at a number of receivers, the Matlab script `EMdataCircle` of Matlab Listing II.77 is called. These data are saved as synthetic data for our inversion methods.
(2) To facilitate the numerical computation of the incident fields, we compute the wave fields pertaining to a number of line sources. They are denoted as G_S. In this chapter, we also need their spatial derivatives, which are denoted as dG_S. We compute their weak forms. Similarly, we compute the weak forms of the Green functions G_R associated with the wave fields measured at the receivers. We also need their spatial derivatives, which are denoted as dG_R. Both types of Green functions are computed with the help of the Matlab function `EGreenSourceReceivers` of Matlab Listing II.78. Next, the incident wave field is computed by using `IncEMwave` of Matlab Listing II.79.
(3) We solve the integral equation for the contrast sources using either function `ITERCGwE` of Matlab Listing II.82 or function `BICGSTABwE` of Matlab Listing II.83. Because the latter converges much faster we prefer the latter one. The integral equation is solved for each source location, indicated by the value q. The forward operator `KopE` and its adjoint `adjKopE` are given by Matlab Listings II.80 and II.81. Note that the function `BICGSTABwE` does not need the adjoints. Furthermore, using the contrast sources, we compute the two electric-field components

Matlab Listing II.76 The `ForwardwE` script to compute synthetic 2D data.

```
clear all; clc; close all; clear workspace;
input = initEM();

% (1) Compute analytically scattered field data -----------------------------
    disp(('')Running EMdataCircle('')); EMdataCircle;

    plotEMcontrast(input); % plot permittivity / permeability contrast

% (2) Compute Green function for sources and receivers --------------------
    pG = EGreenSourceReceivers(input);

% (3) Solve integral equation for contrast source with CGFFT method -------
    E1dataNum = zeros(input.NR,input.NS);
    E2dataNum = zeros(input.NR,input.NS);
    for q = 1 : input.NS
        [E_incq] = IncEMwave(pG,q,input);
         %[w_E]  = ITERCGwE(E_incq,input);
          [w_E]  = BICGSTABwE(E_incq,input);
        [Ercv]   = DOPwE(pG,w_E,input);
        E1dataNum(:,q) = Ercv{1}; E2dataNum(:,q) = Ercv{2};
    end %q_loop

    save EMdataIntEq E1dataNum E2dataNum;

% (4) Plot data at receiver locations -----------------------------------
    figure(9);
       subplot(1,2,1); PlotData(E1data,input);
       title(('')|E_1^{sct}(x_p^R,x_q^S)| = |E_1data_{Bessel}|(''));
       subplot(1,2,2); PlotData(E1dataNum-E1data,input);
       title(('')|E_1data_{Int.Eq.} - E_1data_{Bessel}|(''));
    figure(10);
       subplot(1,2,1); PlotData(E2data,input);
       title(('')|E_2^{sct}(x_p^R,x_q^S)| = |E_2data_{Bessel}|(''));
       subplot(1,2,2); PlotData(E2dataNum-E2data,input);
       title(('')|E_2data_{Int.Eq.} - E_2data_{Bessel}|(''));

  % Compute normalized difference of norms
    error1 = num2str(norm(E1dataNum(:)-E1data(:),1)/norm(E1data(:),1));
    disp([('')error1 =('') error1]);
    error2 = num2str(norm(E2dataNum(:)-E2data(:),1)/norm(E2data(:),1));
    disp([('')error2 =('') error2]);
```

Matlab Listing II.77 The `EMdataCircle` script to compute synthetic 2D data.

```
  input = initEM();
      a = input.a;                                   gam0    = input.gamma_0;
eps_sct = input.eps_sct;                             mu_sct  = input.mu_sct;
gam_sct = gam0 * sqrt(eps_sct*mu_sct);               Z_sct   = sqrt(mu_sct/eps_sct);

% (1) Transform Cartesian coordinates to polar coordinates ----------------
  xR = input.xR;
  xS = input.xS;
  rR = sqrt(xR(1,:).^2 + xR(2,:).^2);   phiR = atan2(xR(2,:),xR(1,:));
  rS = sqrt(xS(1,:).^2 + xS(2,:).^2);   phiS = atan2(xS(2,:),xS(1,:));

% (2) Compute coefficients of Bessel series expansion ----------------------
  arg0 = gam0 * input.a;   args = gam_sct * input.a;
  M = 20;     A = zeros(1,M);                        % increase M for more accuracy
  for m = 1 : M
    Ib0 = besseli(m,arg0);          dIb0 =  besseli(m+1,arg0) + m/arg0 * Ib0;
    Ibs = besseli(m,args);          dIbs =  besseli(m+1,args) + m/args * Ibs;
    Kb0 = besselk(m,arg0);          dKb0 = -besselk(m+1,arg0) + m/arg0 * Kb0;
    denominator = Z_sct * dIbs*Kb0 - dKb0*Ibs;
    A(m)  =     -(Z_sct * dIbs*Ib0 - dIb0*Ibs) / denominator;
  end

% (3) Compute reflected Er field at receivers (data) -----------------------
  E1data = zeros(input.NR,input.NS); E2data =  zeros(input.NR,input.NS);
  for p = 1 : input.NR
    for q = 1:input.NS
      Er = 0;  Ephi = 0;
      for m = 1 : M
         arg0 = gam0*rR(p);  Kb0 =  besselk(m,arg0);
                            dKb0 = -besselk(m+1,arg0) + m./arg0 .* Kb0;
                             KbS =  besselk(m,gam0*rS(q));
           Er = Er   + A(m)*2*m^2.*Kb0.*KbS.*cos(m*(phiR(p)-phiS(q)));
         Ephi = Ephi- A(m)*2*m .*dKb0.*KbS.*sin(m*(phiR(p)-phiS(q)));
      end % m_loop
    Er =            1/(2*pi) * Er./rR(p) ./rS(q);
  Ephi =  gam0 * 1/(2*pi) * Ephi        ./rS(q);
    E1 = cos(phiR(p)) .* Er  - sin(phiR(p)) .* Ephi;
    E2 = sin(phiR(p)) .* Er  + cos(phiR(p)) .* Ephi;
  E1data(p,q) = E1;  E2data(p,q) = E2;
    end % q_loop
  end % p_loop
  save Edata E1data E2data;
```

Matlab Listing II.78 The function `EGreenSourceReceivers` to compute the 2D Green functions for all source and receiver locations.

```
function pG = EGreenSourceReceivers(input)

gam0 = input.gamma_0;
dx    = input.dx;
delta  = (pi)^(-1/2) * dx;            % radius circle with area of dx^2
factor = 2 * besseli(1,gam0*delta) / (gam0*delta);
                                      % factor for weak form if DIS > delta
% Sources
 xS  = input.xS;
   G = cell(1,input.NS);
dG11 = cell(1,input.NS); dG21 = cell(1,input.NS); dG22 = cell(1,input.NS);
for q = 1 : input.NS
      X1  = input.X1 - xS(1,q); X2 = input.X2 - xS(2,q);
      DIS = sqrt(X1.^2 + X2.^2);
       X1 = X1./DIS;  X2 = X2./DIS;
     G{q} = factor * 1/(2*pi) .* besselk(0,gam0*DIS);
       dG = - factor * gam0 .* 1/(2*pi).* besselk(1,gam0*DIS);
  dG11{q} = (2*X1.*X1 - 1) .* (-dG./DIS) + gam0^2 * X1.*X1 .* G{q};
  dG21{q} =  2*X2.*X1      .* (-dG./DIS) + gam0^2 * X2.*X1 .* G{q};
  dG22{q} = (2*X2.*X2 - 1) .* (-dG./DIS) + gam0^2 * X2.*X2 .* G{q};
end %q_loop
pG.GS = G;   pG.dGS11 = dG11;   pG.dGS21 = dG21;   pG.dGS22 = dG22;

% Receivers
 xR  = input.xR;
   G = cell(1,input.NR);
dG11 = cell(1,input.NR); dG21 = cell(1,input.NR); dG22 = cell(1,input.NR);
for p = 1 : input.NR
      X1  = input.X1 - xR(1,p);  X2 = input.X2 - xR(2,p);
      DIS = sqrt(X1.^2 + X2.^2);
       X1 = X1./DIS;  X2 = X2./DIS;
     G{p} = factor * 1/(2*pi) .* besselk(0,gam0*DIS);
       dG = - factor * gam0 .* 1/(2*pi).* besselk(1,gam0*DIS);
  dG11{p} = (2*X1.*X1 - 1) .* (-dG./DIS) + gam0^2 * X1.*X1 .* G{p};
  dG21{p} =  2*X2.*X1      .* (-dG./DIS) + gam0^2 * X2.*X1 .* G{p};
  dG22{p} = (2*X2.*X2 - 1) .* (-dG./DIS) + gam0^2 * X2.*X2 .* G{p};
end % p_loop
pG.GR = G;   pG.dGR11 = dG11;   pG.dGR21 = dG21;   pG.dGR22 = dG22;
```

Matlab Listing II.79 The function `IncEMwave` to compute the incident wave field.

```
function [E_inc] = IncEMwave(pG,q,input)

gam0 = input.gamma_0;      xS = input.xS;
  M1 = xS(1,q) / sqrt(xS(1,q)^2 + xS(2,q)^2);
  M2 = xS(2,q) / sqrt(xS(1,q)^2 + xS(2,q)^2);

E_inc{1} = (-gam0^2 * pG.GS{q} + pG.dGS11{q}) * M1 + pG.dGS21{q} * M2;
E_inc{2} = (-gam0^2 * pG.GS{q} + pG.dGS22{q}) * M2 + pG.dGS21{q} * M1;
```

Matlab Listing II.80 The function `KopE` to compute the operator $\mathcal{K}^E$.

```
function [KwE] = KopE(wE,input)
global nDIM;

KwE = cell(1:nDIM);
for n = 1:nDIM
    KwE{n} = Kop(wE{n},input.FFTG);
end
dummy = graddiv(KwE,input);              % dummy is temporary storage
for n = 1:nDIM
    KwE{n} = KwE{n} - dummy{n} / input.gamma_0^2;
end
```

Matlab Listing II.81 The function `AdjKopE` to compute the adjoint of $\mathcal{K}^E$.

```
function [Kf] = AdjKopE(f,input)
global nDIM

Kf = cell(1:nDIM);
for n = 1:nDIM
    Kf{n} = AdjKop(f{n},input.FFTG);
end
dummy = graddiv(Kf,input);               % dummy is temporary storage
for n = 1:nDIM
    Kf{n} = Kf{n} - dummy{n} / conj(input.gamma_0^2);
end
```

of the scattered wave field at the receiver locations, see function `DOPwE` of Matlab Listing II.84. For convenience of the inverse scattering problem, we also need the adjoint of the data operator, which represents the back-projection of the data from the measurements domain into the object domain. This adjoint is given by `AdjDOPwE` of Matlab Listing II.85. In addition, we have the option to check the accuracy of the adjoint by comparing the definition of the norms in domains

Matlab Listing II.82 The function `ITERCGwE` to compute the contrast sources.

```
function [w_E] = ITERCGwE(E_inc,input)
global nDIM;      CHI_eps = input.CHI_eps;
w_E=cell(1,nDIM);   r_E=cell(1,nDIM);   g_E=cell(1,nDIM);
v_E=cell(1,nDIM);

itmax = 1000; Errcri = input.Errcri;  Norm0 = 0;  it = 0; % initialization
 for n = 1:nDIM
   w_E{n} = zeros(size(E_inc{n}));
   r_E{n} = CHI_eps .* E_inc{n};        Norm0 = Norm0 + norm(r_E{n}(:))^2;
 end
%CheckAdjointE(r_E,input);              % Check once

Error  = 1;     fprintf((')Error =          %g('),Error);
while (it < itmax) && ( Error > Errcri) && (Norm0 > eps)
% determine conjugate gradient direction v
  dummy=cell(1,nDIM);  for n=1:nDIM;  dummy{n}=conj(CHI_eps).*r_E{n}; end
  KE  = AdjKopE(dummy,input);
  AN  = 0;
  for n = 1:nDIM
     g_E{n} =  r_E{n} - KE{n};                 % and window for small CHI_eps
     g_E{n} = (abs(CHI_eps) >= Errcri).*g_E{n};    AN=AN+norm(g_E{n}(:))^2;
  end
  if it == 0
     for n = 1:nDIM;   v_E{n} = g_E{n}; end
  else
     for n = 1:nDIM;   v_E{n} = g_E{n} + AN/AN_1 * v_E{n};   end
  end
% determine step length alpha
  KE =  KopE(v_E,input);
  BN = 0;
  for n = 1:nDIM
      KE{n}= v_E{n} - CHI_eps .* KE{n};          BN = BN + norm(KE{n}(:))^2;
  end
  alpha = AN / BN;
% update contrast sources and AN
  for n = 1:nDIM
      w_E{n} = w_E{n} + alpha * v_E{n};
  end
  AN_1  = AN;
% update residual errors
   Norm = 0;
   for n = 1:nDIM
     r_E{n} = r_E{n} - alpha * KE{n};       Norm = Norm + norm(r_E{n}(:))^2;
   end
   Error = sqrt(Norm / Norm0);         fprintf((')\b\b\b\b\b\b\b\b%6f('),Error);
   it = it+1;
 end % CG iterations

 fprintf((')\b\b\b\b\b\b\b\b%6f\n('),Error);
 disp([(')Number of iterations is (') num2str(it)]);
 if it == itmax; disp([(')itmax reached: err/norm = (') num2str(Error)]); end
```

Matlab Listing II.83 The function `BICGSTABwE` to compute the contrast sources.

```
function [w_E] = BICGSTABwE(E_inc, input)

Data_wE{1}=input.CHI_eps.*E_inc{1};
Data_wE{2}=input.CHI_eps.*E_inc{2};

itmax  = 200;
it     = 0;  % initialization
w_E{1} = zeros(size(E_inc{1}));          w_E{2} = zeros(size(E_inc{2}));
r_E{1} = Data_wE{1};                     r_E{2} = Data_wE{2};
Norm_D = norm(r_E{1}(:))^2 + norm(r_E{2}(:))^2;
eta_D  = 1 / Norm_D;
Error  = 1;           fprintf(('')Error =        %g(''), Error);

while (it < itmax) && ( Error > input.Errcri)
% determine gradient directions
  AN = sum(r_E{1}(:).*conj(Data_wE{1}(:))+r_E{2}(:).*conj(Data_wE{2}(:)));
     if it == 0
      v_E{1} = r_E{1};                    v_E{2} = r_E{2};
     else
      v_E{1} = r_E{1}+(AN/AN_1)*v_E{1};   v_E{2} = r_E{2}+(AN/AN_1)*v_E{2};
     end
% determine step length alpha and update residual error
  KvE    = KopE(v_E, input);
  KvE{1} = v_E{1} - input.CHI_eps .* KvE{1};
  KvE{2} = v_E{2} - input.CHI_eps .* KvE{2};
  BN = sum(KvE{1}(:).*conj(Data_wE{1}(:))+KvE{2}(:).*conj(Data_wE{2}(:)));
  alpha = AN / BN;
  r_E{1} = r_E{1} - alpha * KvE{1};
  r_E{2} = r_E{2} - alpha * KvE{2};
% + succesive overrelaxation (first step of GMRES)
  KrE    = KopE(r_E, input);
  KrE{1} = r_E{1} - input.CHI_eps .* KrE{1};
  KrE{2} = r_E{2} - input.CHI_eps .* KrE{2};
  beta   = (sum(r_E{1}(:).*conj(KrE{1}(:))+r_E{2}(:).*conj(KrE{2}(:)))) ...
           / (norm(KrE{1}(:))^2 + norm(KrE{2}(:))^2);
% update contrast sources and w , v and AN
  w_E{1} = w_E{1} + alpha * v_E{1} + beta * r_E{1};
  w_E{2} = w_E{2} + alpha * v_E{2} + beta * r_E{2};
  v_E{1} = (alpha/beta) * (v_E{1} - beta * KvE{1});
  v_E{2} = (alpha/beta) * (v_E{2} - beta * KvE{2});
  AN_1   = AN;
% update residual errors r_D
  r_E{1} = r_E{1} - beta * KrE{1};
  r_E{2} = r_E{2} - beta * KrE{2};
  Norm_D = norm(r_E{1}(:))^2 + norm(r_E{2}(:))^2;
  Error  = sqrt(eta_D * Norm_D);    % fprintf(('')\b\b\b\b\b\b\b\b%6f(''), Error);
   it = it+1;
 end % CG iterations
 fprintf(('')\b\b\b\b\b\b\b\b%6f\n(''), Error);
 disp([('')Number of BICGSTABwE iterations is ('') num2str(it)]);
 if it == itmax; disp([('')itmax reached: err/norm = ('') num2str(Error)]); end
```

Matlab Listing II.84 The function `DOPwE` to compute synthetic 2D data, for each source location q.

```
function [Ercv] = DOPwE(pG,wE,input)

gam0 = input.gamma_0;        dx = input.dx;

Ercv{1} = zeros(input.NR,1);
Ercv{2} = zeros(input.NR,1);
for p = 1 : input.NR
 Ercv{1}(p,1) = dx^2*sum((gam0^2*pG.GR{p}(:)-pG.dGR11{p}(:)).* wE{1}(:) ...
                                          -pG.dGR21{p}(:) .* wE{2}(:));
 Ercv{2}(p,1) = dx^2*sum((gam0^2*pG.GR{p}(:)-pG.dGR22{p}(:)).* wE{2}(:) ...
                                          -pG.dGR21{p}(:) .* wE{1}(:));
end % p_loop
```

Matlab Listing II.85 The function `AdjDOPwE` to compute synthetic 2D data, for each source location q.

```
function [adjGR] = AdjDOPwE(pG,f,input)

gam0 = input.gamma_0;        dx = input.dx;

adjGR{1} = zeros(input.N1,input.N2);
adjGR{2} = zeros(input.N1,input.N2);
for p = 1 : input.NR
  adjGR{1} = adjGR{1} + conj(gam0^2*pG.GR{p}-pG.dGR11{p}) .* f{1}(p) ...
                                 - conj(pG.dGR21{p}) .* f{2}(p);
  adjGR{2} = adjGR{2} + conj(gam0^2*pG.GR{p}-pG.dGR22{p}) .* f{2}(p) ...
                                 - conj(pG.dGR21{p}) .* f{1}(p);
end % p_loop
adjGR{1} = dx^2 * adjGR{1};
adjGR{2} = dx^2 * adjGR{2};
```

in $\mathbb{D}$ and $\mathbb{M}$. For later convenience, the numerical data obtained from the integral-equation solution are saved.

(4) In order plot the data, we use function `PlotData` of Matlab Listing II.13. To measure the quality of these data, we benchmark it against the analytical ones of step (1). For our electromagnetic example, in Figure 6.3, we present the amplitudes of the two electric-field vectors of the analytical data (left-hand pictures) and the difference between the numerical data obtained from the integral-equation method (right-hand pictures). For both electric-field components, the error is 13%.

Before we discuss the nonlinear inverse scattering problem, we first present the method to create synthetic data, in which we model volume- and interface-contrast sources.

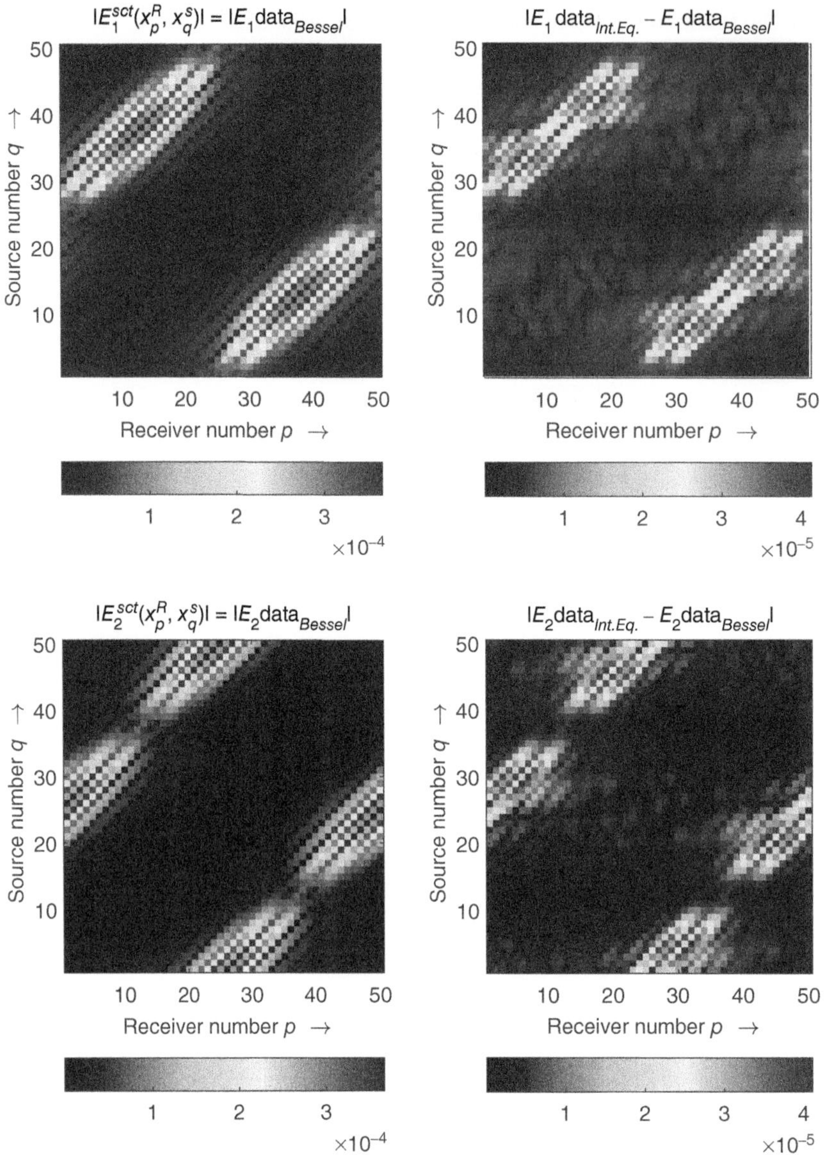

Figure 6.3 Plots of the data amplitudes of the E_1-component (top figures) and the E_2-component (bottom figures) as function of the source and receiver numbers. The relative permittivity is given by $\hat{\varepsilon}_{sct}/\varepsilon_0 = 1.55$. For both field components, the normalized global error in the integral-equation data amounts to 13%.

6.3 Data Modeled with Volume and Interface Contrast Sources

In section, we discuss the computation of synthetic data model with both the volume-contrast sources and the interface-contrast sources. In Chapter 3, we have developed a set of integral equations with volume-contrast sources and interface-contrast sources given by Eqs. (3.125) and (3.126). In 2D, after discretization, these equations are equivalent to

$$\begin{aligned}
\chi^{\varepsilon}_m \langle E^{inc}_1\rangle(\boldsymbol{x}^{(0)}_{T,m}) &= w^{(0)}_{1,m} - \chi^{\varepsilon}_m[\mathcal{K}^{(0,0)}_{1,m} + \partial_1(\mathcal{K}^{(0,1)}_m + \mathcal{K}^{(0,2)}_m)],\\
\chi^{\varepsilon}_m \langle E^{inc}_2\rangle(\boldsymbol{x}^{(0)}_{T,m}) &= w^{(0)}_{2,m} - \chi^{\varepsilon}_m[\mathcal{K}^{(0,0)}_{2,m} + \partial_2(\mathcal{K}^{(0,1)}_m + \mathcal{K}^{(0,2)}_m)],\\
R^{(1)}_m \langle E^{inc}_1\rangle(\boldsymbol{x}^{(1)}_{T,m}) &= \partial w^{(1)}_m - R^{(1)}_m [\mathcal{K}^{(1,0)}_m + \partial_1(\mathcal{K}^{(1,1)}_m + \mathcal{K}^{(1,2)}_m)],\\
R^{(2)}_m \langle E^{inc}_2\rangle(\boldsymbol{x}^{(2)}_{T,m}) &= \partial w^{(2)}_m - R^{(2)}_m [\mathcal{K}^{(2,0)}_m + \partial_2(\mathcal{K}^{(2,1)}_m + \mathcal{K}^{(2,2)}_m)],
\end{aligned} \tag{6.11}$$

where $\chi^{\varepsilon}_m = \chi^{\varepsilon}(\boldsymbol{x}^{(0)}_{T,m})$ and

$$R^{(1)}_m = \frac{\varepsilon(\boldsymbol{x}^{(0)}_{T,m} + \Delta x\, \boldsymbol{i}_1) - \varepsilon(\boldsymbol{x}^{(0)}_{T,m})}{\varepsilon(\boldsymbol{x}^{(0)}_{T,m} + \Delta x\, \boldsymbol{i}_1) + \varepsilon(\boldsymbol{x}^{(0)}_{T,m})}, \quad R^{(2)}_m = \frac{\varepsilon(\boldsymbol{x}^{(0)}_{T,m} + \Delta x\, \boldsymbol{i}_2) - \varepsilon(\boldsymbol{x}^{(0)}_{T,m})}{\varepsilon(\boldsymbol{x}^{(0)}_{T,m} + \Delta x\, \boldsymbol{i}_2) + \varepsilon(\boldsymbol{x}^{(0)}_{T,m})}. \tag{6.12}$$

The diagonal operators are given by

$$\begin{aligned}
\mathcal{K}^{(0,0)}_{1,m}(\boldsymbol{x}^{(0)}_{T,m}) &= \gamma_0^2 \Delta V \sum_{n=1}^{N} \langle\langle G\rangle\rangle(\boldsymbol{x}^{(0)}_{T,m} - \boldsymbol{x}^{(0)}_{T,n})\, \langle w^{(0)}_{1,n}\rangle,\\
\mathcal{K}^{(0,0)}_{2,m}(\boldsymbol{x}^{(0)}_{T,m}) &= \gamma_0^2 \Delta V \sum_{n=1}^{N} \langle\langle G\rangle\rangle(\boldsymbol{x}^{(0)}_{T,m} - \boldsymbol{x}^{(0)}_{T,n})\, \langle w^{(0)}_{2,n}\rangle,\\
\mathcal{K}^{(1,1)}_m(\boldsymbol{x}^{(1)}_{T,m}) &= 2\Delta x \sum_{n=1}^{N} \langle\langle \partial_1 G\rangle\rangle(\boldsymbol{x}^{(0)}_{T,m} - \boldsymbol{x}^{(0)}_{T,n})\, \langle \partial w^{(1)}_n\rangle,\\
\mathcal{K}^{(2,2)}_m(\boldsymbol{x}^{(2)}_{T,m}) &= 2\Delta x \sum_{n=1}^{N} \langle\langle \partial_2 G\rangle\rangle(\boldsymbol{x}^{(0)}_{T,m} - \boldsymbol{x}^{(0)}_{T,n})\, \langle \partial w^{(2)}_n\rangle.
\end{aligned} \tag{6.13}$$

The off-diagonal operators are computed by interpolation and extrapolation of the resulting values of the diagonal operators as

$$\begin{aligned}
\mathcal{K}^{(1,0)}_m &= \mathcal{K}^{(0,0)}_m(\boldsymbol{x}^{(1)}_{T,m}), & \mathcal{K}^{(2,0)}_m &= \mathcal{K}^{(0,0)}_m(\boldsymbol{x}^{(2)}_{T,m}),\\
\mathcal{K}^{(0,1)}_m &= \mathcal{K}^{(1,1)}_m(\boldsymbol{x}^{(0)}_{T,m}), & \mathcal{K}^{(2,1)}_m &= \mathcal{K}^{(1,1)}_m(\boldsymbol{x}^{(2)}_{T,m}),\\
\mathcal{K}^{(0,2)}_m &= \mathcal{K}^{(2,2)}_m(\boldsymbol{x}^{(0)}_{T,m}), & \mathcal{K}^{(1,2)}_m &= \mathcal{K}^{(2,2)}_m(\boldsymbol{x}^{(1)}_{T,m}).
\end{aligned} \tag{6.14}$$

We notice that the system of integral equations listed in Eq. (6.11) is equivalent to the following discrete set of equations:

$$\begin{aligned}
\chi^{\varepsilon}_m \langle E^{inc}_j\rangle(x^{(0)}_m) &= w^{(0)}_{j,m} - \chi^{\varepsilon}_m\, \mathcal{K}^{(0)}_{j,m}, \quad j = 1, 2,\\
R^{(j)}_m \langle E^{inc}_j\rangle(x^{(k)}_m) &= \partial w^{(j)}_m - R^{(j)}_m\, \mathcal{K}^{(j)}_m, \quad j = 1, 2.
\end{aligned} \tag{6.15}$$

The operator $\mathcal{K}^{(0)}_{j,m}$ maps functions defined on the three different grids to functions on the $\boldsymbol{x}^{(0)}_{T,m}$-grid and $\mathcal{K}^{(j)}_m$ maps the functions defined on three different grids to functions on the grid $\boldsymbol{x}^{(j)}_{T,m}$, $j = 1, 2$. They are found as

$$\begin{aligned}
\mathcal{K}^{(0)}_{j,m} &= \mathcal{K}^{(0,0)}_{j,m} + \mathtt{grad}_j \mathcal{K}^{(0,1)}_{j,m} + \mathtt{grad}_j \mathcal{K}^{(0,2)}_{j,m},\\
\mathcal{K}^{(j)}_m &= \mathcal{K}^{(j,0)}_m + \mathtt{grad}_j \mathcal{K}^{(j,1)}_m + \mathtt{grad}_j \mathcal{K}^{(j,2)}_m.
\end{aligned} \tag{6.16}$$

Further, $\mathtt{grad}_j$ is the finite-difference counterpart of the partial derivative with respect to x_j, see Matlab function `gradj` as given by the Matlab Listing I.96.

Matlab Listing II.86 The function `KopEwdw` to compute the volume and interface type operators.

```
function [Kw,Kdw] = KopEwdw(w,dw,input)

% Correction factor for interface integrals
  factor = 2 / (input.gamma_0^2 * input.dx);

  Kw1  = Kop(w{1},input.FFTG);
  Kw2  = Kop(w{2},input.FFTG);
  Kdw1 = factor * Kop(dw{1},input.FFTG);
  Kdw2 = factor * Kop(dw{2},input.FFTG);

% Interpolate the tangential components of Kdw to grid 0
  Kw1 = Kw1 + interpolate(gradj(Kdw1,1,input),0,1,input);
  Kw2 = Kw2 + interpolate(gradj(Kdw2,2,input),0,2,input);

% Extrapolate the normal components of Kdw to grid 0
  Kw{1} = Kw1 + extrapolate(gradj(Kdw2,1,input),0,2,input);
  Kw{2} = Kw2 + extrapolate(gradj(Kdw1,2,input),0,1,input);

% Interpolate the tangential components of Kw to the grid 0
  Kdw{1} = interpolate(Kw{1},1,0,input);
  Kdw{2} = interpolate(Kw{2},2,0,input);
```

Although the Matlab function `KopEwdw` to compute these operators are given in Matlab Listing I.97, a simplified version is given in Matlab Listing II.86. In the latter, we have used that the operators representing the electric-field components tangential to an interface are continuous. The operator $\mathcal{K}_m^{(j)}$ represents the electric-field components on the grid $x_m^{(j)}$ and operator $\mathcal{K}_{j,m}^{(0)}$ represents the jth component of the electric field on the grid $x_m^{(0)}$. Therefore, $\mathcal{K}_m^{(j)}$ follows from a simple linear interpolation of $\mathcal{K}_{j,m}^{(0)}$.

In 2D, the scattered-electric-field components are calculated as

$$\langle E_j^{sct}\rangle(\boldsymbol{x}_T^R) = \sum_{m=1}^{N}\left[\gamma_0{}^2\Delta V\langle G\rangle(\boldsymbol{x}_T^R - \boldsymbol{x}_{T,m}^{(0)})w_{j,m}^{(0)} + 2\Delta A\sum_{k=1}^{2}\partial_j^R\langle G\rangle(\boldsymbol{x}_T^R - \boldsymbol{x}_{T,m}^{(k)})\partial w_m^{(k)}\right]. \tag{6.17}$$

The Matlab function `DOPEwdw` that represents Eq. (6.17) is given in Matlab Listing II.87.

The Matlab script `ForwardwE`, to solve the integral equations and to compute the 2D scattered field both analytically and numerically, is given in Matlab Listing II.88. The following steps are carried out:

(1) For the analytical computation of the electromagnetic wave field scattered from a circular object at a number of receivers, the Matlab script `EMdataCircle` of Matlab Listing II.77 is called.

(2) To compute the weak forms of the Green functions G_S and G_R, associated with the incident wave fields and the wave fields measured at the receivers, we call the function `EGreen-SourceReceivers` of Matlab Listing II.78. Next, the function `initEMcontrastIntf` of Matlab Listing II.89 determines values of interface contrast, while the incident wave field is computed by using `IncEMdwave` of Matlab Listing II.90.

Matlab Listing II.87 The function DOPEwdw to compute synthetic 2D data, for each source location q.

```
function [Ercv] = DOPEwdw(w_E,dwE,input)
gam0 = input.gamma_0;        dx = input.dx;    xR = input.xR;

delta  = (pi)^(-1/2)*dx;                % radius circle with area dx^2
factor = 2 * besseli(1,gam0*delta) / (gam0*delta);

for p = 1 : input.NR
   % Non shifted grid(0)
     X1    = xR(1,p)-input.X1;
     X2    = xR(2,p)-input.X2;
    DIS    = sqrt(X1.^2 + X2.^2);
     G     = factor * 1/(2*pi).* besselk(0,gam0*DIS);
   E1rfl   = gam0^2 * dx^2 * sum( G(:) .* w_E{1}(:) );
   E2rfl   = gam0^2 * dx^2 * sum( G(:) .* w_E{2}(:) );

   % Shifted grid(1)
     X1    = xR(1,p) - input.X1 - dx/2;
     X2    = xR(2,p)-input.X2;
    DIS    = sqrt(X1.^2 + X2.^2);
     X1    = X1./DIS;       X2 = X2./DIS;
    dG     = - factor * gam0 * 1/(2*pi).* besselk(1,gam0*DIS);
   d1_G    =  X1 .* dG;
   d2_G    =  X2 .* dG;
   E1rfl   = E1rfl + 2 * dx * sum(d1_G(:) .* dwE{1}(:));
   E2rfl   = E2rfl + 2 * dx * sum(d2_G(:) .* dwE{1}(:));

   % Shifted grid(2)
     X1    = xR(1,p) - input.X1;
     X2    = xR(2,p) - input.X2 - dx/2;
    DIS    = sqrt(X1.^2 + X2.^2);
      X1   = X1./DIS;       X2 = X2./DIS;
    dG     = - factor * gam0 * (1/(2*pi)) .* besselk(1,gam0*DIS);
   d1_G    =  X1 .* dG;
   d2_G    =  X2 .* dG;
   E1rfl   = E1rfl + 2 * dx * sum(d1_G(:) .* dwE{2}(:));
   E2rfl   = E2rfl + 2 * dx * sum(d2_G(:) .* dwE{2}(:));

   Ercv{1}(p,1) = E1rfl;
   Ercv{2}(p,1) = E2rfl;
end % p_loop
```

(3) Finally, the function BICGSTABwdwE of Matlab Listing II.91 solves the integral equation and the function ForwardwdwE of Matlab Listing II.88 calculates the scattered fields. The results are given in Figure 6.4.
When we compare the results of Figures 6.3 and 6.4, we see that the maximum error in the data is now smaller by a factor of 2. The normalized global error increased from 13% to 16%. The difference between a smooth interface and its stair-case approximation becomes more visible. The extra computation time to include the interface scattering has been paid off.

Matlab Listing II.88 The `ForwardwdwE` script to compute synthetic 2D data, based on volume- and interface-contrast sources.

```
clear all; clc; close all; clear workspace;
input = initEM();

% (1) Compute analytically scattered field data ---------------------------
    disp(('')Running EMdataCircle('')); EMdataCircle;

%     Input interface contrast
    [input] = initEMcontrastIntf(input);

% (2) Compute Green function for sources and receivers -------------------
    pG = EGreenSourceReceivers(input);

% (3) Solve integral equation for contrast source with CGFFT method ------
    E1dataNum = zeros(input.NR,input.NS);
    E2dataNum = zeros(input.NR,input.NS);
    for q = 1 : input.NS
      [E_incq,dEincq] = IncEMdwave(pG,q,input);
      [w_E,dwE]       = BICGSTABwdwE(E_incq,dEincq,input);
      [Ercv]          = DOPEwdw(w_E,dwE,input);
      E1dataNum(:,q)  = Ercv{1};
      E2dataNum(:,q)  = Ercv{2};
    end %q_loop

% (4) Plot data at a number of receivers ---------------------------------
    figure(9);
       subplot(1,2,1);   PlotData(E1data,input);
       title(('')|E_1^{sct}(x_p^R,x_q^S)| = |E_1data_{Bessel}|(''));
       subplot(1,2,2);   PlotData(E1dataNum-E1data,input);
       title(('')|E_1data_{Int.Eq.} - E_1data_{Bessel}|(''));
    figure(10);
       subplot(1,2,1);   PlotData(E2data,input);
       title(('')|E_2^{sct}(x_p^R,x_q^S)| = |E_2data_{Bessel}|(''));
       subplot(1,2,2);   PlotData(E2dataNum-E2data,input);
       title(('')|E_2data_{Int.Eq.} - E_2data_{Bessel}|(''));

  % Compute normalized difference of norms
    error1 = num2str(norm(E1dataNum(:)-E1data(:),1)/norm(E1data(:),1));
    disp([('')error1 =('') error1]);
    error2 = num2str(norm(E2dataNum(:)-E2data(:),1)/norm(E2data(:),1));
    disp([('')error2 =('') error2]);
```

Matlab Listing II.89 The `initEMcontrastIntf` script to initialize the electromagnetic interface contrasts.

```
function [input] = initEMcontrastIntf(input)

R = sqrt(input.X1.^2 + input.X2.^2);
[N1,N2] = size(R);

% Compute permittivity type of (')reflection factors(') ----------------------
   a = input.a;
 eps = input.eps_sct * (R < a) + (R >= a) * 1;

 Rfl{1} = zeros(N1,N2);
 Rfl{2} = zeros(N1,N2);
 Rfl{1}(1:N1-1,:) = (eps(2:N1,:) - eps(1:N1-1,:)) ...
                    ./(eps(2:N1,:) + eps(1:N1-1,:));
 Rfl{2}(:,1:N2-1) = (eps(:,2:N2) - eps(:,1:N2-1)) ...
                    ./(eps(:,2:N2) + eps(:,1:N2-1));
input.Rfl = Rfl;
```

Matlab Listing II.90 The `IncEMdwave` script to compute the 2D incident wave field at the non-shifted grid (0) and shifted grid (1) and (2).

```
function [E_inc,dEinc] = IncEMdwave(pG,q,input)
gam0 = input.gamma_0;     xS = input.xS;

% Non shifted grid (0) -------------------------------------------------
       M1 = xS(1,q) / sqrt(xS(1,q)^2 + xS(2,q)^2);
       M2 = xS(2,q) / sqrt(xS(1,q)^2 + xS(2,q)^2);
 E_inc{1} = (-gam0^2*pG.GS{q} + pG.dGS11{q}) * M1 + pG.dGS21{q} * M2;
 E_inc{2} = (-gam0^2*pG.GS{q} + pG.dGS22{q}) * M2 + pG.dGS21{q} * M1;

% Shifted grid (1) -----------------------------------------------------
 dEinc{1} = 0 * E_inc{1};
        i = 1:input.N1-1;
 dEinc{1}(i,:) = (E_inc{1}(i+1,:) + E_inc{1}(i,:))/2;

% Shifted grid (2) -----------------------------------------------------
 dEinc{2} = 0 * E_inc{2};
        j = 1:input.N2-1;
 dEinc{2}(:,j) = (E_inc{2}(:,j+1) + E_inc{2}(:,j))/2;
```

Matlab Listing II.91 The function `BICGSTABwdwE` to compute the contrast sources.

```
function [w_E,dwE] = BICGSTABwdwE(E_inc,dEinc,input)

        CHI = input.CHI_eps;                    Rfl = input.Rfl;
Data_wE{1} = CHI .* E_inc{1};       Data_dwE{1} = Rfl{1} .* dEinc{1};
Data_wE{2} = CHI .* E_inc{2};       Data_dwE{2} = Rfl{2} .* dEinc{2};

itmax = 200;     it      = 0;      % initialization
w_E{1} = zeros(size(E_inc{1}));              dwE{1} = zeros(size(dEinc{1}));
w_E{2} = zeros(size(E_inc{2}));              dwE{2} = zeros(size(dEinc{2}));
r_E{1} = Data_wE{1};                         drE{1} = Data_dwE{1};
r_E{2} = Data_wE{2};                         drE{2} = Data_dwE{2};
Norm_D = norm(r_E{1}(:))^2 + norm(r_E{2}(:))^2;
eta_D  = 1 / Norm_D;   Error  = 1;     fprintf((')Error =          %g('),Error);

while (it < itmax) && ( Error > input.Errcri) %--------------------------------%
  % determine gradient directions
  AN = sum(r_E{1}(:).*conj(Data_wE{1}(:))+r_E{2}(:).*conj(Data_wE{2}(:)));
  if it == 0;    v_E = r_E;                       dvE = drE;
  else
     v_E{1} = r_E{1}+(AN/AN_1)*v_E{1};    dvE{1} = drE{1}+(AN/AN_1)*dvE{1};
     v_E{2} = r_E{2}+(AN/AN_1)*v_E{2};    dvE{2} = drE{2}+(AN/AN_1)*dvE{2};
  end
  [KvE,KdE] = KopEwdw(v_E,dvE,input); % to determine alpha and beta
  KvE{1} = v_E{1}-CHI.*KvE{1};                KdE{1} = dvE{1}-Rfl{1}.*KdE{1};
  KvE{2} = v_E{2}-CHI.*KvE{2};                KdE{2} = dvE{2}-Rfl{2}.*KdE{2};
  BN = sum(KvE{1}(:).*conj(Data_wE{1}(:))+KvE{2}(:).*conj(Data_wE{2}(:)));
  alpha = AN / BN;    AN_1   = AN;
  r_E{1} = r_E{1} - alpha * KvE{1};         drE{1} = drE{1} - alpha * KdE{1};
  r_E{2} = r_E{2} - alpha * KvE{2};         drE{2} = drE{2} - alpha * KdE{2};
  % + successive overrelaxation (first step of GMRES)
  [KrE,KdrE] = KopEwdw(r_E,drE,input);
  KrE{1} = r_E{1} - CHI .* KrE{1};    KdrE{1} = drE{1} - Rfl{1} .* KdrE{1};
  KrE{2} = r_E{2} - CHI .* KrE{2};    KdrE{2} = drE{2} - Rfl{2} .* KdrE{2};
  beta = (sum(r_E{1}(:).*conj(KrE{1}(:))+r_E{2}(:).*conj(KrE{2}(:)))) ...
         / (norm(KrE{1}(:))^2 +norm(KrE{2}(:))^2);
  %update contrast sources and residual errors
  A = alpha; B = beta;
  w_E{1} = w_E{1}+A*v_E{1}+B*r_E{1};    dwE{1} = dwE{1}+A*dvE{1}+B*drE{1};
  w_E{2} = w_E{2}+A*v_E{2}+B*r_E{2};    dwE{2} = dwE{2}+A*dvE{2}+B*drE{2};
  v_E{1} = (A/B)*(v_E{1} - B*KvE{1});   dvE{1} = (A/B)*(dvE{1} - B*KdE{1});
  v_E{2} = (A/B)*(v_E{2} - B*KvE{2});   dvE{2} = (A/B)*(dvE{2} - B*KdE{2});
  r_E{1} = r_E{1} - B * KrE{1};          drE{1} = drE{1} - B * KdrE{1};
  r_E{2} = r_E{2} - B * KrE{2};          drE{2} = drE{2} - B * KdrE{2};
  Norm_D = norm(r_E{1}(:))^2  + norm(r_E{1}(:))^2 ;
  Error=sqrt(eta_D*Norm_D);   fprintf((')\b\b\b\b\b\b\b\b%6f('),Error);
   it=it+1;
end % CG iterations -----------------------------------------------------------
fprintf((')\b\b\b\b\b\b\b\b%6f\n('),Error);
disp([(')Number of BICGSTABwdwE iterations is (') num2str(it)]);
if it == itmax; disp([(')itmax reached: err/norm = (') num2str(Error)]); end
```

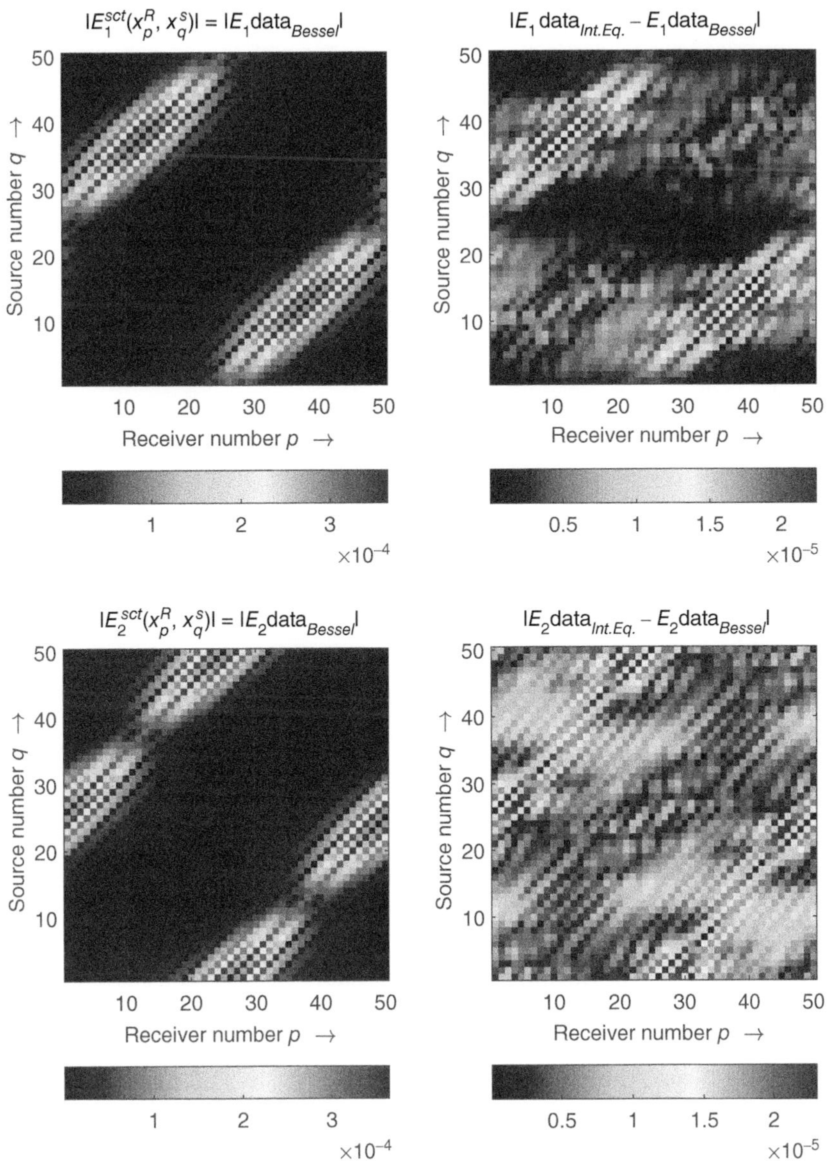

Figure 6.4 Plots of the data amplitudes of the E_1-component (top figures) and the E_2-component (bottom figures) as function of the source and receiver numbers. The relative permittivity is given by $\hat{\varepsilon}_{sct}/\varepsilon_0 = 1.55$. For both field components, the normalized global error in the integral-equation data amounts to 16%.

6.4 Electromagnetic Contrast Source Inversion

Again, we consider only the 2D version and assume a vanishing permeability contrast. The contrast-source inversion method recasts the electromagnetic inversion problem as a minimization of a cost functional, from which the contrast sources $\hat{w}_j^E$ and the permittivity contrast $\hat{\chi}^\varepsilon$ are reconstructed. In our further analysis, we omit the hat symbols. We first introduce the short-hand notations for the unknown wave-field functions $E_{j,q} = E_j(\boldsymbol{x}_m, \boldsymbol{x}_q^S)$ and the known wave-field functions,

$$\begin{aligned} E^{sct}_{\mathrm{M},j,q} &= E^{sct}_{\mathrm{M},j}(\boldsymbol{x}_p^R, \boldsymbol{x}_q^S), \\ E^{inc}_{j,q} &= E^{inc}_{j}(\boldsymbol{x}_m, \boldsymbol{x}_q^S), \end{aligned} \tag{6.18}$$

for $p = 1, 2, \ldots, N^R$ and $q = 1, 2, \ldots, N^S$, where N^R is the total number of receivers and N^S is the total number of sources.

Then, in the present CSI method, the sequences of the contrast sources $w_{j,q}^{E(n)} = w_j^{E(n)}(\boldsymbol{x}_m, \boldsymbol{x}_q^S)$, and the contrast $\chi^{\varepsilon(n)} = \chi^{\varepsilon(n)}(\boldsymbol{x})$, for $n = 1, 2, \ldots$, are iteratively found to minimize the cost functional

$$F^{(n)} = \eta_{\mathrm{M}} F_{\mathrm{M}}(w_{j,q}^{E(n)}) + \eta_{\mathrm{D}}^{(n)} F_{\mathrm{D}}^{E}(w_{j,q}^{E(n)}, \chi^{\varepsilon(n)}), \tag{6.19}$$

where

$$F_{\mathrm{M}} = \sum_{q=1}^{N^S} \sum_{j} \| r_{\mathrm{M},j,q}^{E(n)} \|_{\mathrm{M}}^2, \quad \text{with} \quad \eta_{\mathrm{M}} = \left[\sum_{q=1}^{N^S} \sum_{j} \| E^{sct}_{\mathrm{M},j,q} \|_{\mathrm{M}}^2 \right]^{-1}, \tag{6.20}$$

$$F_{\mathrm{D}}^{E} = \sum_{q=1}^{N^S} \sum_{j} \| r_{\mathrm{D},j,q}^{E(n)} \|_{\mathrm{D}}^2, \quad \text{with} \quad \eta_{\mathrm{D}}^{(n)} = \left[\sum_{q=1}^{N^S} \sum_{j} \| \chi^{\varepsilon(n)} E^{inc}_{j,q} \|_{\mathrm{D}}^2 \right]^{-1}. \tag{6.21}$$

In the present analysis, $\| \cdot \|^2 = \langle \cdot, \cdot \rangle_{\mathrm{M,D}}$ denotes the squared norm for $\boldsymbol{x}_p^R \in \mathbb{M}$ and $\boldsymbol{x} \in \mathbb{D}$, respectively. The residuals are given by

$$\begin{aligned} r_{\mathrm{M},j,q}^{E(n)} &= E^{sct}_{\mathrm{M},j,q} - [G_{\mathrm{M}} w_{j,q}^{E(n)} - \gamma_0^{-2} \partial_j^R \partial_k^R G_{\mathrm{M}} w_{k,q}^{E(n)}], \\ r_{\mathrm{D},j,q}^{E(n)} &= \chi^{\varepsilon} E^{inc}_{j,q} - w_{j,q}^{E(n)} + \chi^{\varepsilon} [G_{\mathrm{D}} w_{j,q}^{E(n)} - \gamma_0^{-2} \partial_j \partial_k G_{\mathrm{D}} w_{k,q}^{E(n)}], \end{aligned} \tag{6.22}$$

where $j = 1, 2$ and $k = 1, 2$. The operators G_{M} and G_{D} are defined in Eqs. (4.21) and (4.22). The first term of Eq. (6.19) measures the residual error $r_{\mathrm{M},j,q}^{E(n)}$ in the measurement domain,, while the second term measures the residual error $r_{\mathrm{D},j,q}^{E(n)}$ in the object domain. The normalization is chosen so that the two terms are equal to one if $w_{j,q}^{E(n)} = 0$. The residual in the first term is a linear function in $w_{j,q}^{E(n)}$, but the residual in the second term is nonlinear in $\chi^{\varepsilon(n)}$ and $w_{j,q}^{E(n)}$. To circumvent this nonlinearity, the CSI method employs an alternate updating scheme for these quantities.

6.4.1 Updating the Contrast Sources

In each iteration, we first update the contrast sources, $w_{j,q}^{E(n)} = w_j^{E(n)}(\boldsymbol{x}_m, \boldsymbol{x}_q^S)$, using the conjugate gradient step,

$$\boxed{w_{j,q}^{E(n)} = w_{j,q}^{E(n-1)} + \alpha_q^{(n)} v_{j,q}^{E(n)},} \tag{6.23}$$

where for each q and n, the parameters $\alpha_q^{(n)}$ are the step lengths and the functions $v_{j,q}^{E(n)}$ are the Polak–Ribière conjugate gradient directions,

$$\boxed{v_{j,q}^{E(1)} = g_{j,q}^{E(1)}, \quad v_{j,q}^{E(n)} = g_{j,q}^{E(n)} + \gamma_q^{(n)} v_{j,q}^{E(n-1)}, \quad n = 2, 3, \ldots,} \tag{6.24}$$

where $g_{j,q}^{E(n)}$ are the gradients of the cost functional $F^{(n)}$ with respect to changes of $w_{j,q}^{E(n-1)}$. The parameter $\gamma_q^{(n)}$ in the Polak–Ribière direction of Eq. (6.24) is defined as

$$\gamma_q^{(n)} = \frac{\sum_j \mathrm{Re}\langle g_{j,q}^{E(n)}, g_{j,q}^{E(n)} - g_{j,q}^{E(n-1)} \rangle_{\mathrm{D}}}{\sum_j \| g_{j,q}^{E(n-1)} \|_{\mathrm{D}}^2}. \tag{6.25}$$

6.4.1.1 Gradient Directions

After discretization of the Green's operators and the norms, similar to the scalar gradient of Eq. (4.37), the gradient of the qth term of the cost functional $F^{(n-1)} = \sum_q F_q^{E(n-1)}$ with respect to the contrast source $\overline{w}_{j,m',q}^{E(n-1)} = \overline{w}_{j,q}^{E(n-1)}(\boldsymbol{x}_{m'}, \boldsymbol{x}_q^S)$ is defined as

$$g_j^{E(n)}(\boldsymbol{x}_{m'}, \boldsymbol{x}_q^S) = -\frac{\partial [F_q^{(n-1)}]}{\partial [\overline{w}_{j,m',q}^{E(n-1)}]}, \tag{6.26}$$

where the qth term of $F^{(n-1)}$ follows from Eqs. (6.19)–(6.22). Hence,

$$g_j^{E(n)}(\boldsymbol{x}_{m'}, \boldsymbol{x}_q^S) = -\eta_{\mathrm{M}} \sum_{p=1}^{N^R} \sum_j \frac{\partial [\overline{r}_{\mathrm{M},j,q}^{E(n-1)}]}{\partial [\overline{w}_{j,m',q}^{E(n-1)}]} r_{\mathrm{M},j,q}^{E(n-1)} - \eta_{\mathrm{D}}^{(n-1)} \sum_{m=1}^{N} \sum_j \frac{\partial [\overline{r}_{\mathrm{D},j,q}^{E(n-1)}]}{\partial [\overline{w}_{j,m',q}^{E(n-1)}]} r_{\mathrm{D},j,q}^{E(n-1)}. \tag{6.27}$$

Carrying out the partial differentiation of the complex conjugate of the residual errors with respect to $\overline{w}_{j,m',q}^{E(n-1)}$, and using Eqs. (6.22), (4.21), and (4.22) we arrive at

$$\begin{aligned} -\frac{\partial [\overline{r}_{\mathrm{M},j,q}^{E(n-1)}]}{\partial [\overline{w}_{j,m',q}^{E(n-1)}]} &= \Delta V [\overline{\gamma}_0^2 \langle \overline{G} \rangle (\boldsymbol{x}_p^R - \boldsymbol{x}_{m'}) - \partial_j^R \partial_k^R \langle \overline{G} \rangle (\boldsymbol{x}_p^R - \boldsymbol{x}_{m'})], \\ -\frac{\partial [\overline{r}_{\mathrm{D},j,q}^{E(n-1)}]}{\partial [\overline{w}_{j,m',q}^{E(n-1)}]} &= \delta_{j,k} \delta_{m,m'} - \overline{\chi}^{\varepsilon(n-1)} \Delta V [\overline{\gamma}_0^2 \langle\langle \overline{G} \rangle\rangle (\boldsymbol{x}_m - \boldsymbol{x}_{m'}) - \partial_j \partial_k \langle\langle \overline{G} \rangle\rangle (\boldsymbol{x}_m - \boldsymbol{x}_{m'})]. \end{aligned} \tag{6.28}$$

Substituting these expressions into Eq. (6.27), we obtain $g_{j,q}^{E(n)} = g_j^{E(n)}(\boldsymbol{x}_m, \boldsymbol{x}_q^S)$ as

$$\begin{aligned} g_{j,q}^{E(n)} &= \eta_{\mathrm{M}} \sum_{p=1}^{N^R} \Delta V\, [\overline{\gamma}_0^2 \langle \overline{G} \rangle (\boldsymbol{x}_p^R - \boldsymbol{x}_{m'}) r_{\mathrm{M},j,q}^{(n-1)} - \partial_j^R \partial_k^R \langle \overline{G} \rangle (\boldsymbol{x}_p^R - \boldsymbol{x}_{m'})\, r_{\mathrm{M},k,q}^{E(n-1)}] \\ &+ \eta_{\mathrm{D}}^{(n-1)} \Bigg[r_{\mathrm{D},j,q}^{Zv(n-1)} - \sum_{m=1}^{N} \Delta V \left[\overline{\gamma}_0^2 \langle \overline{G} \rangle (\boldsymbol{x}_m - \boldsymbol{x}_{m'}) r_{\mathrm{D},j,q}^{E(n-1)} \right. \\ &\qquad \left. - \partial_j \partial_k \langle\langle \overline{G} \rangle\rangle (\boldsymbol{x}_m - \boldsymbol{x}_{m'}) \overline{\chi}^{\varepsilon(n-1)} r_{\mathrm{D},k,q}^{E(n-1)} \right] \Bigg]. \end{aligned} \tag{6.29}$$

Interchanging m' and m, we see that $g_{j,q}^{E(n)} = g_j^{E(n)}(\boldsymbol{x}_m, \boldsymbol{x}_q^S)$ may be written as

$$\begin{aligned} g_{j,q}^{E(n)} &= \eta_{\mathrm{M}} \left[G_{\mathrm{M}}^{\star} \{ r_{\mathrm{M},j,q}^{E(n-1)} \} - \overline{\gamma}_0^{-2} \partial_j^R \partial_k^R [\overline{\gamma}_0^2 G_{\mathrm{M}}^{\star} \{ r_{\mathrm{M},k,q}^{E(n-1)} \}] \right] \\ &+ \eta_{\mathrm{D}}^{(n-1)} \left[r_{\mathrm{D},j,q}^{E(n-1)} - G_{\mathrm{D}}^{\star} \{ \overline{\chi}^{\varepsilon} r_{\mathrm{D},j,q}^{E(n-1)} \} + \overline{\gamma}_0^{-2} \partial_j \partial_k G_{\mathrm{D}}^{\star} \{ \overline{\chi}^{\varepsilon} r_{\mathrm{D},k,q}^{E(n-1)} \} \right]. \end{aligned} \tag{6.30}$$

The adjoints $G_{\mathrm{M}}^{\star}$ and $G_{\mathrm{D}}^{\star}$ are given by Eqs. (4.31) and (4.35).

6.4.1.2 Calculation of the Step Length

The step length $\alpha_q^{(n)}$ is found as minimizer of the functional $F_q^{(n)}$, hence,

$$\alpha_q^{(n)} = \underset{\text{real } \alpha}{\arg\min} \{ F_q^{(n)}(w_{j,q}^{E(n-1)} + \alpha \, v_{j,q}^{E(n)}) \}, \tag{6.31}$$

where $F_q^{(n)}$ as a function of α is given by

$$F_q^{(n)} = \eta_{\mathrm{M}} \sum_j \| r_{\mathrm{M},j,q}^{E(n-1)} - \alpha \, \mathcal{K}_{\mathrm{M},q}^{E} \|_{\mathrm{M}}^2 + \eta_{\mathrm{D}}^{(n-1)} \sum_j \| r_{\mathrm{D},j,q}^{E(n-1)} - \alpha (v_{j,q}^{(n)} - \chi^{\varepsilon(n-1)} \mathcal{K}_{\mathrm{D},j,q}^{E(n)}) \|_{\mathrm{D}}^2, \tag{6.32}$$

in which

$$\mathcal{K}_{\mathrm{M},j,q}^{E(n)} = G_{\mathrm{M}} v_{j,q}^{E(n)} - \gamma_0^{-2} \partial_j \partial_k^R G_{\mathrm{M}} v_{k,q}^{E(n)} \tag{6.33}$$

and

$$\mathcal{K}_{\mathrm{D},j,q}^{E(n)} = G_{\mathrm{D}} v_{j,q}^{E(n)} - \gamma_0^{-2} \partial_j \partial_k G_{\mathrm{D}} v_{k,q}^{E(n)}. \tag{6.34}$$

The real part of the minimizer is found explicitly as

$$\alpha_q^{(n)} = \frac{\mathrm{Re}\left\{ \eta_{\mathrm{M}} \sum_j \langle r_{\mathrm{M},j,q}^{E(n-1)}, \mathcal{K}_{\mathrm{M},j,q}^{E(n)} \rangle + \eta_{\mathrm{D}}^{(n-1)} \sum_j \langle r_{\mathrm{D},j,q}^{E(n-1)}, v_{j,q}^{E(n)} - \chi^{\varepsilon(n-1)} \mathcal{K}_{\mathrm{D},j,q}^{E(n)} \rangle_{\mathrm{D}} \right\}}{\eta_{\mathrm{M}} \sum_j \| \mathcal{K}_{\mathrm{M},j,q}^{E(n)} \|_{\mathrm{M}}^2 + \eta_{\mathrm{D}}^{(n-1)} \sum_j \| v_{j,q}^{E(n)} - \chi^{\varepsilon(n-1)} \mathcal{K}_{\mathrm{D},j,q}^{E(n)} \|_{\mathrm{D}}^2}. \tag{6.35}$$

Using the expression for the gradient of Eq. (6.30), it simplifies to

$$\alpha_q^{(n)} = \frac{\mathrm{Re}\left\{ \sum_j \langle g_{j,q}^{E(n)}, v_{j,q}^{E(n)} \rangle_{\mathrm{D}} \right\}}{\eta_{\mathrm{M}} \sum_j \| \mathcal{K}_{\mathrm{M},j,q}^{E(n)} \|_{\mathrm{M}}^2 + \eta_{\mathrm{D}}^{(n-1)} \sum_j \| v_{j,q}^{E(n)} - \chi^{\varepsilon(n-1)} \mathcal{K}_{\mathrm{D},j,q}^{E(n)} \|_{\mathrm{D}}^2}. \tag{6.36}$$

After each update of the contrast sources, the residual error in the measurement domain becomes

$$\boxed{r_{\mathrm{M},j,q}^{E(n)} = r_{\mathrm{M},j,q}^{E(n-1)} - \alpha_q^{(n)} \mathcal{K}_{\mathrm{M},j,q}^{E(n)}.} \tag{6.37}$$

6.4.2 Updating the Contrast

Using the update for the contrast sources, the fields $E_{j,q}^{(n)} = E_j^{(n)}(\boldsymbol{x}_m, \boldsymbol{x}_q^S)$ are directly obtained as

$$E_{j,q}^{(n)} = E_{j,q}^{(n-1)} + \alpha_q^{(n)} \mathcal{K}_{j,q}^{E(n)}, \quad n = 1, 2, \ldots, \tag{6.38}$$

where $\mathcal{K}_{j,q}^{E(n)}$ is defined by Eq. (6.34). Recall that the position vector $\boldsymbol{x}_m$ and the source location $\boldsymbol{x}^S$ are omitted. Note that $E_{j,q}^{(0)} = E_{j,q}^{inc}$. The update for the contrast is obtained as

$$\boxed{\chi_{csi}^{\varepsilon(n)}(\boldsymbol{x}) = \frac{\sum_{q=1}^{N^S} \sum_j w_j^{E(n)}(\boldsymbol{x}, \boldsymbol{x}_q^S) \, \overline{E}_j^{(n)}(\boldsymbol{x}, \boldsymbol{x}_q^S)}{\sum_{q=1}^{N^S} \sum_j |E_j^{(n)}(\boldsymbol{x}, \boldsymbol{x}_q^S)|^2}.} \tag{6.39}$$

After computation of the contrast update $\chi_{csi}^{\varepsilon(n)}$, the residual errors in the object domain become

$$\boxed{r_{\mathrm{D},j,q}^{E(n)} = \chi_{csi}^{\varepsilon(n)} E_{j,q}^{(n)} - w_{j,q}^{E(n)},} \tag{6.40}$$

which shows that the actual computation of the error from Eq. (6.37) is superfluous. With this result the description of the algorithm is completed, except for the starting values for the contrast sources $w^{E(0)}_{j,q}$.

6.4.3 Initial Estimate

We cannot start with zero contrast sources $w^{E}_{j,q} = 0$, since then $\chi^{\varepsilon(0)} = 0$ and the cost functional $F_{\mathbb{D}}$ is undefined. Therefore, we choose as starting values the contrast sources that only minimize the cost functional

$$F_{\mathbb{M}}(w^{E(0)}_{j,q}) = \sum_{q=1}^{N^S} \sum_j \| E^{sct}_{\mathbb{M},j,q} - \mathcal{K}^{E(0)}_{\mathbb{M},j,q} \|^2_{\mathbb{M}}, \tag{6.41}$$

with

$$\mathcal{K}^{E(0)}_{\mathbb{M},j,q} = G_{\mathbb{M}} w^{E(0)}_{j,q} - \gamma_0^{-2} \partial_j \partial^R_k G_{\mathbb{M}} w^{E(0)}_{k,q}. \tag{6.42}$$

Using the gradient method, we take

$$\boxed{w^{E(0)}_{j,q} = \alpha^{(0)}_q \, g^{E(0)}_{j,q}, \qquad g^{E(0)}_{j,q} = \mathcal{K}^{\star}_{\mathbb{M},j} E^{sct}_{\mathbb{M},j,q},} \tag{6.43}$$

in which

$$\mathcal{K}^{\star}_{\mathbb{M},j} E^{sct}_{\mathbb{M},j,q} = \overline{G}_{\mathbb{M}} E^{sct}_{\mathbb{M},j,q} - \overline{\gamma}_0^{-2} \partial^R_j \partial^R_k \overline{G}_{\mathbb{M}} E^{sct}_{\mathbb{M},k,q}. \tag{6.44}$$

The cost functional is minimized when each term for q is minimized, i.e.

$$\alpha^{(0)}_q = \arg\min_{\alpha} \left\{ \sum_j \| E^{sct}_{\mathbb{M},j,q} - \alpha \, \mathcal{K}^{E(0)}_{\mathbb{M},j,q} \|^2_{\mathbb{M}} \right\}. \tag{6.45}$$

This leads directly to

$$\alpha^{(0)}_q = \frac{\sum_j \langle E^{sct}_{\mathbb{M},j,q}, \mathcal{K}^{E(0)}_{\mathbb{M},j,q} \rangle_{\mathbb{M}}}{\sum_j \| \mathcal{K}^{E(0)}_{\mathbb{M},j,q} \|^2_{\mathbb{M}}}, \tag{6.46}$$

or in simplified form as

$$\boxed{\alpha^{(0)}_q = \frac{\sum_j \| g^{E(0)}_{j,q} \|^2_{\mathbb{D}}}{\sum_j \| \mathcal{K}^{E(0)}_{\mathbb{M},j,q} \|^2_{\mathbb{M}}},} \tag{6.47}$$

where we have used Eqs. (6.42)–(6.44). The gradient $g^{E(0)}_{j,q} = \mathcal{K}^{\star}_{\mathbb{M},j} E^{sct}_{\mathbb{M},jq}$ is the back projection of the data domain $\mathbb{M}$ into the object domain $\mathbb{D}$, and it is a *back propagation* of the electric wave-field data in the object domain.

With this initial estimate $w^{E(0)}_{j,q} = w^{E(0)}_j(\boldsymbol{x}_m, \boldsymbol{x}^S_q)$, the contrast estimates is obtained from

$$\boxed{\chi^{\varepsilon(0)} = \frac{\sum_{q=1}^{N^S} \sum_j w^{E(0)}_{j,q} \, \overline{E}^{inc}_{j,q}}{\sum_{q=1}^{N^S} \sum_j |E^{inc}_{j,q}|^2}.} \tag{6.48}$$

After having discussed the details of the CSI method for the electromagnetic case, in Section 6.4.4, we discuss the regularization procedure.

6.4.4 Updating the Contrast with Multiplicative TV Regularization

To discuss the regularization of the permittivity contrast, we start with the observation that the permittivity contrast, as minimizer of F_{D} with respect to changes of χ^{ε}, is given by

$$\chi_{csi}^{\varepsilon(n)} = \arg\min_{\chi} \left\{ \sum_{q=1}^{N^S} \sum_{j} \| \chi\, E_{j,q}^{(n)} - w_{j,q}^{E(n)} \|_{\mathrm{D}}^2 \right\}, \tag{6.49}$$

or

$$\chi_{csi}^{\varepsilon(n)}(\boldsymbol{x}) = \frac{\sum_{q=1}^{N^S} \sum_j w_{j,q}^{E(n)}(\boldsymbol{x})\, \overline{E}_{j,q}^{(n)}(\boldsymbol{x})}{\sum_{q=1}^{N^S} \sum_j |E_{j,q}^{(n)}(\boldsymbol{x})|^2}. \tag{6.50}$$

Similar to the regularization procedure of the scalar case of Chapter 4, our further analysis simplifies when we consider the minimization of Eq. (6.49) with respect to changes of χ as

$$\chi^{\varepsilon(n)} = \arg\min{}_{\chi} \left\{ \int_{\boldsymbol{x}\in\mathrm{D}} |\chi(\boldsymbol{x}) - \chi_{csi}^{\varepsilon(n)}(\boldsymbol{x})|^2 \,\mathrm{d}V \right\}. \tag{6.51}$$

Hence, the regularization procedure for the permittivity contrast is identical to the one for the wave-speed contrast of Eq. (4.68), we only have to replace the values of $\chi_{csi}^{c(n)}$ by $\chi_{csi}^{\varepsilon(n)}$. This leads to the electromagnetic version of the MRCSI method.

6.4.5 Matlab Codes for the MRCSI Method

In this section, we present the Matlab codes for the electromagnetic version of the present contrast inversion method. The Matlab script `EmMRCSI` is given in Matlab Listing II.93. The following steps are carried out:

(1) We start with the computation of analytical data, by using `EMdataCircle` of Matlab Listing II.77. If these data have been computed with the forward code `ForwardwE` of Matlab Listing II.76 and have been stored in the computer memory, we have the possibility to investigate what the result is of an "Inverse Crime," where both the forward and inverse codes are based on the same computational model. We have the option to load the data from the integral-equation model.
If in the Matlab function `initEM` the value of the "input.Noise" is equal to 1, then there is complex-valued noise added with a standard deviation of $0.2 \times$ mean(|dataCircle|). Adding this so-called computer noise is certainly needed to prevent an inverse crime. However, we are now in the situation that our analytical data and numerical data of the integral-equation method differ about 10–15%, and that we do not need to add extra computer noise.

(2) Similar to the forward code `ForwardwE` of Matlab Listing II.76, we compute the weak forms of the Green functions G_S and G_R and their spatial derivatives with the Matlab function `EGreen-SourceReceivers` of Matlab Listing II.78.
The heart of the MRCSI method is found in the function `ITERMDwE` of the Matlab Listing II.94. In this function, the maximum number of iterations is set to it = 256. Both the initial contrast

Matlab Listing II.92 The function `InitialEstimatewE` to compute an initial estimate of the contrast sources w_j^E and the electric field E_k.

```
function [wE,E,r_M,ErrorM] = ...
                        InitialEstimatewE(eta_M,Einc,pG,E1data,E2data,input)
 NS = input.NS;
 wE  = cell(2,NS);    E  = cell(2,NS);    r_M = cell(2,NS);

% Determine contrast sources by back-projection and minimization of Norm_M
  Norm_M = 0;
  for q = 1 : NS
    dummyq{1} = E1data(:,q);
    dummyq{2} = E2data(:,q);
        g_M     = AdjDOPwE(pG,dummyq,input);
        A       = norm(g_M{1}(:))^2 + norm(g_M{2}(:))^2;
        GRv     = DOPwE(pG,g_M,input);
        B       = norm(GRv{1}(:))^2 + norm(GRv{2}(:))^2;
        alpha   = real(A/B);
      r_M{1,q} = E1data(:,q) - alpha * GRv{1};
      r_M{2,q} = E2data(:,q) - alpha * GRv{2};
       wE{1,q} = alpha * g_M{1};
       wE{2,q} = alpha * g_M{2};
    dummyq{1} = wE{1,q};
    dummyq{2} = wE{2,q};
        KwE     = KopE(dummyq,input);
       Norm_M   = Norm_M + norm(r_M{1,q}(:))^2 + norm(r_M{2,q}(:))^2;
        E{1,q} = Einc{1,q} + KwE{1};
        E{2,q} = Einc{2,q} + KwE{2};
  end %q_loop

 ErrorM = sqrt(eta_M * Norm_M);
```

sources and the initial contrast are estimated with the help of the function `InitialEstimatewE` of Matlab Listing II.92. The contrast sources are determined by back projection. Besides the data operator `DOPwE` of Matlab Listing II.84, we also need its adjoint `AdjDOPwE` of Matlab Listing II.85. The contrast sources follow from minimization of the squared norm on $\mathbb{M}$, viz. ErrorM, after which the contrast follows from Eq. (6.48) with the help of function `UpdateContrastwE` of Matlab Listing II.95. Further, the value of the squared norm on $\mathbb{D}$, viz. ErrorD, is computed. With these initial estimates, we employ the iterative CG method to determine the contrast sources that minimize ErrorM and ErrorD simultaneously. The function `InvITERwE` of Matlab Listing II.96 carries out a single CG iteration, after which the function `UpdateEMContrastwE` of Matlab Listing II.95 determines the updates of the contrast. After the contrast update of Eq. (6.50), the regularization is carried out using the function the function `TVregularizer` of Matlab Listing II.22. Note that regularization is not used in the initial estimate, because we start with an initial estimate of ErrorD = 1. During the iterations, each update of the contrast is shown as an image and a mesh plot on the computer screen. The contrast profiles are saved for a specific number of iterations, cf.

Matlab Listing II.93 The script `EmMRCSI` to reconstruct the contrast χ^ε.

```
clear all; clc; close all; clear workspace;
input = initEM();
% (1) Compute Exact data using anlaytic Bessel function expansion --------
      disp(('')Running EMdataCircle('')); EMdataCircle;
      load Edata E1data E2data;
    % load EMdataIntEq  E1DATAnum  E2DATAnum;
    % E1data = E1DATAnum;  E2data = E2DATAnum;           % Inverse Crime

    % add complex-valued noise to data
      if input.Noise == 1
         E1data = E1data + mean(abs(E1data(:))) * input.Rand;
         E2data = E2data + mean(abs(E2data(:))) * input.Rand;
      end

% (2) Compute Green function for sources and receivers -------------------
      pG = EGreenSourceReceivers(input);

% (3) Apply MRCCSI method and plot inverted permittivity contrast ---------
      Einc = cell(2,input.NS);
      for q = 1 : input.NS
         [E_incq]   = IncEMwave(pG,q,input);
          Einc{1,q} = E_incq{1};    Einc{2,q} = E_incq{2};
      end %q_loop
      [wE] = ITERMDwE(Einc,pG,E1data,E2data,input);

% (4) Compute model data explained by reconstructed contrast sources -----
      E1dataInv= zeros(input.NR,input.NS);
      E2dataInv= zeros(input.NR,input.NS);
      for q = 1 : input.NS
        w_E{1} = wE{1,q};   w_E{2} = wE{2,q};
        [Ercv] = DOPwE(pG,w_E,input);
         E1dataInv(:,q) = Ercv{1};    E2dataInv(:,q) = Ercv{2};
      end % q_loop

    % Plot data and compare to synthetic data
     figure(9);
         subplot(1,2,1);   PlotData(E1data,input);
         title(('')|E_1^{sct}(x_p^R,x_q^S)| = |E_1data_{Bessel}|(''));
         subplot(1,2,2);   PlotData(E1dataInv-E1data,input);
         title(('')|E_1data_{Int.Eq.} - E_1data_{Bessel}|(''));
     figure(10);
         subplot(1,2,1);   PlotData(E2data,input);
         title(('')|E_2^{sct}(x_p^R,x_q^S)| = |E_2data_{Bessel}|(''));
         subplot(1,2,2);   PlotData(E2dataInv-E2data,input);
         title(('')|E_2data_{Int.Eq.} - E_2data_{Bessel}|(''));
   % Compute normalized difference of norms
     error1 = num2str(norm(E1dataInv(:)-E1data(:),1)/norm(E1data(:),1));
     error2 = num2str(norm(E2dataInv(:)-E2data(:),1)/norm(E2data(:),1));
     disp([('')error1 =('') error1]);  disp([('')error2 =('') error2]);

     save MRCSIinv E1dataInv E2dataInv;
```

Matlab Listing II.94 The script `ITERMDwE` to reconstruct the contrast χ^{ε}.

```
function [wE] = ITERMDwE(Einc,pG,E1data,E2data,input)
global nDIM it

   NS = input.NS;
  AN1 = cell(1,NS);
  wEq = cell(2);          vEq = cell(2);         Eq = cell(2);
 g1Eq = cell(2);          rMq = cell(2);        rDq = cell(2);
 g1E  = cell(2,NS);        vE = cell(2,NS);

itmax = 256;
eta_M = 1 / (norm(E1data(:))^2 + norm(E2data(:))^2); % normalization

% (1) Initial estimate of contrast sources
  [wE,E,r_M,ErrorM]= InitialEstimatewE(eta_M,Einc,pG,E1data,E2data,input);
  it = 0;    ErrorD = 1;
  disp([(') Iteration = ('), num2str(it)]);
  [CHI,r_D,eta_D,ErrorD] = UpdateEMContrast(wE,E,Einc,ErrorM,ErrorD,input);
  saveCHI(CHI);
  disp((')----------------------------------------------------------------('));

% (2) Iterative updating of contrast sources and permittivity contrast
  it = 1;
  while (it <= itmax)
    Norm_M = 0;
    for q = 1 : NS
       AN1q = AN1{q};
       for n = 1:nDIM
         wEq{n} =  wE{n,q};    vEq{n} =  vE{n,q};      Eq{n} =   E{n,q};
        g1Eq{n} = g1E{n,q};    rMq{n} = r_M{n,q};     rDq{n} = r_D{n,q};
       end
  [wEq,vEq,Eq,g1Eq,AN1q,rMq] ...
       = InvITERwE(CHI,pG,wEq,vEq,Eq,g1Eq,AN1q,rMq,rDq,eta_M,eta_D,input);
      AN1{q} = AN1q;
     for n = 1:nDIM
         wE{n,q} =  wEq{n};     vE{n,q} = vEq{n};      E{n,q} =  Eq{n};
        g1E{n,q} = g1Eq{n};    r_M{n,q} = rMq{n};
     end
     Norm_M = Norm_M + norm(r_M{1,q}(:))^2 + norm(r_M{2,q}(:))^2;
   end %q_loop
     ErrorM = eta_M * Norm_M;

  disp([(') Iteration = ('), num2str(it)]);
  [CHI,r_D,eta_D,ErrorD]=UpdateEMContrast(wE,E,Einc,ErrorM,ErrorD,input);
  saveCHI(CHI);
  it = it+1;
 end % while
 disp((')----------------------------------------------------------------('));
 Error  = sqrt(ErrorM + ErrorD);
 disp([(')Number of iterations = (') num2str(it)]);
 disp([(') Total Error in M and D = (') num2str(Error)]);
```

Matlab Listing II.95 The script `UpdateEMContrast` to reconstruct contrast χ^{ε}.

```
function [CHI_eps,r_D,eta_D,ErrorD] = ...
                        UpdateEMContrast(wE,E,Einc,ErrorM,ErrorD,input)

% Determine permittivity contrast
  INE_wE = zeros(input.N1,input.N2);      INE_E = zeros(input.N1,input.N2);
  for q = 1 : input.NS
    INE_wE = INE_wE + conj(E{1,q}) .* wE{1,q} + conj(E{2,q}) .* wE{2,q};
    INE_E  = INE_E  + conj(E{1,q}) .*  E{1,q} + conj(E{2,q}) .*  E{2,q};
  end %q_loop
  CHI_eps = (INE_wE) ./ real(INE_E);

  if ErrorD < 1
     CHI_eps = TVregularizer(CHI_eps,ErrorM,ErrorD,input);
  end

% Compute residual error in object domain
  Norm_D = 0;    Norm = 0;     r_D{q} = cell(2,input.NS);
  for q = 1 : input.NS
    r_D{1,q} = CHI_eps .* E{1,q} - wE{1,q};
    r_D{2,q} = CHI_eps .* E{2,q} - wE{2,q};
    Norm_D   = Norm_D + norm(r_D{1,q}(:))^2 + norm(r_D{2,q}(:))^2 ;
    Norm     = Norm   + norm(CHI_eps(:) .* Einc{1,q}(:))^2  ...
                      + norm(CHI_eps(:) .* Einc{2,q}(:))^2;
  end %q_loop
  eta_D = 1 / (Norm);
  ErrorD = eta_D * Norm_D;
  disp([(') ErrorM = ('),num2str(ErrorM), (')    ErrorD  = ('),num2str(ErrorD)]);

% (3) Show intermediate pictures of reconstruction
  x1 = input.X1(:,1);    x2 = input.X2(1,:);
  figure(5); IMAGESC(x1,x2,real(CHI_eps));
  title((')\fontsize{13} Reconstructed Contrast Re[\chi^\epsilon] ('));
  pause(0.1)
```

the script `saveCHI` of Matlab Listing II.51. The script `LoadEImages` of Matlab Listing II.97 displays the contrast profiles at hand.

(3) With the help of the contrast sources solved from the present inversion method, we compute again the scattered field at the receiver locations, see function `DOPEw` of Matlab Listing II.84 and we compare these data with the analytical data used in step (1). By plotting the differences and computing the error being the norm of these differences, we obtain an indication how good the data are explained by the inversion scheme.

In Figure 6.5 (noise-free data) and in Figure 6.6 (noisy data), after a specific number of iterations, we show the reconstructed contrast profiles. The pictures indicated by it = 0 represent the contrast profiles obtained from the initial estimates. We see that the location of the circular boundary can be estimated based on the location of the maximum values of the estimated contrast. Within this range, the estimated contrast has significant negative values. This property is a consequence of the band limitation of the back-projection of the under-sampled data. These maximum contrast values also increase, for an increasing number of iterations. After 16 iterations, the TV regularization becomes

Matlab Listing II.96 The script `InvIterwE` to reconstruct the contrast χ^{ε}.

```
function [wE,vE,E,gE,AN,r_M] ...
        = InvITERwE(CHI,pG,wE,vE,E,g_1E,AN_1,r_M,r_D,eta_M,eta_D,input)
global it
% determine conjugate gradient directions
     g_M     = AdjDOPwE(pG,r_M,input);
   dummy{1}  = conj(CHI) .* r_D{1};
   dummy{2}  = conj(CHI) .* r_D{2};
    [adjKE]  = AdjKopE(dummy,input);
     gE{1}   = eta_M * g_M{1} + eta_D * (r_D{1} - adjKE{1});
     gE{2}   = eta_M * g_M{2} + eta_D * (r_D{2} - adjKE{2});
     AN      = norm(gE{1}(:))^2 + norm(gE{2}(:))^2;
if it > 1
     BN  = sum(gE{1}(:).*conj(g_1E{1}(:))+gE{2}(:).*conj(g_1E{2}(:)));
end
if it == 1
     vE = gE;
else
     vE{1} = gE{1} + real((AN-BN)/AN_1) * vE{1};
     vE{2} = gE{2} + real((AN-BN)/AN_1) * vE{2};
end
% determine step length alpha
    AN_D  = sum(gE{1}(:).*conj(vE{1}(:))+gE{2}(:).*conj(vE{2}(:)));
    Gv_M  = DOPwE(pG,vE,input);
    BN_M  = norm(Gv_M{1}(:))^2 + norm(Gv_M{2}(:))^2;
    [KvE] = KopE(vE,input);
   LvE{1} = vE{1} - CHI .* KvE{1};
   LvE{2} = vE{2} - CHI .* KvE{2};
    BN_D  = norm(LvE{1}(:))^2 + norm(LvE{2}(:))^2;
    alpha = real(AN_D) / (eta_M * BN_M + eta_D * BN_D);
% update contrast source wE and field E
    wE{1} = wE{1}  + alpha * vE{1};
    wE{2} = wE{2}  + alpha * vE{2};
     E{1} = E{1}   + alpha * KvE{1};
     E{2} = E{2}   + alpha * KvE{2};
% update residual error in M
   r_M{1} = r_M{1} - alpha * Gv_M{1}(:);
   r_M{2} = r_M{2} - alpha * Gv_M{2}(:);
```

effective and the spatial variation inside and outside the area with the maximum contrast values is kept to a minimum. We note that, either without noise or with noise, the images of the inverted contrast profiles do not differ very much.

In Figure 6.7 (noise-free data) and in Figure 6.8 (noisy data), we show the differences between the analytical data amplitudes and the inverted data amplitudes obtained from the contrast sources reconstructed after 256 iterations. On the left-hand sides of this figure, we present the amplitudes of the "exact" data using the Bessel series, while on the right-hand side we show the differences between the inverted data and the exact data. Note that the global error between the exact data and the integral-equation data amounts to 13% (see Figure 6.3), while the global error between the analytical data and the inverted data amounts to 0.6% and 11% for noise-free data (see Figure 6.7)

Matlab Listing II.97 The script `loadEimages` to reconstruct the contrast χ^{ε}.

```
input = initEM;      x1 = input.X1(:,1);      x2 = input.X2(1,:);
close all
set(figure,(')Units('),(')centimeters('),(')Position('),[0 -1 20 30]);
sp = subplot(3,3,1);   load Contrast0 CHI_0;    IMAGESC(x1,x2,CHI_0);
     title((')\fontsize{13} it = 0(')); xlabel((')(')); ylabel((')('));
     pos=get(sp,(')Position(')); pos=pos+[0.02 0 0 0]; set(sp,(')Position('),pos);
sp = subplot(3,3,2);   load Contrast1   CHI_1;  IMAGESC(x1,x2,CHI_1);
     title((')\fontsize{13} it = 1('));  xlabel((')(')); ylabel((')('));
     pos=get(sp,(')Position(')); pos=pos+[0 0 0 0];     set(sp,(')Position('),pos);
sp = subplot(3,3,3);   load Contrast4   CHI_4;  IMAGESC(x1,x2,CHI_4);
     title((')\fontsize{13} it = 4('));  xlabel((')(')); ylabel((')('));
     pos=get(sp,(')Position('));pos=pos+[-0.02 0 0 0]; set(sp,(')Position('),pos);

set(figure,(')Units('),(')centimeters('),(')Position('),[0 -1 20 30]);
sp = subplot(3,3,1);   load Contrast8   CHI_8;  IMAGESC(x1,x2,CHI_8);
     title((')\fontsize{13} it = 8('));  xlabel((')(')); ylabel((')('));
     pos=get(sp,(')Position(')); pos=pos+[0.02 0 0 0]; set(sp,(')Position('),pos);
sp = subplot(3,3,2);   load Contrast16  CHI_16; IMAGESC(x1,x2,CHI_16);
     title((')\fontsize{13} it = 16('));  xlabel((')(')); ylabel((')('));
     pos=get(sp,(')Position(')); pos=pos+[0 0 0 0];     set(sp,(')Position('),pos);
sp = subplot(3,3,3);   load Contrast32  CHI_32; IMAGESC(x1,x2,CHI_32);
     title((')\fontsize{13} it = 32('));  xlabel((')(')); ylabel((')('));
     pos=get(sp,(')Position('));pos=pos+[-0.02 0 0 0]; set(sp,(')Position('),pos);

set(figure,(')Units('),(')centimeters('),(')Position('),[0 -1 20 30]);
sp = subplot(3,3,1);  load Contrast64  CHI_64;   IMAGESC(x1,x2,CHI_64);
   title((')\fontsize{13} it = 64('));  xlabel((')(')); ylabel((')('));
     pos=get(sp,(')Position(')); pos=pos+[0.02 0 0 0]; set(sp,(')Position('),pos);
sp = subplot(3,3,2);  load Contrast128 CHI_128; IMAGESC(x1,x2,CHI_128);
   title((')\fontsize{13} it = 128(')); xlabel((')(')); ylabel((')('));
     pos=get(sp,(')Position(')); pos=pos+[0 0 0 0];     set(sp,(')Position('),pos);
sp = subplot(3,3,3);  load Contrast256 CHI_256; IMAGESC(x1,x2,CHI_256);
   title((')\fontsize{13} it = 256(')); xlabel((')(')); ylabel((')('));
     pos=get(sp,(')Position('));pos=pos+[-0.02 0 0 0]; set(sp,(')Position('),pos);
```

and noisy-data (see Figure 6.8), respectively. It is clear that the inverse model for noise-free data explains the analytical data better than our discretized model used in the forward calculations.

To gain more insight into the normalized difference of the modeled scattered-field data with respect to the exact data, the script `CheckInvData` of Matlab Listing II.98 calculates the data errors as follows:

(1) For the nominal contrast values, it calculates data errors using the forward codes based on the wE- and wdwE-models, see first line of Table 6.1.
(2) In the second line of Table 6.1, for the contrast values inverted by the MRCSI method based on the wE-model, we give the error that is not explained by the data.
(3) Again, it calculates the data errors for the inverted contrast values, executing the wE-method or the wdwE-method, see the third line of Table 6.1.

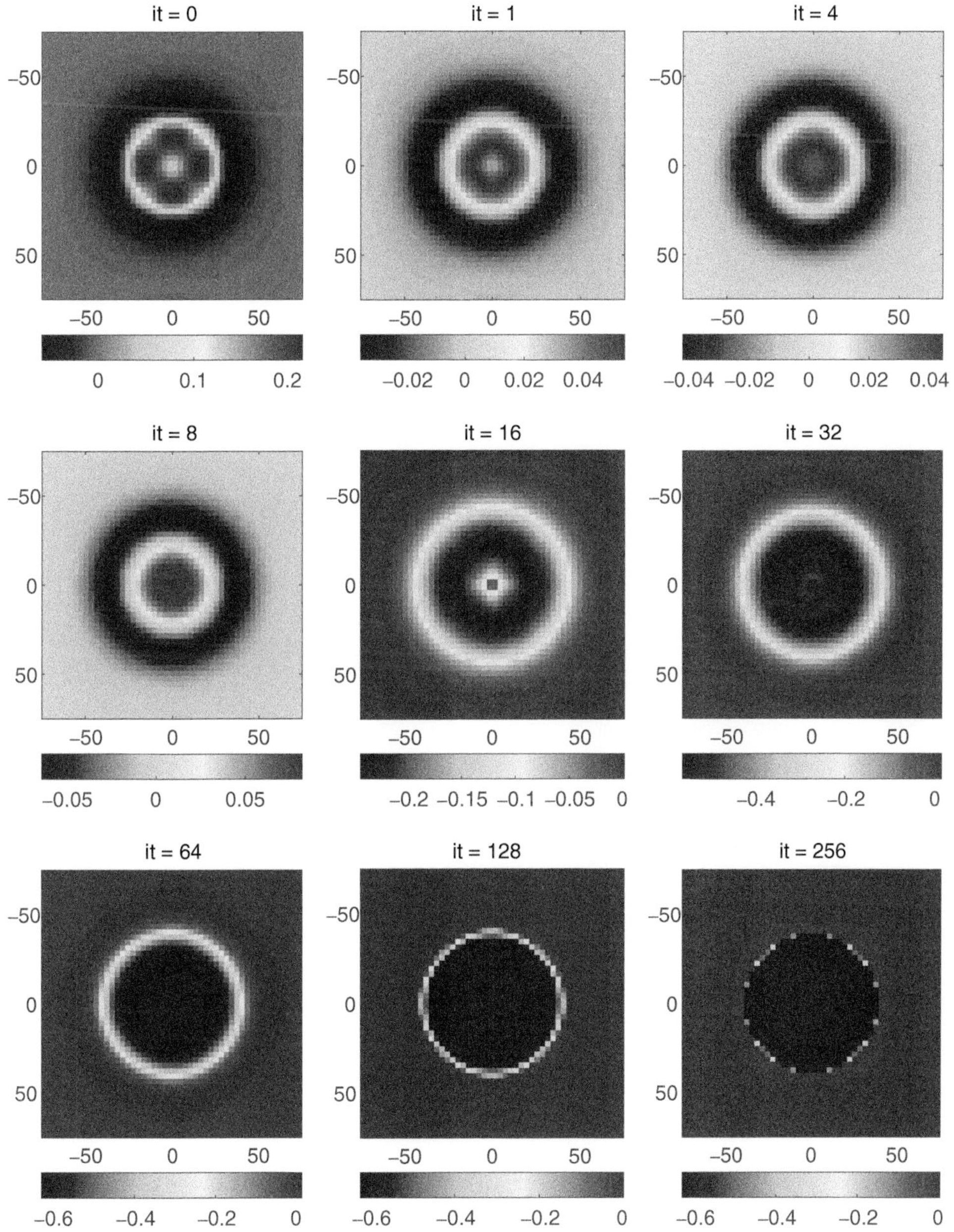

Figure 6.5 Electromagnetic MRCSI method: Image plots of the permittivity contrast reconstruction at different values of the iteration number (it), for **noise-free** data and nominal value of relative permittivity $\hat{\varepsilon}_{sct} = 1.55$.

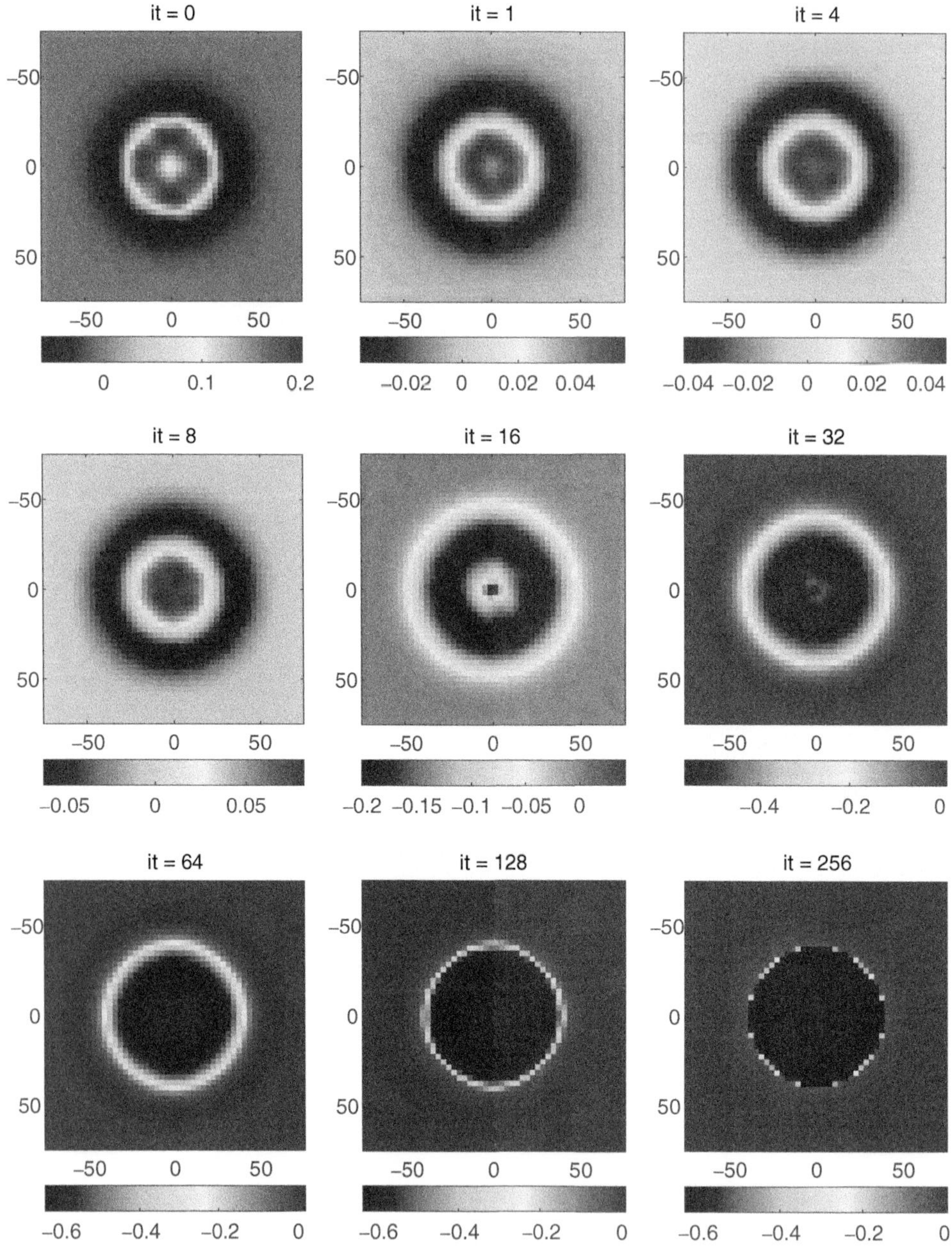

Figure 6.6 Electromagnetic MRCSI method: Image plots of the permittivity contrast reconstruction at different values of the iteration number (it), for **noisy** data and nominal value of relative permittivity $\hat{\varepsilon}_{sct} = 1.55$.

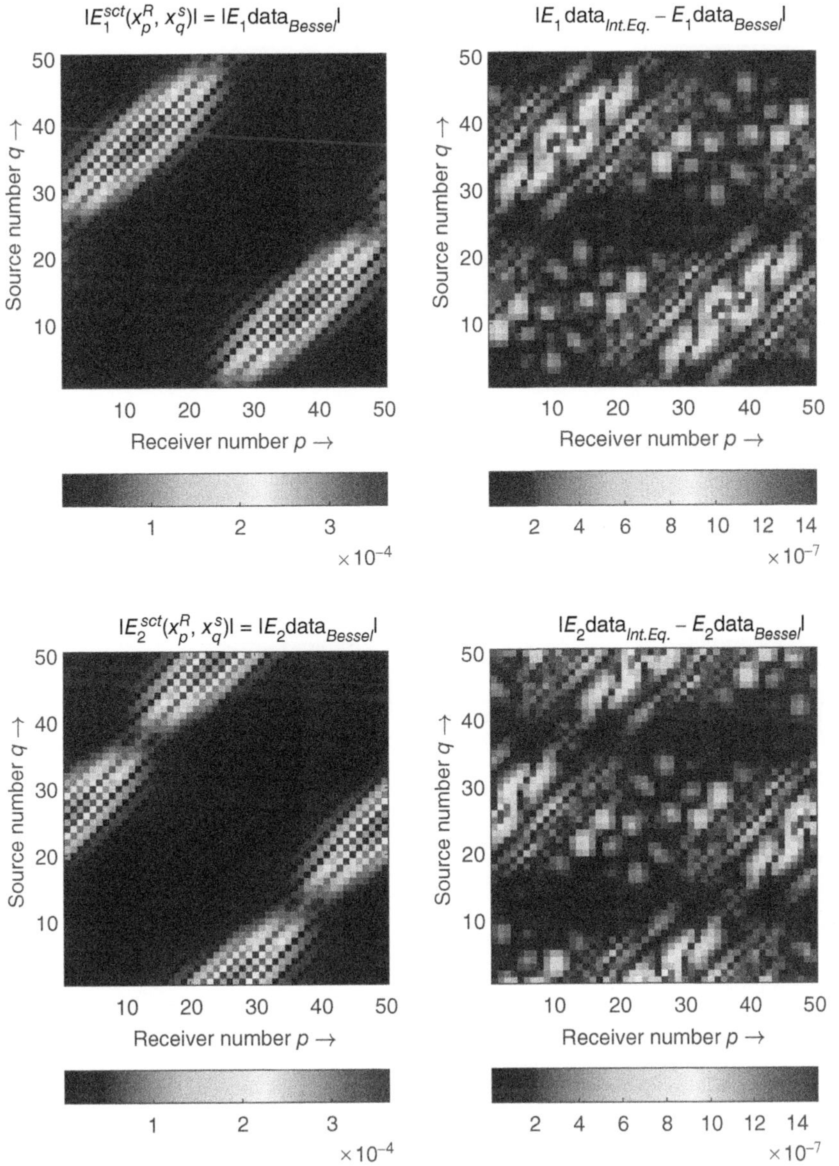

Figure 6.7 Plots of the **noise-free** data amplitudes of the E_1-component (top figures) and the E_2-component (bottom figures) as function of the source and receiver numbers. The relative permittivity is given by $\hat{\varepsilon}_{sct}/\varepsilon_0 = 1.55$. For both field components, the normalized global error in the inverted data amounts to 0.57%.

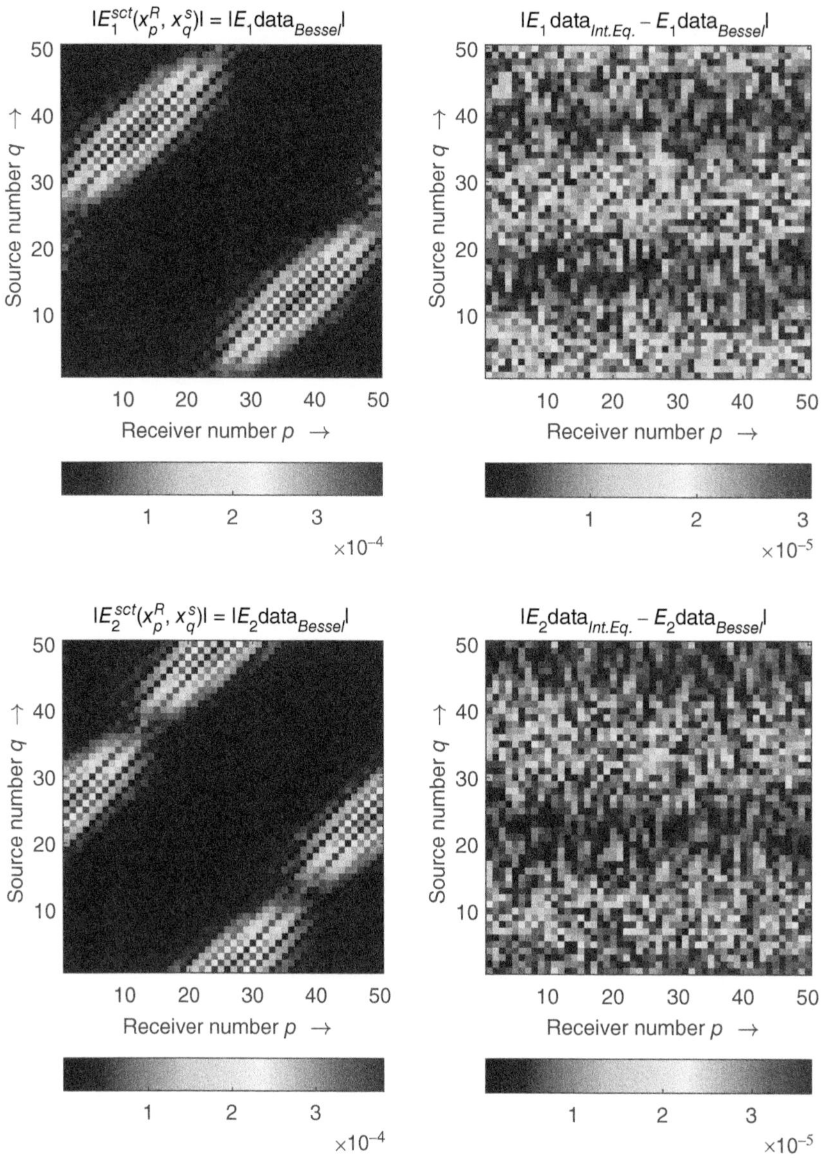

Figure 6.8 Plots of the **noisy** data amplitudes of the E_1-component (top figures) and the E_2-component (bottom figures) as function of the source and receiver numbers. The relative permittivity is given by $\hat{\varepsilon}_{sct}/\varepsilon_0 = 1.55$. For both field components, the normalized global error in the inverted data amounts to 18%.

The figures in this table suggest that the inverted contrast model is closer to the model of the homogeneous circular cylinder than to a stair-case approximation of the curved boundary. Reducing the grid size, will reduce the difference between the different error norms. The errors based on the forward wdwE-model may remain large enough to indicate the difference between a smooth interface and a stair-case type interface. This implies that the wdwE-model can improve the resolution of an inversion scheme.

Matlab Listing II.98 The script `CheckInvData` to estimate the data errors.

```
clear all; clc; close all; clear workspace;
input = initEM();
%     Input interface contrast
      [input] = initEMcontrastIntf(input);
%
%     Load the analytic data and the inverted data from MRCSI ------------
      load Edata E1data E2data;
      load MRCSIinv E1dataInv E2dataInv;
      load Contrast256 CHI_256;
      input.CHI_eps = CHI_256;
      x1 = input.X1(:,1);    x2 = input.X2(1,:);
      figure(5); IMAGESC(x1,x2,real(input.CHI_eps));
      title(('')\fontsize{13} Reconstructed Contrast Re[\chi^\epsilon] (''));

% (1) Compute normalized difference of norms
      error1 = num2str(norm(E1dataInv(:)-E1data(:),1)/norm(E1data(:),1));
      error2 = num2str(norm(E2dataInv(:)-E2data(:),1)/norm(E2data(:),1));
      disp([('')inv ErrorE1 =('') error1]);
      disp([('')inv ErrorE2 =('') error2]);

%     Compute Green function for sources and receivers -------------------
      pG = EGreenSourceReceivers(input);
%     Solve integral equation using wE-method and wdwE-method ------------
      E1DATAwE = zeros(input.NR,input.NS);
      E2DATAwE = zeros(input.NR,input.NS);
      E1DATAwdwE = zeros(input.NR,input.NS);
      E2DATAwdwE = zeros(input.NR,input.NS);
      for q = 1 : input.NS
          [E_incq,dEincq] = IncEMdwave(pG,q,input);
              [w_E]       = BICGSTABwE(E_incq,input);
              [ErSct]     = DOPwE(pG,w_E,input);
          E1DATAwE(:,q)   = ErSct{1};
          E2DATAwE(:,q)   = ErSct{2};
              [w_E,dwE]   = BICGSTABwdwE(E_incq,dEincq,input);
              [ErSct]     = DOPEwdw(w_E,dwE,input);
          E1DATAwdwE(:,q) = ErSct{1};
          E2DATAwdwE(:,q) = ErSct{2};
      end %q_loop

% (2) Compute normalized difference of norms (wE-method)
      error1 = num2str(norm(E1DATAwE(:)-E1data(:),1)/norm(E1data(:),1));
      error2 = num2str(norm(E2DATAwE(:)-E2data(:),1)/norm(E2data(:),1));
      disp([('')wE ErrorE1 =('') error1]);
      disp([('')wE ErrorE2 =('') error2]);

% (3) Compute normalized difference of norms (wdE-method)
      error1 = num2str(norm(E1DATAwdwE(:)-E1data(:),1)/norm(E1data(:),1));
      error2 = num2str(norm(E2DATAwdwE(:)-E2data(:),1)/norm(E2data(:),1));
      disp([('')wdwE Error1 =('') error1]);
      disp([('')wdwE error2 =('') error2]);
```

Table 6.1 Data errors with respect to exact data using Bessel series.

	Forward wE	MRCSI wE	Forward wdwE
Nominal contrast	13%	—	16%
Inverted contrast	—	0.6%	—
Inverted contrast	4.5%	—	26%

6.5 Electromagnetic Gauss–Newton Inversion

In Section 4.9, for scalar wave inversion, we have discussed a Gauss–Newton type of method to determine the finite set of parameters P_k, $k = 1, 2, \dots$, representing a contrast profile $\chi^c(P_k)$. Here, we discuss the Gauss–Newton method for the electromagnetic inverse problem by an iterative minimization of the squared norm of the residual errors in $\mathbb{M}$. In the nth iteration, we define the cost functional as

$$F_{\mathrm{M}}(P_k) = \sum_j \| r^{(n)}_{\mathrm{M},j,q} \|^2_{\mathbb{M}} = \sum_{q=1}^{N^S} \sum_{p=1}^{N^R} \sum_j | r^{(n)}_{\mathrm{M},j,q}(P_k) |^2, \tag{6.52}$$

with

$$r^{(n)}_{\mathrm{M},j,q} = E^{sct}_{\mathrm{M},j,q}(\boldsymbol{x}^R_p, \boldsymbol{x}^S_q) - G_{\mathrm{M}} w^{E(n)}_{j,q}, \tag{6.53}$$

where the data operator $G_{\mathrm{M}} w^{E(n)}_{j,q} = G_{\mathrm{M}} w^{E(n)}_j(\boldsymbol{x}^R_p, \boldsymbol{x}^S_q)$ is written as, cf. Eq. (4.21),

$$G_{\mathrm{M}} w^{E(n)}_{j,q} = \Delta V \sum_{m'=1}^{N} \gamma_0^2 \langle G \rangle (\boldsymbol{x}^R_p - \boldsymbol{x}_{m'})\, w^{E(n)}_j(\boldsymbol{x}_{m'}, \boldsymbol{x}^S_q). \tag{6.54}$$

The contrast source $w^{E(n)}_{j,q} = w^{E(n)}_j(\boldsymbol{x}_m, \boldsymbol{x}^S_q)$ is a solution of the contrast-source equation

$$w^{E(n)}_{j,q} - \chi^{\varepsilon(n-1)}[G_{\mathrm{D}} w^{E(n)}_{j,q} - \gamma_0^{-2} \partial_j \partial_k G_{\mathrm{D}} w^{E(n)}_{k,q}] = \chi^{\varepsilon(n-1)} E^{inc}_{j,q}, \quad q = 1, 2, \dots, N^S, \tag{6.55}$$

where the object operator $G_{\mathrm{D}} w^{E(n)}_{j,q}$ is written as, cf. Eq. (4.22),

$$G_{\mathrm{D}} w^{E(n)}_{j,q} = \Delta V \sum_{m'=1}^{N} \gamma_0^2 \langle\langle G \rangle\rangle (\boldsymbol{x}_m - \boldsymbol{x}_{m'})\, w^{E(n)}_{j,q}(\boldsymbol{x}_{m'}, \boldsymbol{x}^S_{j,q}), \quad m = 1, 2, \dots, N. \tag{6.56}$$

Let us assume that an approximation of the parameter set P_k is known from the previous iteration. We denote this set as $P^{(n-1)}_k$. Then, $\chi^{\varepsilon(n-1)}(\boldsymbol{x}_m)$ is known as well and we solve Eq. (6.55) numerically. The resulting contrast source is substituted in Eq. (6.53). This procedure shows that the residual error is a function of the parameters P_k and can be computed for typical values of P_k. An iterative minimization of the norm of Eq. (6.52) provides the appropriate tool to determine the set of parameters P_k.

Hence, starting with an initial guess $P^{(0)}_k$ for the minimum of the $F_{\mathrm{M}}(P_k)$, the Gauss–Newton method [16] continues with the iterations

$$P^{(n)}_k = P^{(n-1)}_k + \Delta P^{(n)}_k. \tag{6.57}$$

Carrying out a linear Taylor approximation of the residuals $r^{E(n)}_{\mathrm{M},j,q}$ as function of $\Delta P^{(n)}_k$, followed by a minimization of the norm, we arrive at a system of linear equations for $\Delta P^{(n)}_k$, viz.,

$$\sum_{k=1,2,\dots}\sum_{j}\left\langle \frac{\partial r^{E(n)}_{\mathrm{M},j,q}}{\partial P_l}, \frac{\partial r^{E(n)}_{\mathrm{M},j,q}}{\partial P_k}\right\rangle_{\mathrm{M}} \Delta P^{(n)}_k = -\sum_{j}\left\langle \frac{\partial r^{E(n)}_{\mathrm{M},j,q}}{\partial P_l}, r^{E(n)}_{\mathrm{M},j,q}\right\rangle_{\mathrm{M}}, \quad l = 1,\ 2,\dots, \tag{6.58}$$

where the inner product follows from the definition of the norm of Eq. (6.52).

At this point, we remark that analytic derivatives of the residuals with respect to the parameters are not available, and we determine the profile derivatives numerically by defining these derivatives as

$$\frac{\partial r^{E(n)}_{\mathrm{M},j,q}}{\partial P_k} = \frac{r^{E(n)}_{\mathrm{M},j,q}(\boldsymbol{x}, P_{j\neq k}, P_k+\epsilon) - r^{E(n)}_{\mathrm{M},j,q}(\boldsymbol{x}, P_{j\neq k}, P_k)}{\epsilon}, \quad k = 1,\ 2,\dots. \tag{6.59}$$

We consider the numerical example of Section 4.7. Our synthetic data are obtained for a circular cylinder with the radius a and center at the origin, i.e. $x_1^O = 0$ and $x_2^O = 0$. In our numerical procedure to model the data, we use the following parameter representation for the contrast profile:

$$\chi^{\varepsilon}(\boldsymbol{x}) := \frac{1-P_1}{1+\exp[(R-P_2)/s]}, \quad \text{with } R = (x_1^2 + x_2^2)^{\frac{1}{2}}, \tag{6.60}$$

where $P_1 = \varepsilon_{sct}/\varepsilon_0$ and $P_2 = a$ are the real profile parameters to be determined. To investigate the sensitivity of the inversion method at hand, we carry out some numerical experiments using three different mesh grids, viz. $\Delta x = 5, 3,$ and 1 m, respectively. The parameter s determines the sharpness of the profile at the boundary of the circular cylinder. In Section 4.7, we have chosen to normalize this value with respect to the mesh width Δx. We now take $s = 0.01a$ so that the sharpness of the circular cylinder does not depend on the grid size.

6.5.1 Matlab Codes for Gauss–Newton Inversion

In the Matlab script `GaussNewtonwE` of Matlab Listing II.99, we start with the computation of the Green functions, see the function `EGreenSourceReceivers` of Matlab Listing II.78. Subsequently, we call the function `FunDataEMsctCircle` of Matlab Listing II.100 to compute synthetic data using the Bessel series expansion, with input parameters $P_1 = \varepsilon_{sct}/\varepsilon_0$ and $P_2 = a$. The actual Gauss–Newton scheme requires some properly chosen estimates for the input parameters. We have estimated the relative permittivity as $P_1 = 1.05$ and the circle radius as $P_2 = 70$ m, which values are far from the actual values and are assumed to be in a convex convergence domain of the inversion scheme. In each Gauss–Newton iteration, we call the function `FunDataSctIntqwE` of Matlab Listing II.101, which computes a model update for the scattered field. It updates the contrast using Eq. (6.60), solves the integral equations using `BICGSTABwE.m` of Matlab Listing II.83 and computes the scattered field using `DopwE` of Matlab Listing II.84. The Matlab function `FunDataSctIntEq` is used in the Matlab script `GaussNewtonwE` of Matlab Listing II.99 to compute the residual errors between actual data and estimated model data. In addition, this function is used twice to numerically calculate the derivatives with respect to P_1 and P_2. In each Gauss–Newton step, we must start the field computations with $w^E_j = 0$ and we require that the maximum number of iterations in the BICGSTAB method is the same, both for the calculation of the cost functional and for the calculation of the numerical derivatives with

Matlab Listing II.99 The script `GaussNewtonwE` based on the wE-method.

```
clear all; clc; close all; clear workspace
global Pnom P_it
input = initEM(); itmaxGN = 15;      eps = 1e-6;

% Compute Green functions for data operators
  pG = EGreenSourceReceivers(input);

% Computes analytical data for actual parameters
  P = [input.eps_sct, input.a]; EMdata = FunDataEMSctCircle(P,input);

% Start with estimated parameters
  Pnom = [input.eps_sct,input.a];  disp(['Nominal P : ', num2str(Pnom)]);
  P0   = [1.05,          70      ]; disp(['Initial P0: ', num2str(P0)]);
  P_it(1,:) = P0;   % save as global array for plotting

 for it = 1: itmaxGN
 % Compute residual error, construct Jacobian matrix and update P
   P = P_it(it,:);
   data_est     = FunDataSctIntqwE(P,pG,input);
   Residual{1}  = EMdata{1} - data_est{1};
   Residual{2}  = EMdata{2} - data_est{2};
   GNerror      = (norm(Residual{1}(:))+norm(Residual{2}(:))) ...
                 /(norm(EMdata{1}(:))+norm(EMdata{2}(:))) ;
   disp(['            it= ',num2str(it),'  Residual= ',num2str(GNerror)]);
 % Compute numerical derivatives with respect to P
   P(1) = P(1) + eps;  data_plus = FunDataSctIntqwE(P,pG,input);
     dCHI1_P{1} = (data_plus{1} - data_est{1}) /eps;
     dCHI2_P{1} = (data_plus{2} - data_est{2}) /eps;
   P(1) = P(1) - eps;
   P(2) = P(2) + eps;  data_plus = FunDataSctIntqwE(P,pG,input);
    dCHI1_P{2} = (data_plus{1} - data_est{1}) /eps;
    dCHI2_P{2} = (data_plus{2} - data_est{2}) /eps;
   P(2) = P(2) - eps;
 % Compute residual error, construct Jacobian matrix and update P
   b = zeros(1,2); A = zeros(2,2);
   for I = 1:2;
       b(I) = sum(conj(dCHI1_P{I}(:)).*Residual{1}(:)) ...
            + sum(conj(dCHI2_P{I}(:)).*Residual{2}(:));
       for k = 1:2;
           A(I,k) = sum(conj(dCHI1_P{I}(:)).*dCHI1_P{k}(:)) ...
                  + sum(conj(dCHI2_P{I}(:)).*dCHI2_P{k}(:));
       end;
   end;
   DeltaP = real(b/A);             alpha = 1;
   alpha  = 0.618;              %  use Golden ratio for damping
   P = P + alpha * DeltaP ;
   disp(['          P: ', num2str(P)]);
   P_it(it+1,:) = P; % save as global array for plotting
 end; % it_loop

 DisplayParameterswE;
```

Matlab Listing II.100 The function `FunDataEMsctCircle` to compute synthetic data using the Bessel series expansion.

```
function [data_est] = FunDataEMSctCircle(P,input)
global nDIM;

% Connect parameters P to the our actual parameters
  eps_sct = P(1);        a = P(2);

  mu_sct  = input.mu_sct;
  gam0    = input.gamma_0;            gam_sct = gam0 * sqrt(eps_sct*mu_sct);
  Z_sct   = sqrt(mu_sct/eps_sct);

% (1) Transform Cartesian coordinates to polar coordinates ----------------
  xR = input.xR;
  xS = input.xS;
  rR = sqrt(xR(1,:).^2 + xR(2,:).^2);  phiR = atan2(xR(2,:),xR(1,:));
  rS = sqrt(xS(1,:).^2 + xS(2,:).^2);  phiS = atan2(xS(2,:),xS(1,:));

% (2) Compute coefficients of Bessel series expansion ---------------------
  arg0 = gam0 * a;   args = gam_sct * a;
  M = 20;      A = zeros(1,M);                % increase M for more accuracy
  for m = 1 : M;
    Ib0 = besseli(m,arg0);       dIb0 =  besseli(m+1,arg0) + m/arg0 * Ib0;
    Ibs = besseli(m,args);       dIbs =  besseli(m+1,args) + m/args * Ibs;
    Kb0 = besselk(m,arg0);       dKb0 = -besselk(m+1,arg0) + m/arg0 * Kb0;
    denominator = Z_sct * dIbs*Kb0 - dKb0*Ibs;
    A(m)  =      -(Z_sct * dIbs*Ib0 - dIb0*Ibs) / denominator;
  end

% (3) Compute reflected Er field at receivers (data) ----------------------
    data_est = cell(1,nDIM);
  for p = 1 : input.NR;
    for q = 1:input.NS;
      Er = 0;  Ephi = 0;  ZH3 = 0;
      for m = 1 : M;
         arg0 = gam0*rR(p);  Kb0 =  besselk(m,arg0);
                            dKb0 = -besselk(m+1,arg0) + m./arg0 .* Kb0;
                             KbS =  besselk(m,gam0*rS(q));
           Er = Er   + A(m)*2*m^2.*Kb0.*KbS.*cos(m*(phiR(p)-phiS(q)));
         Ephi = Ephi- A(m)*2*m .*dKb0.*KbS.*sin(m*(phiR(p)-phiS(q)));
         ZH3 = ZH3 - A(m)*2*m  .*Kb0.*KbS.*sin(m*(phiR(p)-phiS(q)));
      end % m_loop
    Er =           1/(2*pi) * Er./rR(p)  ./rS(q);
  Ephi =   gam0 * 1/(2*pi) * Ephi        ./rS(q);
  ZH3 = -gam0 * 1/(2*pi) * ZH3          ./rS(q);
    E1 = cos(phiR(p)) .* Er  - sin(phiR(p)) .* Ephi;
    E2 = sin(phiR(p)) .* Er  + cos(phiR(p)) .* Ephi;
  data_est{1}(p,q)  = (E1);
  data_est{2}(p,q)  = (E2);
    end % q_loop
  end % p_loop
```

Matlab Listing II.101 The script `FunDataSctIntqwE` to model data using the wE-method.

```
function [dataE_est] = FunDataSctIntqwE(P,pG,input)

% Update contrast using analytic contrast profile

        s = 0.01 * input.a;
        R = sqrt((input.X1).^2 + (input.X2).^2);
  eps_sct = P(1);

  input.CHI_eps = (1-eps_sct)./(1+exp((R-P(2))/s));

% Estimate data by solving integral equation with BICGSTABwE method

 dataE_est = cell(1,2);
 for q = 1 : input.NS;
   [E_incq,~] = IncEMdwave(pG,q,input);
   [w_E]      = BICGSTABwE(E_incq,input);     % using 3 iterations!
   [Ercv]     = DOPwE(pG,w_E,input);

   dataE_est{1}(:,q) = (Ercv{1});
   dataE_est{2}(:,q) = (Ercv{2});
 end; %q_loop
```

respect to the profile parameters. If these requirements are met, a few iterations will suffice. In our example, we take only three BICGSTAB iterations. Then, the Krylov subspace spanned by the repeated operators, implicitly generated by the BICGSTAB method, has sufficient sensitivity to be used for our inversion purposes. It is not necessary to force the wave-field errors to a very small numerical value.

After constructing the Jacobian and inverting the linear system of equations, we update the parameters P_k as

$$P_k^{(n)} = P_k^{(n-1)} + \alpha \, \Delta P_k^{(n)}, \tag{6.61}$$

in which α is a damping factor. For the original Gauss–Newton scheme α equals one. However, for this value of α, the sum of squares of the residuals may not decrease at every iteration. Note that $\Delta P_k^{(n)}$ is a descent direction, except at the stationary point of the cost functional. In other words, the gradient vector is too long. Reduction of its length will decrease the value of the cost functional. An optimal value for α can be found by using a line-search algorithm, in which α is determined by finding the value that minimizes the cost functional. The drawback is that we need extra solutions of the integral equations in the interval $0 < \alpha < 1$. For our type of problems, we have found that a simple and very effective choice is to reduce the length of the gradient vector by the golden-ratio factor 1.618, viz., $\alpha \approx 1/1.618 \approx 0.618$. Finally, in each iteration, the function `DisplayParameterswE` of Matlab Listing II.102 plots the profile parameters P_k.

For comparison, we also developed a Gauss–Newton inversion scheme, in which we replace the present forward solver based on the volume-contrast sources by the solver based on both volume- and interface-contrast sources. The modifications are straightforward, in fact we only replace the function `BICGSTABwE.m` of Matlab Listing II.83 by `BICGSTABwdwE.m` of Matlab Listing II.91. As a consequence, the Matlab script `GaussNewtonwE` is replaced by the script

Matlab Listing II.102 The script `DisplayParameterswE` to plot the profile parameters, using the forward wE-method.

```
function DisplayParameterswE
global Pnom P_it

set(figure(9), 'Position', [50 50 550 400] );  N = length(P_it(:,1))-1;

subplot(1,2,1)

  plot(0:N,Pnom(1)*ones(N+1,1),':k','Linewidth',2);      hold on;
  plot(0:N, P_it(:,1),'--b','Linewidth',2);
  title('P_1 = eps_{sct}/eps_0','Fontsize',11);
  xlabel('number of iterations','Fontsize',11);
  axis('tight'); axis([ 0 N 1 1.7]);  grid on;

subplot(1,2,2)

  plot(0:N,Pnom(2)*ones(N+1,1),':k','Linewidth',2);      hold on;
  plot(0:N, P_it(:,2),'--b','Linewidth',2);
  title('P_2 = a','Fontsize',11);
  xlabel('number of iterations','Fontsize',11);
  axis('tight'); axis([ 0 N 30 70]);  grid on;
  legend('P nominal ', 'P wE ','Location','NorthEast')
```

`GaussNewtonwdwE` of Matlab Listing II.103, while the functions `FunDataSctIntqwE` and `DisplayParameterswE` are replaced by the functions `FunDataSctIntqwdwE` and `DisplayParameterswdwE`, respectively (see Matlab Listings II.104 and II.105).

In Figures 6.9 and 6.10, we show the reconstructed parameters as a function of the Gauss–Newton iterations. The dotted lines indicate the exact (nominal) values of the profile parameters. The dashed lines represent the reconstructed parameters using the wE-model with volume-contrast sources, while the solid lines represent the reconstructed parameters using the wdwE-model with volume- and interface-contrast sources. In Figure 6.9, we have taken a very rough mesh size ($\Delta x = 5$ m). In the top pictures of this figure, we have used a damping factor $\alpha = 1$. We observe a rather oscillatory behavior of the reconstructed radius a, which means that the inversion scheme lacks some sensitivity to locate the boundary of the scattering object. In the bottom pictures, we have used the golden ratio factor $\alpha = 0.618$, which effectively dampens the oscillations. It is clear that the wdwE-method performs better than the wE-method, although this is at the expense of the increase of computation time. The relative permittivity $\varepsilon_{sct}/\varepsilon_0$ is reconstructed very well. However, for the wE-method, the final value of the reconstructed radius is $a = 39.3$ m. For the wdwE-method, we arrive at $a = 41.3$ m, but it should be taken into account that, for a mesh size of 5 m, the staircase type interface contrast is *a priori* information built in the wdwE-model; it increases the sensitivity, but it limits the accuracy of the approximation. Decreasing the mesh size will improve the reconstruction of a smooth boundary. This is confirmed by Figure 6.10. For $\Delta x = 3$ m (top pictures) we see that the reconstructed value of the radius converged to $a = 39.8$ m for the wE-method, and $a = 40.1$ m for the wdwE-method. Regarding the reconstructed permittivity, there is still a big difference with the wE-method. If we refine the mesh to $\Delta x = 1$ m, we see that the results of the two methods are coming together (bottom pictures).

Matlab Listing II.103 The script GaussNewtonwdwE based on the wdwE-method.

```
clear all; clc; close all; clear workspace
global Pnom P_it
input = initEM(); itmaxGN = 15;     eps = 1e-6;

% Compute Green function for sources and receivers and compute datadata
  pG = EGreenSourceReceivers(input);

% Computes analytical data for actual parameters
  P = [input.eps_sct, input.a]; EMdata = FunDataEMSctCircle(P,input);

% Start with estimated contrast sources
  Pnom = [input.eps_sct,input.a];   disp(['Nominal P : ', num2str(Pnom)]);
  P0   = [1.05,          70    ];   disp(['Initial P0: ', num2str(P0)]);
  P_it(1,:) = P0;  % save as global array for plotting

 for it = 1: itmaxGN
 % Compute residual error, construct Jacobian matrix and update P
   P = P_it(it,:);
   data_est    =  FunDataSctIntqwdwE(P,pG,input);
   Residual{1} = EMdata{1} -  data_est{1};
   Residual{2} = EMdata{2} -  data_est{2};
   GNerror     = (norm(Residual{1}(:))+norm(Residual{2}(:)))...
                /(norm(EMdata{1}(:))+norm(EMdata{2}(:))) ;
   disp(['            it= ',num2str(it),'  Residual= ',num2str(GNerror)]);
 % Compute numerical derivatives with respect to P
   P(1) = P(1) + eps;  data_plus = FunDataSctIntqwdwE(P,pG,input);
     dCHI1_P{1} = (data_plus{1} - data_est{1}) /eps;
     dCHI2_P{1} = (data_plus{2} - data_est{2}) /eps;
   P(1) = P(1) - eps;
   P(2) = P(2) + eps;  data_plus = FunDataSctIntqwdwE(P,pG,input);
    dCHI1_P{2} = (data_plus{1} - data_est{1}) /eps;
    dCHI2_P{2} = (data_plus{2} - data_est{2}) /eps;
   P(2) = P(2) - eps;
 % Compute residual error, construct Jacobian matrix and update P
   b = zeros(1,2); A = zeros(2,2);
   for l = 1:2;
       b(l) = sum(conj(dCHI1_P{l}(:)).*Residual{1}(:))...
            + sum(conj(dCHI2_P{l}(:)).*Residual{2}(:)) ;
       for k = 1:2;
           A(l,k) = sum(conj(dCHI1_P{l}(:)).*dCHI1_P{k}(:))...
                  + sum(conj(dCHI2_P{l}(:)).*dCHI2_P{k}(:)) ;
       end;
   end;
   DeltaP = real(b/A);                alpha = 1;
   alpha  = 0.6180;                 % use Golden ratio for damping
   P = P + alpha * DeltaP ;
   disp(['          P: ', num2str(P)]);
   P_it(it+1,:) = P; % save as global array for plotting
 end; % it_loop

 DisplayParameterswdwE;
```

Matlab Listing II.104 The script `FunDataSctIntqwdwE` to model data using the wdwE-method.

```
function [dataE_est] = FunDataSctIntqwdwE(P,pG,input)

% Update contrast using analytic contrast and determine interface contrast
             s = 0.01 * input.a;
             R = sqrt(input.X1.^2 + input.X2.^2);      [N1,N2] = size(R);
 input.CHI_eps = (1-P(1))./(1+exp((R-P(2))/s));
           eps = 1 - input.CHI_eps;

        Rfl{1} = zeros(N1,N2);          Rfl{2} = zeros(N1,N2);
Rfl{1}(1:N1-1,:) = (eps(2:N1,:) - eps(1:N1-1,:)) ...
                            ./(eps(2:N1,:) + eps(1:N1-1,:));
Rfl{2}(:,1:N2-1) = (eps(:,2:N2) - eps(:,1:N2-1)) ...
                            ./(eps(:,2:N2) + eps(:,1:N2-1));
input.Rfl = Rfl;

% Estimate data by solving integral equation with BICGSTABwE method
dataE_est = cell(1,2);
  for q = 1 : input.NS;
      [E_incq,dEincq] = IncEMdwave(pG,q,input);
      [w_E,dwE] = BICGSTABwdwE(E_incq,dEincq,input); % using 3 iterations!
      [Ercv]    = DOPEwdw(w_E,dwE,input);
      dataE_est{1}(:,q) = (Ercv{1});
      dataE_est{2}(:,q) = (Ercv{2});
  end; %q_loop
```

Matlab Listing II.105 The script `DisplayParameterswdwE` to plot the profile parameters, using the forward wdwE-method.

```
function DisplayParameterswdwE
global Pnom P_it

set(figure(10), 'Position', [50 50 550 400] );  N = length(P_it(:,1))-1;
subplot(1,2,1)
  plot(0:N,Pnom(1)*ones(N+1,1),':k','Linewidth',2);     hold on;
  plot(0:N, P_it(:,1),'-r','Linewidth',2);
  title('P_1 = eps_{sct}/eps_0','Fontsize',11);
  xlabel('number of iterations','Fontsize',11);
  axis('tight'); axis([ 0 N 1 1.7]);  grid on;

subplot(1,2,2)
  plot(0:N,Pnom(2)*ones(N+1,1),':k','Linewidth',2);     hold on;
  plot(0:N, P_it(:,2),'-r','Linewidth',2);
  title('P_2 = a','Fontsize',11);
  xlabel('number of iterations','Fontsize',11);
  axis('tight'); axis([ 0 N 30 70]);  grid on;
  legend('P nominal ', 'P wdwE ','Location','NorthEast')
```

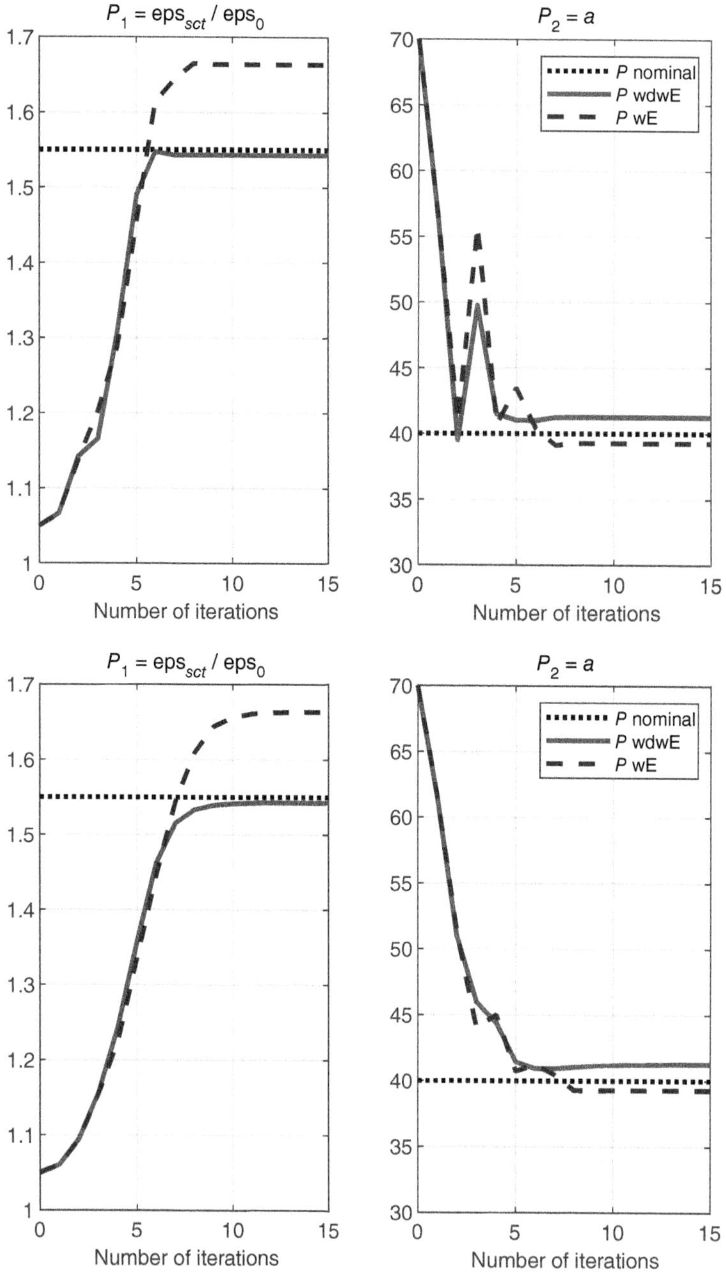

Figure 6.9 Plots of the parameters $\varepsilon_{sct}/\varepsilon_0 = P_1^{(n)}$ and $a = P_2^{(n)}$ as function of the number of Gauss–Newton iterations, for $\Delta x = 5$m, with damping factor $\alpha = 1$ (top pictures) and $\alpha = 0.618$ (bottom pictures), respectively.

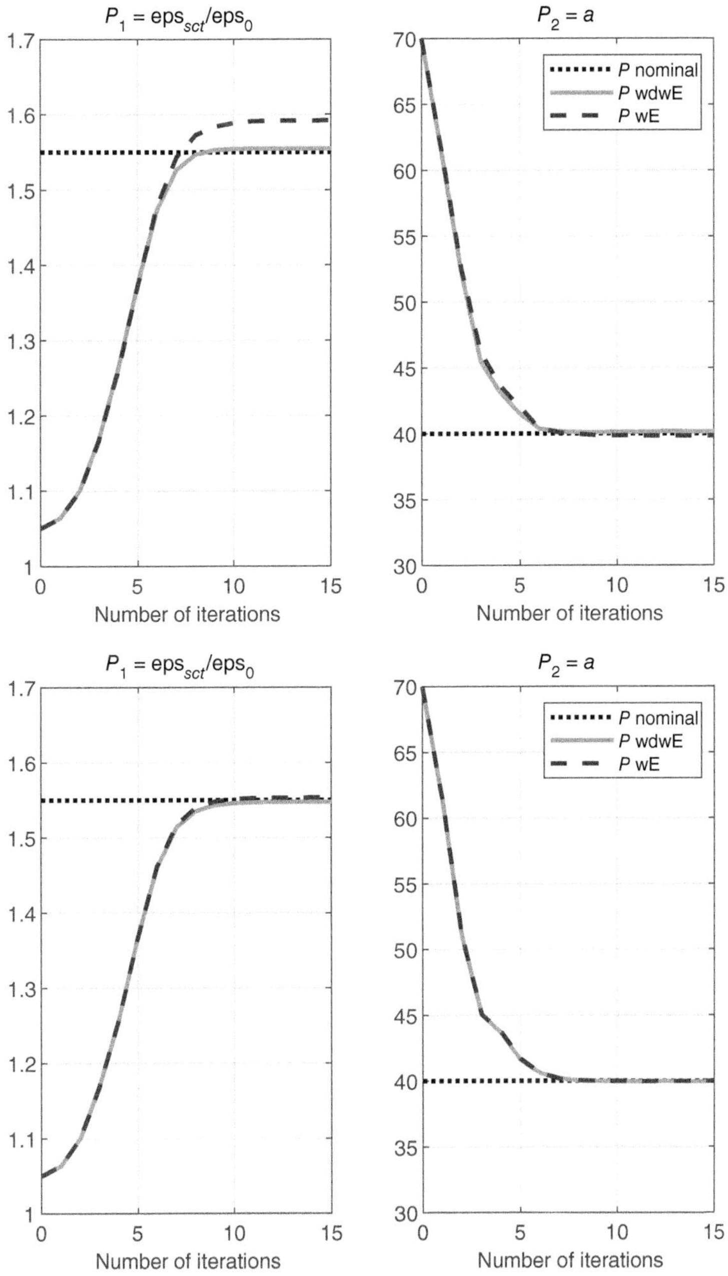

Figure 6.10 Plots of the parameters $\varepsilon_{sct}/\varepsilon_0 = P_1^{(n)}$ and $a = P_2^{(n)}$ as function of the number of Gauss–Newton iterations, with damping factor $\alpha = 0.618$ for $\Delta x = 3$ m, (top pictures) and $\Delta x = 1$ m (bottom pictures), respectively.

6.6 Electromagnetic Defects Metrology

In many technical applications, there is an increasing need for electromagnetic inspection of small defects in a material structure with a known shape and composition. These defects may have occurred during the production process. Further, material aging can cause defects, certainly due to intensive use. Regular inspection by nondestructive testing is then required. If we want to display these defects with electromagnetic waves, we face the essential problem that these defects can be much smaller than a wavelength, while the overall dimensions of the structure are much larger than the wavelength. Nonlinear wave-field inversion, as explained in this chapter, cannot reconstruct the entire structure with simple means. To detect possible defects, a very fine computational grid is required. We then need an enormous amount of computing power, and in addition there is a great danger that the iterative inversion scheme will end up in a local minimum. We cannot *use sledgehummers to crack nuts.*

We propose a linearization of the problem. The ideal structure is known. We denote it as the base model. See our 2D example in Figure 6.11a. It consists of a circular domain with a few inclusions of different material and shape. For this base model, we solve the forward scattering problem and we generate and store synthetic scattering data. For comparison, we compute synthetic data using both the wE-method and the wdwE-method. Let us monitor the structure for changes to the base model. The modified model is referred to as the monitor model. We assume that at the time of measurement, the material differences in the base model are not present anymore. Then, the monitor model is identical to the homogeneous circular cylinder (see Figure 6.11b). For this monitor model we calculate the scattering data analytically using the Bessel series expansion. These data play the role of measured data. In practice, we sometimes miss the phase information of the measured data, but for small differences between the monitor model and the base model, we may use the known phase of the synthetic data computed for the base model. In principle, the difference of the scattered-field data set of the basic model and the data set of the monitor model contains information about the defects, being the differences between the two models. The purpose of this section is to show how the actual differences can be imaged robustly. This so-called defects model is shown in Figure 6.11c.

We start our simulations with a modification of the function `initEMcontrast` to provide a choice of various base models, cf. Matlab Listing II.107. Our base model is constructed by including a few "windows" in the circular domain with a permittivity contrast of $\chi^{\varepsilon} = -0.55$. The permittivity contrast of the windows are equal to $\frac{1}{2}\chi^{\varepsilon} = -0.275$. Assuming that the windows in the monitor model are not present, the defects model consists of a number of small objects with contrast of -0.275. The characteristic spatial dimensions of the windows are between 2 and 20 m. Given the smallest dimension of 2 m, we change the grid size in `initEMgrid` into $\Delta x = 1$ m. Since for our

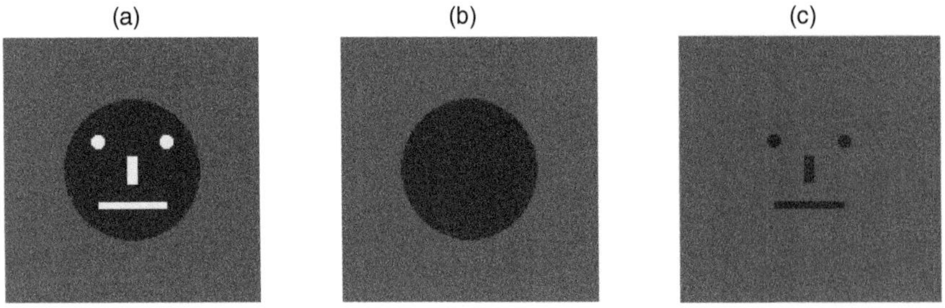

Figure 6.11 Definition of the models in defects metrology. (a) Base model, (b) monitor model, and (c) defects model.

Matlab Listing II.106 The script `Metrology` to compute both analytic data for the monitor model and synthetic data for the base model, using both the wE-method and the wdwE method. These data are processed by focusing functions.

```
clear all; clc; close all; clear workspace;
input = initEM();

% (1) Compute analytically scattered field data for monitor model -------
      disp('Running EMdataCircle'); EMdataCircle;

%     Input interface contrast
      [input] = initEMcontrastIntf(input);

% (2) Compute Green function for sources and receivers ------------------
      pG = EGreenSourceReceivers(input);

% (3) Compute base-model data using either BICGSTABwE or BICGSTABwdwE ---
      E1DATAwE = zeros(input.NR,input.NS);
      E2DATAwE = zeros(input.NR,input.NS);
      for q = 1 : input.NS
          [E_incq,~]      = IncEMdwave(pG,q,input);
              [w_E]       = BICGSTABwE(E_incq,input);
              [ErSct]     = DOPwE(pG,w_E,input);
          E1DATAwE(:,q)   = ErSct{1};
          E2DATAwE(:,q)   = ErSct{2};
      end %q_loop

      E1DATAwdwE = zeros(input.NR,input.NS);
      E2DATAwdwE = zeros(input.NR,input.NS);
      for q = 1 : input.NS
          [E_incq,dEincq] = IncEMdwave(pG,q,input);
              [w_E,dwE]   = BICGSTABwdwE(E_incq,dEincq,input);
              [ErSct]     = DOPEwdw(w_E,dwE,input);
          E1DATAwdwE(:,q) = ErSct{1};
          E2DATAwdwE(:,q) = ErSct{2};
      end %q_loop

% (4) Plot monitor-model and base-model data using wE or wdwE method ----
      [difE1,difE2] = ...
          PlotDataDif(E1data,E2data,E1DATAwE,E2DATAwE,input);
      disp(['difwE1 =' difE1]);
      disp(['difwE2 =' difE2]);

      [difE1,difE2] = ...
          PlotDataDif(E1data,E2data,E1DATAwdwE,E2DATAwdwE,input);
      disp(['difwdwE1 =' difE1]);
      disp(['difwdwE2 =' difE2]);

% (5) Focus the two type of data sets using wE or wdwE method -----------
      Focusing(E1data,E2data,E1DATAwE,  E2DATAwE,  input);
      Focusing(E1data,E2data,E1DATAwdwE,E2DATAwdwE,input);
      PlotDefinitionModels
```

Matlab Listing II.107 The function `initEMcontrast` to select the base model.

```
function [input] = initEMcontrast(input)

input.a   = 40;        % radius circle cylinder
        R = sqrt(input.X1.^2 + input.X2.^2);

input.CHI_eps = (1-input.eps_sct) * (R < input.a);
input.CHI_mu  = (1-input.mu_sct)  * (R < input.a);

% Figures of contrast
x1 = input.X1(:,1);   x2 = input.X2(1,:);
figure(1); IMAGESC(x1,x2,real(input.CHI_eps));
title((')\fontsize{13} \chi^{\epsilon} = 1 - \epsilon_{sct}/ \epsilon_{0}('));
figure(2); mesh(abs(input.CHI_eps),(')edgecolor('), (')k('));
           view(37.5,45); axis xy; axis((')off(')); axis((')tight('))
           axis([1 input.N1 1 input.N2 0 1.5*max(abs(input.CHI_eps(:)))])
figure(3); IMAGESC(x1,x2,real(input.CHI_mu));
title((')\fontsize{13}\chi^{\mu} = 1 - \mu_{sct}/ \mu_{0}('));
figure(4); mesh(abs(input.CHI_mu),(')edgecolor('), (')k('));
           view(37.5,45); axis xy; axis((')off(')); axis((')tight('))
           axis([1 input.N1 1 input.N2 0 10000*max(abs(input.CHI_mu(:)))])
```

frequency of operation the wavelength is equal to 30 m, the question remains whether we are able to detect these small defects.

Next, we run the script `Metrology` of Matlab Listing II.106 to calculate analytic data for the monitor model and synthetic data for the base model. The base models are produced by the Matlab function `initEmcontrust` by selecting $(A_{\text{Eyes}}, A_{\text{Nose}}, A_{\text{Mouth}}) = (1, 0, 0)$, (0,1,0) and (0,0,1), for the left-top image, the left-middle image and the left-bottom image, respectively. We use the wE-method as well as the wdwE-method. It plots the datasets and computes the norms of the data differences, using the function `PlotDataDif` of Matlab Listing II.108. These data are processed by focusing functions, which we explain in the remainder of this section.

In Figure 6.12, for the wE-method, we present some numerical results. For three parts of a "smiley" distribution, the absolute values of the base-model data and the differences between these data and the absolute values of the monitor-model data are presented. The three base models are denoted as the eyes-model, the nose-model, and the mouth-model. We note that it is an almost impossible job to subtract any information of the location of the defects in the spatial domain from differences in data domain (receiver-source domain). A good idea is to project the differences between the monitor-model data and the base-model data from the $(\boldsymbol{x}_p^R, \boldsymbol{x}_q^S)$-space to the (x_1, x_2)-space. To this end, we project both the receiver locations and the source locations to each point (x_1, x_2) of the observational space. We refer to this type of backprojection as a focusing procedure. This procedure is described by the function `divFocus` of Matlab Listing II.109. The receiver-focusing operator is defined as

$$F_p^R(\boldsymbol{x} - \boldsymbol{x}_p^R) = [K_0(\gamma_0|\boldsymbol{x} - \boldsymbol{x}_p^R|)]^{-1}. \tag{6.62}$$

Apart from a constant, it is the inverse of the 2D scalar Green function $G(\boldsymbol{x} - \boldsymbol{x}_p^R)$. In fact, it is a back projection from the receiver point $\boldsymbol{x}_p^R$ to the observation point $\boldsymbol{x}$ in the object domain $\mathbb{D}$. Similarly, the source-focusing operator is defined as

$$F_q^S(\boldsymbol{x} - \boldsymbol{x}_q^S) = [K_0(\gamma_0|\boldsymbol{x} - \boldsymbol{x}_q^S|)]^{-1}. \tag{6.63}$$

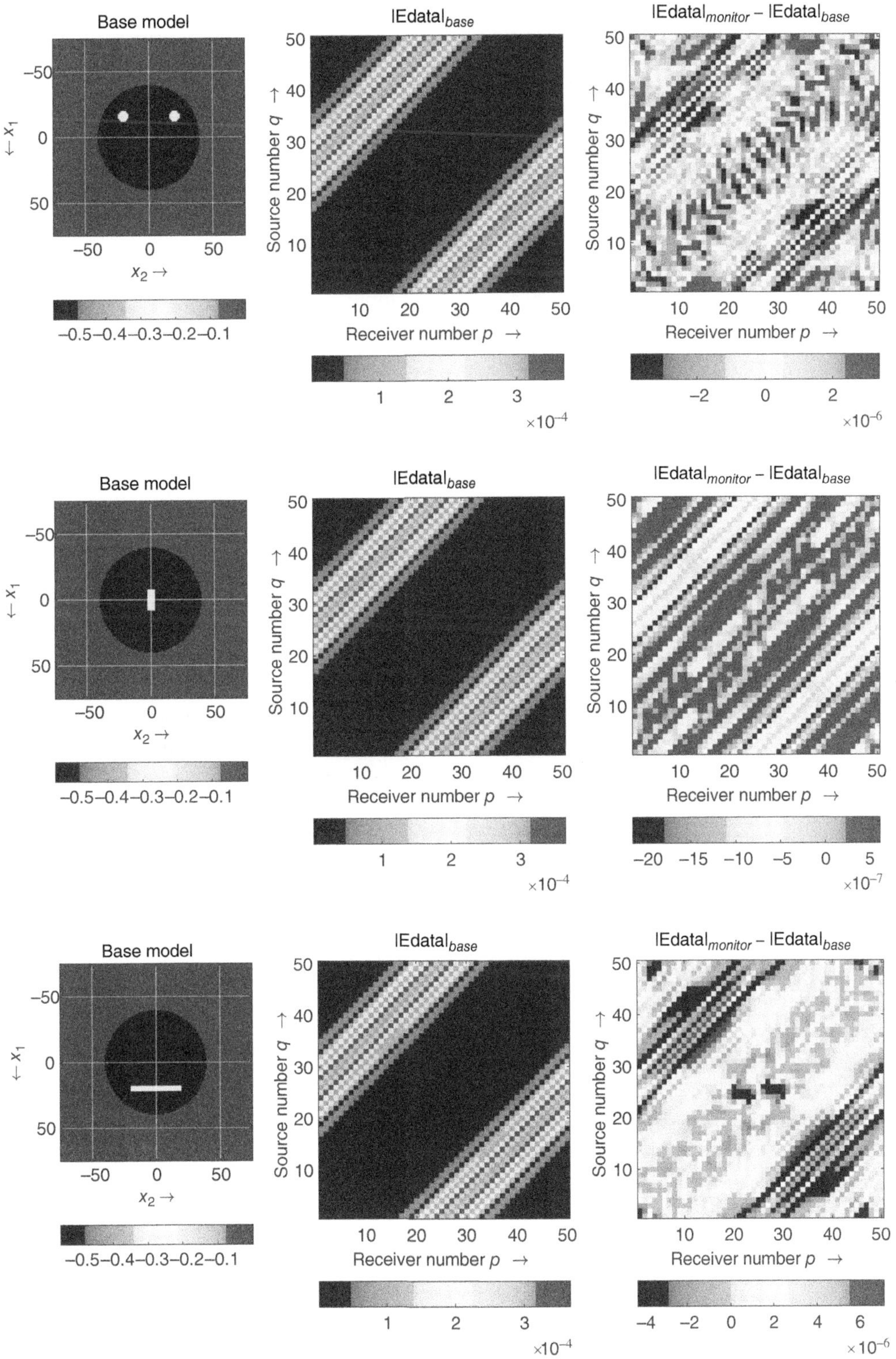

Figure 6.12 Electromagnetic scattered base-model data as function of the receiver and source positions, and the difference between the base-model and the monitor-model data.

Matlab Listing II.108 The function `PlotDataDif` to plot the data obtained by using the wE- and wdwE-methods.

```
function  [difE1 , difE2] =  ...
                 PlotDataDif(E1monitor, E2monitor, E1base, E2base, input)

absEmonitor = sqrt(abs(E1monitor).^2 + abs(E2monitor).^2);
absEbase    = sqrt(abs(E1base).^2 + abs(E2base).^2);

set(figure,'Units','centimeters','Position', [0 0 25 25] );
    sp = subplot(1,3,1);
         IMAGESC(input.X1(:,1),input.X1(:,1),(input.CHI_eps));
         grid on; ax = gca; ax.GridColor = 'w'; ax.GridAlpha =0.5;
         title('\fontsize{12}Base model');
         pos = get(sp,'Position');
         pos = pos+[0.05 0.05 -0.05 -0.057]; set(sp,'Position',pos);
    subplot(1,3,2);
         PlotData(absEbase,input);
         title('\fontsize{12}|Edata|_{base}');
    subplot(1,3,3);
         PlotData(absEmonitor-absEbase,input);
         title('\fontsize{12}|Edata|_{monitor}-|Edata|_{base}');
         colormap(jet.^(1/4));
         h = caxis;  margin = (h(2)-h(1))/4;
         caxis([h(1)+margin; h(2)-margin]);

% Compute normalized difference of norms
   difE1 = num2str(norm(E1monitor(:)-E1base(:),1)/norm(E1monitor(:),1));
   difE2 = num2str(norm(E2monitor(:)-E2base(:),1)/norm(E2monitor(:),1));
```

Next, we focus the two components of the difference in scattered-field data as

$$
\begin{aligned}
F_1(x_1,x_2) &= \sum_{p=1}^{N^R}\sum_{q=1}^{N^S} F_p^R(\boldsymbol{x}-\boldsymbol{x}_p^R)\; F_q^S(\boldsymbol{x}-\boldsymbol{x}_q^S)\; \Delta E_1^{sct}(\boldsymbol{x}_p^R,\boldsymbol{x}_q^S),\\
F_2(x_1,x_2) &= \sum_{p=1}^{N^R}\sum_{q=1}^{N^S} F_p^R(\boldsymbol{x}-\boldsymbol{x}_p^R)\; F_q^S(\boldsymbol{x}-\boldsymbol{x}_q^S)\; \Delta E_2^{sct}(\boldsymbol{x}_p^R,\boldsymbol{x}_q^S).
\end{aligned}
\tag{6.64}
$$

The vector F_j presents the focused electric field. An interesting observation now is that the electric-current density is obtained by taking the divergence of the local electric-flux density. Assuming that the permittivity is constant at the location of the defects, the computation of $\partial_j F_j$ is a measure of the local electric-contrast-current density. At interfaces where the permittivity changes abruptly, the wE-method is expected to produce blurring images, while the wdwE-method is expected to offer a better approach to mapping the local current densities.

6.6.1 Data with Phase Information

First, we assume that every measurement of the scattered field in our monitor model has both amplitude information and phase information. The difference of the scattered electric-field data in

Matlab Listing II.109 The function `divFocus` to compute the focusing operators and to locate the contrast sources by computation the divergence of the focused electric-field vectors.

```
function [DivFocus] = divFocus(data1,data2,input)
  N1 = input.N1; N2 = input.N2;

% Compute receiver focusing operators ------------------------------------
  Focus_R = cell(1,input.NR);
  for p = 1 : input.NR
        X1  = input.X1 - input.xR(1,p);
        X2  = input.X2 - input.xR(2,p);
        DIS = sqrt(X1.^2 + X2.^2);
        Focus_R{p} = 1./ besselk(0,input.gamma_0*DIS);
  end; % p_loop

% Compute source focusing operators --------------------------------------
  Focus_S = cell(1,input.NS);
  for q = 1 : input.NS
        X1 = input.X1 - input.xS(1,q);
        X2 = input.X2 - input.xS(2,q);
       DIS = sqrt(X1.^2 + X2.^2);
       Focus_S{q} = 1./ besselk(0,input.gamma_0*DIS);
  end; % q_loop

% Focus the two components of the electric field data -------------------
  dataFocus{1} = zeros(input.N1,input.N2);
  dataFocus{2} = zeros(input.N1,input.N2);
  for p = 1:input.NR;
    for q = 1 :input.NS;
     dataFocus{1} = dataFocus{1} + data1(p,q) .* Focus_R{p} .* Focus_S{q};
     dataFocus{2} = dataFocus{2} + data2(p,q) .* Focus_R{p} .* Focus_S{q};
    end; % q_loop
  end; % p_loop

% Take divergence of vector (dataFocus{1},dataFocus{2}) ------------------
  DivFocus = DIV(dataFocus,input);
  % Enforce the divergence at the boundary of the test domain to zero
    DivFocus(1,:)  = DivFocus(2,:);     DivFocus(:,1)  = DivFocus(:,2);
    DivFocus(N1,:) = DivFocus(N1-1,:);  DivFocus(:,N2) = DivFocus(:,N2-1);
```

the monitor model and the base model is given as

$$\begin{aligned}\Delta E_1^{sct}(\boldsymbol{x}_p^R,\boldsymbol{x}_q^S) &= E_1^{monitor}(\boldsymbol{x}_p^R,\boldsymbol{x}_q^S) - E_1^{base}(\boldsymbol{x}_p^R,\boldsymbol{x}_q^S),\\ \Delta E_2^{sct}(\boldsymbol{x}_p^R,\boldsymbol{x}_q^S) &= E_2^{monitor}(\boldsymbol{x}_p^R,\boldsymbol{x}_q^S) - E_2^{base}(\boldsymbol{x}_p^R,\boldsymbol{x}_q^S),\end{aligned} \tag{6.65}$$

where $E_j^{monitor}$ is the scattered-electric-field vector in presence of the monitor model and E_j^{base} is the scattered-electric-field vector in presence of the base model. Further, $p = (1, 2, \dots, N^R)$ denotes the receiver number and $q = (1, 2, \dots, N^S)$ denotes the source number. At this point, the function `Focusing` of Matlab Listing II.110 calls the function `divFocus` of Matlab Listing II.109. To determine the difference between the base-model data and the monitor-model data, it is required that the measurement setup in the monitor and base models are identical.

Matlab Listing II.110 The function `Focusing` to focus the data obtained by using the wE- and wdwE-methods.

```
function Focusing(E1monitor,E2monitor,E1base,E2base,input)
x1 = input.X1(:,1);    x2 = input.X2(1,:);

% Plot base model %--------------------------------------------------
figure;
subplot(1,3,1);
  IMAGESC(x1,x2,real(input.CHI_eps));
  grid on; ax = gca; ax.GridColor = 'w'; ax.GridAlpha =0.5;
  title('\fontsize{11}Base model');  xlabel(''); ylabel('');

% Focus E1 and E2 data with phase information %-------------------------
  [dataFocus] = divFocus(E1monitor-E1base,E2monitor-E2base,input);

subplot(1,3,2);
   IMAGESC(x1,x2, abs(dataFocus));
   grid on; ax = gca; ax.GridColor = 'w';  ax.GridAlpha =0.5;
   title('\fontsize{11}Data with phase');
   xlabel(''); ylabel('');
   hold on; phi = 0:.01:2*pi;
   plot(input.a*sin(phi),input.a*cos(phi),'-y','LineWidth',1.5); hold off;
   colormap(jet.^(1/4));
   h = caxis;  margin = (h(2)-h(1))/4;
   caxis([h(1)+ margin; h(2)-margin]);

% Focus Amplitude E1 and E2 data %----------------------------------
  fase1base = E1base./abs(E1base);
  fase2base = E2base./abs(E2base);
  % add fase factors to amplitudes of datawE
    E1monitor = abs(E1monitor) .* fase1base;
    E2monitor = abs(E2monitor) .* fase2base;
    [dataFocus] = divFocus(E1monitor-E1base,E2monitor-E2base,input);

subplot(1,3,3)
   IMAGESC(x1,x2, abs(dataFocus));
   title('\fontsize{11}Amplitude data');
   grid on; ax = gca; ax.GridColor = 'w';  ax.GridAlpha =0.5;
   xlabel(''); ylabel('');
   hold on; phi = 0:.01:2*pi;
   plot(input.a*sin(phi),input.a*cos(phi),'-y','LineWidth',1.5); hold off;
   colormap(jet.^(1/4));
   h = caxis;  margin = (h(2)-h(1))/4;
   caxis([h(1)+ margin; h(2)-margin]);
```

6.6.2 Phaseless Data

Second, we simulate the situation where the monitor data lacks phase information. Then we only have amplitude data. Without phase, such data cannot be focused. In the case that the defects in the contrast profile are small, we can assume that we substitute the known phase of the field in the basic model as a replacement for the missing information of the field in the monitor model. We define the known phase factor of the base-model data as

$$\begin{aligned}\Phi_1(\boldsymbol{x}_p^R,\boldsymbol{x}_q^S) &= \frac{E_1^{base}(\boldsymbol{x}_p^R,\boldsymbol{x}_q^S)}{|E_1^{base}(\boldsymbol{x}_p^R,\boldsymbol{x}_q^S)|},\\ \Phi_2(\boldsymbol{x}_p^R,\boldsymbol{x}_q^S) &= \frac{E_2^{base}(\boldsymbol{x}_p^R,\boldsymbol{x}_q^S)}{|E_2^{base}(\boldsymbol{x}_p^R,\boldsymbol{x}_q^S)|},\end{aligned} \tag{6.66}$$

and complete the amplitude data of the monitor model with these phase factors as

$$\begin{aligned}E_1^{monitor}(\boldsymbol{x}_p^R,\boldsymbol{x}_q^S) &= |E_1^{monitor}(\boldsymbol{x}_p^R,\boldsymbol{x}_q^S)|\ \Phi_1(\boldsymbol{x}_p^R,\boldsymbol{x}_q^S),\\ E_2^{monitor}(\boldsymbol{x}_p^R,\boldsymbol{x}_q^S) &= |E_2^{monitor}(\boldsymbol{x}_p^R,\boldsymbol{x}_q^S)|\ \Phi_2(\boldsymbol{x}_p^R,\boldsymbol{x}_q^S).\end{aligned} \tag{6.67}$$

At this point, the Matlab function `Focusing` calls the function `divFocus` of Matlab Listing II.109.

6.6.3 Focused Data

In Figure 6.13, we present the focused defect images for our eyes-model, nose-model, and mouth-model. For these base models, the scattered-field data are computed using the wE-method. The images using monitor-model data with phase information are displayed in the middle column. The defect locations of the eyes, the nose and the mouth are fairly well positioned, although the eyes and mouth are rather blurred. Furthermore, large artifacts occur in the images of the eyes-model and the mouth-model, namely close to the circular boundary. Band-limited wave phenomena are visible, which is a representation of multiple scattering present in the radial direction of the base model. In the nose-model, the image is very clear because the nose is located far away from the circular boundary. The images using phaseless data are in the right-hand column of Figure 6.13. In these images, the presence of multiscattering in radial direction becomes more dominant. It reduces the resolution of the defects; for example, the eye-model is vaguely visible.

In Figure 6.14, we present the results of our focusing procedure, when the scattered-field data of the base models are computed using the wdwE-method. We now immediately see that the defects are imaged very well. Because the wdwE-method exclusively takes the interface-contrast source into account. We also see that the spatial expansion of the multi-scattering effect decreased considerably. Despite the presence of interaction between the defects close to the circular boundary, the eye model and the mouth model are well resolved. Although the wavelength is 30 m, we can say that the height and width of the nose model are very well reconstructed.

As a final example, we combine our previous basic models into a "smiley" model. When we use the wE-method to calculate the scattered-field data for our smiley model, the upper images

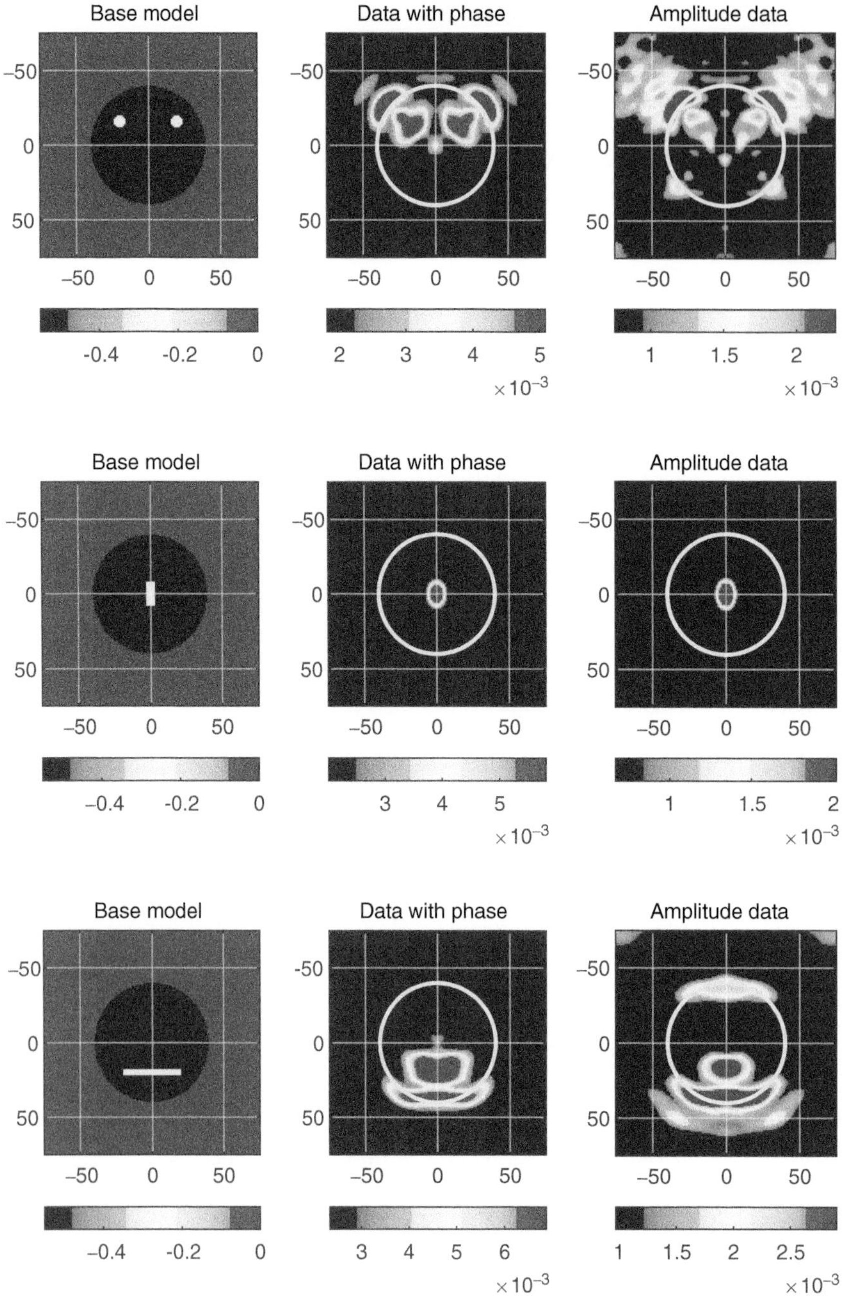

Figure 6.13 Electromagnetic metrology: differences between monitor-model data and base-model data, projected in Cartesian space. The circular line encloses the homogeneous domain of the monitor model. The scattered-field data for the three types of base models are computed using the wE-method.

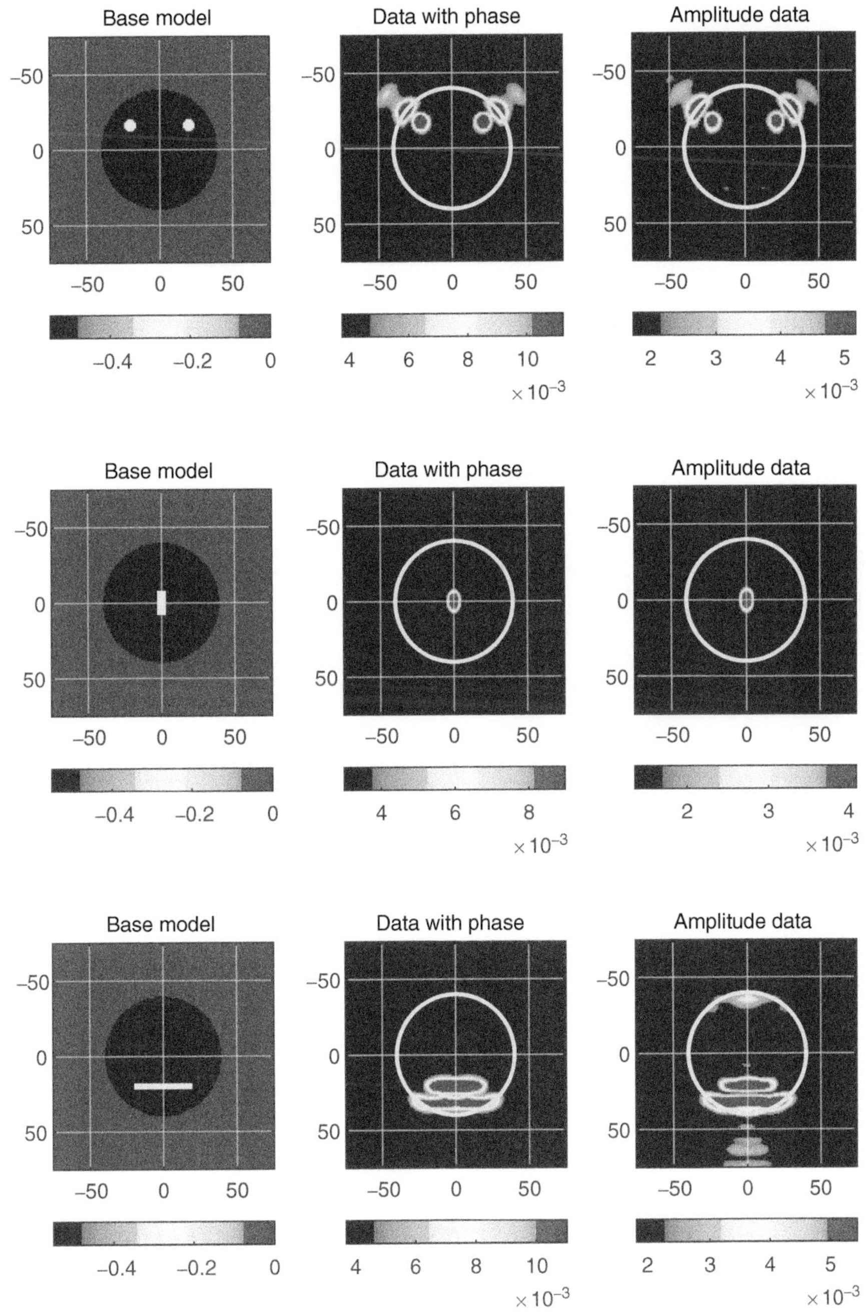

Figure 6.14 Electromagnetic metrology: differences between monitor-model data and base-model data, projected in Cartesian space. The circular line encloses the homogeneous domain of the monitor model. The scattered-field data for the three types of base models are computed using the wdwE-method.

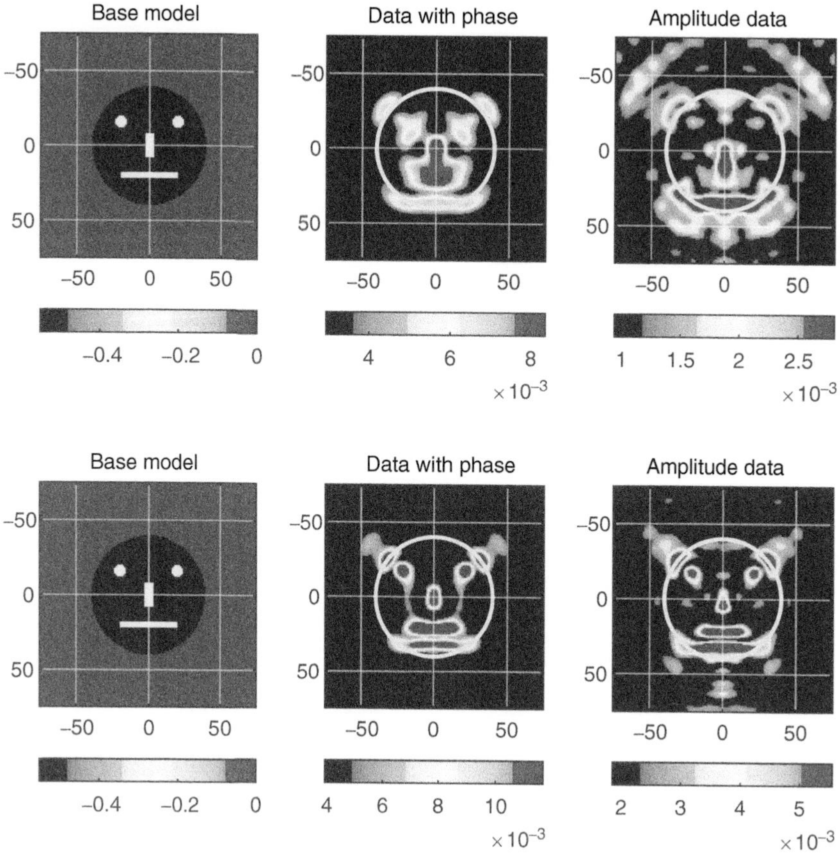

Figure 6.15 Electromagnetic metrology: differences between monitor-model data and base-model data, projected in Cartesian space. The circular line encloses the homogeneous domain of the monitor model. The scattered-field data for the three types of base models are computed using either the wE-method (top pictures) or the wdwE-method (bottom pictures).

of Figure 6.15 represent the results of our focusing procedure. In addition to artifacts due to wave-field interaction with the circular boundary, we also detect interaction between the eyes, nose, and mouth models. The lower images of Figure 6.15 represent the results when we use the wdwE-method. The image quality has improved considerably. The smiley model has been very well reconstructed, even for amplitude data only. The most important artifacts are the occurrence of ears and a beard.

Matlab Scripts

In a companion website (www.wiley.com/go/vandenBerg/ScatteringAlgorithms), all Matlab scripts and functions discussed in this book are stored in a number of folders. To assist the reader of the present book with finding the section in which a specific Matlab script is covered and the correct folder where it is stored, we have made three tables, see Table M.1 for the forward scattering codes, Table M.2 for the inverse scattering codes, and Table M.3 for the forward scattering codes of the canonical objects. The left columns of these tables contain the Matlab scripts. The middle columns show the specific section of this book, while the right columns show the folder where the script is stored. All Matlab functions required to run the Matlab script are also stored in the specific folder.

Table M.1 Matlab scripts of the forward scattering codes, with references to the section and folder locations.

Matlab script	Section	Folder
ForwardCGFFTu.m	Section 1.5	ScalarWavesMfiles
ForwardCGFFTw.m	Section 1.6	ScalarWavesMfiles
ForwardBiCGSTABFFTw.m	Section 1.6.2	ScalarWavesMfiles
TimedomainBiCGSTABFFTw.m	Section 1.7	ScalarWavesMfiles
ForwardCGFFTwpZv.m	Section 2.2	AcousticWavesMfilespZv
ForwardBiCGSTABFFTpZv.m	Section 2.7.2	AcousticWavesMfilespZv
ForwardCGFFTwcdrho.m	Section 2.4	AcousticWavesMfilesPressure
ForwardCGFFTwdw.m	Section 2.7	AcousticWavesMfileswdw
ForwardBiCGSTABFFTwdw.m	Section 2.7.2	AcousticWavesMfileswdw
TimedomainBiCGSTABFFTwdw.m	Section 2.8	AcousticWavesMfileswdw
ForwardCGFFTwE.m	Section 3.2.3	EMWavesMfilesElectricEps
ForwardBiCGSTABFFTwE.m	Section 3.2.3	EMWavesMfilesElectricEps
ForwardBiCGSTABFFTEwdw.m	Section 3.3.3	EMWavesMfilesElectricEpswdw
ForwardBiCGSTABFFTwEH.m	Section 3.4.2	EMWavesMfilesEHepsmu
ForwardBiCGSTABFFTEdw.m	Section 3.5.2	EMWavesMfilesEdw
TimedomainBiCGSTABFFTwE.m	Section 3.6	EMWavesMfilesElectricEps
TimedomainBiCGSTABFFTwdw.m	Section 3.6	EMWavesMfilesElectricEps

Forward and Inverse Scattering Algorithms based on Contrast Source Integral Equations, First Edition. Peter M. van den Berg.

Companion website: www.wiley.com/go/vandenBerg/ScatteringAlgorithms

Table M.2 Matlab scripts of the inverse scattering codes, with references to the section and folder locations.

Matlab script	Section	Folder
Forwardw.m	Section 4.2	InverseScalarWavesMfiles
InverseCSI.m	Section 4.5.5	InverseScalarWavesMfiles
InverseMRCSI.m	Section 4.6.5	InverseScalarWavesMfiles
InverseCSIGaussNewton.m	Section 4.7.2	InverseScalarWavesMfiles
GaussNewtonNumeric.m	Section 4.9.1	InverseScalarWavesGaussNewtonMfiles
GaussNewtonAnalytic.m	Section 4.9.1	InverseScalarWavesGaussNewtonMfiles
ForwardwZv.m	Section 5.2	InverseAcousticWavesZvMfiles
AcMRCSI.m	Section 5.3.5	InverseAcousticWavesZvMfiles
Forwarddw.m	Section 5.4.1	InverseAcousticWavesdwMfiles
MICSI.m	Section 5.5.5	InverseAcousticWavesdwMfiles
ForwardwE.m	Section 6.2	InverseEMwavesElectricEpsMfiles
ForwardwdwE.m	Section 6.3	InverseEMwavesElectricEpsMfiles
EmMRCSI.m	Section 6.4.5	InverseEMwavesElectricEpsMfiles
GaussNewtonwE.m	Section 6.5.1	InverseEMwavesGaussNewtonMfiles
GaussNewtonwdE.m	Section 6.5.1	InverseEMwavesGaussNewtonMfiles
Metrology.m	Section 6.6	InverseEMwavesMetrologyMfiles

Table M.3 Matlab scripts of the forward scattering codes of the canonical objects, with references to the section and folder locations.

Matlab script	Section	Folder
ForwardCanonicalObjects.m	Section 1.C.4	All forward folders Scalar Waves
AcForwardCanonicalObjects.m	Section 2.A.4	All forward folders Acoustic waves
EMForwardCanonicalObjects.m	Section 3.A.3	All forward folders EM waves
DataSctCircle.m	Section 4.2	All folders Scalar wave inversion
AcousticDataSctCircle.m	Section 5.2	All folders Acoustic wave inversion
EMdataCircle.m	Section 6.2	All folders EM wave inversion

References

1 Abramowitz, M. and Stegun, I.A. (1985). *Handbook of Mathematical Functions*, 7e. New York: Dover Publications.

2 Abubakar, A. and Habashy, T.M. (2005). A Green function formulation of the extended Born approximation for three-dimensional electromagnetic modelling. *Wave Motion* 41 (3): 211–27.

3 Abubakar, A., Habashy, T.M., Pan, G., and Li, M.K. (1994). Application of the multiplicative regularized Gauss-Newton algorithm for three dimensional microwave imaging. *IEEE Trans. Antennas Propag.* 60: 2431–2441.

4 Arfken, G. (1985). *Mathematical Methods for Physicists*, 2e. New York: Academic Press.

5 Bleistein, N. (1984). *Mathematical Methods for Wave Phenomena*. New York: Academic Press.

6 Bloemenkamp, R.F., Abubakar, A., and van den Berg, P.M. (2001). Inversion of experimental multi-frequency data using the contrast source inversion method. *Inverse Prob.* 17: 1611–1622.

7 Bojarski, N.N. (1982). Comments on 'Nonuniqeness in inverse source and scattering problems'. *IEEE Trans. Antennas Propag.* 30: 1037–1038.

8 Born, M. and Wolf, E. (1965). *Principles of Optics*. London: Pergamon Press.

9 Charbonnier, P., Blanc-Féraud, L., Aubert, G., and Barlaud, M. (1996). Deterministic edge-preserving regularization in computed imaging. *IEEE Trans. Image Process.* 6: 298–311.

10 Chew, W.C. and Wang, Y.M. (1982). Reconstruction of two-dimensional permittivity distribution using the distorted Born iterative method. *IEEE Trans. Med. Imaging* 9: 218–225.

11 Chew, W.C., Wang, Y.H., Otto, G. et al. (1994). On the inverse source method of solving inverse scattering problems. *Inverse Prob.* 10: 574–553.

12 Colton, D. and Kress, R. (1992). *Inverse Acoustic and Electromagnetic Scattering*. Berlin: Springer-Verlag.

13 Courant, R. and Hilbert, D. (1953). *Methods of Mathematical Physics*, vol. 1. New York: Interscience Publishers.

14 De Hoop, A.T. (1995). *Handbook of Radiation and Scattering of Waves*. London: Academic Press.

15 Devaney, A.J. and Sherman, G.C. (1982). Nonuniqeness in inverse source and scattering problems. *IEEE Trans. Antennas Propag.* 30: 1034–1037.

16 Fletcher, R. (1987). *Practical Methods of Optimization*. Hoboken, NJ: Wiley.

17 Fokkema, J.T. and van den Berg, P.M. (1993). *Seismic Applications of Acoustic Reciprocity*. Amsterdam: Elsevier.

18 Golub, G.H., Heath, M., and Wahba, G. (1979). Generalized cross-validation as a method for choosing a good ridge parameter. *Technometrics* 21: 215–223.

19 Habashy, T.M., Oristaglio, M.L., and de Hoop, A.T. (1994). Simultaneous nonlinear reconstruction of two-dimensional permittivity and conductivity. *Radio Sci.* 29: 1101–1108.

Forward and Inverse Scattering Algorithms based on Contrast Source Integral Equations, First Edition. Peter M. van den Berg.

Companion website: www.wiley.com/go/vandenBerg/ScatteringAlgorithms

20 Hähner, P. (2002). Scattering by media. In: *Scattering*, Two-Volume Set, Chapters 1.2.3 (ed. R. Pike and P. Sabatier), 77. London: Academic Press.

21 Hansen, P.C. (1992). Analysis of discrete ill-posed problems by means of the L-curve. *SIAM Rev.* 34: 561–580.

22 Hestenes, M.R. and Stiefel, E. (1952). Methods of conjugate gradients for solving linear systems. *J. Res. Nat. Bur. Stand.* 49: 409–436.

23 Hönl, H., Maue, A.W., and Westpfahl, K. (1961). Theorie der beugung. In: *Handbuch der Physik* (ed. S. Flugge), 218–573. Berlin: Springer-Verlag.

24 Hopcraft, K.I. and Smith, P.R. (1992). *An Introduction to Electromagnetic Inverse Scattering*. Dordrecht: Springer Science and Business Media.

25 Jackson, J.D. (1975). *Theory of Electromagnetism*, 3e. New York: Wiley.

26 Jones, D.S. (1964). *Classical Electrodynamics*. New York: Pergamon Press.

27 Kantorovich, L.V. and Akilov, G.P. (1958). *Functional Analysis*. New York: Pergamon Press (Translated from the Russian).

28 Kantorovich, L.V. and Krylov, V.I. (1958). *Approximate methods of Higher Analysis*. New York: Noordhoff Interscience (Translated from the Russian).

29 Kittel, C. (2004). *Introduction to Solid State Physics*, 3e. Hoboken, NJ: Wiley.

30 Kleinman, R.E. and van den Berg, P.M. (1991). Iterative methods for solving integral equations. In: *Pier 5, Progress in Electromagnetics Research, Application of Conjugate Gradient Method to Electromagnetics and Signal Analysis*, Chapter 3 (ed. T.K. Sarkar), 67–102. New York: Elsevier.

31 Kress, R. (1989). *Linear Integral Equations, Applied Mathematical Sciences*, vol. 82. Berlin, Heidelberg: Springer-Verlag.

32 Li, L., Zheng, H., and Li, F. (2009). Two-dimensional contrast source inversion method with phaseless data: TM case. *IEEE Trans. Geosci. Remote Sens.* 47: 1719–1736.

33 Maxwell, J.C. (1954). *A Treatise on Electricity & Magnetism*, Two-Volume Set, 3e. New York: Dover Publications.

34 Morozov, V.A. (1984). *Methods for Solving Incorrectly Posed Problems*. New York: Springer-Verlag (Translated from the Russian).

35 Morse, P.M. and Feshbach, H. (1953). *Mathematical Methods of Physics*, Two-Volume Set. New York: McGraw-Hill.

36 Olver, F.W.J., Olde Daalhuis, Olver, F.W.J., Olde Daalhuis et al. (eds.) *NIST Digital Library of Mathematical Functions*. http://dlmf.nist.gov/ Release 1.0.26 of 2020-03-15.

37 Oppenheim, A.V., Willsky, A.S., and Young, I.T. (1983). *Signals and Systems*. London: Prentice Hall.

38 Pike, R. and Sabatier, P. (2002). *Scattering, Scattering and Inverse Scattering in Pure and Applied Sciences*, Two-Volume Set. London: Academic Press.

39 Ralston, A. and Rabinowitz, Ph. (1965). *A First Course in Numerical Analysis*. New York: McGraw-Hill.

40 Rayleigh, J.W.S. (1945). *The Theory of Sound*, Two-Volume Set. New York: Dover Publications.

41 Remis, R.F. and van den Berg, P.M. (2000). On the equivalence of the Newton-Kantorovich and distorted Born methods. *Inverse Prob.* 16: L1–L4.

42 Roger, A. (1981). A Newton-Kantorovich algorithm applied to an electromagnetic inverse. *IEEE Trans. Antennas Propag.* 29: 232–238.

43 Rudin, L.I., Osher, S., and Fatemi, E. (1992). Nonlinear total variation based noise removal algorithms. *Physica D* 60: 259–268.

44 Saad, Y. (2003). *Iterative Methods for Sparse Linear Systems*. Philadelphia, PA: SIAM.

45 Saad, Y. and Schultz, M.H. (1986). GMRES: A generalized minimal residual algorithm for solving nonsymmetric linear systems. *SIAM J. Sci. Stat. Comput.* 7 (3): 856–869.

46 Stone, W. (1982). Comments on "Nonuniqeness in inverse source and scattering problems". *IEEE Trans. Antennas Propag.* 30: 1039.

47 Stratton, J.A. (1941). *Electromagnetic Theory*. New York: McGraw-Hill.

48 Thikhonov, A.N. (1963). Regularization of incorrectly posed problems. *Sov. Math. Dokl.* 4: 1624–1627.

49 Tijhuis, A.G. (1989). Born-type reconstruction of material parameters of an inhomogeneous lossy dielectric slab from reflected-field data. *Wave Motion* 11: 151–173.

50 van den Berg, P.M. (1984). Iterative computational techniques in scattering based upon the integrated square error criterion. *IEEE Trans. Antennas Propag.* AP-32 (10): 1063–1071.

51 van den Berg, P.M. (1991). Iterative schemes based on minimization of uniform error criterion. In: *Pier 5, Progress in Electromagnetics Research, Application of Conjugate Gradient Method to Electromagnetics and Signal Analysis*, Chapter 2 (ed. T.K. Sarkar), 27–65. New York: Elsevier.

52 van den Berg, P.M. and Abubakar, A. (2010). Optical microscopy imaging using the contrast source inversion method. *J. Mod. Opt.* 57 (9): 756–764.

53 van den Berg, P.M. and Haak, K.F.I. (1996). Profile inversion by error in the source type integral equations. In: *Wavefields and Reciprocity* (ed. P.M. van den Berg, H. Blok, and J.T. Fokkema), 87–98. Delft: Delft University Press.

54 van den Berg, P.M. and Kleinman, R.E. (1995). A total variation enhanced modified gradient algorithm for profile reconstruction. *Inverse Prob.* 11: L5–L10.

55 van den Berg, P.M. and Kleinman, R.E. (1997). A contrast source inversion method. *Inverse Prob.* 13: 1607–1620.

56 van den Berg, P.M., van Broekhoven, A.L., and Abubakar, A. (1999). Extended contrast inversion. *Inverse Prob.* 15: 1325–1344.

57 van der Neut, J.R., van den Berg, P.M., Fokkema, J.T., and van Dongen, K.W.A. (2018). Acoustic interface contrast imaging. *Inverse Prob.* 34: 115006 (15pp).

58 van der Vorst, H.A. (1992). Bi-CGSTAB: a fast and smoothly converging variant of BI-CG for the solution of nonsymmetric linear systems. *SIAM J. Sci. Stat. Comput.* 13 (2): 631–644.

59 van Dongen, K.W.A. and Wright, W.M.D. (2007). A full vectorial contrast source inversion scheme for three-dimensional acoustic imaging of both compressibility and density profiles. *J. Acoust. Soc. Am.* 121 (3): 1538–1549.

60 Watson, G.N. (1989). *A Treatise on the Theory of Bessel Functions*. London: Cambridge University Press.

61 Wheeler, G.F. and Crummet, W.P. (1987). The vibrating string controversy. *IEEE Trans. Microwave Theory Tech.* 55 (1): 33–37.

62 Wilcox, C.H. (1959). Spherical means and radiation conditions. *Arch. Ration. Mech. Anal.* 3: 133–148.

63 Zwamborn, A.P.M. and van den Berg, P.M. (1992). The three dimensional weak form of the conjugate gradient FFT method for solving scattering problems. *IEEE Trans. Microwave Theory Tech.* 40 (9): 1757–1766.

Biography

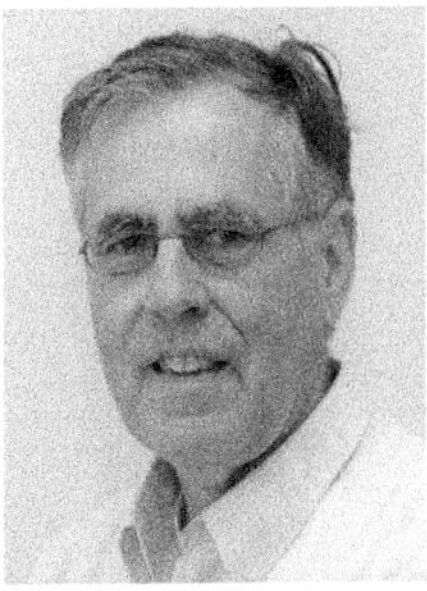

Peter M. van den Berg was born in Rotterdam, The Netherlands, on 11 November 1943. He received the degree in electrical engineering from the Polytechnical School of Rotterdam in 1964, the MSc degree in electrical engineering, and the PhD degree in technical sciences, both from the Delft University of Technology, in 1968, and 1971. From 1967 to 1968, he was employed as a Research Engineer by the Dutch Patent Office. Since 1968, he has been a member of the Scientific Staff of the Electromagnetic Research Group of the Delft University of Technology. He was appointed as a Full Professor in Electromagnetics in the Faculty of Electrical Engineering at the Delft University of Technology in 1981. Since 2004, he is a Research Professor in the Faculty of Applied Sciences of the Delft University. His main research interest is the efficient computation of electromagnetic and acoustic wavefield problems using iterative techniques based on error minimization, the computation of fields in strongly inhomogeneous media, and the use of wave phenomena in seismic data processing. Major interest is in an efficient solution of the nonlinear inverse scattering problem for enhanced imaging in biomedical and geophysical applications. Peter M. van den Berg was a consultant for Shell International, Petroleum GEO-Services, Schlumberger-Doll Research, and ASML. Presently, he is Professor Emeritus in Electromagnetic Theory.

Peter M. van den Berg is (co-)author of 157 publications in international refereed journals and 37 chapters in peer reviewed books, 6 books, 156 international conference papers and 58 international conference abstracts, editorship of 3 special issues of an international refereed journal. He has (co-)supervised 26 successfully completed PhD theses at the Delft University Technology (5 "cum laude"), 2 PhD theses at the University of Delaware, Newark, DA, USA, and 1 PhD thesis at the University of Nijmegen, The Netherlands. He has presented 86 invited lectures, and he is (co)-inventor of 15 patents.

Forward and Inverse Scattering Algorithms based on Contrast Source Integral Equations, First Edition. Peter M. van den Berg.

Companion website: www.wiley.com/go/vandenBerg/ScatteringAlgorithms

Index

a

Acoustic waves
  3D contrast-source integral equations for pressure and particle-velocity 87
  3D contrast-source integral representation for scattered pressure field 87
  cost functional 390
  Matlab Listings
    AdjKOPpZv.m 93
    DIV.m 90
    GRAD.m 90
    graddiv.m 88
    KOPpZv.m 92
  residual error 391
  adjoint operator 90
  BiCGSTAB 160
  canonical configuration
    1D scattering by a slab 170
    3D scattering by a sphere 177
  compressibility contrast 86, 378
  conjugate gradient direction
    Polak-Ribière 391, 417
  Conjugate Gradient method 92, 116, 142
  contrast estimate 394
  contrast sources 85
  contrast update 393
  contrast-source integral equations 378
  contrast-source integral representation 377
  contrast-source inversion 386
  convolution
    spatial derivative Green function 83
  cost functional 416
  Dirac distribution 80
  forward discrete Fourier transform 163
  Green function
    spatial derivative of weak 95
  impulse function 80
  incident particle velocity 83
  incident pressure 83
  integral equation
    for pressure only 86
    for the wave fields 86
    for wave speed and mass density gradient contrast 114
    with interface sources 132, 133
    with volume and interface sources 136
  integral representation
    for wave speed and mass density gradient contrast 114
    scattered particle velocity 85
    scattered pressure 85
  interface contrast 407
  interface contrast estimate 420
  interface contrast sources 405
  interface contrast update 419
  inverse discrete Fourier transform 163
  Kirchhoff approximation 422, 430
  Laplace transformation 80
  lossy media 81
  mass-density contrast 86, 379
  Mass-density Interface Contrast Source Inversion (MICSI) 416
  measurement domain 391, 416
  monopole source 83
  Multiplicative Regularized Contrast Source Inversion (MRCSI) 390
  object domain 391, 416
  particle-velocity contrast source 86, 377
  pressure contrast source 86, 377
  regularization by windowing 420

Forward and Inverse Scattering Algorithms based on Contrast Source Integral Equations, First Edition. Peter M. van den Berg.

Companion website: www.wiley.com/go/vandenBerg/ScatteringAlgorithms

Acoustic waves (*contd.*)
residual error 416
s-domain continuity conditions 81
s-domain wave equation 81
scattered wave field 84
scattering domain 84
step length 392, 418
time-domain continuity conditions 80
time-domain wave equations 79
wave impedance 82
wave speed and mass density contrast 114
wavelet 83
Acoustic waves Forward Matlab Listings
AcForwardCanonicalObjects.m 179
AcousticModifiedContrast.m 120
AcousticSctWavefieldCircle.m 175
AcousticSctWavefieldSlab.m 173
AcousticSctWavefieldSphere.m 178
AdjKOPwcdrho.m 117
AdjKOPwdw.m 145
CheckAdjointpZv.m 102
CheckAdjointwcdrho.m 122
CheckAdjointwdw.m 151
Dopdw.m 157
Dopw.m 156
Dopwcdrho.m 129
DopwpZv.m 110
extrapolate.m 140
ForwardCGFFTwcdrho.m 120
ForwardCGFFTwdw.m 147
ForwardCGFFTwpZv.m 99
Greenfun.m 98
IncAcousticWave.m 100
IncPressurewave.m 149
initAC.m 96
initAcousticContrast.m 99
initAcousticContrastIntf.m 148
initFFTGreenfun.m 97
interpolate.m 139
ITERBiCGSTABwdw.m 162
ITERBiCGSTABwpZv.m 161
ITERCGwcdrho.m 121
ITERCGwdw.m 150
ITERCGwpZv.m 101
KOPwcdrho.m 116
KOPwdw.m 143
plotContrastSourceWcrho.m 123
plotContrastSourcewp.m 103
plotContrastSourcewZv.m 104
plotInterfaceSource.m 152
plotInterfaceSourceWdrho.m 125
plotPressureWavefield.m 108
SnapshotP.m 165
TimedomainBiCGSTABFFTwdw.m 164
Wavelet.m 163
Acoustic waves Inversion Matlab Listings
AcMRCSI.m 396
AcousticDataSctCircle.m 385
AdjDOPMdw.m 414
AdjDOPMwZv.m 390
AdjKOPdw.m 413
AdjKOPwZv.m 389
BICGSTABdw.m 412
BICGSTABwZv.m 388
convert.m 430
dGreenSourcesReceivers.m 386
DOPMdw.m 413
DOPMwZv.m 389
Forwarddw.m 410
ForwardwZv.m 384
Greenfun.m 381
IncPressureWave.m 410
initAC.m 380, 407
initAcousticContrast.m 382, 408
initFFTGreenfun.m 381
InitialEstimatedw.m 422
InitialEstimatewZv.m 395
InvITERdw.m 424
InvITERwZv.m 398
IterCGdw.m 411
ITERCGMDdw.m 423
ITERCGMDwZv.m 397
ITERCGwZv.m 387
KOPdw.m 413
KOPwZv.m 389
loadImages.m 401
loadMeshes.m 402
loadRfl.m 429
MICSI.m 421
RfltoCHI.m 426
saveCHI.m 400
saveRfl.m 427
UpdateContrastdw.m 425
UpdateContrastwZv.m 399

e

Electromagnetic waves
- 2D incident electric field 187
- 2D incident magnetic field 187
- 3D incident electric field 186
- 3D incident magnetic field 186
- 3D integral equation
  - for electric and magnetic fields 190
- cost functional 460
- golden ratio 480
- residual error 460
- base model 486
- BiCGSTAB 209, 222, 243, 258
- canonical configuration
  - 2D scattering by a circular cylinder 284
  - 3D scattering by a sphere 292
- compatibility relations 181
- conjugate gradient direction
  - Polak-Ribière 460
- Conjugate Gradient method 193
- constitutive relations 181
- contrast estimate 463
- contrast source inversion 459
- contrast sources 439
- contrast update 462
- contrast-source integral equations 440
- contrast-source integral representation 439
- Debye potential 279
- defects metrology 486
- defects model 486
- Dirac distribution 182
- EM waves Forward Matlab Listings
  - plotFieldError.m 285
- focused data 493
- forward discrete Fourier transform 269
- Gauss–Newton method 476
- Green function
  - 2D spatial derivatives 187
  - 3D spatial derivatives 186
- impulse function 182
- integral equation
  - 2D E-polarization 191
  - 2D H-polarization 191
  - electric field 191, 212, 251
  - permittivity and permeability contrast 238
  - permittivity and permeability contrast sources 238
  - permittivity contrast source 191
  - volume sources with wave speed and gradient permittivity contrast 212
  - zero wave speed contrast and gradient permeability contrast 251
  - zero wave speed contrast and interface permeability contrast 252
- integral representation
  - electric field 191, 212, 238, 251, 252
  - magnetic field 191, 238
  - permittivity and permeability contrast 238
  - permittivity contrast source 191
  - scattered electric field 189
  - scattered magnetic field 190
- inverse discrete Fourier transform 269
- Laplace transformation 183
- lossy media 184
- measurement domain 460
- Mie series 279
- object domain 460
- permittivity contrast 440
- phaseless data 493
- s-domain continuity conditions 183
- s-domain Maxwell equations 183
- step length 462
- time-domain continuity conditions 182
- time-domain Maxwell equation 181
- wave impedance 185
- wavelet 269

Electromagneticwaves
- wavespeed 185

EM waves Forward Matlab Listings
- AdjKopE.m 193
- CheckAdjointE.m 202
- CURL.m 239
- displayEdata.m 205
- displayHdata.m 206
- DOP3Dmudw.m 267
- Dop3DwE.m 204
- Dop3DwEH.m 248
- DopE3Dwdw.m 234
- DopEwdw.m 233
- DOPmudw.m 266
- DopwE.m 203
- DopwEH.m 247
- Edwvector2matrix.m 260
- EHvector2matrix.m 242

EM waves Forward Matlab Listings (*contd.*)
 EMForwardCanonicalObjects.m 306
 EMincidentTest2D.m 286
 EMincidentTest2Derror.m 288
 EMincidentTest3D.m 299
 EMincidentTest3Derror.m 303
 EMinit.m 196
 EMintCircle.m 289
 EMintSphere.m 304
 EmsctCircle.m 291
 EMsctSphere.m 305
 ForwardBiCGSTABFFTEdw.m 260
 ForwardBiCGSTABFFTEwdw.m 223
 ForwardBiCGSTABFFTwEH.m 243
 ForwardCGFFTwE.m 198
 gradj.m 220
 IncEfield.m 225
 IncEfield3D.m 226
 IncEMsphere.m 300
 IncEMwave.m 200
 IncEtaufield.m 261
 IncEtaufield3D.m 262
 initEMcontrast.m 197
 initEMcontrastIntf.m 224
 initEMgrid.m 197
 ITERBiCGSTABEdw.m 259
 ITERBiCGSTABEwdw.m 227
 ITERBiCGSTABwE.m 210
 ITERBiCGSTABwEH.m 241
 ITERCGwE.m 201
 KopE.m 193
 KopEdw.m 258
 KopEdw3D.m 257
 KopEH.m 240
 KopEwdw.m 221
 plotContrastSourcewE.m 202
 plotContrastSourcewEH.m 245
 plotEMcontrast.m 198
 plotEMInterfaceSourceE.m 229
 plotEMInterfaceSourceEdw.m 263
 SnapshotE.m 273
 TimedomainBiCGSTABFFTEdw.m 274
 TimedomainBiCGSTABFFTwE.m 272
 vector2matrix.m 228
 WaveletE.m 270
EM waves Inversion Matlab Listings
 AdjDOPwE.m 451
 AdjKopE.m 448
 BICGSTABwdwE.m 458
 BICGSTABwE.m 450
 CheckInvData.m 475
 DisplayParameterswdwE.m 483
 DisplayParameterswE.m 481
 divFocus.m 491
 DOPEwdw.m 455
 DOPwE.m 451
 EGreenSourceReceivers.m 447
 EMdataCircle.m 446
 EmMRCSI.m 466
 Focusing.m 492
 ForwardwdwE.m 456
 ForwardwE.m 445
 FunDataEMsctCircle.m 479
 FunDataSctIntqwdwE.m 483
 FunDataSctIntqwE.m 480
 GaussNewtonwdwE.m 482
 GaussNewtonwE.m 478
 IncEMdwave.m 457
 IncEMwave.m 448
 initEM.m 442
 initEMcontrast.m 443, 488
 initEMcontrastIntf.m 457
 initEMgrid.m 443
 InitialEstimatewE.m 465
 InvIterwE.m 469
 ITERCGwE.m 449
 ITERMDwE.m 467
 KopE.m 448
 KopEwdw.m 454
 loadEimages.m 470
 Metrology.m 487
 PlotDataDif.m 490
 UpdateEMContrast.m 468

s

Scalar waves
 L_1-norm 340
 L_2-norm 340
 1D Green function 53
 1D addition theorem 53
 1D domain-integral equation 11
 1D integral representation scattered field 11
 1D spatial Fourier transform 52
 2D Green function 54

in polar coordinates 56
modified Bessel function 55
plane-wave representation 55
2D addition theorem 55
2D domain-integral equation 10
2D integral representation scattered field 10
2D spatial Fourier transform 54
3D Green function 56
in spherical coordinates 58
plane-wave representation 57
3D addition theorem 58, 62
3D integral representation scattered field 8
3D spatial Fourier transform 56
3D domain-integral equation 8
Total Variation (TV) 340
additive regularization 339
adjoint operator 15
amplitude data 368
analytical data 312
BiCGSTAB 42
Born approximation 8, 321
Bromwich integral 43
canonical configuration 58
1D scattering by a slab 58
2D scattering by a circular cylinder 60
3D scattering by a sphere 62
circulant matrix 14
circular convolution 14
conjugate gradient direction
Hesteness-Stiefel 16
Polak-Ribière 325
Conjugate Gradient method (CG) 16
constitutive relation 322
contrast function 8
contrast source 7, 9, 309
Contrast Source Inversion (CSI) 323
contrast update 328
contrast-source integral equation 9, 310
contrast-source integral representation 9, 309
control parameter 340
cost functional 323
data equation 322
data equations 310
data operator 324
Dirac distribution 3
distorted Born approach 322
edge-preserving regularization 340
Euler-Lagrange equation 342
Fast Fourier Transform (FFT) 15
Fermi-Dirac equilibrium distribution 355
forward discrete Fourier transform 44
Fréchet derivative 16, 325, 326
Fréchet derivative derivative 367, 368
Fredholm integral equation 8
Gauss-Newton algorithm 358
Gauss-Newton iteration 358
Gauss-Newton method 322, 355, 369
Green function
1D Green function 10
2D Green function 10
3D Green function 6
strong and weak forms 12
ill-posed equation 322
impulse function 3
incident field line source 10
incident field plane wave 10
incident field point source 6
inner product 15
intensity data 367
Inverse Crime 329
inverse discrete Fourier transform 45
Iterative Born approach 322
Jacobi iteration 344
Laplace transformation 4
lossy media 5
measurement domain 324
measurements domain 310
multiplicative cost functional 341
multiplicative regularization 339
Multiplicative Regularized Contrast Source Inversion (MRCSI) 339
Neumann series 9
Newton-Kantorovich method 322
noise 312
norm 15
normalized cost functional 15
object domain 310, 324
object equation 322
object equations 310
object operator 324
off-centered cylinder 313
operator equation 15
phaseless data 367

Scalar waves (*contd.*)
  propagation coefficient 5
  regularization parameter 339
  residual error 15, 324
  s-domain continuity conditions 4
  s-domain wave equation 4
  scattered wave field 7
  scattering domain 6, 8
  source-type integral equation 322
  standard deviation 312
  state equations 310
  step length 327
  Tikhonov functional 339
  time-domain continuity conditions 3
  time-domain wave equation 3
  wave speed 5
  wave-speed contrast 8
  wavelet 6
  Wavelet.m 45
  weak formulation 11
    circular mean 11
    linear mean 11
    singular kernel 11
    spherical mean 11
Scalar waves Forward Matlab Listings
  AdjKop.m 30
  displayData.m 21, 26
  Dop.m 34
  ForwardBiCGSTABFFTw.m 42
  ForwardCanonicalObjects.m 76, 77
  ForwardCGFFTu.m 24
  ForwardCGw.m 37
  Green.m 23
  IMAGESC.m 21, 24
  IncWave.m 24
  init.m 17
  initContrast.m 21, 24
  initFFTGreen.m 21, 22
  initGrid.m 17, 20
  initSourceReceiver.m 17, 19
  ITERBiCGSTABw.m 43
  ITERCGu.m 24
  ITERCGw.m 38
  Kop.m 30
  Legendre.m 70
  PlotContrastSource.m 32
  plotWavefield.m 21, 25
  SnapshotU.m 47
  TimedomainBiCGSTABFFTw.m 47
  WavefieldIncCircle.m 67
  WavefieldIncSphere.m 70, 73
  WavefieldSctCircle.m 70
  WavefieldSctSlab.m 65
  WavefieldSctSphere.m 70, 76
  WavefieldTotCircle.m 68
  WavefieldTotSlab.m 65
  WavefieldTotSphere.m 70, 74
  Wavelet.m 45
Scalar waves Inversion Matlab Listings
  AdjDopM.m 333
  BICGSTABw.m 319
  CheckAnlNum.m 357
  contrast estimate 329
  DataSctCircle.m 317
  DisplayParameters.m 363
  DopM.m 320
  Forwardw.m 316
  FunDataSctCircle 372
  FunDataSctIntEq 370
  GaussNewtonAnalytic 373
  GaussNewtonCSI.m 362
  GaussNewtonNumeric 371
  Green.m 315
  GreenSourcesReceivers.m 320
  init.m 311
  initContrast.m 315
  initFFTGreen.m 313
  initGrid.m 313
  InitialEstimate.m 332
  initSourceReceiver.m 312
  InverseCSI.m 330
  InverseCSIGaussNewton.m 359
  InverseMRCSI.m 345
  ITERCGMD.m 331
  ITERCGMDGaussNewton.m 360
  ITERCGMDreg.m 346
  ITERCGw.m 318
  PlotData.m 321
  TVregularizer.m 348
  UpdateContrast.m 333
  UpdateContrastGaussNewton.m 361
  UpdateContrastReg.m 347